Differential Equations

Differential Equations

Mark Krusemeyer
CARLETON COLLEGE

Macmillan College Publishing Company
NEW YORK

Maxwell Macmillan Canada
TORONTO

Maxwell Macmillan International
NEW YORK   OXFORD   SINGAPORE   SYDNEY

Editor: Robert W. Pirtle
Production Supervisor: Elaine W. Wetterau
Production Manager: Su Levine
Text Designer: Natasha Sylvester
Cover Designer: Natasha Sylvester
Illustrations: Academy ArtWorks, Inc.

This book was set in Times Roman by York Graphic Services, Inc., and was printed and bound by R. R. Donnelley & Sons Company—Crawfordsville. The cover was printed by Phoenix Color Corp.

Copyright © 1994 by Macmillan College Publishing Company, Inc.

Printed in the United States of America

All rights reserved. No part of this book may be reproduced or transmitted in any form or by any means, electronic or mechanical, including photocopying, recording, or any information storage and retrieval system, without permission in writing from the publisher.

Macmillan College Publishing Company
866 Third Avenue, New York, New York 10022

Macmillan College Publishing Company is part
of the Maxwell Communication Group of Companies.

Maxwell Macmillan Canada, Inc.
1200 Eglinton Avenue East
Suite 200
Don Mills, Ontario M3C 3N1

Library of Congress Cataloging in Publication Data
Krusemeyer, Mark.
   Differential equations / Mark Krusemeyer.
     p. cm.
   Includes index.
   ISBN 0-02-366912-8
    1. Differential equations. I. Title.
QA371.K78 1994
515'.35—dc20

93-63
CIP

Printing: 1 2 3 4 5 6 7 8   Year: 4 5 6 7 8 9 0 1

Preface

As you may suspect from its title, this book is intended as an introduction to differential equations (at an undergraduate level). So are many other books; however, I believe that the style and content of this one will make it particularly useful to many people.

If you have had a course in multivariable calculus and you feel that you can't read this book, then I have failed you. If you are teaching a course from this book and you feel that you have to spend your time explaining to your students what the book is really trying to say, then I have failed you also. And if you fall asleep over the book, well, who knows? Some people need more sleep than others, and some people will find the material less interesting than others. But if you like mathematical ideas or if you like calculations, this book should provide you with plenty of both.

Naturally, if you will be learning the material, there are certain things you'll have to remember, and the summaries at the end of the sections should help with this. Please, though, don't read only the summaries (and/or those examples that look like your homework problems). If you do that, you will have to memorize too much and you will understand too little, because you'll be missing the motivation for what is being done, connections within the subject and with other subjects, and ways of thinking about problems *before* you know how to solve them. If, instead, you actually read the book as well as getting plenty of practice on the exercises, I think that you should end up with real understanding and control of the material, and well prepared for further study.

Meanwhile, if you will be teaching the material, let me assure you that the informal style of the book does not mean that I have given up precision for the sake of vague understanding. I hope that the distinction between proofs and heuristic arguments will be very clear throughout, and that there are ample warnings of logical and computational pitfalls. I also hope that the book will provide a sense of enthusiasm and discovery as well as one of clarity and coherence.

In my own learning and in my own teaching, I've found that going "back and forth" through material—returning for review and more sophisticated understanding to concepts introduced earlier, as well as looking ahead to ideas that will be studied more carefully later—is often very effective. Books can't really be written that way, and this one is no exception; therefore, it's entirely up to you. If you're not used to working "back and forth," this is my one chance to encourage you to give it a try. Don't feel obliged to study *everything* in a section before starting to read the next one; don't feel obliged to cover all ideas in a section before starting to assign some homework from it.

If you are going to be learning from this book in a course, the rest of this preface, which provides mildly technical information about the content of the book before turning to acknowledgments, probably won't be too interesting. If you are planning to use the book for self-study, you might want to keep reading to see what some of your options are.

Although it is possible to teach a year-long course from this book, I expect that a much more common approach will be to choose specific chapters and sections for a semester or quarter course. Nearly every instructor will want to start the course with the material of the first two chapters, but there will be a wide range of opinions as to which other topics should be covered and in what order that should be done. The book offers ample flexibility in this regard; the following diagram indicates the logical dependence between various chapters and sections.

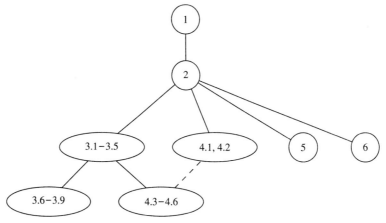

You may notice that Chapter 7 does not appear in the diagram; this chapter, entitled "Appetizers," consists of four relatively unrelated sections, each indicating a possible direction for further study. On the other hand, it is quite feasible to include material from Chapter 7 early in a course. For example, Section 7.2, which introduces numerical methods, can be covered at any point after 1.5. In principle, this is also true of 7.3, but in practice that section, which sketches a proof of the existence and uniqueness theorem, may be too sophisticated to cover at an early stage. Section 7.1 provides an introduction to difference equations, and it can be used as a sequel to 2.5. The final section, 7.4, looks ahead to boundary value problems and separation of variables in partial differential equations, as well as to Fourier series. It might be possible to cover this section after 2.4, but first doing 2.5–2.7 and some of Chapter 5 (at least through 5.3) will provide better motivation.

Let me mention some features from the first six chapters, as well. Chapter 1 starts with a review of relevant material from calculus; in fact, I hope that the discussion of partial derivatives will make the book accessible even to some students who have studied only single-variable calculus. At the end of the chapter, the exercises for Section 1.7 provide practice in recognizing what technique to use for a given first-order (differential) equation. They also introduce some special topics, such as Bernoulli equations and orthogonal trajectories.

To begin Chapter 2, the study of second-order equations is motivated by an introductory section on Newtonian motion. The discussion of this key application resumes in Section 2.7, after the appropriate solution methods have been presented. (Appendix A, at the end of the book, supplies basic information on complex numbers, which is used in Section 2.5 and beyond.)

Chapter 3 provides an extensive and systematic discussion of systems of two first-order equations; several applications are considered in depth. Even if there is no time to go beyond 3.5 or 3.6 in this chapter, students will be introduced to important ideas about the qualitative behavior of solutions, as well as to computational techniques that illustrate the power of linear algebra. (Appendix B supplies elementary background in matrix algebra; the key concepts of eigenvector and eigenvalue are motivated and defined in Section 3.4.) On the other hand, I hope that the discussions of stability and Lyapunov functions in 3.7 and 3.8 will be among the highlights of some courses.

Chapter 4 presents generalizations to higher-order situations of most of the quantitative methods from Chapters 2 and 3. It includes a discussion of matrix exponentials.

The use of power series in our context is motivated in Section 5.1; Section 5.2 then reviews and expands on students' knowledge of power series from calculus, in preparation for the rest of Chapter 5. This chapter features a gradual but thorough discussion of singular points: A brief introduction is followed by a treatment of Euler and Bessel equations as "model" examples, after which the Frobenius method is presented in general.

Chapter 6, on the Laplace transform, features a section on the delta function. If you find this section to be intellectually honest, computationally useful, and accessible to your students, then I have succeeded—and, of course, similar comments apply to other parts of the book.

To the extent that I have succeeded in this long enterprise (a child conceived at the same time as this book might, by now, be starting high school), much credit is due to friends and loved ones as well as students and colleagues. The support of the many people who generously provided encouragement, good advice, and hospitality in a variety of professional and personal circumstances related to the project has been crucial to my seeing it through. I am deeply grateful to these people, as I am grateful to the people listed below for their (particularly tangible) contributions to the book.

The expert typing of most of the manuscript was done by Evelyn Laurent in Princeton, N.J., and Barbara Jenkins in Northfield. Professor Martin Karel at Rutgers University–Camden and Professor Samuel Patterson at Carleton College taught classes using major portions of the typescript and shared their experiences and their students' reactions with me. Marcus Gale, Carleton '93, not only read and commented on the entire typescript and helped me check the answers to the odd-numbered exercises, but also used Mathematica™ to convert my rough sketches into the accurately drawn figures you will find throughout the book.

At Macmillan, production supervisor Elaine Wetterau has tried valiantly, but with charm and patience, to keep me to reasonable deadlines, while allowing me considerable freedom to influence the final form of the typography. Executive editor Robert Pirtle, who took charge of the project a few years ago, has been very helpful and congenial; he arranged for frequent feedback from reviewers. In the most recent "round" of reviews, comments were received from Robert Brown, University of Kansas; Harvey S. Davis, Michigan State University; Laurene V. Fausett, Florida Institute of Technology; Donald Goral, Northern Virginia Community College; Bernard Harris, Northern Illinois University; Gary Roberts, Slippery Rock University; Ralph Showalter, University of Texas; Mo Tavakoli, Chaffey College; and Marvin Zeman, Southern Illinois University. Most comments in this and earlier rounds were helpful and constructive, and often they led directly to substantial changes in the manuscript.

Of course, I take full responsibility for any unclear or misleading passages, errors, and so forth, that remain—and I would really appreciate being told about them! Please write to me at: Department of Mathematics and Computer Science, Carleton College, Northfield, MN 55057.

Mark Krusemeyer

Contents

CHAPTER 1 First-Order Differential Equations   1

        Introduction   *1*
- 1.1 Review of basic facts from calculus   *2*
- 1.2 Separable differential equations   *12*
- 1.3 Applications: Radioactive decay and population growth   *21*
- 1.4 Classification of differential equations   *29*
- 1.5 Initial value problems and integral curves   *34*
- 1.6 Linear first-order equations   *49*
- 1.7 Exact equations and integrating factors   *59*

CHAPTER 2 Second-Order Differential Equations   77

- 2.1 Introduction: Newtonian motion   *77*
- 2.2 General results   *84*
- 2.3 Constant-coefficient equations: Introduction   *98*
- 2.4 Reduction of order   *102*
- 2.5 Constant-coefficient equations (continued)   *107*
- 2.6 Inhomogeneous equations: Introduction   *112*
- 2.7 Mechanical vibration   *119*
- 2.8 Inhomogeneous equations (continued): Variation of constants   *131*

CHAPTER 3 Systems of Two First-Order Differential Equations   143

- 3.1 Problems involving two unknown functions   *143*
- 3.2 Autonomous systems   *155*
- 3.3 Linear systems (elimination method)   *178*
- 3.4 Linear systems (continued); eigenvalues and eigenvectors   *185*
- 3.5 Linear systems (continued): Double or complex conjugate eigenvalues   *200*

	3.6	Phase portraits for linear systems; types of stationary points	210
	3.7	Phase portraits for nonlinear systems; stability	232
	3.8	Lyapunov functions; gradient systems	247
	3.9	The Volterra–Lotka equations and other two-population systems	262

CHAPTER 4 Higher-Order Differential Equations and General Systems — 279

- 4.1 Higher-order differential equations: Introduction   279
- 4.2 Higher-order differential equations (continued)   291
- 4.3 Homogeneous linear systems (autonomous, distinct real eigenvalues)   304
- 4.4 Homogeneous, autonomous linear systems (continued)   322
- 4.5 Homogeneous, autonomous linear systems (using matrix exponentials)   335
- 4.6 General linear systems   354

CHAPTER 5 Power Series Methods — 367

- 5.1 Introduction   367
- 5.2 Power series   376
- 5.3 First case: $x = 0$ an ordinary point   395
- 5.4 Singular points: Introduction   409
- 5.5 Euler equations   417
- 5.6 The Bessel equation   428
- 5.7 Singular points: Conclusion   442

CHAPTER 6 The Laplace Transform — 461

- 6.1 Introduction   461
- 6.2 Transforming initial value problems   478
- 6.3 Step functions   490
- 6.4 The convolution integral   503
- 6.5 The delta "function"   514

CHAPTER 7 Appetizers — 533

Introduction   533
- 7.1 Difference equations (recurrence relations)   533
- 7.2 Numerical methods   545

	7.3	Successive approximation *559*	
	7.4	Partial differential equations *572*	

APPENDIX A Complex Numbers   589

    A.1 Introduction   *589*
    A.2 The complex plane   *592*
    A.3 Power series and complex exponentials   *599*

APPENDIX B Basic Linear Algebra   605

    B.1 Vectors   *605*
    B.2 Matrices and column vectors   *610*
    B.3 Systems of linear equations; inverses of matrices   *615*
    B.4 Determinants   *621*

Answers to Odd-Numbered Exercises   627

Index   681

Differential Equations

CHAPTER 1

First-Order Differential Equations

INTRODUCTION

In many fields in which mathematics is used, such as physics, chemistry, biology, and economics, one encounters "unknown" functions, for which originally no formulas are given. For example, $x(t)$ might be the amount of some substance in a chemical reaction at time t, or it might be the budget deficit of a country at time t. Often one has information (based on experimental evidence or, more likely, on a theoretical model) about such a function, in the form of a statement concerning its rate of change. For example, we will see in Section 1.3 that if $x(t)$ is the amount of a naturally decaying radioactive substance, one can reasonably assume that $\frac{dx}{dt} = Kx$ for some (negative) constant K.

The equation $\frac{dx}{dt} = Kx$ is an example of a differential equation. Solving such an equation means, more or less, finding an explicit formula for all functions $x(t)$ for which the equation is correct. One of the main purposes of this book is to present and motivate various methods for solving specific types of differential equations. Another is to discuss how to obtain information about the unknown functions even in cases in which the differential equations are too difficult to solve.

In this chapter, we will look primarily at differential equations such as $\frac{dx}{dt} = 5x - 2$, $t^2 \frac{dx}{dt} = \sqrt[3]{x+1}$, and $\frac{dx}{dt} = 5tx + t^2$. However, you will also find some preliminary discussion of differential equations involving higher derivatives and/or partial derivatives. We begin with a review of some key ideas from calculus.

1.1 REVIEW OF BASIC FACTS FROM CALCULUS

Let $x(t)$ be a function of one (real) variable t. [When speaking of functions, we'll generally mean functions whose values are real numbers; we do not insist that functions be defined everywhere. For instance, $x(t) = \sqrt{t}$ is acceptable even though this function is not defined for $t < 0$. By the way, you may be more used to having x as the variable and $y = f(x)$ as the function, and this will sometimes be the case. However, our usual notation will be $x(t)$, where in applications t often stands for time.]

We know from calculus that the **derivative** of $x(t)$ can be defined as

$$\frac{dx}{dt} = x'(t) = \lim_{h \to 0} \frac{x(t+h) - x(t)}{h}.$$

Of course, this limit may not exist, and the function $x(t)$ is called **differentiable** when it does; unless otherwise specified, we'll assume that $x(t)$ is differentiable. If the derivative $x'(t)$ is itself differentiable, we can define the second derivative $\dfrac{d^2x}{dt^2} = x''(t)$, and so on.

The definition of the derivative hardly indicates its importance or its uses in actual practice. The following may be the two most important facts about the derivative:

Fact 1 The derivative can be used to find a straight-line approximation (the tangent line) to the graph of the function $x(t)$ near a given point $(t_0, x(t_0))$. Specifically, this **linear approximation** is

$$x(t) \approx x(t_0) + x'(t_0) \cdot (t - t_0) \qquad \text{for } t \text{ near } t_0;$$

the right-hand side of this approximate equation gives the equation of the tangent line at t_0; $x'(t_0)$ is the slope of this tangent line (see Figure 1.1.1).

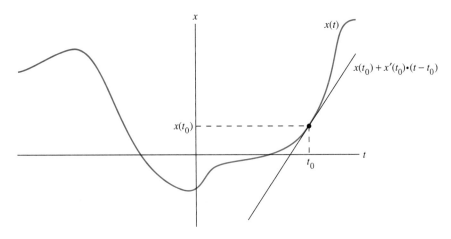

Figure 1.1.1. Tangent line at $(t_0, x(t_0))$.

Fact 2 When the variable t denotes the time, the derivative $\dfrac{dx}{dt}$ equals the **rate of change** of the quantity x with respect to time; more generally, $\dfrac{dx}{dt}$ indicates how rapidly x changes with respect to t.

Fact 1 is used in calculus to investigate the behavior of functions; for instance, if a function $x(t)$ is to have a maximum or a minimum at t_0, the tangent line there must be horizontal, so we must have $x'(t_0) = 0$.

Fact 2 is used in "translating" problems of practical (e.g., physical) origin into mathematical terms. Thus it is used in many applications. Fact 2 also provides intuitive justification for the **chain rule**: If y is a function of x and x is in turn a function of t, so that y is a composite function of t, then $\dfrac{dy}{dt} = \dfrac{dy}{dx}\dfrac{dx}{dt}$.

If two functions $x(t)$ and $y(t)$ have the same derivative, say $f(t)$, everywhere, so that $x'(t) = y'(t) = f(t)$ for all t, then $x(t)$ and $y(t)$ must differ by a constant (see Exercise 4). Therefore, if we know the derivative $f(t)$, this will determine the function itself up to an (arbitrary) constant. The notation $\int f(t)\,dt$ is used for the functions whose derivative is $f(t)$; for example, $\int t^3\,dt = \dfrac{1}{4}t^4 + C$, where C is the arbitrary **constant of integration**.

$\int f(t)\,dt$ is called the **indefinite integral** of $f(t)$, while any specific function whose derivative is $f(t)$ is called an **antiderivative** of $f(t)$.

One way of obtaining an antiderivative of a given function $f(t)$ is to use the **definite integral**, as follows. First recall that for any continuous function f defined on a closed interval $[a, b]$ [1] we can define a number $\int_a^b f(t)\,dt$ (as a limit of Riemann sums). If the function f is nonnegative, this number can be interpreted as the area between the interval $[a, b]$ along the horizontal axis and the graph of f (see Figure 1.1.2).

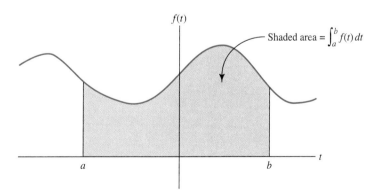

Figure 1.1.2. The definite integral.

---

[1] The **closed interval** $[a, b]$ is the set of all real numbers t with $a \le t \le b$. Similarly, the **open interval** (a, b) consists of all t with $a < t < b$, $(a, b]$ consists of all t with $a < t \le b$, (a, ∞) consists of all t with $t > a$, and so on.

Now if we consider $\int_a^b f(t)\,dt$ as a *function* of the upper bound b, say $x(b) = \int_a^b f(t)\,dt$, then part of the Fundamental Theorem of Calculus states that the derivative of this function x is the given function f, that is, $x'(b) = f(b)$. It is often preferable to have t rather than b as the variable for the function x, but it is considered bad form to write $x(t) = \int_a^t f(t)\,dt$ (using the same letter for the integration variable as for the upper bound). Thus it is customary to use a different "dummy" variable, say s, for the integration. We then have $x(t) = \int_a^t f(s)\,ds$ as an antiderivative of $f(t)$. Note that for this particular antiderivative, $x(a) = 0$. For example, the antiderivative $x(t)$ of e^{-t^2} which satisfies $x(2) = 0$ is given by $x(t) = \int_2^t e^{-s^2}\,ds$. Once again:

The antiderivative $x(t)$ of $f(t)$ with $x(a) = 0$ is given by $x(t) = \int_a^t f(s)\,ds$.

Conversely, instead of using the definite integral to construct antiderivatives, one can often use antiderivatives to compute definite integrals. In fact, the other part of the Fundamental Theorem of Calculus states that if $F(t)$ is any antiderivative of $f(t)$, then $\int_a^b f(t)\,dt = F(b) - F(a)$. For example, since $\frac{1}{4}t^4$ is an antiderivative of t^3, we have

$$\int_2^5 t^3\,dt = \frac{1}{4}\cdot 5^4 - \frac{1}{4}\cdot 2^4 = \frac{609}{4}.$$

Starting in Section 1.7, we will often encounter functions of two variables, usually with x, y as the variables. The graph of such a function, say $z = f(x, y)$, is a surface in 3-space (x,y,z-space). This can make for difficult drawing, and sometimes it is easier to get an idea of the behavior of the function by drawing its **level curves** in the x,y-plane. For each constant C, the level curve of level C is the curve $f(x, y) = C$ along which the function equals C.

EXAMPLE $f(x, y) = x^2 + y^2$.

The graph of this function is the circular paraboloid ("bowl") in 3-space shown in Figure 1.1.3(a), while the level curves in the x,y-plane form the circular pattern in Figure 1.1.3(b). ∎

The relation between the graph of a function and its level curves is the same as the relation between a raised-relief map of a mountain range and the contour lines on a flat map of the same terrain (to the same scale).

Just as the derivative of a function of one variable is used to find a straight-line approximation to its graph, we can use calculus to find a plane approximation, the **tangent plane**, to the graph of $f(x, y)$ near a given point $(x_0, y_0, f(x_0, y_0))$. Just as the tangent line to a plane curve at a given point is determined by its slope, the tangent plane will be determined by two numbers: its slope in the x-direction and its slope in the y-direction. The slope in the x-direction is the slope if y is kept constant, that is, the derivative of the

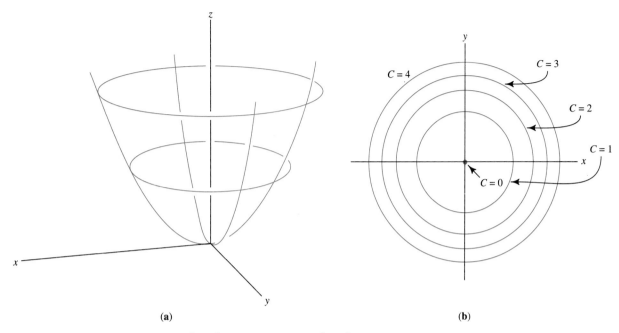

Figure 1.1.3. (a) Graph of $z = x^2 + y^2$. (b) Level curves $x^2 + y^2 = C$.

function $f(x, y_0)$ of x at $x = x_0$. This slope is called the **partial derivative** of f with respect to x (or in the x-direction) at the point (x_0, y_0) and is denoted by $\dfrac{\partial f}{\partial x}(x_0, y_0)$. Thus

$$\frac{\partial f}{\partial x}(x_0, y_0) = \lim_{h \to 0} \frac{f(x_0 + h, y_0) - f(x_0, y_0)}{h}.\ [2]$$

Similarly, the slope of the tangent plane in the y-direction will be the partial derivative of f with respect to y:

$$\frac{\partial f}{\partial y}(x_0, y_0) = \lim_{h \to 0} \frac{f(x_0, y_0 + h) - f(x_0, y_0)}{h}.$$

Once we have the slopes, the approximation of $f(x, y)$ by its tangent plane will be given by

$$z = f(x, y) \approx f(x_0, y_0) + a(x - x_0) + b(y - y_0),$$

where we have abbreviated the slopes by $a = \dfrac{\partial f}{\partial x}(x_0, y_0)$ and $b = \dfrac{\partial f}{\partial y}(x_0, y_0)$. Just as for one variable, the partial derivatives $\dfrac{\partial f}{\partial x}$ and $\dfrac{\partial f}{\partial y}$ are themselves functions (of two variables), since the discussion above is valid for any given x_0, y_0.

[2] As in the one-variable case, we are tacitly assuming that the limit exists.

EXAMPLE 1.1.1 $f(x, y) = x^2 + y^2$.

If we fix $y = y_0$, we have $f(x, y_0) = x^2 + y_0^2$; the derivative of this function of x at x_0 is $\dfrac{\partial f}{\partial x}(x_0, y_0) = 2x_0$. Similarly, $\dfrac{\partial f}{\partial y}(x_0, y_0) = 2y_0$. In practice, given $f(x, y) = x^2 + y^2$, one can find the partial derivatives without using the notation with x_0 and y_0: To find $\dfrac{\partial f}{\partial x}$, differentiate $x^2 + y^2$ with respect to the variable x, considering y as a constant; result: $\dfrac{\partial f}{\partial x} = 2x$. This means that for any point (x_0, y_0), $\dfrac{\partial f}{\partial x}(x_0, y_0) = 2x_0$. Similarly, we have $\dfrac{\partial f}{\partial y} = 2y$. ∎

EXAMPLE 1.1.2 To find the approximation to $f(x, y) = xy^2 + \sin(xy + x) - 2y + 3$ by its tangent plane at $x = y = 0$, we first compute, as in Example 1.1.1,

$$\frac{\partial f}{\partial x} = y^2 + (y + 1)\cos(xy + x) \quad \text{and} \quad \frac{\partial f}{\partial y} = 2xy + x\cos(xy + x) - 2.$$

Substituting $x = y = 0$, we get $f(0, 0) = 3$, $\dfrac{\partial f}{\partial x}(0, 0) = 1$, $\dfrac{\partial f}{\partial y}(0, 0) = -2$, so the approximation is $z = f(x, y) \approx 3 + x - 2y$. ∎

In algebraic terms, the approximation of the graph of $f(x, y)$ by the tangent plane is the approximation of $f(x, y)$ by the *linear* expression $f(x_0, y_0) + a(x - x_0) + b(y - y_0)$, where a and b are again short for the partial derivatives of the function f at (x_0, y_0). However, this approximation will be useful only if we can estimate its accuracy; it turns out that for some functions the "approximation" is actually very misleading.

EXAMPLE 1.1.3 Let the function f be defined by

$$f(x, y) = \begin{cases} \dfrac{1}{xy} & (xy \neq 0) \\ 0 & (x = 0 \text{ or } y = 0). \end{cases}$$

This function is zero along the coordinate axes, but it has very large positive and negative values close to the coordinate axes. Since $f(x, 0) = 0$ for all x, we have $\dfrac{\partial f}{\partial x}(0, 0) = 0$; similarly, $\dfrac{\partial f}{\partial y}(0, 0) = 0$, so the linear approximation to $f(x, y)$ near the point $(0, 0)$ would, according to the above, be the constant function 0. On the other hand, a typical value of f near that point is $f(10^{-6}, 10^{-6}) = 10^{12}$! Obviously, the plane $z = 0$ is not really "tangent" to the graph of f at $(0, 0)$ in any reasonable sense. ∎

SEC. 1.1 REVIEW OF BASIC FACTS FROM CALCULUS

From this example we can see that there is no guarantee in general that we can get much information about the behavior of f near (x_0, y_0) by looking only at how f changes in the x- and y-directions at that point. On the other hand, if (x, y) is near (x_0, y_0), we can go from (x_0, y_0) to (x, y) in two steps, one in the x-direction and one in the y-direction (see Figure 1.1.4).

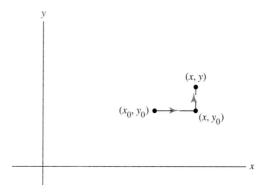

Figure 1.1.4. Moving from (x_0, y_0) to (x, y) following the directions of the coordinate axes.

We can try to use the partial derivatives to see how f will change at each step, as follows. When the point moves from (x_0, y_0) to (x, y_0), y is constant, so the change in f is approximately "the change in x" times "the derivative with respect to x":

$$\underbrace{f(x, y_0) - f(x_0, y_0)}_{\text{change in } f} \approx \frac{\partial f}{\partial x}(x_0, y_0) \cdot \underbrace{(x - x_0)}_{\text{change in } x}. \tag{1}$$

[In the example, $f(10^{-6}, 0) \approx f(0, 0) + \dfrac{\partial f}{\partial x}(0, 0) \cdot 10^{-6}$; in fact, both sides are zero.]

When the point moves from (x, y_0) to (x, y), x is constant, so we have

$$\underbrace{f(x, y) - f(x, y_0)}_{\text{change in } f} \approx \frac{\partial f}{\partial y}(x, y_0) \cdot \underbrace{(y - y_0)}_{\text{change in } y} \tag{2}$$

provided that the partial derivative exists. Note that this partial derivative is taken at (x, y_0) and not at the original point (x_0, y_0). [In the example, the partial derivative $\dfrac{\partial f}{\partial y}(x, y_0)$ does not exist; in fact,

$$f(10^{-6}, y) = \begin{cases} \dfrac{10^6}{y} & (y \neq 0) \\ 0 & (y = 0) \end{cases}$$

and this function is not continuous, let alone differentiable, at $y = 0$.] However, if the partial derivative $\dfrac{\partial f}{\partial y}$ exists and is *continuous* at (x_0, y_0), we can argue that since (x, y_0) is near (x_0, y_0), $\dfrac{\partial f}{\partial y}(x, y_0)$ is approximately equal to $\dfrac{\partial f}{\partial y}(x_0, y_0)$. In that case we can combine approximations (1) and (2) above with

$$\frac{\partial f}{\partial y}(x, y_0) \approx \frac{\partial f}{\partial y}(x_0, y_0) \tag{3}$$

to obtain $f(x, y) - f(x_0, y_0) \approx \dfrac{\partial f}{\partial x}(x_0, y_0) \cdot (x - x_0) + \dfrac{\partial f}{\partial y}(x_0, y_0) \cdot (y - y_0)$ or

$$f(x, y) \approx f(x_0, y_0) + \frac{\partial f}{\partial x}(x_0, y_0) \cdot (x - x_0) + \frac{\partial f}{\partial y}(x_0, y_0) \cdot (y - y_0).$$

This is, once again, the approximation of $f(x, y)$ using the tangent plane.

Although we have still not estimated the error in the approximation above, it now seems reasonable to require that:

1. $\dfrac{\partial f}{\partial x}$ and $\dfrac{\partial f}{\partial y}$ be defined not only at (x_0, y_0), but also at nearby points.

2. $\dfrac{\partial f}{\partial x}$ and $\dfrac{\partial f}{\partial y}$ be continuous at (x_0, y_0).

Under these conditions we may expect to find a reasonable estimate for the error when $f(x, y)$ is replaced by its **linear approximation**

$$f(x_0, y_0) + a(x - x_0) + b(y - y_0)$$

[where $a = \dfrac{\partial f}{\partial x}(x_0, y_0)$, $b = \dfrac{\partial f}{\partial y}(x_0, y_0)$] for (x, y) near (x_0, y_0).

For "standard" functions, conditions 1 and 2 are usually satisfied. It is customary to use the term "tangent plane" only for functions that satisfy these conditions. Such functions, whose partial derivatives are defined and continuous throughout the domain of the function, are called **C^1 functions** (short for "continuously differentiable once functions"). In this book, unless otherwise specified, functions of two variables will be assumed to be C^1.

Just as one can consider higher derivatives for a function of one variable, one can take repeated partial derivatives for a function of two variables. For instance, a function $f(x, y)$ has four "second partials":

$$\frac{\partial^2 f}{\partial x^2} = \frac{\partial}{\partial x}\left(\frac{\partial f}{\partial x}\right), \quad \frac{\partial^2 f}{\partial x \, \partial y} = \frac{\partial}{\partial x}\left(\frac{\partial f}{\partial y}\right), \quad \frac{\partial^2 f}{\partial y \, \partial x} = \frac{\partial}{\partial y}\left(\frac{\partial f}{\partial x}\right), \quad \frac{\partial^2 f}{\partial y^2} = \frac{\partial}{\partial y}\left(\frac{\partial f}{\partial y}\right).$$

EXAMPLE 1.1.4 $f(x, y) = xy^2 + \sin(xy + x) - 2y$.

We have $\dfrac{\partial f}{\partial x} = y^2 + (y + 1)\cos(xy + x)$, so

$$\dfrac{\partial^2 f}{\partial x^2} = -(y+1)^2 \sin(xy+x) \text{ and } \dfrac{\partial^2 f}{\partial y \, \partial x} = 2y + \cos(xy+x) - x(y+1)\sin(xy+x).$$

On the other hand, $\dfrac{\partial f}{\partial y} = 2xy + x \cdot \cos(xy + x) - 2$, so

$$\dfrac{\partial^2 f}{\partial x \, \partial y} = 2y + \cos(xy+x) - x(y+1)\sin(xy+x), \quad \dfrac{\partial^2 f}{\partial y^2} = 2x - x^2 \sin(xy+x). \quad \blacksquare$$

Note that $\dfrac{\partial^2 f}{\partial x \, \partial y} = \dfrac{\partial^2 f}{\partial y \, \partial x}$ in this example; this is true *in general* provided that $\dfrac{\partial^2 f}{\partial x \, \partial y}$ and $\dfrac{\partial^2 f}{\partial y \, \partial x}$ are both continuous.

Now suppose that x and y are, themselves, functions of a variable t, which makes $f(x, y) = f(x(t), y(t))$ a composite function of t.

EXAMPLE 1.1.5 $f(x, y) = x^2 + y^2$, $x(t) = t + \cos t$, $y(t) = \sin t$. Then

$$f(x(t), y(t)) = (t + \cos t)^2 + (\sin t)^2 = t^2 + 2t \cos t + 1. \quad \blacksquare$$

In this situation we can find the derivative of $f(x, y)$ with respect to t from the formula

$$\dfrac{d(f(x, y))}{dt} = \dfrac{\partial f}{\partial x} \dfrac{dx}{dt} + \dfrac{\partial f}{\partial y} \dfrac{dy}{dt}.$$

This formula, which is known as the **chain rule** in two variables, is not too hard to prove. The idea is that when t varies, f changes both because x changes and because y changes. If y were a constant y_0, we would have

$$\dfrac{d(f(x, y_0))}{dt} = \dfrac{d(f(x, y_0))}{dx} \dfrac{dx}{dt} = \dfrac{\partial f}{\partial x} \dfrac{dx}{dt}.$$

The change in f would be due only to the change in x. Since y is actually not constant, we must add the contribution $\dfrac{\partial f}{\partial y} \dfrac{dy}{dt}$, which is due to the change in y, in order to get the rate of change of f. In Example 1.1.5 we have $\dfrac{\partial f}{\partial x} = 2x = 2(t + \cos t)$, $\dfrac{\partial f}{\partial y} = 2y = 2 \sin t$, so

$$\frac{d(f(x,y))}{dt} = \frac{\partial f}{\partial x}\frac{dx}{dt} + \frac{\partial f}{\partial y}\frac{dy}{dt} = 2(t + \cos t)(1 - \sin t) + 2\sin t \cos t$$

$$= 2t + 2\cos t - 2t \cdot \sin t.$$

Of course, this can also be seen directly from $f(x(t), y(t)) = t^2 + 2t \cos t + 1$.

Warning: Remember that there are *two* terms on the right-hand side of the chain rule in two variables!

To see an application of the chain rule, we again consider the level curves $f(x, y) = C$; we will find a formula for the (slope of the) tangent line to such a curve. In first-year calculus this is done by implicit differentiation, but that method works only if the function $f(x, y)$ is known in advance.

EXAMPLE If we know that $f(x, y) = x^2 + y^2$, we can differentiate both sides of $x^2 + y^2 = C$ with respect to x to obtain $2x + 2y\frac{dy}{dx} = 0$, so $\frac{dy}{dx} = -\frac{x}{y}$ is the slope of the tangent line to the level curve at any point where $y \neq 0$. (What happens when $y = 0$?) ∎

We can now use the following general method: Assume that the level curve, or part of it, is the graph of a function $y(x)$. (The same assumption is made in the old method.) Then we have $f(x, y(x)) = C$ along this part of the level curve. Now differentiate both sides of this equation with respect to x; on the left-hand side we must use the chain rule, and the result is $\frac{\partial f}{\partial x} + \frac{\partial f}{\partial y}\frac{dy}{dx} = 0$ or $\frac{dy}{dx} = -\frac{\partial f/\partial x}{\partial f/\partial y}$.

Conclusion. The slope of the level curve of $f(x, y)$ at a given point (x_0, y_0) equals

$$-\frac{\frac{\partial f}{\partial x}(x_0, y_0)}{\frac{\partial f}{\partial y}(x_0, y_0)}.$$

We will use this fact in Section 3.8.

NOTE: This formula can also be obtained from the fact that the gradient vector $\left(\frac{\partial f}{\partial x}, \frac{\partial f}{\partial y}\right)$ at (x_0, y_0) is normal to the level curve there.

EXERCISES

1. Find the straight-line approximation to the graph of $x = t^2 + t$ near $t = 3$.
2. Suppose that $x(t)$ and $y(t)$ are the amounts at time t of two substances that combine in a chemical reaction to produce a third substance. Let $z(t)$ be the amount of the third substance, and suppose that the rate at which the third

substance is produced is proportional to (each of) the amounts of the other two substances. Write down an equation (which will contain an unknown constant of proportionality) expressing this fact.

3. Explain how Fact 2 (p. 3) provides intuitive justification for the (one-variable) chain rule.

4. Suppose that two functions have the same derivative everywhere: $x'(t) = y'(t)$ for all t. Show (or look up the reason in a calculus book) that the functions x and y must differ by a constant.

5. (a) Find the antiderivative $x(t)$ of e^{-t^2} with $x(2) = 15$.
   (b) Show that the antiderivative $x(t)$ of $f(t)$ with $x(a) = k$ is given by
   $$x(t) = k + \int_a^t f(s)\, ds.$$

Sketch the level curves of each of the following functions $f(x, y)$ for $C = 0, \pm 1, \pm 2, \pm 5$.

6. $x^2 - y^2$
7. xy
8. $x + 3y$
9. $x^2 + y$
*10. $x^2 - y^3$

For each of the following functions $f(x, y)$:

(a) Compute $\dfrac{\partial f}{\partial x}$ and $\dfrac{\partial f}{\partial y}$.

(b) Compute the second partial derivatives $\dfrac{\partial^2 f}{\partial x^2}$, $\dfrac{\partial^2 f}{\partial x\, \partial y}$, $\dfrac{\partial^2 f}{\partial y\, \partial x}$, and $\dfrac{\partial^2 f}{\partial y^2}$.

11. $x^2 + \cos y$
12. $\cos(x^2 + y)$
13. $y \cos^2 x$
14. e^{x-2y}

15. Find the tangent plane to the graph of $z = xy - x^2 y^3$ at the point $(1, 1, 0)$.

16. Find the tangent plane to the graph of $z = e^{x-2y}$ at the point where $x = 3$ and $y = 1$.

17. Let $f(x, y) = xy + x^3$, $x(t) = e^{2t} - 1$, $y(t) = e^{2t} + t$. Find $\dfrac{d(f(x, y))}{dt}$ in two ways:
    (a) Using the chain rule in two variables.
    (b) By first finding $f(x(t), y(t))$.

18. (a) Find the equation of the level curve of $f(x, y) = x^2 - y^3$ that passes through $(2, -1)$.
    (b) Find the tangent line at $(2, -1)$ to the level curve in part (a) by using the formula from p. 10 for its slope.
    (c) Check your answer in part (b) using implicit differentiation.

19. Find the equation of the level curve of $f(x, y) = x \cos y$ that passes through the point $(2, \pi/3)$. Then find the tangent line to this curve at $(2, \pi/3)$ by using the formula from p. 10 for its slope.

20. Give a geometric reason that (as stated on p. 10) $\dfrac{dy}{dx} = -\dfrac{x}{y}$ is the slope of the tangent line to $x^2 + y^2 = C$ at any point where $y \neq 0$.

21. (a) Prove, using the chain rule, that $\dfrac{dx}{dy} = \dfrac{1}{dy/dx}$ if x and y are (differentiable) functions *of each other*.

(b) Suppose that y is a function of x. Under what condition on $\dfrac{dy}{dx}$ will x be a function of y?

(c) Illustrate the results of parts (a) and (b) in the following cases:
  (i) $y = x^2$, $x, y > 0$ only.
  (ii) $y = x^2$, no restriction.

FURTHER READING

For a proof that $\dfrac{\partial^2 f}{\partial x\, \partial y} = \dfrac{\partial^2 f}{\partial y\, \partial x}$ provided that both these "mixed partials" are continuous, see Simmons, *Calculus with Analytic Geometry* (McGraw-Hill, 1985), Appendix C.15.

1.2 SEPARABLE DIFFERENTIAL EQUATIONS

A **differential equation** is an equation in which one or more of the derivatives of one or more unknown functions occur; the unknown functions themselves may also occur.[3] These may be functions of one variable or of several variables; accordingly, the derivatives may be ordinary or partial derivatives. If this sounds to you like it opens up an enormous range of possibilities, you're right. We will not study "all" differential equations, and we will give more precise descriptions of the ones we do study. By the way, differential equations are studied because they occur very often, both in mathematics and in its applications. You will see some of these applications during the course of the book.

In this section we consider only the case in which there is one unknown function of one variable.

EXAMPLES
A. $\dfrac{d^2x}{dt^2} + 7x\dfrac{dx}{dt} - x = 55.$

B. $\dfrac{dx}{dt} = 0.$

C. $x' = 3t^2 x.$

In each of these differential equations, x is the unknown function of the variable t. In equation C it would really be better to write $x'(t) = 3t^2 x(t)$, but the shorter notation is more customary. ∎

---

[3] This definition, although standard, is not quite rigorous; however, it will do for our purposes. (See "Further Reading" after the Exercises.)

SEC. 1.2 SEPARABLE DIFFERENTIAL EQUATIONS

A function $x(t)$ that satisfies a differential equation is called a **solution** of that equation. For instance, it turns out that $x(t) = e^{t^3}$ is a solution of equation C. We can check this by computing the derivative: For $x = e^{t^3}$ we have

$$\begin{aligned} x' &= e^{t^3} \cdot 3t^2 \quad \text{(chain rule!)} \\ &= 3t^2 e^{t^3} \\ &= 3t^2 x, \end{aligned}$$

so the function $x(t) = e^{t^3}$ does indeed satisfy equation C.

To **solve** a differential equation means to find *all* the solutions of the equation, that is, all (differentiable) functions for which the equation is true. As we shall see, this is usually too much to hope for. However, certain simple types of differential equations can be solved, and their solution gives some insight into what happens in more general cases.

The simplest type of differential equation, which was touched upon in Section 1.1, is familiar from integral calculus. There one studies (whether one says so or not) equations of the type $\dfrac{dx}{dt} = f(t)$, where f is a given function of t. The solutions to this equation are the antiderivatives of f. $\int f(t)\, dt$ is the general solution; if $F(t)$ is any particular antiderivative of $f(t)$, then $\int f(t)\, dt = F(t) + C$.

EXAMPLE 1.2.1 Solve the differential equation $\dfrac{dx}{dt} = \dfrac{1}{1 + t^2}$.

Solution If $\dfrac{dx}{dt} = \dfrac{1}{1 + t^2}$, then $x = \displaystyle\int \dfrac{1}{1 + t^2}\, dt = \arctan t + C$ [4] for some constant C. ∎

EXAMPLE 1.2.2 Solve the differential equation $\dfrac{dx}{dt} = e^{-t^2}$.

Solution $x = \int e^{-t^2}\, dt$. It has to be admitted that this does not look "done," but it is as far as one can go; it has been shown[5] that it is impossible to write this integral in terms of familiar functions (without using an integral sign). However, it is certainly possible to get some idea of the behavior of specific solutions as functions of t. For example, the solution for which $x(0) = 0$ is $x(t) = \displaystyle\int_0^t e^{-s^2}\, ds$ (as we saw in Section 1.1), and one can find approximate values for $x(t)$ by interpreting the definite integral as an area and using approximation methods for the computation of area. ∎

We now give an example in which we solve a slightly more difficult equation.

---

[4] Many authors use $\sin^{-1}$, $\tan^{-1}$, and so on, for the inverse trigonometric functions; we use arcsin, arctan, and so on, instead.

[5] By the French mathematician Liouville (1809–1882).

EXAMPLE 1.2.3 Solve the differential equation $\dfrac{dx}{dt} = (1 + x^2) \cdot t$.

Solution First rewrite the equation as

$$\frac{1}{1+x^2} \frac{dx}{dt} = t. \tag{1}$$

Now notice that the left-hand side is the derivative, with respect to t, of $\arctan x$:

$$\frac{d}{dt}(\arctan x) = \frac{d(\arctan x)}{dx} \frac{dx}{dt} = \frac{1}{1+x^2} \frac{dx}{dt}.$$

Therefore, our equation can be rewritten $\dfrac{d(\arctan x)}{dt} = t$, and we can integrate this to obtain $\arctan x = \int t\, dt = \tfrac{1}{2}t^2 + C$, which then gives the answer

$$x = \tan\left(\frac{1}{2} t^2 + C\right).$$

As usual, C denotes an arbitrary constant. ∎

Example 1.2.3 may seem a bit contrived. What really happened, and how can one hope to find such a method? First of all, the point of rewriting the equation as (1) was that then the right-hand side depended only on t, while the left-hand side was (a function of x) times (the derivative of x).

Suppose that we have such a situation in general:

$$f(x) \cdot \frac{dx}{dt} = g(t). \tag{2}$$

[In Example 1.2.3, $f(x) = \dfrac{1}{1+x^2}$, and $g(t) = t$.] Then we can integrate with respect to t:

$\int f(x) \cdot \dfrac{dx}{dt}\, dt = \int g(t)\, dt$, and if we change variables in the integration on the left (as in the example, this is nothing but using the chain rule in reverse), we get

$$\int f(x)\, dx = \int g(t)\, dt. \tag{3}$$

If we can actually compute the integrals in this equation, we can then afterward try to find what x can be as a function of t. If we can't compute the integrals, we'll have to accept (3) as the solution to our equation.

We see, then, that any equation of the form (2) can be "solved" by integration. In fact, if you are comfortable with differentials, you can, with some hindsight, abbreviate the process as follows:

$$f(x) \frac{dx}{dt} = g(t). \tag{2}$$

SEC. 1.2 SEPARABLE DIFFERENTIAL EQUATIONS

"Multiply both sides by dt":

$$f(x)\, dx = g(t)\, dt\,.\quad {}^{6}$$

"Integrate both differentials":

$$\int f(x)\, dx = \int g(t)\, dt\,. \qquad (3)$$

A differential equation that is in the form (2): $f(x)\dfrac{dx}{dt} = g(t)$, or that can be rewritten in such a form, is called **separable**, because in the end result x and t have been "separated": In equation (3) the left-hand side involves only x, while the right-hand side involves only t.

EXAMPLE 1.2.4 $x^2 \dfrac{dx}{dt} = 8t^2$.

You might be able to guess one solution right away, without doing any integration; take a minute and look for one. (Give up?[7]) However, we can now do better and find *all* solutions:

$$\int x^2 \frac{dx}{dt}\, dt = \int 8t^2\, dt$$

$$\int x^2\, dx = \int 8t^2\, dt$$

$$\frac{1}{3}x^3 = \frac{8}{3}t^3 + C \quad {}^{8}$$

$$x^3 = 8t^3 + 3C$$

$$x = \sqrt[3]{8t^3 + 3C}\,.$$

(For $C = 0$, we get the solution you may have guessed, $x = 2t$.)

NOTES:

1. To check that the solutions $x = \sqrt[3]{8t^3 + 3C}$ do indeed satisfy $x^2 \dfrac{dx}{dt} = 8t^2$, you can just differentiate them (see Exercise 21). However, this check does not guarantee that the answer is complete; there might be other solutions.

2. Since C is an arbitrary constant, so is $3C$. Therefore, you could simplify the form of the solutions slightly by replacing $3C$ by C, to get $x = \sqrt[3]{8t^3 + C}$. ■

---

[6] At this point, when the equation has been written as $f(x)\, dx = g(t)\, dt$ or $f(x)\, dx - g(t)\, dt = 0$, it is said to be in **differential form**.

[7] How about $x = At$ for a suitable constant A?

[8] C is actually the difference of the two constants arising from the integrals on both sides.

EXAMPLE 1.2.5 $\dfrac{dx}{dt} = \dfrac{t}{x}$.

This equation is not yet in the form $f(x)\dfrac{dx}{dt} = g(t)$, but since it can be rewritten as $x\dfrac{dx}{dt} = t$ $(x \neq 0)$, it is separable and the solutions are given by

$$\int x\,dx = \int t\,dt$$
$$\frac{1}{2}x^2 = \frac{1}{2}t^2 + C$$
$$x = \pm\sqrt{t^2 + 2C}.$$

It is worth making a few observations about this answer. First of all, for each choice of the constant C there are two solutions. However, these are defined only when $t^2 + 2C > 0$. (If $t^2 + 2C < 0$, the square root cannot be taken; $t^2 + 2C = 0$ is also impossible, for this would give $x = 0$, which is impossible in the original equation $\dfrac{dx}{dt} = \dfrac{t}{x}$.) If $C > 0$, then $t^2 + 2C > 0$ for all t, so the solutions with $C > 0$ are defined for all t. If $C \leq 0$, the domain of the solutions is restricted to those values of t for which

$$t^2 + 2C > 0$$
$$t^2 > -2C$$
$$|t| > \sqrt{-2C}.$$

For example, for $C = -1$, we get the solutions $x = \pm\sqrt{t^2 - 2}$, which satisfy $\dfrac{dx}{dt} = \dfrac{t}{x}$ for t in the intervals $(-\infty, -\sqrt{2})$ and $(\sqrt{2}, \infty)$.

In Figure 1.2.1 graphs of the solutions are sketched for various values of C. ∎

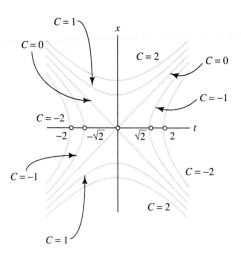

Figure 1.2.1. Solutions of $\dfrac{dx}{dt} = \dfrac{t}{x}$.

EXAMPLE 1.2.6

$x' = e^{x+t}$.

Since $e^{x+t} = e^x e^t$, we can rewrite the equation as $e^{-x} \dfrac{dx}{dt} = e^t$, so it is separable; the solution follows:

$$\int e^{-x}\, dx = \int e^t\, dt$$
$$-e^{-x} = e^t + C$$
$$x^{-x} = -e^t - C$$
$$-x = \log(-e^t - C) \quad ^9$$
$$x = -\log(-e^t - C).$$

NOTE: For certain values of C this solution does not exist; for instance for $C = 0$ we get $x = -\log(-e^t)$, and since $-e^t$ is always negative, this is undefined. In this way we can see that solutions exist only for $C < 0$, and that they are defined only for $t < \log(-C)$ [see Exercise 22(a)]. Figure 1.2.2 shows graphs of the solutions. ∎

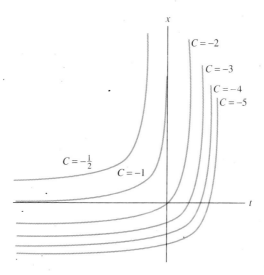

Figure 1.2.2. Solutions of $x' = e^{x+t}$.

You may have noticed that in all the examples we have solved so far, the solutions are "parametrized" by an integration constant C. That is, in these cases one can get all solutions by assigning different values to the constant C. As we will see, this situation often occurs for differential equations that are of "first order," that is, which only contain the first derivative (and no higher derivatives) of the unknown function. (A more complete "dictionary" of various kinds of differential equations will be given in Section 1.4.)

---

[9] In this book, the natural logarithm, to the base e, is denoted by log; so log $e^t = t$. (The "common" logarithm $\log_{10}$ is seldom used in mathematics at this level and beyond.)

Warning: Be sure to introduce the integration constant C immediately after integrating, but before solving for the unknown function! If you forget, you *cannot* salvage the final answer by simply adding "$+C$." For instance, in Example 1.2.6 the "answer" you'd get this way, $x = -\log(-e^t) + C$, is incorrect; as it happens, $-\log(-e^t) + C$ is never even defined.

In our final example, we will not actually be able to solve for the unknown function.

EXAMPLE 1.2.7 $(y^4 - 2)\dfrac{dy}{dx} = x^2$.

As you can see, this time y is the unknown function of x. The equation is separable; in fact, we can integrate immediately:

$$\int (y^4 - 2)\, dy = \int x^2 \, dx$$

$$\frac{1}{5}y^5 - 2y = \frac{1}{3}x^3 + C.$$

On the other hand, we can't solve this for y, so we have to leave the answer this way or in the alternative form

$$\frac{1}{5}y^5 - 2y - \frac{1}{3}x^3 = C.$$

What does this answer mean? Well, it means that *whenever* y is a differentiable function of x such that $\frac{1}{5}y^5 - 2y - \frac{1}{3}x^3 = C$ for some constant C, y will be a solution to our differential equation. (This can be checked by implicit differentiation; see Exercise 23.) See also Exercise 25. Figure 1.2.3 shows curves $\frac{1}{5}y^5 - 2y - \frac{1}{3}x^3 = C$ for various values of C; note

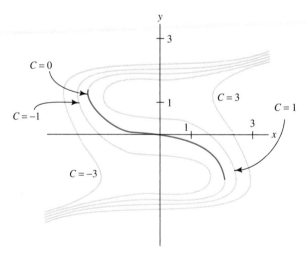

Figure 1.2.3. Solutions of $(y^4 - 2)\dfrac{dy}{dx} = x^2$.

SEC. 1.2 SEPARABLE DIFFERENTIAL EQUATIONS

that one such curve can consist of the graphs of several functions $y(x)$. For example, the "heavy" part of the curve $\frac{1}{5}y^5 - 2y - \frac{1}{3}x^3 = 0$ is the graph of a solution with $y(0) = 0$, and there are two other solutions [with $y(0) \approx 1.8$ and with $y(0) \approx -1.8$; see Exercise 27] whose graphs are part of that same curve. Note that the transitions from one solution to another along the curve occur at points where the tangent line is vertical and thus $\frac{dy}{dx}$ is undefined. ∎

SUMMARY OF KEY CONCEPTS AND TECHNIQUES

Differential equation (not the most general case) (p. 12): An equation involving one or more derivatives of an unknown function of one variable.

Solution of a differential equation (p. 13): A function that satisfies the equation.

Separable differential equation (p. 15): A differential equation that can be written in the form $f(x)\frac{dx}{dt} = g(t)$. To solve such an equation, first integrate both sides to get $\int f(x)\, dx = \int g(t)\, dt$. If possible, compute the integrals, then solve for x.

EXERCISES

For each of the following differential equations:

(a) Solve the equation.
(b) Discuss where the solutions are defined.

1. $x^2 \dfrac{dx}{dt} = t$.

2. $x^2 \dfrac{dx}{dt} = 2t - 1$.

3. $\dfrac{1}{\sqrt{1-x^2}} \dfrac{dx}{dt} = t - 1$.

4. $\dfrac{1}{\sqrt{4-x^2}} \dfrac{dx}{dt} = 3t^2 + 1$.

5. $y' = -y^2 - 1$. (y is a function of x.)

6. $x' = x^2 + 1$. (x is a function of t.)

7. $\dfrac{dx}{dt} = x^2 - t^2 \dfrac{dx}{dt}$.

8. $x^2 \dfrac{dy}{dx} = y^4 - \dfrac{dy}{dx}$.

9. $\dfrac{1}{x} \dfrac{dy}{dx} = e^{-x^2}$ $(x \neq 0)$.

10. $\dfrac{1}{t^2} \dfrac{dx}{dt} = e^{t^3}$ $(t \neq 0)$.

11. $\dfrac{dx}{dt} = \dfrac{1}{x^5}$.

12. $\dfrac{dy}{dx} = -\dfrac{6}{y^3}$.

13. $\dfrac{dx}{dt} = t^2 + t^2 x^2$.

14. $\dfrac{dx}{dt} = (x^2 + 1) \log t$ $(t > 0)$.

15. $x^2 \dfrac{dx}{dt} = \log t$ $(t > 0)$.

16. $x^4 \dfrac{dx}{dt} = 2\cos t + 3$.

17. $\dfrac{dx}{dt} = e^{3x-2t}$.

18. $\dfrac{dy}{dx} = e^{3x+4y}$.

19. $\dfrac{dx}{dt} = -\dfrac{t}{x}$.

20. $\dfrac{dx}{dt} = \dfrac{t+2}{x-1}$.

21. Read the first note following Example 1.2.4, and check by differentiating that the solutions $x = \sqrt[3]{8t^3 + 3C}$ found in that example are correct. Carry out a similar check for Example 1.2.5. It is good practice to do this (time permitting).

22. (See Example 1.2.6.)
    (a) Show that $x = -\log(-e^t - C)$ is defined only for $C < 0$, and then only for $t < \log(-C)$.
    (b) Check that for any $C < 0$, this function is a solution of $x' = e^{x+t}$.

23. (See Example 1.2.7.) Show by implicit differentiation that if $\tfrac{1}{5}y^5 - 2y - \tfrac{1}{3}x^3 = C$, then $(y^4 - 2)\dfrac{dy}{dx} = x^2$.

24. Solve the differential equation $(x^3 + 4)\dfrac{dx}{dt} = \cos t$, and check your answer by implicit differentiation.

25. (a) In Example 1.2.7 and Exercise 23 we've seen that if y is a differentiable function of x such that $\tfrac{1}{5}y^5 - 2y - \tfrac{1}{3}x^3 = C$, then $(y^4 - 2)\dfrac{dy}{dx} = x^2$. Now suppose, instead, that x is a differentiable function of y such that, again, $\tfrac{1}{5}y^5 - 2y - \tfrac{1}{3}x^3 = C$. Show that then $y^4 - 2 = x^2 \dfrac{dx}{dy}$.

    (b) Show that if $y^4 - 2 \neq 0$ and $x^2 \neq 0$, the two equations $(y^4 - 2)\dfrac{dy}{dx} = x^2$ and $y^4 - 2 = x^2 \dfrac{dx}{dy}$ from part (a) are equivalent.

    [Hint: What is the relation between $\dfrac{dy}{dx}$ and $\dfrac{dx}{dy}$?]

    (c) Write the two equations above in differential form (see p. 15, footnote 6). What do you notice?

26. Show that the two equations $f(x)\dfrac{dx}{dt} = g(t)$ and $f(x) = g(t)\dfrac{dt}{dx}$ have the same differential form (see p. 15, footnote 6), and that they are equivalent if $f(x) \neq 0$ and $g(t) \neq 0$.

27. As mentioned on p. 19 and illustrated in Figure 1.2.3, there are solutions of $(y^4 - 2)\dfrac{dy}{dx} = x^2$ with $y(0) \approx 1.8$ and with $y(0) \approx -1.8$ whose graphs are part of the curve $\tfrac{1}{5}y^5 - 2y - \tfrac{1}{3}x^3 = 0$. What are the precise values for $y(0)$ that are being approximated here by ± 1.8?

FURTHER READING

For an example of an equation that fits our description on p. 12 but is not considered a differential equation, see Arnold, *Ordinary Differential Equations* (MIT, 1973), p. 11, footnote.

For more on "functions that cannot be integrated" (such as e^{-t^2}), see Simmons, *Calculus with Analytic Geometry* (McGraw-Hill, 1985), Section 10.8.

1.3 APPLICATIONS: RADIOACTIVE DECAY AND POPULATION GROWTH

Our first application is probably familiar, but because the particular equation it gives rise to is important, we will study it rather carefully.

Consider the decay of a radioactive substance. It is known that the rate of decay of such a substance is proportional to the amount that is present.[10] If we denote the amount of the substance by x and the time by t, this translates to: $\frac{dx}{dt}$ is proportional to x. (Keep in mind that $\frac{dx}{dt}$ is the rate of change of x; see Section 1.1, Fact 2.) In practice, the constant of proportionality may or may not be known; let us call it K. Then we have $\frac{dx}{dt} = K \cdot x$, where K is some negative number. (K is negative because the substance is decaying, which implies that $\frac{dx}{dt} < 0$.) In this equation, x is the unknown function of t. The equation is separable:

$$\frac{dx}{dt} = Kx$$

---

[10] This is thoroughly documented by experimental evidence. It is also intuitively reasonable, since it seems likely that each molecule or atom of the substance has an equal probability, which is independent of what happens to the rest of the substance, of decaying within a given time period.

$$\frac{1}{x}\frac{dx}{dt} = K$$

$$\int \frac{dx}{x} = \int K\,dt$$

$$\log x = Kt + C$$

$$x(t) = e^{Kt} + C.$$

So we find that the amount of the substance at time t is e^{Kt+C}, where K is the (negative) factor of proportionality and C is some constant. In particular, for $t = 0$ we find that $x(0) = e^C$. Therefore, e^C is the amount that is present initially. Using this, we can rewrite the solution as

$$x = e^{Kt+C} = e^{Kt}e^C = e^{Kt}x(0) \quad \text{or} \quad x(t) = x(0)e^{Kt}.$$

Given a specified amount $x(0)$ to start with, the solution will be the one for which $C = \log(x(0))$.

EXAMPLE 1.3.1 Two grams of a radioactive substance are left to decay; it is observed that after a year 1.4 grams are left. What will the amount be after (a total time of) 10 years? What was it after half a year?

Solution Let t denote the time in years, and let x denote the amount of the substance in grams. Then, by the above, $x(t) = x(0)e^{Kt}$. We are not given the constant K, but we do know that $x(0) = 2$, $x(1) = 1.4$. From this we see that $1.4 = x(1) = 2e^K$, and thus $K = \log 0.7$ (≈ -0.357); in particular,

$$x(10) = 2 \cdot e^{10K} = 2 \cdot (0.7)^{10} \approx 0.056$$

and

$$x\left(\frac{1}{2}\right) = 2 \cdot e^{K/2} = 2 \cdot (0.7)^{1/2} \approx 1.673.$$

The amounts are approximately 0.056 and 1.673 grams, respectively. ∎

Let us look again at the equation $\dfrac{dx}{dt} = K \cdot x$. In solving this equation, we used two steps that are questionable. To begin with, we divided by x, which requires that $x \neq 0$. Then we wrote $\int \dfrac{dx}{x} = \log x$, which is legitimate only if $x > 0$. *In our application*, since x represents the amount of the substance under study, these assumptions are reasonable. However, we will encounter the same equation in different situations, so it will be better to solve it in complete generality.

SEC. 1.3 APPLICATIONS: RADIOACTIVE DECAY AND POPULATION GROWTH

Suppose, then, that K is a given constant (positive, negative, or zero) and we wish to solve the equation $\frac{dx}{dt} = K \cdot x$. First we consider the case $x = 0$; clearly, the constant function $x(t) = 0$ is a solution. Then, if $x \neq 0$, we separate the equation:

$$\frac{dx}{dt} = Kx$$

$$\frac{1}{x}\frac{dx}{dt} = K$$

$$\int \frac{dx}{x} = \int K\, dt = Kt + C.$$

Now, if $x > 0$, then $\int \frac{dx}{x} = \log x$, and as earlier we find $x = e^{Kt+C}$. On the other hand, if $x < 0$, then $\int \frac{dx}{x} = \int \frac{d(-x)}{-x} = \log(-x)$, and we get

$$\log(-x) = Kt + C$$
$$x = -e^{Kt+C}.$$

Thus the solutions are

$$\begin{cases} (x > 0) & x(t) = e^{Kt+C} \\ (x = 0) & x(t) = 0 \\ (x < 0) & x(t) = -e^{Kt+C}. \end{cases} \quad (*)$$

NOTES:

1. You may wonder whether there might not be a solution that is positive for some values of t and zero or negative for others. One can see that this is impossible here by using continuity and carefully studying the argument above, or by using the alternative argument given in Exercise 7.

2. Many authors write

$$\int \frac{dx}{x} = \log|x|(+ \text{ constant}) \quad \text{as a shorthand for} \quad \int \frac{dx}{x} = \begin{cases} \log x & (x > 0) \\ \log(-x) & (x < 0). \end{cases}$$

This shorthand has its dangers (see Exercise 22), but it does save space. In the computation above it yields (for $x \neq 0$) $\log|x| = Kt + C$, $|x| = e^{Kt+C}$, and thus $x = \pm e^{Kt+C}$.

Returning to $(*)$, if we look at the values of the solutions for $t = 0$, we find that

$$\begin{cases} (x > 0) & x(0) = e^C \\ (x = 0) & x(0) = 0 \\ (x < 0) & x(0) = -e^C. \end{cases}$$

Therefore, we have
$$x(t) = x(0)e^{Kt}$$
(which we had seen earlier for the case $x > 0$) for *all* the solutions (∗). Also, $x(0)$ can have *any* real value A: To get $A > 0$, take $x > 0$ and $C = \log A$; to get $A = 0$, take $x = 0$; to get $A < 0$, take $x < 0$ and $C = \log(-A)$. [11] In summary, we have:

The general solution of the differential equation $\dfrac{dx}{dt} = Kx$**, where K is a constant, is given by** $x(t) = Ae^{Kt}$**. The arbitrary constant A represents the "initial value"** $x(0)$.

Turning to another application, we now consider the growth of a given population (of humans, animals, or plants). Of course, in reality the number x of individuals comprising the population is always an integer; this integer will be a discontinuous function of time, "jumping" each time a birth or death occurs. (There is a similar phenomenon for the amount of a radioactive substance, considered on the atomic or molecular level!) However, if the population is sufficiently large, one can for practical purposes ignore this fact and pretend that x is a differentiable function of the time.

The growth (or decline) of the population is measured by the **excess birth rate** (the difference between the birth and death rates), sometimes called the "birth rate" for short. This is *not* the rate of change of the number x; it is, rather, that rate of change *divided* by the total size x of the population. (For instance, the excess birth rate might be 11 individuals per 1000 per year.) In other words, $x(t)$ indicates the size of the population as a function of time, and

$$\frac{1}{x}\frac{dx}{dt} = \text{excess birth rate} . \tag{1}$$

One can try to determine the excess birth rate, as a function of time, experimentally; alternatively, one can make a **model**: One can assume that the excess birth rate has a certain form, solve equation (1) using that assumption, and afterward check the predictions made in this way about the behavior of $x(t)$ against reality.

The simplest possible model (and, fortunately, a highly inaccurate one under most circumstances) is the one in which the excess birth rate is taken to be constant:

$$\frac{1}{x}\frac{dx}{dt} = K, \quad K \text{ a (positive) constant} .$$

This equation was considered above, and we found
$$x(t) = x(0) \cdot e^{Kt}$$

---

[11] Of course, you could also check directly that for any constant A, $x(t) = Ae^{Kt}$ is a solution to $\dfrac{dx}{dt} = Kx$.

for the solutions. Here $x(0)$ is the **initial population size**. In other words: A population with a constant excess birth rate will increase exponentially in size. (Of course, if the excess birth rate is *negative*, there will be an exponential decrease.)

Clearly, in practice an exponential increase could only be sustained for a limited time; eventually lack of resources, overcrowding, and similar factors would reduce the excess birth rate.[12] More realistic models, which are still workable, are obtained by assuming that the excess birth rate depends on the population size x but not on anything else. (Naturally, this is not quite true; the excess birth rate will actually depend on external time-dependent circumstances, such as predator population or food supply, as well as on x.) In this way one obtains equations of the form $\frac{1}{x}\frac{dx}{dt} = f(x)$. Regardless of what function $f(x)$ is chosen to represent the excess birth rate, the resulting equation is still separable:

$$\frac{1}{x}\frac{dx}{dt} = f(x)$$

$$\frac{1}{x \cdot f(x)}\frac{dx}{dt} = 1$$

$$\int \frac{dx}{x \cdot f(x)} = \int dt = t + C.$$

One can now try to compute the integral on the left and then try to find x as a function of t. Note that during this computation we assumed that $f(x) \neq 0$. If $f(x) = 0$ for a particular value of x, say $f(x_0) = 0$ with $x_0 > 0$, then the constant function $x(t) = x_0$ will be a solution to our equation $\frac{1}{x}\frac{dx}{dt} = f(x)$ (see Exercise 16). This solution corresponds to a **steady-state** or **equilibrium** population x_0 (which does not actually vary with t).

SUMMARY OF KEY CONCEPTS AND RESULTS

Radioactive decay equation (p. 21): $\frac{dx}{dt} = Kx$, with $K < 0$. This equation is separable.

The same equation for $K > 0$ describes exponential population growth.

General solution of $\frac{dx}{dt} = Kx$ (p. 24): $x(t) = Ae^{Kt}$.   N O T E : $A = x(0)$.

General **population growth equation** (p. 25): $\frac{1}{x}\frac{dx}{dt} = f(x)$, where $f(x)$ represents the **excess birth rate**. This equation is always separable. If $f(x_0) = 0$ with $x_0 > 0$, then x_0 is a **steady-state** or **equilibrium** population.

[12] In fact, this objection applies to any model which predicts that $x(t) \to \infty$ as $t \to \infty$.

EXERCISES

1. Five kilograms of a radioactive substance have decayed to 3 kilograms in 2 years' time. What will the amount be after (a total of) 7 years? How soon will only 1 kilogram be left? When will all of the substance have disappeared?

2. A meteorite was found 50 years ago. At that time, it contained 0.8 gram of a certain radioactive substance; it now contains 0.7 gram of the same substance. If the meteorite landed 500 years ago, how much of the substance did it contain then? When will it no longer contain any of the substance?

3. In half a year, 2 kilograms of a radioactive substance will decay to 1 kilogram. How long will it take 6 kilograms of the same substance to decay to 3 kilograms? And 6 pounds to 3 pounds? (1 pound $\approx$ 0.45 kilogram.)

4. The time required for a radioactive substance to decay to half the original amount is called the **half-life** of the substance. (Half-lives are listed in physical tables; they vary widely between substances, from fractions of a millisecond to millions of years.) Show that the half-life of a given substance is indeed a constant, that is, that it doesn't depend on the original amount of the substance.

5. Find the solution to the differential equation $\dfrac{dx}{dt} = 3x$ that satisfies $x(2) = -5$.

6. Find the solution to $\dfrac{dy}{dx} = -2y$ for which $y(1) = -6$.

7. Suppose that $x(t)$ is a solution to the differential equation $\dfrac{dx}{dt} = Kx$, where K is a constant. Show directly (without using the results of the text) that $x(t)e^{-Kt}$ is a constant, by computing $\dfrac{d}{dt}(x(t)e^{-Kt})$. Call this constant A. Show that $x(t) = Ae^{Kt}$ and that $A = x(0)$.

8. Suppose that someone (the *Enterprise*?) found a "radio-hyperactive" substance, for which the rate of decay was proportional to the *square* of the amount present. Show that the amount of this substance at time t would be given by a function of the form $x = \dfrac{1}{-Kt - C}$ for suitable constants K and C. Show that given a specified amount $x(0)$ to start with, the solution would be the one for which $C = -[x(0)]^{-1}$.

9. The population of the United States in 1900 numbered approximately 76 million; in 1920, approximately 106 million. What would the population have been in the year $1900 + t$ if the excess birth rate were constant? Specifically, in 1940 and in 1960? (Actual figures: 1940, 132 million; 1960, 179 million.)

10. The population of the state of Alabama increased from (approximately) 1000 in 1800 to 9000 in 1810.

SEC. 1.3 APPLICATIONS: RADIOACTIVE DECAY AND POPULATION GROWTH

(a) What would the population have been in 1820, assuming exponential growth? And in the year $1800 + t$?

(b) Surprise: Actual 1820 population, 128,000. What could have happened?

11. Census figures for the city of Apolis and its suburbs show the following populations:

	City	Suburbs
1970	1,200,000	800,000
1980	1,050,000	1,150,000
1990	900,000	1,520,000

(a) Is the suburban population growing exponentially?

(b) Is the city population declining exponentially?

(c) Can you propose a model for what is happening in the Apolis area that is consistent with the given figures?

12. Suppose that the excess birth rate for a certain population is $f(x) = \dfrac{k}{\sqrt{x}}$, where k is a positive constant.

(a) Show that the population will increase "quadratically," with $x = \dfrac{1}{4}(kt + C)^2$ for some constant C.

(b) Suppose that the population was 2500 for $t = 0$ and had doubled (to 5000) for $t = 5$. When will the population have doubled again?

13. A mine produces a radioactive mineral at a constant rate of 25 kilograms per year. This mineral is not used, but stockpiled, and it decays at such a rate that an initial amount of 5 kilograms, if not replenished, would reduce to 3 kilograms in two years. Starting with a zero stockpile, how large will the stockpile be after one year? After t years?
[*Hint:* If x is the amount of the mineral in kilograms and t is the time in years, $\dfrac{dx}{dt} = 25 + kx$. (Explain!) This differential equation is separable; k can be found as in Exercise 1.]

14. An old stockpile of the same radioactive mineral that haunted Exercises 1 and 13 is being depleted (for industrial use) at a constant rate of 4 kilograms per year. In addition, of course, it is decaying naturally. How long will the stockpile last, if it originally contained 100 kilograms?

15. Suppose that the value of your car declines at a rate proportional to its present value. If the car was bought for $11,000 and if a year later it was worth $8500, at what point should you sell the car to a junk dealer who pays a flat $300? Do you think the model we have made here is realistic?

16. Show that if $f(x_0) = 0$ and $x_0 \neq 0$, then $x(t) = x_0$ is a solution to $\dfrac{1}{x}\dfrac{dx}{dt} = f(x)$.

When interest on money in an account is "compounded continuously," the rate of growth of the amount of money is taken to be proportional to the amount. The proportionality constant is called the **interest rate** and is usually expressed as a percentage (per year).

17. (a) Set up a differential equation describing the growth of the amount x of money in an account that earns 5% interest, compounded continuously.
    (b) Suppose that you open the account in part (a) with an initial deposit of $1000. How much money will be in the account two years later?
    (c) By what percentage does the amount of money in the account in part (a) increase each year? [This is called the **annual percentage rate** (APR) of the account.]

18. Bank A offers 6% interest, compounded continuously. Bank B offers 6.125% interest, compounded once a year. Other things being equal, which bank should get your money?

19. Does the answer to Exercise 18 change if the interest paid by bank B is compounded twice a year?

20. Suppose that the excess birth rate for a certain population is $f(x) = 6 + x - x^2$.
    (a) Are there equilibrium values for the population? If so, what are they?
    (b) For x in what intervals will the population increase?
    (c) What do you think will happen to the population as $t \to \infty$? Does the answer depend on the initial size of the population? (*Do not* solve the differential equation!)

*21. The **Verhulst model** of population growth, named after a Belgian statistician who used it to fit census figures, is obtained by taking the excess birth rate function to be of the form $f(x) = K(M - x)$. Here K and M are positive constants. In particular, $f(x)$ is positive for $x < M$, zero for $x = M$, and negative for $x > M$. (Intuitively, M is the population level where growth stops.)
    (a) Find x as a function of t in this model.
    [*Hint*: To perform the integration, use the method of partial fractions from calculus.]
    (b) Show that as $t \to \infty$, x approaches M, without ever reaching M. The one exception occurs when the initial population size $x(0)$ equals M. What happens in that case?

22. Explain what is wrong with the following computation.

$$\int \frac{(2x - 5)\, dx}{x^2 - 5x + 4} = \int \frac{du}{u} \quad \text{(using the substitution } u = x^2 - 5x + 4\text{)}$$
$$= \ln |u| + C$$
$$= \ln |x^2 - 5x + 4| + C;$$

therefore,

$$\int_2^5 \frac{(2x - 5)\, dx}{x^2 - 5x + 4} = \ln |x^2 - 5x + 4| \Big]_{x=2}^5$$
$$= \ln |4| - \ln |-2| = \ln 4 - \ln 2 = \ln 2.$$

FURTHER READING

For an account of how the solutions $x(t) = x(0)e^{Kt}$ to the radioactive decay equation are used in archaeology, see Bowman, *Radiocarbon Dating* (University of California Press/British Museum, 1990).

1.4 CLASSIFICATION OF DIFFERENTIAL EQUATIONS

Now that we have seen examples of particular differential equations and their solutions, we will look briefly at differential equations in general. We'll show how to classify differential equations into various types according to their "appearance." This classification is important because solution methods and properties vary considerably from one type of differential equation to the next. To illustrate the classification, we'll use the following examples.

EXAMPLES

A. $y \dfrac{dy}{dx} - 3xy + x = 0.$

B. $\dfrac{d^2x}{dt^2} + 7x \dfrac{dx}{dt} - x = 55.$

C. $\dfrac{d^2x}{dt^2} + 7x \dfrac{dx}{dt} - x = t^2.$

D. $\dfrac{y'' + y'}{y'' - y^3} = \sqrt{x^2 + 3x + 5}.$

E. $y^{(4)} - 10y^{(3)} + 5y'' - 2y' + y = 1.$ [13] (y is a function of x.)

---

[13] $y^{(3)}$ and $y^{(4)}$ denote the third and fourth derivatives of y, in that order; in general, $f^{(n)}$ denotes the nth derivative of the function f.

F. $\dfrac{du}{dx} = 0$.

G. $\dfrac{\partial u}{\partial x} = 0$. ($u$ is an unknown function of x and y.)

H. $\dfrac{\partial u}{\partial x} + 2\dfrac{\partial u}{\partial y} - u = x^2 y$.

I. $\dfrac{dx}{dt} + \dfrac{dy}{dt} + 2x - y = 0$. ($x$ and y are unknown functions of t.)

J. $\dfrac{\partial^2 u}{\partial x\,\partial y} - \left(\dfrac{\partial u}{\partial t}\right)^2 + 3u = 1$. ($u$ is a function of t, x, and y.) ∎

NOTE: Usually it is clear from the equation what the variable(s) for the unknown function(s) is (or are). Not always, though; see equations E and G! If confusion threatens, it is a good idea to list the variable(s) explicitly.

Now we will start classifying.

Definition   The **order** of a differential equation is the order of the highest derivative (of an unknown function) that occurs in the equation. A differential equation of order n is usually called an nth-order differential equation.

In other words: If only first derivatives (or first partials) occur in the equation, as in equations A, F, G, H, and I above, it is called a *first-order* differential equation. If only first and second derivatives occur, we have a *second-order* differential equation. Examples are equations B, C, D, and J. If higher derivatives occur, we have a higher-order equation, such as the fourth-order differential equation E.

First- and second-order differential equations are not only more easily studied than higher-order ones, but they also turn up more frequently in applications. (This is partly because problems are simplified whenever possible so that mathematicians have a chance.) We will concentrate on first- and second-order differential equations except in Chapter 4.

Definition   An **ordinary** differential equation (ODE) is one in which all the unknown functions are functions of one (and the same) variable. A **partial** differential equation (PDE) is a differential equation that is not an ordinary differential equation.

Equations G, H, and J above are partial differential equations; all the others are ODEs. For a given differential equation, the distinction is usually very clear-cut: If partial derivatives are present, it is a PDE; otherwise, it is an ordinary differential equation.

The use of the word *ordinary* is rooted in tradition, so we are stuck with, and will stick to, it. It would make more sense to speak of differential equations in one variable, as opposed to ("partial") differential equations in several variables. In fact, many, if not most, of the differential equations that arise in applications—for example, those describing heat flow and wave propagation—are partial differential equations. This is to be expected, since there are usually several independent variables in any physical situation. Nevertheless, it is a good idea to study ordinary differential equations first, since ODEs are less complicated and thus their properties are more easily understood. Also, solutions of PDEs are often constructed from the solutions of related ODEs, as we will see in Section 7.4.

Most of this book, then, will be about ODEs, and until Section 7.4 we will simply say "differential equation" instead of "ordinary differential equation." In such a differential equation there may be several unknown functions, as we see in equation I above: The first-order differential equation $\frac{dx}{dt} + \frac{dy}{dt} + 2x - y = 0$ has two unknown functions x and y. Differential equations with several unknown functions, like equations with several unknowns in high-school algebra, usually occur in systems. We will start the study of systems of differential equations in Chapter 3. Until then, we consider only differential equations with one unknown function (of one variable).

There is another important classification to be made in this section. Suppose that in a given nth-order differential equation x is the unknown function of the variable t. Then the equation involves some (or all) of the following: t, x, $\frac{dx}{dt}$, $\frac{d^2x}{dt^2}$, ..., $\frac{d^nx}{dt^n}$, but no other derivatives. By bringing everything to the left-hand side of the equation, it can be rewritten as

$$E\left(t, x, \frac{dx}{dt}, \frac{d^2x}{dt^2}, \ldots, \frac{d^nx}{dt^n}\right) = 0,$$

where E is some expression.

Definition The equation $E(t, x, \frac{dx}{dt}, \frac{d^2x}{dt^2}, \ldots, \frac{d^nx}{dt^n}) = 0$ is called a **linear** differential equation (of order n) if E is a linear expression in the unknown function and its derivatives, that is, if E is linear in x, $\frac{dx}{dt}$, ..., $\frac{d^nx}{dt^n}$. (NOTE: E does *not* have to be linear in the variable, t, as well.) Otherwise, the equation is called **nonlinear**.

In other words, if the differential equation can be rewritten as

$$f_n(t)\frac{d^nx}{dt^n} + f_{n-1}(t)\frac{d^{n-1}x}{dt^{n-1}} + \cdots + f_1(t)\frac{dx}{dt} + f_0(t)x = g(t)$$

for suitable functions $f_n(t)$, $f_{n-1}(t)$, ..., $f_1(t)$, $f_0(t)$, $g(t)$, then it is linear. The functions $f_n(t)$, ..., $f_0(t)$ are called the **coefficients** of the linear differential equation. The function $g(t)$ is often called the **forcing term** (we'll soon see why). In the special case when $g(t) = 0$ (for all t) the linear differential equation is called **homogeneous**[14]; otherwise, it is called **inhomogeneous** or nonhomogeneous.

---

[14] Pronounced *ho-mo-gee-nee-us*; not the same word as for milk!

To get used to these definitions, let's see what happens for first- and second-order differential equations. For a first-order equation, $n = 1$, so a linear first-order equation for the function $x(t)$ looks like

$$f_1(t)\frac{dx}{dt} + f_0(t)x = g(t).$$

For example, $(t^2 - 1)\frac{dx}{dt} + (\sin t + 2)x = 0$ is a homogeneous first-order linear equation, and $(\cos t)\frac{dx}{dt} + 3x + t + 5 = 0$ is an inhomogeneous one. [For the first equation, $f_1(t) = t^2 - 1$, $f_0(t) = \sin t + 2$, $g(t) = 0$, while for the second one, $f_1(t) = \cos t$, $f_0(t) = 3$, $g(t) = -t - 5$.] On the other hand, $x\frac{dx}{dt} + 3x + t + 5 = 0$ is *not* a linear equation, because of the term $x\frac{dx}{dt}$. Similarly, $\frac{dx}{dt} - x^2 = 0$ is not linear, because of the term x^2. In Section 1.6 we will see how to solve *any* linear first-order equation.

For a second-order equation, $n = 2$, so the general form of a linear second-order equation for $x(t)$ is

$$f_2(t)\frac{d^2x}{dt^2} + f_1(t)\frac{dx}{dt} + f_0(t)x = g(t).$$

For instance, $(t - 2)\frac{d^2x}{dt^2} + \frac{dx}{dt} + (t^3 + 1)x + \cos t = 0$ is (inhomogeneous) linear while $\frac{d^2x}{dt^2} + (\frac{dx}{dt})^2 + x = 0$ is nonlinear. We will consider second-order linear equations in Chapter 2. In particular, we will see that if a particle on a spring is subject to an external force $F(t)$, then the position $x(t)$ of the particle satisfies (at least approximately) the equation

$$m\frac{d^2x}{dt^2} + \gamma\frac{dx}{dt} + kx = F(t),$$

where m, γ, and k are constants. Note that this is a linear second-order equation, with $f_2(t) = m$, $f_1(t) = \gamma$, $f_0(t) = k$, $g(t) = F(t)$. The fact that in this case, and similar cases, $g(t)$ is an external force explains why $g(t)$ is called the forcing term.

Of course, when the unknown function has a name other than x, or the variable is not t, the definitions are adjusted accordingly. For instance, $\frac{d^2y}{dx^2} + 7y\frac{dy}{dx} - y = 55$ is a nonlinear second-order equation (because of the term with $y\frac{dy}{dx}$), while $\frac{d^2y}{dx^2} + 7x\frac{dy}{dx} - y = x^2$ is a linear second-order equation, since it is linear in the unknown function and its derivatives.

When making a model of some real-life phenomenon, it is often a good idea to keep the differential equation(s) as close to linear as possible. This is because, as we will see, linear differential equations have special properties that make them easier to solve. To find

SEC. 1.4 CLASSIFICATION OF DIFFERENTIAL EQUATIONS

information about solutions of any nonlinear equations in the model, one can approximate these by linear equations. We will return to this in Chapter 3.

Figure 1.4.1 indicates some of the classifications that we have made so far.

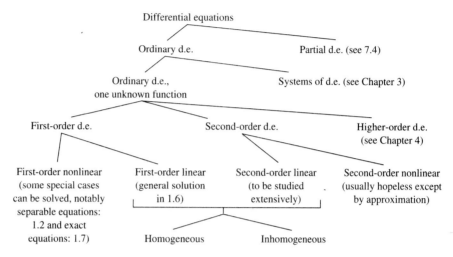

Figure 1.4.1. Classification of differential equations.

SUMMARY OF KEY CONCEPTS

Order of a differential equation (p. 30): Order of the highest derivative that occurs.

Ordinary differential equation (ODE; p. 30): One in which all unknown functions are functions of one (and the same) variable. All other differential equations are **partial** differential equations (PDEs).

nth-order linear differential equation (p. 31):

$$f_n(t)\frac{d^n x}{dt^n} + f_{n-1}(t)\frac{d^{n-1}x}{dt^{n-1}} + \cdots + f_1(t)\frac{dx}{dt} + f_0(t)x = g(t).$$

$f_n(t), \ldots, f_0(t)$ are called the **coefficients**; $g(t)$ is the **forcing term**. The equation is **homogeneous** if $g(t) = 0$ and **inhomogeneous** otherwise.

EXERCISES

For each of the following differential equations, state the order of the equation and whether it is an ODE or a PDE. If it is an ODE, state whether it is linear or not; if it is linear, whether it is homogeneous.

1. $\dfrac{\partial^2 u}{\partial x\, \partial y} + \left(\dfrac{\partial u}{\partial x}\right)^3 = \sin(x + y).$

2. $\dfrac{dx}{dt} - 3x = \cos t.$

3. $\dfrac{dx}{dt} - 3t = \cos x.$

4. $\dfrac{\partial u}{\partial x} + \dfrac{\partial u}{\partial y} = e^u.$

5. $\dfrac{d^2x}{dt^2} + t\dfrac{dx}{dt} + t^2 x = 0.$

6. $\left(\dfrac{dy}{dx}\right)^2 + x\dfrac{dy}{dx} + x^2 y = 0.$

7. $\dfrac{d^2y}{dx^2} + x\dfrac{dy}{dx} + xy^2 = 0.$

8. $\dfrac{d^2x}{dt^2} + x\dfrac{dx}{dt} + xt^2 = 0.$

9. $y'' - 3y' + y = \sin(x + y).$

10. $x^{(3)} - 2x'' + tx' = \sin t.$

11. $x^{(3)} - 2x'' + tx' = 0.$

12. $y^{(3)} - 2y'' - \sin x = 0.$

13. $y^{(3)} - 2y'' = -xy'.$

*14. Show that if $x_1(t)$ and $x_2(t)$ are solutions of the same homogeneous linear equation

$$f_n(t)\dfrac{d^n x}{dt^n} + \cdots + f_1(t)\dfrac{dx}{dt} + f_0(t)x = 0,$$

then for any constants C_1 and C_2, $x(t) = C_1 x_1(t) + C_2 x_2(t)$ is also a solution.

1.5 INITIAL VALUE PROBLEMS AND INTEGRAL CURVES

In this section we discuss first-order differential equations in general, and we will see what should be expected of their solutions. The next section will treat the particular case of linear first-order equations; these can be solved completely.

For now, let's assume that $x(t)$ is the unknown function, so our equations will be of the form $E(t, x, \dfrac{dx}{dt}) = 0$. First of all, it will be desirable to write this in such a way as to "display" the derivative $\dfrac{dx}{dt}$.

EXAMPLES A. $t\dfrac{dx}{dt} = x^3 - t^2 + x\dfrac{dx}{dt}.$ Rewrite this as

$$(t - x)\dfrac{dx}{dt} = x^3 - t^2$$

$$\dfrac{dx}{dt} = \dfrac{x^3 - t^2}{t - x} \qquad (t - x \ne 0).$$

(The case $t - x = 0$ should be considered separately. It yields $x = t$, which is not a solution to the original equation.)

SEC. 1.5 INITIAL VALUE PROBLEMS AND INTEGRAL CURVES

B. $\left(\dfrac{dx}{dt}\right)^2 + x\dfrac{dx}{dt} + 1 = 0.$ There are two possibilities:

$$\dfrac{dx}{dt} = \dfrac{-x + \sqrt{x^2 - 4}}{2} \quad \text{and} \quad \dfrac{dx}{dt} = \dfrac{-x - \sqrt{x^2 - 4}}{2}.$$

Consider these two equations one at a time. (In this case, they are both separable, so they can be solved by the method of Section 1.2.)

C. $\sin\left(\dfrac{dx}{dt}\right) - x\dfrac{dx}{dt} = t^2 - 1.$ This case seems hopeless, since you cannot expect to find an expression for $\dfrac{dx}{dt}$ in terms of x and t. ∎

There is a theorem in analysis, the so-called implicit function theorem, which shows that under reasonable conditions, any equation $E(t, x, \dfrac{dx}{dt}) = 0$ is equivalent to one or more equations of the form $\dfrac{dx}{dt} = f(t, x)$. It may be necessary to treat one or two "exceptional" cases separately (such as $t - x = 0$ in A above).

Unfortunately, the theorem does not indicate a way to find $f(t, x)$ explicitly, so in practice cases such as C above remain hopeless. We will therefore assume that our equation has been converted successfully to the form $\dfrac{dx}{dt} = f(t, x)$. [In A above, we have $f(t, x) = \dfrac{x^3 - t^2}{t - x}$; in B above, there are two equations, with $f(t, x) = \dfrac{-x + \sqrt{x^2 - 4}}{2}$ and $f(t, x) = \dfrac{-x - \sqrt{x^2 - 4}}{2}$, respectively.] In applications where t is the time, this means that the rate of change, $\dfrac{dx}{dt}$, of x is given in terms of x itself and the time t. Suppose that you are also given $x(0)$; that is, you are given what x is to begin with. It is then intuitively reasonable that $x(t)$ should be determined for all t, or at least for all $t \geq 0$ (the future): You know what x is to begin with, and at any given time, once you know what x is, you know how fast x is changing, which should tell you what x will be "at the next instant," and so on.

EXAMPLE Let us return to the first application from Section 1.3. $x(t)$ denotes the amount of a radioactive substance; we have $\dfrac{dx}{dt} = Kx$, K a given negative constant. As we have seen, the solutions to this differential equation are given by $x(t) = Ae^{Kt}$ with $A = x(0)$, so if you are given $x(0)$, you do indeed know what $x(t)$ will be any time in the future [as well as what $x(t)$ was at any time in the past]. Figure 1.5.1 illustrates this. ∎

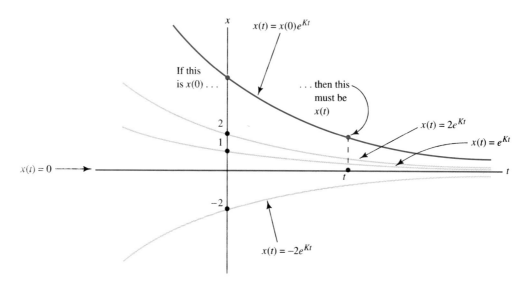

Figure 1.5.1. Solutions of $\frac{dx}{dt} = Kx$ ($K < 0$), the equation describing radioactive decay. Note how $x(0)$ determines $x(t)$ for all t. Only the solutions with $x(0) \geq 0$ have physical meaning.

A problem in which one is given both a differential equation and additional information about the value of the unknown function or its derivatives at specific points is called a **boundary value problem**. The special kind of boundary value problem considered here, where all the information given is for $t = 0$, is called an **initial value problem**. We will see more general cases of such problems later. For now, let us continue to study the case in which we have one **initial condition**, say $x(0) = x_0$, giving $x(0)$, along with the differential equation:

$$\frac{dx}{dt} = f(t, x), \qquad x(0) = x_0.$$

We have said above that it is intuitively reasonable that such an initial value problem should have only one solution. Unfortunately, our intuitive argument, although appealing, is too naive, and in some situations the conclusion is actually wrong.

EXAMPLE 1.5.1 $\frac{dx}{dt} = x^{1/3} (= \sqrt[3]{x}),\ x(0) = 0.$

One solution to this initial value problem can be seen right away: The zero function, $x(t) = 0$, certainly satisfies $\frac{dx}{dt} = x^{1/3}$ and $x(0) = 0$.

SEC. 1.5 INITIAL VALUE PROBLEMS AND INTEGRAL CURVES

To see whether there are any other solutions, we can solve the equation $\frac{dx}{dt} = x^{1/3}$ systematically. It is a separable equation, and for $x \neq 0$ we obtain

$$\int \frac{dx}{x^{1/3}} = \int dt$$

$$\frac{3}{2} x^{2/3} = t + C$$

$$x^2 = \left[\frac{2}{3}(t + C)\right]^3$$

$$x = \pm \left[\frac{2}{3}(t + C)\right]^{3/2}.$$

We see that, just as in Example 1.2.5, there are two solutions here for each choice of the arbitrary constant C. Let's look for a moment at what happens for one particular value of C, say $C = 1$. Our two solutions $x = \pm[\frac{2}{3}(t+1)]^{3/2}$ are only defined for $t \geq -1$. However, for $t = -1$, these solutions are zero, so if we remember that the zero function is also a solution to our equation, we can "piece together" solutions that are defined for all t:

$$x(t) = \begin{cases} 0 & (t \leq -1) \\ \left[\frac{2}{3}(t+1)\right]^{3/2} & (t \geq -1) \end{cases} \quad \text{and} \quad x(t) = \begin{cases} 0 & (t \leq -1) \\ -\left[\frac{2}{3}(t+1)\right]^{3/2} & (t \geq -1) \end{cases}.$$

These "pieced together" solutions are not only continuous, but also differentiable for all t, since the (right) derivative of $[\frac{2}{3}(t+1)]^{3/2}$ at $t = -1$ is zero. Therefore, they really satisfy the equation for all t, including $t = -1$.

There is nothing special about the choice of $C = 1$. Accordingly, we find the solutions

$$x(t) = 0 \, ; \quad x(t) = \begin{cases} 0 & (t \leq -C) \\ \left[\frac{2}{3}(t+C)\right]^{3/2} & (t \geq -C) \end{cases} ; \quad x(t) = \begin{cases} 0 & (t \leq -C) \\ -\left[\frac{2}{3}(t+C)\right]^{3/2} & (t \geq -C) \end{cases}$$

to the equation. The graphs of these solutions for various values of C are shown in Figure 1.5.2.

If we now go back and look at our initial value problem $\frac{dx}{dt} = x^{1/3}$, $x(0) = 0$, we see that it has many solutions—in fact, all the solutions above with $C \leq 0$, as well as the zero solution. Even if we look only near $t = 0$, there are still three solutions that show different behavior immediately after the initial point $t = 0$: the zero solution and the "twin" solutions with $C = 0$ [see Figure 1.5.3(a)]. Note that all three of the graphs have a horizontal tangent line at 0. This shows the fallacy in the naive argument we gave before—knowing only $x(0)$ and the derivative of x when $t = 0$ will *not* tell you what x will be at any future time.

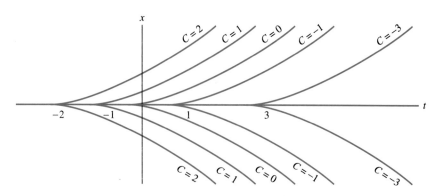

Figure 1.5.2. Solutions of $\dfrac{dx}{dt} = x^{1/3}$. Part of each graph is formed by the part of the t-axis to the left of $-C$. Also, the whole t-axis is the graph of the zero solution.

On the other hand, if we change the initial condition from $x(0) = 0$ to $x(0) = x_0$ for some number $x_0 \neq 0$, the new initial value problem will have just one solution. If $x_0 > 0$, for instance, the only solution to $\dfrac{dx}{dt} = x^{1/3}$, $x(0) = x_0$ is given by

$$x(t) = \begin{cases} 0 & \left(t \leq -\dfrac{3}{2} x_0^{2/3}\right) \\ \left[\dfrac{2}{3}\left(t + \dfrac{3}{2} x_0^{2/3}\right)\right]^{3/2} & \left(t \geq -\dfrac{3}{2} x_0^{2/3}\right) \end{cases}$$

[see Figure 1.5.3(b)]. ∎

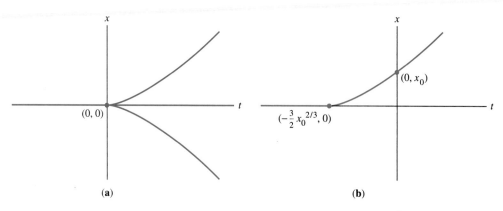

Figure 1.5.3. (a) Three different solutions to the initial value problem $\dfrac{dx}{dt} = x^{1/3}$, $x(0) = 0$. (b) The unique solution to the initial value problem $\dfrac{dx}{dt} = x^{1/3}$, $x(0) = x_0$ ($x_0 > 0$).

Before going back to our general discussion of initial value problems, let us introduce some geometric terminology. The graph of a solution of a differential equation is called an **integral curve** or **solution curve** for that equation. So asking for a solution $x(t)$ to the equation that satisfies $x(0) = x_0$ is the same as asking for an integral curve that goes through the point $(0, x_0)$ in the t,x-plane (see Figure 1.5.4).

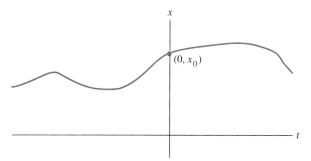

Figure 1.5.4. The graph of a solution to the initial value problem $\dfrac{dx}{dt} = f(t, x)$, $x(0) = x_0$ is the same as an integral curve passing through $(0, x_0)$.

More generally, we can consider any point (t_0, x_0) in the plane and look for integral curves passing through it. This corresponds to the more general initial value problem $\dfrac{dx}{dt} = f(t, x)$, $x(t_0) = x_0$.

The "ideal" situation is the one in which there is exactly one integral curve through each point in the t,x-plane, as in the example of radioactive decay (p. 35; Figure 1.5.1). It turns out that this will be true *provided that f and $\dfrac{\partial f}{\partial x}$, the partial derivative of f with respect to the unknown function x, are defined and continuous* everywhere in the t,x-plane. We will not indicate the reasons for this and for the more general theorem below until Section 7.3, because the proofs are rather difficult and technical and don't contribute greatly to understanding the results.

Warning: Even when there is exactly one integral curve through each point of the t,x-plane, this does *not* imply that the solutions are defined for all t. Example 1.2.3 (p. 14) shows why it doesn't. The equation is $\dfrac{dx}{dt} = (1 + x^2)t$, so we have $f(t, x) = (1 + x^2)t$, and so f and $\dfrac{\partial f}{\partial x}$ are certainly defined and continuous for all x and t. The solutions are $x = \tan(\tfrac{1}{2}t^2 + C)$. Since $\tan \theta$ is defined only when $\theta \neq \dfrac{\pi}{2} + k\pi$ for integers k, none of the solutions are defined for all t. What *is* true is that for each point (t_0, x_0), there is a unique solution whose graph goes through that point. This solution is defined in an interval around t_0. (For a sketch of the integral curves for various values of C, work Exercise 40.)

Now let's go back to the "mysterious" example $\frac{dx}{dt} = x^{1/3}$ that we studied above. In this case, $f(t, x) = x^{1/3}$, $\frac{\partial f}{\partial x} = \frac{1}{3} x^{-2/3}$, so $\frac{\partial f}{\partial x}$ is not defined (let alone continuous) for $x = 0$. This explains why it was possible for us to get several solutions when the initial value $x(0)$ was set to 0. However, we saw that there was only one solution when the initial value was set to some number $x_0 \neq 0$. This again illustrates a general fact:

Theorem **(Existence and Uniqueness Theorem).** If f and $\frac{\partial f}{\partial x}$ are defined and continuous in some rectangle[15] containing (t_0, x_0) in its interior, then there is an integral curve through (t_0, x_0). This integral curve is uniquely determined [that is, no "branching" of the sort shown in Figure 1.5.3(a) occurs] as long as it stays within the rectangle. In particular, the initial value problem $\frac{dx}{dt} = f(t, x), x(t_0) = x_0$ has a unique solution defined for t in some interval containing t_0 in its interior. (See Figure 1.5.5.)

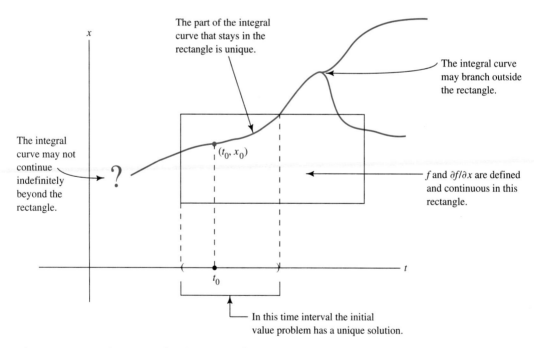

Figure 1.5.5. Existence and uniqueness of an integral curve through (t_0, x_0).

[15] The sides of the rectangle are assumed to be parallel to the coordinate axes.

SEC. 1.5 INITIAL VALUE PROBLEMS AND INTEGRAL CURVES   41

EXAMPLE 1.5.2 $\dfrac{dx}{dt} = -2x^{1/2}$, $x(0) = 4$.

Here $f(t, x) = -2x^{1/2}$ and $\dfrac{\partial f}{\partial x} = -x^{-1/2}$ are defined and continuous for all $x > 0$, so we can certainly find a rectangle around the point $(0, 4)$ in which they are defined and continuous (see Figure 1.5.6).

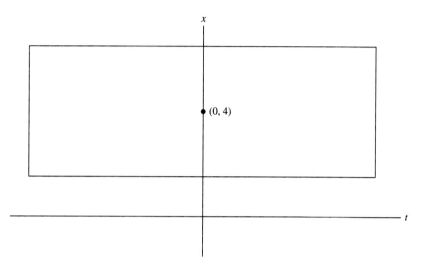

Figure 1.5.6. One rectangle (of many) in which $f(t, x) = -2x^{1/2}$ and $\dfrac{\partial f}{\partial x} = -x^{-1/2}$ are defined and continuous. (See also Exercise 44.)

The Existence and Uniqueness Theorem now guarantees a unique solution to our initial value problem on some interval around $t = 0$. More on this solution in Example 1.5.7. ∎

Often one can use the ideas above to get an indication of the behavior of the solutions of a differential equation when actually solving the equation is too difficult.

EXAMPLE 1.5.3 $\dfrac{dx}{dt} = \sin x + t$.

We can try to solve this equation, but it is not separable or linear, and in fact we are unlikely to succeed with any of the methods we will study. However, since $f(t, x) = \sin x + t$ and $\dfrac{\partial f}{\partial x} = \cos x$ are continuous everywhere, we know that there is a unique integral curve through each point (t_0, x_0) in the plane. What is more, we know the slope of the tangent line to this curve; it is $\sin x_0 + t_0$. In other words, if we draw a (short) line segment with this slope through (t_0, x_0), this segment will be the straight-line approximation to the integral curve near this point. If we do this for enough points (t_0, x_0), we will get a "porcupine" that will give us an idea of what the solution curves must look like. This "porcupine," that is, the collection of line segments of slope $f(t_0, x_0)$ at the various points (t_0, x_0), is called the **direction field** of the differential equation (see Figure 1.5.7). In this particular example we can see, for instance, that any solution will grow without bound as $t \to \infty$ (and as $t \to -\infty$, as well). Also, the pattern of integral curves is periodic in the sense that it will cover itself when shifted vertically through 2π. (Why?) ∎

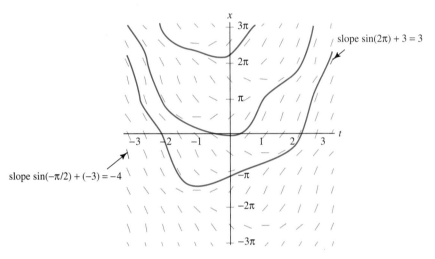

Figure 1.5.7. Direction field and approximate integral curves for the equation $\dfrac{dx}{dt} = \sin(x) + t$.

Naturally, sketching a direction field by hand tends to be a long and tedious job. However, since the job is automatic and repetitious, involving only the computation of $f(t, x)$ at many points, it is well suited for computers. Several commercially available software packages will take a more or less arbitrary differential equation $\dfrac{dx}{dt} = f(t, x)$ and sketch its direction field (using the points of a grid), then trace approximate integral curves from the direction field. Thus one can often get, within seconds, an intuitive idea of the behavior of solutions, whereas actually solving the equation could take hours, days, or forever. In Section 7.2 we will explore related computational approximation methods.

Warning: When sketching a direction field, the line segments should be drawn short enough so that they don't get in each other's way. For instance, if line segments are to be

SEC. 1.5 INITIAL VALUE PROBLEMS AND INTEGRAL CURVES

drawn for $t_0 = 0, 1, 2, \ldots$, each line segment should not extend more than a distance $\frac{1}{4}$ either way in the t-direction from the point (t_0, x_0).

Now suppose that we have an initial value problem for which there is a unique solution and for which we *can* solve the differential equation. To find the solution of the initial value problem, we can first solve the differential equation, and then find the value of the integration constant for which the initial value condition is satisfied. Here are several examples.

EXAMPLE 1.5.4

$\dfrac{dx}{dt} = \dfrac{t}{x}$, $x(0) = -2$.

In Example 1.2.5 we found the general solution to the differential equation, which is $x = \pm\sqrt{t^2 + 2C}$. Since we want $x(0) = -2$, we need $\pm\sqrt{0^2 + 2C} = -2$. In particular, we must have the minus sign on the left-hand side. That is, $-\sqrt{2C} = -2$, which yields $2C = 4$, $C = 2$. So the solution to the initial value problem is

$$x = -\sqrt{t^2 + 4}.$$

Figure 1.5.8, which should be compared to Figure 1.2.1 (p. 16), shows the graph of this solution (as a "heavy" curve), along with other solutions of the differential equation. ∎

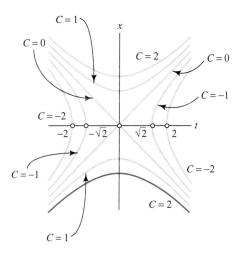

Figure 1.5.8. The solution to the initial value problem $\dfrac{dx}{dt} = \dfrac{t}{x}$, $x(0) = -2$, and other solutions of the differential equation.

EXAMPLE 1.5.5 $x' = e^{x+t}$, $x(1) = 3$.

In Example 1.2.6 we found the general solution $x = -\log(-e^t - C)$. To get $x(1) = 3$, we need

$$-\log(-e - C) = 3$$
$$\log(-e - C) = -3$$
$$-e - C = e^{-3}$$
$$C = -e - e^{-3},$$

so the answer is

$$x = -\log(-e^t + e + e^{-3}).$$

To get an idea of what the graph of this solution looks like, turn to Figure 1.2.2 (p. 17) and imagine the curve for $C = -e - e^{-3}$ (see Exercise 41). ∎

EXAMPLE 1.5.6 $\dfrac{dx}{dt} = (x-2)^2 t^2$, $x(0) = 2$.

This is another separable differential equation, so there is a tendency to go on "automatic pilot":

$$\frac{1}{(x-2)^2}\frac{dx}{dt} = t^2$$

$$\int \frac{dx}{(x-2)^2} = \int t^2\, dt$$

$$\frac{-1}{x-2} = \frac{1}{3}t^3 + C$$

$$x - 2 = \frac{-1}{\frac{1}{3}t^3 + C}$$

$$x = 2 - \frac{1}{\frac{1}{3}t^3 + C}.$$

To get $x(0) = 2$, we need $2 = 2 - \dfrac{1}{C}$, so $\dfrac{1}{C} = 0$, which is impossible! Can you explain what happened?

Since $f(t, x) = (x-2)^2 t^2$ and $\dfrac{\partial f}{\partial x}$ are defined and continuous everywhere, there must be a unique solution to the initial value problem, and there is. We lost the solution when we divided by $(x-2)^2$ and thus assumed that

$$(x-2)^2 \neq 0.$$

SEC. 1.5 INITIAL VALUE PROBLEMS AND INTEGRAL CURVES

If we consider $(x - 2)^2 = 0$, we will find that
$$x = 2$$
is a constant solution to the differential equation, and *this* is the one for which $x(0) = 2$. Of course, we could have found this without first finding the "general" solution above. Note that the true general solution to the differential equation has the slightly awkward form
$$\begin{cases} x = 2 \\ x = 2 - \dfrac{1}{\frac{1}{3}t^3 + C}, \end{cases} \quad C \text{ arbitrary}.$$

(You can think of the special solution $x = 2$ as the limit for $C \to \infty$ of the more general one!) ∎

EXAMPLE 1.5.7   We now return to the initial value problem $\dfrac{dx}{dt} = -2x^{1/2}$, $x(0) = 4$ from Example 1.5.2 (p. 41). Since $x \neq 0$ [at least for t near 0, because $x(0) = 4$], we can divide by $2x^{1/2}$ and integrate:
$$\int \frac{dx}{2x^{1/2}} = -\int dt$$
$$x^{1/2} = -t + C$$
(*) $$x = (-t + C)^2.$$

For $x(0) = 4$ we need $4 = C^2$, so we seem to get two solutions, one with $C = 2$ and one with $C = -2$. However, in Example 1.5.2 we saw that the solution should be unique. What is happening this time?

The trouble is that we lost information at (*), where we went from $x^{1/2} = -t + C$ to $x = (-t + C)^2$. In fact, $x^{1/2} = -t + C$ tells us not only that $x = (-t + C)^2$, but also that $-t + C \geq 0$; that is, $t \leq C$. So the solution with $C = 2$ is valid only for $t \leq 2$, while the solution with $C = -2$ is valid only for $t \leq -2$! Since our initial condition $x(0) = 4$ implies that the solution must be defined for $t = 0$, we cannot accept the solution with $C = -2$. Thus
$$x(t) = (-t + 2)^2 \quad (t \leq 2)$$
is the unique solution to the initial value problem, at least for t near 0. See also Exercise 44. ∎

SUMMARY OF KEY CONCEPTS, RESULTS, AND TECHNIQUES

Initial value problem (p. 36): A differential equation for $x(t)$ together with information about values of x and/or its derivatives for $t = 0$, or more generally for $t = t_0$. So far, we've only considered initial value problems of the form $\frac{dx}{dt} = f(t, x)$, $x(t_0) = x_0$. Such an initial value problem has a unique solution provided that f and $\frac{\partial f}{\partial x}$ are defined and continuous (for a more precise statement, see the Existence and Uniqueness Theorem, p. 40). To find that solution, we can first find the general solution to $\frac{dx}{dt} = f(t, x)$, and then pick the specific solution (usually by choosing C) such that $x(t_0) = x_0$.

Integral curve or **solution curve** (p. 39): The graph of a solution (to a differential equation).

Direction field of $\frac{dx}{dt} = f(t, x)$ (p. 42): A collection of line segments of slope $f(t_0, x_0)$ at various points (t_0, x_0), which can be used to trace approximate integral curves.

EXERCISES

Solve the following initial value problems.

1. $\frac{dx}{dt} = x^2$, $x(0) = 1$.

2. $\frac{dx}{dt} = x^2$, $x(0) = 0$.

3. $\frac{dx}{dt} = \sqrt{1 - x^2}$, $x(0) = \frac{1}{2}$.

4. $\frac{dx}{dt} = x^2 + 1$, $x(0) = 1$.

5. $\frac{dx}{dt} + 3x^2 t = 0$, $x(-1) = 1$.

6. $\frac{dy}{dx} - 3xy^2 = 0$, $y(1) = -1$.

7. $\frac{dx}{dt} = x^2 t$, $x(0) = 2$.

8. $\frac{dx}{dt} = x^2 t$, $x(0) = 0$.

9. $\frac{dx}{dt} = (x + 3)t$, $x(0) = -3$.

10. $\frac{dx}{dt} = (x + 3)t$, $x(0) = 0$.

11. $\frac{dy}{dx} = e^{-x+3y}$, $y(2) = 0$.

12. $\frac{dx}{dt} = e^{2x-t}$, $x(1) = 3$.

13. $x^3 \frac{dx}{dt} = \cos(3t + 1)$, $x(0) = -1$.

14. $x^2 \frac{dx}{dt} = 3 \sin t - 2t$, $x(0) = 2$.

15. $t^2 \frac{dx}{dt} = x - \frac{dx}{dt}$, $x(2) = 0$.

16. $\frac{dx}{dt} = x^3 - t^2 \frac{dx}{dt}$, $x(0) = -5$.

SEC. 1.5 INITIAL VALUE PROBLEMS AND INTEGRAL CURVES

For each of the following initial value problems:

(a) Decide whether you can expect a unique solution, at least on some interval.
(b) Solve the initial value problem.
(c) Sketch several integral curves for the given differential equation. Include (and indicate) the solution(s) to the initial value problem.

17. $\dfrac{dx}{dt} = \dfrac{1}{x}, \quad x(0) = 3.$

18. $\dfrac{dx}{dt} = \dfrac{1}{x}, \quad x(1) = 5.$

19. $\dfrac{dx}{dt} = x - 5, \quad x(1) = 2.$

20. $\dfrac{dx}{dt} = 1 - x, \quad x(2) = 5.$

21. $\dfrac{dx}{dt} = \dfrac{5}{4} x^{1/5}, \quad x(0) = 0.$

22. $\dfrac{dx}{dt} = \dfrac{5}{4} x^{1/5}, \quad x(0) = 2.$

23. $\dfrac{dy}{dx} = y^{2/3}, \quad y(0) = 1.$

24. $\dfrac{dy}{dx} = y^{2/3}, \quad y(0) = 0.$

25. $t \dfrac{dx}{dt} = x, \quad x(0) = 0.$

26. $t \dfrac{dx}{dt} = x, \quad x(0) = 3.$

For each of the following differential equations:

(a) Sketch the direction field in the t,x- or x,y-plane (label the axes!).
(b) Use the direction field to sketch an approximate integral curve through the indicated point. (Don't solve the differential equation yet.)
(c) Solve the initial value problem that gives the integral curve you sketched in part (b), and check that the solution you found is consistent with your sketch.

27. $\dfrac{dx}{dt} = 3; \quad$ point $(0, 1)$.

28. $\dfrac{dy}{dx} = -1; \quad$ point $(1, 2)$.

29. $\dfrac{dy}{dx} = \dfrac{1}{4} xy^2; \quad$ point $(0, -1)$.

30. $\dfrac{dx}{dt} = -\dfrac{1}{4} x^2 t; \quad$ point $(0, 1)$.

For each of the following differential equations, sketch the direction fields in the t,x- or x,y-plane, or have a computer do so. Include enough grid points so that you get an idea of the behavior of the solutions. (Don't try to solve the equations.)

31. $\dfrac{dx}{dt} = \dfrac{1}{2} x - \dfrac{1}{3} t.$

32. $\dfrac{dx}{dt} = \dfrac{1}{3} x + \dfrac{1}{2} t.$

33. $\dfrac{dx}{dt} = xt + 2.$

34. $\dfrac{dy}{dx} = 5 - xy.$

35. $\dfrac{dy}{dx} = x - 2 \sin \pi y.$

36. $\dfrac{dx}{dt} = t + 3 \cos \pi x.$

37. $\dfrac{dy}{dx} = y + \sin \dfrac{\pi x}{4}.$

38. $\dfrac{dy}{dx} = -y + 2 \cos \dfrac{\pi x}{4}.$

39. (See p. 38.) What is the solution to $\dfrac{dx}{dt} = x^{1/3}$, $x(0) = x_0$ if $x_0 < 0$?

40. (See p. 39.) Graph the functions $x = \tan(\tfrac{1}{2}t^2 + C)$ for $C = -\dfrac{\pi}{2}, -\dfrac{\pi}{4}, 0, \dfrac{\pi}{4}, \dfrac{\pi}{2}$ (all in the same graph).

41. (See Example 1.5.5.) Between which of the curves in Figure 1.2.2 (p. 17) would you expect to find the graph of $x = -\log(-e^t - C)$ for $C = -e - e^{-3}$? Is your answer consistent with the initial condition $x(1) = 3$?

42. Find a differential equation whose general solution is
$$\begin{cases} x = 0 \\ x = \dfrac{1}{t^2 + C}, \end{cases} \quad C \text{ arbitrary}.$$

43. Let $x(t)$ denote the size of an insect population; let $x(0) = x_0 > 0$ be the initial population. Assume that the excess birth rate $\dfrac{1}{x}\dfrac{dx}{dt}$ is proportional to x^α, where α is a positive constant. (The case $\alpha = 0$ was considered in Section 1.3.) Let $K > 0$ be the constant of proportionality.
  (a) What initial value problem do we get from this model?
  (b) Solve the initial value problem from part (a) and show that $x(t)$ is defined only for $t < \dfrac{1}{\alpha K x_0^\alpha}$.
  (c) What happens when t approaches $\dfrac{1}{\alpha K x_0^\alpha}$? What would such a "doomsday" result mean in practice, if one were working with this model?

44. (See Examples 1.5.2 and 1.5.7.)
  (a) Show that the initial value problem $\dfrac{dx}{dt} = -2x^{1/2}$, $x(0) = 4$ has a solution that is defined *for all* t.
    [*Hint*: "Piece together" two solutions of the differential equation.]
  (b) Copy Figure 1.5.6 (p. 41) and sketch several integral curves for the equation $\dfrac{dx}{dt} = -2x^{1/2}$ in the figure. In particular, sketch the graph of your solution from part (a).
  (c) Discuss what, if any, branching occurs along these integral curves. How is this related to the rectangle from Figure 1.5.6?

FURTHER READING

To get some of the flavor of the implicit function theorem without studying advanced calculus of several variables, see Burkill, *A First Course in Mathematical Analysis* (Cambridge University Press, 1962), Section 8.9, where an easier case than the one used on p. 35 (specifically, the case with one fewer variable) is proved.

1.6 LINEAR FIRST-ORDER EQUATIONS

We will now show how to solve any linear first-order equation

$$f_1(t)\frac{dx}{dt} + f_0(t)x = g(t) \tag{1}$$

for the unknown function $x(t)$. [Similarly, any equation $f_1(x)\frac{dy}{dx} + f_0(x)y = g(x)$ can be solved for $y(x)$, etc.] As in Section 1.5, we first divide through by $f_1(t)$, in order to "display" the derivative $\frac{dx}{dt}$. If there are values of t for which $f_1(t) = 0$, these must be considered separately. Meanwhile, our equation becomes

$$\frac{dx}{dt} + f(t)x = h(t), \tag{2}$$

where f and h are functions of t given by $f = f_0/f_1$, $h = g/f_1$. In the rest of this section we'll concentrate on solving linear first-order equations of the form (2).

The *homogeneous* case when $h(t) = 0$ [which corresponds to $g(t) = 0$ in equation (1)] turns out to be crucial. In fact, if the equation $\frac{dx}{dt} + f(t)x = h(t)$ is not homogeneous, we first solve the corresponding homogeneous equation $\frac{dx}{dt} + f(t)x = 0$. One example of such a computation occurred in Section 1.3, where we solved $\frac{dx}{dt} - Kx = 0$; the general computation is similar, and you may find it helpful to refer back to Section 1.3 while reading the following discussion.

The homogeneous equation $\frac{dx}{dt} + f(t)x = 0$ is separable, and assuming that $x \neq 0$, we get

$$\frac{1}{x}\frac{dx}{dt} = -f(t)$$

$$\int \frac{dx}{x} = -\int f(t)\,dt$$

$$\begin{matrix}(x>0)\\(x<0)\end{matrix} \quad \left.\begin{matrix}\log x\\ \log(-x)\end{matrix}\right\} = -\int f(t)\,dt$$

$$(x \neq 0) \qquad x = \pm e^{-\int f(t)\,dt}.$$

Note that $\int f(t)\,dt$ is determined only up to a constant. To emphasize this, let's write $\int f(t)\,dt = F(t) + C$, where $F(t)$ is an antiderivative of $f(t)$. Then

$$x = \pm e^{-[F(t)+C]} = \pm e^{-C} \cdot e^{-F(t)}.$$

$\pm e^{-C}$ can be any nonzero constant A; also, we have so far excluded the case $x = 0$, which would correspond to $A = 0$. So we can collect all the solutions of the homogeneous equation in the expression

$$x = Ae^{-F(t)},$$

where A is an arbitrary constant and $\int f(t)\, dt = F(t) + C$.

EXAMPLE For the equation $\dfrac{dx}{dt} - Kx = 0$, we have $f(t) = -K$, $\int f(t)\, dt = -Kt + C$, so we can take $F(t) = -Kt$ and we get the general solution $x = Ae^{Kt}$, as in Section 1.3. ∎

It is sometimes useful to organize things (as in Section 1.3) so that A is the initial value $x(0)$. To do so, we must choose the antiderivative $F(t)$ so it will be zero for $t = 0$; that is, we must take $F(t) = \displaystyle\int_0^t f(s)\, ds$ (see Section 1.1, p. 4). This leads to the alternative expression

$$x(t) = x(0)e^{-\int_0^t f(s)\, ds} \tag{*}$$

for the solutions of the homogeneous equation.

The following example of a homogeneous equation shows the solution method and the two expressions for the general solution.

EXAMPLE 1.6.1 $\dfrac{dx}{dt} + x \sin t = 0.$

Separate the variables:

$$\int \frac{dx}{x} = -\int \sin t\, dt = \cos t + C \quad (x \neq 0), \quad \text{or} \quad x = 0.$$

Integrate: $\left.\begin{array}{l}(x > 0)\\ (x < 0)\end{array}\right\} \begin{array}{l}\log x\\ \log(-x)\end{array}\bigg\} = \cos t + C, \quad \text{or} \quad x = 0.$

Exponentiate: $\qquad\qquad x = \pm e^{\cos t + C} \quad \text{or} \quad x = 0;$ that is,

$$x(t) = Ae^{\cos t},$$

A an arbitrary constant. Note, however, that in this case A is *not* $x(0)$: $x(0) = Ae^{\cos 0} = Ae$, so $A = x(0)/e$. This happened because we used the "simplest" antiderivative, $-\cos t$, of $f(t) = \sin t$, rather than $\displaystyle\int_0^t \sin s\, ds = -\cos t + 1$. Sure enough,

$$x(t) = Ae^{\cos t} = \frac{x(0)}{e}e^{\cos t} = x(0)e^{\cos t - 1},$$

which equals the answer $x(t) = x(0)e^{-\int_0^t \sin s\, ds} = x(0)e^{-(-\cos t + 1)}$ obtained from the general formula (*). ∎

SEC. 1.6 LINEAR FIRST-ORDER EQUATIONS

As you can see in Example 1.6.1, it is typically just as easy to solve linear equations from scratch as to memorize the formulas and work from those. This is equally true in the inhomogeneous case, to which we now return.

To solve $\dfrac{dx}{dt} + f(t)x = h(t)$, which is usually not a separable equation, we will think of this equation as a modified form of the homogeneous equation $\dfrac{dx}{dt} + f(t)x = 0$. Then it seems reasonable that the solutions to our inhomogeneous equation will somehow be "modified forms" of the solutions to the homogeneous equation. Recall that the homogeneous equation has solutions $x(t) = Ae^{-F(t)}$, where $F(t)$ is an antiderivative of $f(t)$. It turns out that the appropriate modification of these solutions is obtained by replacing the arbitrary *constant* A by an unknown *function* $A(t)$, while keeping the factor $e^{-F(t)}$. That is, we will look for solutions of the form

$$x(t) = A(t)e^{-F(t)}$$

to our inhomogeneous equation.

EXAMPLE 1.6.2 If we want to solve $\dfrac{dx}{dt} - 3x = 5$, we can start by finding the solutions $x = Ae^{3t}$ of the homogeneous equation $\dfrac{dx}{dt} - 3x = 0$. [16] We can then look for solutions of the form $x(t) = A(t)e^{3t}$ to our inhomogeneous equation. Note that for $x(t) = A(t)e^{3t}$, the product rule gives $\dfrac{dx}{dt} = \dfrac{dA}{dt}e^{3t} + 3A(t)e^{3t}$. Thus $x(t)$ will be a solution if

$$\underbrace{\frac{dA}{dt}e^{3t} + 3A(t)e^{3t}}_{dx/dt} - \underbrace{3A(t)e^{3t}}_{3x} = 5 .$$

The last two terms on the left cancel, leaving us with

$$\frac{dA}{dt}e^{3t} = 5$$

$$\frac{dA}{dt} = 5e^{-3t} .$$

We can integrate this to find

$$A(t) = \int 5e^{-3t}\, dt = -\frac{5}{3}e^{-3t} + C$$

and, finally,

$$x(t) = A(t)e^{3t} = -\frac{5}{3} + Ce^{3t} . \quad\blacksquare$$

---

[16] The equation $\dfrac{dx}{dt} - 3x = 5$ happens to be separable, as well. See Exercise 2.

Let's find out whether this method will always work. Remember that we are trying to find solutions of the form $x(t) = A(t)e^{-F(t)}$, where $F(t)$ is an antiderivative of $f(t)$, to the inhomogeneous equation $\dfrac{dx}{dt} + f(t)x = h(t)$. By the product rule and the chain rule, the derivative of $x(t) = A(t)e^{-F(t)}$ is

$$\frac{dx}{dt} = \frac{dA}{dt}e^{-F(t)} + A(t)e^{-F(t)} \cdot -F'(t)$$

$$= \frac{dA}{dt}e^{-F(t)} - A(t)e^{-F(t)}f(t).$$

Therefore, $x(t)$ will be a solution if

$$\frac{dA}{dt}e^{-F(t)} - A(t)e^{-F(t)}f(t) + f(t)A(t)e^{-F(t)} = h(t).$$

Once again, the last two terms on the left cancel, leaving us with

$$\frac{dA}{dt}e^{-F(t)} = h(t)$$

$$\frac{dA}{dt} = h(t)e^{F(t)}.$$

This equation for $A(t)$ can be solved directly by integration, and we can then find the solutions $x(t) = A(t)e^{-F(t)}$ to our inhomogeneous equation. In other words, our method will always work!

NOTES:

1. We'll get *all* solutions $x(t)$ this way, because *any* function $x(t)$ can be written in the form $x(t) = A(t)e^{-F(t)}$ simply by choosing $A(t) = x(t)e^{F(t)}$. In other words, when we made the substitution $x(t) = A(t)e^{-F(t)}$ we were not really making any particular assumption about $x(t)$, but only writing $x(t)$ in a form (suggested by the solutions to the homogeneous equation) that made our equation easier to solve.

2. In Example 1.6.2 the solutions $x(t) = -\frac{5}{3} + Ce^{3t}$ to the inhomogeneous equation $\dfrac{dx}{dt} - 3x = 5$ are the same as the solutions $x = Ae^{3t}$ to the homogeneous equation, *except* that $-\frac{5}{3}$ has been added to each of them (and that the arbitrary constant is now called C instead of A). This is a particular case of a general phenomenon: **If one solution of the inhomogeneous equation is known, all others can be obtained by adding the solutions of the homogeneous equation to this one solution**. Exercises 23 and 25 provide reasons for this fact, to which we will return when studying second-order linear equations. In our example, it would have been possible to guess the constant solution $-\frac{5}{3}$ right away and get the general solution from it!

We will give three more examples to illustrate the method, which is known as the method of **variation of constants** (A, which was a constant in the homogeneous case, now varies with t) or **variation of parameters**.

SEC. 1.6 LINEAR FIRST-ORDER EQUATIONS

EXAMPLE 1.6.3 $\dfrac{dx}{dt} + tx = t^2$.

First we solve the homogeneous equation $\dfrac{dx}{dt} + tx = 0$:

$$\frac{1}{x}\frac{dx}{dt} = -t \quad (x \neq 0), \quad \text{or} \quad x = 0$$

$$\int \frac{dx}{x} = -\int t\, dt$$

$$\begin{matrix}(x > 0) \\ (x < 0)\end{matrix} \quad \left.\begin{matrix}\log x \\ \log(-x)\end{matrix}\right\} = -\frac{1}{2}t^2 + C$$

$$x = Ae^{-t^2/2} \quad (x > 0, \ x = 0, \ \text{or} \ x < 0).$$

Now we replace A by $A(t)$ and substitute $x = A(t)e^{-t^2/2}$ in the original equation, to obtain

$$\overbrace{\frac{dA}{dt}e^{-t^2/2} + A(t)e^{-t^2/2}\cdot(-t)}^{dx/dt} + \overbrace{tA(t)e^{-t^2/2}}^{tx} = t^2$$

$$\frac{dA}{dt}e^{-t^2/2} = t^2$$

$$\frac{dA}{dt} = t^2 e^{t^2/2}$$

$$A(t) = \int t^2 e^{t^2/2}\, dt.$$

Like the integral in Example 1.2.2 (p. 13), this integral cannot be written in closed form. (In fact, the integration by parts $\int t^2 e^{t^2/2}\, dt = te^{t^2/2} - \int e^{t^2/2}\, dt$ shows that the two integrals are closely related.) We will have to be satisfied to write

$$x(t) = A(t)e^{-t^2/2},$$

that is,

$$x(t) = e^{-t^2/2}\left(\int t^2 e^{t^2/2}\, dt\right)$$

as the final form of the general solution. ∎

EXAMPLE 1.6.4 $\dfrac{dx}{dt} = 4x + 2t$.

Our standard form for this equation is $\dfrac{dx}{dt} - 4x = 2t$; the general solution of the homogeneous equation $\dfrac{dx}{dt} - 4x = 0$ is $x = Ae^{4t}$. So we substitute $x = A(t)e^{4t}$ in the original equation, and we get

$$\underbrace{\frac{dA}{dt}e^{4t} + A(t)\cdot 4e^{4t}}_{dx/dt} = 4A(t)e^{4t} + 2t$$

$$\frac{dA}{dt}e^{4t} = 2t$$

$$A(t) = \int 2te^{-4t}\,dt.$$

Integrating by parts (using $u = 2t$, $dv = e^{-4t}\,dt$), we find that

$$A(t) = -\frac{1}{2}te^{-4t} - \frac{1}{8}e^{-4t} + C,$$

and the final answer is therefore

$$\begin{aligned}x &= \left(-\frac{1}{2}te^{-4t} - \frac{1}{8}e^{-4t} + C\right)e^{4t} \\ &= -\frac{1}{2}t - \frac{1}{8} + Ce^{4t}.\end{aligned}$$

Figure 1.6.1 shows integral curves for various values of C. ∎

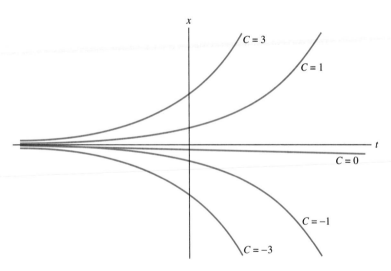

Figure 1.6.1. Solutions of $\dfrac{dx}{dt} = 4x + 2t$.

EXAMPLE 1.6.5

$t\dfrac{dx}{dt} + x = t^2.$

This equation is not in our standard form, so we divide through by t to get

$$\dfrac{dx}{dt} + \dfrac{x}{t} = t \qquad (t \neq 0).$$

Note that if $t = 0$, the original equation yields $x = 0$; in other words, $x(0) = 0$ for any solution that is defined for $t = 0$. We'll come back to this later.

Now we solve the homogeneous equation $\dfrac{dx}{dt} + \dfrac{x}{t} = 0$:

$$\dfrac{dx}{dt} = -\dfrac{x}{t}$$

$$\int \dfrac{dx}{x} = -\int \dfrac{dt}{t} \qquad (x \neq 0)$$

$$\begin{matrix}(x>0)\\(x<0)\end{matrix}\;\;\begin{matrix}\log x\\\log(-x)\end{matrix}\Big\} = \begin{cases}-\log t + C & (t>0)\\-\log(-t) + C & (t<0).\end{cases}$$

Note that we have four cases in all, depending on the signs of x and t.[17] However, once we exponentiate, we simply get

$$x = \dfrac{A}{t},$$

where $A (= 0 \text{ or } \pm e^C)$ is an arbitrary constant. [For example, if $\log(-x) = -\log t + C$, then $-x = e^{-\log t + C} = \dfrac{e^C}{e^{\log t}} = \dfrac{e^C}{t}$, so $x = -\dfrac{e^C}{t}$.]

Thus to solve our original equation we substitute $x = \dfrac{A(t)}{t}$ into it, which yields

$$\dfrac{dA}{dt} = t^2$$

$$A(t) = \dfrac{1}{3}t^3 + C$$

$$x(t) = \dfrac{1}{3}t^2 + \dfrac{C}{t}.$$

Note that of these solutions, only $x = \tfrac{1}{3}t^2$ (the solution with $C = 0$) is defined for $t = 0$. For this solution $x(0) = 0$, as predicted above. ∎

---

[17] As mentioned on p. 23, a common shorthand would be $\log|x| = -\log|t| + C$.

To close this section, let us consider initial value problems based on linear first-order equations. The equation that we have been considering, $\frac{dx}{dt} + f(t)x = h(t)$, can be rewritten $\frac{dx}{dt} = G(t, x)$ with $G(t, x) = h(t) - f(t)x$. By the theorem from Section 1.5 (p. 40), the initial value problem $\frac{dx}{dt} + f(t)x = h(t)$, $x(t_0) = x_0$ will have a unique solution provided that G and $\frac{\partial G}{\partial x} = -f(t)$ are continuous, that is, provided that f and h are continuous functions of t. We can actually strengthen this statement, because we have *solved* the problem: We have seen that the solution has the form $x(t) = A(t)e^{-F(t)}$, where $F(t)$ is such that $\int f(t)\, dt = F(t) + C$ and $A(t)$ satisfies $\frac{dA}{dt} = h(t)e^{F(t)}$. The initial condition will uniquely determine how to choose the integration constant when integrating to find $A(t)$.

Note that this solution is defined *for all t*, that is, for all t for which $f(t)$ and $h(t)$ are defined and continuous (so that the integrations can be performed). For a *linear* first-order equation, then, one cannot have a situation such as for the equation $\frac{dx}{dt} = (1 + x^2)t$ mentioned in Section 1.5 (p. 39), where the solutions are undefined for "unexpected" values of t. In summary, we have the following.

Theorem If $f(t)$ and $h(t)$ are defined and continuous on some interval, then the initial value problem $\frac{dx}{dt} + f(t)x = h(t)$, $x(t_0) = x_0$ has a unique solution, which is defined throughout the interval, whenever t_0 and x_0 are given such that t_0 is in the interior of the interval.

Warning: This theorem applies only to linear first-order equations in our standard form. For example, we saw in Example 1.6.5 that the initial value problem $t\frac{dx}{dt} + x = t^2$, $x(0) = 2$ has no solution.

The theorem above may not seem very important, since we have already seen how to find the solutions. However, in Chapter 2 we'll encounter a similar theorem for linear second-order equations, and for these the solutions are no longer so easily available.

SUMMARY OF KEY RESULTS AND TECHNIQUES

Solving **first-order linear** equations (pp. 49–52):

1. To solve $f_1(t)\frac{dx}{dt} + f_0(t)x = g(t)$, first divide by $f_1(t)$ to obtain the "standard form"
$$\frac{dx}{dt} + f(t)x = h(t).$$

SEC. 1.6 LINEAR FIRST-ORDER EQUATIONS

2. The *homogeneous* equation $\dfrac{dx}{dt} + f(t)x = 0$ is separable, and *its* general solution is

$$x(t) = Ae^{-F(t)} = x(0)e^{-\int_0^t f(s)\,ds},$$

where A is an arbitrary constant and $\int f(t)\,dt = F(t) + C$.

3. **Variation of constants:** To solve $\dfrac{dx}{dt} + f(t)x = h(t)$, replace A in **2** above by $A(t)$ to get $x(t) = A(t)e^{-F(t)}$ and substitute this into the (inhomogeneous) equation. After cancellation, you will have an equation for $\dfrac{dA}{dt}$. Integrate, then multiply by $e^{-F(t)}$ to get $x(t)$.

Finding one solution of the inhomogeneous equation is "enough" (p. 52): If one solution of $\dfrac{dx}{dt} + f(t)x = h(t)$ is known, all others can be obtained by adding the solutions of $\dfrac{dx}{dt} + f(t)x = 0$ to this one solution.

Initial value problem with a linear first-order equation (p. 56): $\dfrac{dx}{dt} + f(t)x = h(t)$, $x(t_0) = x_0$ has a unique solution, defined on any interval with t_0 in its interior on which $f(t)$ and $h(t)$ are defined and continuous.

EXERCISES

1. Check by direct substitution that $x(t) = x_0 e^{-\int_0^t f(s)\,ds}$ is the solution of the equation $\dfrac{dx}{dt} + f(t)x = 0$ for which $x(0) = x_0$.

2. Solve $\dfrac{dx}{dt} - 3x = 5$ as a separable equation, and reconcile your answer with the result of Example 1.6.2 (p. 51).

Solve the following differential equations, and sketch some of their integral curves.

3. $\dfrac{dx}{dt} = -3x + t.$

4. $\dfrac{dx}{dt} = x - 2t.$

5. $x\dfrac{dy}{dx} = 8 - y.$

6. $t\dfrac{dx}{dt} = x + 4.$

7. $x\dfrac{dy}{dx} + 2y = 6x - 3\dfrac{dy}{dx} + 6.$

8. $x\dfrac{dy}{dx} = 3\dfrac{dy}{dx} - 4x + 5y - 3.$

9. $x^2\dfrac{dy}{dx} = 3 - y.$

10. $t^2\dfrac{dx}{dt} = x + 5.$

Solve the following initial value problems.

11. $\dfrac{dx}{dt} = 5x - 2, \quad x(0) = 0.$

12. $\dfrac{dx}{dt} = 2x + 5, \quad x(0) = 0.$

13. $t\dfrac{dx}{dt} = x - 1, \quad x(2) = 3.$

14. $t\dfrac{dx}{dt} = x - 1, \quad x(0) = 0.$

15. $\dfrac{dy}{dx} = e^{2x}y + e^{2x}, \quad y(0) = 2\sqrt{e} - 1.$

16. $\dfrac{dy}{dx} = e^{x}y - e^{x}, \quad y(0) = 3e + 1.$

17. $y\dfrac{dy}{dx} = y^2 + xy, \quad y(0) = 0.$

18. $y\dfrac{dy}{dx} = y^2 + xy, \quad y(0) = 2.$

19. $\dfrac{dx}{dt} = 3t^2x - t^3, \quad x(0) = 0.$ (You can leave an integral in the answer.)

20. $\dfrac{dx}{dt} = t^2x + t^3, \quad x(2) = 0.$ (You can leave an integral in the answer.)

21. $\dfrac{dx}{dt} = tx - 2t^3, \quad x(0) = 1.$

[*Hint:* To find $\int t^3 e^{-t^2/2}\, dt$, use $u = -t^2/2$, followed by integration by parts.]

22. $\dfrac{dx}{dt} = tx + t^3, \quad x(0) = -3.$ (See the hint for Exercise 21.)

*23. In this exercise we'll consider both the homogeneous equation $\dfrac{dx}{dt} + f(t)x = 0$ and the inhomogeneous equation $\dfrac{dx}{dt} + f(t)x = h(t).$

(a) Show directly (i.e., without using how the solutions can be found) that if $x_1(t)$ and $x_2(t)$ are both solutions of the inhomogeneous equation, then their difference $y(t) = x_2(t) - x_1(t)$ is a solution of the homogeneous equation. [*Hint:* What does it mean for $y(t)$ to be such a solution?]

(b) Show that if $x_1(t)$ is a solution of the inhomogeneous equation and $y(t)$ is a solution of the homogeneous equation, then $x_1(t) + y(t)$ is a solution of the inhomogeneous equation.

(c) Show that if one can guess or somehow find *one* solution $x_1(t)$ of the inhomogeneous equation, then one can find *all* solutions of that (inhomogeneous) equation by adding each of the solutions of the *homogeneous* equation to $x_1(t)$ [as in part (b)]. [*Hint:* Suppose that $x_2(t)$ is another solution of the inhomogeneous equation. Use part (a).]

(d) Use part (c) to solve the two inhomogeneous equations $\dfrac{dx}{dt} + 7x = 33$ and $\dfrac{dx}{dt} - 18x = 25.$

[*Hint:* Each of these equations has one *very* simple solution.]

24. Repeat Exercise 13 from Section 1.3 (p. 27), this time treating $\dfrac{dx}{dt} = 25 + kx$ as an inhomogeneous linear equation. [You may want to use Exercise 23(c).]

25. Note that the equation $\dfrac{dA}{dt} = h(t)e^{F(t)}$ (see p. 52) determines $A(t)$ only up to an integration constant. Use this to explain why if one solution of $\dfrac{dx}{dt} + f(t)x = h(t)$ is known, all others can be obtained by adding the solutions of $\dfrac{dx}{dt} + f(t)x = 0$ to this one solution.

1.7 EXACT EQUATIONS AND INTEGRATING FACTORS

In this section we will expand our "repertoire," which so far consists only of the separable and the linear equations, of first-order differential equations that we can solve. As we have seen, "solving" such an equation does not always mean finding an explicit expression for the unknown function. For instance, in Example 1.2.7 (p. 18), y was the unknown function of x, and we had to leave the answer as $\tfrac{1}{5}y^5 - 2y - \tfrac{1}{3}x^3 = C$. This answer could be checked by implicit differentiation; that is, by differentiating both sides we could recover the equation $(y^4 - 2)\dfrac{dy}{dx} = x^2$.

The most general form that such an answer could take, assuming that y was still the unknown function of x, would be $E(x, y) = C$, where E would be some[18] function of two variables. [In the example, $E(x, y) = \tfrac{1}{5}y^5 - 2y - \tfrac{1}{3}x^3$.] Let's see if we can work *backward* and find out *which differential equation(s)* would have $E(x, y) = C$ as their general solution. In other words, we'll start with $E(x, y) = C$ and try to get the differential equation. Since we no longer know which specific function $E(x, y)$ is, we can no longer use implicit differentiation. However, we can still differentiate both sides, and then use the chain rule on the left, as on p. 10:

$$E(x, y) = C$$
$$\dfrac{dE(x, y)}{dx} = 0$$

---

[18] Recall that functions of two variables are assumed to be C^1 (see p. 8).

$$\frac{\partial E}{\partial x}\frac{dx}{dx} + \frac{\partial E}{\partial y}\frac{dy}{dx} = 0$$

$$\frac{\partial E}{\partial x} + \frac{\partial E}{\partial y}\frac{dy}{dx} = 0.$$

A differential equation of this form is called **exact**. In other words, a differential equation

$$M(x, y) + N(x, y)\frac{dy}{dx} = 0 \quad ^{19}$$

is called exact if there exists a function $E(x, y)$ such that

$$\frac{\partial E}{\partial x} = M(x, y), \qquad \frac{\partial E}{\partial y} = N(x, y).$$

In this case, by working in reverse through the steps above, we can see that the general solution to the differential equation is given by $E(x, y) = C$.

EXAMPLE 1.7.1 $2x + y + (x - 2y)\dfrac{dy}{dx} = 0.$

For this equation, $M(x, y) = 2x + y$, $N(x, y) = x - 2y$; the equation is exact if there is a function $E(x, y)$ with $\dfrac{\partial E}{\partial x} = 2x + y$, $\dfrac{\partial E}{\partial y} = x - 2y$. A little experimentation (we'll soon see a systematic method) will show that $E(x, y) = x^2 + xy - y^2$ is such a function, so our differential equation is exact and its general solution is

$$x^2 + xy - y^2 = C.$$

(In this particular example, we could use the quadratic formula to find the possibilities for y as a function of x.) ∎

EXAMPLE 1.7.2 $2x + y + (3x - 2y)\dfrac{dy}{dx} = 0.$

Although this looks very similar to the example above, it is much more difficult to find a function $E(x, y)$ such that $\dfrac{\partial E}{\partial x} = 2x + y$, $\dfrac{\partial E}{\partial y} = 3x - 2y$. In fact, as we'll soon see, it's impossible! ∎

EXAMPLE 1.7.3 $(3x^2 + t)\dfrac{dx}{dt} = e^t - x.$

This time, of course, x is the unknown function of t. Our equation is of the form $M(t, x) + N(t, x)\dfrac{dx}{dt} = 0$ with $M(t, x) = x - e^t$, $N(t, x) = 3x^2 + t$, and it will be exact if

---

[19] Such a differential equation is sometimes written in differential form, as $M(x, y)\,dx + N(x, y)\,dy = 0$ (compare p. 15, footnote 6), in this context. We'll come back to this in Section 3.2.

SEC. 1.7 EXACT EQUATIONS AND INTEGRATING FACTORS

there is a function $E(t, x)$ with $\dfrac{\partial E}{\partial t} = M$, $\dfrac{\partial E}{\partial x} = N$. See Example 1.7.5 below; also see Exercise 1. ∎

Given functions $M(x, y)$ and $N(x, y)$, the problem of finding a function $E(x, y)$ with $\dfrac{\partial E}{\partial x} = M$, $\dfrac{\partial E}{\partial y} = N$ is considered in most books on multivariable calculus. In that context, $(M(x, y), N(x, y))$ is a **vector field** and the function $E(x, y)$ (if it exists) is a **potential function** for this vector field.

If $E(x, y)$ exists, it can be found by starting with the equation $\dfrac{\partial E}{\partial x} = M$ and integrating with respect to x; this yields E up to an integration "constant" $K(y)$ which may depend on y. The second equation, $\dfrac{\partial E}{\partial y} = N$, is then used to find $K(y)$. $K(y)$ will be determined up to an actual constant, which can be fixed arbitrarily [since we need only one function $E(x, y)$].

EXAMPLE In the case of Example 1.7.1, where $M(x, y) = 2x + y$, $N(x, y) = x - 2y$, we have

$$\frac{\partial E}{\partial x} = 2x + y$$

$$E = \int (2x + y)\, dx = x^2 + xy + K(y).$$

Since we must also have $\dfrac{\partial E}{\partial y} = x - 2y$, we get

$$\frac{\partial}{\partial y}[x^2 + xy + K(y)] = x - 2y$$

$$x + K'(y) = x - 2y$$

$$K'(y) = -2y.$$

$K(y) = \int -2y\, dy$ is determined up to a constant; we can take the "easiest" $K(y)$, that is, $K(y) = -y^2$. Then $E(x, y) = x^2 + xy - y^2$, and we again find that the general solution is $x^2 + xy - y^2 = C$. ∎

This method for finding $E(x, y)$ fails if the expression found for $K'(y)$ actually depends on x; in that case, we have a contradiction and there is no function $E(x, y)$ with $\dfrac{\partial E}{\partial x} = M$, $\dfrac{\partial E}{\partial y} = N$.

EXAMPLE In the case of Example 1.7.2, where $M(x, y) = 2x + y$, $N(x, y) = 3x - 2y$, we have, just as in the last example, $E = \int (2x + y) \, dx = x^2 + xy + K(y)$, so we get

$$\frac{\partial}{\partial y}[x^2 + xy + K(y)] = 3x - 2y$$

$$x + K'(y) = 3x - 2y$$

$$K'(y) = 2x - 2y.$$

Since the right-hand side depends on x while the left-hand side doesn't, this is impossible. ∎

Fortunately, there is a test by which one can decide in advance whether a function $E(x, y)$ can be found such that $\dfrac{\partial E}{\partial x} = M$, $\dfrac{\partial E}{\partial y} = N$. It is based on the fact (see p. 9) that for any function E with continuous second partial derivatives, $\dfrac{\partial}{\partial y}\left(\dfrac{\partial E}{\partial x}\right) = \dfrac{\partial}{\partial x}\left(\dfrac{\partial E}{\partial y}\right)$, and thus in our case

$$\frac{\partial M}{\partial y} = \frac{\partial N}{\partial x}.$$

Conversely, it can be shown (see Exercise 32) that if $\dfrac{\partial M}{\partial y} = \dfrac{\partial N}{\partial x}$, a function E with $\dfrac{\partial E}{\partial x} = M$, $\dfrac{\partial E}{\partial y} = N$ really does exist.[20] So we have the

Test for exactness. A differential equation $M(x, y) + N(x, y)\dfrac{dy}{dx} = 0$ is exact if, and only if, $\dfrac{\partial M}{\partial y} = \dfrac{\partial N}{\partial x}$.

EXAMPLE 1.7.4 $(3y^2 - 2x^2 y)\dfrac{dy}{dx} = 2xy^2.$

This equation is not separable or linear; to see whether it is exact, we should write it in the form $M + N\dfrac{dy}{dx} = 0$:

$$-2xy^2 + (3y^2 - 2x^2 y)\frac{dy}{dx} = 0.$$

---

[20] A subtle point: If $M(x, y)$ and $N(x, y)$ are defined in a region of the plane that is not simply connected, it may not be possible to have $E(x, y)$ defined on that entire region. For details, see the discussion of conservative vector fields and Green's theorem in a book on multivariable calculus; for an example, see Exercise 24.

SEC. 1.7 EXACT EQUATIONS AND INTEGRATING FACTORS

Since we have $\frac{\partial}{\partial y}(-2xy^2) = -4xy = \frac{\partial}{\partial x}(3y^2 - 2x^2y)$, the equation is exact. So we look for $E(x, y)$ with $\frac{\partial E}{\partial x} = -2xy^2$, $\frac{\partial E}{\partial y} = 3y^2 - 2x^2y$:

$$\frac{\partial E}{\partial x} = -2xy^2$$

$$E = \int -2xy^2 \, dx = -x^2y^2 + K(y)$$

$$\frac{\partial E}{\partial y} = -2x^2y + K'(y) = 3y^2 - 2x^2y$$

$$K'(y) = 3y^2.$$

$K(y) = y^3$ will do, so we have $E(x, y) = -x^2y^2 + y^3$ and the general solution is given by

$$-x^2y^2 + y^3 = C. \quad \blacksquare$$

EXAMPLE 1.7.5 $x - e^t + (3x^2 + t)\frac{dx}{dt} = 0.$

This is the equation from Example 1.7.3 (p. 60). Since we have $M(t, x) = x - e^t$, $N(t, x) = 3x^2 + t$, we must compare $\frac{\partial M}{\partial x} = 1$ and $\frac{\partial N}{\partial t} = 1$ to see whether the equation is exact. Since they are equal, the equation *is* exact; see Exercise 1. $\quad \blacksquare$

EXAMPLE 1.7.6 $x^2 \frac{dy}{dx} = 3 - y.$

This is Exercise 9 from Section 1.6. Although the equation is linear (and thus can be solved by variation of constants), let's check whether it is exact as well. After rewriting the equation as $(y - 3) + x^2 \frac{dy}{dx} = 0$, we compute $\frac{\partial}{\partial y}(y - 3) = 1$ and $\frac{\partial}{\partial x}(x^2) = 2x$. Since these two partials are not equal, the equation is not exact. $\quad \blacksquare$

Example 1.7.6 may seem surprising. If you did Exercise 9 from Section 1.6, you will have found the answer $y = 3 + Ce^{1/x}$, C an arbitrary constant. This answer can certainly be written as $E(x, y) = C$:

$$y = 3 + Ce^{1/x}$$
$$y - 3 = Ce^{1/x}$$
$$(y - 3)e^{-1/x} = C.$$

Yet we found above that the equation $x^2 \frac{dy}{dx} = 3 - y$ is *not* exact. What is going on? Can you find out before reading further?

To see what is happening, let's start with $(y - 3)e^{-1/x} = C$ and differentiate both sides (as a check, if you like). Using the product rule on the left, we get

$$\frac{dy}{dx} e^{-1/x} + (y - 3)e^{-1/x} \cdot \frac{1}{x^2} = 0$$

or, putting the term with $\frac{dy}{dx}$ last,

$$\frac{1}{x^2} e^{-1/x}(y - 3) + e^{-1/x} \frac{dy}{dx} = 0.$$

This equation is exact, but it is *not the same* as the original equation $(y - 3) + x^2 \frac{dy}{dx} = 0$. Instead, it is a *multiple* of that original equation. Specifically, it is the original equation multiplied by $\frac{1}{x^2} e^{-1/x}$, and it will have the same solutions (at least for $x \neq 0$). Note that this suggests a new method for solving first-order differential equations: Try to find a multiple of the given equation that is exact. In other words, try to find a factor (in the example, $\frac{1}{x^2} e^{-1/x}$), by which the equation can be multiplied to make it exact. Such a factor is called an **integrating factor** for the equation and is usually denoted by μ. (If the equation is exact already, $\mu = 1$ will be an integrating factor.)

How can we find an integrating factor? If the equation $M(x, y) + N(x, y) \frac{dy}{dx} = 0$ is not exact, we want to find a (nonzero) function $\mu(x, y)$ such that

$$\mu(x, y)M(x, y) + \mu(x, y)N(x, y) \frac{dy}{dx} = 0 \quad \text{is exact}.$$

By our test for exactness, this is true when

$$\frac{\partial}{\partial y} (\mu(x, y)M(x, y)) = \frac{\partial}{\partial x} (\mu(x, y)N(x, y)).$$

It can be shown that this partial differential equation for the unknown function μ always has a nonzero solution; more on this below. However, finding such an integrating factor is another matter entirely. If the original equation is separable or linear, it is just as easy to solve by the method of Section 1.2 or 1.6, respectively (although we'll see in Exercises 25 and 26 that our new method can be used also). If the original equation is not separable, linear, or exact, one possible approach is to try to find an integrating factor that depends only on x, or one that depends only on y.

EXAMPLE 1.7.7 Let's return to $(y - 3) + x^2 \frac{dy}{dx} = 0$ (see Example 1.7.6) and try to find an integrating factor without first solving the equation. If $\mu = \mu(x)$ is an integrating factor which depends only on x, then $\mu(x)(y - 3) + \mu(x)x^2 \frac{dy}{dx} = 0$ will be exact, so we must have

$$\frac{\partial}{\partial y}[\mu(x)(y-3)] = \frac{\partial}{\partial x}[\mu(x)x^2]$$

$$\mu(x) = \mu'(x)x^2 + \mu(x) \cdot 2x \quad \text{(using the product rule on the right)}$$

$$\mu(x)(1-2x) = \mu'(x)x^2.$$

This is a separable equation for the unknown $\mu(x)$, and we can solve it:

$$\frac{1-2x}{x^2} = \frac{\mu'(x)}{\mu(x)} = \frac{1}{\mu}\frac{d\mu}{dx}$$

$$\int \frac{1-2x}{x^2} dx = \int \frac{d\mu}{\mu}.$$

Since we are only looking for one integrating factor μ, not for all possible ones, we don't really have to keep track of the signs of x and μ or the integration constant. Thus we get

$$-\frac{1}{x} - 2\log x = \log \mu.$$

Exponentiate:

$$e^{(-1/x - 2\log x)} = \mu$$

$$\mu = \frac{1}{x^2} e^{-1/x}.$$

If we had not already solved the equation $(y-3) + x^2 \frac{dy}{dx} = 0$, we could now multiply it by μ to get the exact equation $\frac{1}{x^2} e^{-1/x}(y-3) + e^{-1/x}\frac{dy}{dx} = 0$ and solve this last equation to get $e^{-1/x}(y-3) = C$. ■

EXAMPLE 1.7.8 $2xy - 2y^2 + (3x^2 - 8xy)\frac{dy}{dx} = 0.$

This equation is not linear or separable; let's see whether it is exact:

$$\frac{\partial}{\partial y}(2xy - 2y^2) = 2x - 4y, \quad \frac{\partial}{\partial x}(3x^2 - 8xy) = 6x - 8y,$$

so it's not exact. If $\mu(x)$ is an integrating factor, then

$$\frac{\partial}{\partial y}[\mu(x)(2xy - 2y^2)] = \frac{\partial}{\partial x}[\mu(x)(3x^2 - 8xy)]$$

$$\mu(x)(2x - 4y) = \mu'(x)(3x^2 - 8xy) + \mu(x)(6x - 8y)$$

$$\mu'(x)(-3x^2 + 8xy) = \mu(x)(4x - 4y)$$

$$\frac{\mu'(x)}{\mu(x)} = \frac{4x - 4y}{-3x^2 + 8xy}.$$

Since the right-hand side depends on y as well as on x, this is impossible: $\dfrac{\mu'(x)}{\mu(x)}$ cannot depend on y. So we must start over; this time, let's look for an integrating factor of the form $\mu(y)$. We want

$$\frac{\partial}{\partial y}[\mu(y)(2xy - 2y^2)] = \frac{\partial}{\partial x}[\mu(y)(3x^2 - 8xy)].$$

This time, we need the product rule on the left:

$$\mu'(y)(2xy - 2y^2) + \mu(y)(2x - 4y) = \mu(y)(6x - 8y)$$

$$\mu'(y)(2xy - 2y^2) = \mu(y)(4x - 4y).$$

This may look as bad as before, but it isn't:

$$\frac{\mu'(y)}{\mu(y)} = \frac{4x - 4y}{2xy - 2y^2} = \frac{4(x - y)}{2y(x - y)} = \frac{2}{y}.$$

Now we have a separable equation for $\mu(y)$; we find easily that $\mu(y) = y^2$ is a solution. Therefore, we multiply the original equation by y^2 and get

$$2xy^3 - 2y^4 + (3x^2y^2 - 8xy^3)\frac{dy}{dx} = 0,$$

which is presumably an exact equation. [It's a good idea to check that it really is: $\dfrac{\partial}{\partial y}(2xy^3 - 2y^4) = 6xy^2 - 8y^3 = \dfrac{\partial}{\partial x}(3x^2y^2 - 8xy^3).$] See Exercise 2 for the rest. ∎

Unfortunately, it is quite unlikely that a "random" differential equation for the unknown function y of x will have an integrating factor of the special form $\mu = \mu(x)$ or $\mu = \mu(y)$. For more on when this does happen, see Exercises 17 and 18. Meanwhile, although finding one may not be feasible, an integrating factor of the more general form $\mu = \mu(x, y)$ does always exist.[21] The intuitive reason for this is that every first-order equation should have a set of solution curves that fill up the plane just as the level curves of a function do, since there should be a unique solution curve through every point (see Section 1.5). Thus there should be a function $E(x, y)$ whose level curves are the solution curves, so that we have $E(x, y) = C$ as a description of the solution curves. Once again, though, finding $\mu(x, y)$ may prove quite difficult. Exercises 27 to 29 deal with specific kinds of first-order equations for which integrating factors are hard to find and specialized "trick" devices are used. Incidentally, if the equation has somehow been solved, one can always look at the answer, put it in the form $E(x, y) = C$ (at least in principle, but usually

---

[21] Note, however, that the integrating factor μ may not be defined on the same region as $M(x, y)$ and $N(x, y)$. For instance, in Example 1.7.7, $M(x, y) = y - 3$ and $N(x, y) = x^2$ are defined everywhere, but $\mu(x, y) = \dfrac{1}{x^2}e^{-1/x}$ is not.

SEC. 1.7 EXACT EQUATIONS AND INTEGRATING FACTORS

this can be done in practice as well), and then differentiate to find out what, *in hindsight*, could have been used as an integrating factor (see Exercise 27).

By now, you may wonder what you should do when presented with some first-order equation without any indication of how it might be solved. A good systematic procedure would be first to check whether the equation is separable, linear, or exact, in that order. If all of these fail, you might look for an integrating factor of either the form $\mu = \mu(x)$ or $\mu = \mu(y)$, or perhaps look through a list of special types of equations such as the ones in Exercises 28 and 29.

Here are a few more examples of solving miscellaneous first-order equations.

EXAMPLE 1.7.9

$(x+1)e^y + x \dfrac{dy}{dx} = 0 \quad (x > 0).$

This equation is separable. We get

$$e^{-y} \frac{dy}{dx} = -\frac{x+1}{x}$$

$$\int e^{-y}\, dy = -\int \left(1 + \frac{1}{x}\right) dx$$

$$-e^{-y} = -(x + \log x) + C \quad \text{(since } x > 0\text{)}$$

$$e^{-y} = x + \log x - C$$

$$-y = \log(x + \log x - C)$$

$$y = -\log(x + \log x - C). \quad \blacksquare$$

EXAMPLE 1.7.10

$y - x^2 + x \dfrac{dy}{dx} = 0.$

This equation is not separable, but it is linear, and we can rewrite it as $\dfrac{dy}{dx} + \dfrac{y}{x} = x$. It can then be solved by variation of constants [see Example 1.6.5 (p. 55), where we had the same equation, except that the unknown function was $x(t)$ instead of $y(x)$].

Answer: $y = \tfrac{1}{3}x^2 + \dfrac{C}{x}.$ $\quad\blacksquare$

EXAMPLE 1.7.11

$(2tx - 1)\dfrac{dx}{dt} = \cos t - x^2.$

Since this equation is neither separable nor linear, rewrite it as

$$x^2 - \cos t + (2tx - 1)\frac{dx}{dt} = 0$$

and check for exactness:

$$\frac{\partial}{\partial x}(x^2 - \cos t) = 2x, \qquad \frac{\partial}{\partial t}(2tx - 1) = 2x,$$

so the equation is exact. We look for a function $E(t, x)$ with partial derivatives $\frac{\partial E}{\partial t} = x^2 - \cos t$, $\frac{\partial E}{\partial x} = 2tx - 1$:

$$\frac{\partial E}{\partial t} = x^2 - \cos t \implies E = \int (x^2 - \cos t)\, dt = x^2 t - \sin t + K(x),$$

which yields $\frac{\partial E}{\partial x} = 2xt + K'(x)$. Thus we must have $2xt + K'(x) = 2tx - 1$, $K'(x) = -1$, and we can take $K(x) = -x$.
Answer: $x^2 t - \sin t - x = C$. ∎

EXAMPLE 1.7.12 $(2x^2 y - 10xy^3)\dfrac{dy}{dx} = 10y^4 - 5xy^2$.

Again, this equation is neither separable nor linear. Rewriting it as

$$5xy^2 - 10y^4 + (2x^2 y - 10xy^3)\frac{dy}{dx} = 0$$

and comparing

$$\frac{\partial}{\partial y}(5xy^2 - 10y^4) = 10xy - 40y^3 \quad \text{to} \quad \frac{\partial}{\partial x}(2x^2 y - 10xy^3) = 4xy - 10y^3$$

reveals the bad news that it isn't exact, either. $\mu(x)$ will be an integrating factor if

$$\frac{\partial}{\partial y}[\mu(x)(5xy^2 - 10y^4)] = \frac{\partial}{\partial x}[\mu(x)(2x^2 y - 10xy^3)]$$

$$\mu(x)(10xy - 40y^3) = \mu'(x)(2x^2 y - 10xy^3) + \mu(x)(4xy - 10y^3)$$

$$\mu(x)(6xy - 30y^3) = \mu'(x)(2x^2 y - 10xy^3)$$

$$\frac{\mu'(x)}{\mu(x)} = \frac{6xy - 30y^3}{2x^2 y - 10xy^3} = \frac{3}{x}.$$

We can indeed find a solution to this separable equation for $\mu(x)$. See Exercise 3 for the rest. ∎

SUMMARY OF KEY CONCEPTS, RESULTS, AND TECHNIQUES

Exact differential equation (p. 60): An equation $M(x, y) + N(x, y)\dfrac{dy}{dx} = 0$ such that there exists a function $E(x, y)$ with $\dfrac{\partial E}{\partial x} = M$, $\dfrac{\partial E}{\partial y} = N$. The general solution to the equation is then given by $E(x, y) = C$.

SEC. 1.7 EXACT EQUATIONS AND INTEGRATING FACTORS

Test for exactness (p. 62): $M(x, y) + N(x, y)\dfrac{dy}{dx} = 0$ is exact if and only if $\dfrac{\partial M}{\partial y} = \dfrac{\partial N}{\partial x}$.

If the equation is exact, $E(x, y)$ can be found by integrating $M(x, y)$ with respect to x, then adjusting the integration "constant" to get $\dfrac{\partial E}{\partial y} = N$.

Integrating factor for a differential equation $M(x, y) + N(x, y)\dfrac{dy}{dx} = 0$ (p. 64):

A nonzero function $\mu(x, y)$ such that the multiple

$$\mu(x, y)M(x, y) + \mu(x, y)N(x, y)\dfrac{dy}{dx} = 0$$

of the original equation is exact. An integrating factor always exists, but unless there is one of the special form $\mu = \mu(x)$ or $\mu = \mu(y)$, it may be difficult to find one.

EXERCISES

1. (a) Find a function $E(t, x)$ with $\dfrac{\partial E}{\partial t} = x - e^t$, $\dfrac{\partial E}{\partial x} = 3x^2 + t$.

    (b) Solve the differential equation $(3x^2 + t)\dfrac{dx}{dt} = e^t - x$ from Example 1.7.3 (p. 60).

2. Finish Example 1.7.8 (p. 65).
3. Finish Example 1.7.12 (p. 68).

For each of the following differential equations, find whether the equation is exact. If so, solve the equation.

4. $e^x + xy^2 + (x^2y - \cos y)\dfrac{dy}{dx} = 0.$

5. $\sin x - x^2y^3 + (e^y - x^3y^2)\dfrac{dy}{dx} = 0.$

6. $(2xy + x^2) + (2xy + y^2)\dfrac{dy}{dx} = 0.$

7. $3x + 2y + (2x - 3)\dfrac{dy}{dx} = 0.$

8. $5y + (5x + y)\dfrac{dy}{dx} = 1.$

9. $3t + 2x + (2t - 3)\dfrac{dx}{dt} = 0.$

10. $5x + (5t + x)\dfrac{dx}{dt} = 1.$

Each of the following differential equations has an integrating factor of the form $\mu(t)$, $\mu(x)$, or $\mu(y)$. Find the integrating factor and solve the given equation or initial value problem.

11. $(1 + 2x)\sin y + (x \cos y)\dfrac{dy}{dx} = 0.$

12. $(1 + 3x) \cos y - (x \sin y) \dfrac{dy}{dx} = 0$.

13. $x - 12t^2x^2 + (3t - 16t^3x) \dfrac{dx}{dt} = 0$.

14. $x^2 + 4tx + (5tx + 8t^2) \dfrac{dx}{dt} = 0$.

15. $6x + 3y^2 + 2 + 2y \dfrac{dy}{dx} = 0, \quad y(0) = 1$.

16. $6x + 2y^2 + 3 + 2y \dfrac{dy}{dx} = 0, \quad y(0) = -1$.

17. Show that the equation $M(x, y) + N(x, y) \dfrac{dy}{dx} = 0$ has an integrating factor of the form $\mu(x)$ if and only if $(\dfrac{\partial M}{\partial y} - \dfrac{\partial N}{\partial x})/N$ is a function of x alone (does not depend on y).

18. Give a test similar to the one in Exercise 17 to determine whether the equation $M(x, y) + N(x, y) \dfrac{dy}{dx} = 0$ has an integrating factor of the form $\mu(y)$, and show that your test works.

19. Consider the differential equation $8y + 3xy^3 + (8x + 4x^2y^2) \dfrac{dy}{dx} = 0$.
    (a) Show that there is no integrating factor of the form $\mu = \mu(x)$.
    (b) Show that there is no integrating factor of the form $\mu = \mu(y)$.
    (c) Show that $\mu(x, y) = xy$ is an integrating factor, and solve the differential equation.

20. Consider the differential equation $6xy + 2y^2 + (3xy + 4x^2) \dfrac{dy}{dx} = 0$.
    (a) Show that there is no integrating factor $\mu = \mu(x)$.
    (b) Show that there is no integrating factor $\mu = \mu(y)$.
    (c) Show that $\mu(x, y) = xy$ is an integrating factor, and solve the differential equation.

21. Show that the equation $M(x, y) + N(x, y) \dfrac{dy}{dx} = 0$ has an integrating factor of the form $\mu(xy)$ (that is, which depends only on the product xy) if and only if $(\dfrac{\partial M}{\partial y} - \dfrac{\partial N}{\partial x})/(xM - yN)$ depends only on xy.

22. Use Exercise 21 to solve $4xy - 3y^3 + (3x^2 - 5xy^2) \dfrac{dy}{dx} = 0$.

23. Use Exercise 21 to solve $3 + 6x^2y^2 - 2xy^3 + (6x^3y - 2x^2y^2 - 1)\dfrac{dy}{dx} = 0$.

24. Let $M(x, y) = \dfrac{y}{x^2 + y^2}$, $N(x, y) = \dfrac{-x}{x^2 + y^2}$. Note that $M(x, y)$ and $N(x, y)$ are defined everywhere except at the origin.
    (a) Show that $\dfrac{\partial M}{\partial y} = \dfrac{\partial N}{\partial x}$.
    (b) Find a function $E(x, y)$ such that $\dfrac{\partial E}{\partial x} = M$, $\dfrac{\partial E}{\partial y} = N$. Where is $E(x, y)$ defined?
    (c) Solve the differential equation $\dfrac{y}{x^2 + y^2} - \dfrac{x}{x^2 + y^2}\dfrac{dy}{dx} = 0$ by using the result of part (b).
    (d) Solve the same equation using a method from earlier in this chapter.

25. Show that any separable equation in the "standard form" $f(x)\dfrac{dx}{dt} = g(t)$ (see p. 15) is exact.

26. Consider the linear first-order equation $\dfrac{dx}{dt} + f(t)x = h(t)$ (see p. 49).
    (a) Show that this equation has an integrating factor that depends only on t, and find a formula for such an integrating factor $\mu(t)$.
    (b) Show that the general solution is $x = \dfrac{C}{\mu(t)} + \dfrac{1}{\mu(t)}\displaystyle\int \mu(t)h(t)\,dt$, and that the first term on the right-hand side is the general solution of the homogeneous equation.

27. (a) Consider the equation $\dfrac{dy}{dx} = \dfrac{2x + 3y}{3x + 2y}$. Introduce a new unknown function $v(x)$ by the substitution $y = vx$, $v = y/x$. Show that v satisfies a separable equation.
    (b) Solve the equation for $v(x)$ you found in part (a), and show that the functions $y(x)$ that are given by $x + y = C(x - y)^5$ satisfy $\dfrac{dy}{dx} = \dfrac{2x + 3y}{3x + 2y}$.
    (c) Use the result of part (b) to find (in hindsight) an integrating factor for the equation $(3x + 2y)\dfrac{dy}{dx} - (2x + 3y) = 0$.

28. (a) Show that for any constants a, b, c, and d, the differential equation
    $$\dfrac{dy}{dx} = \dfrac{cx + dy}{ax + by} = \dfrac{c + dy/x}{a + by/x}$$
    can be solved by the same substitution as in Exercise 27. (Do not actually carry out the computation, once you have found the equation for v and shown how it can be solved.)

(b) Show that, more generally, any differential equation of the form $\dfrac{dy}{dx} = f\left(\dfrac{y}{x}\right)$ can be solved. (Such equations are often known as **homogeneous** differential equations, with all the resulting confusion!)

29. (a) Consider the differential equation $\dfrac{dy}{dx} + xy + (1 - x^2)y^2 = 0$. Show that by using the substitution $v = 1/y$, this equation can be transformed into a linear one. Find the general solution.
[*Hint*: $\int (1 - x^2)e^{-x^2/2}\, dx = xe^{-x^2/2} + C$.]

(b) Show that any differential equation of the form $\dfrac{dy}{dx} + f(x)y + g(x)y^2 = 0$ is transformed into a linear equation by the same substitution $v = 1/y$ (and thus can be solved).

(c) Show that, more generally, any nonlinear differential equation of the form $\dfrac{dy}{dx} + f(x)y + g(x)y^n = 0$ is transformed into a linear equation by the substitution $v = \dfrac{1}{y^{n-1}}$. Nonlinear equations of the special type above are known as **Bernoulli equations** after Jakob Bernoulli (1654–1705), one of a celebrated family of mathematicians.

30. Consider the differential equation $1 + 3x\dfrac{dy}{dx} = 0$.

(a) Show that there is an integrating factor of the form $\mu = \mu(x)$, and solve the equation using this integrating factor.

(b) Show that there is also an integrating factor of the form $\mu = \mu(y)$, and solve the equation using this integrating factor.

(c) Reconcile your answers from parts (a) and (b) with each other.

31. Consider the initial value problem $-xy + (1 - x^2)\dfrac{dy}{dx} = 0$, $y(0) = 2$.

(a) Show that the equation is separable, and solve the initial value problem this way.

(b) Show that there is an integrating factor of the form $\mu = \mu(x)$, and solve the initial value problem this way.

(c) What would have changed in part (a), and in part (b), if the initial condition had been $y(2) = 3$?

*32. Show that if $\dfrac{\partial M}{\partial y} = \dfrac{\partial N}{\partial x}$, there exists a function $E(x, y)$ such that $\dfrac{\partial E}{\partial x} = M$ and $\dfrac{\partial E}{\partial y} = N$. [You may assume that $\dfrac{\partial}{\partial y}\left(\int A(x, y)\, dx\right) = \int \dfrac{\partial A}{\partial y}(x, y)\, dx$ for a function $A(x, y)$ of two variables.]

SEC. 1.7 EXACT EQUATIONS AND INTEGRATING FACTORS

Solve the following first-order equations and initial value problems. (You will have to use various methods from this chapter.)

33. $x^2 - 3y + (-3x + 2e^y - 5)\dfrac{dy}{dx} = 0$.

34. $y - (3x - 1)\dfrac{dy}{dx} = 0$.

35. $y - 1 + 2x\dfrac{dy}{dx} = 0$.

36. $x + \left(2t + \dfrac{1}{x}\right)\dfrac{dx}{dt} = 0$.

37. $2x - \dfrac{3}{t} + t\dfrac{dx}{dt} = 0$.

38. $-3y + 5x - (4y + 3x)\dfrac{dy}{dx} = 0$.

39. $2y - 4x + (2x + 5y)\dfrac{dy}{dx} = 0$.

40. $2y^4 - 6xy + (4xy^3 - 2x^2)\dfrac{dy}{dx} = 0$.

41. $2x^3y - y^2 + (x^4 - 3xy)\dfrac{dy}{dx} = 0$.

42. $(3xy - 2\sin y + 1)\dfrac{dy}{dx} = -\dfrac{3}{2}y^2 + x + 4$, $y(0) = 0$.

43. $(-4xy + 1 - 2y)\dfrac{dy}{dx} = 2y^2 - 3x^2 - 1$, $y(0) = 1$.

44. $2x - 5 + t\dfrac{dx}{dt} = 0$.

45. $2x - 5t + t\dfrac{dx}{dt} = 0$.

46. $3t + 4x - t\dfrac{dx}{dt} = 0$, $x(1) = 2$.

47. $3t + 4 - t\dfrac{dx}{dt} = 0$.

48. $-3y + 5x - \dfrac{dy}{dx} = 0$.

49. $2y - 4x + \dfrac{dy}{dx} = 0$.

50. $1 + (2x - 5y)\dfrac{dy}{dx} = 0.$

51. $1 + (3x + 4y)\dfrac{dy}{dx} = 0.$

52. $2 \sin 3y + (3 \cos 3y)\dfrac{dy}{dx} = 0.$

53. $\cos 2x - \sin 2x \dfrac{dy}{dx} = 0, \quad y\left(\dfrac{\pi}{4}\right) = 1.$

54. $x + t - \sin t + (t - 2 \cos x)\dfrac{dx}{dt} = 0, \quad x(0) = 0.$

55. $\cos t - e^{2t} + 3x^2 - (\sin x - 6xt)\dfrac{dx}{dt} = 0, \quad x(0) = 0.$

56. $-\cos x + y^2 + \left(2xy - \dfrac{1}{1 + y^2}\right)\dfrac{dy}{dx} = 0.$

57. $y + (2x + 4y)\dfrac{dy}{dx} = 0.$

58. $y - (2x + 4y)\dfrac{dy}{dx} = 0.$

59. $(x^2 + 1)y^2 - x\dfrac{dy}{dx} = 0, \quad y(1) = 1.$

60. $y + (y^2 - 3)x^3 \cdot \dfrac{dy}{dx} = 0, \quad y(1) = 1.$

61. Given a function $E(x, y)$ and the set of its level curves $E(x, y) = C$, an **orthogonal trajectory** is defined to be a curve that intersects each of the given curves at right angles. For instance, if $E(x, y) = x^2 + y^2$, the level curves are concentric circles, while straight half-lines (radii) through the origin are orthogonal trajectories. [Sample application: If $E(x, y)$ is the electrical potential at (x, y) in the plane, the field lines of the electric field will be along the orthogonal trajectories.] Show that the orthogonal trajectories are the solution curves of the differential equation $\dfrac{dy}{dx} = \dfrac{\partial E/\partial y}{\partial E/\partial x}.$

[*Hint*: See p. 10 for a formula giving the slope of a level curve at a point.]

Find equations for the orthogonal trajectories (see Exercise 61) for each of the following sets of level curves; sketch a few representative level curves and orthogonal trajectories in each case.

62. $2x - y = C.$

63. $x + 3y = C.$

64. $x^2 + 2y^2 = C$. **65.** $4x^2 + y^2 = C$.
66. $x^2 - y^2 = C$. **67.** $xy = C$.

For each of the following sets of level curves:

(a) Use a computer to sketch several of the level curves (in one figure).
(b) Find equations for the orthogonal trajectories (see Exercise 61).
(c) Use a computer to sketch several of the orthogonal trajectories [preferably as an "overlay" of your figure from part (a)].

68. $x^3 - 3xy^2 = C$. **69.** $6x^2y - 2y^3 + 5y = C$.
70. $x^2 - 4x + y^3 = C$. **71.** $x^3 - y^2 - 2y = C$.

FURTHER READING

For yet another special kind of first-order equation, which is solved by differentiating (!), see Section 4.6, "Clairaut Equations," in a wonderful old-style textbook: Agnew, *Differential Equations*, 2nd ed. (McGraw-Hill, 1960).

CHAPTER 2

Second-Order Differential Equations

2.1 INTRODUCTION: NEWTONIAN MOTION

In this section we consider the motion of a particle along a straight line, which we will call the x-axis. Thus the position of the particle at time t is given by a function $x(t)$. The velocity of the particle is the derivative $x'(t)$ of this function; the acceleration is the second derivative $x''(t)$. [1]

We assume that the particle moves under the influence of a force F, the resultant (total) of all the forces acting on the particle. This force F may depend on x, x', and t; it is positive or negative according to its direction along the axis.

Newton's second law states that force equals mass times acceleration. That is, we can write $x'' = \dfrac{1}{m} F$, where the constant m is the mass of the particle. Since the second derivative, but no higher derivative, of x occurs in this equation, it is a second-order differential equation for x. The difficulty of this equation, and whether it can be solved, depends entirely on what the force F is; one can only hope to solve the equation for relatively simple forces. Here are some examples.

EXAMPLE 2.1.1 Suppose that $F = 0$. This is the case of a "free" particle, which moves without impediment from friction or anything else. In this case $x'' = 0$, so $x' = C$ and $x = Ct + D$ for suitable constants of integration C and D. In other words, the particle moves with constant velocity C from its initial position $x(0) = D$. ∎

---
[1] A different notation, using dots to indicate time derivatives, is often used in texts on mechanics and physics; in this notation, $\dot{x}$ denotes the velocity, $\ddot{x}$ the acceleration.

EXAMPLE 2.1.2 More generally, suppose that $F = F(t)$ depends only on the time (and doesn't depend on x or x'). Then one can (in principle) solve $x'' = \dfrac{1}{m} F(t)$ by simply integrating twice. ∎

EXAMPLE 2.1.3 Suppose that $F = -kx$, where k is some positive constant. This situation occurs (at least approximately) if the particle is attached to a spring and friction is neglected. Here x denotes the displacement of the particle from the equilibrium position (in which the spring is relaxed); see Figure 2.1.1. The fact that F is approximately proportional to x is known as **Hooke's law**; it is only valid when x is small relative to the length of the spring.

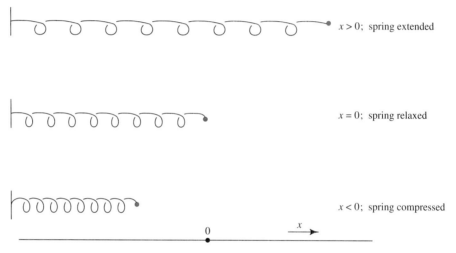

Figure 2.1.1. Possible positions for a particle attached to a spring.

If we do have $F = -kx$, the differential equation describing the motion of the particle becomes $x'' + \dfrac{k}{m} x = 0$. We will show how to solve this equation below. ∎

EXAMPLE 2.1.4 $F = -kx - \gamma x'$, where k and γ are positive constants. This will be the (approximate) expression for the total force if friction is taken into account for the particle attached to the spring in Example 2.1.3. The frictional force is roughly proportional to, and in the opposite direction from, the velocity; this accounts for the term $-\gamma x'$. In this case the differential equation "of motion" (i.e., describing the position x of the moving particle) becomes $mx'' + \gamma x' + kx = 0$. ∎

EXAMPLE 2.1.5 If the spring from the previous examples is hung vertically from the ceiling, there will be an additional force mg on the particle due to gravity. g, the gravitational acceleration, is strictly speaking not constant, but inversely proportional to the square of the distance from the particle to the center of the earth. However, if the spring is not exceptionally long, we can take g to be a constant. The differential equation will be $mx'' + \gamma x' + kx = mg$. ∎

All the examples above give rise to linear second-order equations with constant coefficients. (Bear in mind that this is only because we have chosen to approximate the forces in the actual physical situation.) Later in this chapter we will show how to solve all such equations. We'll also study linear second-order equations whose coefficients are not constant, especially in later chapters. However, we will not always be able to solve those equations.

For now, we will solve the equation $x'' + \dfrac{k}{m}x = 0$ from Example 2.1.3, which describes the motion of a particle on a frictionless horizontal spring. The method used will not apply to more general linear equations with constant coefficients, such as the equation from Example 2.1.4, but it does apply to any equation of the form $x'' - f(x) = 0$. In particular, the same method can, in principle, be used to determine the motion of a particle along the x-axis whenever the force on the particle depends only on its position.

To begin, we shift our emphasis from the unknown function x to the velocity $v = x'$. (Once we've found v, we intend to solve for x.) At first sight this is a dead end, because in terms of v our equation becomes $\dfrac{dv}{dt} + \dfrac{k}{m}x = 0$, and the only way we could express x in terms of v would be as an integral. However, we now make a surprising move. Instead of trying to get rid of x and to get an equation for v as a function of t, we try to get rid of t! That is, we assume that v is a function of x (this will need justification later). By the chain rule, we can then write

$$\frac{dv}{dt} = \frac{dv}{dx}\frac{dx}{dt} = \frac{dv}{dx}v, \quad \text{since } v \text{ is nothing but } \frac{dx}{dt}.$$

This makes our equation $\dfrac{dv}{dx}v + \dfrac{k}{m}x = 0$.

We now have a separable first-order equation for $v(x)$, which we can solve:

$$\int v\, dv = -\int \frac{k}{m}x\, dx$$

$$\frac{1}{2}v^2 = -\frac{k}{2m}x^2 + C$$

$$v = \pm\sqrt{2C - \frac{k}{m}x^2}.$$

Note that these solutions are defined only for $|x| \leq \sqrt{\dfrac{2Cm}{k}}$; in particular, we must have $C \geq 0$.

Now since $v = \dfrac{dx}{dt}$, we get another separable first-order equation (really a pair of them), this time for $x(t)$:

$$\frac{dx}{dt} = \pm\sqrt{2C - \frac{k}{m}x^2}$$

$$\int \frac{dx}{\sqrt{2C - \dfrac{k}{m}x^2}} = \pm\int dt.$$

The left-hand side is a standard integral from calculus. Since $\dfrac{k}{m} > 0$, we get

$$\sqrt{\frac{m}{k}} \arcsin \sqrt{\frac{k}{2mC}}\, x = \pm(t + D),$$

where D is a new integration constant. Finally, we solve for x:

$$\sqrt{\frac{k}{2mC}}\, x = \sin\left(\pm\sqrt{\frac{k}{m}}(t + D)\right)$$

$$x = \pm\sqrt{\frac{2mC}{k}} \sin\left(\sqrt{\frac{k}{m}}(t + D)\right).$$

We can simplify this answer if we note that since $C \geq 0$ is arbitrary, $\pm\sqrt{\dfrac{2mC}{k}}$ can be any constant, as can $\sqrt{\dfrac{k}{m}}D$. This gives us

$$x = C_0 \sin\left(\sqrt{\frac{k}{m}}\, t + D_0\right),$$

where C_0 and D_0 are arbitrary constants.

N O T E : All these solutions are periodic functions of t. This is intuitively quite reasonable: If you stretch out the spring with the particle on it, then let it go, it will oscillate back and forth past the equilibrium position. It does not seem as clear why all the solutions should have the same period $2\pi\sqrt{\dfrac{m}{k}}$, however!

If we can justify our earlier assumption that v is a function of x, then we'll know that we have found the general solution to the equation $x'' + \dfrac{k}{m}x = 0$ above. Of course, we could also substitute $x = C_0 \sin(\sqrt{\dfrac{k}{m}}\, t + D_0)$ into the equation to check that it is a solution (see Exercise 1), but that wouldn't tell us that we have *all* the solutions.

Here is a sketch of a justification. Since v is a function of t, v will be a (composite) function of x provided that t is a function of x. Now x is a function of t, which will have an inverse function on any interval for t where $\dfrac{dx}{dt} > 0$ or where $\dfrac{dx}{dt} < 0$. So t will be an (inverse) function of x if we restrict ourselves to an interval for t where $v = \dfrac{dx}{dt} \neq 0$. How do we know what interval for t to take? We don't (especially since the interval will depend on which solution x we eventually get), but the only way we could really get into trouble would be if v were zero everywhere. But then x would be constant, so x'' would be 0, and since $x'' + \dfrac{k}{m}x = 0$, we'd have $x = 0$. So the only solution that we might have missed by assuming that v is a function of x is the constant solution $x = 0$. However, we did get that solution (for $C_0 = 0$).

As you can see, this argument is a bit involved. It would be preferable if we didn't have to worry about this so much: if there were some way of seeing at the end of the computation that the answer is correct and complete even if some of the intermediate steps are questionable.

In the next section we will see that there is, indeed, a result for second-order equations that tells us what to expect and how to check for completeness of the solutions, especially for second-order linear equations such as the above. Meanwhile, here is an example of finding solutions to a *nonlinear* equation of the form $x'' - f(x) = 0$.

EXAMPLE 2.1.6 $\quad \dfrac{d^2x}{dt^2} = 2x^3 + 2x.$

As on p. 79, we start by considering $v = \dfrac{dx}{dt}$ as an unknown function of x; we then have $\dfrac{d^2x}{dt^2} = \dfrac{dv}{dt} = \dfrac{dv}{dx}\dfrac{dx}{dt} = v\dfrac{dv}{dx}$, so we end up with

$$v\frac{dv}{dx} = 2x^3 + 2x$$

$$\int v\, dv = \int (2x^3 + 2x)\, dx$$

$$\frac{1}{2}v^2 = \frac{1}{2}x^4 + x^2 + C$$

$$v = \pm\sqrt{x^4 + 2x^2 + 2C}.$$

Since $v = \dfrac{dx}{dt}$, this yields

$$\int \frac{dx}{\sqrt{x^4 + 2x^2 + 2C}} = \pm(t + D).$$

Integrals involving the square root of a third- or fourth-degree polynomial, such as the integral we have here, come up quite often, but they usually can't be written in terms of "familiar" functions. (Note that this is unlike the situation for the square root of a *quad-*

ratic polynomial, where we have results such as $\int \frac{dx}{\sqrt{1-x^2}} = \arcsin x + C$.) Such integrals are called **elliptic integrals**,[2] and there is much of importance and mathematical beauty known about them.[3] However, any serious attempt to say more about this would require the theory of functions of a complex variable, and is therefore beyond the scope of this book.

We don't quite have to give up yet, though. Even though we are not able to find $\int \frac{dx}{\sqrt{x^4 + 2x^2 + 2C}}$ for most values of C, there is one particular C for which the integral is quite reasonable. Can you find it?

$C = 0$ is not too bad, but $C = \frac{1}{2}$ is better yet; for $C = \frac{1}{2}$, the square root simply becomes $x^2 + 1$, and we get

$$\int \frac{dx}{x^2 + 1} = \pm(t + D)$$
$$\arctan x = \pm(t + D)$$
$$x = \pm\tan(t + D).$$

Of course, these are by no means *all* the solutions of the original differential equation, but only those for which $C = \frac{1}{2}$. ■

SUMMARY OF KEY RESULTS AND TECHNIQUES

Newton's second law (for particle moving along x-axis; p. 77): $x'' = \frac{1}{m} F$.

Hooke's law (p. 78): For small x (displacement from equilibrium), spring force $= -kx$.

Solving $x'' - f(x) = 0$, where $x(t)$ is the unknown function (p. 79): First use $v = x'$, $x'' = \frac{dv}{dt} = \frac{dv}{dx} v$ to get a separable equation for $v(x)$. Once v is found, setting $v = \frac{dx}{dt}$ yields a separable equation for $x(t)$.

EXERCISES

1. Check by direct substitution that $x = C_0 \sin(\sqrt{\frac{k}{m}} t + D_0)$ really is a solution to $mx'' + kx = 0$.

---
[2] The word *elliptic* alludes to the fact that arc length along an ellipse is given by one such integral.

[3] For instance, just as the inverse function of $\arcsin x = \int \frac{dx}{\sqrt{1-x^2}}$ is the periodic function sin, the inverse functions of certain elliptic integrals are *doubly* periodic functions, with two independent complex periods.

SEC. 2.1 INTRODUCTION: NEWTONIAN MOTION

2. Find the differential equation describing the motion of a particle suspended from a frictionless vertical spring.

3. Rewrite the equation $mx'' + \gamma x' + kx = 0$ from Example 2.1.4 as a first-order differential equation for the velocity $v = x'$ as an unknown function of x. Why does this not look promising for $\gamma \neq 0$?

Find at least one explicit solution (not identically zero) of each of the following differential equations.

[*Hint*: As in Example 2.1.6, finding the general solution is not practical. Since you don't need all the solutions, you can choose the first constant of integration in such a way that the second integration can be carried out.]

4. $\dfrac{d^2x}{dt^2} + \dfrac{1}{4x^3} = 0.$

5. $\dfrac{d^2x}{dt^2} - 80x^{3/5} = 0.$

6. $\dfrac{d^2y}{dx^2} - y^3 = 0.$

7. $\dfrac{d^2x}{dt^2} = \sqrt{x}.$

8. Find a formula (which will involve several indefinite integrations) for the solutions of $\dfrac{d^2x}{dt^2} - f(x) = 0$, where f is an arbitrary (continuous) function.

*9. Find all solutions of $\dfrac{d^2x}{dt^2} - 4x = 0.$

[*Hint*: To integrate $\int \dfrac{dx}{\sqrt{4x^2 + 2C}}$, look at the cases $C > 0$, $C = 0$, $C < 0$ separately and use

$$\int \dfrac{dx}{\sqrt{x^2 + 1}} = \log(x + \sqrt{x^2 + 1}) + D,$$

$$\int \dfrac{dx}{\sqrt{x^2 - 1}} = \begin{cases} \log(x + \sqrt{x^2 - 1}) + D & (x > 1) \\ \log(-x - \sqrt{x^2 - 1}) + D & (x < -1). \end{cases}$$]

*10. Find (do not solve) a second-order equation for $v(x)$ that corresponds to the third-order equation $\dfrac{d^3x}{dt^3} = x^2.$

*11. A rocket is moving in a straight line, directly outward from the center of the earth. Let R be the radius of the earth, let g be the gravitational acceleration on the surface of the earth, and let $x(t)$ be the distance (height) of the rocket from the surface. Assume that the only force on the rocket is the earth's gravity.

  (a) Explain why $x'' = \dfrac{-gR^2}{(x + R)^2}.$

  (b) Find an expression for $v = x'$ as a function of x. (Your answer should involve a constant of integration, C.)

  (c) If the rocket is to "escape" the earth's gravitational field, v must be defined (and positive) for all x. What condition on C does this impose?

(d) Find the **escape velocity** for the rocket: the least velocity with which the rocket could be launched from the surface of the earth and still "escape."

(e) Assume that the rocket was launched, at time $t = 0$, with exactly the escape velocity. Find $x(t)$.

FURTHER READING

To get an idea of what is known about elliptic integrals, see Byrd and Friedman, *Handbook of Elliptic Integrals for Engineers and Scientists*, 2nd ed. (Springer-Verlag, 1971).

2.2 GENERAL RESULTS

In this section we will discuss second-order differential equations in general, just as we discussed first-order equations in Section 1.5. In particular, we'll start studying linear second-order equations, which will occupy us for the rest of this chapter.

As in Section 1.5, we will assume that $x(t)$ is the unknown function; our second-order equations will be of the form

$$E\left(t, x, \frac{dx}{dt}, \frac{d^2x}{dt^2}\right) = 0.$$

This time, we will "display" the highest derivative, $\frac{d^2x}{dt^2}$.

EXAMPLE $x\frac{d^2x}{dt^2} - t\frac{dx}{dt} = x + \frac{d^2x}{dt^2}$. Rewrite this as $\frac{d^2x}{dt^2} = \frac{t}{x-1}\frac{dx}{dt} + \frac{x}{x-1}$. [What happens when $x - 1 = 0$ may have to be considered separately; the constant function $x = 1$ is not a solution, but there might be solutions with $x(t) = 1$ for a particular t.] ∎

As in the case of first-order equations, the implicit function theorem tells us that under reasonable conditions the equation $E(t, x, \frac{dx}{dt}, \frac{d^2x}{dt^2}) = 0$ can be rewritten as one or more equations of the form $\frac{d^2x}{dt^2} = f(t, x, \frac{dx}{dt})$. We will assume for the time being that this has

SEC. 2.2 GENERAL RESULTS

been done. If x denotes the position of a particle moving along an axis, as in Section 2.1, then this means that the acceleration of the particle is given in terms of its velocity, its position, and the time. Since the acceleration is proportional to the total force, this is equivalent to giving the force $F = m \frac{d^2x}{dt^2}$ as a function of t, x, and $\frac{dx}{dt}$. Notice that this is exactly the general situation considered in Section 2.1.

In Section 1.5, when discussing the first-order equation $\frac{dx}{dt} = f(t, x)$, we found that for each choice of $x(0)$, say $x(0) = x_0$, the resulting initial value problem has a unique solution, provided that f and $\frac{\partial f}{\partial x}$ are continuous. Now we have, instead, the second-order equation $\frac{d^2x}{dt^2} = f(t, x, \frac{dx}{dt})$. Should we still expect that there is a unique solution with $x(0) = x_0$ (under reasonable conditions on f)?

EXAMPLE $\frac{d^2x}{dt^2} = -x$. This is a special case ($\frac{k}{m} = 1$) of the equation $x'' + \frac{k}{m}x = 0$ of Example 2.1.3, which was solved later in Section 2.1. The solutions are given (see p. 80) by $x = C_0 \sin(t + D_0)$, with C_0 and D_0 arbitrary constants. If we insist that (for instance) $x(0) = 0$, this means that $C_0 \sin D_0 = 0$. There are many choices of C_0 and D_0 for which this is true; for instance, we can take $D_0 = 0$, and then we'll have $C_0 \sin D_0 = 0$ regardless of what C_0 is. That is, *all* the solutions of the form $x = C_0 \sin t$ will satisfy our initial condition. ∎

From this example we see that we *cannot* expect $\frac{d^2x}{dt^2} = f(t, x, \frac{dx}{dt})$, $x(0) = x_0$ to have a unique solution. We can also see why: In solving the equation, *two* constants of integration were introduced. We cannot expect to solve for both these constants using only one condition on the function x. This can be seen in physical terms, as well: If you know *only* the starting position of a particle (and not, for instance, whether it is moving to the left or to the right and with what speed), you cannot expect to predict where the particle will be in the future.

On the other hand, if you were to know *both* the position *and* the velocity of the particle at $t = 0$, it is reasonable to expect that you might be able to predict its position and velocity in the future: From the position and velocity at any instant you can compute the force on the particle, hence the acceleration of the particle at that instant. This describes how the velocity is changing, which in turn describes how the position is changing.

You may recognize this as the second-order version of the naive argument we gave in Section 1.5 (see p. 35). It is just as naive, and just as wrong (see Exercise 7). Once again, though, there is a grain of truth in it.

We want to consider, then, an initial value problem where along with the equation $\frac{d^2x}{dt^2} = f(t, x, \frac{dx}{dt})$, both x and $\frac{dx}{dt}$ are given for $t = 0$. Let's write $v = \frac{dx}{dt}$ for short, so we have an initial value problem of the form $\frac{d^2x}{dt^2} = f(t, x, v)$, $x(0) = x_0$, $v(0) = v_0$. It turns out that this problem has a unique solution *provided that* f, $\frac{\partial f}{\partial x}$, and $\frac{\partial f}{\partial v}$ are defined and continuous.

EXAMPLE 2.2.1 $(x - 1)\frac{d^2x}{dt^2} + 3x\frac{dx}{dt} + e^x = -t^2$, $x(0) = 0$, $\frac{dx}{dt}\Big|_{t=0} = 2$. [4]

We rewrite the equation as

$$\frac{d^2x}{dt^2} = \frac{-t^2 - 3xv - e^x}{x - 1}.$$

This is in the form given above, with $f(t, x, v) = -\frac{t^2 + 3xv + e^x}{x - 1}$. f, $\frac{\partial f}{\partial x}$, and $\frac{\partial f}{\partial v}$ are certainly defined and continuous *provided* $x \neq 1$. Since the initial conditions given are $x(0) = 0$, $v(0) = 2$, there will be a unique solution (defined for all t in some interval containing 0) to the initial value problem. If, on the other hand, the initial conditions were given as $x(0) = 1$, $v(0) = 2$, there would be no guarantee. Not only that—there would be no solution! (See Exercise 8.) ∎

As in Section 1.5, there is no reason (except convenience) to pose the initial conditions for $t = 0$ in particular; we could do the same for $t = t_0$. The general result is as follows.

Theorem 2.2.1 If f, $\frac{\partial f}{\partial x}$, and $\frac{\partial f}{\partial v}$ are defined and continuous in a rectangular box containing the point (t_0, x_0, v_0) in its interior, then the initial value problem $\frac{d^2x}{dt^2} = f(t, x, v)$, $x(t_0) = x_0$, $v(t_0) = v_0$ has a unique solution defined for t in some interval containing t_0.

This result can be interpreted geometrically by considering "graphs" of solutions in "t, x, v-space," which are curves in this three-dimensional space that play the same role as integral curves (see Section 1.5) in the first-order case. There is also a three-dimensional "direction field" in t, x, v-space. However, curves in three-dimensional space are more difficult, both to draw and to interpret, and we won't pursue this geometric approach. (See Exercises 25 and 26 for simple examples.)

We now turn to the case of *linear* second-order differential equations. These are just about the only ones that one may be able to solve completely by elementary methods.

---

[4] As usual, $\frac{dx}{dt}\Big|_{t=0}$ denotes the value $x'(0)$ of $\frac{dx}{dt}$ for $t = 0$.

From Section 1.4 we know that the general form of a linear second-order equation is

$$f_2(t)\frac{d^2x}{dt^2} + f_1(t)\frac{dx}{dt} + f_0(t)x = g(t).$$

In order to "display" $\frac{d^2x}{dt^2}$, we now divide through by $f_2(t)$. If there are values of t for which $f_2(t) = 0$, these must be considered separately. Meanwhile, we get

$$\frac{d^2x}{dt^2} + p(t)\frac{dx}{dt} + q(t)x = h(t),$$

where $p = \frac{f_1}{f_2}$, $q = \frac{f_0}{f_2}$, and $h = \frac{g}{f_2}$ are certain functions of t. The functions $p(t)$ and $q(t)$ are the **coefficients** of this linear equation; $h(t)$ is sometimes called the **forcing term**, since in physical applications it is often related to an external force.

We can now apply Theorem 2.2.1, since our equation is of the form

$$\frac{d^2x}{dt^2} = f(t, x, v), \quad \text{with} \quad f(t, x, v) = h(t) - p(t)v - q(t)x.$$

Since $\frac{\partial f}{\partial x} = -q(t)$, $\frac{\partial f}{\partial v} = -p(t)$, the conclusions of the theorem apply, provided that p, q, and h are continuous functions of t. In fact, just as in the first-order case (see Section 1.6, p. 56), a stronger result is true: The solution to an initial value problem based on the equation above will be defined for all t for which $p(t)$, $q(t)$, and $h(t)$ are defined and continuous. However, this cannot be proved in the same way as in the first-order case, since in general we won't be able to find the solutions explicitly. We record the result without giving a proof:

Theorem 2.2.2 If $p(t)$, $q(t)$, and $h(t)$ are defined and continuous on some interval,[5] then the initial value problem $\frac{d^2x}{dt^2} + p(t)\frac{dx}{dt} + q(t)x = h(t)$, $x(t_0) = x_0$, $v(t_0) = v_0$ has a unique solution, which is defined throughout the interval, whenever t_0, x_0, and v_0 are given so that t_0 is in the interior of the interval. In particular, if *two* solutions of the linear second-order equation $\frac{d^2x}{dt^2} + p(t)\frac{dx}{dt} + q(t)x = h(t)$ have the same values at t_0 and if their derivatives also are the same at t_0, then those two solutions are *equal* (throughout the interval, not just at t_0).

EXAMPLE The *only* solution of the equation $x'' + t^2x' - x = 0$ that satisfies $x(0) = x'(0) = 0$ is the constant solution $x = 0$. We know this even though we have no idea, as yet, what the other solutions of the equation might be! We also know that if we pick any two numbers x_0 and v_0, there is one and only one solution $x(t)$ such that $x(0) = x_0$ and $x'(0) = v_0$; $x(t)$ will be defined for all t. ∎

[5] In many, if not most, of our applications, the interval will be the whole real line!

We continue to study linear equations of the form $\frac{d^2x}{dt^2} + p(t)\frac{dx}{dt} + q(t)x = h(t)$. Just as for first-order equations (see Section 1.6), the *homogeneous* case $h(t) = 0$ turns out to be crucial. In fact, we'll see in Section 2.8 that there is again a general method, the method of variation of constants, to solve any inhomogeneous equation [with $h(t) \neq 0$], *provided that the corresponding homogeneous equation has been solved.*

The main difficulty, then, lies in solving $\frac{d^2x}{dt^2} + p(t)\frac{dx}{dt} + q(t)x = 0$, the homogeneous equation. (This was easy in the first-order case, because the equation was separable then. Unfortunately, there is no analogue to "separable" for equations of second and higher order.) One solution is obvious: $x = 0$. This solution to a homogeneous linear equation is often called the **trivial** solution, just because it is so obvious.

Fortunately, homogeneous linear equations have special properties that make it unnecessary to find *all* solutions from scratch. In fact, we'll show, in this section and in Section 2.4, that it is actually enough to find *one* nontrivial solution (one solution that is not identically zero) to the equation $x'' + p(t)x' + q(t)x = 0$. All other solutions can then be constructed from that one nontrivial solution!

Let's look at some methods to construct solutions from other solutions.

Warning: These methods work only for homogeneous linear equations.

Method 1: Multiply any solution by a constant. The result will again be a solution. For if x is the original solution and α is the constant, then the new function $y(t) = \alpha x(t)$ will have derivatives $y' = \alpha x'$ and $y'' = \alpha x''$, and so

$$\begin{aligned} y'' + p(t)y' + q(t)y &= \alpha x'' + p(t)\alpha x' + q(t)\alpha x \\ &= \alpha[x'' + p(t)x' + q(t)x] \\ &= \alpha \cdot 0 \quad \text{(because } x \text{ is a solution)} \\ &= 0. \end{aligned}$$

This shows that y is again a solution.

Method 2: Add two solutions. The result will again be a solution (see Exercise 17).

If we start with two solutions $x_1(t)$ and $x_2(t)$ to our equation $x'' + p(t)x' + q(t)x = 0$, we can combine the methods above by first taking separate multiples of $x_1(t)$ and $x_2(t)$ and then adding them. In this way we see that:

For any constants α and β, $\alpha x_1(t) + \beta x_2(t)$ is again a solution.

Methods 1 and 2 are actually special cases of this (for $\beta = 0$ and $\alpha = \beta = 1$, respectively).

EXAMPLE 2.2.2

$x'' - \frac{2}{t}x' + \frac{2}{t^2}x = 0 \quad (t \neq 0).$

It is easy to check (and you should do so) that $x_1(t) = t$ and $x_2(t) = t^2$ are both solutions of this equation. (It's a lot less clear how these solutions were found; eventually, Example

5.4.6 will provide an explanation.) Since the equation is a homogeneous linear one, any function of the form $x(t) = \alpha x_1(t) + \beta x_2(t) = \alpha t + \beta t^2$ will also be a solution, and this is easy to check. (For example, $-3t$, $t - \sqrt{3}\, t^2$, and $-\frac{1}{2}t + 6t^2$ are all solutions.) These are the only solutions we can get from x_1 and x_2 by the methods above; if we keep adding such solutions or multiplying them by constants, the results will have the same form. ∎

A function of the form $x(t) = \alpha x_1(t) + \beta x_2(t)$, where α and β are constants, is called a **linear combination** of the functions x_1 and x_2; α and β are called the coefficients of this linear combination. For the homogeneous linear equation $x'' + p(t)x' + q(t)x = 0$, we have seen that any linear combination of solutions is again a solution. What's more, we will see below that after restricting, if necessary, to a suitable interval for t, *every* solution can be obtained in this way from two particular solutions x_1 and x_2. For instance, in Example 2.2.2 ($x'' - \frac{2}{t}x' + \frac{2}{t^2}x = 0$), we can get every solution as a linear combination of t and t^2, provided that we restrict either to the interval $(0, \infty)$ or to the interval $(-\infty, 0)$. [To see why such a restriction is needed, note that since the equation is undefined for $t = 0$, a function such as

$$x(t) = \begin{cases} -t & (t < 0) \\ 3t - t^2 & (t > 0) \end{cases}$$

is a solution.]

For all this to work, neither of the two solutions x_1 and x_2 should be a constant multiple of the other; in particular, neither of them should be the zero function. Thus, in our example we couldn't have started with the solutions t and $2t$, because their only linear combinations would have been constant multiples of t. However, perhaps surprisingly, this is the *only* restriction on x_1 and x_2. Here is the result, which will be proved below:

Theorem 2.2.3 If both $x_1(t)$ and $x_2(t)$ are solutions of the homogeneous linear second-order equation $\frac{d^2x}{dt^2} + p(t)\frac{dx}{dt} + q(t)x = 0$, where $p(t)$ and $q(t)$ are continuous functions on some interval and $x_1(t)$ and $x_2(t)$ are defined on the same interval, and if neither of x_1 and x_2 is a constant multiple of the other,[6] then all the solutions of the equation on that interval are the linear combinations $\alpha x_1 + \beta x_2$, α and β arbitrary constants.

We'll call $x_1(t)$, $x_2(t)$ a pair of **basic** solutions to the equation if neither of them is a constant multiple of the other. (As we will see in Chapter 4, it is also customary to call them **linearly independent**.)

[6] If x_1 is a constant multiple of x_2, then x_2 is also one of x_1, *except if* $x_1 = 0$. This exceptional case makes for a lot of awkward phrasing of sentences.

EXAMPLE 2.2.3 Let's return to the equation $x'' + x = 0$ (see p. 85 and Section 2.1). We have seen that the solutions are given by $x(t) = C_0 \sin(t + D_0)$, where C_0 and D_0 are arbitrary constants. Now we'll pick two special solutions: Let $x_1(t)$ be the solution with $C_0 = 1$, $D_0 = 0$, so $x_1(t) = \sin t$, and let $x_2(t)$ be the one with $C_0 = 1$, $D_0 = \dfrac{\pi}{2}$, so

$$x_2(t) = \sin\left(t + \frac{\pi}{2}\right) = \cos t.$$

Then neither of x_1 and x_2 is a constant multiple of the other (why?), so Theorem 2.2.3 states that all the solutions are the linear combinations $\alpha \sin t + \beta \cos t$, α, β arbitrary. Sure enough, by the addition formula for sines,

$$C_0 \sin(t + D_0) = C_0 \cos D_0 \cdot \sin t + C_0 \sin D_0 \cdot \cos t$$

is such a linear combination. Note that finding only the two basic solutions $x_1(t) = \sin t$, $x_2(t) = \cos t$ would have been enough to find the general solution! (See also Exercise 28.) ∎

EXAMPLE 2.2.4 Solve the initial value problem

$$x'' - 5x' = 0, \quad x(0) = 2, \quad x'(0) = 15,$$

given that $x_1(t) = 1$ and $x_2(t) = e^{5t}$ are solutions to the differential equation (although they don't satisfy the initial conditions).

Solution By Theorem 2.2.3, the general solution to the differential equation is given by $x(t) = \alpha + \beta e^{5t}$, with α, β arbitrary constants. We need α and β such that $x(0) = 2$ and $x'(0) = 15$. Since $x'(t) = 5\beta e^{5t}$, these conditions yield $\alpha + \beta = 2$, $5\beta = 15$, from which we find $\beta = 3$, $\alpha = -1$. Therefore, we have

$$x(t) = \alpha + \beta e^{5t} = -1 + 3e^{5t}. \quad \blacksquare$$

We now prove Theorem 2.2.3. The proof features a nice application of Theorem 2.2.2, but it is not crucial for understanding the rest of this section, and on first reading you may want to skip ahead to the middle of p. 91.

Let $x_1(t)$ and $x_2(t)$ be solutions, neither of which is a constant multiple of the other, of $\dfrac{d^2x}{dt^2} + p(t)\dfrac{dx}{dt} + q(t)x = 0$, defined on some interval containing t_0, and let $x(t)$ be any solution defined at t_0. We want to show that $x(t)$ is a linear combination $\alpha x_1(t) + \beta x_2(t)$ of the two given solutions. By Theorem 2.2.2, we can do this by showing that for some constants α and β, $x(t)$ and $\alpha x_1(t) + \beta x_2(t)$ are solutions of the same initial value problem (and therefore equal). That is, it's enough to show that we can find α and β such that

$$\begin{cases} x(t_0) = \alpha x_1(t_0) + \beta x_2(t_0) \\ x'(t_0) = \alpha x_1'(t_0) + \beta x_2'(t_0). \end{cases}$$

SEC. 2.2 GENERAL RESULTS

If we put $x_0 = x(t_0)$, $v_0 = x'(t_0)$, this means that we must be able to solve the system of linear equations

$$\begin{cases} \alpha x_1(t_0) + \beta x_2(t_0) = x_0 \\ \alpha x_1'(t_0) + \beta x_2'(t_0) = v_0 \end{cases} \quad (*)$$

for α and β. This can be done (see Appendix B, p. 624) provided that the determinant

$$\begin{vmatrix} x_1(t_0) & x_2(t_0) \\ x_1'(t_0) & x_2'(t_0) \end{vmatrix}$$

is not zero. Of course, this argument will work for any t_0 where the solutions are defined, so all we have to show is that for *some* such t_0,

$$\begin{vmatrix} x_1(t_0) & x_2(t_0) \\ x_1'(t_0) & x_2'(t_0) \end{vmatrix} \neq 0, \quad \text{that is,} \quad x_1(t_0)x_2'(t_0) - x_1'(t_0)x_2(t_0) \neq 0.$$

To see why this should be true, consider the quotient $\dfrac{x_2(t)}{x_1(t)}$ of the two given solutions. This is *not* a constant on our interval (why?); it is a differentiable function, except for those values of t (if any) for which x_1 is zero. By the quotient rule, its derivative is $\dfrac{x_1(t)x_2'(t) - x_1'(t)x_2(t)}{[x_1(t)]^2}$; this is not everywhere zero because the (quotient) function is not constant.[7] So the numerator is not everywhere zero, that is, $x_1(t_0)x_2'(t_0) - x_1'(t_0)x_2(t_0) \neq 0$ for some t_0. For such a t_0, we can solve (*) for α and β. Therefore, all solutions can indeed be found as linear combinations of x_1 and x_2, and the proof of Theorem 2.2.3 is complete.

We have now seen how to find all solutions of the homogeneous equation from two suitable ones; in Section 2.4 we will see how to find them all from just one, with a bit more work. Meanwhile, you may well ask: Given two solutions, how does one know whether or not they are constant multiples of each other? In the examples above this was not really a problem; it is clear that t and t^2, or $\sin t$ and $\cos t$, are not constant multiples of each other. In more complicated cases, though, especially if the solutions are found as integrals or as power series rather than in closed form, it may not be so clear. However, there is a check, which is based on the same idea as our proof of Theorem 2.2.3.

Theorem 2.2.4   If both $x_1(t)$ and $x_2(t)$ are solutions of the equation $x'' + p(t)x' + q(t)x = 0$, defined on some interval on which $p(t)$, $q(t)$ are defined and continuous, and if neither of $x_1(t)$, $x_2(t)$ is a constant multiple of the other, then $\begin{vmatrix} x_1(t) & x_2(t) \\ x_1'(t) & x_2'(t) \end{vmatrix} \neq 0$ for all t in the (interior of the)

---

[7] As you may have noticed (if so, good for you!), this argument is not quite secure, because it is conceivable that $\dfrac{x_2(t)}{x_1(t)}$ would fail to stay constant only for those values of t for which x_1 is zero. See Exercise 31 for a precise argument of this point.

interval. On the other hand, if $x_1(t)$ and $x_2(t)$ are any two differentiable functions on an interval and one of them is a constant multiple of the other, then $\begin{vmatrix} x_1(t) & x_2(t) \\ x_1'(t) & x_2'(t) \end{vmatrix} = 0$ for all t in the interval.

This theorem is proved in Exercise 30. Here is an example of how the theorem can be used.

EXAMPLE 2.2.5 Suppose that we have somehow found the solutions $x_1(t) = e^{-(t-\pi/4)} \cos 2(t - \frac{\pi}{4})$ and $x_2(t) = e^{-t} \sin t \cos t$ to the equation $x'' + 2x' + 5x = 0$. (You can check by direct substitution that they are solutions.) Can we conclude that all solutions are of the form $\alpha e^{-(t-\pi/4)} \cos 2(t - \frac{\pi}{4}) + \beta e^{-t} \sin t \cos t$? Well, only if neither of $x_1(t)$, $x_2(t)$ is a constant multiple of the other. Now the determinant $\begin{vmatrix} x_1(t) & x_2(t) \\ x_1'(t) & x_2'(t) \end{vmatrix}$ equals

$$\begin{vmatrix} e^{-(t-\pi/4)} \cos 2\left(t - \frac{\pi}{4}\right) & e^{-t} \sin t \cos t \\ e^{-(t-\pi/4)}\left[-\cos 2\left(t - \frac{\pi}{4}\right) - 2 \sin 2\left(t - \frac{\pi}{4}\right)\right] & e^{-t}[-\sin t \cos t + \cos^2 t - \sin^2 t] \end{vmatrix}.$$

By Theorem 2.2.4, this determinant is either zero for all t, in which case one of x_1, x_2 is a constant multiple of the other, or it is nonzero for all t, in which case x_1, x_2 form a pair of basic solutions. To find out which, we only have to check one value of t, and we can choose t to make the computation short. For instance, for $t = \frac{\pi}{4}$, the determinant is

$$\begin{vmatrix} 1 & \frac{1}{2} e^{-\pi/4} \\ -1 & -\frac{1}{2} e^{-\pi/4} \end{vmatrix} = 0.$$

So the determinant is *always* zero, and thus x_1 and x_2 (which are obviously not zero) must be constant multiples of each other! (See Exercise 19.) As a result, *not* all solutions will be linear combinations of x_1 and x_2 (see Exercise 20). ∎

The determinant $\begin{vmatrix} x_1(t) & x_2(t) \\ x_1'(t) & x_2'(t) \end{vmatrix}$ that occurs in Theorem 2.2.4 is called the **Wronskian** of the functions x_1 and x_2 (it was named for a Polish mathematician); it is often denoted by $W(x_1, x_2)(t)$ or just $W(x_1, x_2)$.

Warning: If x_1 and x_2 are just any functions, then the Wronskian may very well be zero for some t and nonzero for others. Only if it is known that x_1 and x_2 both satisfy the same equation $x'' + p(t)x' + q(t)x = 0$ does it follow that the Wronskian is either zero for all t where p, q are defined and continuous, or nonzero for all those t. For example, for $x_1(t) = t$, $x_2(t) = t^3 - 2t^2 + 2t$ the Wronskian is

$$W(t, t^3 - 2t^2 + 2t) = \begin{vmatrix} t & t^3 - 2t^2 + 2t \\ 1 & 3t^2 - 4t + 2 \end{vmatrix} = 2t^3 - 2t^2,$$

which is zero both for $t = 0$ and for $t = 1$. This shows that if $x_1(t)$ and $x_2(t)$ would both satisfy an equation $x'' + p(t)x' + q(t)x = 0$, then for $t = 0$ and again for $t = 1$, $p(t)$, $q(t)$, or both would have to be undefined. (In Exercise 21 we will see that there is, indeed, such an equation.)

SUMMARY OF KEY CONCEPTS AND RESULTS

Initial value problem with a second-order equation (p. 85):

$$\frac{d^2x}{dt^2} = f(t, x, v), \quad x(0) = x_0, \quad v(0) = v_0, \quad \text{where} \quad v = \frac{dx}{dt}.$$

Such an initial value problem has a unique solution provided that f, $\dfrac{\partial f}{\partial x}$, and $\dfrac{\partial f}{\partial v}$ are defined and continuous (see Theorem 2.2.1, p. 86, for a more precise statement).

Initial value problem with a second-order linear equation (p. 87):

$$\frac{d^2x}{dt^2} + p(t)\frac{dx}{dt} + q(t)x = h(t), \quad x(t_0) = x_0, \quad v(t_0) = v_0$$

has a unique solution defined on any interval with t_0 in its interior on which $p(t)$, $q(t)$, and $h(t)$ are defined and continuous.

Linear combination of two functions $x_1(t)$, $x_2(t)$: $\alpha x_1(t) + \beta x_2(t)$, where α, β are constants (p. 89).

For a **homogeneous linear** equation, any linear combination of solutions is again a solution (p. 88).

Basic solutions (p. 89): The two functions $x_1(t)$, $x_2(t)$ form a pair of basic solutions to the equation $\dfrac{d^2x}{dt^2} + p(t)\dfrac{dx}{dt} + q(t)x = 0$ if they are solutions, neither of which is a constant multiple of the other. In this situation, *all* the solutions of the equation are linear combinations of x_1 and x_2. (See Theorem 2.2.3, p. 89, for a more precise statement.)

Wronskian (p. 92): $W(x_1, x_2)(t) = \begin{vmatrix} x_1(t) & x_2(t) \\ x_1'(t) & x_2'(t) \end{vmatrix}$. If $x_1(t)$ and $x_2(t)$ are both solutions of $x'' + p(t)x' + q(t)x = 0$ and $p(t)$, $q(t)$ are defined and continuous, then either $W(x_1, x_2)(t) \neq 0$ for all t or $W(x_1, x_2)(t) = 0$ for all t. (See Theorem 2.2.4, p. 91, for a more precise statement.) In the first case $x_1(t)$, $x_2(t)$ form a pair of basic solutions, and in the second case they don't.

EXERCISES

For each of the following initial value problems, decide whether you can expect a unique solution, at least on some interval. Explain your answer.

1. $(x+1)\dfrac{d^2x}{dt^2} + 3\dfrac{dx}{dt} = t, \quad x(0) = -1, \quad \dfrac{dx}{dt}\Big|_{t=0} = 0.$

2. $(x+1)\dfrac{d^2x}{dt^2} + 3\dfrac{dx}{dt} = t, \quad x(0) = 0, \quad \dfrac{dx}{dt}\Big|_{t=0} = -1.$

3. $x'' + 5x' + 4x = 0, \quad x(0) = -1, \quad x'(0) = 3.$

4. $x'' + 5x' + 4x = 0, \quad x(0) = -1.$

5. $\dfrac{d^2x}{dt^2} + \left(\dfrac{dx}{dt}\right)^{1/3} + x = 0, \quad x(0) = 2, \quad \dfrac{dx}{dt}\Big|_{t=0} = 0.$

6. $\dfrac{d^2x}{dt^2} + (t^2-1)\dfrac{dx}{dt} + t^{1/3}x = 0, \quad x(0) = 0, \quad \dfrac{dx}{dt}\Big|_{t=0} = 2.$

7. Consider the initial value problem $x'' = x^{1/3}, x(0) = 0, x'(0) = 0$. Show that in addition to the obvious solution $x = 0$, there are several solutions of the form $x = Ct^3$ (for which values of C?). Does this contradict Theorem 2.2.1 (p. 86)?

8. Show that the initial value problem from Example 2.2.1,

$$(x-1)\dfrac{d^2x}{dt^2} + 3x\dfrac{dx}{dt} + e^x = -t^2, \quad x(0) = 1, \quad \dfrac{dx}{dt}\Big|_{t=0} = 2,$$

has no solution.
[*Hint*: Don't try to solve the differential equation! If there were a solution, what could you say for $t = 0$?] Does this contradict Theorem 2.2.1?

9. (a) Show that $x_1(t) = \dfrac{1}{t}$ and $x_2(t) = 1$ are both solutions of $x'' + 2xx' = 0$.
   (b) Check whether $x_3(t) = x_1(t) + x_2(t) = \dfrac{1}{t} + 1$ is a solution of $x'' + 2xx' = 0$, as well. Does your answer agree with Theorem 2.2.3 (p. 89)? Explain.

10. (a) Show that $x_1(t) = e^t - t$ and $x_2(t) = e^{-t} - t$ are both solutions of $x'' - x = t$.
    (b) Check whether $x_3(t) = x_1(t) - x_2(t) = e^t - e^{-t}$ is a solution of $x'' - x = t$, as well. Does your answer agree with Theorem 2.2.3 (p. 89)? Explain.

Solve each of the following initial value problems, given that $x_1(t)$ and $x_2(t)$ are solutions to the differential equation (although they don't satisfy the initial conditions).

11. $x'' + 4x = 0, \quad x(0) = 1, \quad x'(0) = 3; \quad x_1(t) = \sin 2t, \quad x_2(t) = \cos 2t.$
12. $x'' + 9x = 0, \quad x(0) = 4, \quad x'(0) = -1; \quad x_1(t) = \cos 3t, \quad x_2(t) = \sin 3t.$
13. $x'' - 4x' + 3x = 0, \quad x(0) = 0, \quad x'(0) = 6; \quad x_1(t) = e^t, \quad x_2(t) = e^{3t}.$
14. $x'' + 4x' + 3x = 0, \quad x(0) = 0, \quad x'(0) = -12; \quad x_1(t) = e^{-3t}, \quad x_2(t) = e^{-t}.$

15. (a) Show that $x_1(t) = t^3$ and $x_2(t) = t$ are solutions to the equation
    $t^2 x'' - 3tx' + 3x = 0$.
    (b) Find the general solution to this differential equation for $t > 0$.
    (c) Solve the initial value problem $t^2 x'' - 3tx' + 3x = 0$, $x(1) = 3$, $x'(1) = 5$.
    (d) What happens for $t^2 x'' - 3tx' + 3x = 0$, $x(0) = 0$, $x'(0) = 1$?

16. (a) Show that $x_1(t) = t$ and $x_2(t) = \dfrac{1}{t}$ are solutions to the equation
    $t^2 x'' + tx' - x = 0$.
    (b) Find the general solution to this differential equation for $t > 0$.
    (c) Solve the initial value problem $t^2 x'' + tx' - x = 0$, $x(1) = 1$, $x'(1) = 2$.
    (d) What happens for $t^2 x'' + tx' - x = 0$, $x(0) = 1$, $x'(0) = 2$?

17. (See p. 88.) Show that if x_1 and x_2 are solutions to the equation
    $x'' + p(t)x' + q(t)x = 0$, then so is their sum $x_1 + x_2$.

18. Let $x_1(t)$, $x_2(t)$ be functions, neither of which is a constant multiple of the other.
    (a) Show that if $\alpha x_1(t) + \beta x_2(t) = 0$ for all t, then $\alpha = \beta = 0$.
    (b) Show that if $\alpha x_1(t) + \beta x_2(t) = \gamma x_1(t) + \delta x_2(t)$ for all t, then $\alpha = \gamma$ and $\beta = \delta$. [*Hint*: Subtract.] Note that this shows that all linear combinations of $x_1(t)$, $x_2(t)$ are distinct.

19. (See Example 2.2.5, p. 92.) Show without using the Wronskian that
    $x_1(t) = e^{-(t-\pi/4)} \cos 2(t - \dfrac{\pi}{4})$ is a constant multiple of $x_2(t) = e^{-t} \sin t \cos t$.

20. Explain why in Example 2.2.5, once we know that x_1 and x_2 are constant multiples of each other, we can conclude that not all solutions are of the form $\alpha x_1 + \beta x_2$.

21. Find a differential equation $x'' + p(t)x' + q(t)x = 0$ that has both $x_1(t) = t$ and $x_2(t) = t^3 - 2t^2 + 2t$ as solutions. Where are $p(t)$ and $q(t)$ defined?
    [*Hint*: If $x_1(t)$ is a solution, what equation does that imply for $p(t)$ and $q(t)$? Compare p. 93.]

22. Find a second-order linear homogeneous equation that has both $x_1(t) = e^t$ and $x_2(t) = e^{-t}$ as solutions.

23. Let $x_1(t)$ and $x_2(t)$ be any two twice-differentiable functions. Show that there is a differential equation $x'' + p(t)x' + q(t)x = 0$ that has both $x_1(t)$ and $x_2(t)$ as solutions, provided that the Wronskian of x_1 and x_2 is not identically zero. Where will $p(t)$ and $q(t)$ be defined?

24. Suppose that for the differential equation $x'' + p(t)x' + q(t)x = 0$, $p(t)$ and $q(t)$ defined and continuous at $t = 0$, $x_1(t)$ is the solution that satisfies $x_1(0) = 1$, $x_1'(0) = 0$, while $x_2(t)$ is the solution for which $x_2(0) = 0$, $x_2'(0) = 1$. Find the solution $x(t)$ for which $x(0) = x_0$, $x'(0) = v_0$.

25. Find all solutions of the equation $x'' = 0$; for each solution $x(t)$, find $v = x'$. Then sketch the particular solutions and their derivatives in t, x, v-space for which
    (a) $x(0) = 2$, $x'(0) = -1$; (b) $x(0) = 3$, $x'(0) = 1$.

26. Find all solutions of the equation $x'' = 1$; for each solution $x(t)$, find $v = x'$. Then sketch the particular solutions and their derivatives in t, x, v-space for which
    (a) $x(0) = 0$, $x'(0) = 0$; (b) $x(0) = 2$, $x'(0) = -1$.

27. (a) Show that if x_1 and x_2 are solutions of $x'' + p(t)x' + q(t)x = 0$, their Wronskian $W(x_1, x_2) = x_1 x_2' - x_1' x_2$ satisfies the first-order differential equation $W' = -pW$.
    (b) Solve the equation in part (a) to get an expression for the Wronskian in terms of $\int p(t)\, dt$. [This is a particular case of **Abel's identity**, named after the great Norwegian mathematician N. H. Abel (1802–1829).]

28. (See Example 2.2.3.) Show directly that any linear combination $\alpha \sin t + \beta \cos t$ can be written in the form $C_0 \sin(t + D_0)$ for suitable constants C_0, D_0.

*29. Show directly from Theorem 2.2.2 (p. 87) that the only solutions to $x'' + x = 0$ are given by $x = C_0 \sin(t + D_0)$.
    [*Hint*: Any solution $x(t)$ is defined for all t (why?), so $x(0)$ has some value x_0 and $x'(0)$ has some value v_0. Now show that one of the solutions $C_0 \sin(t + D_0)$ gives these same initial values.]

30. In this exercise we will prove Theorem 2.2.4 (p. 91).
    (a) Show that if $x_1(t) = \alpha x_2(t)$, where α is a constant and x_2 is a differentiable function, then
    $$\begin{vmatrix} x_1(t) & x_2(t) \\ x_1'(t) & x_2'(t) \end{vmatrix} = 0 \quad \text{for all } t.$$
    *(b) Assume that x_1, x_2 is a pair of basic solutions of $x'' + p(t)x' + q(t)x = 0$, defined on some interval, containing t_0 in its interior, on which $p(t)$, $q(t)$ are defined and continuous. By Theorem 2.2.2, the initial value problem $x'' + p(t)x' + q(t)x = 0$, $x(t_0) = x_0$, $x'(t_0) = v_0$ has a unique solution for any choice of x_0, v_0. By Theorem 2.2.3, this unique solution must be of the form $x = \alpha x_1 + \beta x_2$, α and β constants. Use these facts to prove that the Wronskian $W(x_1, x_2)(t) = \begin{vmatrix} x_1(t) & x_2(t) \\ x_1'(t) & x_2'(t) \end{vmatrix}$ is not zero for $t = t_0$.

31. Assume that x_1 and x_2 are solutions of the differential equation $x'' + p(t)x' + q(t)x = 0$ with $p(t)$, $q(t)$ defined and continuous.
    (a) Show that if $\dfrac{x_2(t)}{x_1(t)}$ is constant on *some* interval (with nonempty interior), then x_2 is a constant multiple of x_1 (everywhere, not just on that interval).
    [*Hint*: If $\dfrac{x_2(t)}{x_1(t)} = c$ on some interval, show that $x_2 - cx_1$ and 0 are solutions of the same initial value problem.]
    (b) Show that if neither of x_1, x_2 is a constant multiple of the other, then there is some t_0 for which $x_1(t_0)x_2'(t_0) - x_1'(t_0)x_2(t_0) \neq 0$.
    [*Hint*: x_1 cannot be identically zero (why?), so there must be some interval on which x_1 does not become zero. By part (a), $\dfrac{x_2(t)}{x_1(t)}$ cannot be constant on that interval.]

32. Since the first-order linear equation $\dfrac{dx}{dt} + f(t)x = 0$ has a solution $x = e^{-\int f(t)dt}$ (see Section 1.6, p. 49), it may be tempting to try to find a solution of the same type to the homogeneous second-order linear equation $\dfrac{d^2x}{dt^2} + p(t)\dfrac{dx}{dt} + q(t)x = 0$. Show that if $x = e^{-\int f(t)dt}$ is a solution of this second-order equation, then the unknown function $f(t)$ must satisfy $f'(t) = [f(t)]^2 - p(t)f(t) + q(t)$.

NOTE: An equation such as this, of the form $\dfrac{dy}{dt} = y^2 - p(t)y + q(t)$, is called a **Riccati equation**. Usually, these equations are not easy to solve, and sometimes they are solved by reversing the process above and solving the associated second-order equations.]

Solve the following second-order equations by the method of Exercise 32, that is, by first solving the corresponding Riccati equations.

33. $x'' + x = 0$.

34. $x'' + 4x = 0$.

*__35.__ $x'' - 5x' + 6x = 0$.

*__36.__ Consider the two functions given by $x_1(t) = t^2$, $x_2(t) = \begin{cases} -t^2 & (t \le 0) \\ t^2 & (t \ge 0) \end{cases}$. Show that x_1 and x_2 are differentiable for all t; show that the Wronskian of x_1 and x_2 is identically zero. Neither of x_1, x_2 is a constant multiple of the other (although they are multiples if we restrict either to $t \le 0$ or to $t \ge 0$). Conclude that there is no differential equation $x'' + p(t)x' + q(t)x = 0$ with $p(t)$, $q(t)$ defined and continuous for all t that has $x_1(t)$ and $x_2(t)$ as solutions. Find a differential equation of this form with $p(t)$, $q(t)$ defined for all $t \ne 0$ that does have $x_1(t)$ and $x_2(t)$ as solutions.

FURTHER READING

For a reasonably accessible proof of the "uniqueness" part of Theorem 2.2.2, see Birkhoff and Rota, *Ordinary Differential Equations*, 4th ed. (Wiley, 1989), Chapter 2, Section 4.

2.3 CONSTANT-COEFFICIENT EQUATIONS: INTRODUCTION

For most of the rest of this chapter we consider the case of a linear second-order equation *with constant coefficients*, that is, a differential equation of the form

$$ax'' + bx' + cx = g(t)$$

for the unknown function x of the variable t. Here a, b, and c are constants; a is not zero (otherwise, the equation is of first order; see Section 1.6 for this case). Therefore, we can divide through by a to get

$$x'' + px' + qx = h(t)$$

[with $p = \dfrac{b}{a}$, $q = \dfrac{c}{a}$ constants and $h(t) = \dfrac{1}{a}g(t)$ some function of t]. We will assume that the equation is given in this last form.

As we did for linear first-order equations, we will start with the homogeneous case: $x'' + px' + qx = 0$. Once we know how to solve this equation completely (for any given values of p and q), we will (in Sections 2.6 and 2.8) discuss how to find solutions of the inhomogeneous equation.

So p and q are given constants, and we have the equation $x'' + px' + qx = 0$. How can we find solutions to this (besides the trivial solution $x = 0$)? One approach is to look at the case $q = 0$ first. In this case the equation becomes $x'' + px' = 0$. Since x itself does not occur now, we can think of this as an equation for $v = x'$, namely $v' + pv = 0$. This is a familiar separable equation, and we get

$$v = Ae^{-pt}$$
$$x = \int v\, dt = \int Ae^{-pt}\, dt = -\frac{A}{p}e^{-pt} + C \qquad (p \neq 0).$$

Here A and C are arbitrary constants of integration (you could simplify the answer by replacing $-\dfrac{A}{p}$ by a new constant, B). If $p = 0$, we get $v = A$, $x = At + C$ instead.

Thus we see that for $q = 0$, $p \neq 0$ the solutions we get are constants and exponential functions (and linear combinations of these). Can we expect similar solutions for $q \neq 0$? Well, we certainly can't get nonzero constant solutions if $q \neq 0$ (see Exercise 1). However, exponential functions might perhaps occur as solutions.

With this motivation (or, if you prefer, as a blind guess), let us try to find solutions of the form $x(t) = e^{\lambda t}$, λ a constant,[8] to the equation $x'' + px' + qx = 0$. Differentiating x, we find that $x' = \lambda e^{\lambda t}$, $x'' = \lambda^2 e^{\lambda t}$, so we want

$$\lambda^2 e^{\lambda t} + p\lambda e^{\lambda t} + qe^{\lambda t} = 0$$
$$(\lambda^2 + p\lambda + q)e^{\lambda t} = 0.$$

---
[8] The choice of the Greek letter λ is traditional here.

SEC. 2.3 CONSTANT-COEFFICIENT EQUATIONS: INTRODUCTION

Since the exponential is certainly not zero, this will only happen if $\lambda^2 + p\lambda + q = 0$. Conversely, reversing the steps above shows that:

If λ is a root of the quadratic equation $\lambda^2 + p\lambda + q = 0$, then $x(t) = e^{\lambda t}$ is a solution of $x'' + px' + qx = 0$.

The quadratic equation $\lambda^2 + p\lambda + q = 0$ is called the **characteristic equation** (of the differential equation $x'' + px' + qx = 0$).

As you know from high-school algebra, the roots of $\lambda^2 + p\lambda + q = 0$ are given by $\lambda_{1,2} = \dfrac{-p \pm \sqrt{p^2 - 4q}}{2}$; there are two, one, or no real roots depending on whether $p^2 > 4q$, $p^2 = 4q$, or $p^2 < 4q$. In this section we'll consider the easiest case, $p^2 > 4q$. Thus there will be two distinct real roots, λ_1 and λ_2, and therefore two distinct solutions, $x_1(t) = e^{\lambda_1 t}$ and $x_2(t) = e^{\lambda_2 t}$, to the differential equation. (The other two cases are treated in Sections 2.4 and 2.5.)

Neither of the solutions $x_1(t)$ and $x_2(t)$ is a constant multiple of the other. One way to see this is to compute their Wronskian (see Section 2.2, p. 92):

$$W(e^{\lambda_1 t}, e^{\lambda_2 t}) = \begin{vmatrix} e^{\lambda_1 t} & e^{\lambda_2 t} \\ \lambda_1 e^{\lambda_1 t} & \lambda_2 e^{\lambda_2 t} \end{vmatrix} = (\lambda_2 - \lambda_1) e^{(\lambda_1 + \lambda_2) t},$$

which is nonzero since $\lambda_2 - \lambda_1 \neq 0$ (why?).

We now know that $x_1(t) = e^{\lambda_1 t}$, $x_2(t) = e^{\lambda_2 t}$ is a pair of basic solutions, and by Theorem 2.2.3 (p. 89) the general solution to our differential equation is given by

$$x(t) = \alpha e^{\lambda_1 t} + \beta e^{\lambda_2 t},$$

where α and β are arbitrary constants.

EXAMPLE 2.3.1 $x'' + 3x' + 2x = 0$.

The characteristic equation is $\lambda^2 + 3\lambda + 2 = 0$, which factors as

$$(\lambda + 2)(\lambda + 1) = 0.$$

This has roots $\lambda_1 = -2$, $\lambda_2 = -1$, so we have basic solutions e^{-2t} and e^{-t} to the differential equation. The general solution is $x(t) = \alpha e^{-2t} + \beta e^{-t}$, α, β arbitrary constants. ∎

EXAMPLE 2.3.2 $x'' - 4x' + x = 0$.

The characteristic equation $\lambda^2 - 4\lambda + 1 = 0$ has roots

$$\lambda_{1,2} = \frac{4 \pm \sqrt{16 - 4}}{2} = 2 \pm \sqrt{3},$$

so the general solution of the differential equation is $x(t) = \alpha e^{(2+\sqrt{3})t} + \beta e^{(2-\sqrt{3})t}$. ∎

EXAMPLE 2.3.3 Solve the initial value problem $x'' - 5x' + 6x = 0$, $x(0) = 2$, $x'(0) = -1$.

Solution First we solve the characteristic equation $\lambda^2 - 5\lambda + 6 = 0$. We find $\lambda_1 = 2$, $\lambda_2 = 3$, so our general solution is $x(t) = \alpha e^{2t} + \beta e^{3t}$. Now we have to find α and β such that $x(0) = 2$, $x'(0) = -1$. To begin, we see from $x(t) = \alpha e^{2t} + \beta e^{3t}$ that $x(0) = \alpha + \beta$. Also, differentiating $x(t)$ we get $x'(t) = 2\alpha e^{2t} + 3\beta e^{3t}$, so that $x'(0) = 2\alpha + 3\beta$. Thus we have to find α and β such that $\alpha + \beta = 2$ and $2\alpha + 3\beta = -1$. Elimination yields $\alpha = 7$, $\beta = -5$, so the solution we want is $x(t) = 7e^{2t} - 5e^{3t}$. ∎

EXAMPLE 2.3.4 $x'' - 4x' + 4x = 0$.

In this case, the characteristic equation $\lambda^2 - 4\lambda + 4 = 0$ has a double root, $\lambda = 2$. (Our assumption $p^2 > 4q$ is not valid here!) Therefore, although we know that e^{2t} is a solution of the differential equation, we do not yet know how to find the general solution. This will be done in Section 2.4, using the method of "reduction of order." ∎

EXAMPLE 2.3.5 $x'' + 4x' + 5x = 0$.

For this differential equation, $p^2 < 4q$ ($p = 4, q = 5$), and our assumption is again invalid. We don't know yet how to find any solutions at all in this case; it will be postponed until Section 2.5. ∎

SUMMARY OF KEY CONCEPTS AND RESULTS

Characteristic equation of the constant-coefficient equation $x'' + px' + qx = 0$ (p. 99): $\lambda^2 + p\lambda + q = 0$. If λ is a solution of the characteristic equation, then $x(t) = e^{\lambda t}$ is a solution of the differential equation. If $p^2 > 4q$, the characteristic equation has two distinct real roots λ_1, λ_2. In this case $x_1(t) = e^{\lambda_1 t}$, $x_2(t) = e^{\lambda_2 t}$ is a pair of basic solutions, and $x(t) = \alpha e^{\lambda_1 t} + \beta e^{\lambda_2 t}$ is the general solution, of $x'' + px' + qx = 0$ (p. 99).

EXERCISES

1. Show directly that for $q \neq 0$, the equation $x'' + px' + qx = 0$ has no nontrivial constant solutions.

Solve the following differential equations.

2. $x'' - 4x' + 3x = 0$.

3. $x'' - 3x' = 0$.

4. $\dfrac{d^2y}{dx^2} - 5\dfrac{dy}{dx} + 6y = 0$.

5. $y'' + 2y' - y = 0$. (y is a function of x.)

6. $\dfrac{d^2x}{dt^2} + 3\dfrac{dx}{dt} - 4x = 0$.

Solve the following initial value problems.

7. $3x'' + 9x' - 30x = 0$, $x(0) = 0$, $x'(0) = 1$.
8. $5x'' - 10x' = 0$, $x(0) = 1$, $x'(0) = -1$.
9. $y'' + 8y' - 20y = 0$, $y(0) = -1$, $y'(0) = 2$. (y is a function of x.)
10. $y'' - 9y' + 8y = 0$, $y(2) = 1$, $y'(2) = 0$.
11. $x'' + 4x' + 3x = 0$, $x(3) = 0$, $x'(3) = -5$.
12. $\dfrac{d^2y}{dx^2} + 35167 \dfrac{dy}{dx} + 53782y = 0$, $y(1) = 0$, $y'(1) = 0$.
    [*Hint*: Theorem 2.2.2 (p. 87) will be helpful.]

13. Consider the differential equation $x'' - (2 + r)x' + 2rx = 0$, where r is an arbitrary constant.
    (a) Show that for $r \neq 2$, the general solution is $x(t) = \alpha e^{2t} + \beta e^{rt}$.
    (b) Show that the solution satisfying $x(0) = 0$, $x'(0) = 1$ is $x(t) = \dfrac{e^{rt} - e^{2t}}{r - 2}$
    ($r \neq 2$).
    *(c) Take the limit of the solution in part (b) as $r \to 2$ (use l'Hôpital's rule, with r as the variable!). What function of t do you get? Check whether this function is a solution to the "limit differential equation" $x'' - 4x' + 4x = 0$ obtained for $r = 2$.
    (d) Using part (c), find the general solution of $x'' - 4x' + 4x = 0$.

14. In this problem we outline an alternative method of solution for the equation $x'' + px' + qx = 0$.
    (a) Define a new unknown function $y(t)$ by $y(t) = e^{\mu t}x(t)$, where μ is a constant that we will choose later. Find the second-order differential equation satisfied by $y(t)$.
    [*Hint*: Substitute $x = e^{-\mu t}y$ in the given equation.]
    (b) Show that μ can be chosen so that the term with y' disappears from the new differential equation, which can then (in principle) be solved as in Section 2.1.
    (c) Use the results of Section 2.1 to find the general solution to the equation $x'' + 2x' + 5x = 0$.

2.4 REDUCTION OF ORDER

In this section we assume that one (nontrivial) solution of the homogeneous linear second-order equation $x'' + p(t)x' + q(t)x = 0$ is already known; we will call this known solution $x_1(t)$. Note that the coefficients $p(t)$, $q(t)$ are not assumed to be constant. We now show how to find all other solutions from x_1. As we have seen, it will actually be enough to find one other solution that is not a constant multiple of $x_1(t)$. Surprisingly, this observation indicates a method of attack. Since $x_1(t)$ is not identically zero, a function x that is not a constant multiple of x_1 must be a "variable multiple" of x_1; that is, we can write $x(t) = A(t)x_1(t)$ for some unknown function A. (A may not be defined where the function x_1 is zero.) We will make this substitution and try to solve for the unknown function A rather than for x. Differentiating $x(t) = A(t)x_1(t)$ twice, we get (by the product rule) $x'(t) = A'(t)x_1(t) + A(t)x_1'(t)$, or $x' = A'x_1 + Ax_1'$ for purposes of space conservation, and then

$$x'' = A''x_1 + A'x_1' + A'x_1' + Ax_1'' = A''x_1 + 2A'x_1' + Ax_1''.$$

Substituting into our differential equation yields

$$A''x_1 + 2A'x_1' + Ax_1'' + pA'x_1 + pAx_1' + qAx_1 = 0,$$

which can be rewritten as

$$A''x_1 + A'(2x_1' + px_1) + A(x_1'' + px_1' + qx_1) = 0.$$

However, x_1 is a solution of the original equation, so $x_1'' + px_1' + qx_1 = 0$, and the above simplifies to

$$A''x_1 + A'(2x_1' + px_1) = 0.$$

It is very convenient that the term with A has dropped out, because we'll now be able to think of A' as the unknown function in our last equation. (Keep in mind that x_1', p, and x_1 are all known functions of t.) Is it a miracle? Hardly. After all, we know that if $A(t)$ is constant, that is, if $x(t)$ is a constant multiple of $x_1(t)$, then $x(t)$ will be a solution to the original differential equation. Therefore, the differential equation we found for $A(t)$ *had to* have all the constant functions among its solutions, and this would not have been true if there had been a term with A in addition to the terms with A' and A''.

Enough philosophy. Having found the equation $A''x_1 + A'(2x_1' + px_1) = 0$, we can put $a = A'$, that is, $a(t) = A'(t)$, and get the first-order linear equation

$$x_1 a' + (2x_1' + px_1)a = 0$$

for the new unknown function $a(t)$. This equation is separable, and once we have solved for $a(t)$, we can integrate to find $A(t) = \int a(t)\, dt$, then multiply by the original solution $x_1(t)$ to find the general solution $x(t)$ (that is, to find all possible solutions).

The method above is called the method of **reduction of order**. It is so named because the original linear *second-order* equation is replaced by the linear *first-order* equation for $a(t)$.

SEC. 2.4 REDUCTION OF ORDER

EXAMPLE 2.4.1 Suppose that you are solving the differential equation $x'' + x = 0$ and that you have somehow found the solution $x_1(t) = \sin t$, but no other solutions yet. To find *all* solutions, you can write $x(t) = A(t) \sin t$, find the second derivative $x''(t)$, and then substitute x and x'' into the differential equation:

$$x(t) = A(t) \sin t$$
$$x'(t) = A'(t) \sin t + A(t) \cos t$$
$$x''(t) = A''(t) \sin t + 2A'(t) \cos t - A(t) \sin t,$$

so

$$x'' + x = A'' \sin t + 2A' \cos t - A \sin t + A \sin t = A'' \sin t + 2A' \cos t,$$

which makes your equation

$$A'' \sin t + 2A' \cos t = 0.$$

Note that the term with A has canceled, as promised. You can now proceed to solve for $a(t) = A'(t)$:

$$a' \sin t + 2a \cos t = 0.$$

Separate:

$$\frac{a'}{a} = -2 \frac{\cos t}{\sin t} \quad (\text{or } a = 0).$$

Integrate:

$$\int \frac{da}{a} = -2 \int \frac{\cos t}{\sin t} dt$$

$$\begin{matrix} (a > 0) \\ (a < 0) \end{matrix} \quad \left. \begin{matrix} \log a \\ \log(-a) \end{matrix} \right\} = \begin{cases} -2 \log(\sin t) + C & (\sin t > 0) \\ -2 \log(-\sin t) + C & (\sin t < 0) \end{cases}.$$

Exponentiate:

$$a = \frac{K}{\sin^2 t}.$$

Since $a = A'$, you have to integrate once more to find $A(t)$:

$$A(t) = \int \frac{K}{\sin^2 t} dt = -K \cot t + L.\ ^9$$

Finally, you get

$$x(t) = A(t) \sin t = (-K \cot t + L) \sin t = -K \cos t + L \sin t,$$

where K and L are arbitrary constants, as the general solution to $x'' + x = 0$. (After Sections 2.1 and 2.2, this answer should not be unexpected; note that the minus sign in the end result is optional, since $-K$ is an arbitrary constant just as K is.) ∎

---

[9] Since $\cot t = \dfrac{\cos t}{\sin t}$ has derivative $-\csc^2 t = \dfrac{-1}{\sin^2 t}$.

EXAMPLE 2.4.2 The equation $x^2 y'' - 3xy' + 4y = 0$ for the unknown function $y(x)$ has the solution $y = x^2$ (check this). To find all solutions from this one, we put

$$y = A(x)x^2$$
$$y' = A'x^2 + 2Ax$$
$$y'' = A''x^2 + 4A'x + 2A,$$

and substitute into the original equation to obtain

$$A''x^4 + 4A'x^3 + 2Ax^2 - 3A'x^3 - 6Ax^2 + 4Ax^2 = 0$$

or

$$A''x^4 + A'x^3 = 0.$$

Put $a = A'$:

$$a'x^4 + ax^3 = 0.$$

Assuming for now that $x \neq 0$, we get

$$a' + \frac{a}{x} = 0.$$

The solutions of this separable equation are $a(x) = \dfrac{C}{x}$, C an arbitrary constant, so we get

$$A(x) = \int a(x)\, dx = \begin{cases} C \log x + D & (x > 0) \\ C \log(-x) + D & (x < 0). \end{cases}$$

Therefore, $y = A(x)x^2 = [C \log x + D]x^2$ is the general solution for $x > 0$, while $[C \log(-x) + D]x^2$ is the general solution for $x < 0$. In both cases, C and D are arbitrary constants. We see that $x^2 \log x$ and x^2 form a pair of basic solutions for $x > 0$, while $x^2 \log(-x)$ and x^2 are such a pair for $x < 0$. ∎

As an application of our method, we now return to the constant-coefficient equation $x'' + px' + qx = 0$ (from Section 2.3) in the specific case that $p^2 = 4q$. We saw in Section 2.3 that the characteristic equation $\lambda^2 + p\lambda + q = 0$ has one ("double") real root in this case, namely $\lambda = -\dfrac{p}{2}$. Therefore, we know one solution, $x_1(t) = e^{-\frac{p}{2}t} = e^{-pt/2}$, to the differential equation. To find the general solution, we substitute $x(t) = A(t)e^{-pt/2}$ into the equation $x'' + px' + qx = 0$, which yields

$$A''(t)e^{-pt/2} - p A'(t)e^{-pt/2} + \frac{p^2}{4} A(t)e^{-pt/2}$$
$$+ p A'(t)e^{-pt/2} - \frac{p^2}{2} A(t)e^{-pt/2}$$
$$+ q\, A(t)e^{-pt/2} = 0.$$

SEC. 2.4 REDUCTION OF ORDER

If we remember that $p^2 = 4q$, this simplifies to

$$A''(t)e^{-pt/2} = 0$$
$$A''(t) = 0.$$

So we find that $A(t) = Ct + D$, where C and D are arbitrary constants, and the general solution to our equation is

$$x(t) = (Ct + D)e^{-pt/2}$$
$$= (Ct + D)e^{\lambda t}.$$

Note that $te^{-pt/2}$ and $e^{-pt/2}$ form a pair of basic solutions.

EXAMPLE 2.4.3 Solve the initial value problem

$$x'' - 10x' + 25x = 0, \quad x(0) = 3, \quad x'(0) = 10.$$

Solution Since the characteristic equation $\lambda^2 - 10\lambda + 25 = 0$ has a double root $\lambda = 5$, the general solution to the equation is $x(t) = (Ct + D)e^{5t}$. We now look for C and D such that $x(0) = 3$ and $x'(0) = 10$. Now $x(0) = D$, and since $x'(t) = Ce^{5t} + (Ct + D) \cdot 5e^{5t}$, we find that $x'(0) = C + 5D$. So we need $D = 3$, $C + 5D = 10$, from which we find $C = -5$, $x(t) = (-5t + 3)e^{5t}$. ∎

SUMMARY OF KEY TECHNIQUES AND RESULTS

Reduction of order (p. 102): Given a nontrivial solution $x_1(t)$ of the homogeneous second-order equation $x'' + p(t)x' + q(t)x = 0$, we can get the general solution by substituting $x(t) = A(t)x_1(t)$. This leads to a first-order linear equation for $a(t) = A'(t)$. From $a(t)$, we get $A(t) = \int a(t)\,dt$, and finally $x(t) = A(t)x_1(t)$.

Constant-coefficient equation $x'' + px' + qx = 0$:

If $p^2 > 4q$, the characteristic equation has distinct roots λ_1, λ_2, and $e^{\lambda_1 t}$, $e^{\lambda_2 t}$ is a pair of basic solutions (see Section 2.3).

If $p^2 = 4q$, the characteristic equation has a double root λ, and $e^{\lambda t}$, $te^{\lambda t}$ is a pair of basic solutions; thus $x(t) = \alpha e^{\lambda t} + \beta t e^{\lambda t} = (\alpha + \beta t)e^{\lambda t}$ is the general solution (p. 105).

EXERCISES

In each of the following, a homogeneous linear second-order equation is given, together with one solution. Find the general solution.

1. $t^2 x'' - tx' + x = 0$; $x(t) = t$.
2. $t^2 x'' - 7tx' + 15x = 0$; $x(t) = t^3$.
3. $x'' + 4x = 0$; $x(t) = \cos 2t$.

4. $x'' + 16x = 0$; $x(t) = \sin 4t$.
5. $x'' + 2x' + 2x = 0$; $x(t) = e^{-t} \sin t$.
6. $x'' + 4x' + 5x = 0$; $x(t) = e^{-2t} \cos t$.
7. $\dfrac{d^2y}{dx^2} + 2\dfrac{dy}{dx} + 5y = 0$; $y(x) = \dfrac{\cos 2x}{e^x}$.
8. $y'' + 2y' + 10y = 0$; $y(x) = e^{-x} \sin 3x$.
*9. $t^2 x'' + tx' + x = 0$; $x(t) = \cos(\log t)$ $(t > 0)$.
*10. $f(t)x'' - tx' + x = 0$; $x(t) = t$. [$f(t)$ is some unspecified function.]
*11. $(1 - x^2)y'' - 2xy' + 2y = 0$; $y(x) = x$.

Solve the following initial value problems. (In some cases, a solution of the equation is given, but this solution can't be expected to satisfy the initial conditions.)

12. $x'' + 10x' + 25x = 0$, $x(0) = -1$, $x'(0) = -1$.
13. $x'' - 6x' + 9x = 0$, $x(0) = 2$, $x'(0) = 0$.
14. $x'' + 10x' + 24x = 0$, $x(0) = -1$, $x'(0) = -1$.
15. $x'' - 6x' + 8x = 0$, $x(1) = 2$, $x'(1) = 0$.
16. $\dfrac{d^2y}{dx^2} + 4\dfrac{dy}{dx} + 4y = 0$, $y(5) = 6$, $y'(5) = 0$.
17. $\dfrac{d^2y}{dx^2} - 4\dfrac{dy}{dx} + 4y = 0$, $y(0) = 9$, $y'(0) = -3$.
18. $t^2 x'' - 6tx' + 10x = 0$, $x(1) = 0$, $x'(1) = 1$.
    (One solution is $x = t^2$.)
19. $t^2 x'' - 6tx' + 10x = 0$, $x(0) = 0$, $x'(0) = 1$.
    (One solution is $x = t^2$.)
20. $t^2 x'' + 8tx' + 10x = 0$, $x(1) = 2$, $x'(1) = 0$.
    (One solution is $x = t^{-2}$.)
21. $t^2 x'' + 4tx' - 10x = 0$, $x(-1) = 1$, $x'(-1) = 1$.
    (One solution is $x = t^2$.)
22. $y'' + 4y' + 5y = 0$, $y(0) = 1$, $y'(0) = 1$.
    (One solution is $y = e^{-2x} \sin x$.)
23. $y'' - 4y' + 5y = 0$, $y(0) = 1$, $y'(0) = 1$.
    (One solution is $y = e^{2x} \cos x$.)

24. Explain why in Example 2.4.2, we couldn't expect a pair of basic solutions defined at $x = 0$.

2.5 CONSTANT-COEFFICIENT EQUATIONS (CONTINUED)

We now complete our consideration of the homogeneous constant-coefficient equation $x'' + px' + qx = 0$ by discussing the case $p^2 < 4q$, in which the characteristic equation $\lambda^2 + p\lambda + q = 0$ has no real roots. One rather laborious method of solving the differential equation, using the special case $p = 0$ which was solved in Section 2.1, was presented in Exercise 14 of Section 2.3. We will now present an easier method using complex numbers. For the definition and basic properties of complex numbers, see Appendix A.

If we allow complex numbers, the equation $\lambda^2 + p\lambda + q = 0$ will have two distinct, conjugate roots $-\dfrac{p}{2} \pm \dfrac{i}{2}\sqrt{4q - p^2}$. (We are now assuming that $p^2 < 4q$.) Therefore, if we would allow functions to have complex values for the moment, we might expect that the two functions

$$x_1(t) = e^{[-\frac{p}{2} + \frac{i}{2}\sqrt{4q - p^2}]t} \quad \text{and} \quad x_2(t) = e^{[-\frac{p}{2} - \frac{i}{2}\sqrt{4q - p^2}]t}$$

would be solutions of our differential equation.

However, there are several reasons not to allow our functions to have complex values. One is that in applications, such as the case discussed in Section 2.1 of the particle on the spring, we want only solutions to the equation which have actual physical significance; the position of the particle on the axis is always given by a real number, never by a nonreal complex number. Another reason is that from a mathematical point of view, if one allows functions to have complex values, it seems only logical to let the variable have complex values as well. It turns out that if this is done, there are important changes in the implications of "differentiability" for a function.[10] The study of these phenomena—the theory of functions of a complex variable—is beyond the scope of this book.

We will make the following compromise. *Within computations* we will allow ourselves to use complex numbers whenever convenient. However, we will insist that *in the final answer* to any problem, only real numbers occur. It should also be possible to check the answer without using complex numbers. That is, in principle one could get through this book with the attitude: "Complex numbers don't exist; they're just a trick of the imagination that helps find the right answers." This point of view is not recommended, though!

Next, we will put the idea of a complex-valued solution to a differential equation on a firm footing. Let $z(t)$ be a complex-valued function of the real variable t, that is, a rule which to every real number t associates a complex number $z(t)$. Then we can consider the real-valued ("ordinary") functions $x(t)$, $y(t)$ which are defined by

$$x(t) = \operatorname{Re}[z(t)], \quad y(t) = \operatorname{Im}[z(t)] \quad \text{for all } t.$$

Of course, $z(t) = x(t) + iy(t)$.

---

[10] For example, if z is a complex variable, then the function $f(z) = \bar{z}$, which looks very "nice," is not differentiable.

EXAMPLE 2.5.1 $z(t) = \dfrac{3 - it}{1 + it}$.

If we multiply top and bottom by $1 - it$ to make the denominator a real-valued function, we find that

$$z(t) = \frac{(3 - it)(1 - it)}{(1 + it)(1 - it)} = \frac{3 - t^2 - 4it}{1 + t^2},$$

so

$$x(t) = \frac{3 - t^2}{1 + t^2}, \quad y(t) = \frac{-4t}{1 + t^2}. \quad \blacksquare$$

EXAMPLE 2.5.2 $z(t) = e^{it} + e^{-3it}$.

By Euler's formula (see Appendix A, p. 601) we have $e^{it} = \cos t + i \sin t$ and also $e^{-3it} = \cos 3t - i \sin 3t$, so $x(t) = \text{Re}[z(t)] = \cos t + \cos 3t$, $y(t) = \sin t - \sin 3t$. $\quad \blacksquare$

Definition The complex-valued function $z(t)$ is **differentiable** at t_0 if $x(t)$ and $y(t)$ are differentiable there. Its **derivative** at t_0 is then $z'(t_0) = x'(t_0) + iy'(t_0)$.

Having made this definition, we can talk about complex-valued solutions to a differential equation; these are just differentiable complex-valued functions that satisfy the equation. It can also be shown that all the usual rules of differentiation apply. In particular, if α is a *complex* constant, we have $(\alpha z)' = \alpha z'$ for any complex-valued differentiable function z. Also, the derivative of a sum is the sum of the derivatives: $(z + w)' = z' + w'$ (see Exercise 27). This means that if we have *complex-valued* solutions to a homogeneous linear equation $x'' + p(t)x' + q(t)x = 0$, then the methods of constructing additional solutions that were discussed in Section 2.2 still work: Any linear combination (with complex coefficients) of solutions is again a solution. We will use this fact to get real-valued solutions from complex-valued ones.

Now let us return to the differential equation $x'' + px' + qx = 0$, where p, q are constants and $p^2 < 4q$. We know that the characteristic equation $\lambda^2 + p\lambda + q = 0$ has roots $\lambda = \mu + i\nu$, $\overline{\lambda} = \mu - i\nu$, where we've introduced the abbreviations $\mu = -\dfrac{p}{2}$ and $\nu = \dfrac{1}{2}\sqrt{4q - p^2}$. Now we should check that the functions $x_1(t) = e^{\lambda t}$ and $x_2(t) = e^{\overline{\lambda} t}$ are solutions to the differential equation. Note that we can use Euler's formula (see pp. 601–602) to write these complex exponentials in real and imaginary parts. Thus

$$e^{\lambda t} = e^{(\mu + i\nu)t} = e^{\mu t}e^{i\nu t} = e^{\mu t}(\cos \nu t + i \sin \nu t)$$
$$= e^{\mu t}\cos \nu t + i\, e^{\mu t}\sin \nu t.$$

SEC. 2.5 CONSTANT-COEFFICIENT EQUATIONS (CONTINUED)

It is now easy to check (see Exercise 28) that for $x_1(t) = e^{\lambda t}$, we have $x_1' = \lambda e^{\lambda t}$ and $x_1'' = \lambda^2 e^{\lambda t}$, hence $x_1'' + px_1' + qx_1 = (\lambda^2 + p\lambda + q)e^{\lambda t} = 0$, showing that $x_1(t)$ is indeed a solution. Similarly,

$$x_2(t) = e^{\bar{\lambda} t} = e^{\mu t}(\cos \nu t - i \sin \nu t)$$

is a complex-valued solution.

Note that $x_1(t)$ and $x_2(t)$ are complex conjugates of each other (for any t; we can also say that the functions x_1 and x_2 are complex conjugates). Therefore, the particular linear combinations

$$\frac{1}{2} x_1(t) + \frac{1}{2} x_2(t) = \frac{1}{2} x_1(t) + \frac{1}{2} \overline{x_1(t)} = \text{Re}[x_1(t)]$$

and

$$\frac{1}{2i} x_1(t) - \frac{1}{2i} x_2(t) = \frac{1}{2i} x_1(t) - \frac{1}{2i} \overline{x_1(t)} = \text{Im}[x_1(t)]$$

will be *real-valued* solutions to our equation. Specifically, these are

$$\text{Re}(e^{\lambda t}) = \text{Re}[e^{\mu t}(\cos \nu t + i \sin \nu t)] = e^{\mu t} \cos \nu t$$

and

$$\text{Im}(e^{\lambda t}) = e^{\mu t} \sin \nu t.$$

Neither of these solutions is a constant multiple of the other. (Why?) Therefore, these two real-valued functions form a pair of basic solutions to our equation $x'' + px' + qx = 0$.[11] By Theorem 2.2.3 (p. 89), it follows that the *general* (real) *solution* to our equation is

$$x(t) = \alpha \cdot \text{Re}(e^{\lambda t}) + \beta \cdot \text{Im}(e^{\lambda t}) = e^{\mu t}(\alpha \cos \nu t + \beta \sin \nu t),$$

α, β arbitrary real constants.

EXAMPLE 2.5.3 $x'' + 4x' + 5x = 0$.

The characteristic equation $\lambda^2 + 4\lambda + 5 = 0$ has the complex roots

$$\frac{-4 \pm i\sqrt{20 - 16}}{2} = -2 \pm i.$$

Hence we have complex-valued solutions $e^{(-2+i)t}$ and $e^{(-2-i)t}$ to the differential equation. Since $e^{(-2+i)t} = e^{-2t}e^{it} = e^{-2t}(\cos t + i \sin t)$ and $e^{(-2-i)t} = e^{-2t}(\cos t - i \sin t)$, we can combine these to get the real-valued functions $e^{-2t} \cos t$ and $e^{-2t} \sin t$. This is a pair of basic solutions; the general solution is

$$x(t) = \alpha e^{-2t} \cos t + \beta e^{-2t} \sin t = e^{-2t}(\alpha \cos t + \beta \sin t). \quad \blacksquare$$

---

[11] Since $\text{Re}(e^{\bar{\lambda} t}) = \text{Re}(e^{\lambda t})$ and $\text{Im}(e^{\bar{\lambda} t}) = -\text{Im}(e^{\lambda t})$, we find practically the same basic solutions if we use $\bar{\lambda}$ instead of λ.

EXAMPLE 2.5.4 Solve the initial value problem $x'' + 2x' + 3x = 0$, $x(0) = 1$, $x'(0) = 0$.
We get $\lambda^2 + 2\lambda + 3 = 0$, $\lambda = -1 \pm i\sqrt{2}$. Now

$$e^{(-1+i\sqrt{2})t} = e^{-t}(\cos t\sqrt{2} + i \sin t\sqrt{2}),$$

so

$$\text{Re}[e^{(-1+i\sqrt{2})t}] = e^{-t}\cos t\sqrt{2}, \quad \text{Im}[e^{(-1+i\sqrt{2})t}] = e^{-t}\sin t\sqrt{2}.$$

The general solution is $x(t) = e^{-t}(\alpha \cos t\sqrt{2} + \beta \sin t\sqrt{2})$. In particular, $x(0) = \alpha$. Differentiating $x(t)$, we get

$$x'(t) = -e^{-t}(\alpha \cos t\sqrt{2} + \beta \sin t\sqrt{2}) + \sqrt{2}\,e^{-t}(-\alpha \sin t\sqrt{2} + \beta \cos t\sqrt{2}),$$

so $x'(0) = -\alpha + \sqrt{2}\beta$. Thus from the initial conditions we find the system of equations $\alpha = 1$, $-\alpha + \sqrt{2}\beta = 0$. This gives us $\alpha = 1$, $\beta = \frac{1}{2}\sqrt{2}$, so our final answer is

$$x(t) = e^{-t}(\cos t\sqrt{2} + \frac{1}{2}\sqrt{2} \sin t\sqrt{2}). \quad \blacksquare$$

SUMMARY OF KEY CONCEPTS, TECHNIQUES, AND RESULTS

Complex-valued solution to a differential equation (p. 108): A complex-valued function $z(t) = x(t) + iy(t)$ which satisfies the equation. The derivative of $z(t)$ is defined as $z'(t) = x'(t) + iy'(t)$, and the usual differentiation rules apply.

Constant-coefficient equation $x'' + px' + qx = 0$:
  If $p^2 > 4q$, the characteristic equation has distinct real roots λ_1, λ_2, and $e^{\lambda_1 t}$, $e^{\lambda_2 t}$ is a pair of basic solutions (see Section 2.3).
  If $p^2 = 4q$, the characteristic equation has a double root λ, and $e^{\lambda t}$, $te^{\lambda t}$ is a pair of basic solutions (see Section 2.4).
  If $p^2 < 4q$, the characteristic equation has complex conjugate roots $\lambda = \mu + i\nu$ and $\bar{\lambda} = \mu - i\nu$. $e^{\lambda t} = e^{\mu t}e^{i\nu t} = e^{\mu t}(\cos \nu t + i \sin \nu t)$ and $e^{\bar{\lambda} t} = e^{\mu t}(\cos \nu t - i \sin \nu t)$ are complex-valued functions; $\text{Re}(e^{\lambda t}) = e^{\mu t} \cos \nu t$, $\text{Im}(e^{\lambda t}) = e^{\mu t} \sin \nu t$ is a pair of basic real solutions, so $x(t) = e^{\mu t}(\alpha \cos \nu t + \beta \sin \nu t)$ is the general solution (p. 109).

EXERCISES

Find the real and imaginary parts of the following complex-valued functions.

1. $z = \dfrac{2 + it}{1 - it^2}$.

2. $z = \dfrac{t - 3i}{5 + 2it}$.

3. $z = e^{(3+2i)t}$.

4. $z = e^{(-2+5i)t}$.

5. $z = e^{-4it}$.

6. $z = e^{2-3it}$.

7. $z = (\cos t + i \sin t)^{95}$.
   [*Hint:* Life is short.]

8. $z = (\cos t - i \sin t)^{37}$.

Solve the following differential equations.

9. $x'' - 6x' + 13x = 0$.

10. $x'' + 2x' + 10x = 0$.

11. $x'' + 2x' - 10x = 0$.

12. $x'' + 16x = 0$.

13. $\dfrac{d^2y}{dx^2} + 3y = 0$.

14. $\dfrac{d^2y}{dx^2} + \dfrac{dy}{dx} = 0$.

15. $x'' + 7x' = 0$.

16. $x'' + 6x' + 10x = 0$.

17. $x'' + 6x' + 9x = 0$.

18. $x'' + 6x' + 8x = 0$.

Solve the following initial value problems.

19. $x'' + 6x' + 10x = 0$, $x(0) = 1$, $x'(0) = 3$.

20. $x'' + 6x' + 10x = 0$, $x(2) = 1$, $x'(2) = 3$.

21. $x'' - 2x' + x = 0$, $x(3) = -1$, $x'(3) = 5$.

22. $x'' - 2x' + x = 0$, $x(0) = -1$, $x'(0) = 5$.

23. $x'' - 2x' + 2x = 0$, $x(0) = -1$, $x'(0) = 5$.

24. $x'' - 2x' + 2x = 0$, $x(2) = -1$, $x'(2) = 5$.

25. $x'' - 2x' = 0$, $x(2) = 8$, $x'(2) = 0$.

26. $x'' + 4x' = 0$, $x(0) = 6$, $x'(0) = 1$.

27. (a) Show that if $\alpha = a + bi$ is a complex constant and $z(t) = x(t) + iy(t)$ is a differentiable complex-valued function, then $(\alpha z)' = \alpha z'$.
  (b) Show that $(z + w)' = z' + w'$ for differentiable complex-valued functions $z(t)$, $w(t)$.

28. (See p. 109.) Show that for any complex number $\lambda = \mu + i\nu$, the derivative of the complex-valued function $z(t) = e^{\lambda t}$ is $\lambda e^{\lambda t}$.

29. Consider the differential equation with *complex* coefficients (the only one in this chapter) $x'' + (1 + i)x' + (2 - i)x = 0$.
  (a) Show that $x_1(t) = e^{it}$ is a (complex-valued) solution.
  (b) Show that $\bar{x}_1(t) = e^{-it}$ is *not* a solution and that $\text{Re}[x_1(t)] = \cos t$ is *not* a solution. Does this contradict the theory?
  *(c) Find the general complex-valued solution.
  *(d) Find the general real-valued solution.

FURTHER READING

The implications of differentiability for a function of a complex variable are discussed in any book on complex variables. A particularly thought-provoking one is Pólya and Latta, *Complex Variables* (Wiley, 1974).

2.6 INHOMOGENEOUS EQUATIONS: INTRODUCTION

We will now consider the inhomogeneous linear equation $x'' + px' + qx = h(t)$, where $h(t)$ is a given function of t and the coefficients p, q are constants. In this section, we'll see in a series of examples how a solution to this equation can often be found by educated guessing. Once one solution is found, we will be able to find *all* solutions by a method that was already mentioned in Section 1.6. This method is as follows: First solve the homogeneous equation $x'' + px' + qx = 0$. (We showed how to do this in Sections 2.3 to 2.5.) **Then add the general solution of the homogeneous equation to the one known solution of the inhomogeneous equation; this will yield all solutions of the inhomogeneous equation.** Exercise 37 provides a proof of this fact; we will encounter a more general version as Theorem 2.8.1 (in Section 2.8).

EXAMPLE 2.6.1 Find a solution to the equation $x'' - 4x' + 3x = 17$; then find all the solutions.

Here $h(t) = 17$, a constant; why not look for a constant solution? For a constant solution, $x' = x'' = 0$, and we must have $3x = 17$. Thus we find the constant solution $x = \dfrac{17}{3}$.

To find the general solution, we first solve the homogeneous equation $x'' - 4x' + 3x = 0$. The roots of $\lambda^2 - 4\lambda + 3 = 0$ are $\lambda = 1$ and $\lambda = 3$, so the general solution to the homogeneous equation is $\alpha e^t + \beta e^{3t}$. Therefore, the general solution to our inhomogeneous equation is

$$x(t) = \frac{17}{3} + \alpha e^t + \beta e^{3t}. \quad \blacksquare$$

EXAMPLE 2.6.2 Find a solution to the equation $x'' - 4x' + 3x = \sin t + 2 \cos t$.

Since differentiating $\sin t$ and $\cos t$ yields those "same" functions ($\cos t$ and $-\sin t$), it seems reasonable to look for a solution of the form $x(t) = A \sin t + B \cos t$, where A and B are constants. Then $x' = A \cos t - B \sin t$, $x'' = -A \sin t - B \cos t$. Substituting these in the differential equation, we find

$$(2A + 4B) \sin t + (-4A + 2B) \cos t = \sin t + 2 \cos t,$$

which will be true if $2A + 4B = 1$, $-4A + 2B = 2$. This yields $A = -\dfrac{3}{10}$, $B = \dfrac{2}{5}$, so

$$x(t) = -\frac{3}{10} \sin t + \frac{2}{5} \cos t$$

is a solution of the differential equation. Once again, the general solution would be obtained by adding all solutions of the homogeneous equation to our solution of the inhomogeneous equation:

$$x(t) = -\frac{3}{10} \sin t + \frac{2}{5} \cos t + \alpha e^t + \beta e^{3t}. \quad \blacksquare$$

SEC. 2.6 INHOMOGENEOUS EQUATIONS: INTRODUCTION

EXAMPLE 2.6.3 $x'' - 4x' + 3x = \sin 2t$.

We could try to find a solution of the form $A \sin 2t$, but then the left-hand side of the equation would be $-4A \sin 2t - 8A \cos 2t + 3A \sin 2t$, and the term $-8A \cos 2t$ couldn't cancel against anything. So we should look for a solution of the form $A \sin 2t + B \cos 2t$ instead, just as in Example 2.6.2. This yields the conditions (check these!) $-A + 8B = 1$, $-8A - B = 0$, so we must have $A = -\dfrac{1}{65}$, $B = \dfrac{8}{65}$. The general solution to the inhomogeneous equation will be

$$x(t) = -\frac{1}{65} \sin 2t + \frac{8}{65} \cos 2t + \alpha e^t + \beta e^{3t}.$$ ■

EXAMPLE 2.6.4 $x'' - 4x' + 3x = 5e^t$.

It is tempting to look for a solution of the form Ae^t, but unfortunately all these functions are solutions of the homogeneous equation![12] Because of this, a better idea is to write $x(t) = y(t)e^t$, where y is an unknown function of t, and try to find a solution for y. We have $x' = y'e^t + ye^t$, $x'' = (y'' + 2y' + y)e^t$, so the equation becomes

$$(y'' - 2y')e^t = 5e^t$$
$$y'' - 2y' = 5.$$

We see again that we would have been unable to find a solution with constant y! If we put $y' = z$, we have the first-order equation $z' - 2z = 5$. We could solve this equation completely and eventually find the general solution for x, or we can just take the obvious solution $z = -\dfrac{5}{2}$. Integrating $z = -\dfrac{5}{2}$ gives $y = -\dfrac{5}{2}t + C$, and from this we see that $x(t) = -\dfrac{5}{2}te^t$ is a solution of our original equation. The general solution is therefore

$$x(t) = -\frac{5}{2}te^t + \alpha e^t + \beta e^{3t}.$$

Note that it turned out that $x(t)$ could be taken to be of the form Ate^t (with $A = -\dfrac{5}{2}$). ■

EXAMPLE 2.6.5 $x'' - 2x' + x = 3e^t$.

Again, all functions of the form Ae^t are solutions of the homogeneous equation. In fact, so are all functions of the form $(A + Bt)e^t$. (The characteristic equation $\lambda^2 - 2\lambda + 1 = 0$ has two equal roots $\lambda = 1$; see Section 2.4.) Let us, nevertheless, look for a solution of the

---

[12] Note that if you try $x = Ae^t$ anyway, all the terms on the left will cancel and you'll get $0 = 5e^t$, which is impossible.

form $x(t) = y(t)e^t$. After differentiating twice, substituting, and simplifying, we get

$$y''e^t = 3e^t$$
$$y'' = 3$$
$$y = \frac{3}{2}t^2 + Ct + D,$$

where C and D are constants of integration. Hence the general solution is

$$x(t) = \frac{3}{2}t^2 e^t + Cte^t + De^t;$$

in particular, we have a solution of the form $At^2 e^t$ (with $A = \frac{3}{2}$). ∎

EXAMPLE 2.6.6 Find a solution to $x'' + x = 6e^t$.
In this case we *can* find a solution of the form Ae^t. See Exercise 1. ∎

EXAMPLE 2.6.7 Find a solution to $x'' + x' + x = t^3 - 2t^2$.
If we look for a solution of the form $At^3 + Bt^2$, we will get

$$At^3 + (3A + B)t^2 + (6A + 2B)t + 2B = t^3 - 2t^2,$$

which is impossible. Obviously, more terms must be added to cancel $(6A + 2B)t + 2B$ on the left-hand side. Try $x(t) = At^3 + Bt^2 + Ct + D$, instead. This yields

$$At^3 + (3A + B)t^2 + (6A + 2B + C)t + (2B + C + D) = t^3 - 2t^2.$$

We now equate the coefficients on both sides:

$$\begin{cases} A = 1 \\ 3A + B = -2 \\ 6A + 2B + C = 0 \\ 2B + C + D = 0, \end{cases}$$

and obtain $A = 1$, $B = -5$, $C = 4$, $D = 6$. Thus

$$x(t) = t^3 - 5t^2 + 4t + 6$$

is a solution to our equation (see also Exercise 19). ∎

All the examples above illustrate special cases of a method called the **method of undetermined coefficients.** The idea is to know, or guess, that a solution of a particular form to $x'' + px' + qx = h(t)$ can be found, then to go ahead and find it by substitution. The method can be extended to many common functions $h(t)$; in Section 4.2 we will describe it in the rather general situation that $h(t)$ is any sum of products of polynomials, exponential functions, sines, and cosines. For now we list some special cases, as illustrated by the examples above, in Table 2.6.1.

SEC. 2.6 INHOMOGENEOUS EQUATIONS: INTRODUCTION

Table 2.6.1 Method of undetermined coefficients

$h(t)$	$x(t)$
$Ce^{\alpha t}$	$Ae^{\alpha t}$, but $Ate^{\alpha t}$ if $e^{\alpha t}$ is a solution of the homogeneous equation; $At^2 e^{\alpha t}$ if $te^{\alpha t}$ is also a solution of the homogeneous equation.
$\begin{cases} \sin \alpha t \\ \cos \alpha t \\ C_1 \sin \alpha t + C_2 \cos \alpha t \end{cases}$	$\begin{cases} A \sin \alpha t + B \cos \alpha t, \text{ but } At \sin \alpha t + Bt \cos \alpha t \\ \text{if } \sin \alpha t, \cos \alpha t \text{ are solutions of the homogeneous} \\ \text{equation (see Exercise 34).} \end{cases}$
C	A
$C_0 + C_1 t$	$A_0 + A_1 t$
(etc.)	(etc.)
Any polynomial of degree n	A polynomial $A_0 + A_1 t + \cdots + A_n t^n$ of degree n, but a polynomial of degree $(n + 1)$ if $q = 0$, and a polynomial of degree $(n + 2)$ if also $p = 0$ (see Exercise 36).

Note: To find a solution of the equation $x'' + px' + qx = h(t)$ with $h(t)$ from the left-hand column, try a function of the form $x(t)$ from the right-hand column. The symbols $\alpha, A, B, C, C_0, C_1$, and so on, denote constants.

We conclude this section with two more examples.

EXAMPLE 2.6.8 Find a solution to $x'' - 4x' + 3x = e^{2t} + 4e^{3t} + 3 \sin t - 2 \cos t$.

To get the various terms on the right, we should try to find a solution $x(t)$ that is the sum of corresponding terms (as provided by Table 2.6.1). Specifically, to get e^{2t} on the right, we'll need a term Ae^{2t} in $x(t)$; to get $4e^{3t}$ we'll need a term Bte^{3t} (Be^{3t} won't work, since e^{3t} is a solution to the homogeneous equation[13]); to get $3 \sin t - 2 \cos t$ we'll need terms $C \sin t + D \cos t$. So we try

$$x(t) = Ae^{2t} + Bte^{3t} + C \sin t + D \cos t.$$

This yields

$$Ae^{2t} + 2Be^{3t} + (2C + 4D) \sin t + (2D - 4C) \cos t = e^{2t} + 4e^{3t} + 3 \sin t - 2 \cos t,$$

so $-A = 1$, $2B = 4$, $2C + 4D = 3$, $2D - 4C = -2$, so $A = -1$, $B = 2$, $C = \dfrac{7}{10}$, and $D = \dfrac{2}{5}$; we have a solution

$$x(t) = -e^{2t} + 2te^{3t} + \frac{7}{10} \sin t + \frac{2}{5} \cos t.$$

---

[13] If you hadn't realized this, you'd soon find out by trial and error that a term Be^{3t} wouldn't work.

(The general solution would be $-e^{2t} + 2te^{3t} + \frac{7}{10}\sin t + \frac{2}{5}\cos t + \alpha e^t + \beta e^{3t}$, α, β arbitrary.) ∎

EXAMPLE 2.6.9 Solve the initial value problem $x'' - 3x' + 2x = e^{4t} + 3$, $x(0) = 1$, $x'(0) = 0$.

To solve this problem, we first have to find the *general* solution of the given differential equation. As in previous examples, we do this by finding one specific solution and adding the general solution of the homogeneous equation to it. The homogeneous equation $x'' - 3x' + 2x = 0$ has general solution $\alpha e^t + \beta e^{2t}$. Since the terms e^{4t} and 3 on the right-hand side of our (inhomogeneous) equation aren't solutions of the homogeneous equation, there will be no complications: We can expect to find a solution of the form $x(t) = Ae^{4t} + B$. [**Warning:** This specific solution will *not* yet satisfy the initial conditions $x(0) = 1$, $x'(0) = 0$.]

For $x = Ae^{4t} + B$, we have $x' = 4Ae^{4t}$, $x'' = 16Ae^{4t}$. Thus $x'' - 3x' + 2x = 6Ae^{4t} + 2B$ and we need $6A = 1$, $2B = 3$, so $A = \frac{1}{6}$, $B = \frac{3}{2}$. Therefore, $\frac{1}{6}e^{4t} + \frac{3}{2}$ is one solution to our equation and

$$x(t) = \frac{1}{6}e^{4t} + \frac{3}{2} + \alpha e^t + \beta e^{2t}$$

is the general solution. *Now* we must find the constants α, β such that $x(0) = 1$, $x'(0) = 0$. $x(0) = 1$ yields $\frac{1}{6} + \frac{3}{2} + \alpha + \beta = 1$; we have $x'(t) = \frac{2}{3}e^{4t} + \alpha e^t + 2\beta e^{2t}$, so $x'(0) = 0$ yields $\frac{2}{3} + \alpha + 2\beta = 0$. Solving the equations $\frac{1}{6} + \frac{3}{2} + \alpha + \beta = 1$, $\frac{2}{3} + \alpha + 2\beta = 0$, we get $\alpha = -\frac{2}{3}$, $\beta = 0$, so the answer is

$$x(t) = \frac{1}{6}e^{4t} + \frac{3}{2} - \frac{2}{3}e^t. \quad ∎$$

N O T E : In Chapter 6 we will present a completely different approach to the solution of a differential equation of the form $x'' + px' + qx = h(t)$. This approach—the use of the Laplace transform—will be especially fast for initial value problems.

SUMMARY OF KEY TECHNIQUES AND RESULTS

Method of **undetermined coefficients**: To find one solution of the inhomogeneous constant-coefficient equation $x'' + px' + qx = h(t)$, make an educated guess as to the form of the solution and substitute a general function of that form for $x(t)$ in the equation. For specific suggestions, see Table 2.6.1, p. 115. Once one solution has been found, the general solution can be obtained by adding the general solution of the *homogeneous* equation to that one solution (p. 112).

SEC. 2.6 INHOMOGENEOUS EQUATIONS: INTRODUCTION

EXERCISES

Find one solution to each of the following differential equations.

1. $x'' + x = 6e^t$.
2. $x'' + 4x = 2e^{3t}$.
3. $x'' - 3x' + 2x = \sin t + 2 \cos t$.
4. $x'' - 5x' + 6x = 3 \sin t - \cos t$.
5. $x'' - 3x' + 2x = -e^{2t}$.
6. $x'' - 5x' + 6x = 4e^{3t}$.
7. $x'' + 9x' - x = 17 \sin t$.
8. $x'' - 2x' + x = 5 \cos t$.
9. $x'' - x' - x = t^2 + t - 5$.
10. $x'' + 2x' - x = 3t^2 - t - 1$.
11. $x'' + 4x' + 6x = 19$.
12. $x'' - 3x' - 2x = -24$.
13. $\dfrac{d^2y}{dx^2} - 7\dfrac{dy}{dx} + 6y = e^x - 12$.
14. $\dfrac{d^2y}{dx^2} + 5\dfrac{dy}{dx} + 4y = e^{-x} + 16$.
15. $\dfrac{d^2y}{dx^2} + 4y = 3 \sin 5x - \cos 5x$.
16. $\dfrac{d^2y}{dx^2} + 16y = \sin 2x + 3 \cos 2x$.
17. $x'' + 2x' + 5x = e^{2t} - 10 \cos t$.
18. $x'' + 4x' + 5x = e^{3t} + 16 \sin t$.

19. Find the general solution to the equation $x'' + x' + x = t^3 - 2t^2$ of Example 2.6.7.

Solve the following initial value problems.

20. $x'' - 4x' + 3x = t + 1$, $x(0) = 0$, $x'(0) = 0$.
21. $x'' + 4x' + 3x = 1 - t$, $x(0) = 0$, $x'(0) = 0$.
22. $x'' - 3x' + 2x = e^t$, $x(0) = 1$, $x'(0) = 2$.
23. $x'' - 3x' + 2x = e^t$, $x(0) = -1$, $x'(0) = 0$.
24. $\dfrac{d^2y}{dx^2} - 5\dfrac{dy}{dx} + 4y = -17 \sin x$, $y(0) = 0$, $y'(0) = 1$.
25. $\dfrac{d^2y}{dx^2} - 5\dfrac{dy}{dx} + 4y = 17 \cos x$, $y(0) = \dfrac{1}{2}$, $y'(0) = \dfrac{5}{2}$.
26. $x'' + 3x' + 2x = e^{3t}$, $x(0) = 1$, $x'(0) = -1$.
27. $x'' - 3x' + 2x = 10e^{-3t}$, $x(0) = 0$, $x'(0) = 0$.
28. $x'' + 2x' + x = 2e^{-t}$, $x(0) = 0$, $x'(0) = 0$.
29. $x'' + 4x' + 4x = e^{-2t}$, $x(0) = 3$, $x'(0) = -1$.
30. $\dfrac{d^2y}{dx^2} - 3\dfrac{dy}{dx} + 2y = e^{2x} - 6e^{-x}$, $y(0) = 0$, $y'(0) = 0$.
31. $\dfrac{d^2y}{dx^2} - 3\dfrac{dy}{dx} + 2y = e^x - 6e^{-2x}$, $y(0) = 0$, $y'(0) = 0$.
32. $x'' - 3x' + 2x = 5 \sin t + e^{-t}$, $x(0) = 0$, $x'(0) = 1$.

33. $x'' + 3x' + 2x = 5\cos t - e^t$, $x(0) = 0$, $x'(0) = 1$.

34. **(a)** Let $\alpha \neq 0$ be a constant. Show that there is never a solution of the form $x(t) = A\sin\alpha t + B\cos\alpha t$ to the equation $x'' + \alpha^2 x = C\sin\alpha t + D\cos\alpha t$, where C and D are constants, unless $C = D = 0$.
(b) Find a solution of the form $x(t) = At\sin 3t + Bt\cos 3t$ to the equation $x'' + 9x = -3\sin 3t$.

35. Solve the initial value problem $\dfrac{d^2y}{dx^2} + 4y = 3\sin 2x - \cos 2x$, $y(0) = 5$, $y'(0) = 1$.

36. **(a)** Show that there is no solution of the form $x(t) = A_0 + A_1 t + \cdots + A_n t^n$ to the equation $x'' + px' = t^n$, where p is constant and n is a positive integer.
(b) Solve the equation $x'' + 3x' = t^2$.
(c) Solve the initial value problem $x'' - x' = 4t^3 - 3t^2 - 14t + 1$, $x(0) = 0$, $x'(0) = -1$.

37. In this exercise we consider the inhomogeneous equation $x'' + px' + qx = h(t)$ and the related homogeneous equation $x'' + px' + qx = 0$.
(a) Show that if $x_1(t)$ and $x_2(t)$ are solutions of the inhomogeneous equation, then $y(t) = x_2(t) - x_1(t)$ is a solution of the homogeneous equation.
[*Hint*: What does it mean for $y(t)$ to be such a solution?]
(b) Show that if $x(t)$ is a solution of the inhomogeneous equation and $y(t)$ is a solution to the homogeneous equation, then $x(t) + y(t)$ is a solution to the inhomogeneous equation.
(c) Show that if one can guess or otherwise find one solution $x_1(t)$ of the inhomogeneous equation, one can then find *all* solutions to that equation by adding each of the solutions of the homogeneous equation to $x_1(t)$ [as in part (b)].
[*Hint*: Suppose that $x_2(t)$ is another solution of the inhomogeneous equation. Use part (a).]

2.7 MECHANICAL VIBRATION

We are now in a position to solve the various equations we found in Section 2.1 for the motion of a particle attached to a spring. We will also study a more general case, in which the particle is "forced" to vibrate.

Recall that if the spring is suspended from the ceiling, the forces on the particle are the gravitational force mg (that is, the weight of the particle), the spring force $-kx$, and the frictional force $-\gamma x'$ (see Figure 2.7.1). Here x denotes the displacement of the particle downward from the position where the spring is relaxed. Then the total force on the particle is $-\gamma x' - kx + mg$, and from Newton's second law ($F = mx''$) we get the equation

$$mx'' + \gamma x' + kx = mg$$

describing the motion of the particle. This is an inhomogeneous second-order linear equation; we see immediately that there is a constant solution $x = \dfrac{mg}{k}$. This solution corresponds to the equilibrium position of the particle: If the spring is stretched by $\dfrac{mg}{k}$, the upward spring force $k \cdot \dfrac{mg}{k}$ will exactly cancel the downward gravitational force mg. Therefore, it is possible for the particle to be at rest with $x = \dfrac{mg}{k}$. As we have seen in Section 2.6, we can now get the general solution to our equation by solving the homogeneous equation $mx'' + \gamma x' + kx = 0$ and adding $\dfrac{mg}{k}$ to all its solutions.

Note that this same homogeneous equation governs the motion of the particle if the spring is in a horizontal plane, so that gravity does not have to be taken into account (see

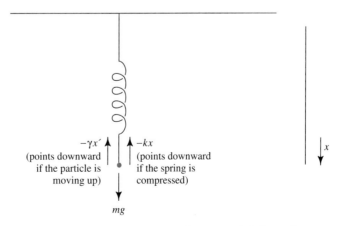

Figure 2.7.1. Forces on a particle suspended from the ceiling by a spring.

Example 2.1.4, p. 78). Another way of putting this is that the only influence gravity has on the motion of the particle is to change the equilibrium position; *relative to the equilibrium position* the motion is the same.

We will now discuss the motion of the particle relative to the equilibrium position, that is, the behavior of the solutions to $mx'' + \gamma x' + kx = 0$ for various values of k, γ, and m. If we write the equation as $x'' + \dfrac{\gamma}{m} x' + \dfrac{k}{m} x = 0$, it is in the form considered in Sections 2.3 to 2.5. In particular, the behavior of the solutions will depend on whether $(\dfrac{\gamma}{m})^2$ is greater than, equal to, or less than $\dfrac{4k}{m}$. Since γ is a *nonnegative* constant, these cases correspond to

$$\gamma > 2\sqrt{km}, \quad \gamma = 2\sqrt{km}, \quad \gamma < 2\sqrt{km},$$

respectively. They are referred to as the **overdamped**, **critically damped**, and **underdamped** cases, in that order. (The term *damping* refers to the introduction of a frictional force, which may be a "natural" result of the motion of the particle, for example through the air, or an artificial restraining force. Our equation is valid in either case—provided that the frictional force is proportional to, and in the opposite direction from, the velocity.)

Let us see what happens if we start with $\gamma = 0$ (no friction) and then increase γ more and more, so that the motion will become first critically damped and then overdamped, after having been underdamped.

Case 1 $\gamma < 2\sqrt{km}$ (**underdamped** case).

In this case there are no real roots of the characteristic equation $\lambda^2 + \dfrac{\gamma}{m} \lambda + \dfrac{k}{m} = 0$; if we let $\lambda = -\dfrac{\gamma}{2m} + \dfrac{i}{2m}\sqrt{4km - \gamma^2}$ and $\bar{\lambda}$ be the complex roots, then we have complex-valued solutions $e^{\lambda t}$ and $e^{\bar{\lambda} t}$, from which we derive the pair of basic real-valued solutions

$$\operatorname{Re}(e^{\lambda t}) = e^{-\gamma t/2m} \cos(\sqrt{4km - \gamma^2} \cdot \dfrac{t}{2m}), \quad \operatorname{Im}(e^{\lambda t}) = e^{-\gamma t/2m} \sin(\sqrt{4km - \gamma^2} \cdot \dfrac{t}{2m});$$

the general solution is a linear combination of these two with arbitrary coefficients C_1, C_2.

An important special case is $\gamma = 0$, the **undamped** case; for $\gamma = 0$, our basic solutions become $\cos \sqrt{\dfrac{k}{m}} t$ and $\sin \sqrt{\dfrac{k}{m}} t$. We had already solved this case in Section 2.1; there we found the expression $C_0 \sin(\sqrt{\dfrac{k}{m}} t + D_0)$ for the general solution of $m'' + kx = 0$. Just as we saw in Example 2.2.3 for $\dfrac{k}{m} = 1$, we can go back and forth between the two ways of expressing the solution, since

$$C_0 \sin\left(\sqrt{\dfrac{k}{m}} t + D_0\right) = C_0 \cos D_0 \cdot \sin \sqrt{\dfrac{k}{m}} t + C_0 \sin D_0 \cdot \cos \sqrt{\dfrac{k}{m}} t.$$

The advantage of expressing the general solution as $C_0 \sin(\sqrt{\frac{k}{m}} t + D_0)$ rather than as a linear combination $C_1 \sin \sqrt{\frac{k}{m}} t + C_2 \cos \sqrt{\frac{k}{m}} t$ is that we see clearly that each solution is simply a sine function with the phase shifted, that is, with t replaced by t plus a constant. The motion $x(t) = C_0 \sin(\sqrt{\frac{k}{m}} t + D_0)$ is periodic with period $2\pi \sqrt{\frac{m}{k}}$ and amplitude $|C_0|$ (see Figure 2.7.2). In other words, the particle oscillates around the equilibrium position, with $|C_0|$ the farthest distance it attains from equilibrium; the vibration goes on forever, with $2\pi \sqrt{\frac{m}{k}}$ being the time needed for one complete oscillation.

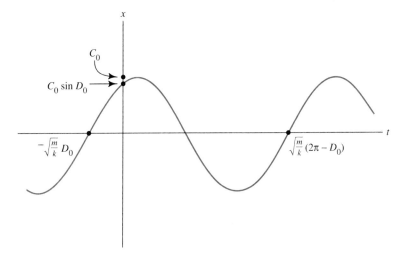

Figure 2.7.2. Undamped vibration $(C_0 > 0, 0 < D_0 < 2\pi)$.

Now let us go back to the general underdamped case, and assume that $\gamma > 0$. Just as we could write any linear combination of $\cos \sqrt{\frac{k}{m}} t$, $\sin \sqrt{\frac{k}{m}} t$ as $C_0 \sin(\sqrt{\frac{k}{m}} t + D_0)$, we can write any linear combination of the basic solutions $e^{-\gamma t/2m} \cos \omega t$ and $e^{-\gamma t/2m} \sin \omega t$ as $x(t) = C_0 e^{-\gamma t/2m} \sin(\omega t + D_0)$. (Here we have abbreviated the constant $\frac{1}{2m} \sqrt{4km - \gamma^2}$ by ω.) This function is a product of an exponential function and a sine function, and this is apparent in its behavior: Like the sine function, it oscillates, but its amplitude is not constant; like the exponential function $e^{-\gamma t/2m}$, it tends to zero when $t \to \infty$, but the decrease is not monotonic. Since $|\sin(\omega t + D_0)| \leq 1$ for all t, the graph of $x(t)$ lies between the graphs of $C_0 e^{-\gamma t/2m}$ and $-C_0 e^{-\gamma t/2m}$ (see Figure 2.7.3, p. 122). The vibration still goes on forever, but the displacements get successively smaller, and for t large enough the vibration will be imperceptible; the particle will appear to be at rest.

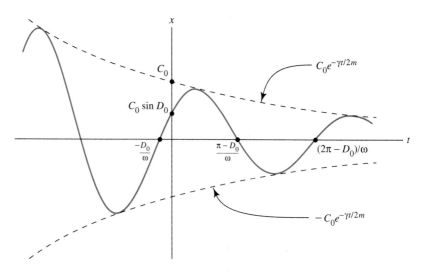

Figure 2.7.3. The underdamped case ($C_0 > 0, 0 < D_0 < 2\pi$).

Note that the "period" of the motion,

$$\frac{2\pi}{\omega} = \frac{4\pi m}{\sqrt{4km - \gamma^2}} = 2\pi \sqrt{\frac{m}{k}} \cdot \frac{1}{\sqrt{1 - \frac{\gamma^2}{4km}}},$$

is an increasing function of γ. That is, if friction is increased, the particle will vibrate more slowly. In fact, as γ increases toward the critically damped case $\gamma = 2\sqrt{km}$, $1 - \frac{\gamma^2}{4km}$ approaches zero, and so the "period" grows without bound.

Case 2 $\gamma = 2\sqrt{km}$ (**critical** damping).

If friction is increased until the friction coefficient γ reaches $2\sqrt{km}$, there is a fundamental change in the behavior of the solutions: Their oscillatory character disappears completely. To see this, note that for $\gamma = 2\sqrt{km}$ the characteristic equation

$$\lambda^2 + \frac{\gamma}{m}\lambda + \frac{k}{m} = 0 \quad \text{has two equal roots} \quad \lambda_{1,2} = -\frac{\gamma}{2m} = -\sqrt{\frac{k}{m}};$$

therefore, the general solution of $x'' + \frac{\gamma}{m}x' + \frac{k}{m}x = 0$ will be

$$x(t) = C_1 e^{-(\sqrt{k/m})t} + C_2 t e^{-(\sqrt{k/m})t} = (C_1 + C_2 t)e^{-(\sqrt{k/m})t}.$$

Any of these solutions approaches zero as $t \to \infty$, so the motion "dies out" just as in the underdamped case, but it is no longer oscillatory: $x(t)$ is zero only for $t = -\frac{C_1}{C_2}$. [If

SEC. 2.7 MECHANICAL VIBRATION

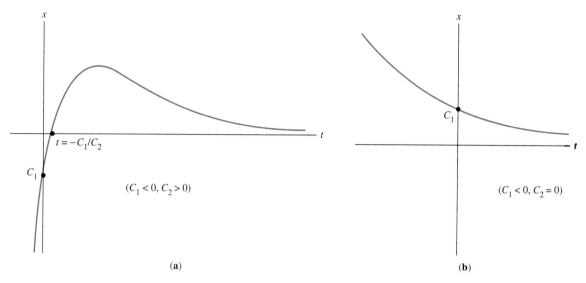

Figure 2.7.4. The critically damped case.

$C_2 = 0$, $x(t)$ is never zero at all.] Depending on the actual values of C_1 and C_2, the graph of $x(t)$ may look, for instance, like one of the graphs in Figure 2.7.4.

Case 3 $\gamma > 2\sqrt{km}$ (**overdamped** case).

In this case the characteristic equation $\lambda^2 + \dfrac{\gamma}{m}\lambda + \dfrac{k}{m} = 0$ has distinct real roots $\lambda_{1,2} = -\dfrac{\gamma}{2m} \pm \dfrac{1}{2m}\sqrt{\gamma^2 - 4km}$, and the differential equation has $x(t) = C_1 e^{\lambda_1 t} + C_2 e^{\lambda_2 t}$ as its general solution. λ_1 and λ_2 are both negative (see Exercise 3), so we still have $x(t) \to 0$ as $t \to \infty$. The graphs of the solutions are not unlike the graphs in the critically damped case; again, there is at most one value of t for which $x(t) = 0$. That is, the particle will be in the equilibrium position at most once; after that, the particle will go to one side of the equilibrium position and eventually approach equilibrium from that side, without (theoretically) ever reaching it.

We will now consider a somewhat more general situation, in which there is, in addition to the spring force and the frictional force, an external force $F(t)$ on the particle; we assume that this external force depends only on the time. In this case the displacement $x(t)$ of the particle from its equilibrium position will satisfy the differential equation

$$mx'' + \gamma x' + kx = F(t).$$

In the next section we will discuss a method, the method of variation of constants, by which such an equation can be solved for any function $F(t)$. Meanwhile, we'll consider the special case in which $F(t)$ is a *periodic* function of the time, more particularly

$$F(t) = F_0 \cos \Omega t, \quad \Omega > 0. \quad [14]$$

As we saw in Section 2.6, we can find a solution of the equation

$$mx'' + \gamma x' + kx = F_0 \cos \Omega t$$

by substituting $x(t) = C_1 \cos \Omega t + C_2 \sin \Omega t$ and solving for the constants C_1 and C_2, *provided* $\cos \Omega t$ and $\sin \Omega t$ don't satisfy the homogeneous equation $mx'' + \gamma x' + kx = 0$. Under this assumption, a straightforward computation (see Exercise 12) yields

$$C_1 = \frac{k - m\Omega^2}{(k - m\Omega^2)^2 + \gamma^2 \Omega^2} F_0, \quad C_2 = \frac{\gamma \Omega}{(k - m\Omega^2)^2 + \gamma^2 \Omega^2} F_0. \quad (*)$$

Just as in the discussion of the underdamped case without external force, we can write $C_1 \cos \Omega t + C_2 \sin \Omega t$ in the form $C_0 \cos(\Omega t - B_0)$ for suitable constants C_0 and B_0.[15] The general solution is then found by adding the general solution of the homogeneous equation—underdamped, critically damped, or overdamped, as the case may be—to this one.

Note that as $t \to \infty$, all solutions of the homogeneous equation approach zero (except in the undamped case $\gamma = 0$), as we saw above. As a result, *every solution of the inhomogeneous equation* approaches our particular solution $C_0 \cos(\Omega t - B_0)$. This means that after sufficient time has elapsed, the particle is "forced" to vibrate with the frequency[16] corresponding to the external force: The motion is approximately periodic with period $\dfrac{2\pi}{\Omega}$.

However, this need not be true in the case $\gamma = 0$, and we will now study this case, the undamped case, separately.

You may also have noticed that we haven't yet studied the exceptional case in which $\cos \Omega t$ and $\sin \Omega t$ satisfy the homogeneous equation. It turns out that we again have $\gamma = 0$ in that case (see Exercise 17).

In the undamped case $\gamma = 0$, the following simplifications and special phenomena occur. The equation becomes

$$mx'' + kx = F_0 \cos \Omega t$$

and $x(t) = C_1 \cos \Omega t$, with $C_1 = \dfrac{F_0}{k - m\Omega^2}$, is a solution to this. [Just substitute $\gamma = 0$ in

---

[14] The reason that it is customary to prefer $F(t) = F_0 \cos \Omega t$ over $F(t) = F_0 \sin \Omega t$ in setting up the equation (there is no difference except that time is shifted by $\dfrac{\pi}{2\Omega}$) is that, this way, $F(0) = F_0$, which is easy to remember. $\Omega > 0$ is also for convenience; $\cos(-\Omega t) = \cos \Omega t$, so there is no real difference between the cases $\Omega > 0$ and $\Omega < 0$.

[15] Again, the sign of B_0 and the choice of cos rather than sin are convenient rather than essential (see Exercise 13).

[16] By definition, the frequency of the periodic function $F_0 \cos \Omega t$ is $\dfrac{\Omega}{2\pi}$ (the number of periods per unit time).

SEC. 2.7 MECHANICAL VIBRATION

(∗).] However, if $\cos \Omega t$ is a solution to the homogeneous equation, then $k - m\Omega^2 = 0$ (and vice versa), and obviously C_1 is undefined. Note that the condition $k - m\Omega^2 = 0$ works out to

$$\Omega = \sqrt{\frac{k}{m}} = \omega.\ ^{17}$$

This means that the period $\frac{2\pi}{\Omega}$ of the external force equals the period $\frac{2\pi}{\omega}$ that we found earlier for the solutions to the homogeneous equation. In this case we must look, instead, for a solution of the form

$$x(t) = At \cos \Omega t + Bt \sin \Omega t$$

(see Section 2.6). A straightforward computation (see Exercise 17) then yields $A = 0$, $B = \dfrac{F_0}{2m\Omega}$. The solution

$$x(t) = \frac{F_0}{2m\Omega} t \sin \Omega t$$

obtained in this way behaves unexpectedly as $t \to \infty$: It oscillates, but with *linearly increasing amplitude* (see Figure 2.7.5). This phenomenon is called **resonance**. Of course, in practice the amplitude of the oscillations can't really increase without bound, because of the physical limitations of the spring. We have already noted that our expression for the spring force is only approximate. Still, resonance phenomena do occur, and this is the

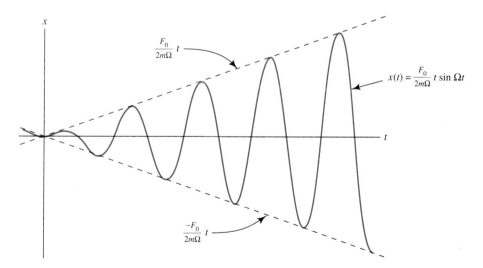

Figure 2.7.5. Resonance: undamped case, $\Omega = \omega$.

---

[17] This abbreviation is consistent with our use of ω on p. 121, since for $\gamma = 0$, $\dfrac{1}{2m}\sqrt{4km - \gamma^2} = \sqrt{\dfrac{k}{m}}$.

easiest possible model that gives some mathematical idea why. Intuitively the reason is that the particle "wants to vibrate" with period $\dfrac{2\pi}{\omega}$ anyway; if an external force with the same period is applied, this reinforces the tendency of the particle to do so.

Resonance phenomena have important practical consequences—both fortunate and unfortunate ones. For example, resonance makes it possible to tune in to a specific radio frequency. In fact, in Section 3.1 we'll study an electrical circuit that gives rise to an inhomogeneous equation similar to $mx'' + \gamma x' + kx = F_0 \cos \Omega t$, so that we'll be able to find a condition for electrical resonance in that circuit.

On the other hand, there have been several disastrous cases in which bridges have collapsed because a periodic external force (with resonance frequency) resulted from the passage of marching soldiers or from eddy effects due to wind (see "Further Reading" following the Exercises).

Finally, let us investigate the case $\gamma = 0$, $\Omega \neq \omega$: the undamped case without resonance. We have a solution $\dfrac{F_0}{k - m\Omega^2} \cos \Omega t$ to the inhomogeneous equation

$$mx'' + kx = F_0 \cos \Omega t;$$

the general solution will be

$$x(t) = A \cos \omega t + B \sin \omega t + \frac{F_0}{k - m\Omega^2} \cos \Omega t, \quad A, B \text{ arbitrary}.$$

Let's see, in particular, what happens if the particle is at rest for $t = 0$. (If the force is applied starting then, our solution will hold on the interval $t \geq 0$.) We have $x(0) = 0$, $x'(0) = 0$. Solving for A and B, we get $A + \dfrac{F_0}{k - m\Omega^2} = 0$, $B = 0$. So the solution will be

$$x(t) = \frac{F_0}{k - m\Omega^2} (\cos \Omega t - \cos \omega t).$$

To study the behavior of this function, it's convenient to use the trigonometric identity $\cos \phi - \cos \psi = -2 \sin \dfrac{\phi + \psi}{2} \sin \dfrac{\phi - \psi}{2}$ (see Exercise 8), which yields

$$x(t) = \frac{2F_0}{k - m\Omega^2} \sin \frac{1}{2}(\omega + \Omega)t \sin \frac{1}{2}(\omega - \Omega)t.$$

This function is the product of two oscillating functions. The faster oscillation has period $\dfrac{4\pi}{\omega + \Omega}$, and the function can be thought of as an "oscillation" with this period, whose amplitude itself oscillates with the larger period $\dfrac{4\pi}{|\omega - \Omega|}$. This will be especially obvious if Ω is close to ω (without being equal). In this case the smaller period is close to the natural period $\dfrac{2\pi}{\omega}$ of the system, while the other period is very much larger.

SEC. 2.7 MECHANICAL VIBRATION

A similar phenomenon occurs for acoustical, rather than mechanical, vibrations when two instruments that are almost, but not quite, at the same pitch (that is, frequency of vibration) are sounded together. This creates a "beating" sensation, which is unpleasant to the ear, as the volume of sound pulses regularly. Hence the word **beats**, which has also come to be used for other than acoustical vibrations (see Figure 2.7.6).

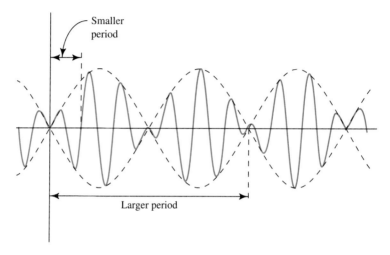

Figure 2.7.6. Beats: undamped case, $\Omega \neq \omega$.

SUMMARY OF KEY CONCEPTS AND RESULTS

Equation of motion for particle on vertical spring, no external force (p. 119): $mx'' + \gamma x' + kx = mg$.

Equilibrium position (p. 119): $x = \dfrac{mg}{h}$.

Equation of motion as above, but *relative to equilibrium position* (p. 120): $mx'' + \gamma x' + kx = 0$.

Undamped case (p. 120): $\gamma = 0$. Solutions periodic with period $2\pi \sqrt{\dfrac{m}{k}}$.

Underdamped case (p. 120): $\gamma < 2\sqrt{km}$. Solutions oscillate with "period" which increases with γ; $x(t) \to 0$ as $t \to \infty$, provided that $\gamma > 0$ (p. 121).

Critically damped case (p. 122): $\gamma = 2\sqrt{km}$. Solutions do not oscillate; $x(t) \to 0$ as $t \to \infty$.

Overdamped case (p. 123): $\gamma > 2\sqrt{km}$. Behavior of solutions fairly similar to critically damped case.

Equation of motion as above, but with periodic forcing term (pp. 123–124): $mx'' + \gamma x' + kx = F_0 \cos \Omega t$. For $\gamma > 0$, all solutions approach a particular solution $C_0 \cos(\Omega t - B_0)$ as $t \to \infty$: The particle is "forced" to vibrate with period $\dfrac{2\pi}{\Omega}$ (p. 124).

Resonance (p. 125): If $\gamma = 0$ (no friction) and $\Omega = \sqrt{\dfrac{k}{m}}$ (the period of the external force equals the "normal" period of the particle), a solution is found that oscillates with increasing amplitude.

Beats (pp. 126–127): If $\gamma = 0$ and $\Omega \neq \sqrt{\dfrac{k}{m}}$, a solution is found that combines a slower and a faster oscillation; this effect is particularly noticeable if Ω is close to $\sqrt{\dfrac{k}{m}}$.

EXERCISES

1. The position of a particle on a spring (without external force) is given by
$$x(t) = e^{-3t} \cos\left(5t - \frac{\pi}{2}\right).$$
   (a) Is the motion of the particle underdamped, critically damped, or overdamped?
   (b) What are the initial ($t = 0$) position and velocity of the particle?
   (c) Given $m = 2$, find the spring constant k and the damping constant γ.
   (d) If m is changed to $m = 10$, how would we have to change k and γ to produce the same motion?

2. The position of a particle on a spring (without external force) is given by
$$x(t) = 5e^{-t} - 2e^{-7t}.$$
   (a) Is the motion underdamped, critically damped, or overdamped?
   (b) What are the initial position and velocity of the particle?
   (c) Given $m = 5$, find the spring constant k and the damping constant γ.
   (d) If m is changed to $m = 10$, how would we have to change k and γ to produce the same motion?

3. (See p. 123.) Show that if $\gamma > 2\sqrt{km}$, then both roots
$$\lambda_{1,2} = -\frac{\gamma}{2m} \pm \frac{1}{2m}\sqrt{\gamma^2 - 4km}$$
of the characteristic equation are negative (k, m are positive constants).

4. (See p. 123.) Assuming that k and m are positive constants, sketch the graph of $x(t) = (C_1 + C_2 t)e^{-(\sqrt{k/m})t}$ for all possible signs of the constants C_1 and C_2 (positive, negative, or zero).

5. In one figure, sketch the graphs of the following functions:
$$e^{-t} - 2e^{-3t};\ 2e^{-t} - e^{-3t};\ e^{-t} + e^{-3t};\ -e^{-t} - 2e^{-3t};\ -e^{-t} + 2e^{-3t}.$$

Find a second-order differential equation to which all these functions are solutions. If this equation describes the motion of a particle on a spring, is the motion underdamped, overdamped, or critically damped?

6. (See p. 121.) Suppose that a vibration is considered imperceptible at the point at which its amplitude has decreased to 1% of its original amplitude. Find a formula for the amount of time it will take a vibration to become imperceptible in the underdamped case ($\gamma < 2\sqrt{km}$). What happens as $\gamma \to 0$?

7. In the text, we assumed that $k > 0$ in the equation $mx'' + \gamma x' + kx = mg$. Solve the equation in the case $k = 0$. (This corresponds to the case where there is no spring: The particle is falling, subject only to friction.) Find, in particular, the solution corresponding to the initial conditions $x(0) = 0$, $x'(0) = 0$ (particle starting from rest). Sketch the behavior of this solution as a function of t.

8. Prove the identity $\cos \phi - \cos \psi = -2 \sin \dfrac{\phi + \psi}{2} \sin \dfrac{\phi - \psi}{2}$.

    [*Hint*: $\phi = \dfrac{\phi + \psi}{2} + \dfrac{\phi - \psi}{2}$.]

9. As two musicians are tuning, they find that their instruments are combining to produce beats, with a slow oscillation of 4 Hz (four periods per second) and a fast oscillation of 440 Hz. What frequencies (in periods per second) are being produced by the two instruments separately?

10. Sketch the graph of the function $\sin 2t \sin 4t$ for $0 \leq t \leq 2\pi$.

11. Sketch the graph of the function $\sin t \sin 6t$ for $0 \leq t \leq 2\pi$.

12. Show that if $x_1(t) = C_1 \cos \Omega t + C_2 \sin \Omega t$ satisfies the equation $mx'' + \gamma x' + kx = F_0 \cos \Omega t$, $\Omega > 0$, then

$$C_1 = \frac{k - m\Omega^2}{(k - m\Omega^2)^2 + \gamma^2 \Omega^2} F_0, \quad C_2 = \frac{\gamma \Omega}{(k - m\Omega^2)^2 + \gamma^2 \Omega^2} F_0.$$

13. Show that $C_1 \cos \Omega t + C_2 \sin \Omega t = C_0 \cos(\Omega t - B_0)$ exactly when C_0, B_0 are polar coordinates for the point in the plane whose rectangular coordinates are (C_1, C_2).

Use Exercise 13 to write each of the following functions in the form $C_0 \cos(\Omega t - B_0)$.

14. $2 \cos 3t + 2\sqrt{3} \sin 3t$

15. $\cos 4t - \sin 4t$

16. (a) Use the results of Exercises 12 and 13 to show that when the periodic external force $F_0 \cos \Omega t$ is applied to a particle on a spring, the amplitude of the resulting oscillation will (in the long run) be $\dfrac{F_0}{\sqrt{(k - m\Omega^2)^2 + \gamma^2 \Omega^2}}$.

    (b) Show that if $\gamma^2 \geq 2km$, this amplitude is a decreasing function of Ω (assuming that $\Omega > 0$ and given F_0, m, k, and γ).

(c) Show, on the other hand, that if $\gamma^2 < 2km$, the amplitude is as large as possible for $\Omega = \dfrac{1}{2m}\sqrt{4km - 2\gamma^2}$. Note that, provided $\gamma > 0$, this value for Ω is *less* than the constant ω from p. 121; thus the "resonance frequency" $\dfrac{\Omega}{2\pi}$ for which the particle responds "best" to the external force is less than the "natural frequency" $\dfrac{\omega}{2\pi}$ of the particle's damped oscillations without an external force. Can you explain this?
[*Hint*: Think about pushing someone on a swing.]

17. (a) Show that if $\Omega > 0$ and if $\cos \Omega t$, $\sin \Omega t$ satisfy the homogeneous equation $mx'' + \gamma x' + kx = 0$, then $\gamma = 0$ and $\Omega = \sqrt{\dfrac{k}{m}}$.

    (b) Show that in this case, the unique solution of the form $At \cos \Omega t + Bt \sin \Omega t$ to the inhomogeneous equation $mx'' + \gamma x' + kx = F_0 \cos \Omega t$ is $\dfrac{F_0}{2m\Omega} t \sin \Omega t$.

    (c) Discuss the behavior of the *other* solutions to the inhomogeneous equation as $t \to \infty$ in this case ($\gamma = 0$, $\Omega = \sqrt{\dfrac{k}{m}}$).

18. Two particles are suspended from identical springs. The first particle is twice as heavy as the second. Intuitively, which of the particles would you expect to see vibrate more quickly? Confirm this by assuming that $\gamma = 0$ and finding (precisely) how the resonance frequencies (see p. 124, footnote 16) of the two particles compare.

*19. In this exercise we consider the overdamped case of a particle attached to a spring.

    (a) Show that $\gamma - \sqrt{\gamma^2 - 4km} = \dfrac{4km}{\gamma + \sqrt{\gamma^2 - 4km}}$. (This is just algebra.)

    (b) Show that when γ is substantially larger than $2\sqrt{km}$, one of the roots of the characteristic equation $\lambda^2 + \dfrac{\gamma}{m}\lambda + \dfrac{k}{m} = 0$ is approximately $-\dfrac{k}{\gamma}$.

    (c) Assume that k and m are fixed but that the damping constant γ is variable. Explain why your result from part (b) is evidence that for given $x(0)$ and $x'(0)$, $x(t)$ will usually approach zero *more slowly* (as $t \to \infty$) for larger $\gamma > 2\sqrt{km}$. This may seem paradoxical at first; can you give an intuitive explanation? Note that this phenomenon explains the term *over*damped!

FURTHER READING

For a tragicomic account of the 1940 collapse of the Tacoma Narrows Bridge (due to resonance), see Braun, *Differential Equations and Their Applications*, 3rd ed. (Springer-Verlag, 1982), Section 2.6.1.

2.8 INHOMOGENEOUS EQUATIONS (CONTINUED): VARIATION OF CONSTANTS

In this section we continue to study inhomogeneous second-order linear equations. We'll no longer assume that they have constant coefficients, and so our equations will have the form

$$x'' + p(t)x' + q(t)x = h(t),$$

where $p(t)$, $q(t)$, and $h(t)$ are given functions. In Section 2.6 we saw that in case p and q are constant, it is enough to find one solution to such an equation; this is because the homogeneous equation can be solved in this case. Although in general it may not be possible to solve the homogeneous equation, the same relationship between solutions of the homogeneous and inhomogeneous equations continues to hold, as stated in the following theorem.

Theorem 2.8.1 Given the inhomogeneous linear equation

$$x'' + p(t)x' + q(t)x = h(t)$$

and the corresponding homogeneous equation

$$x'' + p(t)x' + q(t)x = 0,$$

the general solution of the inhomogeneous equation can be obtained by adding any *particular* solution of the *inhomogeneous* equation to the *general* solution of the *homogeneous* equation.[18]

That is, if we add a solution of the homogeneous equation to one of the inhomogeneous equation, we get another solution of the inhomogeneous equation, and *all* solutions of the inhomogeneous equation can be obtained in this way from just one of them. For the proof, which is relatively straightforward, see Exercise 41. For a more general result, the **superposition principle**, which shows how to get the solutions if the forcing term $h(t)$ is a linear combination of simpler functions $h_1(t)$, $h_2(t)$, see Exercise 42.

Warning: Make sure that the equation is linear! For instance, $x'' - x^2 = -1$ "looks like" an inhomogeneous equation. One solution is $x = 1$, while $x = \dfrac{6}{t^2}$ is a solution to $x'' - x^2 = 0$. However, if we add these two, we get $x = 1 + \dfrac{6}{t^2}$; this is *not* a solution to $x'' - x^2 = -1$ (see Exercise 33). The explanation is that the equation is not linear, because of the presence of the term x^2.

---

[18] In this context, the solutions of the homogeneous equation, which are added to the **particular solution** of the inhomogeneous equation, are sometimes called the **complementary solutions**.

In the rest of this section we will assume that the solutions to the homogeneous equation $x'' + p(t)x' + q(t)x = 0$ are known. (This is true, in particular, if p and q are constants, but we will encounter other cases as well.) As we have seen, we can pick any two basic solutions $x_1(t)$ and $x_2(t)$ of the homogeneous equation, and then the general solution of the homogeneous equation will be $\alpha x_1(t) + \beta x_2(t)$, α and β arbitrary constants. We will assume that this has been done.[19]

Now we wish to solve the inhomogeneous equation

$$x'' + p(t)x' + q(t)x = h(t).$$

If we can somehow guess one solution, we are done, as we saw above. If not, the following method, although laborious, will *always* work. We will look for a solution of the form

$$x(t) = \alpha(t)x_1(t) + \beta(t)x_2(t),$$

where $x_1(t)$ and $x_2(t)$ are still the basic solutions of the homogeneous equation and α and β are now unknown functions.[20] Note that this is analogous to the approach used in Section 1.6 for first-order equations.

When we substitute the expression above for $x(t)$ into the inhomogeneous equation, we will obviously get a (rather involved) differential equation for the unknown functions α and β. We cannot expect to solve for both α and β from just one equation. Therefore, we can try to choose a second equation for α and β and make α and β satisfy both equations at the same time. We will choose this second equation so as to make the computation easy!

Let's see what happens when we differentiate:

$$x = \alpha x_1 + \beta x_2$$
$$x' = \alpha' x_1 + \alpha x_1' + \beta' x_2 + \beta x_2';$$

rearrange:

$$x' = \alpha' x_1 + \beta' x_2 + \alpha x_1' + \beta x_2'.$$

Now we will choose our second equation to be

$$\alpha' x_1 + \beta' x_2 = 0.$$

This has the advantage of simplifying x'; also, it means that when we differentiate x' to get x'', at least we will not get terms with α'' and β''. In fact, we will have

$$x' = \alpha x_1' + \beta x_2'$$
$$x'' = \alpha' x_1' + \beta' x_2' + \alpha x_1'' + \beta x_2'' \quad \text{(after rearranging)}.$$

Substituting all this into our inhomogeneous equation and combining terms, we find

$$\alpha' x_1' + \beta' x_2' + \alpha(x_1'' + px_1' + qx_1) + \beta(x_2'' + px_2' + qx_2) = h.$$

---

[19] It is actually enough to know one nonzero solution $x_1(t)$ of the homogeneous equation; a second basic solution $x_2(t)$ can then be found by reduction of order (see Section 2.4 and Exercises 35 to 38; also see Exercise 39).

[20] It will be possible to write solutions in this form in different ways. For instance, if $x_1(t) = \sin t$, $x_2(t) = \cos t$, then $tx_1(t) + (t^2 + \sin t)x_2(t) = (t + \cos t)x_1(t) + t^2 x_2(t)$.

SEC. 2.8 INHOMOGENEOUS EQUATIONS (CONTINUED): VARIATION OF CONSTANTS

Remember, though, that x_1 and x_2 are solutions of the homogeneous equation, so we have $x_1'' + px_1' + qx_1 = x_2'' + px_2' + qx_2 = 0$, and the equation above simplifies to

$$\alpha' x_1' + \beta' x_2' = h.$$

So we get the following system of differential equations for α and β:

$$\begin{cases} \alpha' x_1 + \beta' x_2 = 0 \\ \alpha' x_1' + \beta' x_2' = h. \end{cases} \quad (*)$$

Actually, this is a system of linear equations in α' and β', and so we can simply solve for α' and β', provided that the determinant $\begin{vmatrix} x_1 & x_2 \\ x_1' & x_2' \end{vmatrix}$ is not zero. But this determinant is exactly the Wronskian—the determinant that occurred in Theorem 2.2.4! Therefore, the determinant will not be zero, and we will be able to solve for α' and β'. Integration will then yield α and β. Finally, we'll have $x = \alpha x_1 + \beta x_2$.

EXAMPLE 2.8.1 Solve the equation $x'' - x = t^2$ by this method.

First, we solve $x'' - x = 0$; the characteristic equation is $\lambda^2 - 1 = 0$, so $\lambda = \pm 1$ and we get $x_1 = e^t$, $x_2 = e^{-t}$ as a pair of basic solutions of the homogeneous equation.

Thus we will substitute $x(t) = \alpha(t)e^t + \beta(t)e^{-t}$. First, however, we differentiate this to obtain (after rearranging) $x' = \alpha' e^t + \beta' e^{-t} + \alpha e^t - \beta e^{-t}$, and we require that

$$\alpha' e^t + \beta' e^{-t} = 0. \quad (1)$$

This leaves us with $x' = \alpha e^t - \beta e^{-t}$. Differentiating again, we get

$$x'' = \alpha' e^t - \beta' e^{-t} + \alpha e^t + \beta e^{-t}.$$

Now we substitute x'' and x into the given differential equation, which yields

$$\alpha' e^t - \beta' e^{-t} + \alpha e^t + \beta e^{-t} - \alpha e^t - \beta e^{-t} = t^2$$

or

$$\alpha' e^t - \beta' e^{-t} = t^2. \quad (2)$$

Combining (1) and (2), we get

$$\alpha' e^t = \frac{1}{2} t^2, \qquad \beta' e^{-t} = -\frac{1}{2} t^2,$$

so

$$\alpha' = \frac{1}{2} t^2 e^{-t}, \qquad \beta' = -\frac{1}{2} t^2 e^t,$$

$$\alpha = \frac{1}{2} \int t^2 e^{-t}\, dt, \qquad \beta = -\frac{1}{2} \int t^2 e^t\, dt.$$

Using integration by parts twice for each integral leads to

$$\alpha = \frac{1}{2}(-t^2 - 2t - 2)e^{-t} + C, \qquad \beta = -\frac{1}{2}(t^2 - 2t + 2)e^t + D.$$

Finally, we have

$$x(t) = \alpha \cdot e^t + \beta \cdot e^{-t} = \frac{1}{2}(-t^2 - 2t - 2) + Ce^t - \frac{1}{2}(t^2 - 2t + 2) + De^{-t}$$

$$x(t) = -t^2 - 2 + Ce^t + De^{-t}.$$

We could also have found this answer by the method of undetermined coefficients (from Section 2.6); in fact, that would have been easier. However, our new method is much more general, because it doesn't depend on being able to make an educated guess about what a solution should look like. ∎

EXAMPLE 2.8.2 Solve $x'' - x = h(t)$, where $h(t)$ is *any* given (continuous) function.

As in the special case from Example 2.8.1, we substitute $x(t) = \alpha(t)e^t + \beta(t)e^{-t}$, and we require that

$$\alpha' e^t + \beta' e^{-t} = 0. \tag{1}$$

The given equation then yields

$$\alpha' e^t - \beta' e^{-t} = h(t). \tag{2}$$

Solving (1) and (2) for α' and β', we get

$$\alpha' = \frac{1}{2} h(t)e^{-t}, \qquad \beta' = -\frac{1}{2} h(t)e^t$$

$$\alpha = \int \frac{1}{2} h(t)e^{-t}\, dt, \qquad \beta = -\int \frac{1}{2} h(t)e^t\, dt$$

$$x(t) = e^t \int \frac{1}{2} h(t)e^{-t}\, dt - e^{-t} \int \frac{1}{2} h(t)e^t\, dt.$$

Notice that the integrals are determined only up to integration constants; adding such constants, say C and D, will add $Ce^t - De^{-t}$, a solution of the homogeneous equation, to $x(t)$. ∎

We'll give two more examples below, but first a few general comments on our new method, which is known as the method of **variation of constants** or **variation of parameters**. This method was introduced by the French mathematician Lagrange in 1774, only about a century after calculus was invented!

It is possible to find a formula for the general solution $x(t)$ in terms of $x_1(t)$, $x_2(t)$, and $h(t)$ (see Exercise 34), but it's hardly possible to remember this formula. In practice, it's easier to go through our procedure, substituting $x = \alpha x_1 + \beta x_2$, in any given problem. You could also decide to memorize equations (∗) (p. 133). If you do so, beware of the following!

Warning: The simultaneous equations

$$\begin{cases} \alpha' x_1 + \beta' x_2 = 0 \\ \alpha' x_1' + \beta' x_2' = h(t) \end{cases} \tag{∗}$$

apply *only* if $h(t)$ is the right-hand side of the inhomogeneous equation

$$x'' + p(t)x' + q(t)x = h(t)$$

in that *(standard) form*. For instance, for the equation $t^2 x'' - 3tx' + 8x = 5t$ the correct $h(t)$ is not $5t$ but $\dfrac{5t}{t^2} = \dfrac{5}{t}$, since the equation first has to be divided by t^2 to make the coefficient of x'' equal 1.

EXAMPLE 2.8.3 Solve the equation $t^2 x'' - 7tx' + 15x = t^5$, given that t^5 and t^3 form a pair of basic solutions for the homogeneous equation.

We look for a solution $x(t) = \alpha(t)t^5 + \beta(t)t^3$ to the inhomogeneous equation. This yields $x' = \alpha' t^5 + \beta' t^3 + 5\alpha t^4 + 3\beta t^2$, and we require that

$$\alpha' t^5 + \beta' t^3 = 0. \tag{1}$$

We then have

$$x' = 5\alpha t^4 + 3\beta t^2$$
$$x'' = 5\alpha' t^4 + 3\beta' t^2 + 20\alpha t^3 + 6\beta t.$$

Substituting these expressions into the given equation and simplifying, we get

$$5\alpha' t^6 + 3\beta' t^4 = t^5. \tag{2}$$

If we had started from (∗), we would have obtained (2) in the form $5\alpha' t^4 + 3\beta' t^2 = t^3$, since the differential equation in standard form is $x'' - \dfrac{7}{t}x' + \dfrac{15}{t^2}x = t^3$. Either way, from (1) and (2) we get

$$\alpha' = \frac{1}{2t}, \quad \beta' = -\frac{1}{2}t,$$

so

$$\alpha = \int \frac{dt}{2t} = \begin{cases} \dfrac{1}{2}\log t & (t > 0) \\ \dfrac{1}{2}\log(-t) & (t < 0) \end{cases} + C \quad \text{and} \quad \beta = \int -\frac{1}{2}t\, dt = -\frac{1}{4}t^2 + D.$$

Thus for $t > 0$ the general solution is

$$x(t) = \left[\frac{1}{2}\log t + C\right]t^5 + \left[-\frac{1}{4}t^2 + D\right]t^3,$$

while for $t < 0$ we have

$$x(t) = \left[\frac{1}{2}\log(-t) + C\right]t^5 + \left[-\frac{1}{4}t^2 + D\right]t^3.$$

NOTES:

1. These answers can be simplified a bit; for instance, for $t > 0$,

$$x(t) = \frac{1}{2}t^5 \log t + \left(C - \frac{1}{4}\right)t^5 + Dt^3$$

$$= \frac{1}{2}t^5 \log t + C_1 t^5 + Dt^3,$$

where $C_1 = C - \frac{1}{4}$ and D are arbitrary constants. If we could somehow have guessed the solution $\frac{1}{2}t^5 \log t$, we could have saved a lot of work!

2. The solutions can be defined for $t = 0$ (see Exercise 43). ∎

EXAMPLE 2.8.4 Solve the initial value problem

$$\frac{d^2y}{dx^2} - 3\frac{dy}{dx} + 2y = e^{5x} \cos 3x, \quad y(0) = -\frac{149}{150}, \quad y'(0) = \frac{13}{75}.$$

The general solution to the homogeneous equation is easily found to be $y = \alpha e^x + \beta e^{2x}$. As we'll see in Section 4.2, we could find a particular solution to our inhomogeneous equation by using undetermined coefficients: There will be such a solution of the form $y = e^{5x}(A \cos 3x + B \sin 3x)$. However, let's use variation of constants instead. We put

$$y(x) = \alpha(x)e^x + \beta(x)e^{2x},$$

$$\frac{dy}{dx} = \alpha' e^x + \beta' e^{2x} + \alpha e^x + 2\beta e^{2x};$$

we require that

$$\alpha' e^x + \beta' e^{2x} = 0. \tag{1}$$

Then

$$\frac{dy}{dx} = \alpha e^x + 2\beta e^{2x},$$

$$\frac{d^2y}{dx^2} = \alpha' e^x + 2\beta' e^{2x} + \alpha e^x + 4\beta e^{2x}.$$

Substituting in the inhomogeneous equation, we get

$$\alpha' e^x + 2\beta' e^{2x} = e^{5x} \cos 3x. \tag{2}$$

From (1) and (2) we get

$$\alpha' = -e^{4x} \cos 3x, \qquad \beta' = e^{3x} \cos 3x$$

$$\alpha = -\int e^{4x} \cos 3x \, dx, \qquad \beta = \int e^{3x} \cos 3x \, dx.$$

In calculus courses, these integrals are usually found by integrating by parts twice. A more efficient method uses the fact that cos $3x$ is the real part of e^{3ix}, together with integration of a complex-valued exponential function. For instance,

$$\int e^{4x} \cos 3x \, dx = \int e^{4x} \cdot \text{Re}(e^{3ix}) \, dx$$

$$= \int \text{Re}[e^{(4+3i)x}] \, dx$$

$$= \text{Re}\left[\int e^{(4+3i)x} \, dx\right]$$

$$= \text{Re}\left[\frac{e^{(4+3i)x}}{4+3i} + C_1\right]$$

$$= \text{Re}\left[\frac{e^{4x}(\cos 3x + i \sin 3x)}{4+3i} + C_1\right]$$

$$= \text{Re}\left[\frac{e^{4x}(\cos 3x + i \sin 3x)(4-3i)}{25} + C_1\right]$$

$$= \frac{1}{25}e^{4x}(4 \cos 3x + 3 \sin 3x) + C,$$

where C is the real part of the complex integration constant C_1. Similarly, $\int e^{3x} \cos 3x \, dx = \frac{1}{6}e^{3x}(\cos 3x + \sin 3x) + D$. Thus we have

$$y = \alpha e^x + \beta e^{2x}$$

$$= -\left[\frac{1}{25}e^{4x}(4 \cos 3x + 3 \sin 3x) + C\right]e^x + \left[\frac{1}{6}e^{3x}(\cos 3x + \sin 3x) + D\right]e^{2x}$$

$$= \frac{1}{150}e^{5x} \cos 3x + \frac{7}{150}e^{5x} \sin 3x + Ce^x + De^{2x};$$

to finish the problem, we have to find the constants C, D such that $y(0) = -\frac{149}{150}$ and $y'(0) = \frac{13}{75}$ (see Exercise 1). ∎

SUMMARY OF KEY RESULTS AND TECHNIQUES

Finding one solution of an inhomogeneous equation can be enough (p. 131): If we have one solution $x_0(t)$ of the inhomogeneous linear equation $x'' + p(t)x' + q(t)x = h(t)$, then we can get all of them by adding the general solution of the *homogeneous* equation $x'' + p(t)x' + q(t)x = 0$ to that particular solution $x_0(t)$.

Method of **variation of constants** (p. 132): To solve $x'' + p(t)x' + q(t)x = h(t)$, start with a pair of basic solutions $x_1(t)$, $x_2(t)$ to the *homogeneous* equation, and substitute $x(t) = \alpha(t)x_1(t) + \beta(t)x_2(t)$. Requiring $\alpha' x_1 + \beta' x_2 = 0$ simplifies the computation and leads to the simultaneous equations

$$\begin{cases} \alpha' x_1 + \beta' x_2 = 0 \\ \alpha' x_1' + \beta' x_2' = h(t) \end{cases} \text{for } \alpha', \beta'.$$

After solving for α', β', integrate to get α, β and substitute back into $x = \alpha x_1 + \beta x_2$ to get the final answer.

EXERCISES

1. Finish Example 2.8.4.

Solve the following differential equations and initial value problems.

2. $t^2 x'' - 3tx' + 3x = t^{5/2}$. ($t^3$ and t form a pair of basic solutions for the homogeneous equation.)
3. $t^2 x'' - 3tx' + 3x = t$. (See Exercise 2.)
4. $x'' - 6x' + 8x = \sqrt{t}\, e^t$. (Don't try to evaluate the integrals.)
5. $x'' + 6x' + 8x = \sqrt{t}\, e^{-t}$. (Don't try to evaluate the integrals.)
6. $\dfrac{d^2 y}{dx^2} - 2 \dfrac{dy}{dx} + y = e^x / x$, $y(1) = 2$, $y'(1) = 0$.
7. $x'' - 6x' + 9x = t^{-1/2} e^{3t}$, $x(0) = 1$, $x'(0) = 0$.
8. $x'' - 4x' + 4x = t^{4/3} e^{2t}$, $x(0) = 1$, $x'(0) = 0$.
9. $x'' + 2x' + x = e^{-t}/t^2$, $x(1) = 1$, $x'(1) = 1$.
10. $x'' - 4x' + 3x = 2\sin t - \cos t$.
11. $x'' + 5x' + 4x = 34 \cos t$.
12. $\dfrac{d^2 y}{dx^2} + 4y = \dfrac{1}{\cos^3 2x}$, $y(0) = 1$, $y'(0) = 0$.
13. $x'' + 16x = \dfrac{1}{\sin^3 4t}$, $x(\dfrac{\pi}{8}) = 1$, $x'(\dfrac{\pi}{8}) = 0$.
14. $x'' - 7x = 2$.
15. $\dfrac{d^2 y}{dx^2} - 11y = 3x + 5$.
16. $x'' + 2x' + x = te^{-t/2}$.
17. $x'' - 2x' + x = 2te^{t/2}$.
18. $x'' + x = \dfrac{1}{\cos t}$. (You may assume that $-\dfrac{\pi}{2} < t < \dfrac{\pi}{2}$, so $\cos t > 0$.)

SEC. 2.8 INHOMOGENEOUS EQUATIONS (CONTINUED): VARIATION OF CONSTANTS

19. $\dfrac{d^2y}{dx^2} + y = \dfrac{1}{\sin x} \quad (0 < x < \pi).$

20. $t^2 x'' + 4tx' + 2x = \dfrac{2}{t}, \quad x(1) = 1, \quad x'(1) = 0.$ (t^{-1} and t^{-2} form a pair of basic solutions for the homogeneous equation.)

21. $t^2 x'' + 4tx' + 2x = t^{3/2}, \quad x(1) = 1, \quad x'(1) = 0.$ (See Exercise 20.)

22. $\dfrac{d^2y}{dx^2} - 4\dfrac{dy}{dx} + 3y = e^{2x} \cos x.$

23. $x'' + 5x' + 4x = e^{-3t} \cos t.$

24. $x^2 \dfrac{d^2y}{dx^2} + 3x \dfrac{dy}{dx} - 8y = 4x - \dfrac{1}{x^4}.$ (x^2 and x^{-4} form a pair of basic solutions for the homogeneous equation.)

25. $x^2 \dfrac{d^2y}{dx^2} + 3x \dfrac{dy}{dx} - 8y = 6x^2 - 3.$ (See Exercise 24.)

Find formulas for the solutions of the following differential equations.

26. $x'' - 6x' + 8x = h(t).$
27. $x'' + 6x' + 8x = h(t).$
28. $\dfrac{d^2y}{dx^2} + 4y = h(x).$
29. $x'' + 16x = h(t).$
30. $t^2 x'' - 3tx' + 3x = h(t).$ (See Exercise 2.)
31. $x^2 \dfrac{d^2y}{dx^2} + 3x \dfrac{dy}{dx} - 8y = h(x).$ (See Exercise 24.)
32. $x'' = h(t).$

33. (a) Check that $x = 1$ is a solution to $x'' - x^2 = -1$ and that $x = \dfrac{6}{t^2}$ is a solution to $x'' - x^2 = 0.$

(b) Check that, nevertheless, $x = 1 + \dfrac{6}{t^2}$ is not a solution to $x'' - x^2 = -1.$

(c) How would you (in principle) find the general solution to $x'' - x^2 = -1$?

34. Let $x_1(t), x_2(t)$ be a pair of basic solutions for the differential equation $x'' + p(t)x' + q(t)x = 0.$

(a) Show that $x(t) = \alpha(t)x_1(t) + \beta(t)x_2(t)$ will be a solution of

$$x'' + p(t)x' + q(t)x = h(t) \quad \text{provided} \quad \alpha' = \dfrac{-h(t)x_1(t)}{W(t)}, \quad \beta' = \dfrac{h(t)x_2(t)}{W(t)},$$

where $W(t)$ denotes the Wronskian $\begin{vmatrix} x_1(t) & x_2(t) \\ x_1'(t) & x_2'(t) \end{vmatrix}.$

(b) Show that the general solution of the inhomogeneous equation is

$$x(t) = x_2(t) \cdot \int \frac{h(t)x_1(t)}{W(t)} dt - x_1(t) \cdot \int \frac{h(t)x_2(t)}{W(t)} dt.$$

35. (a) Check that $y = e^{-x^2}$ is a solution to $xy'' - y' - 4x^3y = 0$.
    (b) Find the general solution of the equation in part (a). (Use reduction of order.)
    (c) Find the general solution to $xy'' - y' - 4x^3y = x^5 + x^3$.

36. (a) Check that $x = t^3$ is a solution to $t^2x'' - 5tx' + 9x = 0$.
    (b) Find the general solution of the equation in part (a).
    (c) Find the solution to the initial value problem

$$t^2x'' - 5tx' + 9x = 2t^3 - 1, \quad x(1) = 0, \quad x'(1) = 1.$$

*37. (a) Check that $x = t$ is a solution to

$$(\sin t - t \cos t)x'' - (t \sin t)x' + (\sin t)x = 0.$$

(b) Find the general solution of the equation in part (a).

[*Hint*: $\dfrac{2t \cos t + t^2 \sin t - 2 \sin t}{t(\sin t - t \cos t)} = -\dfrac{2}{t} + \dfrac{t \sin t}{\sin t - t \cos t}$; the second term on the right can be integrated using $u = \sin t - t \cos t$.]

(c) Find a formula for the general solution of

$$(\sin t - t \cos t)x'' - (t \sin t)x' + (\sin t)x = h(t).$$

*38. (a) Show that $x = e^t$ is a solution to

$$(\cos t - \sin t)x'' + (2 \sin t)x' + (-\sin t - \cos t)x = 0.$$

(b) Find the general solution of the differential equation in part (a).

[*Hint*: $\dfrac{-2 \cos t}{\cos t - \sin t} = -1 + \dfrac{-\cos t - \sin t}{\cos t - \sin t}$; the second term on the right can be integrated using $u = \cos t - \sin t$.]

(c) Find the solution to the initial value problem

$$(\cos t - \sin t)x'' + (2 \sin t)x' + (-\sin t - \cos t)x = 3e^{-t}(1 - 2 \sin t \cos t),$$
$$x(0) = 0, \quad x'(0) = 0.$$

[*Hint*: $1 - 2 \sin t \cos t = (\cos t - \sin t)^2$. (Why?)]

*39. Suppose that a nonzero solution $x_1(t)$ to the homogeneous linear equation $x'' + p(t)x' + q(t)x = 0$ is known. Show that the method of reduction of order can be applied directly to the *inhomogeneous* equation $x'' + p(t)x' + q(t)x = h(t)$, by showing that if you substitute $x(t) = A(t)x_1(t)$ in the inhomogeneous equation, you will get an inhomogeneous *first-order* linear equation for $a(t) = A'(t)$.

40. Use the method of Exercise 39 to solve $x'' - 4x' + 4x = t$, using only the fact that $x_1(t) = e^{2t}$ is a solution to the homogeneous equation.

41. In this exercise we consider the inhomogeneous equation

$$x'' + p(t)x' + q(t)x = h(t)$$

and the related homogeneous equation

$$x'' + p(t)x' + q(t)x = 0.$$

(a) Show that if $x(t)$ and $x_0(t)$ are solutions of the inhomogeneous equation, then $y(t) = x(t) - x_0(t)$ is a solution of the homogeneous equation.

(b) Show that if $x_0(t)$ is a solution of the inhomogeneous equation and $y(t)$ is a solution of the homogeneous equation, then $x_0(t) + y(t)$ is a solution of the inhomogeneous equation.

(c) Show that if one can guess or somehow find *one* solution $x_0(t)$ of the inhomogeneous equation, then one can find *all* solutions of that (inhomogeneous) equation by adding each of the solutions of the *homogeneous* equation to $x_0(t)$ [as in part (b)]. Note that this proves Theorem 2.8.1 (p. 131).

[*Hint*: Suppose that $x(t)$ is another solution of the inhomogeneous equation. Use part (a).]

42. Suppose that $h(t) = Ah_1(t) + Bh_2(t)$, where A and B are constants.

(a) Show that if $x_1(t)$ is a solution of $x'' + p(t)x' + q(t)x = h_1(t)$ and $x_2(t)$ is a solution of $x'' + p(t)x' + q(t)x = h_2(t)$, then $Ax_1(t) + Bx_2(t)$ is a solution of $x'' + p(t)x' + q(t)x = h(t)$.

*(b) Show that unless $A = B = 0$, *every* solution of $x'' + p(t)x' + q(t)x = h(t)$ can be obtained by the procedure described in part (a). You may assume that t is restricted, if necessary, to some interval on which $p(t)$, $q(t)$, $h_1(t)$, $h_2(t)$, and $h(t)$ are defined and continuous, so that solutions really exist.

[*Hint*: Once you get one solution, what else do you need?]

*43. (a) Show that one can define a function $f(t)$, which is continuous for all t, such that $f(t) = \frac{1}{2}t^5 \log t$ for $t > 0$ and $f(t) = \frac{1}{2}t^5 \log(-t)$ for $t < 0$.

(b) Show that the function $f(t)$ defined in part (a) is actually differentiable for all t.

(c) (See Example 2.8.3.) Show that for *any* constants E, F, G, H, there is a solution to $t^2 x'' - 7tx' + 15x = t^5$ which is defined for all t, and for which

$$x(t) = \begin{cases} \left[\frac{1}{2}\log t + E\right]t^5 + \left[-\frac{1}{4}t^2 + F\right]t^3 & (t > 0) \\ \left[\frac{1}{2}\log(-t) + G\right]t^5 + \left[-\frac{1}{4}t^2 + H\right]t^3 & (t < 0). \end{cases}$$

(d) What are the initial values $x(0)$, $x'(0)$ for various solutions? Does this contradict Theorem 2.2.2 (p. 87)?

CHAPTER 3

Systems of Two First-Order Differential Equations

3.1 PROBLEMS INVOLVING TWO UNKNOWN FUNCTIONS

In this chapter we will study differential equations that contain two unknown functions x and y of (one and) the same variable t. In applications, t will usually represent the time.

Since we cannot expect to solve for both unknown functions x and y if only one equation involving them is given, a second equation relating x and y will be needed.

EXAMPLE 3.1.1 In calculus one considers "related rate problems," in which the rate of change of one quantity is given and the rate of change of a related quantity is to be found. For instance, suppose that a conical tank is filling up with water at the constant rate of 8 ft^3 per minute, and we want to find out how fast the depth of the water in the tank is increasing. If $h(t)$ is the depth and $r(t)$ is the "surface radius" of the water in the tank (Figure 3.1.1, p. 144), then by the formula for the volume of a cone we know that $\frac{d}{dt}\left(\frac{1}{3}\pi r^2 h\right) = 8$. This gives us the differential equation $\frac{2}{3}\pi r h \frac{dr}{dt} + \frac{1}{3}\pi r^2 \frac{dh}{dt} = 8$. If we also know the shape of the tank, that will provide a second equation relating r and h. In a typical calculus problem, we would then find $\frac{dh}{dt}$ at a specific instant (say, when the water is 5 feet deep). However, we can do better and actually find the function $h(t)$ (see Exercise 14). ■

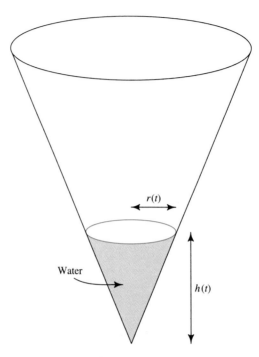

Figure 3.1.1. The conical tank of Example 3.1.1.

EXAMPLE 3.1.2 In Section 2.1 we considered the motion of a particle on a spring. Neglecting friction, we found the equation $x'' + \frac{k}{m} x = 0$ for the position $x(t)$ of the particle. To solve this equation, we introduced a second unknown function $v(t) = x'(t)$, the velocity. This gave us the differential equation $\frac{dv}{dt} + \frac{k}{m} x = 0$ for the two unknown functions, together with the equation $v = \frac{dx}{dt}$. (We then proceeded to use a "trick," writing $\frac{dv}{dt} = \frac{dv}{dx} \frac{dx}{dt} = v \frac{dv}{dx}$.)

More generally, if a particle moves along the x-axis under the influence of a force $F(x, x', t)$, we have the equation $x'' = \frac{1}{m} F(x, x', t)$ (see Section 2.1). If we again introduce the second unknown function $v = x'$, this will give us the first-order equation $v' = \frac{1}{m} F(x, v, t)$ involving v and x. Therefore, we have the system

$$\begin{cases} x' = v \\ v' = \frac{1}{m} F(x, v, t) \end{cases}$$

of two differential equations for the unknown functions x and v. ∎

SEC. 3.1 PROBLEMS INVOLVING TWO UNKNOWN FUNCTIONS

EXAMPLE 3.1.3   Similarly, any second-order linear differential equation $x'' + p(t)x' + q(t)x = h(t)$ can be rewritten as a pair of equations

$$\begin{cases} x' = v \\ v' = -p(t)v - q(t)x + h(t) \end{cases}$$

for the unknown functions x and v. However, this is not to say that this system is necessarily easier to solve than the original equation; in fact, we will see that some systems are solved by the reverse of this process, eliminating all but one of the unknown functions to get a higher-order equation in the remaining one. On the other hand, qualitative information is often easier to get for first-order systems than for the corresponding higher-order equations. ■

EXAMPLE 3.1.4   In Section 1.3 we considered the growth of a single population; in particular, a model was discussed in which the excess birth rate is taken to be constant, leading to the equation $\frac{1}{x}\frac{dx}{dt} = K$ for the size $x(t)$ of the population.

Now we will look at two populations sharing the same habitat. Such a situation can give rise to **competition**, if both populations use the same limited resources (food supply, etc.); to **predation**, if one population feeds on the other; or—in nature, this tends to be the least obvious case—to **symbiosis**, when the presence of each population is beneficial to the other. For the moment we will consider only the case of predation, and we will make a model for this case. In doing so, we'll assume that the species eaten (the **prey**) makes up an essential part of the diet of the eating species (the **predators**). For example, the habitat might be a cornfield, the prey species might be field mice, and the predator species, (wild) cats.

Let $x(t)$ be the size of the predator ("cat") population, while $y(t)$ is the size of the prey ("mouse") population. Just as in Section 1.3, $\frac{1}{x}\frac{dx}{dt}$ and $\frac{1}{y}\frac{dy}{dt}$ will be the respective excess birth rates for the two species.

First consider the prey birth rate $\frac{1}{y}\frac{dy}{dt}$. If there were no predators, we would be back in the case of Section 1.3, and according to our previous model, this rate would then be a constant. Now that there are predators, we have to expect that the excess birth rate of the prey will depend on the number of predators: The more predators there are, the more prey die (are eaten), so $\frac{1}{y}\frac{dy}{dt}$ will be a decreasing function of x. The simplest possible equation that describes such a situation is

$$\frac{1}{y}\frac{dy}{dt} = K - Lx,$$

where K and L are positive constants. Once again, if $x = 0$ (no predators) we have the old equation $\frac{1}{y}\frac{dy}{dt} = K$ describing the growth of the prey population.

Now let's look at the predators. Here the situation is different, since an essential part of the predators' diet is made up of the prey; that is, if there were no prey, the predators would not survive. In other words, we should expect the excess birth rate $\frac{1}{x}\frac{dx}{dt}$ to be *negative* for $y = 0$ and to be an increasing function of y. The simplest form for such a function is $-M + Py$, where M and P are positive constants. This leads to the equation

$$\frac{1}{x}\frac{dx}{dt} = -M + Py.$$

Combining this with the equation found above for the prey birth rate and clearing denominators, we find the following pair of differential equations:

$$\begin{cases} \dfrac{dx}{dt} = -Mx + Pxy \\ \dfrac{dy}{dt} = Ky - Lxy, \end{cases} \tag{$*$}$$

where K, L, M, and P are positive constants. These equations are known as the **Volterra–Lotka equations**; Vito Volterra (1860–1940), a mathematician, and Alfred Lotka (1880–1949), a biophysicist, proposed the model above.

It is easy to raise objections to this model. For instance, although the predators will thrive if the numbers of prey increase, a point of satiation might be expected (as in an all-you-can-eat restaurant). In other words, it might be more realistic to assume that the predator birth rate increases with y until some maximum is reached and then stays at that maximum even if y increases further. Also, the objections that apply to the "exponential" one-species model from Section 1.3 (and that led, for instance, to the Verhulst model; see Section 1.3, Exercise 21) are valid here, as well. Therefore, we should not expect the equations above to be satisfied in any practical "cats and mice" situation. What we *are* entitled to hope is that a mathematical solution to the equations will give us a qualitative idea of how the functions $x(t)$ and $y(t)$ behave in practice, for instance as t becomes large.[1] We will return to the consideration of equations ($*$) later in the chapter, especially in Section 3.9. ∎

EXAMPLE 3.1.5

We consider an electrical circuit consisting of four elements that are connected in series (i.e., "one after the other"). First we'll describe the elements separately.

A **resistor** is a piece of metal wire, or of another conducting material. If there is a difference in voltage (electrical potential) between the two ends of the resistor, then an electrical current will pass through the resistor. Conversely, if a current passes through the resistor, this will give rise to a voltage difference between the two ends. **Ohm's law** says that in many situations, the current will be proportional to this voltage drop. If this is so, the resistor is called **linear**. The constant R such that $V_R = R \cdot I$, where I is the current and

---

[1] It was pointed out in Section 1.3 that the exponential one-species model gave rise to the unreasonable prediction of unlimited growth as $t \to \infty$. However, the two-species model is better in this respect, as we will see.

SEC. 3.1 PROBLEMS INVOLVING TWO UNKNOWN FUNCTIONS

V_R is the voltage drop across the resistor, is then called the **resistance**. The symbol for a linear resistor of resistance R is $\underset{R}{-\!\!\!\!\bigwedge\!\!\bigwedge\!\!\bigwedge\!\!-}$.

A **capacitor** consists of a pair of parallel metal plates separated by air (or by another nonconducting material). If an electric current flows from one plate through the rest of the circuit to the other plate, there will be a resulting build-up of electrostatic charge on the plates, which will cause a difference in electrical potential between them. This difference, the voltage drop V_C across the capacitor, can be assumed to be proportional to the charge Q on the plates; the constant C such that $V_C = \dfrac{Q}{C}$ is called the **capacitance** of the capacitor. Since the charge on the plates is built up by the transfer of electrons by the current, the rate of change of the charge is the current itself: We have $I = \dfrac{dQ}{dt}$. As a result, we find $I = C\dfrac{dV_C}{dt}$ as the relation between the current through the circuit and the voltage drop across the capacitor. Alternatively, we can write $V_C = \dfrac{1}{C}Q = \dfrac{1}{C}\int I\,dt$; this will determine V_C up to a constant of integration if I is given as a function of t. The symbol for a (linear) capacitor of capacitance C is $\underset{C}{-|\!|-}$.

An **inductor** is a metal coil; when a nonconstant electrical current passes through such a coil, the interaction between the different loops of the coil gives rise to a voltage drop across the coil. We may assume that this voltage drop is proportional to the rate of change of the current, that is, $V_L = L\dfrac{dI}{dt}$, where V_L is the voltage drop across the inductor and L is a constant, the **inductance**. The notation for a (linear) inductor of inductance L is $\underset{L}{-\text{\small{OOO}}-}$.

The fourth element in our circuit will be a **generator**, that is, a machine that produces a certain voltage $E(t)$, which may depend on the time. This voltage is produced independently from the rest of the circuit and is (in theory) not influenced by anything that takes place in the rest of the circuit. The generator will be denoted by $\underset{E(t)}{-\bigcirc-}$.

Using only the four elements described above, one can assemble many different electrical networks; however, for the moment we consider only the circuit shown in Figure 3.1.2 (on p. 148). This is, for obvious reasons, called an *RLC* circuit.

It is a basic fact of electrical theory, known as **Kirchhoff's second law**, that the sum of the voltage changes encountered in going around the circuit must be zero.[2] Therefore, the

---

[2] Note that if this were not the case, it would be impossible to define electrical potential at a point, since the potential (or voltage) difference between two points would depend on the path taken between those two points. Kirchhoff's laws are named after the German physicist Kirchhoff (1824–1887). His first law states that at any junction in a circuit, the total incoming current must equal the total outgoing current.

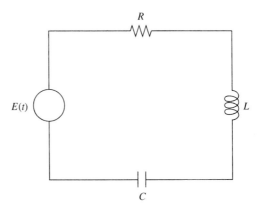

Figure 3.1.2. An *RLC* circuit.

voltage $E(t)$ produced by the generator must be equal to the sum of the voltage drops across the other three elements:

$$E(t) = V_R + V_L + V_C = RI + L\frac{dI}{dt} + V_C.$$

Since, on the other hand, we have $I = C\dfrac{dV_C}{dt}$, we have a pair of differential equations for the unknown functions I and V_C of t:

$$\begin{cases} RI + L\dfrac{dI}{dt} + V_C = E(t) \\ I = C\dfrac{dV_C}{dt}. \end{cases}$$

If we want to display the derivatives, we can rewrite these equations as

$$\begin{cases} \dfrac{dI}{dt} = -\dfrac{R}{L}I - \dfrac{1}{L}V_C + \dfrac{E(t)}{L} \\ \dfrac{dV_C}{dt} = \dfrac{1}{C}I. \end{cases}$$

Once again, R, L, and C are constants and $E(t)$ is a known function.

Our system of differential equations can be solved by eliminating I. In fact, substituting $I = C\dfrac{dV_C}{dt}$ in the first equation yields

$$LC\frac{d^2V_C}{dt^2} + RC\frac{dV_C}{dt} + V_C = E(t).$$

SEC. 3.1 PROBLEMS INVOLVING TWO UNKNOWN FUNCTIONS

This inhomogeneous second-order equation for V_C, or the equivalent equation

$$L\frac{d^2Q}{dt^2} + R\frac{dQ}{dt} + \frac{1}{C}Q = E(t)$$

for the unknown charge on the capacitor plates, can be solved by the methods of Chapter 2. In fact, there is a strong analogy between this last equation and the equation (which we considered in Section 2.7) $mx'' + \gamma x' + kx = F(t)$ for a vibrating particle on a spring. See Exercise 15. ∎

In this chapter we will study pairs of differential equations such as the ones found in Examples 3.1.4 and 3.1.5. Usually, one can solve only the simplest such systems of two equations, whether by elimination as in Example 3.1.5 or by a different method. However, we'll also see how to get qualitative information about the solutions of more difficult systems. Later, in Chapter 4, we'll study the more general case of n differential equations involving n unknown functions.

Throughout the rest of this chapter we will assume, unless stated otherwise, that x and y are the unknown functions of the variable t and that we are given a system of differential equations of the form

$$\begin{cases} \dfrac{dx}{dt} = f(t, x, y) \\ \dfrac{dy}{dt} = g(t, x, y) \end{cases}$$

for these unknown functions. [In Example 3.1.4, we have $f(t, x, y) = -Mx + Pxy$ and $g(t, x, y) = Ky - Lxy$, while in Example 3.1.5, $f(t, I, V_C) = -\dfrac{R}{L}I - \dfrac{1}{L}V_C + \dfrac{E(t)}{L}$, $g(t, I, V_C) = \dfrac{1}{C}I$. In general, f and g can be any given functions.] If t is the time, this means that we are given the rates of change of x and y in terms of x, y, and t. If we are also given $x(0)$ and $y(0)$, we might expect (as in the case of one unknown function, discussed in Section 1.5) that $x(t)$ and $y(t)$ would be determined for all t. Note that $x(0)$ and $y(0)$ must *both* be specified and that we expect to find *both* $x(t)$ and $y(t)$ once this is done. It is therefore convenient to call a *pair of functions* $(x(t), y(t))$ such that

$$\begin{cases} \dfrac{dx}{dt} = f(t, x, y) \\ \dfrac{dy}{dt} = g(t, x, y) \end{cases}$$

a solution of this system of differential equations. The graph of such a solution in t, x, y-space, that is, the set of points $(t, x(t), y(t))$, is called an integral curve of the system. It can be shown that just as in Section 1.5, there is exactly one integral curve through any point

(t_0, x_0, y_0), provided that f, g, and the partial derivatives $\dfrac{\partial f}{\partial x}, \dfrac{\partial f}{\partial y}, \dfrac{\partial g}{\partial x}, \dfrac{\partial g}{\partial y}$ of f and g with respect to the unknown functions are defined and continuous. In other words, under these assumptions, an initial value problem of the form

$$\begin{cases} \dfrac{dx}{dt} = f(t, x, y), & x(t_0) = x_0 \\ \dfrac{dy}{dt} = g(t, x, y), & y(t_0) = y_0 \end{cases}$$

has a unique solution.

Warning: Just as in Section 1.5, the solution may not be defined for all t, even though the equations look perfectly harmless. It may happen that $x(t)$ or $y(t)$ becomes infinite as t approaches some finite value. Of course, if this happens in an application, there may be a reason for alarm or to refine the model being used. (Compare Section 1.5, Exercise 43.)

EXAMPLES

A. In Example 3.1.4, given an initial predator population $x(0)$ and an initial prey population $y(0)$, the equations

$$\begin{cases} \dfrac{dx}{dt} = -Mx + Pxy \\ \dfrac{dy}{dt} = Ky - Lxy \end{cases}$$

will uniquely determine the functions $x(t)$, $y(t)$ which describe the subsequent growth (or decline) of the populations.

B. In Example 3.1.5, the capacitor voltage $V_C(0)$ and the current $I(0)$ for $t = 0$ will determine $V_C(t)$ and $I(t)$, *provided* that the generator voltage $E(t)$ is a (known) continuous function of t. This condition is not always met in practice; for instance, if the voltage is drawn from a 6-volt battery that is switched on at time $t = 1$, then

$$E(t) = \begin{cases} 0 & (t < 1) \\ 6 & (t > 1). \end{cases}$$

We will see in Chapter 6 how cases of this sort can be studied either by separate consideration of the equations "before" and "after" the discontinuity or by using the Laplace transform. ∎

SUMMARY OF KEY CONCEPTS

Volterra–Lotka equations (p. 146): $\dfrac{dx}{dt} = -Mx + Pxy$, $\dfrac{dy}{dt} = Ky - Lxy$, K, L, M, P positive constants. Here x and y are the predator and prey populations, respectively.

SEC. 3.1 PROBLEMS INVOLVING TWO UNKNOWN FUNCTIONS

Relations between current and voltage (pp. 146–147):
For a (linear) resistor: $V = RI$.
For a capacitor: $V = \dfrac{Q}{C}$, $I = \dfrac{dQ}{dt} = C\dfrac{dV}{dt}$.
For an inductor: $V = L\dfrac{dI}{dt}$.

Solution of a system $\dfrac{dx}{dt} = f(t, x, y)$, $\dfrac{dy}{dt} = g(t, x, y)$ (p. 149): A pair of functions $(x(t), y(t))$ that satisfy both equations. Provided that f, g, $\dfrac{\partial f}{\partial x}$, $\dfrac{\partial f}{\partial y}$, $\dfrac{\partial g}{\partial x}$, $\dfrac{\partial g}{\partial y}$ are defined and continuous, there is a unique solution with $x(t_0) = x_0$, $y(t_0) = y_0$.

EXERCISES

For each of the following second-order equations, write down an equivalent system of two first-order equations.

1. $x'' - 3tx' + (\sin t)x = t^2$.
2. $\dfrac{d^2y}{dx^2} + 5x\dfrac{dy}{dx} - x^2 y = 0$.
3. $x'' + 4(x')^2 - 6xx' = 17$.
4. $x'' - 5x^2 x' + 7 \sin x = 0$.

5. Consider two species sharing the same habitat. Assume that one of the species preys on the other "for sport" but that the prey are not essential to the predators' diet. (Big game hunting?) Make a model for this situation. Can you solve the differential equations?

6. Consider two species that are in competition with each other and share the same habitat. Propose a model that will describe the growth (or decline) of the two populations.

7. Consider two species in a symbiotic relationship. Assume that each species, although helpful to the other, is not essential to the other's survival. Also assume that the resources available to each species are unlimited. Make a model for this situation.

8. Redo Exercise 7, but this time assume that each species is crucial to the other's survival.

9. Suppose that three species of fish, A, B, and C, coexist in a pond. The diet of species A consists partly of species B and partly of insects. The diet of species B consists almost exclusively of species C. The diet of species C consists

exclusively of insects. Make a model describing the growth (or decline) of the three populations:
(a) assuming an unlimited supply of insects;
(b) assuming a limited supply of insects.

10. Assume that $x \neq 0$. Show that the Volterra–Lotka system

$$\begin{cases} \dfrac{dx}{dt} = -Mx + Pxy \\ \dfrac{dy}{dt} = Ky - Lxy \end{cases}$$

is equivalent to the system

$$\begin{cases} y = \dfrac{1}{Px}\dfrac{dx}{dt} + \dfrac{M}{P} \\ \dfrac{d^2x}{dt^2} - \dfrac{1}{x}\left(\dfrac{dx}{dt}\right)^2 + (Lx - K)\dfrac{dx}{dt} + (LMx^2 - KMx) = 0 \end{cases}.$$

Note that y has now been eliminated, but that the resulting equation for x is highly nonlinear.

11. Consider the electrical network shown in Figure 3.1.3.
(a) Justify the system of equations

$$\begin{cases} L\dfrac{dI_2}{dt} = E(t) - RI \\ I = I_1 + I_2 \\ C\dfrac{dV_C}{dt} = I_1 \\ V_C = L\dfrac{dI_2}{dt} \end{cases}$$

for the currents and capacitor voltage in the network.

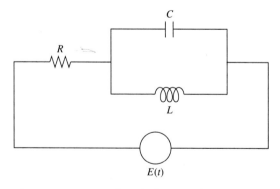

Figure 3.1.3. An electrical network.

(b) Show how the system in part (a) could be solved by elimination. (Don't actually solve it, but do show how the system reduces to a single second-order equation.)

12. Consider the electrical network shown in Figure 3.1.4.
   (a) Find a system of equations for the currents and capacitor voltage in the network.
   (b) Show how your system could be solved by elimination. [You may assume that $E(t)$ is differentiable.]

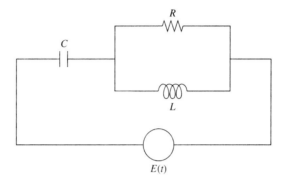

Figure 3.1.4. Another electrical network.

*13. Find a system of two differential equations for the two capacitor voltages in the network shown in Figure 3.1.5.
[*Hint*: First set up equations involving currents and voltages, then eliminate the currents.]

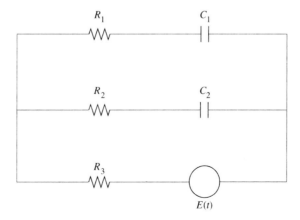

Figure 3.1.5. One more electrical network.

14. Suppose that in Example 3.1.1, the tank is 30 feet high and the top of the tank is 20 feet across.

   (a) Find $\dfrac{dh}{dt}$ when the water is 5 feet deep.

   (b) Assuming that the tank started to fill up at time $t = 0$ (minutes), find $h(t)$.
   [*Hint*: First find a separable differential equation for $h(t)$.]

   (c) Using your answer from part (b), find out when the water is 5 feet deep, and then confirm your answer from part (a).

15. (a) Explain why, in Example 3.1.5, the equations

$$LC\frac{d^2V_C}{dt^2} + RC\frac{dV_C}{dt} + V_C = E(t) \quad \text{and} \quad L\frac{d^2Q}{dt^2} + R\frac{dQ}{dt} + \frac{1}{C}Q = E(t)$$

are equivalent.

   (b) If we think of Q as analogous to the displacement x of a particle on a spring, what are the analogues of the mass m, the spring constant k, and the damping constant γ?

   (c) Suppose that $E(t) = 0$. For what values of L, R, and C will the "electrical vibration" be (i) underdamped, (ii) critically damped, and (iii) overdamped?

   (d) Suppose that $R = 0$ and $E(t) = E_0 \cos \Omega t$. For what value of Ω will resonance occur in the circuit?

16. Show that the initial value problem

$$y\frac{dx}{dt} = y + 3t, \qquad x(0) = 1,$$

$$(x-1)\frac{dy}{dt} = 3x - e^t, \qquad y(0) = 0$$

has no solutions. Does this contradict the statement on p. 150 about initial value problems having unique solutions?

FURTHER READING

For an explanation of how inductors work, see a physics textbook such as Halliday and Resnick, *Fundamentals of Physics*, 3rd ed. (Wiley, 1988).

3.2 AUTONOMOUS SYSTEMS

At the heart of much of scientific thought is the assumption that experiments may be repeated at will with identical results. For instance, if two chemical substances react to form a third, then this reaction will take place in the same way if the experiment is repeated a hundred years later, provided that the amounts of the substances and all external factors (temperature, presence of catalysts, etc.) are the same. As a result, when differential equations are used to make a mathematical model of an experiment, we will not expect these equations to depend explicitly on the time unless time-dependent external factors are taken into account. In the situation we introduced in Section 3.1, this means that our system of equations

$$\begin{cases} \dfrac{dx}{dt} = f(t, x, y) \\ \dfrac{dy}{dt} = g(t, x, y) \end{cases}$$

will actually have the form

$$\begin{cases} \dfrac{dx}{dt} = f(x, y) \\ \dfrac{dy}{dt} = g(x, y) \end{cases}$$

unless external influences which depend on t are taken into account.

For instance, in Example 3.1.4, the equations

$$\begin{cases} \dfrac{dx}{dt} = -Mx + Pxy \\ \dfrac{dy}{dt} = Ky - Lxy \end{cases} \quad \text{(see p. 146)}$$

do not include t explicitly. If we start with a certain number of cats and mice, we expect their populations to grow, or decline, regardless of whether the experiment is done now or next year. However, if external circumstances change (for instance, if the cornfield is plowed under), we may want to take this into account in our model. This might be done, for instance, by replacing the constants K, L, M, and P by suitable functions of the time.

In Example 3.1.5, the first equation in the system

$$\begin{cases} \dfrac{dI}{dt} = -\dfrac{R}{L} \cdot I - \dfrac{1}{L} \cdot V_C + \dfrac{E(t)}{L} \\ \dfrac{dV_C}{dt} = \dfrac{1}{C} \cdot I \end{cases} \quad \text{(see p. 148)}$$

does depend explicitly on t. This is because of the presence of the generator; the generator voltage $E(t)$ has to be considered an external factor. For instance, if

$$E(t) = \begin{cases} 0 & (t < 1) \\ 6 & (t > 1), \end{cases}$$

then the time $t = 1$ is no longer an arbitrary one; it is now *the* moment at which the power supply is switched on. If we now start, say, with $I = 0$, $V_C = 0$ for $t = 0$, we will expect nothing to happen in the circuit until $t = 1$, since currents and voltages do not occur spontaneously. On the other hand, if we "repeat" the experiment starting with $I = 0$, $V_C = 0$ for $t = 2$, then we should expect I and V_C to change immediately from these initial values. Of course, this is not a true repeat of the experiment, since an external factor, the generator voltage, is different.

NOTE: In some situations, it might be possible to construct a more elaborate model in which $E(t)$ is not considered a known function, but instead E itself is found from a differential equation. If so, there might be a system of equations for I, V_C, and E (and possibly other unknown functions) that did not depend explicitly on t.

A system of differential equations such as

$$\begin{cases} \dfrac{dx}{dt} = f(x, y) \\ \dfrac{dy}{dt} = g(x, y), \end{cases}$$

in which all derivatives are with respect to t but the equations do not depend explicitly on the variable t, is called **autonomous**.[3] Although we will eventually (in Section 4.6) discuss the solution of one particular kind of nonautonomous system, which includes the system from Example 3.1.5, we'll usually deal only with autonomous systems. These have several special properties that make it easier to analyze their solutions; also, they occur frequently in applications, for the reasons mentioned above.

In this chapter our systems will have the form $\dfrac{dx}{dt} = f(x, y)$, $\dfrac{dy}{dt} = g(x, y)$; we'll always assume that f and g are C^1, so that f, g, and their partial derivatives are defined and continuous (see p. 8).[4] A solution of such a system is a pair of functions $(x(t), y(t))$ satisfying both equations; as we saw in Section 3.1, there is a unique solution satisfying the initial conditions $x(t_0) = x_0$, $y(t_0) = y_0$.

It is often useful, although it may seem artificial at first, to think of x and y as the coordinates in an x,y-plane. A solution $(x(t), y(t))$ then describes a parametric curve in this

---

[3] Note the name, which derives from the kind of practical situation modeled by such a system.

[4] On the other hand, we *will* consider situations in which f and g are C^1 functions defined only in some region of the x,y-plane.

SEC. 3.2 AUTONOMOUS SYSTEMS

plane: For each value of the parameter t a point of the plane is given. Such a parametric curve, the set of the points $(x(t), y(t))$ for all possible values of the parameter t, is called a **trajectory** or a **path** of the autonomous system.[5] If we think of t as the time, then $(x(t), y(t))$ is a moving point (in the plane) that follows the trajectory; the initial conditions indicate that at time t_0 the point is at (x_0, y_0): $(x(t_0), y(t_0)) = (x_0, y_0)$.

If the moving point starts from the same position (x_0, y_0) but at a different time, say at $t = 0$, it will follow the same trajectory. The reason is that when $(x(t), y(t))$ is a solution to the system, $(x(t + t_0), y(t + t_0))$ is a solution also (see Exercise 60). This last solution is at (x_0, y_0) for $t = 0$. Its path is the same, since as t varies arbitrarily, $t + t_0$ does also, so that the set of *all* points $(x(t + t_0), y(t + t_0))$ is the same as the set of *all* points $(x(t), y(t))$. We can conclude that there is exactly one trajectory through each point (x_0, y_0) in the region where f and g are C^1. In particular, different trajectories cannot intersect (if they did, there would be several trajectories through the intersection point).

EXAMPLE The solution of the system $\dfrac{dx}{dt} = \dfrac{1}{4}y$, $\dfrac{dy}{dt} = x$ satisfying the initial conditions $x(1) = \sqrt{e}$, $y(1) = 2\sqrt{e}$ is $x(t) = e^{t/2}$, $y(t) = 2e^{t/2}$. (Check this; we'll see later in this chapter how such a solution can be found.) The corresponding trajectory is the set of all points $(e^{t/2}, 2e^{t/2})$ in the x, y-plane, that is, the part of the line $y = 2x$ with $x > 0$ [Figure 3.2.1(a), p. 158]. If we change the initial conditions to $x(0) = \sqrt{e}$, $y(0) = 2\sqrt{e}$, we must replace t by $t + 1$ in the solution, which yields the new solution $x(t) = e^{(t+1)/2}$, $y(t) = 2e^{(t+1)/2}$. (Check that this is indeed a solution to the system and that it satisfies the new initial conditions.) The trajectory is still the same; the moving point reaches each position along the trajectory, in particular the "initial position" $(\sqrt{e}, 2\sqrt{e})$, one unit of time earlier [see Figure 3.2.1(b)]. ∎

Since $\dfrac{dx}{dt}$ and $\dfrac{dy}{dt}$ are the components of the velocity vector[6] of the moving point $(x(t), y(t))$, the equations

$$\begin{cases} \dfrac{dx}{dt} = f(x, y) \\ \dfrac{dy}{dt} = g(x, y) \end{cases}$$

indicate what the velocity of the moving point is when it reaches a particular position (x, y). Note, again, that this does not depend on the time when that position is reached.

---

[5] Note that the trajectories are the projections of the *integral curves* introduced in Section 3.1—the sets of points $(t, x(t), y(t))$—onto the x,y-plane.

[6] See Appendix B, p. 607.

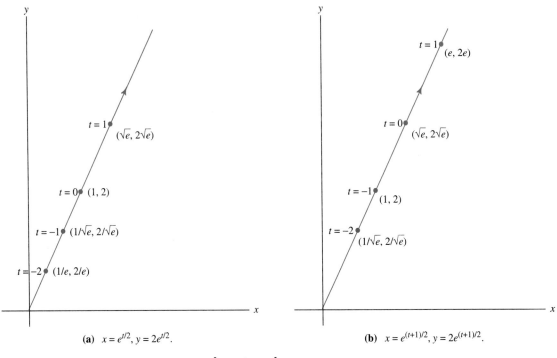

Figure 3.2.1. Solutions to the system $\dfrac{dx}{dt} = \dfrac{1}{4}y$, $\dfrac{dy}{dt} = x$: (a) with initial conditions $x(1) = \sqrt{e}$, $y(1) = 2\sqrt{e}$; (b) with initial conditions $x(0) = \sqrt{e}$, $y(0) = 2\sqrt{e}$.

EXAMPLE 3.2.1

$$\dfrac{dx}{dt} = -y, \quad \dfrac{dy}{dt} = x.$$

For this system of equations, the velocity vector of the moving point will be $(-y, x)$ when the point is at (x, y). For instance, if at some particular time we have $x = 2$, $y = 3$, then the moving point will have the velocity $(-3, 2)$, so it will be traveling with speed $\sqrt{13}$ in the direction indicated in Figure 3.2.2.

N O T E : The arrows in Figure 3.2.2 were actually shortened (as suggested on p. 42 in a similar situation) so that they wouldn't interfere with each other. That is, instead of drawing the velocity vector $(-y, x)$ at position (x, y), we drew $(-ky, kx)$ for some constant $k < 1$. □

Just as in this example, we can always sketch the velocity vector $(f(x, y), g(x, y))$ of the moving point (or a suitably scaled-down multiple) for various positions (x, y) in the plane. The collection of these vectors is called the **direction field** of the system. Sketching the direction field, which can be done by hand or—preferably—by computer, will give some idea of what the trajectories might look like (without having to solve the equations!) For

SEC. 3.2 AUTONOMOUS SYSTEMS

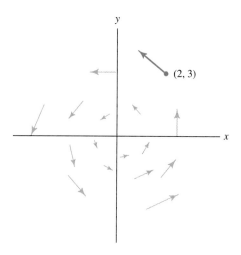

Figure 3.2.2. Directions of travel of a moving point $(x(t), y(t))$ whose motion is described by $\dfrac{dx}{dt} = -y$, $\dfrac{dy}{dt} = x$.

instance, in Figure 3.2.2 we can see that the moving point has a tendency to turn around the origin in the counterclockwise direction. However, we can't expect to find out whether the trajectories are closed curves or curves spiraling inward or outward without explicit computations (see Figure 3.2.3, p. 160).

We'll soon see that for the system $\dfrac{dx}{dt} = -y, \dfrac{dy}{dt} = x$ of Example 3.2.1, Figure 3.2.3(a) is correct; in fact, the trajectories are circles. The vectors drawn in Figure 3.2.2 are tangent to these circles, and the relation between Figure 3.2.2 and Figure 3.2.3(a) is therefore the same as the relation between the direction field and the solution curves for a first-order differential equation, as shown in Figure 1.5.7.[7] (Note, however, that we now have x and y, rather than t and x, along the axes.) We will make this more explicit below.

A sketch such as Figure 3.2.3(a), which indicates the actual behavior of the trajectories of the system, is called a **phase portrait** or **phase diagram** for the system. [Again, note the difference between the phase portrait, as shown in Figure 3.2.3(a), and the direction field that approximates it, shown in Figure 3.2.2.] Note also that the direction of travel along the trajectories is indicated. Phase portraits are an important qualitative tool for the study of autonomous systems.

We continue our study of the system $\dfrac{dx}{dt} = f(x, y), \dfrac{dy}{dt} = g(x, y)$. If we assume that $\dfrac{dx}{dt} \neq 0$, [that is, $f(x, y) \neq 0$], so that t is an (inverse) function of x and $y(t)$ is a composite

---

[7] Since the direction field of a first-order equation consists of line segments rather than vectors (see Section 1.5), it does not have exactly the same form as the direction field of a system.

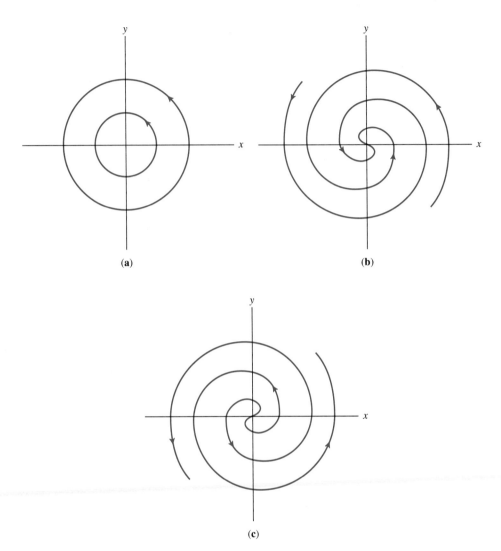

Figure 3.2.3. Which of these sets of trajectories corresponds to Figure 3.2.2?

function of x,[8] then we have $\dfrac{dy}{dx} = \dfrac{dy}{dt}\dfrac{dt}{dx} = \dfrac{dy}{dt} \bigg/ \dfrac{dx}{dt}$, so

$$\frac{dy}{dx} = \frac{g(x, y)}{f(x, y)}. \tag{1}$$

Note that this is a first-order equation, from which t has disappeared completely, for the function y of x.

[8] Compare the discussion on pp. 80–81.

EXAMPLE 3.2.1 *(continued)*

For the system we considered earlier, we have $\dfrac{dx}{dt} = -y$, $\dfrac{dy}{dt} = x \Rightarrow \dfrac{dy}{dx} = -\dfrac{x}{y}$. The direction field of this equation "is" the field shown in Figure 3.2.2 (more precisely, the part of Figure 3.2.2 where $y \neq 0$, with the arrows replaced by line segments). Since the equation $\dfrac{dy}{dx} = -\dfrac{x}{y}$ is separable, we can solve it; we get $y = \pm\sqrt{2C - x^2}$ (compare Example 1.2.5). The solution curves are semicircles, as shown in Figure 3.2.4(a). We can conclude that the trajectories of our system $\dfrac{dx}{dt} = -y$, $\dfrac{dy}{dt} = x$ will follow these semicircles as long as $y \neq 0$. □

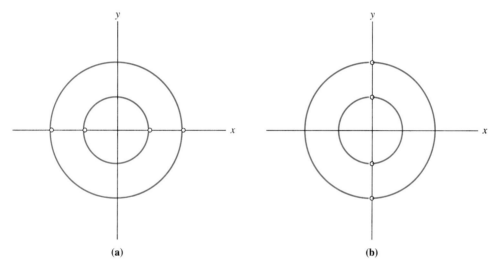

Figure 3.2.4. (a) Solutions of $\dfrac{dy}{dx} = -\dfrac{x}{y}$ $(y \neq 0)$; (b) Solutions of $\dfrac{dx}{dt} = -\dfrac{y}{x}$ $(x \neq 0)$.

Now suppose we consider x as a function of y, instead. To do so, we must have $\dfrac{dy}{dt} \neq 0$, that is, $g(x, y) \neq 0$, and then we obtain

$$\frac{dx}{dy} = \frac{f(x, y)}{g(x, y)}. \tag{2}$$

For the system of Example 3.2.1, this yields $\dfrac{dx}{dy} = -\dfrac{y}{x}$, $x = \pm\sqrt{2C - y^2}$, and we get semicircles cut off by the y-axis [Figure 3.2.4(b)].

If we are at a point where $f(x, y)$ and $g(x, y)$ are *both* nonzero, then it doesn't matter whether we consider y as a function of x or vice versa; equations (1) and (2) are then equivalent. (Remember that by the chain rule, $\dfrac{dy}{dx} \cdot \dfrac{dx}{dy} = 1$.) To illustrate this, they are often presented in **differential form** as

$$g(x, y)\, dx - f(x, y)\, dy = 0. \tag{3}$$

Equation (3) can be interpreted as follows: If y is a function of x, "divide by dx" to get $g(x, y) - f(x, y) \frac{dy}{dx} = 0$ or $\frac{dy}{dx} = \frac{g(x, y)}{f(x, y)}$ as the equation for y; if x is a function of y, "divide by dy" in similar fashion. The advantage of (3) is that x and y play similar roles and that there is no need to distinguish between the different cases when writing down the equation.

EXAMPLE 3.2.1 (concluded)

For the system $\frac{dx}{dt} = -y$, $\frac{dy}{dt} = x$, we have the equation $x\, dx + y\, dy = 0$ in differential form. The solution curves are obtained by "patching together" the semicircles from Figure 3.2.4(a) and (b); they are circles around the origin, as claimed previously. Note, however, that this means only that the trajectories of the system $\frac{dx}{dt} = -y$, $\frac{dy}{dt} = x$ will *follow* these circles—it does not guarantee that the moving point goes all the way around the circles (although it does so *in this case*; compare Exercises 56 and 57). It also doesn't indicate whether the point is moving clockwise or counterclockwise. To get this information, we can refer to the direction field (Figure 3.2.2) or consider the signs of $\frac{dx}{dt}, \frac{dy}{dt}$ in different quadrants.[9] Either way, we'll find that the motion is counterclockwise, and we end up with the phase portrait shown in Figure 3.2.3(a), as promised on p. 159. ∎

Here is another example of sketching the phase portrait of a system using the first-order equation relating x and y.

EXAMPLE 3.2.2

$\frac{dx}{dt} = y^2 + 1$, $\frac{dy}{dt} = y(y^2 + 1)$.

Since $y^2 + 1$ is never zero, y will be a function of x everywhere, and we get

$$\frac{dy}{dx} = \frac{dy/dt}{dx/dt} = \frac{y(y^2 + 1)}{y^2 + 1} = y.$$

This equation has solutions $y = Ae^x$, so the trajectories follow the curves $y = Ae^x$. Since $\frac{dx}{dt} > 0$ everywhere (why?), the curves are followed from left to right. See Figure 3.2.5. ∎

---

[9] For instance, in the first quadrant $\frac{dx}{dt} = -y < 0$, $\frac{dy}{dt} = x > 0$, so the point is moving to the left and upward.

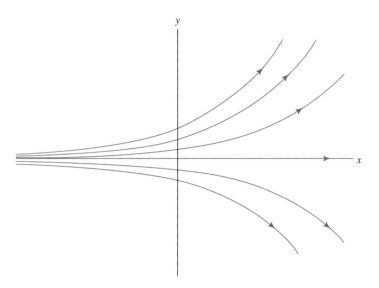

Figure 3.2.5. Phase portrait for $\dfrac{dx}{dt} = y^2 + 1$, $\dfrac{dy}{dt} = y(y^2 + 1)$.

So far we've seen that for a system $\dfrac{dx}{dt} = f(x, y)$, $\dfrac{dy}{dt} = g(x, y)$ we have $\dfrac{dy}{dx} = \dfrac{g(x, y)}{f(x, y)}$ where $f(x, y) \neq 0$, $\dfrac{dx}{dy} = \dfrac{f(x, y)}{g(x, y)}$ where $g(x, y) \neq 0$, and thus $g(x, y)\, dx - f(x, y)\, dy = 0$ in any case. However, we should consider what happens at a point where $f(x, y)$ and $g(x, y)$ are *both* zero, so that neither x nor y is a function of the other. Let (x_0, y_0) be such a point, that is, assume $f(x_0, y_0) = g(x_0, y_0) = 0$. In this case $\dfrac{dx}{dt} = \dfrac{dy}{dt} = 0$ at (x_0, y_0): The "moving" point is standing still at the position (x_0, y_0)! But then the moving point will always stay at (x_0, y_0), *and it will always have been there*. For in this case $x = x_0$, $y = y_0$ is a (constant) solution to our system $\dfrac{dx}{dt} = f(x, y)$, $\dfrac{dy}{dt} = g(x, y)$, and by the uniqueness of solutions, any solution $(x(t), y(t))$ that is equal to (x_0, y_0) for some time t must be this constant solution. In other words, the moving point can neither "enter" nor "leave" (x_0, y_0); it is either always or never at that position. [However, if it is never at that position, the moving point could still approach (x_0, y_0) as a limit, either for $t \to \infty$ or for $t \to -\infty$.] In particular, the point (x_0, y_0) constitutes a complete trajectory.

In this situation, that is, when $f(x_0, y_0) = g(x_0, y_0) = 0$, the point (x_0, y_0) is called a **stationary point** (other names are **critical point** and **equilibrium point**) of the system $\dfrac{dx}{dt} = f(x, y)$, $\dfrac{dy}{dt} = g(x, y)$.

As we'll see, stationary points are particularly important and can be worth finding even when the rest of the phase portrait is difficult to draw precisely.

EXAMPLE 3.2.3 Our system $\frac{dx}{dt} = -y$, $\frac{dy}{dt} = x$ from Example 3.2.1 has a stationary point at the origin $(0, 0)$. ∎

EXAMPLE 3.2.4 Consider the system $\frac{dx}{dt} = x^2 - 2xy - x$, $\frac{dy}{dt} = xy + y^2 - 3y$. This system will have a stationary point at (x_0, y_0) if

$$x_0^2 - 2x_0 y_0 - x_0 = x_0 y_0 + y_0^2 - 3y_0 = 0.$$

In practice, when finding the stationary points, the subscript "0" is dropped while one solves the simultaneous equations, here

$$x^2 - 2xy - x = 0, \qquad xy + y^2 - 3y = 0.$$

These particular equations can be solved by factoring each of them to get

$$x(x - 2y - 1) = 0, \qquad (x + y - 3)y = 0.$$

Since a product such as $x(x - 2y - 1)$ is zero if at least one of its two factors x and $x - 2y - 1$ is zero, there are four possibilities in all:

1. $x = 0$, $x + y - 3 = 0$.
2. $x = 0$, $y = 0$.
3. $x - 2y - 1 = 0$, $x + y - 3 = 0$.
4. $x - 2y - 1 = 0$, $y = 0$.

Solving each of these systems of equations separately yields $(0, 3)$, $(0, 0)$, $\left(\frac{7}{3}, \frac{2}{3}\right)$, and $(1, 0)$, so there are four stationary points in all.

NOTE: Sketching the phase portrait would be considerably more difficult than in the previous examples. For one thing, the first-order equation for y as a function of x is $\frac{dy}{dx} = \frac{dy/dt}{dx/dt} = \frac{xy + y^2 - 3y}{x^2 - 2xy - x}$, and it is not clear how to solve this. Meanwhile, sketching the direction field by finding the vectors $(x^2 - 2xy - x, xy + y^2 - 3y)$ at various positions (x, y) is realistic only if a computer is available, but then it can lead to interesting observations and questions about the phase portrait; see Figure 3.2.6 and Exercises 63 and 64. ∎

EXAMPLE 3.2.5 The system $\frac{dx}{dt} = -3x + 6xy$, $\frac{dy}{dt} = 2y - xy$ has stationary points [at (x, y)] where $-3x + 6xy = 2y - xy = 0$. From this we see, using a method similar to that of Example 3.2.4, that stationary points occur at $(0, 0)$ and $\left(2, \frac{1}{2}\right)$, and nowhere else. (This is a particular case of the Volterra–Lotka system from Section 3.1. See Exercise 55.) ∎

SEC. 3.2 AUTONOMOUS SYSTEMS

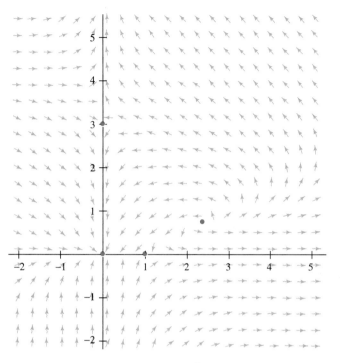

Figure 3.2.6. Direction field for $\dfrac{dx}{dt} = x^2 - 2xy - x$, $\dfrac{dy}{dt} = xy + y^2 - 3y$.

EXAMPLE 3.2.6

$\dfrac{dx}{dt} = x - y^2$, $\dfrac{dy}{dt} = y + x^2$.

To find the stationary points of this system, we must solve $x - y^2 = 0$, $y + x^2 = 0$. These equations don't factor, but we can eliminate x (or y): $x - y^2 = 0 \Rightarrow x = y^2$; substituting this into $y + x^2 = 0$ yields $y + y^4 = 0 \Rightarrow y(y^3 + 1) = 0$, so $y = 0$ or $y = -1$. Since $x = y^2$, the stationary points are $(0, 0)$ and $(1, -1)$. ∎

EXAMPLE 3.2.7

In calculus a function $V(x, y)$ of two variables is said to have a **critical point** at (x_0, y_0) if both its partial derivatives are zero there. At such a point, the tangent plane to the graph of V (see pp. 4–5) will be horizontal. If the function V has a (local) maximum or minimum[10] at (x_0, y_0), it will have a critical point there; however, other types of critical point also exist. In Section 3.8 we will classify critical points of V using the autonomous system

---

[10] As you might expect, the function V is said to have a **local minimum** at (x_0, y_0) if its value there is less than or equal to its value at any nearby point, that is, if there exists a circle of some radius $r_0 > 0$ around (x_0, y_0) such that $V(x_0, y_0) \leq V(x, y)$ for all points (x, y) inside the circle. For the definition of **local maximum**, replace $V(x_0, y_0) \leq V(x, y)$ by $V(x_0, y_0) \geq V(x, y)$.

$$\frac{dx}{dt} = -\frac{\partial V}{\partial x}, \quad \frac{dy}{dt} = -\frac{\partial V}{\partial y}.$$ (V is assumed to be known, so that $\frac{\partial V}{\partial x}$ and $\frac{\partial V}{\partial y}$ are known functions of x and y.) Notice that this system has stationary points exactly where V has critical points, that is, where $\frac{\partial V}{\partial x} = \frac{\partial V}{\partial y} = 0$. A system

$$\frac{dx}{dt} = -\frac{\partial V}{\partial x}, \quad \frac{dy}{dt} = -\frac{\partial V}{\partial y}$$

of this type is called a **gradient system**.[11] ■

EXAMPLE 3.2.8 Consider a particle of mass m moving along the x-axis. Recall from Example 3.1.2 (p. 144) that we have the system $x' = v$, $v' = \frac{1}{m} F(x, v, t)$, where x and the velocity v are the unknown functions. Now assume that the system is autonomous, that is, that the force is a function of x and v alone:

$$x' = v, \quad v' = \frac{1}{m} F(x, v).$$

A solution to this system is a parametric curve $(x(t), v(t))$ in the x,v-plane. [If you refer to $(x(t), v(t))$ as a "moving point," remember that it is *not* the particle, whose motion takes place along the x-axis.] A stationary point will occur when $v = \frac{1}{m} F(x, v) = 0$, that is, when $v = F = 0$. This means that the particle is standing still (on the x-axis) and that there is no force on it (which would make it begin to move). In this situation the particle is said to be in **equilibrium**. ■

When studying the qualitative behavior of a system, it is useful to identify the stationary points and to study the behavior of the trajectories near these points. One reason is that in applications (not only in the case described in Example 3.2.8) stationary points usually correspond to "equilibrium" situations, which are often desirable. However, it is impossible in practice to keep an experiment in exact equilibrium—there will always be small disturbances from the outside, to say nothing of the "disturbance" created by the fact that the equations in the mathematical model are only approximately correct. This raises questions such as the following: If the moving point $(x(t), y(t))$ starts near, but not at, a stationary point (because it has been displaced from the stationary point by a disturbance), will it then remain near the stationary point, perhaps even approaching it in the limit as

---

[11] To understand the name, note that the system requires that the velocity vector $\left(\frac{dx}{dt}, \frac{dy}{dt}\right)$ be opposite at all times to the gradient $\left(\frac{\partial V}{\partial x}, \frac{\partial V}{\partial y}\right)$ of V. The minus signs are traditional but not essential, since V could be replaced by $-V$.

SEC. 3.2 AUTONOMOUS SYSTEMS

$t \to \infty$, or will it go elsewhere? In the first case the equilibrium is "stable"; otherwise, it is "unstable." We will study such questions, and give more precise definitions of "stable" and "unstable," later in the chapter. Fortunately, these questions can usually be tackled by studying the **linear approximation** to the system near the stationary point; this is obtained by replacing $f(x, y)$ and $g(x, y)$ by their linear approximations (see p. 8) near the stationary point.

EXAMPLE The system $\dfrac{dx}{dt} = -3x + 6xy$, $\dfrac{dy}{dt} = 2y - xy$ of Example 3.2.5 has the linear approximation $\dfrac{dx}{dt} = -3x$, $\dfrac{dy}{dt} = 2y$ near the stationary point $(0, 0)$. ∎

In general, if the system $\dfrac{dx}{dt} = f(x, y)$, $\dfrac{dy}{dt} = g(x, y)$ has a stationary point at (x_0, y_0), then its linear approximation near that stationary point is

$$\frac{dx}{dt} = a(x - x_0) + b(y - y_0), \qquad \frac{dy}{dt} = c(x - x_0) + d(y - y_0),$$

where

$$a = \frac{\partial f}{\partial x}(x_0, y_0), \qquad b = \frac{\partial f}{\partial y}(x_0, y_0), \qquad c = \frac{\partial g}{\partial x}(x_0, y_0), \qquad d = \frac{\partial g}{\partial y}(x_0, y_0).$$

The technique of linear approximation makes the study of linear systems, that is, systems for which $f(x, y)$ and $g(x, y)$ are linear functions, very important, and we'll study such systems in Sections 3.3 to 3.6. For now, however, we return to the study of a general system $\dfrac{dx}{dt} = f(x, y)$, $\dfrac{dy}{dt} = g(x, y)$ by way of the associated first-order equation $g(x, y) \, dx - f(x, y) \, dy = 0$. As we've seen in Section 1.7,[12] such an equation always has an integrating factor $\mu(x, y)$, and thus its solutions can be written in the form $E(x, y) = C$. In particular, the trajectories of the system $\dfrac{dx}{dt} = f(x, y)$, $\dfrac{dy}{dt} = g(x, y)$ will follow the level curves $E(x, y) = C$. In other words, the function $E(x, y)$ **will be constant along each trajectory**. A function $E(x, y)$ with this property (and which is not constant throughout the plane) is called a **first integral** of the system.[13] As you might expect, knowing a first integral is often very helpful in drawing the phase portrait, but finding a first integral can be very difficult in practice. Here are some examples.

---

[12] In Section 1.7, $g(x, y)$, $-f(x, y)$ were called $M(x, y)$, $N(x, y)$ respectively; it was assumed that y is a function of x, and so the equation was written as $M(x, y) + N(x, y) \dfrac{dy}{dx} = 0$.

[13] There will be many first integrals for any one system. See Exercise 52.

EXAMPLE 3.2.9 In Example 3.2.2 we saw that for the system $\frac{dx}{dt} = y^2 + 1$, $\frac{dy}{dt} = y(y^2 + 1)$, the trajectories follow the curves $y = Ce^x$. Since this can be rewritten as $ye^{-x} = C$, the function ye^{-x} is a first integral for the system. ∎

EXAMPLE 3.2.10 For the system $\frac{dx}{dt} = y$, $\frac{dy}{dt} = x$, the associated first-order equation $x\,dx - y\,dy = 0$ is separable. Solving it yields

$$\int x\,dx = \int y\,dy$$
$$\frac{1}{2}x^2 = \frac{1}{2}y^2 + C$$
$$x^2 - y^2 = 2C.$$

Thus $E(x, y) = x^2 - y^2$ is a first integral for the system; the trajectories follow the hyperbolas $x^2 - y^2 = C$ (where the original constant C has been replaced by $\frac{1}{2}C$). Note that the trajectories can't be the complete hyperbolas, since each hyperbola (for $C \neq 0$) has two distinct branches. On the degenerate hyperbola $x^2 - y^2 = 0$ (i.e., the pair of lines $y = \pm x$) there are actually five trajectories: the stationary point $(0, 0)$ and the four half-lines to either side of this point (see Figure 3.2.7). See also Exercise 53. ∎

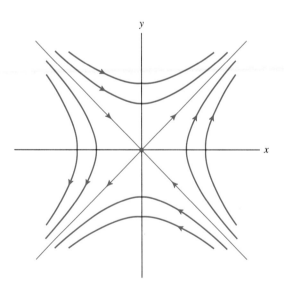

Figure 3.2.7. Phase portrait for $\frac{dx}{dt} = y$, $\frac{dy}{dt} = x$. The trajectories lie on the hyperbolas $x^2 - y^2 = C$.

SEC. 3.2 AUTONOMOUS SYSTEMS

In physical applications, knowledge of a first integral can sometimes lead to a "conservation principle" of some sort. Here is a classic example.

EXAMPLE 3.2.11 Consider a particle moving along an axis as in Example 3.2.8 (p. 166), and suppose further that the force depends only on x. We then have the system $\dfrac{dx}{dt} = v$, $\dfrac{dv}{dt} = \dfrac{1}{m} F(x)$ with associated first-order equation $\dfrac{1}{m} F(x)\, dx - v\, dv = 0$. As in Example 3.2.10, we find

$$\int mv\, dv = \int F(x)\, dx$$

$$\frac{1}{2} mv^2 = \int F(x)\, dx.$$

Of course, the integral on the right "hides" an integration constant. Now recall from physics that the **potential energy** of the particle is defined to be the *negative* integral of the force, that is, the work that was done to get the particle where it is despite the force. Again, this is determined up to a constant of integration; we can, for instance, declare the potential energy to be zero for $x = 0$, and we then have

$$P(x) = -\int_0^x F(s)\, ds \quad ^{14}$$

as the potential energy function. This allows us to rewrite the equation above as

$$\frac{1}{2} mv^2 = -P(x) + C \quad \text{or} \quad P(x) + \frac{1}{2} mv^2 = C.$$

In particular, $E(x, v) = P(x) + \dfrac{1}{2} mv^2$ is a first integral for the system. Since $\dfrac{1}{2} mv^2$ is the **kinetic energy** of the particle, $E(x, v)$ represents the **total** energy: **The total energy of the particle is constant throughout its motion**. This is a special case of a basic conservation law in physics. ∎

To check whether a *given* function $E(x, y)$ is a first integral for a given system, there is no need to solve the system or even the associated equation $g(x, y)\, dx - f(x, y)\, dy = 0$. Instead, one can compute the rate of change of $E(x, y)$, as the moving point $(x(t), y(t))$ follows the trajectories, directly from the chain rule (see p. 9):

$$\frac{d(E(x, y))}{dt} = \frac{\partial E}{\partial x} \frac{dx}{dt} + \frac{\partial E}{\partial y} \frac{dy}{dt}$$

$$= \frac{\partial E}{\partial x} f(x, y) + \frac{\partial E}{\partial y} g(x, y).$$

---

[14] s is a dummy variable, as on p. 4.

EXAMPLE 3.2.12 If you suspected on the basis of the direction field in Figure 3.2.2 that $E(x, y) = x^2 + y^2$ might be a first integral for the system $\frac{dx}{dt} = -y$, $\frac{dy}{dt} = x$, you could compute

$$\frac{d(E(x, y))}{dt} = \frac{\partial E}{\partial x}\frac{dx}{dt} + \frac{\partial E}{\partial y}\frac{dy}{dt}$$
$$= 2x \cdot (-y) + 2y \cdot x = 0$$

to confirm your suspicions. ∎

This idea is of limited use for now, although it is helpful in checking the results of computations, because it's usually hard to come up with a first integral without solving the first-order equation associated to the system. However, a slight extension of the idea, in Section 3.8, will lead to the important concept of a Lyapunov function.

We now conclude this section with another example of finding a first integral.

EXAMPLE 3.2.13 Find a first integral for the system $\frac{dx}{dt} = 2xy - x^2$, $\frac{dy}{dt} = 3xy - 2y^2$, and sketch the phase portrait.

The first-order equation $(3xy - 2y^2)\, dx + (x^2 - 2xy)\, dy = 0$ is not separable or linear. Neither is it exact, since $\frac{\partial}{\partial y}(3xy - 2y^2) = 3x - 2y$ and $\frac{\partial}{\partial x}(x^2 - 2xy) = 2x - 2y$ are not equal. If there is an integrating factor of the special form $\mu = \mu(x)$, then

$$\frac{\partial}{\partial y}[\mu(x)(3xy - 2y^2)] = \frac{\partial}{\partial x}[\mu(x)(x^2 - 2xy)].$$

It turns out [see Exercise 58(a)] that this does work, for $\mu(x) = x$. Multiplying our equation by x yields the exact equation

$$(3x^2y - 2xy^2)\, dx + (x^3 - 2x^2y)\, dy = 0,$$

which has solutions $x^3y - x^2y^2 = C$ (why?). Thus

$$E(x, y) = x^3y - x^2y^2$$

is a first integral for the system.

To sketch the curves $x^3y - x^2y^2 = C$, note that in the special case $C = 0$ the "curve" consists of the three lines $x = 0$, $y = 0$, $y = x$ (see Figure 3.2.8). For $C \neq 0$ we have $x \neq 0$ and the quadratic formula yields $y = \dfrac{x^2 \pm \sqrt{x^4 - 4C}}{2x}$. See Exercise 58 for the rest. ∎

SEC. 3.2 AUTONOMOUS SYSTEMS

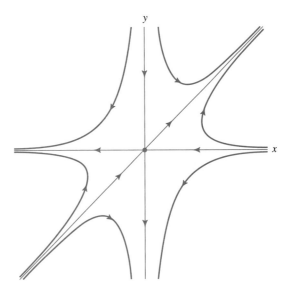

Figure 3.2.8. Phase portrait for $\dfrac{dx}{dt} = 2xy - x^2$, $\dfrac{dy}{dt} = 3xy - 2y^2$.

SUMMARY OF KEY CONCEPTS, RESULTS, AND TECHNIQUES

Autonomous system with unknown functions $x(t)$, $y(t)$ (p. 156):

$$\frac{dx}{dt} = f(x, y), \quad \frac{dy}{dt} = g(x, y). \quad (f, g \text{ are assumed to be } C^1 \text{ functions.})$$

Trajectory of an autonomous system (p. 157): The parametric curve described by a solution $(x(t), y(t))$. Through each point (x_0, y_0) in the plane (in a region where f, g are C^1 functions), there is a unique trajectory of the system $\dfrac{dx}{dt} = f(x, y)$, $\dfrac{dy}{dt} = g(x, y)$.

Phase portrait of an autonomous system (p. 159): Sketch of the trajectories in the x, y-plane, showing the direction of travel.

Direction field of $\dfrac{dx}{dt} = f(x, y)$, $\dfrac{dy}{dt} = g(x, y)$ (p. 158): The collection of vectors $(f(x, y), g(x, y))$ (or suitably scaled-down multiples) for various points (x, y).

First-order equation relating x and y: For the system $\dfrac{dx}{dt} = f(x, y)$, $\dfrac{dy}{dt} = g(x, y)$, we have

$$\frac{dy}{dx} = \frac{g(x, y)}{f(x, y)} \text{ where } f(x, y) \neq 0 \text{ (p. 160) and } \frac{dx}{dy} = \frac{f(x, y)}{g(x, y)} \text{ where } g(x, y) \neq 0. \text{ In all}$$

cases we have the first-order equation **in differential form** $g(x, y)\, dx - f(x, y)\, dy = 0$. (p. 162) If (any of) the first-order equation(s) can be solved, this can be very helpful in sketching the phase portrait of the original system.

Stationary point (critical point, equilibrium point) of $\dfrac{dx}{dt} = f(x, y)$, $\dfrac{dy}{dt} = g(x, y)$ (p. 163): A point (x_0, y_0) for which $f(x_0, y_0) = g(x_0, y_0) = 0$. For a stationary point (x_0, y_0), the system has a constant solution $x(t) = x_0$, $y(t) = y_0$, so the point constitutes a complete trajectory.

Linear approximation to the system $\dfrac{dx}{dt} = f(x, y)$, $\dfrac{dy}{dt} = g(x, y)$ near a given stationary point (x_0, y_0) (p. 167):

$$\frac{dx}{dt} = a(x - x_0) + b(y - y_0), \qquad \frac{dy}{dt} = c(x - x_0) + d(y - y_0),$$

where

$$a = \frac{\partial f}{\partial x}(x_0, y_0), \qquad b = \frac{\partial f}{\partial y}(x_0, y_0), \qquad c = \frac{\partial g}{\partial x}(x_0, y_0), \qquad d = \frac{\partial g}{\partial y}(x_0, y_0).$$

First integral of $\dfrac{dx}{dt} = f(x, y)$, $\dfrac{dy}{dt} = g(x, y)$ (p. 167): A nonconstant function $E(x, y)$ which is constant along each trajectory, so that

$$\frac{d(E(x, y))}{dt} = \frac{\partial E}{\partial x}\frac{dx}{dt} + \frac{\partial E}{\partial y}\frac{dy}{dt} = 0.$$

Usually found by solving $g(x, y)\, dx - f(x, y)\, dy = 0$ to get $E(x, y) = C$.

EXERCISES

Sketch the direction field (or get a computer to do it for you) for each of the following systems.

1. $\dfrac{dx}{dt} = y$, $\dfrac{dy}{dt} = -x$.
2. $\dfrac{dx}{dt} = 3x + 2y$, $\dfrac{dy}{dt} = 2x + 3y$.
3. $\dfrac{dx}{dt} = x - y$, $\dfrac{dy}{dt} = 5x - y$.
4. $\dfrac{dx}{dt} = 2x$, $\dfrac{dy}{dt} = 3y$.
5. $\dfrac{dx}{dt} = x + y$, $\dfrac{dy}{dt} = -x$.
6. $\dfrac{dx}{dt} = -x + y$, $\dfrac{dy}{dt} = -x$.
*7. $\dfrac{dx}{dt} = y - x^3$, $\dfrac{dy}{dt} = -x + y^3$.
*8. $\dfrac{dx}{dt} = -3x + 6xy$, $\dfrac{dy}{dt} = 2y - xy$.

Find all stationary points of the following systems.

9. $\dfrac{dx}{dt} = y + x^2$, $\dfrac{dy}{dt} = -x + y^2$.
10. $\dfrac{dx}{dt} = y - x^2$, $\dfrac{dy}{dt} = -x + y^2$.

SEC. 3.2 AUTONOMOUS SYSTEMS

11. $\dfrac{dx}{dt} = y + x^3, \quad \dfrac{dy}{dt} = -x + y^3.$ **12.** $\dfrac{dx}{dt} = y - x^3, \quad \dfrac{dy}{dt} = -x + y^3.$

For each of the following systems:
(a) Sketch the direction field.
(b) Find all stationary points.
(c) Solve the first-order equation relating x and y that corresponds to the system.
(d) Sketch the phase portrait, using the information from parts (a), (b), and (c). Make sure to indicate the directions of travel along the trajectories.

13. $\dfrac{dx}{dt} = -2y, \quad \dfrac{dy}{dt} = x.$ **14.** $\dfrac{dx}{dt} = 4y, \quad \dfrac{dy}{dt} = -x.$

15. $\dfrac{dx}{dt} = 2y - 6, \quad \dfrac{dy}{dt} = -2x.$ **16.** $\dfrac{dx}{dt} = -2y, \quad \dfrac{dy}{dt} = 2x + 4.$

*17. $\dfrac{dx}{dt} = -2y(y+1), \quad \dfrac{dy}{dt} = x(y+1).$ ***18.** $\dfrac{dx}{dt} = 4y(y-3), \quad \dfrac{dy}{dt} = -x(y-3).$

Write the following differential equations in differential form. Also write the "equivalent" differential equation for x as an unknown function of y, and discuss where in the plane the two equations really are equivalent.

19. $\dfrac{dy}{dx} = (1 + x^2)y.$ **20.** $(\sin^2 x + 1)\dfrac{dy}{dx} = e^y + 2.$

21. $\dfrac{dy}{dx} - y = 2x + 3.$ **22.** $3\dfrac{dx}{dy} - 2x^2 = y^2 - 1.$

Find the linear approximations to the following systems near $(0, 0)$. (Check that this is a stationary point in each case.)

23. $\dfrac{dx}{dt} = 3x + 7xy - 5y^2, \quad \dfrac{dy}{dt} = y + \sin x.$

24. $\dfrac{dx}{dt} = e^x - 1, \quad \dfrac{dy}{dt} = 3x - xy.$

25. $\dfrac{dx}{dt} = 2x + y + xy - x^3, \quad \dfrac{dy}{dt} = xy + 2x^3 - y\sin y.$

26. $\dfrac{dx}{dt} = -y + xy, \quad \dfrac{dy}{dt} = x(y - 4).$

For each of the following systems:
(a) Find all stationary points.
(b) Find the linear approximation near each of the stationary points.

27. $\dfrac{dx}{dt} = xy - 1, \quad \dfrac{dy}{dt} = 4x - y.$ **28.** $\dfrac{dx}{dt} = 4xy - 1, \quad \dfrac{dy}{dt} = x - y.$

29. $\dfrac{dx}{dt} = 2x - 3y - 5, \quad \dfrac{dy}{dt} = 4x + y - 3.$

30. $\dfrac{dx}{dt} = 4x - 6y - 2, \quad \dfrac{dy}{dt} = -3x + 5y + 1.$

Find first integrals for each of the following systems.

31. $\dfrac{dx}{dt} = x + x^2, \quad \dfrac{dy}{dt} = -y - xy.$

32. $\dfrac{dx}{dt} = \cos x, \quad \dfrac{dy}{dt} = y \sin x.$

33. $\dfrac{dx}{dt} = y, \quad \dfrac{dy}{dt} = y^2.$

34. $\dfrac{dx}{dt} = -\dfrac{1}{2}x, \quad \dfrac{dy}{dt} = x + y.$

35. $\dfrac{dx}{dt} = \alpha x + \beta y, \quad \dfrac{dy}{dt} = \gamma x - \alpha y,$ where α, β, γ are arbitrary constants.

36. $\dfrac{dx}{dt} = x \cos y, \quad \dfrac{dy}{dt} = -\sin y + x^2.$

37. $\dfrac{dx}{dt} = 2x + 3y^2 - e^y, \quad \dfrac{dy}{dt} = \sin x - 6x^2 - 2y.$

38. $\dfrac{dx}{dt} = x^2 + y^2, \quad \dfrac{dy}{dt} = x^3 - 2xy.$

39. $\dfrac{dx}{dt} = \cos x - \cos y, \quad \dfrac{dy}{dt} = y \sin x + \cos x.$

40. $\dfrac{dx}{dt} = 12y - 3x^2 - 5x^4 y^2, \quad \dfrac{dy}{dt} = 2xy + 4x^3 y^3.$

41. $\dfrac{dx}{dt} = -2xy - 4x^3 y^3, \quad \dfrac{dy}{dt} = 5x^2 y^4 + 3y^2 - 5x^2.$

For each of the following systems:

(a) Find a first integral.
(b) Sketch the phase portrait. (Remember to indicate the stationary points, if any, and the directions of travel along the trajectories.)

42. $\dfrac{dx}{dt} = -4, \quad \dfrac{dy}{dt} = 2x.$

43. $\dfrac{dx}{dt} = 2y, \quad \dfrac{dy}{dt} = 3.$

44. $\dfrac{dx}{dt} = 5x, \quad \dfrac{dy}{dt} = -5y.$

45. $\dfrac{dx}{dt} = -3x, \quad \dfrac{dy}{dt} = 3y.$

46. $\dfrac{dx}{dt} = -2y - 6, \quad \dfrac{dy}{dt} = 2x.$

47. $\dfrac{dx}{dt} = 2y, \quad \dfrac{dy}{dt} = 4 - 2x.$

48. $\dfrac{dx}{dt} = 2y, \quad \dfrac{dy}{dt} = 12x^2.$

49. $\dfrac{dx}{dt} = -2y, \quad \dfrac{dy}{dt} = -3x^2.$

50. $\dfrac{dx}{dt} = \dfrac{2}{x^2+1}, \quad \dfrac{dy}{dt} = \dfrac{4xy}{(x^2+1)^2}.$ **51.** $\dfrac{dx}{dt} = \dfrac{-1}{x^2+1}, \quad \dfrac{dy}{dt} = \dfrac{-2xy}{(x^2+1)^2}.$

52. (a) Show that the system $\dfrac{dx}{dt} = 1, \dfrac{dy}{dt} = \dfrac{1}{x}$ ($x > 0$) has both $V_1(x, y) = xe^{-y}$ and $V_2(x, y) = y - \log x$ as first integrals.
[*Hint*: Use the method of Example 3.2.12.]
(b) Can you explain the connection between $V_1(x, y)$ and $V_2(x, y)$?

53. Show that the directions of travel in Figure 3.2.7 (p. 168) are indicated correctly. Discuss the connection between Figure 3.2.7 and Figure 1.2.1 (p. 16). (Note that t and x in the latter correspond to x and y in the former.)

54. Explain why it is not practical to draw a phase portrait for a nonautonomous system $\dfrac{dx}{dt} = f(t, x, y), \dfrac{dy}{dt} = g(t, x, y).$

55. Consider the Volterra–Lotka system $\dfrac{dx}{dt} = -Mx + Pxy, \dfrac{dy}{dt} = Ky - Lxy$ from Example 3.1.4.
(a) Find all stationary points of the system.
(b) Show that the differential equation $\dfrac{dy}{dx} = \dfrac{Ky - Lxy}{-Mx + Pxy}$ associated with the system is separable.
(c) Find a first integral for the system.

56. Show that for the system $\dfrac{dx}{dt} = -y, \dfrac{dy}{dt} = x$ the speed of the moving point on a trajectory equals the radius of the trajectory. Conclude that every solution $(x(t), y(t))$ is periodic with period 2π.

57. Consider the system $\dfrac{dx}{dt} = -y(x + 1), \dfrac{dy}{dt} = x(x + 1).$
(a) Find all stationary points of the system.
(b) Show that the corresponding first-order equation for x and y is $x\,dx + y\,dy = 0$, just as in Example 3.2.1.
(c) Show that the trajectories follow circles around the origin.
(d) Show that the solution satisfying $x(0) = 1, y(0) = 0$ is
$$x(t) = \dfrac{1 - t^2}{1 + t^2}, \quad y(t) = \dfrac{2t}{1 + t^2}.$$
Draw the corresponding points on the trajectory for $t = -10, t = -1, t = 0, t = 1, t = 10$. What happens as $t \to \infty$? As $t \to -\infty$?

58. In this exercise we complete Example 3.2.13.
(a) Solve $\dfrac{\partial}{\partial y}[\mu(x)(3xy - 2y^2)] = \dfrac{\partial}{\partial x}[\mu(x)(x^2 - 2xy)]$ to find an integrating factor.

*(b) Sketch the curve $y = \dfrac{x^2 + \sqrt{x^4 - 4C}}{2x}$ for $C = 1$; in the same diagram,

sketch $y = \dfrac{x^2 - \sqrt{x^4 - 4C}}{2x}$ for $C = 1$.

*(c) As part (b), but for $C = -1$.

*(d) Check that your results from parts (b) and (c) are consistent with Figure 3.2.8, and explain why the directions of travel indicated in that figure are correct.

59. Solve the initial value problem
$$\dfrac{dx}{dt} = 3xy - 2x - y, \quad \dfrac{dy}{dt} = y^2 - 5xy^3 + 4y^6, \qquad x(0) = 1, \quad y(0) = 1.$$
[*Hint*: Only a confirmed optimist would look for the general solution of the system.]

60. Show that if $(x(t), y(t))$ is a solution of the system $\dfrac{dx}{dt} = f(x, y), \dfrac{dy}{dt} = g(x, y)$, then $(x(t + t_0), y(t + t_0))$ is a solution as well, where t_0 is an arbitrary constant. [*Hint*: Chain rule in one variable!]

61. (a) Show that if $(x(t), y(t))$ is a solution of the system $\dfrac{dx}{dt} = f(x, y), \dfrac{dy}{dt} = g(x, y)$,

then $(x(-t), y(-t))$ is a solution of $\dfrac{dx}{dt} = -f(x, y), \dfrac{dy}{dt} = -g(x, y)$.

(b) Show that the two systems in part (a) have the same trajectories, but that these are traversed in opposite directions.

(c) Explain why the system $\dfrac{dx}{dt} = -f(x, y), \dfrac{dy}{dt} = -g(x, y)$ is said to be obtained

by **time reversal** from the system $\dfrac{dx}{dt} = f(x, y), \dfrac{dy}{dt} = g(x, y)$.

*62. (a) Let $(x(t), y(t))$ be a solution of the system $\dfrac{dx}{dt} = f(x, y), \dfrac{dy}{dt} = g(x, y)$. Find a related system to which $(x(2t), y(2t))$ is a solution.

(b) How are the trajectories of the two systems in part (a) related? How are the speeds with which the "moving points" follow the trajectories related?

63. (See Example 3.2.4.) For each of the following pairs of initial conditions, use Figure 3.2.6 to predict what will happen as $t \to \infty$ to the solution $(x(t), y(t))$ of the system $\dfrac{dx}{dt} = x^2 - 2xy - x, \dfrac{dy}{dt} = xy + y^2 - 3y$ which satisfies those conditions.

(a) $x(0) = 4, \quad y(0) = 3$.
(b) $x(0) = 2, \quad y(0) = 1.5$.
(c) $x(0) = 1, \quad y(0) = -2$.
(d) $x(0) = 0, \quad y(0) = 2$.
(e) $x(0) = -2, \quad y(0) = 4$.
(f) $x(0) = -2, \quad y(0) = -2$.

SEC. 3.2 AUTONOMOUS SYSTEMS

*64. Figure 3.2.6 suggests that at least four different things can happen as $t \to \infty$ to a trajectory of the system $\dfrac{dx}{dt} = x^2 - 2xy - x$, $\dfrac{dy}{dt} = xy + y^2 - 3y$ which starts near, but not at, the stationary point $\left(\dfrac{7}{3}, \dfrac{2}{3}\right)$. Describe four such possibilities. Two of the possibilities seem rather unlikely. Which ones?

For each of the following systems:

(a) Find all stationary points.
(b) Use a computer to sketch the direction field.
(c) Use the direction field to predict the behavior as $t \to \infty$ of the solution that satisfies the given initial conditions.

65. $\dfrac{dx}{dt} = xy$, $\dfrac{dy}{dt} = (y - x - 2)(y + x - 3)$; $x(0) = 7$, $y(0) = 0$.

66. $\dfrac{dx}{dt} = xy - x^2$, $\dfrac{dy}{dt} = y^2 + xy - 6y$; $x(0) = 3$, $y(0) = 5$.

67. $\dfrac{dx}{dt} = y - 1$, $\dfrac{dy}{dt} = x^2 - y^3$; $x(0) = 1$, $y(0) = -1$.

68. $\dfrac{dx}{dt} = y - 1$, $\dfrac{dy}{dt} = x^2 - y^3$; $x(0) = 2$, $y(0) = -1$.

*69. $\dfrac{dx}{dt} = 3xy - y^2$, $\dfrac{dy}{dt} = (x - 2)^2 - y^2$; $x(0) = 3$, $y(0) = -1$.

*70. $\dfrac{dx}{dt} = 3xy - y^2$, $\dfrac{dy}{dt} = (x - 2)^2 - y^2$; $x(0) = 3$, $y(0) = -3$.

3.3 LINEAR SYSTEMS (ELIMINATION METHOD)

A general (perhaps not autonomous) system of two differential equations $\frac{dx}{dt} = f(t, x, y)$, $\frac{dy}{dt} = g(t, x, y)$ is called **linear** if f and g are linear in the unknown functions x and y (but not necessarily in t). Note that this definition corresponds exactly to the definition of "linear" for a single differential equation, as given in Section 1.4, p. 31.

EXAMPLE $\quad \frac{dx}{dt} = t^2 x + (\sin t) y - 5, \; \frac{dy}{dt} = x + 2y + t^3$ is a linear nonautonomous system. ∎

Now let's restrict ourselves once again to the autonomous case, in which f and g do not depend explicitly on t. If a linear system is autonomous, it will be of the form

$$\frac{dx}{dt} = ax + by + k_1, \qquad \frac{dy}{dt} = cx + dy + k_2,$$

where a, b, c, d, k_1, k_2 are constants. Such a system will have a stationary point at (x_0, y_0) if $ax_0 + by_0 + k_1 = cx_0 + dy_0 + k_2 = 0$. Usually, the equations $ax + by + k_1 = 0$, $cx + dy + k_2 = 0$ will describe two intersecting lines in the x,y-plane, so there will be exactly one stationary point (x_0, y_0): the intersection point of the lines. However, there are a few exceptional cases: If the lines are parallel, there will be no stationary points at all; if the lines coincide, the whole line will consist of stationary points. Finally, there are the cases in which $a = b = 0$, or $c = d = 0$, or both, which means one, or both, of the lines is degenerate. In all these exceptional cases the system is relatively easy to solve (see Exercises 13 and 14). We will therefore assume that there is a unique stationary point (x_0, y_0). The system can then be simplified by introducing a change of coordinates, as follows. Consider the unknown functions $X = x - x_0$, $Y = y - y_0$ instead of x, y; note that these new unknown functions are zero at the stationary point. Now we find a system of differential equations for the new unknown functions, starting with

$$\begin{aligned}
\frac{dX}{dt} &= \frac{d(x - x_0)}{dt} \\
&= \frac{dx}{dt} \quad \text{(since } x_0 \text{ is constant)} \\
&= ax + by + k_1 \\
&= ax + by - ax_0 - by_0 \quad \text{(since } ax_0 + by_0 + k_1 = 0\text{)} \\
&= a(x - x_0) + b(y - y_0) \\
&= aX + bY.
\end{aligned}$$

SEC. 3.3 LINEAR SYSTEMS (ELIMINATION METHOD)

Similarly, we find (see Exercise 11) that $\frac{dY}{dt} = cX + dY$, and so the new system is

$$\begin{cases} \dfrac{dX}{dt} = aX + bY \\ \dfrac{dY}{dt} = cX + dY. \end{cases}$$

In other words, we have gotten rid of the constants k_1 and k_2 by shifting the stationary point to the origin.

EXAMPLE 3.3.1 The system $\frac{dx}{dt} = 3x + 2y + 4$, $\frac{dy}{dt} = 2x + 3y + 1$ has a stationary point at $(-2, 1)$. In terms of the new coordinates $x + 2$ and $y - 1$, the first equation becomes

$$\frac{d(x+2)}{dt} = \frac{dx}{dt} = 3x + 2y + 4 = 3(x+2) + 2(y-1).$$

Similarly,

$$\frac{d(y-1)}{dt} = 2(x+2) + 3(y-1).$$

Hence if we write $X = x + 2$, $Y = y - 1$, we'll have the new system

$$\begin{cases} \dfrac{dX}{dt} = 3X + 2Y \\ \dfrac{dY}{dt} = 2X + 3Y. \end{cases}$$

We'll return to this example in Note 2 after Example 3.3.2. □

By the above, it will be enough to be able to solve a *homogeneous* system (that is, one for which $k_1 = k_2 = 0$), such as

$$\begin{cases} \dfrac{dx}{dt} = ax + by \\ \dfrac{dy}{dt} = cx + dy. \end{cases}$$

There are several methods by which this linear homogeneous system can be solved. One method starts by solving the equation $(cx + dy) - (ax + by)\frac{dy}{dx} = 0$ for the trajectories, using the "trick" substitution $y = vx$, as outlined in Exercises 27 and 28 of Section 1.7. Once the trajectories are found, y is (in principle) known as a function of x, so at that point

$\frac{dx}{dt} = ax + by$ can be written as a separable equation for $x = x(t)$. This method is not recommended, since the computations tend to be long and messy.

We will present two other methods of solution for the system above. Both of these can be generalized to similar systems with more than two unknown functions, but the method which will be introduced in the next section is faster in the case of these larger systems; it is also better suited to qualitative analysis. On the other hand, the **elimination method**, which we now present, is the quickest way to actually solve the system

$$\begin{cases} \dfrac{dx}{dt} = ax + by \\ \dfrac{dy}{dt} = cx + dy. \end{cases}$$

The idea of the elimination method is to eliminate one of the unknown functions, converting the system into a single *second-order* equation for the other unknown function. This last equation is then solved using the methods of Chapter 2.

EXAMPLE 3.3.2

$\dfrac{dx}{dt} = 3x + 2y$, $\dfrac{dy}{dt} = 2x + 3y$.

We can use the first equation to write y in terms of x and $\dfrac{dx}{dt}$: $y = \dfrac{1}{2}\dfrac{dx}{dt} - \dfrac{3}{2}x$.

Therefore, $\dfrac{dy}{dt} = \dfrac{1}{2}\dfrac{d^2x}{dt^2} - \dfrac{3}{2}\dfrac{dx}{dt}$. Substituting these expressions for y and $\dfrac{dy}{dt}$ into the second equation of the system, we get

$$\frac{1}{2}\frac{d^2x}{dt^2} - \frac{3}{2}\frac{dx}{dt} = 2x + \frac{3}{2}\frac{dx}{dt} - \frac{9}{2}x$$

$$\frac{d^2x}{dt^2} - 6\frac{dx}{dt} + 5x = 0.$$

This is a second-order equation (with constant coefficients) for x, from which y has been eliminated. Its characteristic equation is $\lambda^2 - 6\lambda + 5 = 0$, with roots $\lambda = 1$ and $\lambda = 5$, so its general solution is $x(t) = \alpha e^t + \beta e^{5t}$. Substituting this back into $y = \dfrac{1}{2}\dfrac{dx}{dt} - \dfrac{3}{2}x$, we get $y(t) = -\alpha e^t + \beta e^{5t}$. Thus the general solution to the system is

$$\begin{cases} x(t) = \alpha e^t + \beta e^{5t} \\ y(t) = -\alpha e^t + \beta e^{5t}, \end{cases}$$

where α and β are arbitrary constants.

NOTES:

1. If you look at the answer for a while, you may notice that $x(t) + y(t) = 2\beta e^{5t}$ and $x(t) - y(t) = 2\alpha e^t$ are actually "easier" functions than $x(t)$ and $y(t)$ themselves. They also satisfy easier differential equations, as we'll see in the next section.

2. We can now finish Example 3.3.1 (p. 179). In the notation of that example, we have found that $X = \alpha e^t + \beta e^{5t}$, $Y = -\alpha e^t + \beta e^{5t}$, and so the general solution to

$$\frac{dx}{dt} = 3x + 2y + 4, \quad \frac{dy}{dt} = 2x + 3y + 1$$

is given by

$$x = X - 2 = \alpha e^t + \beta e^{5t} - 2, \quad y = -\alpha e^t + \beta e^{5t} + 1. \quad \blacksquare$$

EXAMPLE 3.3.3 Find the solution to

$$\begin{cases} \dfrac{dx}{dt} = x - y \\ \dfrac{dy}{dt} = 5x - y \end{cases}$$

that satisfies the initial conditions $x(0) = 1$, $y(0) = 0$.

This time the first equation yields $y = -\dfrac{dx}{dt} + x$, so we get $\dfrac{dy}{dt} = -\dfrac{d^2x}{dt^2} + \dfrac{dx}{dt}$ and $\dfrac{d^2x}{dt^2} + 4x = 0$. We have *complex-valued* solutions e^{2it} and e^{-2it} to this equation (see Section 2.5); the general real-valued solution is $x(t) = \alpha \cos 2t + \beta \sin 2t$. From this we find that the general solution of our system is

$$\begin{cases} x(t) = \alpha \cos 2t + \beta \sin 2t \\ y(t) = (\alpha - 2\beta) \cos 2t + (\beta + 2\alpha) \sin 2t. \end{cases}$$

We must now choose α and β in such a way that $x(0) = 1$, $y(0) = 0$. This yields $\alpha = 1$, $\alpha - 2\beta = 0$, so $\alpha = 1$, $\beta = \dfrac{1}{2}$, and the required solution is

$$\begin{cases} x(t) = \cos 2t + \dfrac{1}{2} \sin 2t \\ y(t) = \dfrac{5}{2} \sin 2t. \end{cases} \quad \blacksquare$$

EXAMPLE 3.3.4 $\dfrac{dx}{dt} = 4x + y - 3, \quad \dfrac{dy}{dt} = -x + 2y + 12$.

The stationary point is at the intersection of the lines $4x + y = 3$ and $x = 2y + 12$, which turns out to be at $(2, -5)$. So we'll introduce the new coordinates $X = x - 2$, $Y = y + 5$, and the system then becomes

$$\frac{dX}{dt} = 4X + Y, \quad \frac{dY}{dt} = -X + 2Y.$$

Now we get $Y = \dfrac{dX}{dt} - 4X$ from the first equation, and substituting this into the second equation yields

$$\dfrac{d^2X}{dt^2} - 4\dfrac{dX}{dt} = -X + 2\dfrac{dX}{dt} - 8X$$

or

$$\dfrac{d^2X}{dt^2} - 6\dfrac{dX}{dt} + 9X = 0.$$

We find

$$X = (\alpha + \beta t)e^{3t} \quad \text{(see Section 2.4)},$$

so

$$Y = \dfrac{dX}{dt} - 4X = (\beta - \alpha - \beta t)e^{3t}$$

and finally

$$x = (\alpha + \beta t)e^{3t} + 2, \quad y = (\beta - \alpha - \beta t)e^{3t} - 5, \quad \alpha, \beta \text{ arbitrary}. \quad \blacksquare$$

EXAMPLE 3.3.5 $\quad \dfrac{dx}{dt} = 2x, \quad \dfrac{dy}{dt} = 3x - y.$

In this case we can't use the first equation to write y in terms of x and $\dfrac{dx}{dt}$, since y is missing from the first equation! On the other hand, this means that y has already been eliminated: We can solve the first equation directly for x and then substitute the result in the second equation to get a first-order equation for y (see Exercise 5). Alternatively, we could use the second equation to write x in terms of y and $\dfrac{dy}{dt}$ and substitute the result into the first equation. $\quad \blacksquare$

SUMMARY OF KEY CONCEPTS AND TECHNIQUES

Linear system (p. 178): $\dfrac{dx}{dt} = f(t, x, y)$, $\dfrac{dy}{dt} = g(t, x, y)$ is a linear system if f and g are linear in x and y (but not necessarily in t). If such a system is autonomous, it will have the form $\dfrac{dx}{dt} = ax + by + k_1$, $\dfrac{dy}{dt} = cx + dy + k_2$. The intersection point (x_0, y_0) of the lines $ax + by + k_1 = 0$, $cx + dy + k_2 = 0$ will be a stationary point (if it exists). Introducing the new unknown functions $X = x - x_0$, $Y = y - y_0$ simplifies the system to $\dfrac{dX}{dt} = aX + bY$, $\dfrac{dY}{dt} = cX + dY$.

SEC. 3.3 LINEAR SYSTEMS (ELIMINATION METHOD)

Elimination method (p. 180): Use one of the two given equations to express one of the unknown functions (y, say) in terms of the other (x) and its derivative $\left(\dfrac{dx}{dt}\right)$, then substitute this expression into the other equation to obtain a second-order equation (for x).

EXERCISES

Solve the following systems and initial value problems.

1. $\begin{cases} \dfrac{dx}{dt} = 4x + 3y \\ \dfrac{dy}{dt} = 3x + 4y. \end{cases}$

2. $\begin{cases} \dfrac{dx}{dt} = 2x - y \\ \dfrac{dy}{dt} = -x + 2y. \end{cases}$

3. $\begin{cases} \dfrac{dx}{dt} = -y + 5 \\ \dfrac{dy}{dt} = 4x - 8. \end{cases}$

4. $\begin{cases} \dfrac{dx}{dt} = x + 2y, \quad x(0) = 0, \\ \dfrac{dy}{dt} = 3x + 6y, \quad y(0) = 1. \end{cases}$

5. $\begin{cases} \dfrac{dx}{dt} = 2x, \quad x(0) = 1, \\ \dfrac{dy}{dt} = 3x - y, \quad y(0) = 3. \end{cases}$

6. $\begin{cases} \dfrac{dx}{dt} = 3x + 4y + 7 \\ \dfrac{dy}{dt} = -4x + 3y - 1. \end{cases}$

7. $\begin{cases} \dfrac{dx}{dt} = x + y + 3, \quad x(0) = 5, \\ \dfrac{dy}{dt} = y - 1, \quad y(0) = 2. \end{cases}$

8. $\begin{cases} \dfrac{dx}{dt} = x + y \\ \dfrac{dy}{dt} = -x + 3y. \end{cases}$

9. $\begin{cases} \dfrac{dx}{dt} = x + y \\ \dfrac{dy}{dt} = 3y. \end{cases}$

10. $\begin{cases} \dfrac{dx}{dt} = 55x + 374y, \quad x(0) = 0, \\ \dfrac{dy}{dt} = 17x - 1923y, \quad y(0) = 0. \end{cases}$

11. Show that if (x_0, y_0) is a stationary point for the system $\dfrac{dx}{dt} = ax + by + k_1$, $\dfrac{dy}{dt} = cx + dy + k_2$, then for $X = x - x_0$, $Y = y - y_0$ we have $\dfrac{dX}{dt} = cX + dY$.

12. Show that the characteristic equation of the second-order differential equation obtained by elimination from the system
$$\begin{cases} \dfrac{dx}{dt} = ax + by \\ \dfrac{dy}{dt} = cx + dy \end{cases}$$

is
$$\lambda^2 - (a+d)\lambda + (ad-bc) = 0.$$

13. Consider the linear autonomous system $\dfrac{dx}{dt} = ax + by + k_1$, $\dfrac{dy}{dt} = cx + dy + k_2$, and suppose that the line $ax + by + k_1 = 0$ is degenerate; that is, suppose that $a = b = 0$. Explain how the system can be solved by first solving for x.

14. Consider the linear autonomous system $\dfrac{dx}{dt} = ax + by + k_1$, $\dfrac{dy}{dt} = cx + dy + k_2$, and suppose that the lines $ax + by + k_1 = 0$ and $cx + dy + k_2 = 0$ are parallel (or identical) and nondegenerate.
   (a) Assuming that a and b are both nonzero, show that there exist constants q, r such that $\dfrac{d(y - qx)}{dt} = r$.

   [*Hint*: First show that because the lines are parallel, $\dfrac{c}{a} = \dfrac{d}{b}$; let $q = \dfrac{c}{a}$.]

   (b) Show that the conclusion of part (a) is still correct if a or b is zero.
   (c) Explain how the system can be solved.

Use the methods of Exercises 13 and 14, or ad hoc methods of your choice, to solve the following systems.

15. $\begin{cases} \dfrac{dx}{dt} = 2 \\ \dfrac{dy}{dt} = x + y - 4. \end{cases}$

16. $\begin{cases} \dfrac{dx}{dt} = -x + 5y - 6 \\ \dfrac{dy}{dt} = -1. \end{cases}$

17. $\begin{cases} \dfrac{dx}{dt} = x + y + 7 \\ \dfrac{dy}{dt} = 2x + 2y + 14. \end{cases}$

18. $\begin{cases} \dfrac{dx}{dt} = 2x - 4y + 6 \\ \dfrac{dy}{dt} = 3x - 6y + 8. \end{cases}$

19. $\begin{cases} \dfrac{dx}{dt} = x + y - 1 \\ \dfrac{dy}{dt} = 3x + 3y + 5. \end{cases}$

20. $\begin{cases} \dfrac{dx}{dt} = 4x + 2y - 1 \\ \dfrac{dy}{dt} = 2x + y + 1. \end{cases}$

3.4 LINEAR SYSTEMS (CONTINUED); EIGENVALUES AND EIGENVECTORS

In this section we continue our study of systems of the form

$$\frac{dx}{dt} = ax + by, \qquad \frac{dy}{dt} = cx + dy,$$

where a, b, c, and d are constants. We'll see how such systems can be simplified, and then solved, by suitable changes of coordinates.

EXAMPLE 3.4.1

$\dfrac{dx}{dt} = 3x + 2y, \quad \dfrac{dy}{dt} = 2x + 3y.$

This is the system from Example 3.3.2; we saw there (see p. 180, Note 1) that the general solution has a simpler form when it is expressed in terms of the functions $x + y$ and $x - y$ rather than x and y. With hindsight, then, we can try to solve the system by introducing new unknown functions $X = x + y$ and $Y = x - y$. For these new functions we get the equations

$$\frac{dX}{dt} = \frac{d(x+y)}{dt} = \frac{dx}{dt} + \frac{dy}{dt} = (3x + 2y) + (2x + 3y) = 5x + 5y = 5X$$

and

$$\frac{dY}{dt} = (3x + 2y) - (2x + 3y) = x - y = Y.$$

Note that the new system $\dfrac{dX}{dt} = 5X$, $\dfrac{dY}{dt} = Y$ can be solved right away, since the two new equations are **uncoupled** (that is, independent of each other), with each equation involving only one unknown function. We find $X = Ce^{5t}$, $Y = De^{t}$, where C and D are arbitrary constants. Therefore, we have $x + y = Ce^{5t}$, $x - y = De^{t}$, and solving for x and y yields

$$x = \frac{1}{2}(Ce^{5t} + De^{t}), \qquad y = \frac{1}{2}(Ce^{5t} - De^{t})$$

as the general solution to our system.[15]

Geometrically, introducing the new unknown functions X, Y amounts to a linear change of coordinates in the x,y-plane. The (new) X-axis is given by $Y = 0$, that is, $y = x$; the Y-axis is given by $y = -x$. As shown in Figure 3.4.1 (p. 186), the positive direction along the X-axis is the direction in which $X = x + y$ increases, and similarly for the Y-axis. ∎

---

[15] Note that this solution isn't quite in the form we found in Section 3.3; the constants C, D are related to the constants α, β from Section 3.3 by $C = 2\beta$, $D = 2\alpha$.

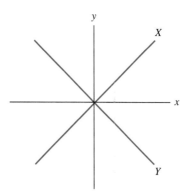

Figure 3.4.1. New axes (in color) after the change of coordinates $X = x + y$, $Y = x - y$.

While the particular change of coordinates $X = x + y$, $Y = x - y$ works well in the example above, we should expect that it usually won't. To solve a system of the general form $\dfrac{dx}{dt} = ax + by$, $\dfrac{dy}{dt} = cx + dy$ we will presumably need a more general change of coordinates, say with $X = px + qy$, $Y = rx + sy$. In particular, there is no reason to expect the X- and Y-axes (given by $Y = 0$ and $X = 0$, respectively) to end up perpendicular to each other, as they did in Example 3.4.1.

Our plan is, then, given a system $\dfrac{dx}{dt} = ax + by$, $\dfrac{dy}{dt} = cx + dy$, to try to find new coordinates $X = px + qy$, $Y = rx + sy$ such that the differential equations for these new unknown functions are easier than the original equations. Ideally, we would like $\dfrac{dX}{dt}$ to be a constant multiple of X, while $\dfrac{dY}{dt}$ is a constant multiple of Y. (In Example 3.4.1, we had $\dfrac{dX}{dt} = 5X$, $\dfrac{dY}{dt} = Y$.) If we succeed in this, then we can immediately solve for X and Y, and finally we can reverse the change of coordinates to get the original unknown functions x and y.

If we do take $X = px + qy$, where p and q are constants, we find

$$\frac{dX}{dt} = p\frac{dx}{dt} + q\frac{dy}{dt} = p(ax + by) + q(cx + dy) = (ap + cq)x + (bp + dq)y.$$

For this to be a constant multiple of X, say λX, we must have

$$ap + cq = \lambda p, \qquad bp + dq = \lambda q.$$

SEC. 3.4 LINEAR SYSTEMS (CONTINUED); EIGENVALUES AND EIGENVECTORS

These simultaneous equations for p and q can be rewritten as

$$\begin{cases} (a - \lambda)p + cq = 0 \\ bp + (d - \lambda)q = 0. \end{cases}$$

Of course, one solution to these is $p = q = 0$. However, this trivial solution is useless, since it yields $X = 0$: The "new coordinate" would be identically zero! In order for there to be other solutions for p and q, we must have $\det \begin{bmatrix} a - \lambda & c \\ b & d - \lambda \end{bmatrix} = 0$ (see Appendix B, p. 624), which works out to $(a - \lambda)(d - \lambda) - bc = 0$ or

$$\lambda^2 - (a + d)\lambda + (ad - bc) = 0.$$

This quadratic equation for λ is called the **characteristic equation** of our linear system. It is the same as the characteristic equation found by the elimination method (see Section 3.3, Exercise 12). If λ is a root of the characteristic equation, then we can find a "new coordinate" $X = px + qy$ which satisfies an equation of the form $\dfrac{dX}{dt} = \lambda X$. Thus if the characteristic equation has *two* distinct roots, we can find two new coordinates X and Y corresponding to the roots and solve the system by using these new coordinates. For instance, for the system $\dfrac{dx}{dt} = 3x + 2y$, $\dfrac{dy}{dt} = 2x + 3y$ from Example 3.4.1, the characteristic equation is $\lambda^2 - 6\lambda + 5 = 0$, which has roots $\lambda = 5$ and $\lambda = 1$. We could have seen from this that there would be new coordinates X and Y satisfying $\dfrac{dX}{dt} = 5X$, $\dfrac{dY}{dt} = Y$.

Here is an example of our method as explained so far. However, the method will be improved soon, so don't take this example as a model for efficient computation.

EXAMPLE 3.4.2 Find the general solution of the system

$$\frac{dx}{dt} = 2y, \qquad \frac{dy}{dt} = -x + 3y.$$

Solution (to be "streamlined" later): Here $a = 0$, $b = 2$, $c = -1$, $d = 3$, so the characteristic equation will be $\lambda^2 - 3\lambda + 2 = 0$. It has roots $\lambda = 1$ and $\lambda = 2$, so there must be new coordinates X and Y such that $\dfrac{dX}{dt} = X$, $\dfrac{dY}{dt} = 2Y$. Let's find them: If $X = px + qy$, then $\dfrac{dX}{dt} = p(2y) + q(-x + 3y) = -qx + (2p + 3q)y$. So to have $\dfrac{dX}{dt} = X$ means that $-q = p$ and $2p + 3q = q$. These two equations for p and q are equivalent. We can choose, for instance, $p = 1$, $q = -1$, which will give $X = x - y$ as our new coordinate, or we can choose any (nonzero) multiple of this coordinate. Similarly, if we let $Y = rx + sy$, then $\dfrac{dY}{dt} = -sx + (2r + 3s)y$, so we need $-s = 2r$, $2r + 3s = 2s$ in order for $\dfrac{dY}{dt} = 2Y$ to be the equation for Y. These equations for r and s are again equivalent; if we take $r = 1$, for example, we will have $s = -2$ and $Y = x - 2y$. Thus we can choose the change of coordinates $X = x - y$, $Y = x - 2y$. In terms of these new coordinates, the system becomes $\frac{dX}{dt} = X$, $\frac{dY}{dt} = 2Y$, so we have $X = Ce^t$, $Y = De^{2t}$ (C, D arbitrary). Returning to the original coordinates, this means that $x - y = Ce^t$, $x - 2y = De^{2t}$; solving for x and y, we finally get

$$x = 2Ce^t - De^{2t}, \quad y = Ce^t - De^{2t}$$

as our general solution. ∎

Computations such as the above can be made more efficient once we realize the surprising fact that it is unnecessary to actually find the new coordinates X and Y (explicitly)! What we really want to know about the change of coordinates is how to do it *in reverse*. After all, if the characteristic equation has distinct roots λ_1, λ_2 and X, Y are corresponding new coordinates, we already know that we'll get

$$\frac{dX}{dt} = \lambda_1 X, \quad \frac{dY}{dt} = \lambda_2 Y \quad \text{and thus} \quad X = Ce^{\lambda_1 t}, Y = De^{\lambda_2 t}.$$

What we have to find out is how to recover x, y from this information. So instead of looking for the numbers p, q, r, s such that $X = px + qy$, $Y = rx + sy$ will serve as the new coordinates, it will be more practical to look right away for the numbers P, Q, R, S such that in the end $x = PX + QY$, $y = RX + SY$. (In Example 3.4.2, to undo the change of coordinates $X = x - y$, $Y = x - 2y$, we need to know that $x = 2X - Y$, $y = X - Y$.)

For the rest of this discussion, it is convenient to use some notation and terminology from linear algebra (see Appendix B). Specifically, we can think of (x, y) and (X, Y) as unknown vectors, which depend on t and which are transformed into each other by the matrices $\begin{bmatrix} p & q \\ r & s \end{bmatrix}$ and $\begin{bmatrix} P & Q \\ R & S \end{bmatrix}$:

$$\begin{bmatrix} X \\ Y \end{bmatrix} = \begin{bmatrix} p & q \\ r & s \end{bmatrix} \begin{bmatrix} x \\ y \end{bmatrix}, \quad \begin{bmatrix} x \\ y \end{bmatrix} = \begin{bmatrix} P & Q \\ R & S \end{bmatrix} \begin{bmatrix} X \\ Y \end{bmatrix}.$$

In particular, these matrices are inverses of each other; we'll write $\mathbf{T} = \begin{bmatrix} P & Q \\ R & S \end{bmatrix}$ for the matrix we want to find, so that $\mathbf{T}^{-1} = \begin{bmatrix} p & q \\ r & s \end{bmatrix}$. (In Example 3.4.2, we had $\mathbf{T} = \begin{bmatrix} 2 & -1 \\ 1 & -1 \end{bmatrix}$, $\mathbf{T}^{-1} = \begin{bmatrix} 1 & -1 \\ 1 & -2 \end{bmatrix}$.)

SEC. 3.4 LINEAR SYSTEMS (CONTINUED); EIGENVALUES AND EIGENVECTORS

Meanwhile, the unknown vector[16] (x, y) has the derivative $\left(\dfrac{dx}{dt}, \dfrac{dy}{dt}\right)$, and our system of equations

$$\frac{dx}{dt} = ax + by, \quad \frac{dy}{dt} = cx + dy$$

can be rewritten as

$$\frac{d}{dt}\begin{bmatrix} x \\ y \end{bmatrix} = \begin{bmatrix} a & b \\ c & d \end{bmatrix}\begin{bmatrix} x \\ y \end{bmatrix}.$$

We'll put $\mathbf{A} = \begin{bmatrix} a & b \\ c & d \end{bmatrix}$, so that we get the vector differential equation

$$\frac{d}{dt}\begin{bmatrix} x \\ y \end{bmatrix} = \mathbf{A}\begin{bmatrix} x \\ y \end{bmatrix}.$$

(For the system $\dfrac{dx}{dt} = 2y$, $\dfrac{dy}{dt} = -x + 3y$ from Example 3.4.2, we have $\mathbf{A} = \begin{bmatrix} 0 & 2 \\ -1 & 3 \end{bmatrix}$.)

When we make the change of coordinates to $\begin{bmatrix} X \\ Y \end{bmatrix} = \mathbf{T}^{-1}\begin{bmatrix} x \\ y \end{bmatrix}$, the derivative undergoes the same change:

$$\frac{d}{dt}\begin{bmatrix} X \\ Y \end{bmatrix} = \mathbf{T}^{-1}\frac{d}{dt}\begin{bmatrix} x \\ y \end{bmatrix} \quad \text{(see Exercise 20)},$$

so we have

$$\frac{d}{dt}\begin{bmatrix} X \\ Y \end{bmatrix} = \mathbf{T}^{-1}\mathbf{A}\begin{bmatrix} x \\ y \end{bmatrix} = \mathbf{T}^{-1}\mathbf{A}\mathbf{T}\begin{bmatrix} X \\ Y \end{bmatrix}$$

as the vector differential equation in the new coordinates. Note that the matrix in the new vector differential equation is $\mathbf{T}^{-1}\mathbf{A}\mathbf{T}$. Remember that we want this last equation to be equivalent to $\dfrac{dX}{dt} = \lambda_1 X$, $\dfrac{dY}{dt} = \lambda_2 Y$. That is, we want to get

$$\frac{d}{dt}\begin{bmatrix} X \\ Y \end{bmatrix} = \begin{bmatrix} \lambda_1 & 0 \\ 0 & \lambda_2 \end{bmatrix}\begin{bmatrix} X \\ Y \end{bmatrix}.$$

So we see that we should look for a transformation matrix $\mathbf{T}$ such that

$$\mathbf{T}^{-1}\mathbf{A}\mathbf{T} = \begin{bmatrix} \lambda_1 & 0 \\ 0 & \lambda_2 \end{bmatrix}.$$

Multiplying both sides by $\mathbf{T}$ on the left yields

$$\mathbf{A}\mathbf{T} = \mathbf{T}\begin{bmatrix} \lambda_1 & 0 \\ 0 & \lambda_2 \end{bmatrix},$$

---

[16] Since (x, y) depends on t, you may prefer to call (x, y) the unknown *vector function*.

that is,

$$\mathbf{A}\begin{bmatrix} P & Q \\ R & S \end{bmatrix} = \begin{bmatrix} P & Q \\ R & S \end{bmatrix}\begin{bmatrix} \lambda_1 & 0 \\ 0 & \lambda_2 \end{bmatrix}$$

or

$$\mathbf{A}\begin{bmatrix} P \vdots Q \\ R \vdots S \end{bmatrix} = \begin{bmatrix} \lambda_1 P \vdots \lambda_2 Q \\ \lambda_1 R \vdots \lambda_2 S \end{bmatrix}$$

(the dotted vertical lines are for reference in the next step). Another way to put this is that

$$\mathbf{A}\begin{bmatrix} P \\ R \end{bmatrix} = \begin{bmatrix} \lambda_1 P \\ \lambda_1 R \end{bmatrix} = \lambda_1 \cdot \begin{bmatrix} P \\ R \end{bmatrix} \quad \text{(left of the dotted lines)}$$

and

$$\mathbf{A}\begin{bmatrix} Q \\ S \end{bmatrix} = \begin{bmatrix} \lambda_2 Q \\ \lambda_2 S \end{bmatrix} = \lambda_2 \cdot \begin{bmatrix} Q \\ S \end{bmatrix} \quad \text{(right of the dotted lines)}.$$

Thus the vectors (P, R) and (Q, S) should have the special property that the matrix $\mathbf{A}$ transforms them into multiples of themselves. [In Example 3.4.2, $(P, R) = (2, 1)$, $(Q, S) = (-1, -1)$, and we have

$$\begin{bmatrix} 0 & 2 \\ -1 & 3 \end{bmatrix}\begin{bmatrix} 2 \\ 1 \end{bmatrix} = \begin{bmatrix} 2 \\ 1 \end{bmatrix} = 1 \cdot \begin{bmatrix} 2 \\ 1 \end{bmatrix}, \quad \begin{bmatrix} 0 & 2 \\ -1 & 3 \end{bmatrix}\begin{bmatrix} -1 \\ -1 \end{bmatrix} = \begin{bmatrix} -2 \\ -2 \end{bmatrix} = 2 \cdot \begin{bmatrix} -1 \\ -1 \end{bmatrix}.]$$

Vectors with this property are called **eigenvectors**, and the corresponding constants λ are called the **eigenvalues**, of the matrix $\mathbf{A}$. There is one exception: Since $\mathbf{A}\begin{bmatrix} 0 \\ 0 \end{bmatrix} = \lambda \cdot \begin{bmatrix} 0 \\ 0 \end{bmatrix}$ for any constant λ, and since the vector $(0, 0)$ is useless for our purposes, we exclude this case. This leaves us with the following.

Definition The vector (ξ_1, ξ_2) is called an **eigenvector** of the matrix $\mathbf{A}$ for the **eigenvalue** λ if the vector is not the zero vector and

$$\mathbf{A}\begin{bmatrix} \xi_1 \\ \xi_2 \end{bmatrix} = \lambda \cdot \begin{bmatrix} \xi_1 \\ \xi_2 \end{bmatrix}.$$

NOTE: Any number λ, including 0, can occur as an eigenvalue.

We can see directly that the eigenvalues of $\mathbf{A}$ are just the roots of the characteristic equation. For if $\begin{bmatrix} a & b \\ c & d \end{bmatrix}\begin{bmatrix} \xi_1 \\ \xi_2 \end{bmatrix} = \lambda \cdot \begin{bmatrix} \xi_1 \\ \xi_2 \end{bmatrix}$, then, comparing coordinates, we have

$$\begin{cases} a\xi_1 + b\xi_2 = \lambda\xi_1 \\ c\xi_1 + d\xi_2 = \lambda\xi_2. \end{cases}$$

SEC. 3.4 LINEAR SYSTEMS (CONTINUED); EIGENVALUES AND EIGENVECTORS

These simultaneous equations for ξ_1, ξ_2 can be rewritten as

$$\begin{cases} (a - \lambda)\xi_1 + b\xi_2 = 0 \\ c\xi_1 + (d - \lambda)\xi_2 = 0; \end{cases} \qquad (*)$$

they have a nontrivial solution (a solution besides $\xi_1 = \xi_2 = 0$) only when $\det\begin{bmatrix} a - \lambda & b \\ c & d - \lambda \end{bmatrix} = 0$ (see Appendix B, p. 624). This yields $(a - \lambda)(d - \lambda) - bc = 0$ or $\lambda^2 - (a + d)\lambda + (ad - bc) = 0$, the characteristic equation.[17]

To find an eigenvector (ξ_1, ξ_2) for an eigenvalue λ, we simply solve the simultaneous equations $(*)$. We have seen above that if there are two distinct (real) eigenvalues λ_1, λ_2, that is, two distinct roots of the characteristic equation, then we can take $\mathbf{T} = \begin{bmatrix} P & Q \\ R & S \end{bmatrix}$ as the transformation matrix, where (P, R) and (Q, S) are eigenvectors for λ_1, λ_2, respectively.[18] So **the eigenvectors**[19] **of A are used as the columns of the transformation matrix T**; we then have

$$\mathbf{T}^{-1}\mathbf{A}\mathbf{T} = \begin{bmatrix} \lambda_1 & 0 \\ 0 & \lambda_2 \end{bmatrix}, \qquad \begin{bmatrix} x \\ y \end{bmatrix} = \mathbf{T}\begin{bmatrix} X \\ Y \end{bmatrix} = \mathbf{T}\begin{bmatrix} Ce^{\lambda_1 t} \\ De^{\lambda_2 t} \end{bmatrix}.$$

NOTE: A matrix of the form $\begin{bmatrix} \lambda_1 & 0 \\ 0 & \lambda_2 \end{bmatrix}$ is called a **diagonal** matrix; finding a matrix $\mathbf{T}$ such that $\mathbf{T}^{-1}\mathbf{A}\mathbf{T}$ is diagonal is called **diagonalizing** the matrix $\mathbf{A}$. The matrix $\mathbf{A}$ is **diagonalizable** if this can be done; in Section 3.5 we'll see cases in which $\mathbf{A}$ is not diagonalizable.

EXAMPLE 3.4.3 Find the eigenvalues and eigenvectors of the matrix $\mathbf{A} = \begin{bmatrix} 11 & -3 \\ 36 & -10 \end{bmatrix}$. Then solve the system $\dfrac{dx}{dt} = 11x - 3y$, $\dfrac{dy}{dt} = 36x - 10y$.

The eigenvalues are the solutions of the characteristic equation

$$\det\begin{bmatrix} 11 - \lambda & -3 \\ 36 & -10 - \lambda \end{bmatrix} = 0$$

$$(11 - \lambda)(-10 - \lambda) + 108 = 0$$

$$\lambda^2 - \lambda - 2 = 0$$

$$(\lambda - 2)(\lambda + 1) = 0,$$

so we have $\lambda_1 = 2$ and $\lambda_2 = -1$.

---

[17] Note that $\begin{bmatrix} a - \lambda & b \\ c & d - \lambda \end{bmatrix}$ is actually the *transpose* of the matrix $\begin{bmatrix} a - \lambda & c \\ b & d - \lambda \end{bmatrix}$ that came up on p. 187. However, its determinant is the same.

[18] Actually, before making this statement we should have checked that the matrix $\mathbf{T}$ always has an inverse, so that it can be used as a transformation matrix (see Exercise 25).

[19] The phrase "*the* eigenvectors" here is sloppy (although often used this way): If (P, R) is an eigenvector, so is any nonzero multiple of (P, R). Replacing (P, R) by the multiple $k \cdot (P, R)$ has the effect of replacing the "new" coordinate X by $\dfrac{1}{k} X$.

(NOTE: The order in which the eigenvalues are listed is random. Once the order is chosen, though, it should be maintained consistently throughout the problem. The eigenvectors—to be found below—should be put in the columns of the transformation matrix in the same order.)

An eigenvector (ξ_1, ξ_2) for the eigenvalue 2 is a nonzero vector such that

$$\begin{bmatrix} 11 & -3 \\ 36 & -10 \end{bmatrix} \begin{bmatrix} \xi_1 \\ \xi_2 \end{bmatrix} = 2 \cdot \begin{bmatrix} \xi_1 \\ \xi_2 \end{bmatrix} = \begin{bmatrix} 2\xi_1 \\ 2\xi_2 \end{bmatrix}$$

or

$$\begin{cases} 11\xi_1 - 3\xi_2 = 2\xi_1 \\ 36\xi_1 - 10\xi_2 = 2\xi_2. \end{cases}$$

These equations are satisfied exactly when $\xi_2 = 3\xi_1$ (check this). So any vector of the form $(\xi_1, 3\xi_1) = \xi_1 \cdot (1, 3)$ with $\xi_1 \neq 0$ is an eigenvector for $\lambda = 2$. If only one eigenvector is needed, we can take $(1, 3)$.

Similarly, the eigenvectors (ξ_1, ξ_2) for the eigenvalue -1 satisfy

$$\begin{cases} 11\xi_1 - 3\xi_2 = -\xi_1 \\ 36\xi_1 - 10\xi_2 = -\xi_2, \end{cases}$$

that is, $\xi_2 = 4\xi_1$. So the eigenvectors for $\lambda = -1$ are the nonzero multiples of the vector $(1, 4)$.

To solve the system, remember that there is a coordinate change $\begin{bmatrix} X \\ Y \end{bmatrix} = \begin{bmatrix} p & q \\ r & s \end{bmatrix} \begin{bmatrix} x \\ y \end{bmatrix}$, $\begin{bmatrix} x \\ y \end{bmatrix} = \begin{bmatrix} P & Q \\ R & S \end{bmatrix} \begin{bmatrix} X \\ Y \end{bmatrix}$ which transforms the system to $\dfrac{dX}{dt} = 2X$, $\dfrac{dY}{dt} = -Y$, since 2 and -1 are the eigenvalues. We then have $X = Ce^{2t}$, $Y = De^{-t}$. The columns $\begin{bmatrix} P \\ R \end{bmatrix}$ and $\begin{bmatrix} Q \\ S \end{bmatrix}$ of the matrix **T** can be chosen to be $\begin{bmatrix} 1 \\ 3 \end{bmatrix}$ and $\begin{bmatrix} 1 \\ 4 \end{bmatrix}$ (the eigenvectors found above). So we get

$$\begin{bmatrix} x \\ y \end{bmatrix} = \begin{bmatrix} 1 & 1 \\ 3 & 4 \end{bmatrix} \begin{bmatrix} Ce^{2t} \\ De^{-t} \end{bmatrix} = \begin{bmatrix} Ce^{2t} + De^{-t} \\ 3Ce^{2t} + 4De^{-t} \end{bmatrix},$$

that is,

$$\begin{cases} x = Ce^{2t} + De^{-t} \\ y = 3Ce^{2t} + 4De^{-t}, \end{cases}$$

as the general solution of our system.

NOTE: By the definition of addition and scalar multiplication of vectors (see Appendix B, Section 1), we can think of the general solution vector (x, y) as built up from two basic solutions, as follows:

$$(x, y) = (Ce^{2t} + De^{-t}, 3Ce^{2t} + 4De^{-t}) = C \cdot (e^{2t}, 3e^{2t}) + D \cdot (e^{-t}, 4e^{-t}).$$

Here $(e^{2t}, 3e^{2t})$ is the solution with $C = 1$, $D = 0$, while $(e^{-t}, 4e^{-t})$ is the one with $C = 0$, $D = 1$. ∎

EXAMPLE 3.4.4 Solve the system $\frac{dx}{dt} = 2y$, $\frac{dy}{dt} = -x + 3y$ (from Example 3.4.2, p. 187) using eigenvectors.

In this case we have $\mathbf{A} = \begin{bmatrix} 0 & 2 \\ -1 & 3 \end{bmatrix}$, so the eigenvalues are the roots of

$$\det \begin{bmatrix} -\lambda & 2 \\ -1 & 3-\lambda \end{bmatrix} = 0$$

$$\lambda^2 - 3\lambda + 2 = 0.$$

Thus we have $\lambda = 1$, $\lambda = 2$. To find an eigenvector for $\lambda = 1$, we look for (ξ_1, ξ_2) such that $\begin{bmatrix} 0 & 2 \\ -1 & 3 \end{bmatrix} \begin{bmatrix} \xi_1 \\ \xi_2 \end{bmatrix} = \begin{bmatrix} \xi_1 \\ \xi_2 \end{bmatrix}$, which yields $2\xi_2 = \xi_1$; for instance, we can take $(2, 1)$. Similarly, $(1, 1)$ is an eigenvector for $\lambda = 2$ (check this). So we can take $\mathbf{T} = \begin{bmatrix} 2 & 1 \\ 1 & 1 \end{bmatrix}$ as the transformation matrix to act on $\begin{bmatrix} X \\ Y \end{bmatrix} = \begin{bmatrix} Ce^t \\ De^{2t} \end{bmatrix}$. The general solution is thus found to be

$$\begin{bmatrix} x \\ y \end{bmatrix} = \begin{bmatrix} 2 & 1 \\ 1 & 1 \end{bmatrix} \begin{bmatrix} Ce^t \\ De^{2t} \end{bmatrix}, \quad \text{that is,} \quad \begin{cases} x = 2Ce^t + De^{2t} \\ y = Ce^t + De^{2t}. \end{cases}$$

NOTE: Once again, the general solution can be written as

$$(x, y) = C \cdot (2e^t, e^t) + D \cdot (e^{2t}, e^{2t}).$$

In fact, one can introduce "basic solutions" for systems much as in Section 2.2 for second-order equations. We will see this again in the next chapter (see also Exercise 22). Meanwhile, note that $(2e^t, e^t) = e^t \cdot (2, 1)$ and $(e^{2t}, e^{2t}) = e^{2t} \cdot (1, 1)$: Our special solutions for $C = 1$, $D = 0$ and for $C = 0$, $D = 1$ are obtained by multiplying the eigenvector for λ by $e^{\lambda t}$, for $\lambda = 1$ and $\lambda = 2$, respectively. This also works in general [see Exercise 22(b)]. □

Even though in the streamlined method above we no longer find the new coordinates X and Y in terms of x and y, it turns out that it is still easy to find the coordinate *axes*. This will be very useful in sketching the phase portrait of a given linear system, as we'll see in the examples below and in Section 3.6. In fact, we'll now see that the eigenvectors indicate where the new axes are; specifically, they are located along the new axes. This is because $x = P$, $Y = R$ corresponds to $X = 1$, $Y = 0$: $\begin{bmatrix} P \\ R \end{bmatrix} = \begin{bmatrix} P & Q \\ R & S \end{bmatrix} \begin{bmatrix} 1 \\ 0 \end{bmatrix}$. Therefore, the first eigenvector (P, R) is the "unit vector" along the (new) X-axis. (Since the new axes may not be perpendicular, though, there is no notion of "length" in the new coordinate system—so the concept of "unit vector" is not all that significant.) Similarly, the vector (Q, S)—the second eigenvector—has new coordinates $X = 0$, $Y = 1$, so it lies along the (new) Y-axis.

EXAMPLE 3.4.4 (continued)

Sketch the phase portrait of the system

$$\frac{dx}{dt} = 2y, \quad \frac{dy}{dt} = -x + 3y.$$

We have already seen (on p. 193) that there are eigenvectors (2, 1) and (1, 1) for $\lambda = 1$ and $\lambda = 2$, respectively. Therefore, we have $X = Ce^t$, $Y = De^{2t}$, with the vector (2, 1) along the X-axis and (1, 1) along the Y-axis. To see what the parametric curves look like, note that $X^2 = C^2 e^{2t}$ and $Y = De^{2t}$ are proportional. Thus the trajectories will follow curves of the form $Y = KX^2$ (where $K = D/C^2$ is constant). An exception occurs when $C = 0$; in this case, $X = 0$, $Y = De^{2t}$, so the trajectory lies along the Y-axis; it is the positive Y-axis if $D > 0$, the stationary point at the origin if $D = 0$, and the negative Y-axis if $D < 0$. Each curve $Y = KX^2$ contains two trajectories, one with positive X (or C) and one with negative X, in addition to the stationary point. In particular, for $K = 0$, the positive and negative X-axis are trajectories. All trajectories (except the stationary point) approach infinity as $t \to \infty$, since $e^t \to \infty$, $e^{2t} \to \infty$ as $t \to \infty$. All trajectories "originate" at the stationary point in the sense that they approach the origin as $t \to -\infty$. (Remember, though, that a nonconstant trajectory can never reach a stationary point.)

In Figure 3.4.2(a) the trajectories are drawn as they would appear if the X- and Y-axes were perpendicular. To get the true phase portrait, this sketch must be "distorted," to account for the actual location of the axes [Figure 3.4.2(b) and (c)]. See Exercise 23. ∎

EXAMPLE 3.4.5

Sketch the phase portrait of the system

$$\frac{dx}{dt} = -5x + 4y, \quad \frac{dy}{dt} = 8x - y.$$

Here we have $\mathbf{A} = \begin{bmatrix} -5 & 4 \\ 8 & -1 \end{bmatrix}$, and we find $\lambda = 3$ and $\lambda = -9$ as the eigenvalues [see Exercise 6(a)]. (1, 2) is an eigenvector for $\lambda = 3$, while (1, −1) is an eigenvector for $\lambda = -9$ [see Exercise 6(b)]. Thus we have $X = Ce^{3t}$, $Y = De^{-9t}$ with (1, 2) along the X-axis and (1, −1) along the Y-axis. Now note that $X^3 Y = C^3 D =$ constant, so the trajectories follow curves of the form $X^3 Y = K$. [20] In particular, for $K = 0$ we get trajectories along the X- and Y-axes. Figure 3.4.3(a) on p. 196 shows the appearance of the trajectories if the X- and Y-axes are drawn to be perpendicular, and Figure 3.4.3(b) shows the actual phase portrait (see also Exercise 10). ∎

In the next section we'll deal with the situation in which the characteristic equation $\det \begin{bmatrix} a - \lambda & b \\ c & d - \lambda \end{bmatrix} = 0$ does not have two distinct real roots, that is, in which $\mathbf{A}$ has at most one real eigenvalue.

---

[20] In other words, $X^3 Y$ is a *first integral* for the system! (See Section 3.2.)

SEC. 3.4 LINEAR SYSTEMS (CONTINUED); EIGENVALUES AND EIGENVECTORS

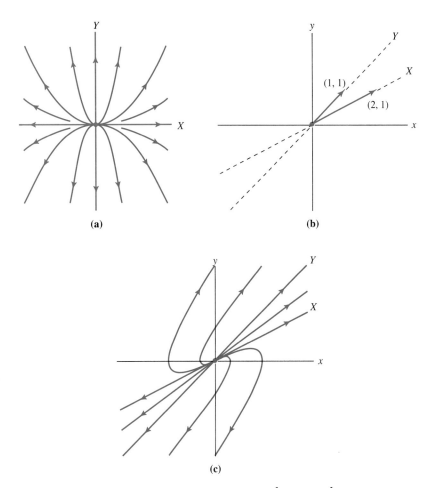

Figure 3.4.2. Drawing the phase portrait for $\dfrac{dx}{dt} = 2y$, $\dfrac{dy}{dt} = -x + 3y$: (a) new coordinates drawn along perpendicular axes; (b) locating the X- and Y-axes using eigenvectors; (c) actual phase portrait.

SUMMARY OF KEY CONCEPTS, RESULTS, AND TECHNIQUES

Vector differential equation (p. 189): The system $\dfrac{dx}{dt} = ax + by$, $\dfrac{dy}{dt} = cx + dy$ can be rewritten as $\dfrac{d}{dt}\begin{bmatrix} x \\ y \end{bmatrix} = \mathbf{A}\begin{bmatrix} x \\ y \end{bmatrix}$, where $\mathbf{A} = \begin{bmatrix} a & b \\ c & d \end{bmatrix}$.

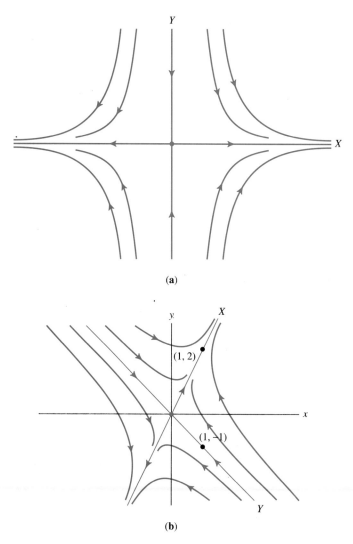

Figure 3.4.3. Drawing the phase portrait for $\dfrac{dx}{dt} = -5x + 4y$, $\dfrac{dy}{dt} = 8x - y$: (a) new coordinates drawn along perpendicular axes; (b) actual phase portrait. (See Example 3.4.5, p. 194.)

Change of coordinates (pp. 188–189): If the 2×2 **transformation matrix T** has an inverse, we can change coordinates from x, y to X, Y such that $\begin{bmatrix} x \\ y \end{bmatrix} = \mathbf{T}\begin{bmatrix} X \\ Y \end{bmatrix}$, $\begin{bmatrix} X \\ Y \end{bmatrix} = \mathbf{T}^{-1}\begin{bmatrix} x \\ y \end{bmatrix}$. Under this change of coordinates, the vector differential equation $\dfrac{d}{dt}\begin{bmatrix} x \\ y \end{bmatrix} = \mathbf{A}\begin{bmatrix} x \\ y \end{bmatrix}$ becomes $\dfrac{d}{dt}\begin{bmatrix} X \\ Y \end{bmatrix} = \mathbf{T}^{-1}\mathbf{A}\mathbf{T}\begin{bmatrix} X \\ Y \end{bmatrix}$.

SEC. 3.4 LINEAR SYSTEMS (CONTINUED); EIGENVALUES AND EIGENVECTORS

Eigenvector of a 2×2 matrix $\mathbf{A} = \begin{bmatrix} a & b \\ c & d \end{bmatrix}$ (p. 190): A nonzero vector (ξ_1, ξ_2) such that $\mathbf{A}\begin{bmatrix} \xi_1 \\ \xi_2 \end{bmatrix} = \lambda \cdot \begin{bmatrix} \xi_1 \\ \xi_2 \end{bmatrix}$ for some constant λ. The constants λ that occur in this way are called the **eigenvalues** of $\mathbf{A}$; they are the roots of the **characteristic equation** $\det \begin{bmatrix} a - \lambda & b \\ c & d - \lambda \end{bmatrix} = 0$. To find the eigenvectors of $\mathbf{A}$, first find the eigenvalues, then solve $\mathbf{A}\begin{bmatrix} \xi_1 \\ \xi_2 \end{bmatrix} = \lambda \cdot \begin{bmatrix} \xi_1 \\ \xi_2 \end{bmatrix}$, or equivalently $\begin{cases} (a - \lambda)\xi_1 + b\xi_2 = 0 \\ c\xi_1 + (d - \lambda)\xi_2 = 0 \end{cases}$, for each eigenvalue λ separately.

Solving $\dfrac{d}{dt}\begin{bmatrix} x \\ y \end{bmatrix} = \mathbf{A}\begin{bmatrix} x \\ y \end{bmatrix}$ using eigenvectors (pp. 189–191): If the characteristic equation has two distinct roots λ_1, λ_2 (for the other cases, see Section 3.5), we can find an eigenvector (P, R) for λ_1 and an eigenvector (Q, S) for λ_2. Then using the transformation matrix $\mathbf{T} = \begin{bmatrix} P & Q \\ R & S \end{bmatrix}$ whose columns are these eigenvectors, we have

$$\mathbf{T}^{-1}\mathbf{A}\mathbf{T} = \begin{bmatrix} \lambda_1 & 0 \\ 0 & \lambda_2 \end{bmatrix}:$$

$\mathbf{T}^{-1}\mathbf{A}\mathbf{T}$ is a **diagonal** matrix with the eigenvalues as its diagonal entries. Therefore, the differential equations for the new coordinates are $\dfrac{dX}{dt} = \lambda_1 X$, $\dfrac{dY}{dt} = \lambda_2 Y$, and we get

$$X = Ce^{\lambda_1 t}, \qquad Y = De^{\lambda_2 t}, \qquad \begin{bmatrix} x \\ y \end{bmatrix} = \mathbf{T}\begin{bmatrix} X \\ Y \end{bmatrix} = \begin{bmatrix} P & Q \\ R & S \end{bmatrix}\begin{bmatrix} Ce^{\lambda_1 t} \\ De^{\lambda_2 t} \end{bmatrix}.$$

Also, $\begin{bmatrix} P \\ R \end{bmatrix}$ is along the X-axis and $\begin{bmatrix} Q \\ S \end{bmatrix}$ is along the Y-axis. To draw the phase portrait of the system $\dfrac{d}{dt}\begin{bmatrix} x \\ y \end{bmatrix} = \mathbf{A}\begin{bmatrix} x \\ y \end{bmatrix}$, first draw X and Y along perpendicular axes, then "distort" this sketch to indicate the actual locations of the X- and Y-axes (pp. 193–194).

EXERCISES

For each of the following matrices $\mathbf{A}$, find:

(a) The eigenvalues.
(b) The eigenvectors.
(c) If possible, a matrix $\mathbf{T}$ such that $\mathbf{T}^{-1}\mathbf{A}\mathbf{T} = \begin{bmatrix} \lambda_1 & 0 \\ 0 & \lambda_2 \end{bmatrix}$ for suitable numbers λ_1, λ_2.

1. $\begin{bmatrix} 2 & 5 \\ 0 & 3 \end{bmatrix}$ 2. $\begin{bmatrix} -1 & 0 \\ 1 & -4 \end{bmatrix}$ 3. $\begin{bmatrix} 2 & 1 \\ 1 & 2 \end{bmatrix}$

4. $\begin{bmatrix} \frac{1}{2} & \frac{1}{2} \\ \frac{1}{2} & \frac{1}{2} \end{bmatrix}$ 5. $\begin{bmatrix} 3 & 1 \\ 6 & -2 \end{bmatrix}$ 6. $\begin{bmatrix} -5 & 4 \\ 8 & -1 \end{bmatrix}$

7. $\begin{bmatrix} 1 & 2 \\ 0 & 0 \end{bmatrix}$ 8. $\begin{bmatrix} 0 & 1 \\ -2 & 0 \end{bmatrix}$ 9. $\begin{bmatrix} 1 & 3 \\ -3 & 1 \end{bmatrix}$

10. How do we know the direction of travel along the curves $X^3Y = K$ in Figure 3.4.3(a)?

For each of the following systems and initial value problems:

(a) Solve the system or initial value problem using eigenvectors.
(b) Sketch the phase portrait of the system.

11. $\dfrac{dx}{dt} = 2x + y$, $\dfrac{dy}{dt} = x + 2y$. (Exercise 3 will be helpful.)

12. $\dfrac{dx}{dt} = 3x - 2y$, $\dfrac{dy}{dt} = 2x - 2y$.

13. $\dfrac{dx}{dt} = -2x - y$, $\dfrac{dy}{dt} = -4y$.

14. $\dfrac{dx}{dt} = 4y$, $\dfrac{dy}{dt} = -\dfrac{1}{2}x + 3y$, $x(0) = 8$, $y(0) = -1$.

15. $\dfrac{dx}{dt} = -5x + 8y$, $\dfrac{dy}{dt} = 4x - y$, $x(0) = 0$, $y(0) = 15$.

16. $\dfrac{dx}{dt} = -x$, $\dfrac{dy}{dt} = x + 4y$.

17. $\dfrac{dx}{dt} = 15x - 48y$, $\dfrac{dy}{dt} = 4x - 13y$.

18. $\dfrac{dx}{dt} = -5x - 4y + 9$, $\dfrac{dy}{dt} = 6x + 5y - 11$.
    [*Hint*: First change coordinates so the stationary point is at the origin.]

19. $\dfrac{dx}{dt} = 10x + 12y + 8$, $\dfrac{dy}{dt} = -x + 2y - 4$.

20. Show that if $\begin{bmatrix} X \\ Y \end{bmatrix} = \begin{bmatrix} p & q \\ r & s \end{bmatrix} \begin{bmatrix} x \\ y \end{bmatrix}$ with constants p, q, r, s, then

$\dfrac{d}{dt}\begin{bmatrix} X \\ Y \end{bmatrix} = \begin{bmatrix} p & q \\ r & s \end{bmatrix} \dfrac{d}{dt}\begin{bmatrix} x \\ y \end{bmatrix}$.

21. Show that the solutions found on p. 188 (in Example 3.4.2) and on p. 193 (in Example 3.4.4) for the system $\dfrac{dx}{dt} = 2y$, $\dfrac{dy}{dt} = -x + 3y$ are equivalent.

22. Consider the system $\dfrac{dx}{dt} = ax + by$, $\dfrac{dy}{dt} = cx + dy$.

   (a) Show that if $(x_1(t), y_1(t))$ and $(x_2(t), y_2(t))$ are solutions of the system, then so is any linear combination
   $$C \cdot (x_1, y_1) + D \cdot (x_2, y_2) = (Cx_1 + Dx_2, Cy_1 + Dy_2),$$
   where C and D are constants.

   (b) Show that if (ξ_1, ξ_2) is an eigenvector of $\mathbf{A} = \begin{bmatrix} a & b \\ c & d \end{bmatrix}$ for the eigenvalue λ, then $e^{\lambda t}(\xi_1, \xi_2) = (e^{\lambda t}\xi_1, e^{\lambda t}\xi_2)$ is a solution of the system.

   (c) Note that by part (b), if there are eigenvectors (P, R) and (Q, S) for the (distinct) eigenvalues λ_1 and λ_2, then $(e^{\lambda_1 t} \cdot P, e^{\lambda_1 t} \cdot R)$ and $(e^{\lambda_2 t} \cdot Q, e^{\lambda_2 t} \cdot S)$ will be solutions; hence by part (a),
   $$(Ce^{\lambda_1 t} \cdot P + De^{\lambda_2 t} \cdot Q, Ce^{\lambda_1 t} \cdot R + De^{\lambda_2 t} \cdot S)$$
   will be a solution. Check that this solution is the same as the general solution $\begin{bmatrix} x \\ y \end{bmatrix} = \begin{bmatrix} P & Q \\ R & S \end{bmatrix} \begin{bmatrix} X \\ Y \end{bmatrix}$ found in the text.

23. Sketch the curves $Y = KX^2$ in the x,y-plane, where $Y = x - 2y$, $X = x - y$, for $K = 0, \pm\dfrac{1}{2}, \pm 1, \pm 2$. [Compare Figure 3.4.2(c), p. 195.]

*24. Show that in the situation of the text, the vectors (p, q) and (r, s) which form the rows of $\mathbf{T}^{-1}$ are eigenvectors of $\mathbf{A}^T$, where $\mathbf{A}^T = \begin{bmatrix} a & c \\ b & d \end{bmatrix}$ is the transpose of the matrix $\mathbf{A}$.

*25. Show that if (P, R) and (Q, S) are eigenvectors of $\mathbf{A} = \begin{bmatrix} a & b \\ c & d \end{bmatrix}$ for the eigenvalues λ_1 and λ_2 respectively, with $\lambda_1 \neq \lambda_2$, then the matrix $\mathbf{T} = \begin{bmatrix} P & Q \\ R & S \end{bmatrix}$ always has an inverse.
   [*Hint*: If not, det $\mathbf{T} = 0$ and one of the two vectors (P, R) and (Q, S) will be a multiple of the other.]

26. Show that if $\mathbf{A} = \begin{bmatrix} a & b \\ c & d \end{bmatrix}$ has distinct eigenvalues λ_1 and λ_2, then their product $\lambda_1 \lambda_2$ equals det $\mathbf{A}$.
   [*Hint*: The characteristic equation must factor as $(\lambda - \lambda_1)(\lambda - \lambda_2)$.]

3.5 LINEAR SYSTEMS (CONTINUED): DOUBLE OR COMPLEX CONJUGATE EIGENVALUES

As we have seen, the characteristic equation of the system $\frac{dx}{dt} = ax + by$, $\frac{dy}{dt} = cx + dy$ works out to $\lambda^2 - (a + d)\lambda + (ad - bc) = 0$. Since this is a quadratic equation for λ, it has either two distinct real roots, a double real root, or a pair of (complex conjugate) nonreal roots. We will now consider the latter two cases.

If the characteristic equation has a double root $\lambda_1 = \lambda_2 = \lambda$, there will usually be only one eigenvector (up to multiplication by a nonzero constant), so that our method from Section 3.4 will fail for lack of a transformation matrix $\mathbf{T}$ for which $\mathbf{T}^{-1}\mathbf{A}\mathbf{T} = \begin{bmatrix} \lambda & 0 \\ 0 & \lambda \end{bmatrix}$.[21]

EXAMPLE 3.5.1

$\frac{dx}{dt} = -3x + y$, $\frac{dy}{dt} = -x - y$.

The matrix $\mathbf{A} = \begin{bmatrix} -3 & 1 \\ -1 & -1 \end{bmatrix}$ has characteristic equation

$$(-3 - \lambda)(-1 - \lambda) + 1 = 0$$
$$\lambda^2 + 4\lambda + 4 = 0,$$

so $\lambda = -2$ is a double root (or **double eigenvalue**). If (ξ_1, ξ_2) is an eigenvector for $\lambda = -2$, then

$$\begin{bmatrix} -3 & 1 \\ -1 & -1 \end{bmatrix} \begin{bmatrix} \xi_1 \\ \xi_2 \end{bmatrix} = -2 \begin{bmatrix} \xi_1 \\ \xi_2 \end{bmatrix},$$

so

$$\begin{cases} -3\xi_1 + \xi_2 = -2\xi_1 \\ -\xi_1 - \xi_2 = -2\xi_2, \end{cases}$$

which yields $\xi_1 = \xi_2$. Hence the only eigenvectors are the nonzero multiples of $(1, 1)$. □

Despite this difficulty, we'll still be able to solve the system by a variation of the method from Section 3.4. First of all, it can be shown [see Exercise 28(b)] that if the *first* column of the transformation matrix $\mathbf{T}$ is taken to be an eigenvector (which is still possible) and the second column is chosen *arbitrarily* (since we don't have another eigenvector), then provided $\mathbf{T}$ is invertible, the new matrix $\mathbf{T}^{-1}\mathbf{A}\mathbf{T}$ will have the form $\begin{bmatrix} \lambda & \sigma \\ 0 & \lambda \end{bmatrix}$

---

[21] The only exception occurs when $a = d = \lambda$, $b = c = 0$ (see Exercise 29). In this case every nonzero vector is an eigenvector. What's more, the system then has the form $\frac{dx}{dt} = \lambda x$, $\frac{dy}{dt} = \lambda y$, so we see right away that the general solution is $x = Ce^{\lambda t}$, $y = De^{\lambda t}$.

for some constant σ.[22] This means that in the new coordinates, the system becomes

$$\begin{cases} \dfrac{dX}{dt} = \lambda X + \sigma Y \\ \dfrac{dY}{dt} = \lambda Y. \end{cases}$$

The second of these equations yields $Y = Ce^{\lambda t}$; one can then solve the resulting equation $\dfrac{dX}{dt} = \lambda X + \sigma Ce^{\lambda t}$ for X by variation of constants (see Section 1.6). Finally, the matrix $\mathbf{T}$ is applied to $\begin{bmatrix} X \\ Y \end{bmatrix}$ to find the solution $\begin{bmatrix} x \\ y \end{bmatrix}$ in the original coordinates.

EXAMPLE 3.5.1 (continued) For the system $\dfrac{dx}{dt} = -3x + y$, $\dfrac{dy}{dt} = -x - y$ considered above, we found that $(1, 1)$ is an eigenvector of $\mathbf{A}$, so we can take any matrix $\mathbf{T}$ of the form $\mathbf{T} = \begin{bmatrix} 1 & \cdots \\ 1 & \cdots \end{bmatrix}$, provided that $\mathbf{T}$ is invertible. For instance, if we take $\mathbf{T} = \begin{bmatrix} 1 & 0 \\ 1 & 1 \end{bmatrix}$, then $\mathbf{T}^{-1} = \begin{bmatrix} 1 & 0 \\ -1 & 1 \end{bmatrix}$ and

$$\mathbf{T}^{-1}\mathbf{A}\mathbf{T} = \begin{bmatrix} 1 & 0 \\ -1 & 1 \end{bmatrix}\begin{bmatrix} -3 & 1 \\ -1 & -1 \end{bmatrix}\begin{bmatrix} 1 & 0 \\ 1 & 1 \end{bmatrix} = \begin{bmatrix} -3 & 1 \\ 2 & -2 \end{bmatrix}\begin{bmatrix} 1 & 0 \\ 1 & 1 \end{bmatrix} = \begin{bmatrix} -2 & 1 \\ 0 & -2 \end{bmatrix}.$$

So if we use the change of coordinates $\begin{bmatrix} X \\ Y \end{bmatrix} = \begin{bmatrix} 1 & 0 \\ -1 & 1 \end{bmatrix}\begin{bmatrix} x \\ y \end{bmatrix}$, then

$$\dfrac{dX}{dt} = -2X + Y, \qquad \dfrac{dY}{dt} = -2Y.$$

The second equation yields $Y = Ce^{-2t}$, hence $\dfrac{dX}{dt} = -2X + Ce^{-2t}$. The general solution to this last equation is $X = (Ct + D)e^{-2t}$ (see Exercise 24). Therefore, we get

$$\begin{bmatrix} x \\ y \end{bmatrix} = \begin{bmatrix} 1 & 0 \\ 1 & 1 \end{bmatrix}\begin{bmatrix} (Ct + D)e^{-2t} \\ Ce^{-2t} \end{bmatrix} = \begin{bmatrix} (Ct + D)e^{-2t} \\ (Ct + C + D)e^{-2t} \end{bmatrix},$$

so the general solution to our original system is

$$\begin{cases} x = (Ct + D)e^{-2t} \\ y = (Ct + C + D)e^{-2t} \end{cases}, \quad C \text{ and } D \text{ arbitrary constants.}$$

NOTE: As in Section 3.4, we have basic solutions, this time

$$(x, y) = (t \cdot e^{-2t}, (t + 1)e^{-2t}) \quad \text{and} \quad (x, y) = (e^{-2t}, e^{-2t}),$$

---

[22] It turns out that one can actually arrange to get $\sigma = 1$, unless $\mathbf{A} = \begin{bmatrix} \lambda & 0 \\ 0 & \lambda \end{bmatrix}$, the exceptional case from the preceding footnote. However, for us, arranging to have $\sigma = 1$ won't be worth the trouble.

of which every solution is a linear combination. The second basic solution can be written as $e^{-2t}(1, 1)$, so it is again of the form $e^{\lambda t}(\xi_1, \xi_2)$, where (ξ_1, ξ_2) is the eigenvector for the eigenvalue λ. Compare Section 3.4, Exercise 22(b). ∎

Now we consider the final case, in which there are no real eigenvalues. Instead, the characteristic equation $\lambda^2 - (a + d)\lambda + (ad - bc) = 0$ will have two complex roots, which are complex conjugates of each other. Just as in Section 2.5, these roots will be denoted by $\lambda = \mu + i\nu$ and $\bar{\lambda} = \mu - i\nu$. [Here we have $\mu = \frac{1}{2}(a + d)$, $\nu = \frac{1}{2}\sqrt{4(ad - bc) - (a + d)^2}$; in particular, $\nu > 0$]. If we allow vectors to have complex coordinates, then there will be an eigenvector for each of these complex eigenvalues; if (ξ_1, ξ_2) is an eigenvector for λ, the complex conjugate vector $(\bar{\xi}_1, \bar{\xi}_2)$ will be an eigenvector for $\bar{\lambda}$ (see Exercise 26).

EXAMPLE 3.5.2 Find the (complex) eigenvalues and eigenvectors of $\mathbf{A} = \begin{bmatrix} 3 & 4 \\ 2 & -1 \end{bmatrix}$.

The characteristic equation is $\lambda^2 - 2\lambda + 5 = 0$ (check this!), so we get the eigenvalues $\lambda, \bar{\lambda} = \dfrac{2 \pm i\sqrt{20 - 4}}{2} = 1 \pm 2i$. To find an eigenvector for $\lambda = 1 + 2i$, we look for (ξ_1, ξ_2) such that

$$\begin{bmatrix} 3 & -4 \\ 2 & -1 \end{bmatrix} \begin{bmatrix} \xi_1 \\ \xi_2 \end{bmatrix} = (1 + 2i) \begin{bmatrix} \xi_1 \\ \xi_2 \end{bmatrix}$$

or

$$\begin{cases} 3\xi_1 - 4\xi_2 = (1 + 2i)\xi_1 \\ 2\xi_1 - \xi_2 = (1 + 2i)\xi_2; \end{cases}$$

both equations reduce to $\xi_1 = (1 + i)\xi_2$, so $(1 + i, 1)$ is an eigenvector for $\lambda = 1 + 2i$. A similar computation can be made for $\bar{\lambda} = 1 - 2i$, but this is unnecessary: the conjugate vector $(\overline{1 + i}, \bar{1}) = (1 - i, 1)$ is an eigenvector for $\bar{\lambda}$. ∎

Once the (complex) eigenvectors have been found, the method of Section 3.4 can be used to find the general complex-valued solution to the system. That is, we can use the coordinate transformation $\mathbf{T} = \begin{bmatrix} \xi_1 & \bar{\xi}_1 \\ \xi_2 & \bar{\xi}_2 \end{bmatrix}$; in terms of the new coordinates given by $\begin{bmatrix} X \\ Y \end{bmatrix} = \mathbf{T}^{-1} \begin{bmatrix} x \\ y \end{bmatrix}$, the system will become

$$\begin{cases} \dfrac{dX}{dt} = \lambda X \\ \dfrac{dY}{dt} = \bar{\lambda} Y; \end{cases}$$

the new matrix will be $\mathbf{T}^{-1}\mathbf{AT} = \begin{bmatrix} \lambda & 0 \\ 0 & \bar{\lambda} \end{bmatrix}$. The general complex-valued solution is then given by $\begin{bmatrix} X \\ Y \end{bmatrix} = \begin{bmatrix} Ce^{\lambda t} \\ De^{\bar{\lambda} t} \end{bmatrix}$, where C and D are *complex* constants; in terms of the original coordinates, we have

$$\begin{bmatrix} x \\ y \end{bmatrix} = \mathbf{T} \begin{bmatrix} X \\ Y \end{bmatrix} = \begin{bmatrix} \xi_1 & \bar{\xi}_1 \\ \xi_2 & \bar{\xi}_2 \end{bmatrix} \begin{bmatrix} Ce^{\lambda t} \\ De^{\bar{\lambda} t} \end{bmatrix}$$

or

$$\begin{cases} x = \xi_1 Ce^{\lambda t} + \bar{\xi}_1 De^{\bar{\lambda} t} \\ y = \xi_2 Ce^{\lambda t} + \bar{\xi}_2 De^{\bar{\lambda} t}. \end{cases}$$

EXAMPLE 3.5.3 Consider the system $\dfrac{dx}{dt} = 3x - 4y$, $\dfrac{dy}{dt} = 2x - y$. For the matrix $\mathbf{A} = \begin{bmatrix} 3 & -4 \\ 2 & -1 \end{bmatrix}$, we found $\lambda = 1 + 2i$ and $(\xi_1, \xi_2) = (1 + i, 1)$ in Example 3.5.2. Therefore, the general complex-valued solution to the system will be given by

$$\begin{bmatrix} x \\ y \end{bmatrix} = \begin{bmatrix} 1+i & 1-i \\ 1 & 1 \end{bmatrix} \begin{bmatrix} Ce^{(1+2i)t} \\ De^{(1-2i)t} \end{bmatrix}.$$

That is,

$$x = (1+i)Ce^{(1+2i)t} + (1-i)De^{(1-2i)t}, \qquad y = Ce^{(1+2i)t} + De^{(1-2i)t},$$

where C and D are arbitrary complex constants. □

Among the complex-valued solutions, we should look for the real-valued ones; these are the ones with $x = \bar{x}$, $y = \bar{y}$. It can be shown that they are also precisely the ones for which the constants C and D are related by $D = \bar{C}$ (see Exercise 27). For these solutions we have

$$x = \xi_1 Ce^{\lambda t} + \bar{\xi}_1 \overline{C} e^{\bar{\lambda} t} = 2 \operatorname{Re}(\xi_1 Ce^{\lambda t}), \quad \text{and similarly} \quad y = 2 \operatorname{Re}(\xi_2 Ce^{\lambda t}).$$

Remember that ξ_1, ξ_2, and C are all *complex* numbers; since C is an arbitrary complex constant, we can replace C by $\dfrac{1}{2}C$. This will simplify our formulas slightly, to

$$x = \operatorname{Re}(\xi_1 Ce^{\lambda t}), \qquad y = \operatorname{Re}(\xi_2 Ce^{\lambda t}).$$

If we write

$$e^{\lambda t} = e^{(\mu+i\nu)t} = e^{\mu t}e^{i\nu t} = e^{\mu t}(\cos \nu t + i \sin \nu t),$$

we see that the functions x and y will be linear combinations of $e^{\mu t} \cos \nu t$ and $e^{\mu t} \sin \nu t$.

EXAMPLE 3.5.3 (continued) For the system $\dfrac{dx}{dt} = 3x - 4y$, $\dfrac{dy}{dt} = 2x - y$ considered above, we get

$$x = \operatorname{Re}((1+i)Ce^{(1+2i)t}), \qquad y = \operatorname{Re}(Ce^{(1+2i)t}).$$

To get more explicit expressions for x and y, we can set $C = C_1 + iC_2$. We then have

$$\begin{aligned}x &= \text{Re}[(1+i)(C_1 + iC_2)e^{(1+2i)t}]\\&= \text{Re}[(1+i)(C_1 + iC_2)e^t(\cos 2t + i \sin 2t)]\\&= \text{Re}[e^t((C_1 - C_2) + i(C_1 + C_2))(\cos 2t + i \sin 2t)]\\&= e^t[(C_1 - C_2)\cos 2t - (C_1 + C_2)\sin 2t]\end{aligned}$$

and

$$\begin{aligned}y &= \text{Re}[(C_1 + iC_2)e^t(\cos 2t + i \sin 2t)]\\&= e^t(C_1 \cos 2t - C_2 \sin 2t).\end{aligned}$$

So the general (real-valued) solution to the system is

$$\begin{cases} x = e^t[(C_1 - C_2)\cos 2t - (C_1 + C_2)\sin 2t] \\ y = e^t(C_1 \cos 2t - C_2 \sin 2t), \end{cases}$$

C_1 and C_2 arbitrary (real) constants.

NOTE: As in Section 3.4, every solution is a linear combination of two basic solutions, namely

$$(x, y) = (e^t(\cos 2t - \sin 2t), e^t \cos 2t) \quad (\text{for } C_1 = 1, C_2 = 0)$$

and

$$(x, y) = (e^t(\cos 2t + \sin 2t), e^t \sin 2t) \quad (\text{for } C_1 = 0, C_2 = -1).$$

See Exercise 30. ∎

EXAMPLE 3.5.4 In Example 3.2.1 we considered the system $\dfrac{dx}{dt} = -y$, $\dfrac{dy}{dt} = x$. We now solve this system, using complex eigenvectors.

The matrix $\mathbf{A} = \begin{bmatrix} 0 & -1 \\ 1 & 0 \end{bmatrix}$ has eigenvalues $\lambda = i$, $\bar\lambda = -i$; an eigenvector for λ is $(\xi_1, \xi_2) = (1, -i)$ (see Exercise 2). Therefore, writing $C = C_1 + iC_2$, we obtain

$$\begin{aligned}x &= \text{Re}(\xi_1 Ce^{\lambda t}) = \text{Re}[(C_1 + iC_2)e^{it}]\\&= C_1 \cos t - C_2 \sin t,\\y &= \text{Re}(\xi_2 Ce^{\lambda t}) = \text{Re}[-i(C_1 + iC_2)e^{it}]\\&= \text{Re}[(C_2 - iC_1)e^{it}]\\&= C_1 \sin t + C_2 \cos t.\end{aligned}$$

Note that

$$\begin{aligned}x^2 + y^2 &= (C_1 \cos t - C_2 \sin t)^2 + (C_1 \sin t + C_2 \cos t)^2\\&= C_1^2 + C_2^2 = \text{constant},\end{aligned}$$

which agrees with our earlier result that the trajectories of this particular system are circles with center at the origin. ∎

SEC. 3.5 LINEAR SYSTEMS (CONTINUED): DOUBLE OR COMPLEX CONJUGATE EIGENVALUES

EXAMPLE 3.5.5 Solve the system

$$\frac{dx}{dt} = x + 5y, \qquad \frac{dy}{dt} = -2x + 3y.$$

Here we have $\mathbf{A} = \begin{bmatrix} 1 & 5 \\ -2 & 3 \end{bmatrix}$, so the characteristic equation is $(1 - \lambda)(3 - \lambda) + 10 = 0$ or $\lambda^2 - 4\lambda + 13 = 0$, and we find $\lambda, \overline{\lambda} = 2 \pm 3i$. To find an eigenvector for $\lambda = 2 + 3i$, we look for (ξ_1, ξ_2) with

$$\begin{cases} \xi_1 + 5\xi_2 = (2 + 3i)\xi_1 \\ -2\xi_1 + 3\xi_2 = (2 + 3i)\xi_2. \end{cases}$$

When we try to simplify these equations we get

$$(-1 - 3i)\xi_1 + 5\xi_2 = 0, \quad -2\xi_1 + (1 - 3i)\xi_2 = 0,$$

which may look disconcerting at first. However, these last two equations really are equivalent to each other, since $(-1 - 3i)\xi_1 + 5\xi_2 = 0$ yields

$$\xi_1 = \frac{5}{1 + 3i}\xi_2 = \frac{5(1 - 3i)}{(1 + 3i)(1 - 3i)}\xi_2 = \frac{1 - 3i}{2}\xi_2, \quad {}^{23}$$

as does $-2\xi_1 + (1 - 3i)\xi_2 = 0$. To avoid fractions, we can, for instance, take $\xi_2 = 2$ and get $(\xi_1, \xi_2) = (1 - 3i, 2)$ as one eigenvector for λ. Finally, we get

$$x = \text{Re}[(1 - 3i)(C_1 + iC_2)e^{(2+3i)t}], \qquad y = \text{Re}[2(C_1 + iC_2)e^{(2+3i)t}];$$

see Exercise 10 for the rest. ∎

EXAMPLE 3.5.6 Solve the initial value problem

$$\frac{dx}{dt} = -5x - 2y, \quad \frac{dy}{dt} = 18x + 7y, \quad x(0) = -2, \quad y(0) = 11.$$

This time, $\mathbf{A} = \begin{bmatrix} -5 & -2 \\ 18 & 7 \end{bmatrix}$ has characteristic equation $\lambda^2 - 2\lambda + 1 = 0$, so we have a double eigenvalue $\lambda = 1$. The eigenvectors are the nonzero multiples of $(1, -3)$ (see Exercise 9), so we can use any transformation matrix $\mathbf{T}$ with first column $\begin{bmatrix} 1 \\ -3 \end{bmatrix}$. Let's take $\mathbf{T} = \begin{bmatrix} 1 & 0 \\ -3 & 1 \end{bmatrix}$. We then find $\mathbf{T}^{-1} = \begin{bmatrix} 1 & 0 \\ 3 & 1 \end{bmatrix}$ and

$$\mathbf{T}^{-1}\mathbf{A}\mathbf{T} = \begin{bmatrix} 1 & 0 \\ 3 & 1 \end{bmatrix}\begin{bmatrix} -5 & -2 \\ 18 & 7 \end{bmatrix}\begin{bmatrix} 1 & 0 \\ -3 & 1 \end{bmatrix} = \begin{bmatrix} 1 & -2 \\ 0 & 1 \end{bmatrix}.$$

---
[23] For details on arithmetic with complex numbers, see Appendix A, Section A.1.

Thus for the new coordinates X, Y with $\begin{bmatrix} X \\ Y \end{bmatrix} = T^{-1} \begin{bmatrix} x \\ y \end{bmatrix}$, we have

$$\frac{dX}{dt} = X - 2Y, \qquad \frac{dY}{dt} = Y.$$

The second of these equations yields $Y = Ce^t$, and the first equation then becomes

$$\frac{dX}{dt} = X - 2Ce^t.$$

The homogeneous equation $\frac{dX}{dt} = X$ has general solution $X = Ae^t$, so we now put $X = A(t)e^t$ (variation of constants) and obtain

$$\frac{dA}{dt} e^t + A(t)e^t = A(t)e^t - 2Ce^t$$

$$\frac{dA}{dt} = -2C$$

$$A(t) = -2Ct + D,$$

from which we get $X = (-2Ct + D)e^t$. Now we go back to the old coordinates:

$$\begin{bmatrix} x \\ y \end{bmatrix} = \mathbf{T} \begin{bmatrix} X \\ Y \end{bmatrix} = \begin{bmatrix} 1 & 0 \\ -3 & 1 \end{bmatrix} \begin{bmatrix} (-2Ct + D)e^t \\ Ce^t \end{bmatrix},$$

in other words,

$$x = (-2Ct + D)e^t, \qquad y = (6Ct + C - 3D)e^t.$$

In particular, $x(0) = D$, $y(0) = C - 3D$, so from the given initial conditions we have $D = -2$, $C - 3D = 11$. That is, $C = 5$, $D = -2$, and the final answer is

$$x = (-10t - 2)e^t, \qquad y = (30t + 11)e^t. \quad \blacksquare$$

SUMMARY OF KEY RESULTS AND TECHNIQUES

Solving $\frac{d}{dt}\begin{bmatrix} x \\ y \end{bmatrix} = \mathbf{A}\begin{bmatrix} x \\ y \end{bmatrix}$ using eigenvectors (continued): If the characteristic equation has a double root λ (a **double eigenvalue**), we can use any invertible transformation matrix $\mathbf{T}$ whose first column is an eigenvector for λ to get $\mathbf{T}^{-1}\mathbf{A}\mathbf{T} = \begin{bmatrix} \lambda & \sigma \\ 0 & \lambda \end{bmatrix}$. We can then solve the new system $\frac{dX}{dt} = \lambda X + \sigma Y$, $\frac{dY}{dt} = \lambda Y$ by noting that $Y = Ce^{\lambda t}$ and then using variation of constants for the first equation (p. 201).

SEC. 3.5 LINEAR SYSTEMS (CONTINUED): DOUBLE OR COMPLEX CONJUGATE EIGENVALUES

If the characteristic equation has complex roots $\lambda, \bar{\lambda} = \mu \pm i\nu$, we can get the general complex-valued solution as in Section 3.4 (using complex eigenvectors) (pp. 202–203). If (ξ_1, ξ_2) is a complex eigenvector for λ, then the general *real-valued* solution is given by $x = \operatorname{Re}(\xi_1 C e^{\lambda t})$, $y = \operatorname{Re}(\xi_2 C e^{\lambda t})$, where $C = C_1 + i C_2$ is an arbitrary complex constant (p. 203).

EXERCISES

Find all (real or complex) eigenvalues and eigenvectors of the following matrices.

1. $\begin{bmatrix} 0 & 1 \\ 0 & 0 \end{bmatrix}$
2. $\begin{bmatrix} 0 & -1 \\ 1 & 0 \end{bmatrix}$
3. $\begin{bmatrix} 0 & 3 \\ -1 & 0 \end{bmatrix}$
4. $\begin{bmatrix} 13 & 0 \\ 0 & 13 \end{bmatrix}$
5. $\begin{bmatrix} 1 & 1 \\ -3 & 1 \end{bmatrix}$
6. $\begin{bmatrix} -1 & -5 \\ 2 & -3 \end{bmatrix}$
7. $\begin{bmatrix} 5 & 1 \\ -1 & 3 \end{bmatrix}$
8. $\begin{bmatrix} -1 & \frac{1}{2} \\ -\frac{1}{2} & 0 \end{bmatrix}$
9. $\begin{bmatrix} -1 & -2 \\ 2 & 3 \end{bmatrix}$

10. Finish Example 3.5.5.

Use eigenvectors to solve the following systems of differential equations and initial value problems.

11. $\dfrac{dx}{dt} = 5x - y$, $\dfrac{dy}{dt} = x + 3y$.

12. $\dfrac{dx}{dt} = -2x + y$, $\dfrac{dy}{dt} = -x - 4y$.

13. $\dfrac{dx}{dt} = 2y$, $\dfrac{dy}{dt} = -8x$.

14. $\dfrac{dx}{dt} = -x + 3y$, $\dfrac{dy}{dt} = -3x - y$, $x(0) = 5$, $y(0) = -1$.

15. $\dfrac{dx}{dt} = -x + \dfrac{1}{2}y$, $\dfrac{dy}{dt} = -\dfrac{1}{2}x$, $x(0) = 2$, $y(0) = 4$.

16. $\dfrac{dx}{dt} = y$, $\dfrac{dy}{dt} = 0$.

17. $\dfrac{dx}{dt} = x - 5y$, $\dfrac{dy}{dt} = x + 3y + 8$.
[*Hint:* First change coordinates so that the stationary point is at the origin.]

18. $\dfrac{dx}{dt} = -3x + 4y + 6, \quad \dfrac{dy}{dt} = -2x - 7y + 4.$

19. $\dfrac{dx}{dt} = 7x + 3y + 6, \quad \dfrac{dy}{dt} = -3x + y + 2.$

20. $\dfrac{dx}{dt} = x - 2y, \quad \dfrac{dy}{dt} = 2x - 3y + 1.$

21. $\dfrac{dx}{dt} = -x + 4y, \quad \dfrac{dy}{dt} = -5x + 3y, \quad x(0) = 0, \quad y(0) = 5.$

22. $\dfrac{dx}{dt} = 2x + 2y, \quad \dfrac{dy}{dt} = -x + 4y, \quad x(0) = 1, \quad y(0) = 3.$

23. $\dfrac{dx}{dt} = -3y, \quad \dfrac{dy}{dt} = 3x.$

24. Solve the equation $\dfrac{dX}{dt} = -2X + Ce^{-2t}$ (see p. 201).

25. Show that the system $\dfrac{dX}{dt} = \lambda X + \sigma Y, \quad \dfrac{dY}{dt} = \lambda Y$ has general solution
$X = (\sigma Ct + D)e^{\lambda t}, \ Y = Ce^{\lambda t}, \ C, D$ arbitrary.

26. Let $\mathbf{A} = \begin{bmatrix} a & b \\ c & d \end{bmatrix}$ be a matrix with real entries that has conjugate eigenvalues λ and $\bar{\lambda}$. Show that if (ξ_1, ξ_2) is an eigenvector for λ, then $(\bar{\xi}_1, \bar{\xi}_2)$ is an eigenvector for $\bar{\lambda}$.

27. (a) Show that if $D = \bar{C}$, then the solution
$$x = \xi_1 Ce^{\lambda t} + \bar{\xi}_1 De^{\bar{\lambda} t}, \ y = \xi_2 Ce^{\lambda t} + \bar{\xi}_2 De^{\bar{\lambda} t}$$
is real-valued.

*(b) Show, conversely, that if $x = \xi_1 Ce^{\lambda t} + \bar{\xi}_1 De^{\bar{\lambda} t}$ is real-valued, then $D = \bar{C}$ or $\xi_1 = 0$.

*(c) Show that if the solution in part (a) is real-valued, then $D = \bar{C}$.
[*Hint*: Is it possible that both $\xi_1 = 0$ and $\xi_2 = 0$?]

*28. (a) Show that if (P, R) is an eigenvector of the matrix $\mathbf{A} = \begin{bmatrix} a & b \\ c & d \end{bmatrix}$ for the eigenvalue λ, and if (Q, S) is any other vector such that $\begin{bmatrix} P & Q \\ R & S \end{bmatrix} = \mathbf{T}$ has an inverse, then $\mathbf{T}^{-1}\mathbf{AT}$ has the form $\begin{bmatrix} \lambda & \cdots \\ 0 & \cdots \end{bmatrix}$, where dots stand for unspecified entries.

[*Hint*: Find $\mathbf{T}^{-1}\mathbf{AT}\begin{bmatrix} 1 \\ 0 \end{bmatrix}$.]

*(b) Now assume that **A** has a *double* eigenvalue $\lambda \neq 0$. Show that in the situation of part (a), $\mathbf{T}^{-1}\mathbf{A}\mathbf{T}$ has the form $\begin{bmatrix} \lambda & \sigma \\ 0 & \lambda \end{bmatrix}$.

[*Hint*: Consider $\det(\mathbf{T}^{-1}\mathbf{A}\mathbf{T})$ and use the idea of Exercise 26 from Section 3.4.]

NOTE: This result is still valid for $\lambda = 0$, but it is slightly harder to prove in that case.

*29. Show that if $\mathbf{A} = \begin{bmatrix} a & b \\ c & d \end{bmatrix}$ has a double eigenvalue λ and more than one eigenvector (up to multiplication by a nonzero constant), then $\mathbf{A} = \begin{bmatrix} \lambda & 0 \\ 0 & \lambda \end{bmatrix}$.

30. (a) Show that the real basic solutions we found in the Note after Example 3.5.3 (on p. 204) are the real and imaginary parts of $e^{(1+2i)t}(1+i, 1)$, the special complex-valued solution obtained by multiplying the eigenvector for $\lambda = 1 + 2i$ by $e^{\lambda t}$.

*(b) Show that whenever $\mathbf{A} = \begin{bmatrix} a & b \\ c & d \end{bmatrix}$ has complex eigenvalues $\lambda, \bar{\lambda} = \mu \pm i\nu$ (with $\nu > 0$), with (ξ_1, ξ_2) an eigenvector for λ, then $\operatorname{Re}[e^{\lambda t}(\xi_1, \xi_2)]$ and $\operatorname{Im}[e^{\lambda t}(\xi_1, \xi_2)]$ form a pair of basic solutions for the system $\dfrac{dx}{dt} = ax + by$, $\dfrac{dy}{dt} = cx + dy$.

3.6 PHASE PORTRAITS FOR LINEAR SYSTEMS; TYPES OF STATIONARY POINTS

In this section and the next, we return (as announced in Section 3.2) to the study of the qualitative behavior of an autonomous system near a stationary point. We'll assume that the system is in the form

$$\begin{cases} \dfrac{dx}{dt} = f(x, y) \\ \dfrac{dy}{dt} = g(x, y), \end{cases} \quad (*)$$

with C^1 functions f and g, and that (x_0, y_0) is a stationary point, in other words, that $f(x_0, y_0) = g(x_0, y_0) = 0$. In this situation the system has as its **linear approximation** (see p. 167) for (x, y) near (x_0, y_0):

$$\begin{cases} \dfrac{dx}{dt} = a(x - x_0) + b(y - y_0) \\ \dfrac{dy}{dt} = c(x - x_0) + d(y - y_0), \end{cases} \quad (**)$$

where we have used the abbreviations

$$a = \frac{\partial f}{\partial x}(x_0, y_0), \quad b = \frac{\partial f}{\partial y}(x_0, y_0), \quad c = \frac{\partial g}{\partial x}(x_0, y_0), \quad d = \frac{\partial g}{\partial y}(x_0, y_0).$$

As we have seen in Sections 3.3 to 3.5, the linear system (**) can be solved by first changing the unknown functions to $x - x_0$, $y - y_0$ and then solving the resulting homogeneous system, whether by the elimination method or by using one more change of coordinates. Although the solutions we get in this way usually won't be solutions to our original system (*) [since (**) is only an approximation to (*)], their qualitative behavior will in most cases provide insight into the behavior of solutions to (*). Accordingly, we will start by systematically studying the phase portraits of linear systems [such as (**)]. While doing so, we'll assume that the stationary point has been shifted to the origin, so that the system has the form

$$\frac{d}{dt}\begin{bmatrix} x \\ y \end{bmatrix} = \mathbf{A}\begin{bmatrix} x \\ y \end{bmatrix}, \text{ with } \mathbf{A} = \begin{bmatrix} a & b \\ c & d \end{bmatrix}.$$

Two such phase portraits were already drawn in Section 3.4 (see Figures 3.4.2 and 3.4.3). In each of these cases, the matrix $\mathbf{A}$ had two distinct real eigenvalues λ_1 and λ_2. If this is so and we choose eigenvectors (P, R) and (Q, S) for λ_1 and λ_2 respectively, as was done in Section 3.4, then we'll get $X = Ce^{\lambda_1 t}$, $Y = De^{\lambda_2 t}$ for the new coordinates X, Y such that $\begin{bmatrix} x \\ y \end{bmatrix} = \begin{bmatrix} P & Q \\ R & S \end{bmatrix}\begin{bmatrix} X \\ Y \end{bmatrix}$. In particular, for $D = 0$ there will be trajectories along the X-axis (the positive X-axis if $C > 0$, etc.), while for $C = 0$ there will be trajectories along the Y-axis. The directions of travel for these trajectories will depend on the signs of λ_1 (for

SEC. 3.6 PHASE PORTRAITS FOR LINEAR SYSTEMS; TYPES OF STATIONARY POINTS

$D = 0$) and λ_2, respectively; they will be toward the origin for negative eigenvalues and away from the origin for positive eigenvalues. All other trajectories follow curves, whose equations can be found by eliminating t from $X = Ce^{\lambda_1 t}$, $Y = De^{\lambda_2 t}$ as follows:

$$\frac{X}{C} = e^{\lambda_1 t}, \quad \frac{Y}{D} = e^{\lambda_2 t},$$

so

$$\left(\frac{X}{C}\right)^{\lambda_2} = e^{\lambda_1 \lambda_2 t} = \left(\frac{Y}{D}\right)^{\lambda_1}. \tag{1}$$

If we want to simplify this further, there is a slight technical complication: we may not be able to write $\left(\dfrac{X}{C}\right)^{\lambda_2} = \dfrac{X^{\lambda_2}}{C^{\lambda_2}}$, since X^{λ_2} may not be defined. (If λ_2 is not an integer, then $X^{\lambda_2} = e^{\lambda_2 \log X}$ is usually defined for $X > 0$ only.) The easiest way around this is to look first at the first quadrant of the X,Y-plane, that is, at the region where $X > 0$, $Y > 0$. The trajectories in the other quadrants can eventually be found from the trajectories in the first quadrant by reflections in the X- and Y-axes (see Exercise 23).

In the rest of this discussion we'll assume where necessary that $X > 0$, $Y > 0$. Equation (1) then becomes

$$\frac{X^{\lambda_2}}{C^{\lambda_2}} = \frac{Y^{\lambda_1}}{D^{\lambda_1}},$$

which gives us

$$Y^{\lambda_1} = \frac{D^{\lambda_1}}{C^{\lambda_2}} X^{\lambda_2}$$

or (except if $\lambda_1 = 0$)

$$Y = KX^{\lambda_2/\lambda_1}, \quad \text{where} \quad K = \frac{D}{C^{\lambda_2/\lambda_1}} \text{ is a new constant}.$$

The appearance of these curves and the directions of travel along them (and along their reflections in the other quadrants) will depend on the signs and relative size of λ_1 and λ_2, and we'll now list the various possible cases. Remember that we are assuming, for now, that the eigenvalues λ_1 and λ_2 are *real* and *distinct*.

Case I $\lambda_1 > 0$, $\lambda_2 > 0$.

Subcase 1: $\lambda_1 < \lambda_2$. Since λ_1 and λ_2 are positive, the trajectories "start" at the stationary point (that is, $X = Ce^{\lambda_1 t}$, $Y = De^{\lambda_2 t}$ both approach zero as $t \to -\infty$) and go to infinity as $t \to \infty$. Since $\dfrac{\lambda_2}{\lambda_1} > 1$, Y goes to infinity "faster" than X does. Also, the trajectories all "leave" the stationary point in the X-direction, except for the special trajectories along the Y-axis (see Exercise 24). Figure 3.6.1(i) (p. 213) shows the shape of the "phase portrait" in the new coordinates.

NOTE: Figure 3.4.2(a) is a special case of this.

Subcase 2: $\lambda_1 > \lambda_2$. This is just subcase 1 with the roles of X and Y reversed. See Figure 3.6.1(ii).

Case II $\lambda_1 > 0$, $\lambda_2 < 0$. In this case $\dfrac{\lambda_2}{\lambda_1}$ is negative and the curves $Y = KX^{\lambda_2/\lambda_1}$ are "generalized hyperbolas" (ordinary hyperbolas if $\lambda_2/\lambda_1 = -1$). Except for the special trajectories along the axes, $X = Ce^{\lambda_1 t}$ goes to infinity as $t \to \infty$ while Y goes to infinity as $t \to -\infty$. See Figure 3.6.1(iii).

NOTE: Figure 3.4.3(a) is a special case of this.

Case III $\lambda_1 < 0$, $\lambda_2 > 0$. This is case II with the roles of X and Y reversed. See Figure 3.6.1(iv).

Case IV $\lambda_1 < 0$, $\lambda_2 < 0$. In this case $\dfrac{\lambda_2}{\lambda_1}$ is positive, and since $\dfrac{\lambda_2}{\lambda_1} = \dfrac{-\lambda_2}{-\lambda_1}$, the trajectories follow the same curves as if the eigenvalues were $-\lambda_1$ and $-\lambda_2$ (this would be Case I, since $-\lambda_1 > 0$ and $-\lambda_2 > 0$). However, the direction of travel is now *toward* the stationary point. There are again two subcases, which are shown in Figure 3.6.1(v) and (vi). We can also see the connection between Case IV and Case I by noting that $Ce^{\lambda_1 t} = Ce^{(-\lambda_1)(-t)}$, $De^{\lambda_2 t} = De^{(-\lambda_2)(-t)}$. Thus if we replace λ_1 by $-\lambda_1$ and λ_2 by $-\lambda_2$, we'll get the same solutions, except with t replaced by $-t$. This is known as **time reversal**; the trajectories follow the same curves, but in the opposite direction.[24] Similarly, Case III can be obtained by time reversal from Case II.

Case V $\lambda_1 = 0$, $\lambda_2 > 0$. In this case we have $X = C$, $Y = De^{\lambda_2 t}$, so X is constant and Y goes to infinity as $t \to \infty$, provided that $D \neq 0$. For $D = 0$ we find infinitely many *constant* solutions, that is, stationary points; the line $Y = 0$ consists entirely of stationary points. See Figure 3.6.1(vii).

Case VI $\lambda_1 = 0$, $\lambda_2 < 0$. As Case V, except that $Y \to 0$ as $t \to \infty$; see Figure 3.6.1(viii). (This case can be obtained by time reversal from Case V.)

Case VII $\lambda_1 > 0$, $\lambda_2 = 0$. Case V with X, Y reversed; see Figure 3.6.1(ix).

Case VIII $\lambda_1 < 0$, $\lambda_2 = 0$. Case VI with X, Y reversed; see Figure 3.6.1(x).

Note that to get the actual phase portraits corresponding to Figure 3.6.1, the location of the eigenvectors (in other words, of the new coordinate axes) in the x,y-plane has to be taken into account in each case, as was done in Figures 3.4.2(c) and 3.4.3(b). The actual phase portraits will therefore have "distorted" versions of the shapes in Figure 3.6.1. Also, if the stationary point was originally not at the origin, the actual phase portrait will be shifted. (We'll return to this below; see Examples 3.6.1 and 3.6.3.)

---

[24] For a more general case of time reversal, see Section 3.2, Exercise 61.

SEC. 3.6 PHASE PORTRAITS FOR LINEAR SYSTEMS; TYPES OF STATIONARY POINTS 213

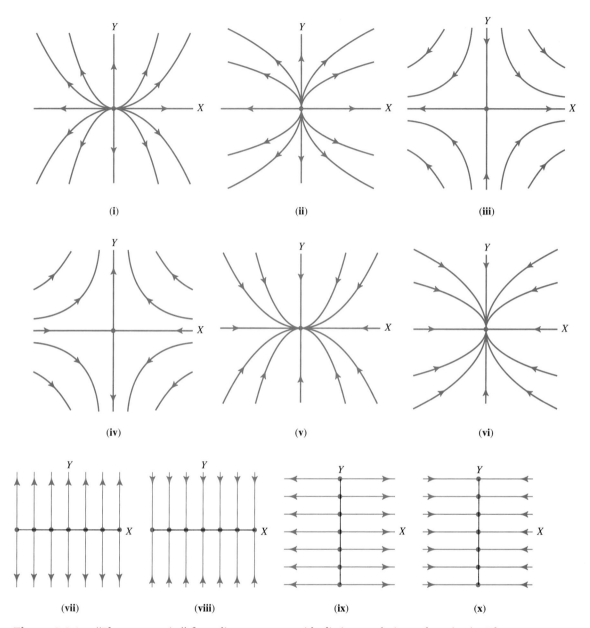

Figure 3.6.1. "Phase portraits" for a linear system with distinct real eigenvalues λ_1, λ_2. The new coordinates have been drawn along perpendicular axes in each case.
(i) $\lambda_1 > 0$, $\lambda_2 > 0$; $\lambda_1 < \lambda_2$. (ii) $\lambda_1 > 0$, $\lambda_2 > 0$; $\lambda_1 > \lambda_2$. (iii) $\lambda_1 > 0$, $\lambda_2 < 0$. (iv) $\lambda_1 < 0$, $\lambda_2 > 0$.
(v) $\lambda_1 < 0$, $\lambda_2 < 0$; $|\lambda_1| < |\lambda_2|$. (vi) $\lambda_1 < 0$, $\lambda_2 < 0$; $|\lambda_1| > |\lambda_2|$. (vii) $\lambda_1 = 0$, $\lambda_2 > 0$.
(viii) $\lambda_1 = 0$, $\lambda_2 < 0$. (ix) $\lambda_1 > 0$, $\lambda_2 = 0$. (x) $\lambda_1 < 0$, $\lambda_2 = 0$.

Similar configurations of the trajectories near a stationary point will also frequently occur for nonlinear systems. Indeed, stationary points are classified by the appearance of the nearby trajectories. Unfortunately, the names used in this classification vary from one author to the next! *We* will call a stationary point a **stable node** if the trajectories nearby[25] look like (a distorted version of) Figure 3.6.1(v) or (vi). An **unstable node** corresponds to Figure 3.6.1(i) or (ii), while a stationary point is called a **saddle point** if the trajectories behave as in Figure 3.6.1(iii) or (iv). Note that for a stable node all nearby trajectories approach the stationary point as $t \to \infty$; for an unstable node they are traversed in the other direction, so they "originate at" the stationary point (as $t \to -\infty$). For a saddle point most trajectories leave the vicinity of the stationary point as $t \to \infty$, but two special trajectories approach the stationary point as $t \to \infty$. We won't give any special names to the configurations in Figure 3.6.1(vii) to (x); in practice these occur less often, since it is relatively unusual that an eigenvalue will be exactly zero. More on the classification of stationary points below—so far we've only done the case that λ_1, λ_2 are real and distinct!

Now suppose that the matrix $\mathbf{A} = \begin{bmatrix} a & b \\ c & d \end{bmatrix}$ has a *double* eigenvalue λ. We have seen in Section 3.5 that such a matrix usually has only "one" eigenvector (up to constant multiples), and that after a change of coordinates the system will have the form

$$\frac{dX}{dt} = \lambda X + \sigma Y, \qquad \frac{dY}{dt} = \lambda Y.$$

This yields $Y = Ce^{\lambda t}$, $X = (\sigma Ct + D)e^{\lambda t}$ (see Section 3.5, Exercise 25). Because of the term with $te^{\lambda t}$ that occurs in X, it is now a little more difficult both to find equations for the trajectories and to graph them (see Exercise 25 of this section). The resulting "phase portraits" (with the new coordinates drawn along perpendicular axes) are shown in Figure 3.6.2. Note the special case $\sigma = 0$; this is the exceptional case in which $\mathbf{A} = \begin{bmatrix} \lambda & 0 \\ 0 & \lambda \end{bmatrix}$ and any nonzero vector is an eigenvector. Accordingly, the trajectories *all* lie along straight lines in this case; in fact, if $Y = Ce^{\lambda t}$, $X = De^{\lambda t}$, then $DY = CX$.

A stationary point will be called an **improper node** if the nearby trajectories look like Figure 3.6.2(i), (iii), (vii), or (ix). [The improper node is *stable* for Figure 3.6.2(vii) and (ix), *unstable* for Figure 3.6.2(i) and (iii).] The configuration in Figure 3.6.2(viii) is called a **stable node**, along with the ones (for distinct eigenvalues) in Figure 3.6.1(v) and (vi) (on p. 213). This is not as confusing as it looks, because Figure 3.6.2(viii) is what one might intuitively expect as the transition between Figure 3.6.1(v) and Figure 3.6.1(vi). Similarly, Figure 3.6.2(ii) shows an **unstable node**; it can be thought of as a transition from Figure 3.6.1(i) to Figure 3.6.1(ii). No special names are given to the remaining cases in Figure 3.6.2.

Finally, we have to consider the case in which $\mathbf{A} = \begin{bmatrix} a & b \\ c & d \end{bmatrix}$ has (complex) conjugate eigenvalues $\lambda = \mu + i\nu$, $\bar{\lambda} = \mu - i\nu$ with corresponding eigenvectors (ξ_1, ξ_2), $(\bar{\xi}_1, \bar{\xi}_2)$. In

---

[25] We include the word *nearby* so that the definition will remain valid for nonlinear systems—to be studied in the next section.

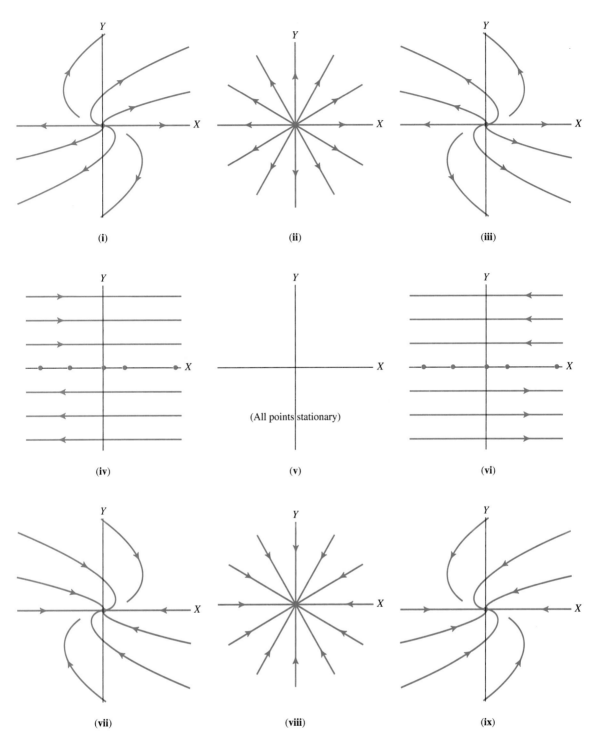

Figure 3.6.2. "Phase portraits" for a linear system with a double eigenvalue λ. The new coordinates X, Y for which $\dfrac{dX}{dt} = \lambda X + \sigma Y$, $\dfrac{dY}{dt} = \lambda Y$ have been drawn along perpendicular axes.
(i) $\lambda > 0$, $\sigma > 0$. (ii) $\lambda > 0$, $\sigma = 0$. (iii) $\lambda > 0$, $\sigma < 0$. (iv) $\lambda = 0$, $\sigma > 0$. (v) $\lambda = 0$, $\sigma = 0$.
(vi) $\lambda = 0$, $\sigma < 0$. (vii) $\lambda < 0$, $\sigma > 0$. (viii) $\lambda < 0$, $\sigma = 0$. (ix) $\lambda < 0$, $\sigma < 0$.

this case it looks at first sight as if the new coordinates $\begin{bmatrix} X \\ Y \end{bmatrix} = \mathbf{T}^{-1} \begin{bmatrix} x \\ y \end{bmatrix}$ introduced in Section 3.5 (where $\mathbf{T} = \begin{bmatrix} \xi_1 & \bar{\xi}_1 \\ \xi_2 & \bar{\xi}_2 \end{bmatrix}$; see p. 202) can't be used to draw the phase portrait, since X, Y are complex (not real) even for real-valued solutions x, y. On the other hand, if we give up on X, Y entirely and try to eliminate t from x, y in order to find equations for the trajectories, messy computations result. [Compare Example 3.5.3, pp. 203–204, where we found $x = e^t((C_1 - C_2)\cos 2t - (C_1 + C_2)\sin 2t)$, $y = e^t(C_1 \cos 2t - C_2 \sin 2t)$ — try eliminating t!]

The trick is to change X and Y into real-valued functions by taking their real and imaginary parts. This looks like it might produce four unrelated functions $\operatorname{Re}(X), \operatorname{Im}(X), \operatorname{Re}(Y),$ and $\operatorname{Im}(Y)$. However, we know from Section 3.5 that the real-valued solutions are those for which $X = Ce^{\lambda t}$, $Y = \overline{C}e^{\bar{\lambda} t}$.[26] Therefore, X and Y are complex conjugates, so $\operatorname{Re}(Y) = \operatorname{Re}(X), \operatorname{Im}(Y) = -\operatorname{Im}(X)$, and

$$\begin{bmatrix} X \\ Y \end{bmatrix} = \begin{bmatrix} \operatorname{Re}(X) + i\operatorname{Im}(X) \\ \operatorname{Re}(X) - i\operatorname{Im}(X) \end{bmatrix} = \begin{bmatrix} 1 & i \\ 1 & -i \end{bmatrix} \begin{bmatrix} \operatorname{Re}(X) \\ \operatorname{Im}(X) \end{bmatrix}$$

for real-valued solutions x, y. This will enable us to rewrite the system

$$\frac{dX}{dt} = \lambda X, \qquad \frac{dY}{dt} = \bar{\lambda} Y$$

in terms of $\operatorname{Re}(X)$ and $\operatorname{Im}(X)$; if we put

$$X_1 = \operatorname{Re}(X), \qquad X_2 = \operatorname{Im}(X)$$

for short, the result will be

$$\frac{dX_1}{dt} = \mu X_1 - \nu X_2, \qquad \frac{dX_2}{dt} = \nu X_1 + \mu X_2$$

(see Exercise 26; reminder: μ and ν are the real and imaginary parts of the eigenvalue λ, and $\nu > 0$).

EXAMPLES A. In Example 3.5.3, the original system

$$\frac{dx}{dt} = 3x - 4y, \qquad \frac{dy}{dt} = 2x - y$$

can be rewritten first as $\dfrac{dX}{dt} = (1 + 2i)X, \quad \dfrac{dY}{dt} = (1 - 2i)Y$ and then as

$$\frac{dX_1}{dt} = X_1 - 2X_2, \qquad \frac{dX_2}{dt} = 2X_1 + X_2.$$

---

[26] Recall that on p. 203 we found that $X = Ce^{\lambda t}$, $Y = De^{\bar{\lambda} t}$, and $D = \overline{C}$ for real-valued solutions.

B. The system $\frac{dx}{dt} = -y$, $\frac{dy}{dt} = x$ from Example 3.5.4 was already in this form (with $\mu = 0$, $\nu = 1$). ∎

The solutions of our newest system

$$\frac{dX_1}{dt} = \mu X_1 - \nu X_2, \qquad \frac{dX_2}{dt} = \nu X_1 + \mu X_2$$

can be found directly, but it's easier to note that since $X_1 = \mathrm{Re}(X)$ and $X_2 = \mathrm{Im}(X)$, we have

$$X_1 = \mathrm{Re}(Ce^{\lambda t}) = \mathrm{Re}[(C_1 + iC_2)e^{\mu t}(\cos \nu t + i \sin \nu t)]$$
$$X_1 = e^{\mu t}(C_1 \cos \nu t - C_2 \sin \nu t)$$

and similarly,

$$X_2 = e^{\mu t}(C_1 \sin \nu t + C_2 \cos \nu t).$$

Although it looks almost as formidable to eliminate t from these functions as from x, y, it isn't really, because

$$X_1^2 + X_2^2 = e^{2\mu t}(C_1 \cos \nu t - C_2 \sin \nu t)^2 + e^{2\mu t}(C_1 \sin \nu t + C_2 \cos \nu t)^2$$
$$= (C_1^2 + C_2^2)e^{2\mu t}.$$

The rest of the elimination is discussed in Exercise 27, but we can already see at this point what the trajectories will look like. The easiest case is when $\mu = 0$; in this case we see that $X_1^2 + X_2^2 = C_1^2 + C_2^2$ is *constant* along the trajectories, so these are circles in the X_1, X_2-plane; see Figure 3.6.3(i), p. 218. (We had already seen this for Example 3.5.4, p. 204.) In the actual phase portrait, that is, in the x, y-plane, the trajectories may be "distorted" circles, that is, ellipses. Since ν is positive, the circles *in the X_1, X_2-plane* will be traversed counterclockwise (check this!). However, in the x, y-plane the ellipses may be traversed clockwise or counterclockwise, depending on the location of the positive X_1- and X_2-axes; see below.

If we pass from $\mu = 0$ to $\mu > 0$ (but leave ν the same), both X_1 and X_2 are multiplied by $e^{\mu t}$, which is a function that increases from zero (as $t \to -\infty$) to infinity (as $t \to \infty$). As a result, each trajectory in the X_1, X_2-plane, instead of circling the origin at a constant distance, circles the origin at an ever-increasing distance and spirals out to infinity. [See Figure 3.6.3(ii), where, to avoid cluttering the diagram, only one spiral trajectory has been drawn.] Again, the trajectories will, in general, be "distorted" to elliptical spirals in the x, y-plane; this is why we couldn't have found their behavior by looking at $x^2 + y^2$ instead of $X_1^2 + X_2^2$.

For $\mu < 0$ we again get spirals, this time spiraling in toward the stationary point (which is reached in the limit as $t \to \infty$). See Figure 3.6.3(iii).

In configurations such as Figure 3.6.3(i), the stationary point is called a **center**; a center is "surrounded" by trajectories that are closed curves.[27] Figure 3.6.3(ii) and (iii) show

---

[27] This definition is somewhat tricky to make precise for nonlinear systems, because not *all* the trajectories near the center need to be closed curves. (See "Further Reading" following the Exercises.)

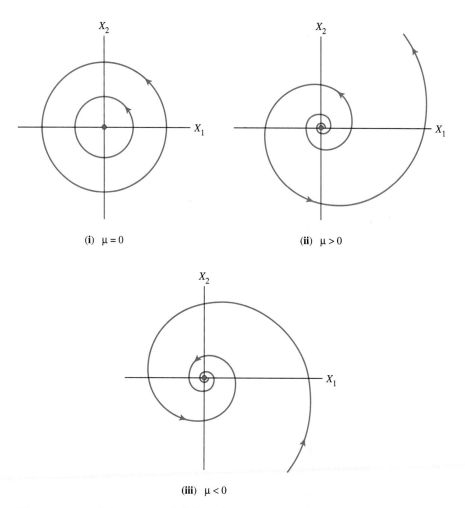

Figure 3.6.3. "Phase portraits" for a linear system with complex conjugate eigenvalues $\nu \pm i\nu$. $X_1 = \text{Re}(X)$, $X_2 = \text{Im}(X)$ have been drawn along perpendicular axes. In the actual phase portraits, the "distorted" circles may be traversed in either direction, and the spirals may turn in either direction.

spiral points. The spiral point in Figure 3.6.3(iii), for which all nearby trajectories approach the stationary point as $t \to \infty$, is called **stable**, while the spiral point in Figure 3.6.3(ii) is **unstable**.

Finally, to sketch the actual phase portrait in the x,y-plane, we have to find the X_1- and X_2-axes there. Combining

$$\begin{bmatrix} x \\ y \end{bmatrix} = \begin{bmatrix} \xi_1 & \bar{\xi}_1 \\ \xi_2 & \bar{\xi}_2 \end{bmatrix} \begin{bmatrix} X \\ Y \end{bmatrix} \text{ (from p. 203)} \quad \text{with} \quad \begin{bmatrix} X \\ Y \end{bmatrix} = \begin{bmatrix} 1 & i \\ 1 & -i \end{bmatrix} \begin{bmatrix} X_1 \\ X_2 \end{bmatrix} \text{ (from p. 216)},$$

SEC. 3.6 PHASE PORTRAITS FOR LINEAR SYSTEMS; TYPES OF STATIONARY POINTS

we get

$$\begin{bmatrix} x \\ y \end{bmatrix} = \begin{bmatrix} \xi_1 & \bar{\xi}_1 \\ \xi_2 & \bar{\xi}_2 \end{bmatrix} \begin{bmatrix} 1 & i \\ 1 & -i \end{bmatrix} \begin{bmatrix} X_1 \\ X_2 \end{bmatrix}$$
$$= \begin{bmatrix} \xi_1 + \bar{\xi}_1 & i(\xi_1 - \bar{\xi}_1) \\ \xi_2 + \bar{\xi}_2 & i(\xi_2 - \bar{\xi}_2) \end{bmatrix} \begin{bmatrix} X_1 \\ X_2 \end{bmatrix}$$
$$= \begin{bmatrix} 2\operatorname{Re}(\xi_1) & -2\operatorname{Im}(\xi_1) \\ 2\operatorname{Re}(\xi_2) & -2\operatorname{Im}(\xi_2) \end{bmatrix} \begin{bmatrix} X_1 \\ X_2 \end{bmatrix}.$$

In particular, for $X_1 = 1$, $X_2 = 0$ we get

$$(x, y) = (2\operatorname{Re}(\xi_1), 2\operatorname{Re}(\xi_2)),$$

so the positive X_1-axis is in the direction of $(\operatorname{Re}(\xi_1), \operatorname{Re}(\xi_2))$, *the real part of the eigenvector* (ξ_1, ξ_2). Similarly, we see (Exercise 21) that the positive X_2-axis is in the direction *opposite* to the imaginary part of the eigenvector.

EXAMPLE 3.6.1 Sketch the phase portrait of the linear system

$$\frac{dx}{dt} = -5x + 2y - 1, \qquad \frac{dy}{dt} = -20x + 7y - 6.$$

First of all, the stationary point occurs for $-5x + 2y - 1 = -20x + 7y - 6 = 0$, that is, at $x = -1$, $y = -2$, so we adopt $x + 1$ and $y + 2$ as new unknown functions and obtain the homogeneous system

$$\begin{cases} \dfrac{d(x+1)}{dt} = -5(x+1) + 2(y+2) \\ \dfrac{d(y+2)}{dt} = -20(x+1) + 7(y+2). \end{cases}$$

The eigenvalues of the matrix $\mathbf{A} = \begin{bmatrix} -5 & 2 \\ -20 & 7 \end{bmatrix}$ are $\lambda, \bar{\lambda} = 1 \pm 2i$, and an eigenvector for $\lambda = 1 + 2i$ is $(\xi_1, \xi_2) = (1, 3 + i)$. (Check these statements!)

Since $\lambda = 1 + 2i$, we have $\mu = 1$, $\nu = 2$, and the "phase portrait" in the X_1, X_2-plane looks like Figure 3.6.3(ii); the stationary point is an unstable spiral point. To find the actual phase portrait (in the x, y-plane), note that the vector $(\operatorname{Re}(\xi_1), \operatorname{Re}(\xi_2)) = (1, 3)$ gives the direction of the positive X_1-axis while $(\operatorname{Im}(\xi_1), \operatorname{Im}(\xi_2)) = (0, 1)$ does the same for the *negative* X_2-axis; the positive X_2-axis, then, is in the direction of $(0, -1)$. Since these axes go through the stationary point $(-1, -2)$, they are located as shown in Figure 3.6.4(a), p. 220. In particular, as shown in Figure 3.6.4(b), the trajectories, which are elliptical spirals in the x, y-plane, are traversed *clockwise* (from the positive X_1-direction to the positive X_2-direction). They are traversed *outward* since $\mu > 0$. ∎

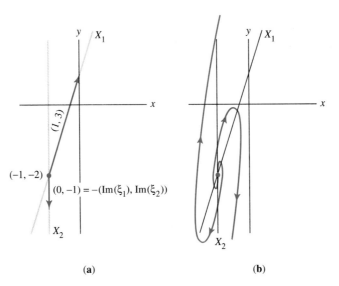

Figure 3.6.4. (a) Location of X_1- and X_2-axes. (b) Phase portrait for $\dfrac{dx}{dt} = -5x + 2y - 1$, $\dfrac{dy}{dt} = -20x + 7y - 6$.

EXAMPLE 3.6.2 Sketch the phase portrait of the system

$$\frac{dx}{dt} = -10x + 8y, \qquad \frac{dy}{dt} = -18x + 14y.$$

This time, the stationary point is at the origin. The matrix $\mathbf{A} = \begin{bmatrix} -10 & 8 \\ -18 & 14 \end{bmatrix}$ has a double eigenvalue $\lambda = 2$, and the "only" eigenvector is $(2, 3)$. (Check these statements!) We can take $\mathbf{T} = \begin{bmatrix} 2 & 0 \\ 3 & 1 \end{bmatrix}$, and we then find $\mathbf{T}^{-1} = \begin{bmatrix} \frac{1}{2} & 0 \\ -\frac{3}{2} & 1 \end{bmatrix}$, $\mathbf{T}^{-1}\mathbf{A}\mathbf{T} = \begin{bmatrix} 2 & 4 \\ 0 & 2 \end{bmatrix}$.

So we have new coordinates X, Y for which

$$\frac{dX}{dt} = 2X + 4Y, \qquad \frac{dY}{dt} = 2Y,$$

and we are in the case shown in Figure 3.6.2(i). To get the actual phase portrait in the x,y-plane, note that $\begin{bmatrix} x \\ y \end{bmatrix} = \begin{bmatrix} 2 & 0 \\ 3 & 1 \end{bmatrix}\begin{bmatrix} X \\ Y \end{bmatrix}$ and that the X- and Y-axes are in the directions given by $X = 1, Y = 0$ and $X = 0, Y = 1$, respectively. That is, the X-axis is in the direction of the eigenvector $(2, 3)$, while the Y-axis is in the direction of $(0, 1)$; see Figure

SEC. 3.6 PHASE PORTRAITS FOR LINEAR SYSTEMS; TYPES OF STATIONARY POINTS 221

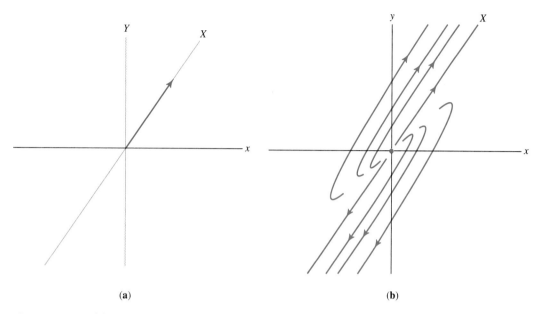

(a)   (b)

Figure 3.6.5. (a) Location of X- and Y-axes.
(b) Phase portrait for $\dfrac{dx}{dt} = -10x + 8y$, $\dfrac{dy}{dt} = -18x + 14y$.

3.6.5(a). Combining this information with Figure 3.6.2(i) yields the phase portrait in Figure 3.6.5(b).

NOTE: Once we know that $\lambda = 2$ is a double eigenvalue (and that we're not in the exceptional case where $\mathbf{A} = \begin{bmatrix} 2 & 0 \\ 0 & 2 \end{bmatrix}$), we know that we'll have the situation of either Figure 3.6.2(i) or Figure 3.6.2(iii). Also, the X-axis will be in the direction of the eigenvector $(2, 3)$. Instead of choosing $\mathbf{T}$, computing $\mathbf{T}^{-1}\mathbf{AT}$, and locating the Y-axis to decide in which direction the trajectories are bending, you could also decide between Figure 3.6.5(b) and Figure 3.6.6 by considering the direction field. For instance, along the x-axis we have $y = 0$, so $\dfrac{dx}{dt} = -10x$, $\dfrac{dy}{dt} = -18x$ there. Since the vectors $(-10x, -18x)$ along the positive x-axis point downward and to the left, Figure 3.6.5(b) must be correct, and Figure 3.6.6 (p. 222) is incorrect. ∎

EXAMPLE 3.6.3   Sketch the phase portrait of the system

$$\frac{dx}{dt} = -4x - y + 11, \qquad \frac{dy}{dt} = -4x - 4y + 8.$$

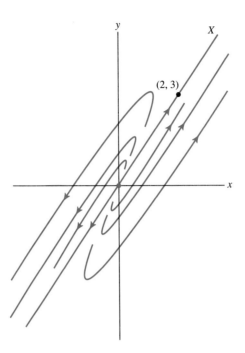

Figure 3.6.6. *Incorrect* phase portrait for $\dfrac{dx}{dt} = -10x + 8y$, $\dfrac{dy}{dt} = -18x + 14y$.

The stationary point is at $(3, -1)$. The matrix $\mathbf{A} = \begin{bmatrix} -4 & -1 \\ -4 & -4 \end{bmatrix}$ has eigenvalues -2 and -6, so we are in Case IV from p. 212 and we have a stable node. If we take the eigenvalues in the order $\lambda_1 = -2$, $\lambda_2 = -6$, then we have $|\lambda_1| < |\lambda_2|$, so the trajectories approach the stationary point from the X-direction[28] [see Figure 3.6.1(v)]. That direction is given by the eigenvector $(1, -2)$ for $\lambda_1 = -2$, while the Y-direction is given by $(1, 2)$. See Figure 3.6.7. ∎

EXAMPLE 3.6.4 Sketch the phase portrait of the system

$$\frac{dx}{dt} = 10x + 4y, \quad \frac{dy}{dt} = 15x + 6y.$$

The matrix $\mathbf{A} = \begin{bmatrix} 10 & 4 \\ 15 & 6 \end{bmatrix}$ has eigenvalues $\lambda_1 = 0$ and $\lambda_2 = 16$, so we are in the case shown in Figure 3.6.1(vii), and we just have to find "the" eigenvectors $(2, -5)$ and $(2, 3)$ for λ_1, λ_2, respectively. See Figure 3.6.8. ∎

---

[28] Except for the special trajectories along the Y-axis.

SEC. 3.6 PHASE PORTRAITS FOR LINEAR SYSTEMS; TYPES OF STATIONARY POINTS 223

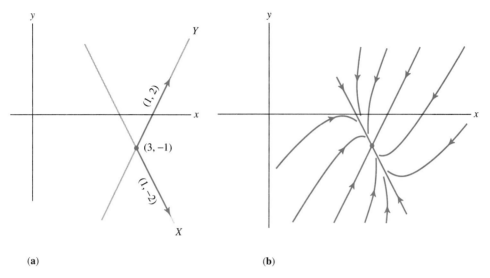

(a) **(b)**

Figure 3.6.7. (a) Location of X- and Y-axes.
(b) Phase portrait for $\dfrac{dx}{dt} = -4x - y + 11$, $\dfrac{dy}{dt} = -4x - 4y + 8$.

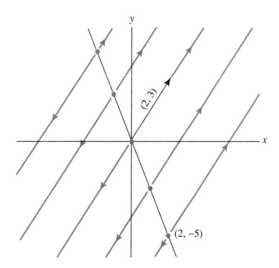

Figure 3.6.8. Phase portrait for
$\dfrac{dx}{dt} = 10x + 4y$, $\dfrac{dy}{dt} = 15x + 6y$.

EXAMPLE 3.6.5 Sketch the phase portrait of the system

$$\frac{dx}{dt} = -3x - y, \qquad \frac{dy}{dt} = 13x + y.$$

The matrix $\mathbf{A} = \begin{bmatrix} -3 & -1 \\ 13 & 1 \end{bmatrix}$ has eigenvalues $\lambda, \bar{\lambda} = -1 \pm 3i$, and as an eigenvector for $\lambda = -1 + 3i$, we find $(\xi_1, \xi_2) = (1, -2 - 3i)$. We are in the situation of Figure 3.6.3(iii) (since $\mu = -1 < 0$), so we have a stable spiral point. The spirals will turn from the positive X_1-axis to the positive X_2-axis[29]; the positive X_1-axis is in the direction of $(\text{Re}(\xi_1), \text{Re}(\xi_2)) = (1, -2)$, while the positive X_2-axis is in the direction of $-(\text{Im}(\xi_1), \text{Im}(\xi_2)) = (0, 3)$. See Figure 3.6.9. ∎

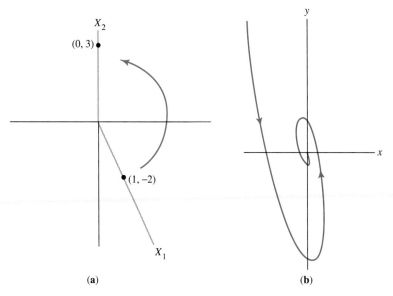

Figure 3.6.9. (a) Direction of spiraling.
(b) Phase portrait for $\dfrac{dx}{dt} = -3x - y$, $\dfrac{dy}{dt} = 13x + y$.

---

[29] Remember that for this to work, $\nu = \text{Im}(\lambda)$ must be positive, so we could *not* have chosen $\lambda = -1 - 3i$.

SEC. 3.6 PHASE PORTRAITS FOR LINEAR SYSTEMS; TYPES OF STATIONARY POINTS

SUMMARY OF KEY CONCEPTS, RESULTS, AND TECHNIQUES

For a linear system $\dfrac{d}{dt}\begin{bmatrix} x \\ y \end{bmatrix} = \mathbf{A}\begin{bmatrix} x \\ y \end{bmatrix}$, the following basic types of stationary points can occur (at the origin). In each case the diagram shows an "ideal" phase portrait, from which the actual phase portrait is obtained by "distortion" to take the true axis directions into account; more on this below. The eigenvalues of $\mathbf{A}$ are denoted by λ_1, λ_2 if they are real and by $\lambda = \mu + i\nu$, $\bar{\lambda} = \mu - i\nu$ (with $\nu > 0$) otherwise.

Unstable node:

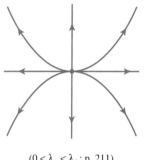 or 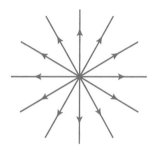 or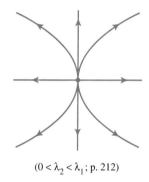

$(0 < \lambda_1 < \lambda_2;\ \text{p. 211})$ $\left(\mathbf{A} = \begin{bmatrix} \lambda_1 & 0 \\ 0 & \lambda_1 \end{bmatrix},\ \lambda_1 > 0;\ \text{p. 214}\right)$ $(0 < \lambda_2 < \lambda_1;\ \text{p. 212})$

Stable node:

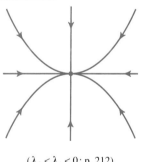 or 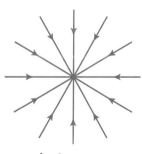 or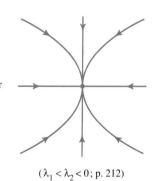

$(\lambda_2 < \lambda_1 < 0;\ \text{p. 212})$ $\left(\mathbf{A} = \begin{bmatrix} \lambda_1 & 0 \\ 0 & \lambda_1 \end{bmatrix},\ \lambda_1 < 0;\ \text{p. 214}\right)$ $(\lambda_1 < \lambda_2 < 0;\ \text{p. 212})$

Improper node (p. 214):

(unstable)

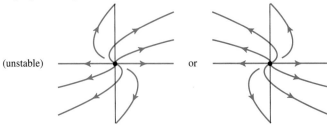

$\left(\lambda_1 = \lambda_2 > 0,\ \mathbf{A} \neq \begin{bmatrix} \lambda_1 & 0 \\ 0 & \lambda_1 \end{bmatrix}\right)$

(stable)

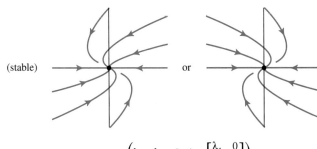

$\left(\lambda_1 = \lambda_2 < 0,\ \mathbf{A} \neq \begin{bmatrix} \lambda_1 & 0 \\ 0 & \lambda_1 \end{bmatrix}\right)$

Saddle point (p. 214):

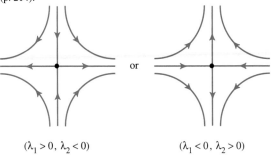

$(\lambda_1 > 0,\ \lambda_2 < 0)$ $(\lambda_1 < 0,\ \lambda_2 > 0)$

Spiral point (pp. 217–218):

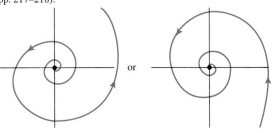

$(\lambda = \mu + i\nu,\ \mu > 0,\ \nu > 0,\ \text{unstable})$ $(\lambda = \mu + i\nu,\ \mu < 0,\ \nu > 0,\ \text{stable})$

SEC. 3.6 PHASE PORTRAITS FOR LINEAR SYSTEMS; TYPES OF STATIONARY POINTS

Center (p. 217):

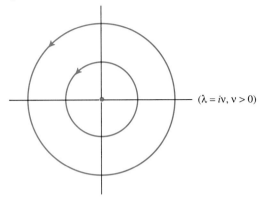

$(\lambda = i\nu, \nu > 0)$

Other cases:

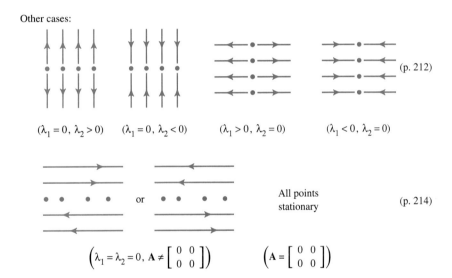

$(\lambda_1 = 0, \lambda_2 > 0)$ $(\lambda_1 = 0, \lambda_2 < 0)$ $(\lambda_1 > 0, \lambda_2 = 0)$ $(\lambda_1 < 0, \lambda_2 = 0)$ (p. 212)

or   All points stationary   (p. 214)

$\left(\lambda_1 = \lambda_2 = 0, \mathbf{A} \neq \begin{bmatrix} 0 & 0 \\ 0 & 0 \end{bmatrix}\right)$ $\left(\mathbf{A} = \begin{bmatrix} 0 & 0 \\ 0 & 0 \end{bmatrix}\right)$

In the diagrams above, when λ_1, λ_2 are real, the horizontal axis corresponds to λ_1; that is, in the actual phase portrait it has the direction of an eigenvector for λ_1. If also $\lambda_2 \neq \lambda_1$, then the vertical axis has the direction of an eigenvector for λ_2. If $\lambda = \mu + i\nu$ (with $\nu > 0$), the horizontal axis has the direction of the real part of an eigenvector for λ, while the vertical axis has the direction *opposite* the imaginary part. Depending on these directions, the elliptical spirals or ellipses forming the phase portrait in the case of a spiral point or center, respectively, may turn out to be described either counterclockwise (as in the "ideal" cases shown) or clockwise.

If an inhomogeneous linear system $\dfrac{dx}{dt} = ax + by + k_1$, $\dfrac{dy}{dt} = cx + dy + k_2$ has a stationary point at (x_0, y_0), then its phase portrait will be as above with $\mathbf{A} = \begin{bmatrix} a & b \\ c & d \end{bmatrix}$, but with the "origin" shifted to the stationary point.

EXERCISES

For each of the following linear systems:

(a) Indicate whether the stationary point is a node, a spiral point, an improper node, a center, a saddle point, or none of these. (If it is a node, a spiral point, or an improper node, indicate whether it is stable or unstable.)

(b) Use eigenvectors to sketch the phase portrait. Make sure to indicate the directions of travel along the trajectories.

1. $\dfrac{dx}{dt} = -5x + 2y + 3$, $\dfrac{dy}{dt} = -5x + y + 4$.

2. $\dfrac{dx}{dt} = -11x - 15y + 3$, $\dfrac{dy}{dt} = 10x + 14y - 2$.

3. $\dfrac{dx}{dt} = 12x - 5y - 19$, $\dfrac{dy}{dt} = 30x - 13y - 47$.

4. $\dfrac{dx}{dt} = 2y + 6$, $\dfrac{dy}{dt} = -5x + 6y - 2$.

5. $\dfrac{dx}{dt} = 3x - y$, $\dfrac{dy}{dt} = 4x - y + 2$.

6. $\dfrac{dx}{dt} = -3x - y + 5$, $\dfrac{dy}{dt} = 4x + y - 7$.

7. $\dfrac{dx}{dt} = 5x - 12y + 17$, $\dfrac{dy}{dt} = x - 2y + 3$.

8. $\dfrac{dx}{dt} = 3y - 6$, $\dfrac{dy}{dt} = -2x - 5y + 4$.

9. $\dfrac{dx}{dt} = 2x - 4y$, $\dfrac{dy}{dt} = x - 2y$.

10. $\dfrac{dx}{dt} = -2x + 8$, $\dfrac{dy}{dt} = -2y - 4$.

11. $\dfrac{dx}{dt} = 7x + 10y + 3$, $\dfrac{dy}{dt} = -5x - 7y + 1$.

12. $\dfrac{dx}{dt} = 5x - 2y + 2$, $\dfrac{dy}{dt} = 5x - y + 6$.

13. $\dfrac{dx}{dt} = -2y - 2$, $\dfrac{dy}{dt} = 5x - 6y - 1$.

14. $\dfrac{dx}{dt} = 9x - 18y$, $\dfrac{dy}{dt} = 3x - 6y$.

15. $\dfrac{dx}{dt} = -9x + 18y$, $\dfrac{dy}{dt} = -3x + 6y$.

For each of the following linear systems, use eigenvectors to (a) solve the system and (b) sketch the phase portrait.

16. $\dfrac{dx}{dt} = 8x - y - 3, \quad \dfrac{dy}{dt} = -4x + 8y + 24.$

17. $\dfrac{dx}{dt} = 2x + 4y, \quad \dfrac{dy}{dt} = x + 2y.$

18. $\dfrac{dx}{dt} = -6x + 2y + 14, \quad \dfrac{dy}{dt} = 2x - 6y - 10.$

19. $\dfrac{dx}{dt} = 3x - y - 7, \quad \dfrac{dy}{dt} = -x + 3y + 5.$

20. How are the answers to Exercises 18 and 19 related? Why did this happen?

21. Explain why, in the situation of p. 219, the positive X_2-axis is in the direction opposite to the imaginary part of the eigenvector.

22. Show that if the matrix $\mathbf{A}$ has distinct real eigenvalues λ_1, λ_2, then the system
$$\frac{d}{dt}\begin{bmatrix} x \\ y \end{bmatrix} = \mathbf{A}\begin{bmatrix} x \\ y \end{bmatrix}$$
has $Y^{\lambda_1} X^{-\lambda_2}$ as a first integral (for $X > 0$, $Y > 0$), where the new coordinates X, Y correspond to λ_1, λ_2, respectively.

23. Let $X = Ce^{\lambda_1 t}$, $Y = De^{\lambda_2 t}$ be the general solution, in the new coordinates, of a system with two distinct real eigenvalues λ_1 and λ_2 (see pp. 210–211).
    (a) Show that the trajectories lie in the various quadrants of the X, Y-plane as follows:

    $C > 0, D > 0$: first quadrant;
    $C < 0, D > 0$: second quadrant;
    $C < 0, D < 0$: third quadrant;
    $C > 0, D < 0$: fourth quadrant.

    (b) Show that the trajectories in the second quadrant can be found from those in the first quadrant by reflection in the Y-axis.
    [*Hint*: Don't eliminate t.]
    (c) Show that the trajectories in the third and fourth quadrants can be found from the others by reflection in the X-axis.

24. (a) Show that if K and L are constants with $L > 1$, then the graph of $Y = KX^L$ in the x, y-plane has a horizontal tangent line at $X = 0$.
    (b) Show that in Case I, Subcase 1, on p. 211, all trajectories that do not lie along the Y-axis "leave" the stationary point in the X-direction.

25. Consider the system $\dfrac{dX}{dt} = \lambda X + \sigma Y$, $\dfrac{dY}{dt} = \lambda Y$ (see p. 214), which has the general solution $Y = Ce^{\lambda t}$, $X = (\sigma Ct + D)e^{\lambda t}$.
    (a) Show that if $\lambda = 0$, all points along the X-axis are stationary points. Show that in this case the phase portrait has the form indicated in Figure 3.6.2(iv), (v), or (vi), according to whether $\sigma > 0$, $\sigma = 0$, or $\sigma < 0$.

(b) From now on, assume that $\lambda \neq 0$. Show that there are two trajectories along the X-axis besides the stationary point, and that they are traversed toward the stationary point if $\lambda < 0$ and away from the stationary point if $\lambda > 0$.

(c) Show that the trajectories with $Y > 0$ are of the form $X = \dfrac{\sigma}{\lambda} Y \log Y + BY$,

where $B = \dfrac{D}{C} - \dfrac{\sigma}{\lambda} \log C$, while the trajectories with $Y < 0$ are of the form

$X = \dfrac{\sigma}{\lambda} Y \log(-Y) + BY$, where $B = \dfrac{D}{C} - \dfrac{\sigma}{\lambda} \log(-C)$.

[*Hint*: Eliminate t by writing t in terms of Y.]

(d) Show that if $\sigma = 0$, the phase portrait has the form indicated in Figure 3.6.2(ii) or (viii), according to whether $\lambda > 0$ or $\lambda < 0$.

Now assume that $\sigma \neq 0$, so we can write

$$\begin{cases} X = \dfrac{\sigma}{\lambda}(Y \log Y + AY) & (Y > 0) \\ X = \dfrac{\sigma}{\lambda}(Y \log(-Y) + AY) & (Y < 0) \end{cases} \quad \text{with } A = \dfrac{\lambda}{\sigma} B.$$

(e) Sketch the graph of $X = Y \log Y + AY$ for $A = -1, 0, 1$. (Use graph-sketching techniques from calculus.)

(f) Show that the phase portrait has the form indicated in Figure 3.6.2(i), (iii), (vii), or (ix), depending on whether λ and σ are positive or negative.

26. In this exercise we'll rewrite the system $\dfrac{dX}{dt} = \lambda X$, $\dfrac{dY}{dt} = \bar{\lambda} Y$ (with $\lambda = \mu + i\nu$, $Y = \bar{X}$; see p. 216) in terms of $X_1 = \operatorname{Re}(X)$ and $X_2 = \operatorname{Im}(X)$.

(a) Find a matrix **B** such that $\begin{bmatrix} X_1 \\ X_2 \end{bmatrix} = \mathbf{B} \begin{bmatrix} X \\ Y \end{bmatrix}$.

[*Hint*: We know from p. 216 that $\begin{bmatrix} X \\ Y \end{bmatrix} = \begin{bmatrix} 1 & i \\ 1 & -i \end{bmatrix} \begin{bmatrix} X_1 \\ X_2 \end{bmatrix}$.]

(b) Explain why $\dfrac{d}{dt}\begin{bmatrix} X_1 \\ X_2 \end{bmatrix} = \mathbf{B} \begin{bmatrix} \lambda & 0 \\ 0 & \bar{\lambda} \end{bmatrix} \begin{bmatrix} 1 & i \\ 1 & -i \end{bmatrix} \begin{bmatrix} X_1 \\ X_2 \end{bmatrix}$.

(c) Show that $\dfrac{dX_1}{dt} = \mu X_1 - \nu X_2$, $\dfrac{dX_2}{dt} = \nu X_1 + \mu X_2$, as claimed on p. 216.

27. (See p. 217.)

(a) Show that t can be eliminated from the equations

$$\begin{cases} X_1 = e^{\mu t}(C_1 \cos \nu t - C_2 \sin \nu t) \\ X_2 = e^{\mu t}(C_1 \sin \nu t + C_2 \cos \nu t) \end{cases}$$

to get equations for the trajectories. (Don't try to simplify the result!)
The equations found in part (a) become more manageable if polar coordinates are used in the X_1, X_2-plane, which can be thought of as the complex plane (see Appendix A, p. 595, for a discussion of polar coordinates in the complex plane).

SEC. 3.6 PHASE PORTRAITS FOR LINEAR SYSTEMS; TYPES OF STATIONARY POINTS

Specifically, let
$$X = X_1 + iX_2 = R(\cos\phi + i\sin\phi) \quad \text{and} \quad C = C_1 + iC_2 = \rho(\cos\alpha + i\sin\alpha).$$
Note that $R = R(t)$ and $\phi = \phi(t)$ are new unknown functions with $R = \sqrt{X_1^2 + X_2^2}$, $\tan\phi = \dfrac{X_2}{X_1}$, while ρ and α are new constants. Now show the following:

(b) $R = \rho e^{\mu t}$.
(c) $X_1 = R\cos(\nu t + \alpha)$, $X_2 = R\sin(\nu t + \alpha)$.
(d) $\phi = \nu t + \alpha$ (up to multiples of 2π).
(e) $R = \rho e^{\mu(\phi - \alpha)/\nu}$. This is a polar equation of a logarithmic spiral in the X_1, X_2-plane (a circle, if $\mu = 0$).

28. Consider the inhomogeneous linear system $\dfrac{dx}{dt} = x$, $\dfrac{dy}{dt} = x + 1$.

    (a) Show that the system has *no* stationary points.
    (b) Find the general solution.
    (c) Sketch the phase portrait.

For each of the following linear systems, the coefficients depend on a parameter k. If you have a computer program available that will sketch the phase portrait of the system, have it do so for various values of k. You should see qualitatively different behavior for values of k in different ranges (for instance, the origin might be a spiral point for $k < -1$ and a node for $k > -1$). Try to find the values of k for which transitions occur (such as $k = -1$ in the hypothetical situation just mentioned), and describe what transitions you observe. Then use the methods of this section to show that your observations are correct.

29. $\dfrac{dx}{dt} = (10 + k)x + 17y$, $\dfrac{dy}{dt} = -5x + (k - 12)y$.

30. $\dfrac{dx}{dt} = (k - 3)x - 3y$, $\dfrac{dy}{dt} = 6x + (k + 6)y$.

31. $\dfrac{dx}{dt} = (3 - 4k)x + (16k - 9)y$, $\dfrac{dy}{dt} = (1 - k)x + (4k - 3)y$.

32. $\dfrac{dx}{dt} = (k + 4)x - (k + 2)y$, $\dfrac{dy}{dt} = (k + 8)x - (k + 4)y$.

33. $\dfrac{dx}{dt} = (5 - 2k)x + (5 - k)y$, $\dfrac{dy}{dt} = (4k - 9)x + (2k - 9)y$.

34. $\dfrac{dx}{dt} = kx - ky$, $\dfrac{dy}{dt} = (k - 7)x + (6 - k)y$.

FURTHER READING

For a precise definition of "center," see Sánchez, *Ordinary Differential Equations and Stability Theory: An Introduction* (Freeman, 1968), p. 86.

3.7 PHASE PORTRAITS FOR NONLINEAR SYSTEMS; STABILITY

For a nonlinear system of the form $\frac{dx}{dt} = f(x, y)$, $\frac{dy}{dt} = g(x, y)$, it will usually not be possible to find the general solution. Although an integrating factor for the equation $g(x, y)\, dx - f(x, y)\, dy = 0$ always exists, it is another matter to find such an integrating factor, or a first integral for the system. Thus it may not even be possible to find equations for the trajectories of the system. This means that the phase portrait can only be drawn approximately.

Of course, you could try to sketch the phase portrait by first plotting the direction field $(f(x, y), g(x, y))$ at various points (x, y), as discussed in Section 3.2 (compare Figure 3.2.2). This method is similar in spirit to that of sketching the graph of a function by plotting more or less arbitrary points. The disadvantage is the same—important features may not show up if the points happen not to have been chosen well. Even with the help of a computer, which makes it practicable to plot the direction field at hundreds or even thousands of points, key features of the phase portrait can be missed if the ranges of x and y are not chosen well. Naturally, if you don't know what to expect, you may not make good choices!

A more systematic approach starts by finding the stationary points and then examining the behavior near each individual stationary point, using the linear approximation to the system there. In fact, in applications one is often concerned only with the behavior near one particular stationary point (which represents a desired equilibrium situation). The bad news[30] is that in exceptional cases, the actual phase portrait near a stationary point can be quite unlike that of the linear approximation.

EXAMPLE 3.7.1 The system $\frac{dx}{dt} = x^2$, $\frac{dy}{dt} = y$ has a stationary point at the origin; the linear approximation there to the system is $\frac{dx}{dt} = 0$, $\frac{dy}{dt} = y$. This linear system has matrix $\mathbf{A} = \begin{bmatrix} 0 & 0 \\ 0 & 1 \end{bmatrix}$, with eigenvalues $\lambda = 0$ and $\lambda = 1$; as we saw in Section 3.6, its phase portrait has the form shown in Figure 3.7.1(a). (The "old" coordinates x and y are also the "new" coordinates, since $\mathbf{A}$ is a diagonal matrix.) On the other hand, we can actually solve the original system (see Exercise 11) and we then find that its phase portrait is as shown in Figure 3.7.1(b). Clearly, the two phase portraits are quite dissimilar. In particular, the phase portrait of the linear approximation shows a whole line of stationary points, due to $\lambda = 0$ being an eigenvalue for $\mathbf{A}$. ∎

As was pointed out on p. 214, the case in which an eigenvalue is zero is relatively rare. Since it creates some complications (as seen in the example above; see also Exercises 12 to 14), we'll avoid this situation from now on. That is, when we consider the nonlinear system

$$\frac{dx}{dt} = f(x, y), \quad \frac{dy}{dt} = g(x, y)$$

---
[30] Good news will follow. See p. 234.

SEC. 3.7 PHASE PORTRAITS FOR NONLINEAR SYSTEMS; STABILITY

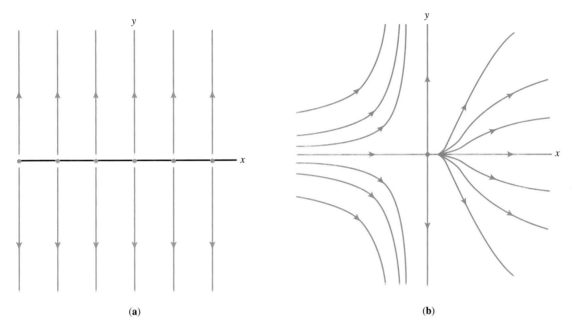

Figure 3.7.1. Phase portraits for (a) the linear approximation $\frac{dx}{dt} = 0$, $\frac{dy}{dt} = y$; (b) the system $\frac{dx}{dt} = x^2$, $\frac{dy}{dt} = y$.

and its linear approximation[31]

$$\frac{dx}{dt} = a(x - x_0) + b(y - y_0), \quad \frac{dy}{dt} = c(x - x_0) + d(y - y_0)$$

near a stationary point (x_0, y_0), we'll assume that the matrix $\mathbf{A} = \begin{bmatrix} a & b \\ c & d \end{bmatrix}$ has no eigenvalue zero. In other words (see Exercise 15), we'll assume that $\det \mathbf{A} = ad - bc \neq 0$.

NOTE: Other textbooks often impose what seem to be additional restrictions, in particular:

A. A restriction to ensure that the linear approximation near (x_0, y_0) is a "good" one.

B. The restriction that the stationary point (x_0, y_0) must be **isolated**, that is, that there must be some circle around this point which does not contain any other stationary points of the system.

However, it can be shown that under our blanket assumption (see p. 156) that f and g are C^1 functions, restriction A is automatic, while B follows from our assumption that $ad - bc \neq 0$.

---
[31] As usual, we'll use the abbreviations $a = \frac{\partial f}{\partial x}(x_0, y_0)$, $b = \frac{\partial f}{\partial y}(x_0, y_0)$, and so on; see p. 167.

Even with the restriction $ad - bc \neq 0$, it's still possible for the phase portrait of a system near a stationary point to be essentially different from that of the linear approximation to the system, as the next example shows.

EXAMPLE 3.7.2 Consider the three systems

$$\frac{dx}{dt} = -2y, \quad \frac{dy}{dt} = 2x; \tag{1}$$

$$\frac{dx}{dt} = -2y + x^3, \quad \frac{dy}{dt} = 2x + y^3; \tag{2}$$

$$\frac{dx}{dt} = -2y - x^3, \quad \frac{dy}{dt} = 2x - y^3. \tag{3}$$

Each of these systems has a stationary point at the origin; system (1), which is linear, is the linear approximation to all three systems there. This linear system has as its matrix $\mathbf{A} = \begin{bmatrix} 0 & -2 \\ 2 & 0 \end{bmatrix}$, with eigenvalues $\lambda = \pm 2i$; as we saw in Section 3.6, its phase portrait has a center at the origin [Figure 3.7.2(1)]. On the other hand, we'll see later (Section 3.8, Example 3.8.1 and Exercise 21) that near the origin, the trajectories of system (2) spiral outward from the origin, while the trajectories of (3) spiral inward toward the origin [Figure 3.7.2(2) and (3)]. Apparently, the behavior of the trajectories in these cases depends on the presence and sign of the nonlinear terms x^3, y^3. ∎

From the point of view of many applications, the behavior of system (2) in the example above is particularly disconcerting. If the stationary point represents a desirable "equilibrium" situation, then a slight departure from this situation will cause the "moving point" to follow a trajectory spiraling outward and thus the system will move farther and farther from equilibrium—the stationary point is "unstable." On the other hand, the linear approximation (1) shows a "stable" stationary point: A slight departure from the origin here will put the moving point on a trajectory that circles closely around the origin. System (3) shows even "better" behavior: Not only is the stationary point at the origin stable, but any slight departure from the origin is "corrected" in the long run as the moving point spirals back in and approaches the origin as $t \to \infty$. The stationary point for system (3) is said to be "asymptotically stable."

Clearly, if a system with an unstable stationary point and a system with an asymptotically stable stationary point have the same linear approximation [like systems (2) and (3) in the example], this linear approximation is almost useless in predicting the behavior of the system. The good news (finally!) is that, assuming that $ad - bc \neq 0$, this can *only* happen if the linear approximation has a center. In all other cases (if the linear approximation has a proper or improper node, a saddle point, or a spiral point) the stability of the stationary point for the system will be the same as for its linear approximation.

To make this precise, we'll have to give the long-awaited definitions of *stable* and *unstable*, as well as a definition of *asymptotically stable*. These definitions are rather

SEC. 3.7 PHASE PORTRAITS FOR NONLINEAR SYSTEMS; STABILITY

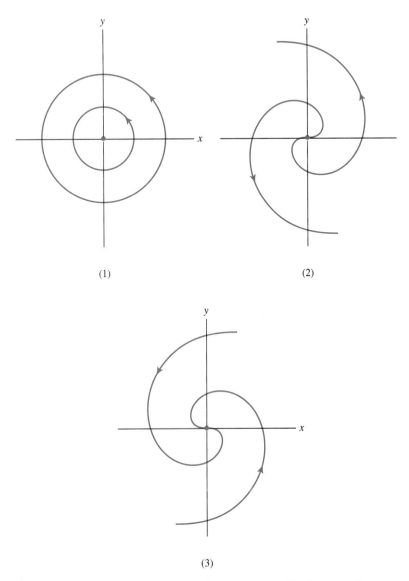

Figure 3.7.2. Phase portraits for three systems with the same linear approximation $\dfrac{dx}{dt} = -2y$, $\dfrac{dy}{dt} = 2x$ near their common stationary point $(0, 0)$.

subtle (which is why they've been postponed so far). Perhaps the best way to appreciate them is to try to devise them oneself. For instance, how should we define "stable" for a stationary point P? We could try starting with the idea that "once a trajectory is close enough to P, it stays close enough." "Close enough" can be interpreted as "inside a circle of small enough radius" or, equivalently, "within a small enough distance." Would

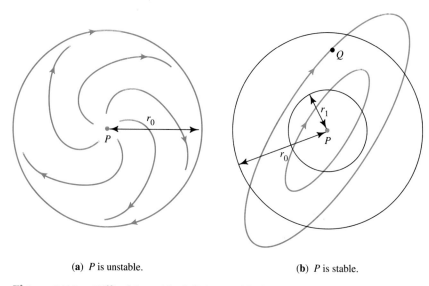

(a) *P* is unstable.   (b) *P* is stable.

Figure 3.7.3. Difficulties with defining *stable* for a stationary point *P*.

it be possible for any given distance to be "small enough"? Figure 3.7.3(a) shows a problem: In the situation of this figure, any trajectory through a point inside the circle of radius r_0 with center at P will remain inside that circle, yet the trajectories spiral outward from P and we wouldn't want to call this stationary point stable. To resolve this problem, we might notice that the trajectories *don't* stay inside smaller circles around P. We could then try insisting that for P to be stable, any trajectory through a point inside *any* circle with center at P should stay inside that circle. There would once again be a problem, though, as shown in Figure 3.7.3(b). In this figure, P is a center (the system could even be linear, in which case the trajectories would be ellipses with P at their center; see p. 217), and thus P should be stable according to any successful definition of stability. However, the figure shows that although Q is inside the circle of radius r_0 and center P, the trajectory through Q does not stay inside that circle. Note that we can, nevertheless, arrange for a (roughly) elliptical trajectory to stay inside the circle of radius r_0 by insisting that the trajectory go through a point inside a *smaller* circle (of radius r_1, shown in the figure) around P. This is the key idea behind the precise definition, which is as follows.

Definition The stationary point P is called **stable** if for every radius $r_0 > 0$ we can find another (possibly smaller) radius $r_1 > 0$ such that any trajectory that passes through a point inside the circle with center P and radius r_1 will thereafter stay inside the circle with center P and radius r_0. A stationary point that is not stable is called **unstable**.

As mentioned above, we will also give a definition of *asymptotically stable* for a stationary point P. The idea here is that P should be stable and that trajectories which start close enough to P should approach P as $t \to \infty$. More precisely:

SEC. 3.7 PHASE PORTRAITS FOR NONLINEAR SYSTEMS; STABILITY

Definition The stationary point P is called **asymptotically stable**[32] if it is stable *and* if, also, there exists a radius $r > 0$ such that every trajectory which passes through a point inside the circle with center P and radius r will approach P as $t \to \infty$.

N O T E : Although it may seem plausible that the existence of such a radius $r > 0$ would automatically imply that P is stable, this is not true; for an example, see Birkhoff and Rota, *Ordinary Differential Equations*, 4th ed. (Wiley, 1989), p. 130. (The terminology used in this reference is different from ours; Birkhoff and Rota refer to their example as "asymptotically stable" but not "strictly stable.")

Now let's see what these definitions give us in the case of a *linear* system

$$\begin{cases} \dfrac{dx}{dt} = a(x - x_0) + b(y - y_0) \\ \dfrac{dy}{dt} = c(x - x_0) + d(y - y_0) \end{cases} \quad \text{with } ad - bc \neq 0.$$

If the eigenvalues λ_1, λ_2 of the matrix $\mathbf{A} = \begin{bmatrix} a & b \\ c & d \end{bmatrix}$ are real and distinct, Figure 3.6.1 (p. 213) shows the possible phase portraits. Since $ad - bc \neq 0$, λ_1 and λ_2 are nonzero (see Exercise 15), so we only have to consider cases (i) to (vi) of Figure 3.6.1, which are redrawn in Figure 3.7.4(i) to (vi) (on p. 238) for convenience. It's easy to see that in cases (i) to (iv) the stationary point is unstable, because one can start arbitrarily close to the stationary point and follow a trajectory that goes out to infinity along one of the "new" axes. These cases are precisely the ones in which at least one of the eigenvalues λ_1, λ_2 is positive. In cases (v) and (vi), on the other hand, all trajectories lead "inward"[33] toward the stationary point, which is therefore stable and even asymptotically stable.[34] These are the cases in which both eigenvalues are negative.

Turning to the case of a double eigenvalue $\lambda \neq 0$, we see from Figure 3.6.2 (p. 215) that the pattern is similar: The stationary point is unstable if $\lambda > 0$ and asymptotically stable if $\lambda < 0$. Finally, we have the case in which $\mathbf{A}$ has complex conjugate eigenvalues $\lambda = \mu + i\nu$, $\overline{\lambda} = \mu - i\nu$. Figure 3.6.3 (p. 218) illustrates that the stationary point is unstable if $\mu > 0$ and asymptotically stable if $\mu < 0$. In the "transitional" case $\mu = 0$, the stationary point is a center, which is stable but not asymptotically stable.

Note that in all cases, the signs of the real parts of the eigenvalues (which are the signs of the eigenvalues themselves if these are real) determine stability. We can summarize our results for linear systems as follows.

---

[32] Sometimes "strictly stable." See the Note below.

[33] The word *inward* should not be taken too literally. See Exercise 17.

[34] Although we now have precise definitions of stability and asymptotic stability, we will use pictorial arguments rather than rigorous proofs when checking whether a stationary point has these properties.

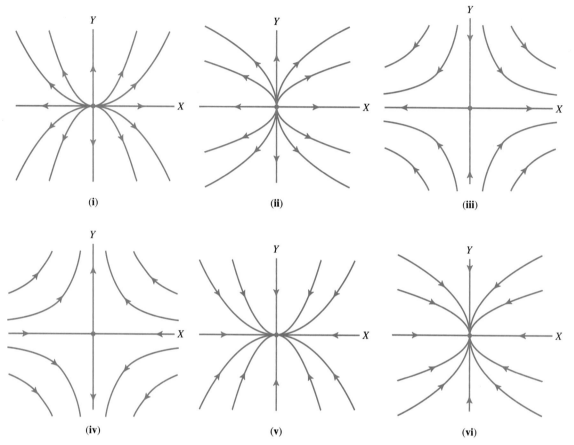

Figure 3.7.4. "Phase portraits" for a linear system with distinct, real, nonzero eigenvalues λ_1, λ_2. The new coordinates have been drawn along perpendicular axes.
(i) $\lambda_1 > 0, \lambda_2 > 0$; $\lambda_1 < \lambda_2$. (ii) $\lambda_1 > 0, \lambda_2 > 0$; $\lambda_1 > \lambda_2$. (iii) $\lambda_1 > 0, \lambda_2 < 0$.
(iv) $\lambda_1 < 0, \lambda_2 > 0$. (v) $\lambda_1 < 0, \lambda_2 < 0$; $|\lambda_1| < |\lambda_2|$. (vi) $\lambda_1 < 0, \lambda_2 < 0$; $|\lambda_1| > |\lambda_2|$.

Theorem 3.7.1 (**Stability for Linear Systems**). The stationary point (x_0, y_0) of the system

$$\frac{dx}{dt} = a(x - x_0) + b(y - y_0), \quad \frac{dy}{dt} = c(x - x_0) + d(y - y_0) \quad \text{with} \quad ad - bc \neq 0$$

is asymptotically stable if the real parts of the eigenvalues of $\mathbf{A} = \begin{bmatrix} a & b \\ c & d \end{bmatrix}$ are both negative, and is unstable if at least one of those real parts is positive. If the eigenvalues are purely imaginary, the linear system has a center and the stationary point is stable, but not asymptotically stable.

As announced earlier, for nonlinear systems stability of a stationary point will be determined by the stability of the linear approximation to the system there, provided that $ad - bc \neq 0$ and that the linear approximation does not have a center. Thus we have:

SEC. 3.7 PHASE PORTRAITS FOR NONLINEAR SYSTEMS; STABILITY

Theorem 3.7.2 (**Stability for Nonlinear Systems**). Let (x_0, y_0) be a stationary point of the autonomous system

$$\frac{dx}{dt} = f(x, y), \quad \frac{dy}{dt} = g(x, y)$$

and let

$$\frac{dx}{dt} = a(x - x_0) + b(y - y_0), \quad \frac{dy}{dt} = c(x - x_0) + d(y - y_0)$$

be the linear approximation to the system near (x_0, y_0). Assume that $ad - bc \neq 0$. Then the given stationary point is asymptotically stable if the real parts of the eigenvalues of $\mathbf{A} = \begin{bmatrix} a & b \\ c & d \end{bmatrix}$ are both negative and is unstable if at least one of those real parts is positive.

EXAMPLE 3.7.3 Consider the system

$$\begin{cases} \dfrac{dx}{dt} = -y + (x - y)(x^2 + y^2 - 1) \\ \dfrac{dy}{dt} = x + (x + y)(x^2 + y^2 - 1). \end{cases}$$

A stationary point of this system occurs whenever both

$$-y + (x - y)(x^2 + y^2 - 1) = 0 \quad \text{and} \quad x + (x + y)(x^2 + y^2 - 1) = 0.$$

Although there is a general method, the method of **resultants**, by which one can in principle solve simultaneous polynomial equations such as the above (see ''Further Reading'' following the Exercises), we won't present this method. Often a bit of ingenuity will provide a faster way, while if the equations are not polynomial equations, one has to fall back on ingenuity anyway. In our case we can write

$$y = (x - y)(x^2 + y^2 - 1) \quad \text{and} \quad x = -(x + y)(x^2 + y^2 - 1),$$

then ''cross-multiply'' to get

$$-y(x + y)(x^2 + y^2 - 1) = x(x - y)(x^2 + y^2 - 1).$$

We now distinguish two cases: Either $x^2 + y^2 - 1 = 0$, or we can divide by $x^2 + y^2 - 1$ to obtain $-y(x + y) = x(x - y)$. If $x^2 + y^2 - 1 = 0$, our original equations yield $y = 0$ and $x = 0$, respectively, so $x^2 + y^2 - 1 = -1$, a contradiction. This leaves us with the case $-y(x + y) = x(x - y)$, which simplifies to $-y^2 = x^2$. However, this is only possible for $x = y = 0$. (Why?) Thus the origin, which is indeed a stationary point of our system, is the only one.

The linear approximation to the system near $(0, 0)$ is $\dfrac{dx}{dt} = -x$, $\dfrac{dy}{dt} = -y$ [see Exercise 18(a)]. Since both eigenvalues of $\mathbf{A} = \begin{bmatrix} -1 & 0 \\ 0 & -1 \end{bmatrix}$ are negative (in fact, there is a double eigenvalue $\lambda = -1$), we can use our theorem to conclude that the origin is an asymptoti-

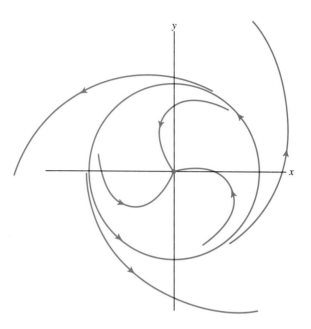

Figure 3.7.5. Phase portrait of the system
$$\frac{dx}{dt} = -y + (x - y)(x^2 + y^2 - 1),$$
$$\frac{dy}{dt} = x + (x + y)(x^2 + y^2 - 1).$$

cally stable stationary point. However, note that this does *not* imply that *all* trajectories approach the origin as $t \to \infty$, only that trajectories which start close enough to the origin do. In fact, the phase portrait shows a closed trajectory, or **cycle**, the circle $x^2 + y^2 = 1$ (see Figure 3.7.5). All trajectories inside the cycle approach the origin as $t \to \infty$, while trajectories outside the cycle spiral outward to infinity [see Exercise 18(d)]. ■

In general, the region of the plane in which trajectories approach a particular asymptotically stable stationary point is called the **attractor basin** of that point. In the example above, the attractor basin of $(0, 0)$ is the region inside the circle $x^2 + y^2 = 1$, that is, the open disk given by the inequality $x^2 + y^2 < 1$. While finding the attractor basin, or at least getting some idea of the extent of this region, is quite important in applications, the linear approximation is no help for this. In fact, the attractor basin for the stationary point of the *linear approximation* will (assuming the point to be asymptotically stable) always be the whole plane. The method of Lyapunov functions, which will be presented in the next section, can sometimes help determine the attractor basin as well as stability for a stationary point.

Here is one more example in which linear approximation is used to investigate the behavior of the system near each stationary point.

EXAMPLE 3.7.4

The system

$$\begin{cases} \dfrac{dx}{dt} = x - 3y - (x - y)^3 \\ \dfrac{dy}{dt} = -3x + y + (x - y)^3 \end{cases}$$

has stationary points where $x - 3y = (x - y)^3$ and $3x - y = (x - y)^3$. Combining these equations yields $x - 3y = 3x - y$ and thus $y = -x$; if we substitute this back into the equation $x - 3y = (x - y)^3$, we get $4x = (2x)^3 = 8x^3$, from which $x = 0$ or $x = \pm\dfrac{1}{\sqrt{2}}$. Thus there are three stationary points, at $(0, 0)$, $(\dfrac{1}{\sqrt{2}}, -\dfrac{1}{\sqrt{2}})$, and $(-\dfrac{1}{\sqrt{2}}, \dfrac{1}{\sqrt{2}})$. The linear approximations to the system near the stationary points are

$$\dfrac{dx}{dt} = x - 3y, \quad \dfrac{dy}{dt} = -3x + y,$$

$$\dfrac{dx}{dt} = -5\left(x - \dfrac{1}{\sqrt{2}}\right) + 3\left(y + \dfrac{1}{\sqrt{2}}\right), \quad \dfrac{dy}{dt} = 3\left(x - \dfrac{1}{\sqrt{2}}\right) - 5\left(y + \dfrac{1}{\sqrt{2}}\right), \text{ and}$$

$$\dfrac{dx}{dt} = -5\left(x + \dfrac{1}{\sqrt{2}}\right) + 3\left(y - \dfrac{1}{\sqrt{2}}\right), \quad \dfrac{dy}{dt} = 3\left(x + \dfrac{1}{\sqrt{2}}\right) - 5\left(y - \dfrac{1}{\sqrt{2}}\right)$$

respectively [see Exercise 19(a)]. Thus for the last two stationary points we have the same matrix $\mathbf{A} = \begin{bmatrix} -5 & 3 \\ 3 & -5 \end{bmatrix}$; since this matrix has negative eigenvalues [Exercise 19(b)], the stationary points $(\pm\dfrac{1}{\sqrt{2}}, \mp\dfrac{1}{\sqrt{2}})$ are asymptotically stable. For the stationary point at the origin, on the other hand, we have $\mathbf{A} = \begin{bmatrix} 1 & -3 \\ -3 & 1 \end{bmatrix}$, one of whose eigenvalues is positive, so the linear approximation there has a saddle point and $(0, 0)$ is an unstable stationary point. The actual phase portrait is shown in Figure 3.7.6 (on p. 242). Note the trajectories along the line $y = x$, which approach the saddle point at the origin as $t \to \infty$ [see Exercise 19(c)]. The line separates the attractor basins of the stable nodes $(-\dfrac{1}{\sqrt{2}}, \dfrac{1}{\sqrt{2}})$ (above the line) and $(\dfrac{1}{\sqrt{2}}, -\dfrac{1}{\sqrt{2}})$ (below the line). ∎

Just as in the example above, it turns out in general that if the linear approximation near a stationary point to a system has a saddle point there, then so does the system itself. In particular, there will be two special trajectories that approach the saddle point (from opposite directions) as $t \to \infty$. In the example, these trajectories were along the straight line $y = x$, but usually they would be along curves. Such a curve is called a **separatrix**; as

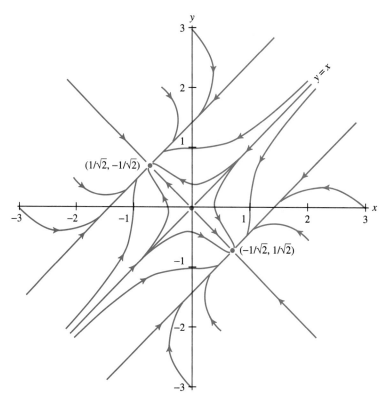

Figure 3.7.6. Phase portrait of the system
$$\frac{dx}{dt} = x - 3y - (x - y)^3, \quad \frac{dy}{dt} = -3x + y + (x - y)^3.$$

in the example, it separates trajectories to either side of it with markedly different long-term behavior (see Figure 3.7.7).

The resemblance of the phase portrait of a nonlinear system near a stationary point to the phase portrait of its linear approximation there extends beyond the case of a saddle point. In fact, provided that $ad - bc \neq 0$ and the matrix $\mathbf{A} = \begin{bmatrix} a & b \\ c & d \end{bmatrix}$ has *distinct* eigenvalues that are not purely imaginary, the stationary points of the system and its linear approximation will always be of the same type (saddle point, stable node, unstable node, stable spiral point, or unstable spiral point). The following exceptions are not covered by this.

1. $ad - bc = 0$: At least one eigenvalue 0. See Example 3.7.1 (p. 232) for an illustration of how the phase portraits of the system and its linear approximation may differ.

2. The linear approximation has a center (and the eigenvalues of $\mathbf{A}$ are purely imaginary). See Example 3.7.2 (p. 234).

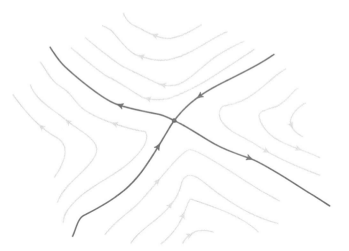

Figure 3.7.7. Separatrices (shown by heavy lines) near a saddle point.

3. The eigenvalues of **A** are equal and nonzero. In this case, although the *stability* of the stationary point for the system and its linear approximation will be the same, the *type* may not be. Specifically, the linear approximation will always have a proper or improper node, while the actual system may have a spiral point (see Exercise 20).

SUMMARY OF KEY CONCEPTS, RESULTS, AND TECHNIQUES

Isolated (stationary point; p. 233): The stationary point P of a system is called isolated if there is some circle around P inside which there are no other stationary points of the system. This is automatic if $\det \mathbf{A} \neq 0$ for the matrix **A** of the linear approximation to the system near P.

Stable (stationary point; p. 236): P is called stable if for every circle around P we can find a second circle such that any trajectory which passes through a point inside the second circle will thereafter stay inside the original circle. Otherwise, P is **unstable**.

Asymptotically stable (stationary point; p. 237): P is called asymptotically stable if P is stable *and* there exists $r > 0$ such that every trajectory which passes within a distance r of P will approach P as $t \to \infty$.

Provided that $\det \mathbf{A} \neq 0$, the stationary point of a linear system with 2×2 matrix **A** is unstable if at least one of the eigenvalues of **A** has positive real part, has a (stable) center if the eigenvalues are purely imaginary, and is asymptotically stable if both eigenvalues have negative real part (p. 238). *Except in the case of a center*, these results also apply to a stationary point of a nonlinear system whose linear approximation near that point has matrix **A** (p. 239).

In fact, if the linear approximation has a saddle point, a node (with $\lambda_1 \neq \lambda_2$), or a spiral point, then so does the nonlinear system (p. 242).

Attractor basin of an asymptotically stable stationary point (p. 240): The region of the plane in which the trajectories approach that point as $t \to \infty$.

Separatrix (p. 241): Curve along which a trajectory approaches a saddle point as $t \to \infty$.

EXERCISES

For each of the following systems:

(a) Find all stationary points.
(b) Use the linear approximation near each stationary point to get as much information as you can about that stationary point.
(c) Sketch a plausible phase portrait for the system.

[For Exercises 1 to 4, part (a) and some of part (b) appeared earlier as Exercises 27 to 30 of Section 3.2.]

1. $\dfrac{dx}{dt} = xy - 1$, $\dfrac{dy}{dt} = 4x - y$.

2. $\dfrac{dx}{dt} = 4xy - 1$, $\dfrac{dy}{dt} = x - y$.

3. $\dfrac{dx}{dt} = 2x - 3y - 5$, $\dfrac{dy}{dt} = 4x + y - 3$.

4. $\dfrac{dx}{dt} = 4x - 6y - 2$, $\dfrac{dy}{dt} = -3x + 5y + 1$.

5. $\dfrac{dx}{dt} = x^2 + 3xy - 4x$, $\dfrac{dy}{dt} = 2xy - 6y^2 + 4y$.

6. $\dfrac{dx}{dt} = x^2 - 5xy + 4x$, $\dfrac{dy}{dt} = -2xy + 3y^2 - y$.

7. $\dfrac{dx}{dt} = y + x^2$, $\dfrac{dy}{dt} = -x + y^2$.

8. $\dfrac{dx}{dt} = y - x^2$, $\dfrac{dy}{dt} = -x + y^2$.

9. $\dfrac{dx}{dt} = y - x^3$, $\dfrac{dy}{dt} = -x + y^3$.

10. $\dfrac{dx}{dt} = y + x^3$, $\dfrac{dy}{dt} = -x + y^3$.

11. (a) Find the general solution of the system $\dfrac{dx}{dt} = x^2$, $\dfrac{dy}{dt} = y$.

(b) Show that the phase portrait of this system is as shown in Figure 3.7.1(b).

Each of the following systems has a stationary point at $(0, 0)$, and its linear approximation near this stationary point is $\dfrac{dx}{dt} = 0$, $\dfrac{dy}{dt} = y$. Sketch the phase portrait of each system, e·r by finding the general solution and then eliminating t or by finding the equations of the trajectories by first writing $\dfrac{dy}{dx}$ in terms of x and y.

12. $\dfrac{dx}{dt} = x^3$, $\dfrac{dy}{dt} = y$.   13. $\dfrac{dx}{dt} = y^2$, $\dfrac{dy}{dt} = y$.   14. $\dfrac{dx}{dt} = xy$, $\dfrac{dy}{dt} = y$.

15. Show that the matrix $\mathbf{A} = \begin{bmatrix} a & b \\ c & d \end{bmatrix}$ has $\lambda = 0$ as an eigenvalue if and only if $ad - bc = 0$.

16. Show that a necessary and sufficient condition for the *linear* system

$$\dfrac{dx}{dt} = a(x - x_0) + b(y - y_0), \quad \dfrac{dy}{dt} = c(x - x_0) + d(y - y_0)$$

to have (x_0, y_0) as its *only* stationary point is that $ad - bc \neq 0$.

17. (a) Show that for the linear system $\dfrac{dx}{dt} = -2y$, $\dfrac{dy}{dt} = x - 3y$ the stationary point is asymptotically stable.

(b) Show that for any solution $(x(t), y(t))$ of the system,

$$\dfrac{d(x^2 + y^2)}{dt} = -2xy - 6y^2.$$

(c) Show that at the point $(-4, 1)$, the trajectory through that point is actually moving "outward" from the origin in the sense that the distance from a point on the trajectory to the origin is increasing.

(d) Sketch the phase portrait of the system and reconcile your findings from parts (a) and (c).

18. (a) Show that the linear approximation to the system

$$\dfrac{dx}{dt} = -y + (x - y)(x^2 + y^2 - 1), \quad \dfrac{dy}{dt} = x + (x + y)(x^2 + y^2 - 1),$$

near its stationary point $(0, 0)$ is $\dfrac{dx}{dt} = -x$, $\dfrac{dy}{dt} = -y$.

(b) Show that for any solution $(x(t), y(t))$ of the system,
$$\frac{d(x^2 + y^2)}{dt} = 2(x^2 + y^2)(x^2 + y^2 - 1).$$

(c) Show that the unit circle $x^2 + y^2 = 1$ is one trajectory of the system.
[*Hint*: A familiar parametrization of the unit circle gives a solution to the system.]

(d) Show that trajectories outside the unit circle move away from the origin, while trajectories inside the unit circle approach the origin.
[*Hint*: What is the connection between $x^2 + y^2$ and the distance to the origin? Use part (b).]

19. (a) Show that the linear approximations to the system
$$\frac{dx}{dt} = x - 3y - (x - y)^3, \qquad \frac{dy}{dt} = -3x + y + (x - y)^3$$
near its stationary points are those given on p. 241.

(b) Find the eigenvalues for the matrices $\begin{bmatrix} -5 & 3 \\ 3 & -5 \end{bmatrix}$ and $\begin{bmatrix} 1 & -3 \\ -3 & 1 \end{bmatrix}$.

(c) Show that if $y = x$ for one point on a trajectory, then the whole trajectory lies on the line $y = x$. Show that the directions of motion along this line are those indicated in Figure 3.7.6.

*(d) Find all other straight lines in the plane along which trajectories lie, and investigate the directions of motion along these lines.

*20. Consider the system $\dfrac{dx}{dt} = f(x, y)$, $\dfrac{dy}{dt} = g(x, y)$, where the functions f and g are defined as follows: $f(0, 0) = g(0, 0) = 0$, and for $(x, y) \neq (0, 0)$,
$$f(x, y) = -x + \frac{y}{\log(x^2 + y^2)}, \qquad g(x, y) = -y - \frac{x}{\log(x^2 + y^2)}.$$

(a) Show that f and g are C^1 functions everywhere inside the circle $x^2 + y^2 = 1$ (that is, at any point in the open disk $x^2 + y^2 < 1$).

(b) Show that the linear approximation to the system near the stationary point $(0, 0)$ is $\dfrac{dx}{dt} = -x$, $\dfrac{dy}{dt} = -y$ and thus has a stable node.

(c) If r, θ denote the usual polar coordinates, so that $x = r \cos \theta$, $y = r \sin \theta$, $r = \sqrt{x^2 + y^2}$, $\tan \theta = \dfrac{y}{x}$, show that for any solution $(x(t), y(t))$ of the system we have $\dfrac{dr}{dt} = -r$ and $\dfrac{d\theta}{dt} = \dfrac{-1}{2 \log r}$.

SEC. 3.8 LYAPUNOV FUNCTIONS; GRADIENT SYSTEMS

(d) By finding [from part (c)] and integrating an expression for $\dfrac{d\theta}{dr}$, show that for any trajectory inside the unit circle there is some constant C such that
$$\theta = \frac{1}{2}\log(-\log r) + C \text{ along the trajectory.}$$

(e) Show that the origin is a spiral point of the system.

FURTHER READING

For the method of resultants, see Chapter X in Mostowski and Stark, *Introduction to Higher Algebra* (Macmillan, 1964).

3.8 LYAPUNOV FUNCTIONS; GRADIENT SYSTEMS

As we have seen, one can often determine the "local" qualitative behavior (near a stationary point) of an autonomous system by studying its linear approximation. However, there are exceptions (such as the case when the linear approximation has a center); also, the "local" information may not be helpful in answering "global" (large-scale) questions, for instance in finding the size of the attractor basin for an asymptotically stable stationary point. For both these reasons, the method of Lyapunov functions,[35] which is the topic of the first part of this section, is often useful. When using this method, one investigates the rates of change of various functions along the trajectories of the system and tries to find a suitable function that decreases along every trajectory.[36] For instance, if the distance to a stationary point is always decreasing, then that stationary point must surely be stable!

EXAMPLE 3.8.1 Let's reconsider the system
$$\frac{dx}{dt} = -2y - x^3, \qquad \frac{dy}{dt} = 2x - y^3,$$

---

[35] After the Russian mathematician Lyapunov (also spelled Liapunov, Liapunoff, etc.; 1857–1918).

[36] Compare this with trying to find a first integral, which is *constant* along every trajectory.

which was system (3) in Example 3.7.2. The linear approximation to this system near the stationary point $(0, 0)$ is

$$\frac{dx}{dt} = -2y, \qquad \frac{dy}{dt} = 2x,$$

which has a center there. This leaves the question of stability completely open. However, let's see how the distance to $(0, 0)$ changes along the trajectories. This distance is, of course, $\sqrt{x^2 + y^2}$, but since we're mainly concerned with whether the distance is increasing or decreasing, we can use $x^2 + y^2$, the square of the distance, instead. If we put $V(x, y) = x^2 + y^2$, then the rate of change of this function along the trajectories will be (by the chain rule; see p. 9):

$$\frac{d(V(x, y))}{dt} = \frac{\partial V}{\partial x}\frac{dx}{dt} + \frac{\partial V}{\partial y}\frac{dy}{dt} = 2x(-2y - x^3) + 2y(2x - y^3) = -2x^4 - 2y^4$$
$$= -2(x^4 + y^4).$$

In particular, for any point (x, y) except the origin, $V(x, y)$ will be decreasing along the trajectory there, since $x^4 + y^4 > 0$. Therefore, the distance to the origin will also be decreasing, and thus $(0, 0)$ is a stable stationary point. As we'll see below, this point is actually asymptotically stable.

By contrast, a similar computation for

$$\frac{dx}{dt} = -2y + x^3, \qquad \frac{dy}{dt} = 2x + y^3,$$

which was system (2) in Example 3.7.2, will show that the distance to the origin is *increasing* along trajectories of this system; thus the origin is an unstable stationary point.[37] ■

Of course, for most systems things will not go as smoothly as in the two cases above. Even changing our system slightly can lead to difficulties, as in the next example.

EXAMPLE 3.8.2

$$\frac{dx}{dt} = y - x^3, \quad \frac{dy}{dt} = -2x - y^3.$$

Once again, the linear approximation near the stationary point $(0, 0)$ has a center. For $V(x, y) = x^2 + y^2$, we have

$$\frac{d(V(x, y))}{dt} = 2x\frac{dx}{dt} + 2y\frac{dy}{dt} = 2x(y - x^3) + 2y(-2x - y^3) = -2x^4 - 2xy - 2y^4.$$

---

[37] Although it is intuitively reasonable, our conclusion that the origin must be unstable isn't as obvious as it may seem. One could imagine a situation in which trajectories spiraled outward, but in such a way that any trajectory which would go through a point at any distance r, say, to the origin would stay within distance $2r$ of the origin. In such a situation the origin would be stable (see the definition on p. 236). For a precise argument showing that it isn't, see Exercise 21.

SEC. 3.8 LYAPUNOV FUNCTIONS; GRADIENT SYSTEMS

Unfortunately, this is not *always* negative; for instance, for $x = -\frac{1}{10}$, $y = \frac{1}{10}$ we have $-2x^4 - 2xy - 2y^4 = \frac{-2}{10^4} + \frac{2}{10^2} - \frac{2}{10^4} = 0.0196$, and so the trajectory through the point $(-\frac{1}{10}, \frac{1}{10})$ proceeds from that point to other points that are farther from the origin. True, $-2x^4 - 2xy - 2y^4$ is negative for "most" points (x, y) (including all points in the first and third quadrants). Thus it would be reasonable to expect that as the trajectories "circle" the origin, they tend to come inward and that therefore the origin will be stable, or even asymptotically stable. However, proving this using our current methods would be tricky, and this is where Lyapunov's idea comes in. The idea is to *replace* (when necessary) $x^2 + y^2$ by a different function $V(x, y)$, which has similar properties but which *does* decrease everywhere along the trajectories. ∎

For simplicity, we'll assume for now that the stationary point whose stability we're studying is at the origin. (This can always be arranged by shifting the coordinate axes.)

Definition The function[38] $V(x, y)$ is called a **Lyapunov function** for the autonomous system $\frac{dx}{dt} = f(x, y)$, $\frac{dy}{dt} = g(x, y)$ [which has a stationary point at $(0, 0)$] if:

(a) $\dfrac{d(V(x, y))}{dt} = \dfrac{\partial V}{\partial x} \cdot f(x, y) + \dfrac{\partial V}{\partial y} \cdot g(x, y) \leq 0$ for all points[39] $(x, y) \neq (0, 0)$

and

(b) $V(0, 0) = 0$, $V(x, y) > 0$ for all points[39] $(x, y) \neq (0, 0)$.

NOTES:

1. Because of the condition $V(0, 0) = 0$ just before it, the second condition in (b) can be rephrased as: $V(x, y) > V(0, 0)$ for all points $(x, y) \neq (0, 0)$. Once this is done, it is no longer really essential to require that $V(0, 0) = 0$. This is because we can always add a constant to the function V without affecting either of the conditions $\dfrac{d(V(x, y))}{dt} \leq 0$ and $V(x, y) > V(0, 0)$ for points $(x, y) \neq (0, 0)$. By adjusting this constant, we can then arrange to have $V(0, 0) = 0$. Note that the condition that $V(x, y) > V(0, 0)$ for all points $(x, y) \neq (0, 0)$ implies, in particular, that V has a minimum at $(0, 0)$. It would not be enough simply to require that V has a minimum, however, because this would leave open the possibility that $V(x, y) = V(0, 0)$ for some point $(x, y) \neq (0, 0)$. Allowing this possibility would cause trouble later on (see Exercise 31).

---

[38] Remember (see p. 8) that all functions of two variables are assumed to be C^1, unless otherwise specified.

[39] Actually, we're starting here with the easiest case of the definition. As we'll see later, there can be a restriction on which points $(x, y) \neq (0, 0)$ are considered; the phrase "for all points" is then modified.

2. A function V satisfying both conditions in (b): $V(0, 0) = 0$, $V(x, y) > 0$ for all points $(x, y) \neq (0, 0)$, is called **positive definite**.

3. The condition in (a) says that V does not increase anywhere along the trajectories. An equally reasonable condition would be that V *decrease* everywhere along the trajectories (except at the stationary point). A function V for which $\dfrac{d(V(x, y))}{dt} < 0$ for all points $(x, y) \neq (0, 0)$ and which satisfies the conditions in (b) is called a **strict Lyapunov function**.

The easiest case of **Lyapunov's theorem** is as follows:

Theorem 3.8.1 If the system $\dfrac{dx}{dt} = f(x, y)$, $\dfrac{dy}{dt} = g(x, y)$ has a Lyapunov function, then the stationary point $(0, 0)$ is stable. If the system has a strict Lyapunov function, then $(0, 0)$ is asymptotically stable.

EXAMPLE 3.8.3 In Example 3.8.1 we saw that for $V(x, y) = x^2 + y^2$, we have $\dfrac{d(V(x, y))}{dt} = -2(x^4 + y^4)$, which is negative for all points $(x, y) \neq (0, 0)$. V also satisfies the conditions in (b) of the definition: $V(0, 0) = 0$ and $V(x, y) > 0$ for all points $(x, y) \neq (0, 0)$. So V is a strict Lyapunov function for the system $\dfrac{dx}{dt} = -2y - x^3$, $\dfrac{dy}{dt} = 2x - y^3$. Therefore, the stationary point $(0, 0)$ of this system is asymptotically stable. ∎

EXAMPLE 3.8.4 Let's return to the system $\dfrac{dx}{dt} = y - x^3$, $\dfrac{dy}{dt} = -2x - y^3$ of Example 3.8.2. Our previous work showed that $x^2 + y^2$ is *not* a Lyapunov function for this system, so to apply the theorem, we need to find a different Lyapunov function. The easiest functions that satisfy the conditions in (b) for a Lyapunov function are of the form $V(x, y) = ax^2 + by^2$ with constants $a, b > 0$. By multiplying by a positive constant (specifically, $\dfrac{1}{a}$), such a function can be assumed to have the form $V(x, y) = x^2 + cy^2$ with $c > 0$. Since for such a function we do have $V(0, 0) = 0$ and $V(x, y) > 0$ for all $(x, y) \neq (0, 0)$, it's enough if we can choose c such that $\dfrac{d(V(x, y))}{dt} = \dfrac{d}{dt}(x^2 + cy^2) \leq 0$ for all points $(x, y) \neq (0, 0)$. (If we can make the inequality strict, we'll have a strict Lyapunov function.) Well,

$$\frac{d}{dt}(x^2 + cy^2) = 2x\frac{dx}{dt} + 2cy\frac{dy}{dt}$$
$$= 2x(y - x^3) + 2cy(-2x - y^3)$$
$$= -2x^4 + (2 - 4c)xy - 2cy^4.$$

SEC. 3.8 LYAPUNOV FUNCTIONS; GRADIENT SYSTEMS

So we're looking for a value of c for which $-2x^4 + (2 - 4c)xy - 2cy^4$ is never positive. Since the "cross term" $(2 - 4c)xy$ is the only one that could ever be positive, it seems natural to make this term zero by choosing c such that $2 - 4c = 0$, that is, $c = \frac{1}{2}$. For $c = \frac{1}{2}$ we get

$$\frac{d}{dt}\left(x^2 + \frac{1}{2}y^2\right) = -2x^4 - y^4 < 0 \qquad \text{for all } (x, y) \neq (0, 0).$$

Since we already know that $V(x, y) = x^2 + \frac{1}{2}y^2$ satisfies the conditions in (b) for a Lyapunov function, V is a strict Lyapunov function and thus the origin is asymptotically stable. ∎

Of course, it would be too much to expect always to be able to find a Lyapunov function of the form $x^2 + cy^2$. Finding a Lyapunov function is an "art," rather than a "science": There is no particular method that is always reliable. In some situations, especially applications to physics, an "energy function" will work, since the energy of an autonomous physical system is either constant (when all energy is conserved) or decreasing (when some energy is dissipated, for instance in the form of heat due to friction).

EXAMPLE 3.8.5 Consider a nonlinear spring for which friction is taken into account. As in Section 2.1, the differential equation for the displacement x from equilibrium will be $mx'' + \gamma x' = F(x)$ with positive constants m, γ. Here $F(x)$ is the spring force, which will be positive when x is negative and vice versa. As in Example 3.1.2 (p. 144), we can replace this equation by the equivalent system

$$\frac{dx}{dt} = v, \qquad \frac{dv}{dt} = \frac{1}{m}F(x) - \frac{\gamma}{m}v.$$

The stationary point $x = 0$, $v = 0$ of this system corresponds to the equilibrium of the spring. We now consider the **mechanical energy** $E(x, v)$ of the spring, which is the sum of the kinetic energy $\frac{1}{2}mv^2$ and the potential energy $P(x) = -\int_0^x F(\xi)\, d\xi$ (compare Example 3.2.11, p. 169):

$$E(x, v) = \frac{1}{2}mv^2 + P(x).$$

Note that $E(0, 0) = 0$. Since $F(\xi)$ is positive for ξ negative and vice versa, $P(x)$ will be positive for all $x \neq 0$ [see Exercise 24(a)]. Therefore [Exercise 24(b)], $E(x, v) > 0$ for all points $(x, v) \neq (0, 0)$, so E is positive definite, that is, the conditions in (b) for a Lyapunov function are satisfied by E. As to the rate of change of E along the trajectories, we get

$$\frac{d(E(x, v))}{dt} = mv\frac{dv}{dt} + P'(x)\frac{dx}{dt}$$

$$= mv\left(\frac{1}{m}F(x) - \frac{\gamma}{m}v\right) - F(x)v$$

$$= -\gamma v^2.$$

In particular, we have $\dfrac{d(E(x, v))}{dt} \leq 0$ for all x, v, so $E(x, v)$ is a Lyapunov function and the equilibrium of the spring is stable. Notice that we can draw this conclusion despite the fact that we have very little chance of solving the equation $mx'' + \gamma x' = F(x)$ in general. ∎

A proof of Lyapunov's theorem is beyond the scope of this book (see ''Further Reading'' following the Exercises). While the theorem is intuitively reasonable, one can imagine difficulties that might arise. For instance, $V(x, y)$ might be a strict Lyapunov function but decrease more and more slowly along each trajectory, so that the trajectories would not necessarily approach the stationary point as $t \to \infty$.

So far, we've considered only ''global'' Lyapunov functions, that is, functions that have the required properties for all points $(x, y) \neq (0, 0)$. However, often one cannot expect such functions to exist, even when the stationary point is, in fact, asymptotically stable. For instance, in Example 3.7.3 (p. 239) we saw that the system

$$\frac{dx}{dt} = -y + (x - y)(x^2 + y^2 - 1), \qquad \frac{dy}{dt} = x + (x + y)(x^2 + y^2 - 1)$$

has an asymptotically stable stationary point at the origin, but that only trajectories inside the circle $x^2 + y^2 = 1$ approach the origin as $t \to \infty$. Accordingly, if we try to use $V(x, y) = x^2 + y^2$ as a Lyapunov function, we find that

$$\frac{d(V(x, y))}{dt} = 2x\frac{dx}{dt} + 2y\frac{dy}{dt} = 2(x^2 + y^2)(x^2 + y^2 - 1),$$

which is negative for $0 < x^2 + y^2 < 1$ but positive for $x^2 + y^2 > 1$.

Fortunately, Lyapunov's theorem remains valid if we restrict the function $V(x, y)$ to a domain[40] D containing the origin. That is, if $V(0, 0) = 0$ and if $\dfrac{d(V(x, y))}{dt} \leq 0$ and $V(x, y) > 0$ for all points $(x, y) \neq (0, 0)$ in D, then the origin is stable; if the first inequality can be strengthened [still for all points $(x, y) \neq (0, 0)$ in D] to $\dfrac{d(V(x, y))}{dt} < 0$, then the origin is asymptotically stable.

---

[40] In the technical sense, that is, an open and connected set. (A set is **open** if for *each* point in the set, the set contains a disk of positive radius around that point.)

For instance, in the case from Example 3.7.3 mentioned above, we could take D to be the region inside the circle $x^2 + y^2 = 1$. By our earlier computation, $V(x, y) = x^2 + y^2$ would then be a strict Lyapunov function on D, and this shows that the origin is asymptotically stable.

In general, the existence of a strict Lyapunov function on D does *not* imply that *every* trajectory which contains a point of D will approach the origin. The reason is that trajectories may leave D, either at some finite time or in the limit as $t \to \infty$, as the following example shows.

EXAMPLE 3.8.6 Consider the system $\dfrac{dx}{dt} = x(x-2)$, $\dfrac{dy}{dt} = -y$, which does have the origin as a stationary point. If we want to try using $V(x, y) = x^2 + y^2$ as a Lyapunov function, we first compute $\dfrac{dV}{dt} = 2x \dfrac{dx}{dt} + 2y \dfrac{dy}{dt} = 2x^2(x-2) - 2y^2$. Since $\dfrac{dV}{dt} < 0$ when $y^2 > x^2(x-2)$, it is natural to let D be the set of all points (x, y) for which $y^2 > x^2(x-2)$, together with the origin. This set, which is a domain, is shaded in Figure 3.8.1 [see Exercise 22(a)]. V will be a strict Lyapunov function on D, and thus the origin is an asymptotically stable stationary point. [We could also have seen this using the linear approximation; see Exercise 22(b).] However, there are trajectories in D that don't approach the origin as $t \to \infty$. In fact, there are two such trajectories along the vertical line $x = 2$ [see Exercise 22(c)]. ∎

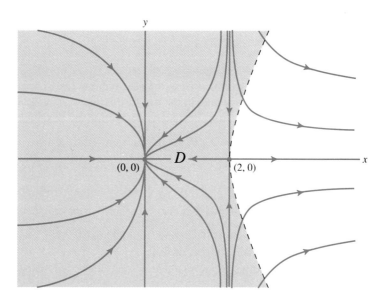

Figure 3.8.1. Phase portrait of the system $\dfrac{dx}{dt} = x(x-2)$, $\dfrac{dy}{dt} = -y$. Although $V(x, y) = x^2 + y^2$ is a strict Lyapunov function on the shaded region D, not all trajectories through points in D approach $(0, 0)$ as $t \to \infty$.

Despite such problems, it is often possible to use a (strict) Lyapunov function to get information about the size of the attractor basin of an asymptotically stable critical point (see "Further Reading" following the Exercises).

Of course, to make any of this work, one first has to come up with a Lyapunov function. Very often, except in situations for which there is a natural "energy" function, the best candidates are **quadratic forms**: functions of the form

$$V(x, y) = k_1 x^2 + k_2 xy + k_3 y^2$$

with constants k_1, k_2, k_3. Such a function is positive definite if $k_1 > 0$ and $k_2^2 < 4k_1 k_3$ (see Exercise 23). Note that the functions $V(x, y) = x^2 + cy^2$ considered earlier are of this form.

To conclude our brief discussion of Lyapunov functions, we now state Lyapunov's theorem in the more general case of a stationary point at (x_0, y_0).

Theorem 3.8.2 Let (x_0, y_0) be a stationary point for the autonomous system $\dfrac{dx}{dt} = f(x, y)$, $\dfrac{dy}{dt} = g(x, y)$, and let D be a domain[41] containing (x_0, y_0). Suppose that a function (*Lyapunov function*) $V(x, y)$, defined on D, exists such that:

(a) $\dfrac{d(V(x, y))}{dt} = \dfrac{\partial V}{\partial x} f(x, y) + \dfrac{\partial V}{\partial y} g(x, y) \leq 0$ for all points $(x, y) \neq (x_0, y_0)$ in D and

(b) $V(x, y) > V(x_0, y_0)$ for all points $(x, y) \neq (x_0, y_0)$ in D.

Then the stationary point (x_0, y_0) is stable; if the inequality in (a) is actually a strict inequality, the stationary point is asymptotically stable.

This theorem will be used in the next section, for a Lyapunov function of a somewhat unusual type. Meanwhile, in the rest of this section, we'll take a longer look at **gradient systems**, which we first encountered in Section 3.2 (Example 3.2.7, pp. 165–166). Recall that a gradient system has the form

$$\frac{dx}{dt} = -\frac{\partial V}{\partial x}, \quad \frac{dy}{dt} = -\frac{\partial V}{\partial y},$$

where V is a function of two variables. For our general theory to apply, $-\dfrac{\partial V}{\partial x}$ and $-\dfrac{\partial V}{\partial y}$ should be C^1 functions. Therefore, V should be a C^2 function; that is, all second partials of V should be defined and continuous.

The trajectories of the gradient system above will satisfy the equation

$$\frac{dy}{dx} = \frac{dy/dt}{dx/dt} = \frac{\partial V/\partial y}{\partial V/\partial x} \quad \left(\text{where } \frac{\partial V}{\partial x} \neq 0\right).$$

---

[41] That is, an open and connected set. (See footnote 40 on p. 252.)

SEC. 3.8 LYAPUNOV FUNCTIONS; GRADIENT SYSTEMS

In other words, at any point (x_0, y_0) the slope of the trajectory through that point will be $\dfrac{\dfrac{\partial V}{\partial y}(x_0, y_0)}{\dfrac{\partial V}{\partial x}(x_0, y_0)}$. On the other hand, we've seen in Section 1.1 (on p. 10) that the slope of the level curve of V at (x_0, y_0) is $-\dfrac{\dfrac{\partial V}{\partial x}(x_0, y_0)}{\dfrac{\partial V}{\partial y}(x_0, y_0)}$. Since the product of the two slopes is -1, we can conclude that the trajectory through any point is *perpendicular* to the level curve through that point.[42] One can check (see Exercise 29) that this remains true for points where either $\dfrac{\partial V}{\partial x}$ or $\dfrac{\partial V}{\partial y}$ is zero, but not both. Note that at any point where $\dfrac{\partial V}{\partial x} = \dfrac{\partial V}{\partial y} = 0$, the system has a stationary point and the function V has a critical point.

Of course, there will be two (opposite) directions perpendicular to the level curve through any point. To see which of these is the direction of the trajectory, let's see how the function V changes along the trajectory. We get

$$\frac{dV}{dt} = \frac{\partial V}{\partial x}\frac{dx}{dt} + \frac{\partial V}{\partial y}\frac{dy}{dt} = -\left(\frac{\partial V}{\partial x}\right)^2 - \left(\frac{\partial V}{\partial y}\right)^2.$$

Therefore, $\dfrac{dV}{dt} < 0$ (except for points where $\dfrac{\partial V}{\partial x} = \dfrac{\partial V}{\partial y} = 0$).[43] So the trajectories are followed in the direction of *decreasing V*.

EXAMPLE 3.8.7 Let $V(x, y) = 2x^2 + y^2$. The level curves of V are concentric ellipses centered at the origin; larger values of V correspond to larger ellipses. Thus the trajectories of the gradient system

$$\frac{dx}{dt} = -4x, \qquad \frac{dy}{dt} = -2y$$

will be along curves perpendicular to the ellipses, and these curves will be followed inward (toward the origin; see Figure 3.8.2, p. 256). In fact, most of the trajectories lie on parabolas (see Exercise 28). ∎

Example 3.8.7 is unusually simple in several respects. The gradient system is linear, and its only stationary point is at the origin. As for the function $V(x, y)$, its critical point at the origin is clearly a minimum. We will now look at critical points of functions $V(x, y)$ in general, and classify them using gradient systems.

---

[42] That is, the trajectories are **orthogonal trajectories** for the level curves. See Section 1.7, Exercise 61.

[43] However, V is not necessarily a Lyapunov function, even if the origin is a stationary point (see Exercise 32).

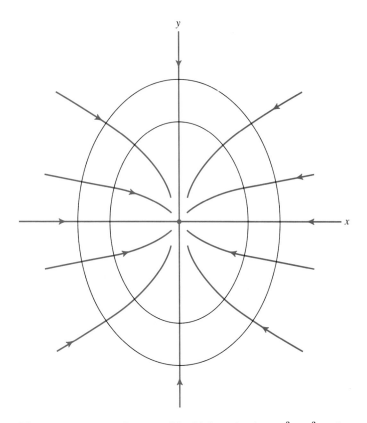

Figure 3.8.2. Level curves (black) for $V(x, y) = 2x^2 + y^2$ and trajectories (in color) for the gradient system $\dfrac{dx}{dt} = -\dfrac{\partial V}{\partial x}$, $\dfrac{dy}{dt} = -\dfrac{\partial V}{\partial y}$.

Suppose that (x_0, y_0) is a critical point of $V(x, y)$, and thus a stationary point of the system $\dfrac{dx}{dt} = -\dfrac{\partial V}{\partial x}$, $\dfrac{dy}{dt} = -\dfrac{\partial V}{\partial y}$. To study this stationary point, we use the linear approximation (see p. 167)

$$\frac{dx}{dt} = a(x - x_0) + b(y - y_0), \qquad \frac{dy}{dt} = c(x - x_0) + d(y - y_0),$$

where the coefficients a, b, c, d are the partial derivatives of $-\dfrac{\partial V}{\partial x}$ and $-\dfrac{\partial V}{\partial y}$, evaluated at (x_0, y_0). For instance, $\dfrac{\partial}{\partial x}\left(-\dfrac{\partial V}{\partial x}\right) = -\dfrac{\partial^2 V}{\partial x^2}$, so $a = -\dfrac{\partial^2 V}{\partial x^2}(x_0, y_0)$; similarly, we have $b = -\dfrac{\partial^2 V}{\partial y \, \partial x}(x_0, y_0)$, $c = -\dfrac{\partial^2 V}{\partial x \, \partial y}(x_0, y_0)$, $d = -\dfrac{\partial^2 V}{\partial y^2}(x_0, y_0)$. Since the mixed second partials of the C^2 function V are equal (see p. 9), we have $b = c$. Thus the matrix of the (linear) approximating system is of the form $\mathbf{A} = \begin{bmatrix} a & b \\ b & d \end{bmatrix}$.

As we've seen in Section 3.6, we can classify the stationary point of this linear system once we know the eigenvalues of $\mathbf{A}$. These are the solutions of the characteristic equation $(a - \lambda)(d - \lambda) - b^2 = 0$ (see p. 191), and after a little algebra we find them to be
$$\lambda_{1,2} = \frac{a + d \pm \sqrt{(a - d)^2 + 4b^2}}{2}$$
[see Exercise 26(a)]. Since $(a - d)^2 + 4b^2$ is never negative, the eigenvalues are always real.[44] If both eigenvalues are negative, the stationary point is a stable node; this happens when $ad > b^2$ and $a < 0, d < 0$ [see Exercise 26(d)]. Similarly, if $ad > b^2$ and $a > 0, d > 0$, then both eigenvalues are positive, and the stationary point is an unstable node. On the other hand, if $ad < b^2$, then λ_1 and λ_2 have opposite signs [see Exercise 26(b)] and so we have a saddle point. In all these cases, the stationary point of the original system $\frac{dx}{dt} = -\frac{\partial V}{\partial x}, \frac{dy}{dt} = -\frac{\partial V}{\partial y}$ is of the same type as that of the linear approximation[45] (see p. 242). Let's consider each case separately.

Case 1 $ad > b^2, a < 0, d < 0$. Since in this case the stationary point (x_0, y_0) is a stable node[45], all nearby trajectories of the gradient system approach (x_0, y_0) as $t \to \infty$. We saw earlier that V decreases along the trajectories, so for all points (x, y) sufficiently near (x_0, y_0), except for (x_0, y_0) itself, we must have $V(x, y) > V(x_0, y_0)$ (since the trajectory through (x, y) approaches (x_0, y_0) and V is continuous and decreases along that trajectory).[46] Therefore, V has a *local minimum* at (x_0, y_0).

Case 2 $ad > b^2, a > 0, d > 0$. We now have an unstable node[45], with all nearby trajectories "originating at" (x_0, y_0) (as $t \to -\infty$). By an argument similar to the above, we see that V has a *local maximum* at (x_0, y_0).

Case 3 $ad < b^2$. Since (x_0, y_0) is now a saddle point, there are two trajectories (separatrices) approaching (x_0, y_0) as $t \to \infty$ (see Figure 3.7.7, p. 243). On the other hand, there are two trajectories "originating at" (x_0, y_0). In particular, there are points (x, y) arbitrarily close to (x_0, y_0) for which $V(x, y) > V(x_0, y_0)$, but there are also such arbitrarily close points with $V(x, y) < V(x_0, y_0)$. It follows that V has neither a maximum nor a minimum at (x_0, y_0); in fact, it will not be surprising that V has a *saddle point* there.

If we remember that $a = -\frac{\partial^2 V}{\partial x^2}(x_0, y_0)$ and so on, we can summarize the three cases above in the following test, which is usually found without proof in books on calculus.

---

[44] This is a special case of a general result from linear algebra: The eigenvalues of a real **symmetric** matrix (a matrix that remains the same if its rows are replaced by its columns and vice versa) are always real.

[45] Provided the eigenvalues are distinct. [See Exercise 26(e) for the case $\lambda_1 = \lambda_2$.]

[46] Actually, V decreases except for points where $\frac{\partial V}{\partial x} = \frac{\partial V}{\partial y} = 0$, but these are stationary points of the system. Thus the trajectories through such points consist of the single points themselves. In particular, (x_0, y_0) is the only such point "nearby," since nearby trajectories must approach (x_0, y_0).

Theorem 3.8.3 (Second Derivative Test). Let V be a C^2 function of two variables, and let (x_0, y_0) be a critical point of V. To classify this critical point, evaluate $\dfrac{\partial^2 V}{\partial x^2}$, $\dfrac{\partial^2 V}{\partial x\, \partial y} = \dfrac{\partial^2 V}{\partial y\, \partial x}$, and $\dfrac{\partial^2 V}{\partial y^2}$ there. If (still at the critical point) $\dfrac{\partial^2 V}{\partial x^2} \cdot \dfrac{\partial^2 V}{\partial y^2} - \left(\dfrac{\partial^2 V}{\partial x\, \partial y}\right)^2 < 0$, then the point is a saddle point. If $\dfrac{\partial^2 V}{\partial x^2} \cdot \dfrac{\partial^2 V}{\partial y^2} - \left(\dfrac{\partial^2 V}{\partial x\, \partial y}\right)^2 > 0$, then the point is a local extremum: specifically, a local minimum if $\dfrac{\partial^2 V}{\partial x^2} > 0$ [and $\dfrac{\partial^2 V}{\partial y^2} > 0$; see Exercise 27(a)] and a local maximum if $\dfrac{\partial^2 V}{\partial x^2} < 0$ [and $\dfrac{\partial^2 V}{\partial y^2} < 0$].

Note that the test does not give any information if $\dfrac{\partial^2 V}{\partial x^2} \cdot \dfrac{\partial^2 V}{\partial y^2} - \left(\dfrac{\partial^2 V}{\partial x\, \partial y}\right)^2 = 0$. This corresponds to the case of an eigenvalue 0 [see Exercise 27(b)], in which we have seen that the phase portrait of a system may differ considerably from that of its linear approximation!

SUMMARY OF KEY CONCEPTS AND RESULTS

Positive definite (p. 250): A function $V(x, y)$ is positive definite if $V(0, 0) = 0$ and $V(x, y) > 0$ for all $(x, y) \neq (0, 0)$.

Lyapunov function, *special case* (p. 249): Let the system $\dfrac{dx}{dt} = f(x, y)$, $\dfrac{dy}{dt} = g(x, y)$ have a stationary point at $(0, 0)$. Then $V(x, y)$ is a Lyapunov function for this system if V is positive definite and $\dfrac{d(V(x, y))}{dt} = \dfrac{\partial V}{\partial x} f(x, y) + \dfrac{\partial V}{\partial y} g(x, y) \leq 0$ for all (x, y). A Lyapunov function V is **strict** if $\dfrac{d(V(x, y))}{dt} < 0$ for all $(x, y) \neq (0, 0)$.

Lyapunov's theorem, *special case* (p. 250): The stationary point at $(0, 0)$ is stable if the system has a Lyapunov function, and asymptotically stable if it has a strict Lyapunov function. Good candidates for a Lyapunov function include functions of the form $x^2 + cy^2$ and "energy" functions (p. 251).

Lyapunov function, *general case* (p. 254): Let the system $\dfrac{dx}{dt} = f(x, y)$, $\dfrac{dy}{dt} = g(x, y)$ have a stationary point at (x_0, y_0), and let D be a domain containing (x_0, y_0). A function $V(x, y)$ defined on D is a Lyapunov function if $\dfrac{d(V(x, y))}{dt} \leq 0$ for all (x, y) in D and $V(x, y) > V(x_0, y_0)$ for all $(x, y) \neq (x_0, y_0)$ in D. A Lyapunov function V is **strict** if $\dfrac{d(V(x, y))}{dt} < 0$ for all $(x, y) \neq (x_0, y_0)$ in D.

SEC. 3.8 LYAPUNOV FUNCTIONS; GRADIENT SYSTEMS

Lyapunov's theorem, *general case* (p. 254): The stationary point at (x_0, y_0) is stable if the system has a Lyapunov function on D, and asymptotically stable if it has a strict Lyapunov function on D. [However, in the latter case, it is not necessarily true that all of D is in the attractor basin of (x_0, y_0); see Example 3.8.6.]

Gradient system (pp. 166, 254): A system of the form $\dfrac{dx}{dt} = -\dfrac{\partial V}{\partial x}, \dfrac{dy}{dt} = -\dfrac{\partial V}{\partial y}$, where V is a C^2 function. The trajectories of a gradient system are perpendicular to the level curves of V and are followed in the direction of decreasing V (p. 255). A gradient system has stationary points exactly where V has critical points. Classifying these stationary points leads to the Second Derivative Test from multivariable calculus (p. 258).

EXERCISES

Find Lyapunov functions for the following systems and draw an appropriate conclusion about stability in each case.

[*Hint*: Look for functions of the form $V(x, y) = x^2 + cy^2$.]

1. $\dfrac{dx}{dt} = 4y, \quad \dfrac{dy}{dt} = -x.$

2. $\dfrac{dx}{dt} = -y, \quad \dfrac{dy}{dt} = 4x.$

3. $\dfrac{dx}{dt} = -y - 2x^3, \quad \dfrac{dy}{dt} = 3x - y^3.$

4. $\dfrac{dx}{dt} = 4y - x^5, \quad \dfrac{dy}{dt} = -x.$

5. $\dfrac{dx}{dt} = 2y - (3x - y)(4 - x^2 - y^2), \quad \dfrac{dy}{dt} = -2x - (x + 5y)(4 - x^2 - y^2).$

6. $\dfrac{dx}{dt} = 2y + (x - 2y)(x^2 + 2y^2 - 1), \quad \dfrac{dy}{dt} = -x + (x + 2y)(x^2 + 2y^2 - 1).$

In each of the following cases, explain why $V(x, y)$ is *not* a Lyapunov function for the given system.

7. $V(x, y) = x^2 + y^2, \quad \dfrac{dx}{dt} = -y + 2x^3, \quad \dfrac{dy}{dt} = x - y^3.$

8. $V(x, y) = x^2 + y^2, \quad \dfrac{dx}{dt} = -y - 2x^3, \quad \dfrac{dy}{dt} = 3x - y^3.$

9. $V(x, y) = x^2 + y^3, \quad \dfrac{dx}{dt} = -x^3, \quad \dfrac{dy}{dt} = -y^2.$

10. $V(x, y) = x^5 + y^2, \quad \dfrac{dx}{dt} = -x^2, \quad \dfrac{dy}{dt} = -y^3.$

Which of the following functions are positive definite? Of those that are not, which could possibly be used as a Lyapunov function on some domain D containing $(0, 0)$?

11. $x^2 + y^3$

12. $x^2 + y^4$

13. $x^2 + 2xy + 2y^2$
    [*Hint*: Complete the square.]
14. $x^2 - 2xy + 2y^2$
15. $x^2 + 2xy + y^2$
16. $x^2 - 3xy + y^2$
17. $4(x^2 + y^2) - (x^2 + y^2)^2$
18. $(x^2 + y^2)(x^2 + y^2 - 2)$
19. $\cos(x^2 + y^2)$
20. $\sin(x^2 + y^2)$

21. Consider (as on p. 248) the system $\dfrac{dx}{dt} = -2y + x^3$, $\dfrac{dy}{dt} = 2x + y^3$ and the function $V(x, y) = x^2 + y^2$.

   (a) Show that $\dfrac{d(V(x, y))}{dt} = 2(x^4 + y^4)$.

   (b) Show that $\dfrac{d(V(x, y))}{dt} \geq [V(x, y)]^2$.

   *(c) Show that each trajectory (except for the stationary point) reaches arbitrarily large distances from the origin as $t \to \infty$.

   [*Hint*: If the distance to the origin for $t = 0$ is r, then $\dfrac{d(V(x, y))}{dt} \geq r^4$ for $t \geq 0$ (why?), and hence (explain!) $V(x, y) \geq r^2 + r^4 t$.]

   (d) Show that the origin is unstable.

22. (a) Show that in Figure 3.8.1, the set D consisting of all points (x, y) for which $y^2 > x^2(x - 2)$, together with the origin, is drawn correctly.

   (b) Use the linear approximation to the system $\dfrac{dx}{dt} = x(x - 2)$, $\dfrac{dy}{dt} = -y$ near the origin to show that the stationary point $(0, 0)$ is asymptotically stable.

   (c) Show that in addition to the stationary point $(2, 0)$, there are two trajectories of the system along the line $x = 2$.

   (d) Show that the phase portrait of the system is as shown in Figure 3.8.1.

*23. (a) Show that if $k_1 > 0$ and $k_2^2 < 4k_1 k_3$, the function
$$V(x, y) = k_1 x^2 + k_2 xy + k_3 y^2$$
is positive definite.
[*Hint*: Complete the square.]

   (b) Show that, conversely, if $k_1 x^2 + k_2 xy + k_3 y^2$ is positive definite, then $k_1 > 0$ and $k_2^2 < 4k_1 k_3$.

24. Consider the situation of Example 3.8.5 (p. 251).
   (a) Show that $P(x) > 0$ for all $x \neq 0$.
   (b) Show that $E(x, v) > 0$ for all points $(x, v) \neq (0, 0)$.
   (c) Can we use the Lyapunov function $E(x, v)$ to show that the equilibrium of the spring is *asymptotically* stable? Explain.

25. (a) Show that in Example 3.8.5 (p. 251), the linear approximation to the system

$$\frac{dx}{dt} = v, \quad \frac{dv}{dt} = \frac{1}{m}F(x) - \frac{\gamma}{m}v \quad \text{is} \quad \frac{dx}{dt} = v, \quad \frac{dv}{dt} = -\frac{k}{m}x - \frac{\gamma}{m}v,$$

where $k = -F'(0)$. [The notation is chosen to correspond to the linear case, for which $F(x) = -kx$.]

(b) Use part (a) to show that if $k > 0$, the equilibrium of the spring is asymptotically stable.

26. (a) Show that the eigenvalues of the matrix $\mathbf{A} = \begin{bmatrix} a & b \\ b & d \end{bmatrix}$ are

$$\lambda_{1,2} = \frac{a + d \pm \sqrt{(a-d)^2 + 4b^2}}{2}.$$

(b) Show that λ_1, λ_2 have the same sign (positive or negative) if $ad > b^2$ and that λ_1, λ_2 have opposite signs if $ad < b^2$. What happens if $ad = b^2$?

(c) Show that if $ad > b^2$, then a and d are either both negative or both positive.

(d) Show that λ_1, λ_2 are both negative if $ad > b^2$ and $a < 0, d < 0$ and that λ_1, λ_2 are both positive if $ad > b^2$ and $a > 0, d > 0$.

*(e) Show that λ_1, λ_2 are *equal* if and only if $a = d$ and $b = 0$. Can you adapt the arguments given in Case 1 and Case 2 on p. 257 to this case?

27. (a) Explain why the wording of the Second Derivative Test can include the phrase " ... a local minimum if $\frac{\partial^2 V}{\partial x^2} > 0$ " rather than " ... a local minimum if $\frac{\partial^2 V}{\partial x^2} > 0$ and $\frac{\partial^2 V}{\partial y^2} > 0$ " (see p. 258 for the context).

(b) Show that if $\frac{\partial^2 V}{\partial x^2} \cdot \frac{\partial^2 V}{\partial y^2} - \left(\frac{\partial^2 V}{\partial x \, \partial y}\right)^2 = 0$, the matrix of the linear approximation to the gradient system $\frac{dx}{dt} = -\frac{\partial V}{\partial x}, \frac{dy}{dt} = -\frac{\partial V}{\partial y}$ has an eigenvalue zero.

28. Show that most of the trajectories of the gradient system $\frac{dx}{dt} = -4x, \frac{dy}{dt} = -2y$ lie on parabolas. What special trajectories are there?

29. Let V be a C^2 function.

(a) Show that if either $\frac{\partial V}{\partial x}(x_0, y_0)$ or $\frac{\partial V}{\partial y}(x_0, y_0)$ is zero, but not both, then the level curve of V through (x_0, y_0) and the trajectory of the gradient system $\frac{dx}{dt} = -\frac{\partial V}{\partial x}, \frac{dy}{dt} = -\frac{\partial V}{\partial y}$ through (x_0, y_0) are perpendicular there.

(b) Why is this no longer true if $\frac{\partial V}{\partial x}(x_0, y_0) = \frac{\partial V}{\partial y}(x_0, y_0) = 0$?

30. Show that if the function $E(x, y)$ is a first integral for the gradient system

$$\frac{dx}{dt} = -\frac{\partial V}{\partial x}, \frac{dy}{dt} = -\frac{\partial V}{\partial y},$$

then, conversely, $V(x, y)$ is a first integral for

$$\frac{dx}{dt} = -\frac{\partial E}{\partial x}, \quad \frac{dy}{dt} = -\frac{\partial E}{\partial y}.$$

31. Explain why it would not be a good idea to replace (b) in the definition of Lyapunov function by: $V(0, 0) = 0$ and V has a minimum at $(0, 0)$. (See Note 1, p. 249.)

32. Suppose that V has a critical point at the origin. What is needed for V to be a Lyapunov function for the system $\dfrac{dx}{dt} = -\dfrac{\partial V}{\partial x}, \ \dfrac{dy}{dt} = -\dfrac{\partial V}{\partial y}$?

FURTHER READING

For more on Lyapunov functions and gradient systems, including a proof of Lyapunov's theorem and a theorem on the size of the attractor basin, see Hirsch and Smale, *Differential Equations, Dynamical Systems, and Linear Algebra* (Academic Press, 1974), Chapter 9.

3.9 THE VOLTERRA–LOTKA EQUATIONS AND OTHER TWO-POPULATION SYSTEMS

In this section we return to the study of systems of differential equations which arise from models describing the growth and/or decline of two populations that share the same habitat, and we'll apply the qualitative techniques developed in this chapter to these systems.

Recall that in Example 3.1.4, we developed a model to describe the population sizes of predators (x) and prey (y) sharing the same habitat. This led to the Volterra–Lotka equations

$$\frac{dx}{dt} = -Mx + Pxy, \quad \frac{dy}{dt} = Ky - Lxy,$$

in which K, L, M, and P are positive constants.

To begin the study of this system, we can find the stationary points and (at least for specific values of K, L, M, P) sketch the direction field. It turns out that there are two stationary points, one at the origin and one at $(\frac{K}{L}, \frac{M}{P})$ [see p. 175, Exercise 55(a)]. Figure 3.9.1 shows the direction field in the case $K = 2$, $L = 1$, $M = 2$, $P = 3$. Since only situations with $x \geq 0$ and $y \geq 0$ are meaningful in our application, the direction field has only been drawn in the first quadrant.

We can guess from this sketch that the origin is a saddle point, with a special trajectory along the x-axis approaching the origin as $t \to \infty$ and one along the y-axis "originating at"

SEC. 3.9 THE VOLTERRA–LOTKA EQUATIONS AND OTHER TWO-POPULATION SYSTEMS

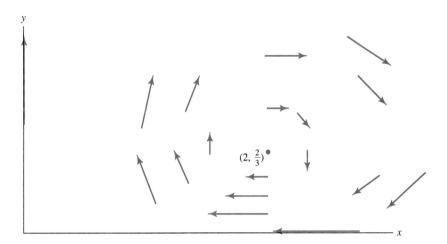

Figure 3.9.1. Direction field for the Volterra–Lotka system
$\frac{dx}{dt} = -2x + 3xy$, $\frac{dy}{dt} = 2y - xy$.

$(0, 0)$ as $t \to -\infty$ and approaching infinity as $t \to \infty$. This is indeed the case (see Exercise 2). However, the special trajectory along the y-axis is not entirely realistic; as pointed out on p. 25, we cannot expect unlimited growth for the prey population, even in the absence of predators!

It is more interesting to see what our model might predict as the long-term behavior of a trajectory starting at a point (x, y) for which both x and y are positive. From the sketch we see that the trajectories will have a tendency to turn in a clockwise direction around the stationary point $(\frac{K}{L}, \frac{M}{P})$, but it's not clear whether they spiral inward, spiral outward, or follow closed curves (compare Figure 3.2.3, p. 160). To study the stability of the stationary point, it's convenient to first shift it to the origin by introducing the new coordinates $X = x - \frac{K}{L}$, $Y = y - \frac{M}{P}$. In terms of X and Y, the system can be rewritten as

$$\frac{dX}{dt} = \frac{PK}{L}Y + PXY, \qquad \frac{dY}{dt} = -\frac{LM}{P}X - LXY$$

[see Exercise 3(a)]. Near our stationary point $X = 0$, $Y = 0$ this has linear approximation $\frac{dX}{dt} = \frac{PK}{L}Y$, $\frac{dY}{dt} = -\frac{LM}{P}X$, and we have $\mathbf{A} = \begin{bmatrix} 0 & \frac{PK}{L} \\ -\frac{LM}{P} & 0 \end{bmatrix}$. This matrix has a pair of purely imaginary eigenvalues $\lambda, \bar{\lambda} = \pm i\sqrt{KM}$ [see Exercise 3(b)], so the *linear approximation* has a center at $X = 0$, $Y = 0$. Unfortunately, as we've seen (compare Figure

3.7.2, p. 235), this is a special case, in which we can't be sure yet of the stability of the stationary point of the actual system.

A natural next step is to try to find a Lyapunov function for the system

$$\frac{dX}{dt} = \frac{PK}{L}Y + PXY, \quad \frac{dY}{dt} = -\frac{LM}{P}X - LXY.$$

If we try the easiest possible form (see p. 250) of such a function, that is, if we try $V(X, Y) = X^2 + cY^2$ with $c > 0$, we get

$$\frac{dV}{dt} = 2X\left(\frac{PK}{L}Y + PXY\right) + 2cY\left(-\frac{LM}{P}X - LXY\right)$$

$$= \left(2\frac{PK}{L} - 2c\frac{LM}{P}\right)XY + 2PX^2Y - 2cLXY^2.$$

However, there is no choice of c that would make this positive, or even nonnegative, for all X, Y close enough to zero (see Exercise 16). Other "simple" candidates to be Lyapunov functions give similarly disappointing results; the reason for this will become clear below.

Fortunately, we can analyze the behavior of the system by solving the differential equation for the trajectories, which turns out to be separable! The computation can be done using either the original coordinates x, y or the new ones X, Y (see Exercise 4). Here is the version using x and y:

$$\frac{dy}{dx} = \frac{dy/dt}{dx/dt} = \frac{Ky - Lxy}{-Mx + Pxy} = \frac{(K - Lx)y}{x(-M + Py)}$$

$$\frac{-M + Py}{y}\frac{dy}{dx} = \frac{K - Lx}{x}$$

$$\int \frac{-M + Py}{y}dy = \int \frac{K - Lx}{x}dx$$

$$-M \log y + Py = K \log x - Lx + C \quad \text{(assuming that } x > 0, y > 0\text{)}$$
$$Lx + Py - K \log x - M \log y = C.$$

At this point we have found a *first integral* $E(x, y) = Lx + Py - K \log x - M \log y$ (defined for $x > 0$, $y > 0$) for the system. Using the Second Derivative Test (see p. 258), it is easy to show (Exercise 6) that E has a local minimum at the stationary point $(\frac{K}{L}, \frac{M}{P})$. In fact, our proof of the Second Derivative Test actually shows slightly more (on p. 257): For all points (x, y) sufficiently near $(\frac{K}{L}, \frac{M}{P})$, except for $(\frac{K}{L}, \frac{M}{P})$ itself, we have $E(x, y) > E(\frac{K}{L}, \frac{M}{P})$. On the other hand, since E is constant along the trajectories, we have $\frac{d(E(x, y))}{dt} = 0$ for all these points. This shows that, in the sense of Theorem 3.8.2

(p. 254), E is a Lyapunov function! We can conclude that the stationary point $(\frac{K}{L}, \frac{M}{P})$ is stable; we will soon see that it is a center.

To get a better understanding of the phase portrait, we can study the level curves of E, along which the trajectories lie, more closely. Since E has a local minimum at $(\frac{K}{L}, \frac{M}{P})$, we might expect nearby level curves to be closed; how about level curves through points in the first quadrant that are "far away" from this stationary point? One way to investigate this is to see how E varies if one follows straight lines through the stationary point (these are *not* trajectories). For example, along the vertical line $x = \frac{K}{L}$ we have

$$E(x, y) = E\left(\frac{K}{L}, y\right) = K + Py - K \log \frac{K}{L} - M \log y$$
$$= Py - M \log y + \text{constant}.$$

This function of y has a local minimum (as expected) at $y = \frac{M}{P}$, but what's more, it is decreasing for *all* y with $0 < y < \frac{M}{P}$, increasing for *all* y with $y > \frac{M}{P}$, and it grows without bound as $y \to \infty$ and as y decreases toward 0 [see Figure 3.9.2, p. 266 and Exercise 7(a)]. Therefore, the function E takes on each value greater than $E(\frac{K}{L}, \frac{M}{P})$ exactly twice along the vertical line $x = \frac{K}{L}$,[47] once on either side of the stationary point. It is not too hard to show [see Exercise 7(b) and (c)] that E behaves similarly along any line[47] through $(\frac{K}{L}, \frac{M}{P})$. We can conclude the following:

1. There are no points (x, y) (inside the first quadrant) where $E(x, y) < E(\frac{K}{L}, \frac{M}{P})$.

2. The only point inside the first quadrant where $E(x, y) = E(\frac{K}{L}, \frac{M}{P})$ is the stationary point.

3. For every value C with $C > E(\frac{K}{L}, \frac{M}{P})$, the level curve $E(x, y) = C$ intersects each line through $(\frac{K}{L}, \frac{M}{P})$ exactly twice in the first quadrant, once on either side of the stationary point.

---

[47] More precisely, along the part of the line inside the first quadrant.

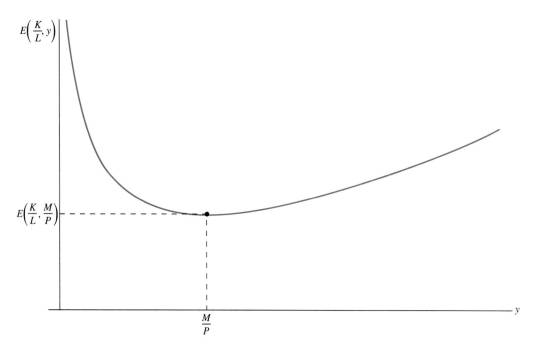

Figure 3.9.2. $E(x, y)$ along the vertical line $x = \dfrac{K}{L}$.

It follows (as illustrated by Figure 3.9.3; we won't present a formal proof) that the level curves [for $C > E(\dfrac{K}{L}, \dfrac{M}{P})$] are *closed*.

Now we know that *all* the trajectories inside the first quadrant (except for the stationary point) follow *closed* curves $Lx + Py - K \log x - M \log y = C$. The phase portrait of the system is drawn in Figure 3.9.4. Note that unless we somehow have exactly the equilibrium (stationary) populations $x = \dfrac{K}{L}$ for the predators and $y = \dfrac{M}{P}$ for the prey, the two populations will cycle, as one of the trajectories is followed round and round the stationary point.[48]

We can see intuitively what is happening in each part of such a cycle. For instance, suppose that the trajectory follows the curve *ABCD* shown by a heavy line in Figure 3.9.4. As the moving point (along the trajectory) arrives at *A*, the prey population (y) reaches its peak and thus life is ideal for the predators. As a result, the predator population (x) increases, and beyond *A* increased predation starts making inroads on the prey. As a result, prey becomes less abundant (y decreases), and the predator population will rise more slowly, reaching its maximum at *B*. See Exercise 1.

---

[48] At first sight, it might not be clear why the trajectory would have to go around the entire level curve; for instance, one could imagine the moving point going more and more slowly along the level curve so that it never got all the way around. However, it can be shown that in such a case, the level curve would have to contain a stationary point (which would be the limit of the moving point as $t \to \infty$).

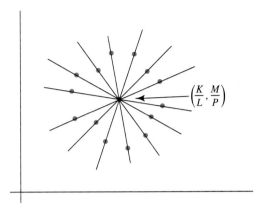

Figure 3.9.3. A level curve $E(x, y) = C$ in the first quadrant which meets each line through $\left(\dfrac{K}{L}, \dfrac{M}{P}\right)$ exactly once on either side of that point is a closed curve.

Population cycles of this sort have indeed been observed in several predator–prey situations, starting with Volterra's 1926 study of fish populations in the Adriatic. The fact that our model is clearly a bit simplistic makes it all the more impressive that the model succeeds in predicting qualitative features of such interactions!

By the way, in retrospect we can see why a Lyapunov function was hard to find. Since such a function must be either decreasing or constant along the trajectories and since the

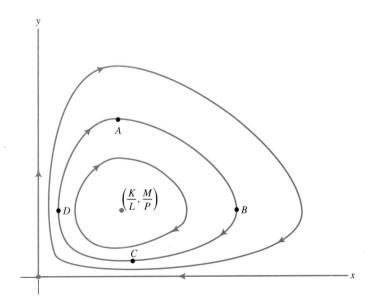

Figure 3.9.4. Phase portrait for the Volterra–Lotka system.

trajectories follow closed curves, any such function must be *constant* along the trajectories. (If not, for example if in Figure 3.9.4 the function would decrease along the trajectory from A to B, a contradiction would arise when the trajectory eventually returned to A.) In other words, in this situation (where the trajectories are closed curves) only a first integral can be a Lyapunov function.

So far, we've looked only at a predator–prey situation; what happens when, instead, two species are in competition with each other? Exercise 6 of Section 3.1 asked for a model of this situation; probably the simplest possible model is as follows. Let x and y be the population sizes for the two species. Since the species are competing for limited resources, the excess birth rate of each species will be a decreasing function of the population size of the other. Taking the easiest possible form for these functions that will allow each species to grow in the absence of the other, we get

$$\frac{1}{x}\frac{dx}{dt} = M - Py, \qquad \frac{1}{y}\frac{dy}{dt} = K - Lx$$

or

$$\frac{dx}{dt} = Mx - Pxy, \qquad \frac{dy}{dt} = Ky - Lxy,$$

where K, L, M, and P are positive constants. Something you might think about before reading further: Does this seem to be a realistic model? Do you expect it to predict the qualitative behavior of the functions x, y successfully?

Since our new system looks very similar to the Volterra–Lotka system (only the sign of $\frac{dx}{dt}$ has changed) we can expect the computations to be somewhat similar as well. There are again two stationary points, the origin and $(\frac{K}{L}, \frac{M}{P})$. However, this time the origin is an unstable node, while $(\frac{K}{L}, \frac{M}{P})$ is a saddle point (see Exercise 5). The equation for the trajectories is again separable, and we find

$$E(x, y) = Lx - Py - K \log x + M \log y \qquad (x > 0, y > 0)$$

as a first integral. However, this time the level curves of E aren't closed, and they are somewhat harder to draw. Although it can certainly be done, we will instead use a different approach to find out what will happen to the two populations in the long run.

To begin with, we sketch the direction field of the system; Figure 3.9.5(a) shows the result in the specific case $K = 2$, $L = 1$, $M = 2$, $P = 3$. It appears that as $t \to \infty$, most trajectories will head toward infinity either near the positive x-axis or near the positive y-axis; we can argue this more precisely as follows. The horizontal line $y = \frac{M}{P}$ and the vertical line $x = \frac{K}{L}$ divide the first quadrant into four regions, as shown in Figure 3.9.5(b).

SEC. 3.9 THE VOLTERRA–LOTKA EQUATIONS AND OTHER TWO-POPULATION SYSTEMS

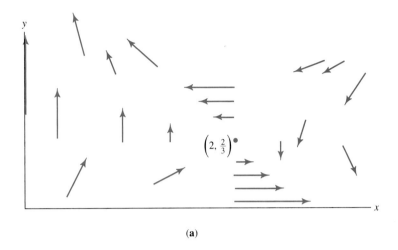

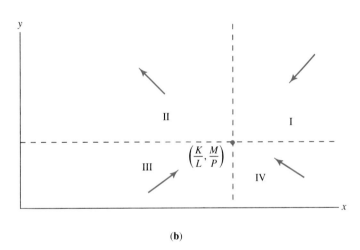

Figure 3.9.5. (a) Direction field for the system $\dfrac{dx}{dt} = 2x - 3xy$, $\dfrac{dy}{dt} = 2y - xy$. (b) Division of the first quadrant into regions.

In region I, which is defined by the inequalities $x > \dfrac{K}{L}$, $y > \dfrac{M}{P}$, we have $\dfrac{dx}{dt} < 0$ and $\dfrac{dy}{dt} < 0$ [see Exercise 9(a)], so the trajectories in this region lead to the left and downward. Similarly, all trajectories in region II lead to the left and upward, and so on [Exercise 9(b)]; this is indicated schematically in Figure 3.9.5(b) by one arrow in each region.

What will happen to a trajectory that "starts" in region I? Exactly one such trajectory, along a separatrix, will approach the saddle point $(\dfrac{K}{L}, \dfrac{M}{P})$ as $t \to \infty$ (compare Figure

3.7.7, p. 243). The others will "veer off" either above the stationary point, crossing the vertical line $x = \dfrac{K}{L}$ into region II, or to the right of the stationary point, crossing $y = \dfrac{M}{P}$ into region IV. Similarly, except for one trajectory along a separatrix, every trajectory starting in region III will end up either in region II or in region IV. On the other hand, once a trajectory is in region II or region IV, it will stay in that same region.

Let's consider a trajectory in region II. A point moving along such a trajectory has to keep moving to the left and upward, so it would seem that it either has to approach some point on the positive y-axis or go upward to infinity. It can be shown that this is indeed so, and that because there are no stationary points on the positive y-axis, it actually has to go upward to infinity (see footnote 48, on p. 266, for a sketch of a similar argument). On the other hand, it's not hard to see (Exercise 10) that if x is decreasing and $y \to \infty$ along a curve $Lx - Py - K \log x + M \log y =$ constant, then x is approaching 0. So the trajectories in region II approach the positive y-axis as they go upward to infinity.

Similarly, the trajectories in region IV approach the positive x-axis as they go to infinity (to the right). We now see that the phase portrait must look essentially like Figure 3.9.6. This means, though, that our model is predicting that unless the initial values for x and y are exactly such that $(x(t), y(t))$ will follow a separatrix, which is highly unlikely, one of the two species will die out in the long run. Even if $(x(t), y(t))$ starts out along a separatrix or at the stationary point, a small one-time change in x or y, for instance due to a disease or unusual weather, would be enough to set one of the two species on the road to eventual extinction.

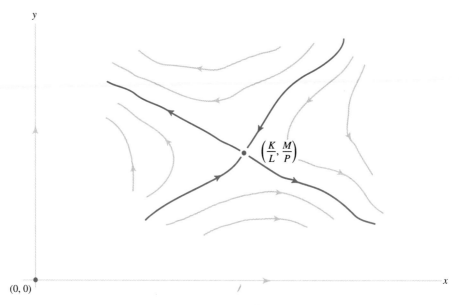

Figure 3.9.6. Phase portrait for the system $\dfrac{dx}{dt} = Mx - Pxy$, $\dfrac{dy}{dt} = Ky - Lxy$.

Based on actual experience, these predictions seem unrealistic. Although there are certainly cases of competition in nature in which one species "drives out" another which subsequently becomes extinct, there are also cases in which competing species coexist indefinitely. What went wrong?

In setting up our model, we took the excess birth rate of each species to be a decreasing function of the population size of *the other* due to competition for limited resources. However, the same limitations on resources should make the excess birth rate also be a decreasing function of the population size of *the species itself*. Thus a more realistic model would be given by equations such as

$$\begin{cases} \dfrac{1}{x}\dfrac{dx}{dt} = A_1 - B_1 x - C_1 y \\ \dfrac{1}{y}\dfrac{dy}{dt} = A_2 - C_2 x - B_2 y \end{cases}$$

with positive constants $A_1, A_2, B_1, B_2, C_1, C_2$; the constants B_1, B_2 measure the inhibiting effect of each species on its own growth, while C_1 and C_2 describe the inhibiting effect of the competition. The resulting system

$$\begin{cases} \dfrac{dx}{dt} = x(A_1 - B_1 x - C_1 y) \\ \dfrac{dy}{dt} = y(A_2 - C_2 x - B_2 y) \end{cases}$$

has four stationary points,[49] at

$$(0,0), \quad \left(0, \frac{A_2}{B_2}\right), \quad \left(\frac{A_1}{B_1}, 0\right), \quad \text{and} \quad \left(\frac{A_1 B_2 - A_2 C_1}{B_1 B_2 - C_1 C_2}, \frac{B_1 A_2 - C_2 A_1}{B_1 B_2 - C_1 C_2}\right).$$

However, depending on the values of the constants A_1, A_2, and so on, the fourth stationary point may or may not be in the first quadrant (and, of course, it has no significance in our application if it isn't). We can see this geometrically, because the fourth point is the intersection point of the two lines $B_1 x + C_1 y = A_1$ and $C_2 x + B_2 y = A_2$; there are four cases, as shown in Figure 3.9.7 (p. 272), depending on the configuration of these lines.

In each case, the lines divide the first quadrant into regions analogous to regions I to IV in Figure 3.9.5(b). Within each region, $\dfrac{dx}{dt}$ and $\dfrac{dy}{dt}$ have fixed signs, so either all trajectories within the region lead to the left or they all lead to the right, and similarly for upward or downward. Figure 3.9.7 indicates these directions schematically.

While we'll leave detailed analysis of the stationary points for Exercises 11 to 14, we can make some informed guesses on the basis of Figure 3.9.7. For example, in case (a) we

---

[49] Provided that $B_1 B_2 \neq C_1 C_2$; otherwise, there are three, or possibly infinitely many.

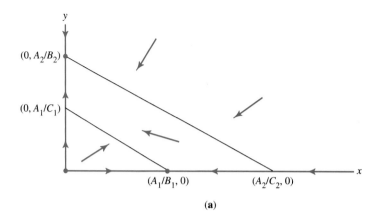

(a)

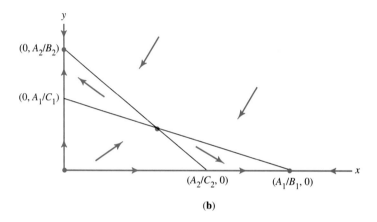

(b)

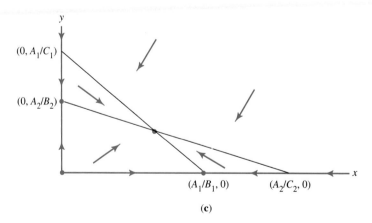

(c)

Figure 3.9.7. Possible configurations in the first quadrant for the lines $B_1 x + C_1 y = A_1$, $C_2 x + B_2 y = A_2$, and schematic directions of the trajectories in each region and along the axes.

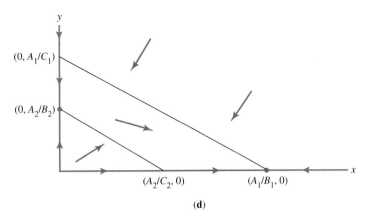

(d)

Figure 3.9.7 (Continued)

expect $(0, 0)$ to be an unstable node, $(\frac{A_1}{B_1}, 0)$ to be a saddle point, and $(0, \frac{A_2}{B_2})$ to be a stable node. Furthermore, it looks like in this case all trajectories except the ones along the x-axis approach $(0, \frac{A_2}{B_2})$ as $t \to \infty$. In other words, if in case (a) the second species is represented at all, it will "win" the competition, with y approaching the limiting value $\frac{A_2}{B_2}$ as the first species becomes extinct. Similarly, in case (d) the second species will become extinct, and x will approach its limiting value $\frac{A_1}{B_1}$.

In case (b), the fourth stationary point is a saddle point, so only the two trajectories along the separatrices approach this "state of coexistence" as $t \to \infty$. All other trajectories will approach one of $(\frac{A_1}{B_1}, 0)$ and $(0, \frac{A_2}{B_2})$, which are both stable nodes in this case. The system $\frac{dx}{dt} = Mx - Pxy$, $\frac{dy}{dt} = Ky - Lxy$ from our first (inadequate) model for competing species can be considered as an "extreme case" of (b) for which $B_1 = B_2 = 0$, so that the two stable nodes are "off at infinity" (and the names of the constants A_1, A_2, C_1, C_2 are changed to M, K, P, L, respectively).

In case (c), on the other hand, the fourth stationary point is a stable node, and except for the special trajectories along the axes, all trajectories will approach this point as $t \to \infty$. In this case, then, the two competing species do coexist in the long run. See Exercise 15.

Incidentally, note that for our improved model the differential equation for the trajectories, $\frac{dy}{dx} = \frac{y(A_2 - C_2x - B_2y)}{x(A_1 - B_1x - C_1y)}$, is no longer separable. We cannot expect to find explicit equations for the trajectories, and yet we've seen that the qualitative behavior of the system is quite predictable!

SUMMARY OF KEY RESULTS AND TECHNIQUES

For the Volterra–Lotka equations $\frac{dx}{dt} = -Mx + Pxy$, $\frac{dy}{dt} = Ky - Lxy$, the trajectories (for $x > 0$, $y > 0$) follow the closed curves $Lx + Py - K\log x - M\log y = C$ around the stationary point $(\frac{K}{L}, \frac{M}{P})$, which is a center (pp. 265–266).

In sketching phase portraits, it can be useful to divide the plane into regions according to the signs of $\frac{dx}{dt}$ and $\frac{dy}{dt}$ (pp. 268–270).

The system $\frac{dx}{dt} = x(A_1 - B_1 x - C_1 y)$, $\frac{dy}{dt} = y(A_2 - C_2 x - B_2 y)$ for population sizes x, y of competing species leads to reasonable predictions (pp. 271–273), whereas the simplified system with $B_1 = B_2 = 0$ is, in general, inadequate.

EXERCISES

1. Give an intuitive explanation of what is happening along each part of the cycle *ABCD* in Figure 3.9.4 (an explanation for part *AB* is given on p. 266).

2. (a) Show that the origin is a saddle point for the Volterra–Lotka system $\frac{dx}{dt} = -Mx + Pxy$, $\frac{dy}{dt} = Ky - Lxy$. (Use the linear approximation.)
   (b) Find all solutions of the system with $x = 0$; also with $y = 0$. Show that the special trajectories along the axes are followed as described on pp. 262–263.

3. (a) Show that the change of coordinates given by $X = x - \frac{K}{L}$, $Y = y - \frac{M}{P}$ transforms the Volterra–Lotka system into
$$\frac{dX}{dt} = \frac{PK}{L} Y + PXY, \qquad \frac{dY}{dt} = -\frac{LM}{P} X - LXY.$$
   (b) Show that the linear approximation to this system near $X = 0$, $Y = 0$ has a matrix with eigenvalues $\pm i\sqrt{KM}$.

4. (a) Solve the differential equation for the trajectories of the system
$$\frac{dX}{dt} = \frac{PK}{L} Y + PXY, \qquad \frac{dY}{dt} = -\frac{LM}{P} X - LXY.$$
   (b) Reconcile your answer with the answer found on p. 264 using the "old" coordinates x, y.

SEC. 3.9 THE VOLTERRA–LOTKA EQUATIONS AND OTHER TWO-POPULATION SYSTEMS

5. (a) Show that the system $\frac{dx}{dt} = Mx - Pxy$, $\frac{dy}{dt} = Ky - Lxy$ has exactly two stationary points, at $(0, 0)$ and at $(\frac{K}{L}, \frac{M}{P})$.
   (b) Classify these stationary points.

6. Show that the function given by $E(x, y) = Lx + Py - K \log x - M \log y$ has a local minimum at $(\frac{K}{L}, \frac{M}{P})$.

7. (a) Show that the graph of $E(\frac{K}{L}, y) = Py - M \log y + A$ (A is some constant) is as shown in Figure 3.9.2.
   (b) Show that $E(x, y) = Lx + Py - K \log x - M \log y$ varies in a similar way along the horizontal line through $(\frac{K}{L}, \frac{M}{P})$.
   *(c) Show that $E(x, y)$ varies in a similar way along *any* line through $(\frac{K}{L}, \frac{M}{P})$.
   [*Hint*: Unless the line is vertical, it has an equation of the form
   $$y - \frac{M}{P} = m\left(x - \frac{K}{L}\right).]$$

8. A biologist using the Volterra–Lotka model observes the following population figures at various points of a cycle: $x = 50$, $y = 2000$; $x = 80$, $y = 1200$; $x = 55$, $y = 800$; $x = 30$, $y = 1100$. What would she expect the equilibrium populations to be? (A calculator or computer will be helpful.)

9. (a) Show that in region I shown in Figure 3.9.5(b), $\frac{dx}{dt} < 0$ and $\frac{dy}{dt} < 0$, so all trajectories lead to the left and downward.
   (b) Give similar arguments for regions II, III, and IV.

10. Show that if x is decreasing and $y \to \infty$ along a curve
    $$Lx - Py - K \log x + M \log y = \text{constant} \quad (K, L, M, P > 0),$$
    then $x \to 0$.

11. (a) Show that in Figure 3.9.7, we have case (a) if and only if $A_1 C_2 < A_2 B_1$ and $A_1 B_2 < A_2 C_1$.
    (b) Assuming that $A_1 C_2 < A_2 B_1$ and $A_1 B_2 < A_2 C_1$, classify the three stationary points along the axes (using linear approximations to the system).

12. (a) Give inequalities, similar to the ones in Exercise 11(a), that characterize case (d) in Figure 3.9.7.
    (b) Classify the three stationary points along the axes in this case (using linear approximation).

13. (a) Give inequalities, similar to the ones in Exercise 11(a), that characterize case (b) in Figure 3.9.7.
    (b) Classify the three stationary points along the axes in this case (using linear approximation).
    (c) Show that the linear approximation to the system near the fourth stationary point has matrix $\begin{bmatrix} -B_1\bar{x} & -C_1\bar{x} \\ -C_2\bar{y} & -B_2\bar{y} \end{bmatrix}$, where

    $$\bar{x} = \frac{A_1 B_2 - A_2 C_1}{B_1 B_2 - C_1 C_2}, \quad \bar{y} = \frac{B_1 A_2 - C_2 A_1}{B_1 B_2 - C_1 C_2}$$

    are the coordinates of the stationary point.
    (d) Show that the product of the eigenvalues of this matrix is $(B_1 B_2 - C_1 C_2)\bar{x}\bar{y}$.
    [*Hint*: If the eigenvalues are λ_1 and λ_2 and the characteristic equation is $\lambda^2 + p\lambda + q = 0$, then since $\lambda^2 + p\lambda + q = (\lambda - \lambda_1)(\lambda - \lambda_2)$, we have $\lambda_1 \lambda_2 = q$.]
    (e) Show that in case (b) in Figure 3.9.7, the fourth stationary point is a saddle point.
    [*Hint*: First show, using the results of part (a), that $B_1 B_2 - C_1 C_2 < 0$.]

14. (a) Give inequalities, similar to the ones in Exercise 11(a), that characterize case (c) in Figure 3.9.7.
    (b) Classify the three stationary points along the axes in this case (using linear approximation).
    (c) Using the result of Exercise 13(c), show that the eigenvalues of the matrix for the linear approximation near the fourth stationary point are given by

    $$\lambda = \frac{-(B_1\bar{x} + B_2\bar{y}) \pm \sqrt{(B_1\bar{x} - B_2\bar{y})^2 + 4C_1 C_2 \bar{x}\bar{y}}}{2},$$

    and explain why this shows that the eigenvalues are real.
    *(d) Using the results of Exercise 13(d) and of (c) of this exercise, show that in case (c) in Figure 3.9.7, the fourth stationary point is a stable node.
    [*Hint*: First show, using the inequalities from (a), that $B_1 B_2 - C_1 C_2 > 0$.]

15. (a) Using the results of Exercises 13 and 14, show that in our model for competing species peaceful coexistence will occur if and only if $A_1 B_2 > A_2 C_1$ and $A_2 B_1 > A_1 C_2$.
    (b) Show that if A_1 and A_2 are considered variable, peaceful coexistence can occur if and only if $B_1 B_2 > C_1 C_2$.

16. Show (assuming that K, L, M, P are positive constants) that there is no number $c > 0$ such that $(2\frac{PK}{L} - 2c - \frac{LM}{P})XY + 2PX^2 Y - 2cLXY^2 \geq 0$ for all X, Y close enough to zero.

17. Consider the system $\frac{dx}{dt} = -y + x^3$, $\frac{dy}{dt} = x - y^3$.
    (a) Find all stationary points; show that there are two saddle points and that at the origin, the linear approximation has a center.

(b) Show that $V(x, y) = x^2 + cy^2$ is *not* a Lyapunov function for the system (for any $c > 0$).

(c) Sketch the curves $-y + x^3 = 0$ and $x - y^3 = 0$ and indicate how these curves divide the plane into regions. For each region, find the signs of $\dfrac{dx}{dt}$ and $\dfrac{dy}{dt}$.

(d) Use the results of parts (a), (b), and (c) to sketch a plausible phase portrait for the system.

*(e) Show that the system has two special trajectories along the hyperbola $xy = -1$.

[*Hint*: First show that $\dfrac{d(xy)}{dt} = 0$ if $xy = -1$.]

*(f) Show that $E(x, y) = \dfrac{x^2 + y^2}{(1 + xy)^2}$ is a first integral for the system.

*(g) Show that the curve $E(x, y) = \dfrac{1}{2}$ consists of two hyperbolas.

[*Hint*: First show that $E(x, y) = \dfrac{1}{2}$ is equivalent to $2(x - y)^2 = (1 - xy)^2$.]

Where are these hyperbolas in relation to the stationary points? Use this information to sketch an improved version of the phase portrait.

FURTHER READING

For more examples of interesting nonlinear systems, with suggested computer experiments to investigate their behavior, see Koçak, *Differential and Difference Equations Through Computer Experiments*, 2nd ed. (Springer-Verlag, 1989).

CHAPTER 4

Higher-Order Differential Equations and General Systems

4.1 HIGHER-ORDER DIFFERENTIAL EQUATIONS: INTRODUCTION

Although we have seen that many of the differential equations encountered in practice are first- and second-order equations, there are important applications in which higher-order equations occur. For instance, if a uniform elastic beam is put along the x-axis and subjected to a vertical load $p(x)$ (which may not be constant along the beam), then the deflection y of the beam from the horizontal will satisfy the equation $k\dfrac{d^4y}{dx^4} = p(x)$, where k is a constant that depends on the physical properties of the beam.

This particular equation can be solved immediately by four successive integrations, but other problems concerning elasticity and vibrations of mechanical systems give rise to more complicated fourth-order equations. Thus the treatment of higher-order equations is not only of theoretical interest. (See also ''Further Reading'' following the Exercises.)

Fortunately, much of the theory for second-order equations can be extended to the higher-order case (although the notation tends to be a bit more unpleasant). Accordingly, we'll summarize the theory fairly quickly, while paying particular attention to new difficulties that show up. It may be a good idea to review the corresponding topics in Chapter 2 as you go along!

Recall that an nth-order differential equation involves the nth derivative $\dfrac{d^nx}{dt^n}$ of some unknown function $x(t)$; by the implicit function theorem, it is (under reasonable conditions) possible to rewrite the equation as one or more equations of the ''explicit'' form $\dfrac{d^nx}{dt^n} = f\left(t, x, \dfrac{dx}{dt}, \ldots, \dfrac{d^{n-1}x}{dt^{n-1}}\right)$, and we'll assume that this has been done. An **initial**

value problem consists of such an equation $\dfrac{d^n x}{dt^n} = f\left(t, x, \dfrac{dx}{dt}, \ldots, \dfrac{d^{n-1}x}{dt^{n-1}}\right)$ together with initial conditions

$$x(t_0) = x_0, \quad \dfrac{dx}{dt}\bigg|_{t=t_0} = v_0, \quad \dfrac{d^2 x}{dt^2}\bigg|_{t=t_0} = a_0, \quad \ldots, \quad \dfrac{d^{n-1}x}{dt^{n-1}}\bigg|_{t=t_0} = z_0.$$

Such an initial value problem will have a unique solution (defined for all t in some interval containing t_0) provided that f and its partial derivatives with respect to $x, \dfrac{dx}{dt}, \ldots, \dfrac{d^{n-1}x}{dt^{n-1}}$ are defined and continuous.[1]

EXAMPLE The initial value problem

$$\dfrac{d^3 x}{dt^3} = \dfrac{tx - \dfrac{dx}{dt}}{1 + \dfrac{d^2 x}{dt^2}}, \quad x(0) = 0, \quad \dfrac{dx}{dt}\bigg|_{t=0} = 1, \quad \dfrac{d^2 x}{dt^2}\bigg|_{t=0} = 1$$

will have a unique solution (finding it is another matter!). If the last condition is changed to $\dfrac{d^2 x}{dt^2}\bigg|_{t=0} = -1$, there will no longer be a solution. (Why?) ∎

A *linear n*th-order differential equation

$$\dfrac{d^n x}{dt^n} + p_{n-1}(t)\dfrac{d^{n-1}x}{dt^{n-1}} + \cdots + p_1(t)\dfrac{dx}{dt} + p_0(t)x = h(t)$$

will have solutions that are defined whenever $p_{n-1}(t), \ldots, p_1(t), p_0(t)$, and $h(t)$ are defined and continuous. Just as for second-order equations, we will consider the homogeneous case first; in the next section we'll study the methods of variation of constants and of undetermined coefficients, which can be used in the inhomogeneous case.

For a homogeneous nth-order linear equation

$$\dfrac{d^n x}{dt^n} + p_{n-1}(t)\dfrac{d^{n-1}x}{dt^{n-1}} + \cdots + p_1(t)\dfrac{dx}{dt} + p_0(t)x = 0$$

we would like to show that there are n "basic" solutions $x_1(t), x_2(t), \ldots, x_n(t)$ such that every solution $x(t)$ is a **linear combination** $x(t) = \alpha_1 x_1(t) + \alpha_2 x_2(t) + \cdots + \alpha_n x_n(t)$ of these (with constants $\alpha_1, \alpha_2, \ldots, \alpha_n$, called the coefficients of the linear combination). Conversely, it is easy to show that any linear combination of solutions is again a solution (see Exercise 13).

---

[1] More precisely, defined and continuous in a rectangular box containing $(t_0, x_0, v_0, a_0, \ldots, z_0)$ in its interior. Compare the Existence and Uniqueness Theorem in the first-order case (see p. 40).

SEC. 4.1 HIGHER-ORDER DIFFERENTIAL EQUATIONS: INTRODUCTION

EXAMPLE 4.1.1

The homogeneous third-order equation

$$\frac{d^3x}{dt^3} - \frac{5}{t}\frac{d^2x}{dt^2} + \frac{12}{t^2}\frac{dx}{dt} - \frac{12}{t^3}x = 0$$

has solutions $x_1(t) = t$, $x_2(t) = t^3$, $x_3(t) = t^4$. (Check this!) Given this information, we'd like to conclude that the general solution is $x(t) = \alpha_1 t + \alpha_2 t^3 + \alpha_3 t^4$, with $\alpha_1, \alpha_2, \alpha_3$ arbitrary constants. We will return to this example; note for future reference that the coefficients of the differential equation are defined and continuous *except at* $t = 0$. ☐

Just as for a second-order equation, we can't quite expect *any* n solutions $x_1(t), x_2(t)$, ..., $x_n(t)$ to form a set of basic solutions to the equation. Remember that for a second-order equation, two solutions can only form a pair of basic solutions if neither of them is a constant multiple of the other. In our case, n solutions $x_1(t), x_2(t), \ldots, x_n(t)$ can only form a set of basic solutions if none of them is a linear combination of the others.

This property of functions[2] is called linear independence. That is, the functions $x_1, \ldots, x_n$ are called **linearly independent** if none of them is a linear combination of the others. For instance, in the example above $x_1(t) = t$, $x_2(t) = t^3$, and $x_3(t) = t^4$ are linearly independent, because t is not a linear combination of t^3 and t^4, t^3 is not a linear combination of t and t^4, and t^4 is not one of t and t^3. As you can see, checking linear independence in this way is tedious; it is more efficient to check the following alternative criterion, which is justified in Exercise 15.

Test for linear independence. The functions[2] $x_1, x_2, \ldots, x_n$ are linearly independent if, and only if, the *only* way a linear combination $\alpha_1 x_1 + \alpha_2 x_2 + \cdots + \alpha_n x_n$ can be the zero function is for all the constants $\alpha_1, \alpha_2, \ldots, \alpha_n$ to equal zero.

EXAMPLE 4.1.1 (continued)

For the functions $x_1(t) = t$, $x_2(t) = t^3$, $x_3(t) = t^4$ above, the linear combination $\alpha_1 x_1(t) + \alpha_2 x_2(t) + \alpha_3 x_3(t) = \alpha_1 t + \alpha_2 t^3 + \alpha_3 t^4$ can be the zero function only if the coefficients α_1, α_2, and α_3 are all zero, so x_1, x_2, x_3 are linearly independent. ☐

EXAMPLE 4.1.2

The functions y_1, y_2, y_3, y_4 given by $y_1(t) = t^2$, $y_2(t) = t^3$, $y_3(t) = -2t^2 - 3t^3$, $y_4(t) = t^4$ are **linearly dependent** (that is, not linearly independent), since we can take the linear combination $2y_1 + 3y_2 + y_3 + 0 \cdot y_4$, and this will be the zero function even though $\alpha_1 = 2, \alpha_2 = 3, \alpha_3 = 1, \alpha_4 = 0$ are not all zero. (According to the definition, since the functions y_1, y_2, y_3, y_4 are not linearly independent, one of them should be a linear combination of the others. See Exercise 14.) ■

---

[2] Or, more generally, vectors in a vector space; see textbooks on linear algebra.

Let us assume for now that we have n linearly independent solutions $x_1(t), \ldots, x_n(t)$ to the homogeneous equation

$$\frac{d^n x}{dt^n} + p_{n-1}(t)\frac{d^{n-1} x}{dt^{n-1}} + \cdots + p_1(t)\frac{dx}{dt} + p_0(t)x = 0,$$

on an interval where the coefficients $p_{n-1}(t), \ldots, p_0(t)$ are defined and continuous. To show that every solution $x(t)$ is a linear combination of $x_1(t), \ldots, x_n(t)$, it is enough to show that $x(t)$ is a solution to the same initial value problem as some linear combination $\alpha_1 x_1(t) + \cdots + \alpha_n x_n(t)$. That is, we want to be able to find numbers $\alpha_1, \ldots, \alpha_n$ such that

$$\begin{cases} \alpha_1 x_1(t_0) & + \cdots + \alpha_n x_n(t_0) & = x(t_0) \\ \alpha_1 x_1'(t_0) & + \cdots + \alpha_n x_n'(t_0) & = x'(t_0) \\ \alpha_1 x_1''(t_0) & + \cdots + \alpha_n x_n''(t_0) & = x''(t_0) \\ & \vdots & \\ \alpha_1 x_1^{(n-1)}(t_0) & + \cdots + \alpha_n x_n^{(n-1)}(t_0) & = x^{(n-1)}(t_0). \end{cases}$$

This is possible provided that the determinant

$$\begin{vmatrix} x_1(t_0) & \cdots & x_n(t_0) \\ x_1'(t_0) & \cdots & x_n'(t_0) \\ \vdots & & \vdots \\ x_1^{(n-1)}(t_0) & \cdots & x_n^{(n-1)}(t_0) \end{vmatrix}$$

is not zero (see Appendix B, p. 624); as in the case of a second-order equation, it is enough to show that this is true for *some* value of t_0. The determinant above is called the **Wronskian** (evaluated at t_0) of the n functions $x_1, \ldots, x_n$, and it is denoted by $W(x_1, \ldots, x_n)(t_0)$. It can be shown that if the Wronskian $W(x_1, \ldots, x_n)$ were identically zero, the functions would be linearly dependent. The proof of this is a bit more involved than in the case of two functions; the hard part is to show that the functions would be linearly dependent on some interval, and then it follows by uniqueness that they would be linearly dependent everywhere. For details, see Exercises 18 and 19, or the paper "Why does the Wronskian work?", *American Mathematical Monthly* 95(1988), p. 46.

We can conclude, then, that **any n linearly independent solutions** $x_1(t), x_2(t), \ldots, x_n(t)$ **form a set of basic solutions** in the sense that all the solutions of the homogeneous linear equation are the linear combinations $\alpha_1 x_1(t) + \alpha_2 x_2(t) + \cdots + \alpha_n x_n(t)$ of these. Note that this is an analogue to Theorem 2.2.3 (p. 89).

There is also an analogue to Theorem 2.2.4 (see Exercise 17 for a proof): If $x_1(t), \ldots, x_n(t)$ are any n functions which are $(n-1)$ times differentiable and linearly dependent, then their Wronskian

$$W(x_1, \ldots, x_n)(t) = \begin{vmatrix} x_1(t) & \cdots & x_n(t) \\ \vdots & & \vdots \\ x_1^{(n-1)}(t) & \cdots & x_n^{(n-1)}(t) \end{vmatrix}$$

is zero for all t. If, on the other hand, $x_1(t), \ldots, x_n(t)$ are linearly independent solutions of the equation $\dfrac{d^n x}{dt^n} + p_{n-1}(t)\dfrac{d^{n-1} x}{dt^{n-1}} + \cdots + p_1(t)\dfrac{dx}{dt} + p_0(t)x = 0$, then their Wronskian is never zero as long as $p_0(t), p_1(t), \ldots, p_{n-1}(t)$ are defined and continuous.

EXAMPLE 4.1.3   For the functions $x_1(t) = t$, $x_2(t) = t^3$, $x_3(t) = t^4$ considered in Example 4.1.1, the Wronskian is

$$W(x_1, x_2, x_3)(t) = \begin{vmatrix} t & t^3 & t^4 \\ 1 & 3t^2 & 4t^3 \\ 0 & 6t & 12t^2 \end{vmatrix} = t(36t^4 - 24t^4) = 12t^5.$$

This is never zero except for $t = 0$, the value of t where the coefficients $p_2(t) = -\dfrac{5}{t}$, $p_1(t) = \dfrac{12}{t^2}$, $p_0(t) = -\dfrac{12}{t^3}$ of the differential equation satisfied by x_1, x_2, x_3 are undefined (as was noted on p. 281).

On the other hand, for the four functions $y_1(t) = t^2$, $y_2(t) = t^3$, $y_3(t) = -2t^2 - 3t^3$, $y_4(t) = t^4$ from Example 4.1.2, the Wronskian is

$$W(y_1, y_2, y_3, y_4)(t) = \begin{vmatrix} t^2 & t^3 & -2t^2 - 3t^3 & t^4 \\ 2t & 3t^2 & -4t - 9t^2 & 4t^3 \\ 2 & 6t & -4 - 18t & 12t^2 \\ 0 & 6 & -18 & 24t \end{vmatrix},$$

which is zero for all t [see Exercise 14(c)]. ∎

As in the case of a second-order equation, the easiest type of nth-order equation is homogeneous with constant coefficients. Such an equation looks like

$$\frac{d^n x}{dt^n} + p_{n-1} \frac{d^{n-1} x}{dt^{n-1}} + \cdots + p_1 \frac{dx}{dt} + p_0 x = 0$$

with constants $p_0, p_1, \ldots, p_{n-1}$. We can look for solutions of the form $x(t) = e^{\lambda t}$, and we find that $e^{\lambda t}$ is indeed a solution provided that

$$\lambda^n + p_{n-1} \lambda^{n-1} + \cdots + p_1 \lambda + p_0 = 0.$$

Therefore, if this polynomial equation (called the **characteristic equation**) has n distinct solutions (roots) $\lambda_1, \ldots, \lambda_n$, then we will have n solutions $e^{\lambda_1 t}, e^{\lambda_2 t}, \ldots, e^{\lambda_n t}$ of our differential equation. To see whether these are linearly independent, we can compute their Wronskian, which is

$$W(e^{\lambda_1 t}, \ldots, e^{\lambda_n t}) = \begin{vmatrix} e^{\lambda_1 t} & e^{\lambda_2 t} & \cdots & e^{\lambda_n t} \\ \lambda_1 e^{\lambda_1 t} & \lambda_2 e^{\lambda_2 t} & \cdots & \lambda_n e^{\lambda_n t} \\ \lambda_1^2 e^{\lambda_1 t} & \lambda_2^2 e^{\lambda_2 t} & \cdots & \lambda_n^2 e^{\lambda_n t} \\ \vdots & \vdots & & \vdots \\ \lambda_1^{n-1} e^{\lambda_1 t} & \lambda_2^{n-1} e^{\lambda_2 t} & \cdots & \lambda_n^{n-1} e^{\lambda_n t} \end{vmatrix}$$

$$= e^{\lambda_1 t} \cdot e^{\lambda_2 t} \cdot \ldots \cdot e^{\lambda_n t} \cdot \begin{vmatrix} 1 & 1 & \cdots & 1 \\ \lambda_1 & \lambda_2 & \cdots & \lambda_n \\ \lambda_1^2 & \lambda_2^2 & \cdots & \lambda_n^2 \\ \vdots & \vdots & & \vdots \\ \lambda_1^n & \lambda_2^n & \cdots & \lambda_n^n \end{vmatrix}.$$

Here we have factored $e^{\lambda_1 t}, e^{\lambda_2 t}, \ldots, e^{\lambda_n t}$ out of the first, second, ..., nth column, respectively. Since the exponentials are never zero, the Wronskian will be nonzero (and thus the functions will be linearly independent) provided that the determinant

$$\begin{vmatrix} 1 & \cdots & 1 \\ \lambda_1 & \cdots & \lambda_n \\ \vdots & & \vdots \\ \lambda_1^{n-1} & \cdots & \lambda_n^{n-1} \end{vmatrix}$$

is nonzero. This determinant is of a famous type; it is called a **Vandermonde determinant**. In Exercise 19 of Appendix B, Section B.4, it is shown that this Vandermonde determinant equals the product $(\lambda_2 - \lambda_1)(\lambda_3 - \lambda_1)(\lambda_3 - \lambda_2) \cdots$ of all differences $\lambda_j - \lambda_i$ with $j > i$ and $1 \le i \le n, 1 \le j \le n$. Therefore, the determinant is nonzero provided that none of the factors $\lambda_j - \lambda_i$ is zero, that is, provided that the numbers $\lambda_1, \ldots, \lambda_n$ are distinct. Thus we have found:

If the characteristic equation $\lambda^n + p_{n-1}\lambda^{n-1} + \cdots + p_1 \lambda + p_0 = 0$ **has n distinct roots** $\lambda_1, \ldots, \lambda_n$, **then the differential equation**

$$\frac{d^n x}{dt^n} + p_{n-1}\frac{d^{n-1}x}{dt^{n-1}} + \cdots + p_1\frac{dx}{dt} + p_0 x = 0 \qquad \text{(with constant coefficients)}$$

has n basic solutions $e^{\lambda_1 t}, \ldots, e^{\lambda_n t}$ **and general solution**

$$x(t) = \alpha_1 e^{\lambda_1 t} + \cdots + \alpha_n e^{\lambda_n t}, \qquad \alpha_1, \ldots, \alpha_n \text{ arbitrary constants.}$$

EXAMPLE 4.1.4 Solve the initial value problem

$$x^{(3)} + 2x'' - 13x' + 10x = 0, \qquad x(0) = 0, \quad x'(0) = -9, \quad x''(0) = 15.$$

The characteristic equation is $\lambda^3 + 2\lambda^2 - 13\lambda + 10 = 0$. If we notice that $\lambda = 1$ is a root of this equation, we can factor out $(\lambda - 1)$ and get

$$(\lambda - 1)(\lambda^2 + 3\lambda - 10) = 0$$
$$(\lambda - 1)(\lambda - 2)(\lambda + 5) = 0.$$

Thus we have $\lambda_1 = 1, \lambda_2 = 2, \lambda_3 = -5$ and the general solution is

$$x(t) = \alpha_1 e^t + \alpha_2 e^{2t} + \alpha_3 e^{-5t}.$$

Now we need to find $\alpha_1, \alpha_2, \alpha_3$ such that the initial conditions $x(0) = 0, x'(0) = -9, x''(0) = 15$ are satisfied. This yields the system of linear equations

$$\begin{cases} \alpha_1 + \alpha_2 + \alpha_3 = 0 \\ \alpha_1 + 2\alpha_2 - 5\alpha_3 = -9 \\ \alpha_1 + 4\alpha_2 + 25\alpha_3 = 15. \end{cases}$$

Solving these (by elimination or row reduction), we get $\alpha_1 = 2, \alpha_2 = -3, \alpha_3 = 1$, so the solution to our initial value problem is

$$x(t) = 2e^t - 3e^{2t} + e^{-5t}. \qquad \blacksquare$$

Note that even for a third-degree (constant-coefficient) differential equation, the characteristic equation $\lambda^3 + p_2\lambda^2 + p_1\lambda + p_0 = 0$ is usually hard to solve.[3] For $n \geq 5$ there is, by a celebrated theorem of Abel and Galois, *no* general algebraic method for the solution of nth-degree polynomial equations. However, the roots of the characteristic equation can be approximated arbitrarily closely, for instance using Newton's method.

Additional complications arise if the characteristic equation has complex or multiple (repeated) roots. Complex (nonreal) roots come in pairs $\lambda, \bar{\lambda}$ [see Appendix A, Section A.2, Exercise 21(c)], just as for second-order equations. If $\lambda = \mu + i\nu$, $\bar{\lambda} = \mu - i\nu$ is such a pair of roots, then $e^{\lambda t}$ and $e^{\bar{\lambda} t}$ are complex-valued solutions to the differential equation, which can be combined to give the real-valued solutions

$$\frac{1}{2}(e^{\lambda t} + e^{\bar{\lambda} t}) = \text{Re}(e^{\lambda t}) = e^{\mu t} \cos \nu t$$

and

$$\frac{1}{2i}(e^{\lambda t} - e^{\bar{\lambda} t}) = \text{Im}(e^{\lambda t}) = e^{\mu t} \sin \nu t.$$

Provided that all the roots of the characteristic equation are distinct, it can be shown that the various functions $e^{\mu t} \cos \nu t$, $e^{\mu t} \sin \nu t$ (for complex roots $\lambda = \mu + i\nu$, $\bar{\lambda}$) and $e^{\lambda t}$ (for real roots λ) are linearly independent, so we will have n basic solutions that can be combined to give the general solution.

EXAMPLE 4.1.5 $x^{(4)} + x'' - 10x' = 0$.

The characteristic equation $\lambda^4 + \lambda^2 - 10\lambda = 0$ factors:

$$\lambda(\lambda^3 + \lambda - 10) = 0$$
$$\lambda(\lambda - 2)(\lambda^2 + 2\lambda + 5) = 0.$$

Since $\lambda^2 + 2\lambda + 5 = 0$ has complex roots $\lambda = -1 \pm 2i$, we get $\lambda_1 = 0$, $\lambda_2 = 2$ as two real roots and $\lambda_3 = -1 + 2i$, $\bar{\lambda}_3 = -1 - 2i$ as a pair of complex roots to the characteristic equation. The corresponding basic solutions are

$$e^{\lambda_1 t} = 1, \quad e^{\lambda_2 t} = e^{2t}, \quad \text{Re}(e^{\lambda_3 t}) = e^{-t} \cos 2t, \quad \text{Im}(e^{\lambda_3 t}) = e^{-t} \sin 2t,$$

so the general (real) solution to the differential equation is

$$x(t) = \alpha_1 + \alpha_2 e^{2t} + \alpha_3 e^{-t} \cos 2t + \alpha_4 e^{-t} \sin 2t, \qquad \alpha_1, \alpha_2, \alpha_3, \alpha_4 \text{ arbitrary.} \quad \blacksquare$$

Finally, we have the case of multiple roots. Recall (from Section 2.4) that for a second-order constant coefficient equation $x'' + px' + qx = 0$ for which λ is a double root of $\lambda^2 + p\lambda + q = 0$, a pair of basic solutions is given by $e^{\lambda t}, te^{\lambda t}$. It can be shown in general[4]

---

[3] In fact, although a solution method has been known since the sixteenth century (see "Further Reading" following the Exercises), even professional mathematicians are not always familiar with it!

[4] This is shown in one particular case in Exercise 16.

that if λ is a multiple root that occurs m times in the characteristic equation, then along with the basic solution $e^{\lambda t}$ the differential equation also has basic solutions $te^{\lambda t}, t^2 e^{\lambda t}, \ldots,$ $t^{m-1}e^{\lambda t}$. If λ is a multiple *complex* root, we can take real and imaginary parts of these solutions. In all cases we will end up with n (linearly independent) basic solutions in all for an nth-order differential equation.

EXAMPLE 4.1.6 $x^{(3)} - 6x'' + 12x' - 8x = 0$.

The characteristic equation can be rewritten as $(\lambda - 2)^3 = 0$, so there is a triple root $\lambda = 2$. Thus we have basic solutions $e^{2t}, te^{2t}, t^2 e^{2t}$ and general solution

$$x(t) = \alpha_1 e^{2t} + \alpha_2 te^{2t} + \alpha_3 t^2 e^{2t}$$
$$= (\alpha_1 + \alpha_2 t + \alpha_3 t^2)e^{2t}, \qquad \alpha_1, \alpha_2, \alpha_3 \text{ arbitrary.} \quad \blacksquare$$

EXAMPLE 4.1.7 $x^{(4)} + 2x'' + x = 0$.

The characteristic equation is

$$\lambda^4 + 2\lambda^2 + 1 = 0$$
$$(\lambda^2 + 1)^2 = 0$$
$$(\lambda - i)^2(\lambda + i)^2 = 0,$$

so we have a pair of double complex roots $\lambda = i, \overline{\lambda} = -i$. Along with the usual basic solutions $\text{Re}(e^{\lambda t}) = \cos t$, $\text{Im}(e^{\lambda t}) = \sin t$ we therefore also have $\text{Re}(te^{\lambda t}) = t \cos t$, $\text{Im}(te^{\lambda t}) = t \sin t$, and the general solution is

$$x(t) = \alpha_1 \cos t + \alpha_2 \sin t + \alpha_3 t \cos t + \alpha_4 t \sin t, \qquad \alpha_1, \alpha_2, \alpha_3, \alpha_4 \text{ arbitrary.} \quad \blacksquare$$

NOTE: In practice this complication (of multiple roots) is rare.

SUMMARY OF KEY CONCEPTS, RESULTS, AND TECHNIQUES

Existence and uniqueness of solutions: An initial value problem

$$\frac{d^n x}{dt^n} = f\left(t, x, \frac{dx}{dt}, \ldots, \frac{d^{n-1}x}{dt^{n-1}}\right), \quad x(t_0) = x_0, \quad \left.\frac{dx}{dt}\right|_{t=t_0} = v_0, \ldots, \left.\frac{d^{n-1}x}{dt^{n-1}}\right|_{t=t_0} = z_0$$

has a unique solution provided f and its partial derivatives with respect to $x, \frac{dx}{dt}, \ldots,$ $\frac{d^{n-1}x}{dt^{n-1}}$ are defined and continuous. (See p. 280 for a more precise statement.)

Linear combination of functions $x_1(t), \ldots, x_n(t)$ (p. 280): Any function of the form $x(t) = \alpha_1 x_1(t) + \cdots + \alpha_n x_n(t)$, where the coefficients $\alpha_1, \ldots, \alpha_n$ are constants.

Linearly independent (p. 281): The functions $x_1, \ldots, x_n$ are linearly independent if none of them is a linear combination of the others, or, equivalently, if the only way for $\alpha_1 x_1 + \cdots + \alpha_n x_n$ to be identically zero is for $\alpha_1, \ldots, \alpha_n$ all to be zero.

Basic solutions (p. 282): Any n linearly independent solutions $x_1, \ldots, x_n$ of a *homogeneous linear* differential equation

$$\frac{d^n x}{dt^n} + p_{n-1}(t)\frac{d^{n-1}x}{dt^{n-1}} + \cdots + p_1(t)\frac{dx}{dt} + p_0(t)x = 0$$

form a set of basic solutions. That is, *any* solution $x(t)$ can be written as a linear combination of $x_1(t), \ldots, x_n(t)$.

Wronskian (p. 282):

$$W(x_1, \ldots, x_n)(t) = \begin{vmatrix} x_1(t) & \cdots & x_n(t) \\ x_1'(t) & \cdots & x_n'(t) \\ \vdots & & \vdots \\ x_1^{(n-1)}(t) & \cdots & x_n^{(n-1)}(t) \end{vmatrix}.$$

If $x_1(t), \ldots, x_n(t)$ are all solutions of

$$\frac{d^n x}{dt^n} + p_{n-1}(t)\frac{d^{n-1}x}{dt^{n-1}} + \cdots + p_1(t)\frac{dx}{dt} + p_0(t)x = 0$$

and $p_0(t), \ldots, p_{n-1}(t)$ are defined and continuous, then either $W(x_1, \ldots, x_n)(t) \neq 0$ for all t, and $x_1, \ldots, x_n$ form a set of basic solutions, or $W(x_1, \ldots, x_n)(t) = 0$ for all t, and they don't.

Constant-coefficient homogeneous equation

$$\frac{d^n x}{dt^n} + p_{n-1}\frac{d^{n-1}x}{dt^{n-1}} + \cdots + p_1\frac{dx}{dt} + p_0 x = 0:$$

The **characteristic equation** (p. 283) $\lambda^n + p_{n-1}\lambda^{n-1} + \cdots + p_1\lambda + p_0 = 0$ will have n roots, some of which may be repeated or complex. If λ is a real root, then $e^{\lambda t}$ is a solution of the differential equation (p. 283). If $\lambda = \mu + i\nu$, $\bar\lambda = \mu - i\nu$ form a pair of complex conjugate roots, then $\text{Re}(e^{\lambda t}) = e^{\mu t}\cos\nu t$ and $\text{Im}(e^{\lambda t}) = e^{\mu t}\sin\nu t$ are solutions (p. 285). If λ is an m-fold root, then $te^{\lambda t}, \ldots, t^{m-1}e^{\lambda t}$ are solutions, and if λ is a multiple *complex* root, we can take real and imaginary parts of these solutions (p. 286). Combining these cases appropriately, we can always find n basic solutions for the (constant-coefficient, homogeneous) differential equation, provided that we can solve its characteristic equation.

EXERCISES

Solve the following differential equations and initial value problems.

1. $\dfrac{d^3 x}{dt^3} - \dfrac{d^2 x}{dt^2} + 4\dfrac{dx}{dt} - 4x = 0.$

2. $\dfrac{d^3 y}{dx^3} - \dfrac{d^2 y}{dx^2} + 9\dfrac{dy}{dx} - 9y = 0.$

3. $\dfrac{d^3y}{dx^3} - 3\dfrac{d^2y}{dx^2} + \dfrac{dy}{dx} - 3y = 0.$
   [Hint: $y = e^{3x}$ is one solution.]

4. $\dfrac{d^3x}{dt^3} - 5\dfrac{d^2x}{dt^2} + \dfrac{dx}{dt} - 5x = 0.$
   [Hint: $x = e^{5t}$ is one solution.]

5. $\dfrac{d^4x}{dt^4} + 8\dfrac{d^2x}{dt^2} + 16x = 0.$

6. $\dfrac{d^4x}{dt^4} - 5\dfrac{d^2x}{dt^2} + 4x = 0,\quad x(0) = 7,\quad x'(0) = -5,\quad x''(0) = 13,\quad x^{(3)}(0) = -29.$

7. $x^{(3)} - 5x'' + 5x' - 4x = 0.$ (x is a function of t.)

8. $x^{(3)} + 5x'' + 3x' - 9x = 0.$

9. $\dfrac{d^4y}{dx^4} - 10\dfrac{d^2y}{dx^2} + 9y = 0,\quad y(0) = -1,\quad y'(0) = -1,\quad y''(0) = 23,$
   $y^{(3)}(0) = -25.$

10. $\dfrac{d^4y}{dx^4} + 6\dfrac{d^2y}{dx^2} + 9y = 0.$

11. Find a linear differential equation with constant coefficients that has $\sin 2t$ and te^{3t} as solutions.

12. Find a linear differential equation with constant coefficients that has e^{-t} and $t \sin t$ as solutions.

13. Show that if $x_1(t), \ldots, x_n(t)$ are solutions of the homogeneous linear equation
    $$\dfrac{d^n x}{dt^n} + p_{n-1}(t)\dfrac{d^{n-1}x}{dt^{n-1}} + \cdots + p_1(t)\dfrac{dx}{dt} + p_0(t)x = 0,$$ then so is any linear combination $\alpha_1 x_1(t) + \cdots + \alpha_n x_n(t)$.

14. Consider the functions y_1, y_2, y_3, y_4 given by
    $$y_1(t) = t^2,\quad y_2(t) = t^3,\quad y_3(t) = -2t^2 - 3t^3,\quad y_4(t) = t^4$$
    (see Examples 4.1.2 and 4.1.3).
    (a) Is y_4 a linear combination of $y_1, y_2,$ and y_3? Why?
    (b) Find one of the functions that can be written as a linear combination of the others, and write it that way.
    (c) Show directly that $W(y_1, y_2, y_3, y_4)(t) = 0$ for all t.

15. (a) Assume that none of the functions $x_1, \ldots, x_n$ is a linear combination of the others. Show that the only way a linear combination $\alpha_1 x_1 + \alpha_2 x_2 + \cdots + \alpha_n x_n$ can be the zero function is for all the coefficients $\alpha_1, \ldots, \alpha_n$ to equal zero. [Hint: What would happen if, for instance, α_1 was not zero?]
    (b) Conversely, assume that the only way a linear combination of the functions $x_1, \ldots, x_n$ can be the zero function is for all the coefficients to equal zero. Show that none of the functions is a linear combination of the others.

[*Hint:* What would happen if, for instance, x_1 was a linear combination of the other functions?]

16. **(a)** Show that for the differential equation $\dfrac{d^3x}{dt^3} - 3c\dfrac{d^2x}{dt^2} + 3c^2\dfrac{dx}{dt} - c^3x = 0$
    (where c is any constant), the characteristic equation has $\lambda = c$ as a triple root; in particular, e^{ct} is one solution to the differential equation.
    (b) In the situation of part (a), introduce a new unknown function $A(t)$ with $x(t) = e^{ct}A(t)$. (*Note:* This is the method of reduction of order from Section 2.4.) Show that the differential equation for $A(t)$ becomes $\dfrac{d^3A}{dt^3} = 0$. Solve for $A(t)$ and $x(t)$.
    *****(c)** Show that for the differential equation
    $$\frac{d^m x}{dt^m} - mc\frac{d^{m-1}x}{dt^{m-1}} + \frac{m(m-1)}{2}c^2\frac{d^{m-2}x}{dt^{m-2}} + \cdots + \binom{m}{k}(-c)^k\frac{d^{m-k}x}{dt^{m-k}} + \cdots$$
    $$+ m(-c)^{m-1}\frac{dx}{dt} + (-c)^m x = 0,$$
    the characteristic equation has $\lambda = c$ as an m-fold root. [In part (a) you did the special case $m = 3$.]
    *****(d)** In the situation of part (c), introduce a new unknown function $A(t)$ with $x(t) = e^{ct}A(t)$. Solve for $A(t)$ and $x(t)$. [You should get $e^{ct}, te^{ct}, \ldots, t^{m-1}e^{ct}$ as a set of basic solutions for $x(t)$.]

*17. (Compare Section 2.2, Exercise 30.)
    (a) Show that if $x_1(t), \ldots, x_n(t)$ are linearly *dependent* [and $(n-1)$ times differentiable] functions, then their Wronskian $W(x_1, \ldots, x_n)(t)$ is zero for all t.
    (b) Assume, on the other hand, that $x_1(t), \ldots, x_n(t)$ form a set of basic solutions of $x^{(n)} + p_{n-1}(t)x^{n-1} + \cdots + p_1(t)x' + p_0(t)x = 0$, and that $p_0(t), p_1(t), \ldots, p_{n-1}(t)$ are defined and continuous at t_0. Show that $W(x_1, \ldots, x_n)(t_0) \neq 0$.

**18. In this exercise we show that if the Wronskian $W(x_1, \ldots, x_n)$ is identically zero, then there is some interval (with nonempty interior) on which $x_1, \ldots, x_n$ are linearly dependent. [For an example of two functions with $W(x_1, x_2) \equiv 0$ for which x_1, x_2 are not linearly dependent if you don't restrict to some interval, see Section 2.2, Exercise 36.]
    (a) Show that for any function $g(t)$,
    $$W(gx_1, gx_2, \ldots, gx_n) = g^n \, W(x_1, x_2, \ldots, x_n).$$
    [*Hint:* Write out the determinant on the left-hand side and use row reduction.]
    (b) Show that $W(x_1, x_2, \ldots, x_n) = x_1^n \, W\!\left(1, \dfrac{x_2}{x_1}, \ldots, \dfrac{x_n}{x_1}\right)$.
    [*Hint:* Take $g = \dfrac{1}{x_1}$ in part (a).]

(c) Show that $W(x_1, x_2, \ldots, x_n) = x_1^n W\left(\left(\dfrac{x_2}{x_1}\right)', \left(\dfrac{x_3}{x_1}\right)', \ldots, \left(\dfrac{x_n}{x_1}\right)'\right)$.

(d) Show by induction on n that if $W(x_1, \ldots, x_n) \equiv 0$, then there is some interval on which $x_1, \ldots, x_n$ are linearly dependent.

19. Show that if $x_1(t), \ldots, x_n(t)$ are solutions of

$$\frac{d^n x}{dt^n} + p_{n-1}(t)\frac{d^{n-1} x}{dt^{n-1}} + \cdots + p_1(t)\frac{dx}{dt} + p_0(t)x = 0$$

with $p_0(t), p_1(t), \ldots, p_{n-1}(t)$ defined and continuous and if $x_1, \ldots, x_n$ are linearly dependent on some interval (with nonempty interior), then $x_1, \ldots, x_n$ are linearly dependent everywhere.
[*Hint*: Compare Section 2.2, Exercise 31(a).]

FURTHER READING

For an example showing how a third-order differential equation might come up in economics, see Section 3.5 in McCann, *Introduction to Ordinary Differential Equations* (Harcourt Brace Jovanovich, 1982).

For a fascinating account of the discovery of solution methods for third- and fourth-degree polynomial equations, see van der Waerden, *A History of Algebra* (Springer-Verlag, 1985).

4.2 HIGHER-ORDER DIFFERENTIAL EQUATIONS (CONTINUED)

As promised (threatened?), we now return to the case of an inhomogeneous linear nth-order equation

$$\frac{d^n x}{dt^n} + p_{n-1}(t)\frac{d^{n-1}x}{dt^{n-1}} + \cdots + p_1(t)\frac{dx}{dt} + p_0(t)x = h(t).$$

Assuming that the corresponding homogeneous equation can be solved, it's enough to find one solution to our inhomogeneous equation; all other solutions can then be found by adding all solutions of the homogeneous equation to that one "particular" solution. (This is the nth-order analogue of Theorem 2.8.1. See Exercise 32.) We'll assume that the solutions to the homogeneous equation are known, and that $x_1(t), x_2(t), \ldots, x_n(t)$ form a set of basic solutions to the homogeneous equation. Thus $x_1(t), x_2(t), \ldots, x_n(t)$ are linearly independent and the general solution to the homogeneous equation is given by $x(t) = \alpha_1 x_1(t) + \cdots + \alpha_n x_n(t)$, $\alpha_1, \ldots, \alpha_n$ arbitrary constants.

If the coefficients $p_{n-1}(t), \ldots, p_1(t), p_0(t)$ of our inhomogeneous equation are especially simple, in particular if they are constant, we may be able to guess at the form of a solution to the inhomogeneous equation, as we did in Section 2.6. We'll return to this later in this section. If this idea doesn't work, the method of variation of constants from Section 2.8 can be extended to our situation. To apply this method, we look for a solution of the form

$$x(t) = \alpha_1(t)x_1(t) + \cdots + \alpha_n(t)x_n(t)$$

to our inhomogeneous equation, where $\alpha_1, \ldots, \alpha_n$ are now unknown functions of t. (In Section 2.8 we did the case $n = 2$.) When we substitute this $x(t)$ into the inhomogeneous equation, we'll get one (complicated) differential equation for the n unknown functions $\alpha_1, \ldots, \alpha_n$. To solve for n unknown functions, we really should have a system of n equations, so we can try to impose $(n - 1)$ additional conditions on the functions $\alpha_1, \ldots, \alpha_n$. As in the special case $n = 2$, these conditions will be chosen so as to make the computation as easy as possible.

When we differentiate $x = \alpha_1 x_1 + \cdots + \alpha_n x_n$, we get (after rearranging)

$$x' = \alpha_1' x_1 + \cdots + \alpha_n' x_n + \alpha_1 x_1' + \cdots + \alpha_n x_n'.$$

Our first additional condition will be

$$\alpha_1' x_1 + \cdots + \alpha_n' x_n = 0.$$

This will leave us with $x' = \alpha_1 x_1' + \cdots + \alpha_n x_n'$; if we differentiate again, we end up with $x'' = \alpha_1' x_1' + \cdots + \alpha_n' x_n' + \alpha_1 x_1'' + \cdots + \alpha_n x_n''$. Unless $n = 2$ (in which case we've already used up the only additional condition we're entitled to; see Section 2.8 for this case) our second additional condition is

$$\alpha_1' x_1' + \cdots + \alpha_n' x_n' = 0.$$

This will simplify x'' just as the first condition simplified x', and we get

$$x'' = \alpha_1 x_1'' + \cdots + \alpha_n x_n''.$$

We continue in this way through the $(n-1)$st derivative, which will look like
$$x^{(n-1)} = \alpha_1' x_1^{(n-2)} + \cdots + \alpha_n' x_n^{(n-2)} + \alpha_1 x_1^{(n-1)} + \cdots + \alpha_n x_n^{(n-1)};$$
our last additional condition,
$$\alpha_1' x_1^{(n-2)} + \cdots + \alpha_n' x_n^{(n-2)} = 0,$$
will leave us with the simplified expression
$$x^{(n-1)} = \alpha_1 x_1^{(n-1)} + \cdots + \alpha_n x_n^{(n-1)},$$
from which we get
$$x^{(n)} = \alpha_1' x_1^{(n-1)} + \cdots + \alpha_n' x_n^{(n-1)} + \alpha_1 x_1^{(n)} + \cdots + \alpha_n x_n^{(n)}.$$
We now substitute $x = \alpha_1 x_1 + \cdots + \alpha_n x_n$ and our simplified expressions for $x', x'', \ldots, x^{(n-1)}$, as well as the expression above for $x^{(n)}$, into the differential equation
$$x^{(n)} + p_{n-1} x^{(n-1)} + \cdots + p_1 x' + p_0 x = h.$$
When we come up for air a few minutes later after recombining terms, we'll have
$$\alpha_1' x_1^{(n-1)} + \cdots + \alpha_n' x_n^{(n-1)} + \alpha_1 [x_1^{(n)} + p_{n-1} x_1^{(n-1)} + \cdots + p_1 x_1' + p_0 x_1]$$
$$+ \cdots$$
$$+ \alpha_n [x_n^{(n)} + p_{n-1} x_n^{(n-1)} + \cdots + p_1 x_n' + p_0 x_n] = h.$$
However, because the functions $x_1, \ldots, x_n$ are each solutions to the homogeneous equation $x^{(n)} + p_{n-1} x^{(n-1)} + \cdots + p_1 x' + p_0 x = 0$, each of the expressions in square brackets is zero. So we end up with
$$\alpha_1' x_1^{(n-1)} + \cdots + \alpha_n' x_n^{(n-1)} = h$$
along with the $(n-1)$ additional conditions that we imposed earlier. This gives us the system
$$\begin{cases} \alpha_1' x_1 + \cdots + \alpha_n' x_n = 0 \\ \alpha_1' x_1' + \cdots + \alpha_n' x_n' = 0 \\ \vdots \\ \alpha_1' x_1^{(n-2)} + \cdots + \alpha_n' x_n^{(n-2)} = 0 \\ \alpha_1' x_1^{(n-1)} + \cdots + \alpha_n' x_n^{(n-1)} = h \end{cases}$$
for the unknown functions $\alpha_1, \ldots, \alpha_n$ of t (although really only the derivatives $\alpha_1', \ldots, \alpha_n'$ occur). This is the nth-order analogue of the system
$$\begin{cases} \alpha' x_1 + \beta' x_2 = 0 \\ \alpha' x_1' + \beta' x_2' = h \end{cases}$$
that we found in Section 2.8. As in the case $n = 2$, the system is actually a *linear* system of equations in $\alpha_1', \ldots, \alpha_n'$, and its determinant
$$\begin{vmatrix} x_1 & \cdots & x_n \\ x_1' & \cdots & x_n' \\ \vdots & & \vdots \\ x_1^{(n-1)} & \cdots & x_n^{(n-1)} \end{vmatrix}$$

is exactly the Wronskian $W(x_1, \ldots, x_n)$ studied in Section 4.1. Therefore, the determinant will not be zero, and we will be able to solve for $\alpha'_1, \ldots, \alpha'_n$. These derivatives can then be integrated to find $\alpha_1, \ldots, \alpha_n$, and our solution method of **variation of constants** is complete.

EXAMPLE 4.2.1

Solve the equation $x^{(3)} + 2x'' - 13x' + 10x = h(t)$, where h is a given function.

In Example 4.1.4 (p. 284) we saw that the general solution to the homogeneous equation is $\alpha_1 e^t + \alpha_2 e^{2t} + \alpha_3 e^{-5t}$. We can take $x_1 = e^t, x_2 = e^{2t}, x_3 = e^{-5t}$ as basic solutions and look for a solution of the form $x(t) = \alpha_1(t)e^t + \alpha_2(t)e^{2t} + \alpha_3(t)e^{-5t}$ to our inhomogeneous equation. Differentiating and rearranging, we get

$$x' = \alpha'_1 e^t + \alpha'_2 e^{2t} + \alpha'_3 e^{-5t} + \alpha_1 e^t + 2\alpha_2 e^{2t} - 5\alpha_3 e^{-5t}.$$

Our first additional condition will be

$$\alpha'_1 e^t + \alpha'_2 e^{2t} + \alpha'_3 e^{-5t} = 0, \tag{1}$$

which leaves us with $x' = \alpha_1 e^t + 2\alpha_2 e^{2t} - 5\alpha_3 e^{-5t}$. Differentiating and rearranging again yields

$$x'' = \alpha'_1 e^t + 2\alpha'_2 e^{2t} - 5\alpha'_3 e^{-5t} + \alpha_1 e^t + 4\alpha_2 e^{2t} + 25\alpha_3 e^{-5t};$$

the second additional condition is

$$\alpha'_1 e^t + 2\alpha'_2 e^{2t} - 5\alpha'_3 e^{-5t} = 0. \tag{2}$$

We now have $x'' = \alpha_1 e^t + 4\alpha_2 e^{2t} + 25\alpha_3 e^{-5t}$, from which

$$x^{(3)} = \alpha'_1 e^t + 4\alpha'_2 e^{2t} + 25\alpha'_3 e^{-5t} + \alpha_1 e^t + 8\alpha_2 e^{2t} - 125\alpha_3 e^{-5t}.$$

Substituting this, along with our previous expressions for x' and x'', into the original equation, we get

$$\begin{aligned}
\alpha'_1 e^t + 4\alpha'_2 e^{2t} + 25\alpha'_3 e^{-5t} + \alpha_1 e^t + 8\alpha_2 e^{2t} - 125\alpha_3 e^{-5t} &\quad \leftarrow x^{(3)} \\
+ \, 2(\alpha_1 e^t + 4\alpha_2 e^{2t} + 25\alpha_3 e^{-5t}) &\quad \leftarrow 2x'' \\
- \, 13(\alpha_1 e^t + 2\alpha_2 e^{2t} - 5\alpha_3 e^{-5t}) &\quad \leftarrow -13x' \\
+ \, 10(\alpha_1 e^t + \alpha_2 e^{2t} + \alpha_3 e^{-5t}) &\quad \leftarrow 10x \\
= h(t). &
\end{aligned}$$

Everything cancels except for $\alpha'_1 e^t + 4\alpha'_2 e^{2t} + 25\alpha'_3 e^{-5t} = h(t)$. Combining this with (1) and (2), we have

$$\begin{cases} \alpha'_1 e^t + \alpha'_2 e^{2t} + \alpha'_3 e^{-5t} = 0 \\ \alpha'_1 e^t + 2\alpha'_2 e^{2t} - 5\alpha'_3 e^{-5t} = 0 \\ \alpha'_1 e^t + 4\alpha'_2 e^{2t} + 25\alpha'_3 e^{-5t} = h(t). \end{cases}$$

Solving for $\alpha'_1, \alpha'_2,$ and α'_3 eventually gets us

$$\alpha'_1 = -\frac{1}{6} h(t) e^{-t}, \qquad \alpha'_2 = \frac{1}{7} h(t) e^{-2t}, \qquad \alpha'_3 = \frac{1}{42} h(t) e^{5t},$$

so

$$\alpha_1 = \int -\frac{1}{6} h(t)e^{-t}\, dt, \quad \alpha_2 = \int \frac{1}{7} h(t)e^{-2t}\, dt, \quad \alpha_3 = \int \frac{1}{42} h(t)e^{5t}\, dt,$$

from which

$$x(t) = \left(\int -\frac{1}{6} h(t)e^{-t}\, dt\right)e^{t} + \left(\int \frac{1}{7} h(t)e^{-2t}\, dt\right)e^{2t} + \left(\int \frac{1}{42} h(t)e^{5t}\, dt\right)e^{-5t}$$

$$= -\frac{1}{6} e^{t} \int h(t)e^{-t}\, dt + \frac{1}{7} e^{2t} \int h(t)e^{-2t}\, dt + \frac{1}{42} e^{-5t} \int h(t)e^{5t}\, dt.$$

As in Example 2.8.2, the integrals are determined only up to integration constants; adding such constants will add solutions of the homogeneous equation to $x(t)$. ∎

EXAMPLE 4.2.2 Solve the initial value problem

$$t^3 x^{(3)} - 2t^2 x'' + 3tx' - 3x = 9t^4 + 2t^2, \quad x(1) = 0, \quad x'(1) = 4, \quad x''(1) = 7,$$

given that $x_1(t) = t$, $x_2(t) = t \log t$ $(t > 0)$ and $x_3(t) = t^3$ are all solutions of the homogeneous equation.[5]

Since the given functions x_1, x_2, x_3 are linearly independent (which can be seen directly or by taking their Wronskian; see Exercise 33), they form a set of basic solutions for the homogeneous equation. Thus we look for a solution of the form

$$x(t) = \alpha_1(t)t + \alpha_2(t) t \log t + \alpha_3(t) t^3$$

to our inhomogeneous equation. This yields

$$x' = \alpha_1' t + \alpha_2' t \log t + \alpha_3' t^3 + \alpha_1 + \alpha_2(\log t + 1) + 3\alpha_3 t^2,$$

so we let

$$\alpha_1' t + \alpha_2' t \log t + \alpha_3' t^3 = 0 \tag{1}$$

and then have

$$x' = \alpha_1 + \alpha_2(\log t + 1) + 3\alpha_3 t^2,$$

$$x'' = \alpha_1' + \alpha_2'(\log t + 1) + 3\alpha_3' t^2 + \alpha_2 \cdot \frac{1}{t} + 6\alpha_3 t.$$

Our second additional condition is

$$\alpha_1' + \alpha_2'(\log t + 1) + 3\alpha_3' t^2 = 0, \tag{2}$$

---

[5] In Chapter 5 we'll see a method (presented there for second-order equations) by which these solutions could be found; see Section 5.5, Exercise 10.

SEC. 4.2 HIGHER-ORDER DIFFERENTIAL EQUATIONS (CONTINUED)

and we then have $x'' = \alpha_2 \cdot \frac{1}{t} + 6\alpha_3 t$, $x^{(3)} = \alpha_2' \cdot \frac{1}{t} + 6\alpha_3' t + \alpha_2 \cdot -\frac{1}{t^2} + 6\alpha_3$. The original equation becomes

$$t^3 \left[\alpha_2' \cdot \frac{1}{t} + 6\alpha_3' t + \alpha_2 \cdot -\frac{1}{t^2} + 6\alpha_3 \right] - 2t^2 \left[\alpha_2 \cdot \frac{1}{t} + 6\alpha_3 t \right]$$
$$+ 3t[\alpha_1 + \alpha_2(\log t + 1) + 3\alpha_3 t^2] - 3[\alpha_1 t + \alpha_2 t \log t + \alpha_3 t^3] = 9t^4 + 2t^2.$$

All terms with $\alpha_1, \alpha_2, \alpha_3$ cancel, as they should, and we have $\alpha_2' t^2 + 6\alpha_3' t^4 = 9t^4 + 2t^2$ along with the equations $\alpha_1' t + \alpha_2' t \log t + \alpha_3' t^3 = 0$, $\alpha_1' + \alpha_2'(\log t + 1) + 3\alpha_3' t^2 = 0$ from (1) and (2) above. Now we solve for $\alpha_1', \alpha_2', \alpha_3'$ and integrate these (see Exercise 1). The result will be

$$x = t^4 - 2t^2 + C_1 t + C_2 t \log t + C_3 t^3,$$

where C_1, C_2, and C_3 are integration constants. It should now be routine to solve for C_1, C_2, and C_3 from the given initial conditions (again, see Exercise 1). ∎

Warning: The warning from Section 2.8 about the variation of constants method applies here, as well. If you decide to write down the system

$$\begin{cases} \alpha_1' x_1 + \cdots + \alpha_n' x_n = 0 \\ \vdots \\ \alpha_1' x_1^{(n-2)} + \cdots + \alpha_n' x_n^{(n-2)} = 0 \\ \alpha_1' x_1^{(n-1)} + \cdots + \alpha_n' x_n^{(n-1)} = h(t) \end{cases}$$

immediately, first make sure that $h(t)$ is the right-hand side of the given equation *in standard form*. That is, the left-hand side should look like $x^{(n)} + p_{n-1}(t)x^{(n-1)} + \cdots$. For instance, in Example 4.2.2, the equation $t^3 x^{(3)} - 2t^2 x'' + \cdots = 9t^4 + 2t^2$ is not in standard form, and you would have to divide both sides by t^3 first.

Let's reconsider that example, which involves the differential equation

$$t^3 x^{(3)} - 2t^2 x'' + 3tx' - 3x = 9t^4 + 2t^2.$$

In practice you might well be able to guess that this equation will have a solution of type $x = At^4 + Bt^3 + Ct^2 + Dt + E$ for suitable constants A through E; since t and t^3 were given to be solutions of the homogeneous equation, this implies that $x = At^4 + Ct^2 + E$ is a solution. At this point you can try substituting $x = At^4 + Ct^2 + E$ into the equation and seeing whether you can find values of A, C, E for which you do indeed have a solution:

$$x = At^4 + Ct^2 + E$$
$$x' = 4At^3 + 2Ct$$
$$x'' = 12At^2 + 2C$$
$$x^{(3)} = 24At,$$

so

$$t^3 x^{(3)} - 2t^2 x'' + 3tx' - 3x$$
$$= t^3(24At) - 2t^2(12At^2 + 2C) + 3t(4At^3 + 2Ct) - 3(At^4 + Ct^2 + E)$$
$$= 9At^4 - Ct^2 - 3E,$$

so we need $9At^4 - Ct^2 - 3E = 9t^4 + 2t^2$ or $A = 1$, $C = -2$, $E = 0$. This yields a solution $x = t^4 - 2t^2$, which combines with the solutions of the homogeneous equation to yield the general solution $x = t^4 - 2t^2 + C_1 t + C_2 t \log t + C_3 t^3$, as before. Note that this method, which involves the "educated guessing" of the form of one solution, is nothing but the **method of undetermined coefficients**, of which we have already seen several cases in Section 2.6. This method usually does not work very well for equations that don't have constant coefficients (the example above is an exception); for such equations, it tends to be difficult to guess the form that a solution might take. However, for constant-coefficient equations of type $x^{(n)} + p_{n-1} x^{(n-1)} + \cdots + p_1 x' + p_0 x = h(t)$, the method of undetermined coefficients is often useful.

We will now see how one can find a solution of such an equation (and thus, provided that one can solve the characteristic equation $\lambda^n + p_{n-1} \lambda^{n-1} + \cdots + p_1 \lambda + p_0 = 0$, find all solutions) in case $h(t)$ is any sum of products of polynomials, exponential functions, sines, and cosines. If sines and cosines are involved, one can either work with them directly (this was done in several examples in Section 2.6) or express them in terms of complex exponentials (see Appendix A, p. 602).

EXAMPLE 4.2.3   If $h(t) = (t^3 - t) e^{2t} \sin t \cos 3t$, we can use

$$\sin t = \frac{1}{2i} (e^{it} - e^{-it}), \quad \cos 3t = \frac{1}{2} (e^{3it} + e^{-3it})$$

to get

$$\begin{aligned} h(t) &= \frac{1}{2i} \cdot \frac{1}{2} (t^3 - t) e^{2t} (e^{it} - e^{-it})(e^{3it} + e^{-3it}) \\ &= \frac{(t^3 - t)}{4i} e^{2t} (e^{4it} - e^{2it} + e^{-2it} - e^{-4it}) \\ &= \frac{(t^3 - t)}{4i} [e^{(4i+2)t} - e^{(2i+2)t} + e^{(2-2i)t} - e^{(2-4i)t}]. \quad \blacksquare \end{aligned}$$

Although in particular cases it may be more practical to work with the sines and cosines directly, we'll assume for the general discussion that they have been rewritten as in Example 4.2.3. This means that $h(t)$ is a sum of products of polynomials and complex exponential functions. In fact, by multiplying out these products, $h(t)$ will become a sum of terms of the form $Ct^m e^{\gamma t}$, where for each term C and γ are complex constants and $m \geq 0$ is an integer exponent. [In Example 4.2.3, the first such term is $\frac{1}{4i} t^3 e^{(4i+2)t}$, for which $C = \frac{1}{4i}$, $m = 3$, and $\gamma = 4i + 2$.] If $h(t)$ is a real-valued function, these terms will form complex conjugate pairs $Ct^m e^{\gamma t}$, $\overline{C} t^m e^{\overline{\gamma} t}$; each such pair makes a real-valued contribution $Ct^m e^{\gamma t} + \overline{C} t^m e^{\overline{\gamma} t} = 2 \operatorname{Re}(Ct^m e^{\gamma t})$ to $h(t)$.

SEC. 4.2 HIGHER-ORDER DIFFERENTIAL EQUATIONS (CONTINUED)

To find a solution of the equation $x^{(n)} + p_{n-1}x^{(n-1)} + \cdots + p_1 x' + p_0 x = h(t)$ where $h(t)$ is a sum of various terms as described above, we can deal with the terms one at a time, then add up solutions we get for each term separately. That is, we have the following result.

Theorem **(Superposition Principle).** If $h(t) = h_1(t) + \cdots + h_k(t)$ and if for each i with $1 \leq i \leq k$, the function $x_i(t)$ is a solution of $x^{(n)} + p_{n-1}x^{(n-1)} + \cdots + p_1 x' + p_0 x = h_i(t)$, then $x(t) = x_1(t) + \cdots + x_k(t)$ is a solution of $x^{(n)} + p_{n-1}x^{(n-1)} + \cdots + p_1 x' + p_0 x = h(t)$.

The proof of this theorem is straightforward (see Exercise 34); it does not require the coefficients $p_0, \ldots, p_{n-1}$ to be constants. Note that if two terms in $h(t)$ are complex conjugates, say $h_1(t)$ and $\overline{h_1(t)}$, then a solution $x_1(t)$ to

$$x^{(n)} + p_{n-1}x^{(n-1)} + \cdots + p_1 x' + p_0 x = h_1(t)$$

will automatically supply us with a solution $\overline{x_1(t)}$ to the "conjugate" equation with $\overline{h_1(t)}$ on the right (see Exercise 31). Together, $x_1(t)$ and $\overline{x_1(t)}$ will make a real-valued contribution to $x(t)$.

For instance, in the situation of Example 4.2.3, the first term $h_1(t) = \dfrac{t^3}{4i} e^{(4i+2)t}$ has complex conjugate $-\dfrac{t^3}{4i} e^{(2-4i)t}$, which is also a term in $h(t)$. In fact, in this situation it will be enough to find solutions $x_1(t), x_2(t), x_3(t), x_4(t)$ corresponding to the four terms $\dfrac{t^3}{4i} e^{(4i+2)t}, -\dfrac{t}{4i} e^{(4i+2)t}, -\dfrac{t^3}{4i} e^{(2i+2)t}, \dfrac{t}{4i} e^{(2i+2)t}$, respectively;

$$x(t) = x_1(t) + \overline{x_1(t)} + x_2(t) + \overline{x_2(t)} + x_3(t) + \overline{x_3(t)} + x_4(t) + \overline{x_4(t)}$$
$$= 2\,\mathrm{Re}[x_1(t) + x_2(t) + x_3(t) + x_4(t)]$$

will then give us the solution we need.

So how do we get a solution corresponding to a single term in $h(t)$? Well, to find a solution of $x^{(n)} + p_{n-1}x^{(n-1)} + \cdots + p_1 x' + p_0 x = Ct^m e^{\gamma t}$, we can try a function of the form $x(t) = (A_m t^m + \cdots + A_1 t + A_0) e^{\gamma t}$ with constants $A_0, \ldots, A_m$, that is, a polynomial of degree m multiplied by the same exponential that occurs on the right-hand side. This will work unless $e^{\gamma t}$ is a solution of the homogeneous equation. From Section 4.1 we know that this exceptional case occurs exactly if γ is a root of the characteristic equation $\lambda^n + p_{n-1}\lambda^{n-1} + \cdots + p_1 \lambda + p_0 = 0$. If so, the degree of the polynomial factor in $x(t)$ must be raised—to degree $(m+1)$ if γ is a simple root of the characteristic equation, to degree $(m+2)$ if γ is a double root, and so on. Although proofs of all this are a bit beyond our scope, it should be clear that taking derivatives of a function of the form above,

$$(A_m t^m + \cdots + A_1 t + A_0) e^{\gamma t},$$

will result in similar expressions. It is then reasonable to hope that the combination $x^{(n)} + p_{n-1}x^{(n-1)} + \cdots + p_1 x' + p_0 x$ of these expressions can be made to equal $Ct^m e^{\gamma t}$ by choosing the constants $A_0, A_1, \ldots, A_m$ suitably. This hope will fail if the terms with $t^m e^{\gamma t}$ on the left all cancel, and it turns out that this will happen exactly if $e^{\gamma t}$ is a solution of the homogeneous equation.

Here are a few more examples using the method of undetermined coefficients. (See Section 2.6 for more examples.)

EXAMPLE 4.2.4 Find a solution of $x^{(3)} + x = 2te^{-t}$.

The right-hand side has only one term, with $m = 1$, so we can try $x = (At + B)e^{-t}$. Differentiating three times and simplifying after each differentiation, we get

$$x' = [-At + (A - B)]e^{-t}$$
$$x'' = [At + (B - 2A)]e^{-t}$$
$$x^{(3)} = [-At + (3A - B)]e^{-t}.$$

However, we then find that

$$x^{(3)} + x = [-At + (3A - B)]e^{-t} + (At + B)e^{-t} = 3Ae^{-t},$$

so we would need $3Ae^{-t} = 2te^{-t}$, which is impossible. (**Warning**: You *can't* take $A = \frac{2}{3}t$. Why not?) What went wrong is that -1 is a (simple) root of the characteristic equation $\lambda^3 + 1 = 0$, so e^{-t} is a solution of the homogeneous equation. Therefore, we should raise the degree of the polynomial by 1 and try $x = (At^2 + Bt)e^{-t}$.[6] This time we get

$$x' = [-At^2 + (2A - B)t + B]e^{-t}$$
$$x'' = [At^2 + (B - 4A)t + (2A - 2B)]e^{-t}$$
$$x^{(3)} = [-At^2 + (6A - B)t + (3B - 6A)]e^{-t},$$

and we end up with $x^{(3)} + x = [6At + (3B - 6A)]e^{-t}$. Since this must equal $2te^{-t}$, we need $6A = 2$, $3B - 6A = 0 \Rightarrow A = \frac{1}{3}$, $B = \frac{2}{3}$, and we see that

$$x = \left(\frac{1}{3}t^2 + \frac{2}{3}t\right)e^{-t}$$

is a solution to our equation. (See Exercise 2 for the general solution.) ∎

EXAMPLE 4.2.5 Solve the initial value problem

$$x'' - 4x = \sin t \sin 2t, \qquad x(0) = -\frac{4}{65}, \qquad x'(0) = 1.$$

---

[6] It's not necessary to include a term Ce^{-t}, since this would be a solution of the homogeneous equation anyway.

First we find one solution to the given inhomogeneous equation. We start by rewriting the right-hand side:

$$\sin t \sin 2t = \frac{1}{2i}(e^{it} - e^{-it}) \cdot \frac{1}{2i}(e^{2it} - e^{-2it})$$

$$= -\frac{1}{4}(e^{3it} + e^{-3it} - e^{it} - e^{-it}).$$

Now we need to find solutions to $x'' - 4x = -\frac{1}{4}e^{3it}$ and $x'' - 4x = \frac{1}{4}e^{it}$ (adding these solutions and their complex conjugates will give a solution to our inhomogeneous equation). For the first of these, we can take $x_1 = Ce^{3it}$, $x_1' = 3iCe^{3it}$, $x_1'' = -9Ce^{3it}$, and we get $-9C - 4C = -\frac{1}{4} \Rightarrow C_1 = \frac{1}{52}$, $x_1 = \frac{1}{52}e^{3it}$. Similarly, $x_2 = -\frac{1}{20}e^{it}$ is a solution to $x'' - 4x = \frac{1}{4}e^{it}$. Thus,

$$x_1 + \bar{x}_1 + x_2 + \bar{x}_2 = 2\,\mathrm{Re}(x_1 + x_2)$$

$$= 2\,\mathrm{Re}\left(\frac{1}{52}e^{3it} - \frac{1}{20}e^{it}\right)$$

$$= \frac{1}{26}\cos 3t - \frac{1}{10}\cos t$$

is one solution to the inhomogeneous equation. See Exercise 3 for the rest of the computation. ∎

EXAMPLE 4.2.6 Find the general solution of

$$x'' + 4x = (8t^2 - 2)e^{2t} + e^t \sin 2t + 4e^t \cos 2t.$$

The characteristic equation is $\lambda^2 + 4 = 0$, so we get $\lambda = \pm 2i$ and $\cos 2t$, $\sin 2t$ form a pair of basic solutions to the homogeneous equation. To get a solution for the inhomogeneous equation, we can use the superposition principle (p. 297) and consider separate equations

$$x'' + 4x = (8t^2 - 2)e^{2t} \quad \text{and} \quad x'' + 4x = e^t \sin 2t + 4e^t \cos 2t.$$

For the first of these, we can take $x_1 = (At^2 + Bt + C)e^{2t}$ and solve for A, B, C. [It would have been possible to separate $(8t^2 - 2)e^{2t}$ further into terms $8t^2e^{2t}$ and $-2e^{2t}$, but this would end up being more work.] See Exercise 4(a) for details; the result will be

$$x_1 = (t^2 - t)e^{2t}.$$

For the second equation $x'' + 4x = e^t \sin 2t + 4e^t \cos 2t$, we could rewrite the right-hand side in terms of complex exponentials (as in Example 4.2.5). However, we can also take

$x_2 = De^t \sin 2t + Ee^t \cos 2t = e^t(D \sin 2t + E \cos 2t)$, since the derivative of such an expression is a similar one:

$$x_2' = e^t(D \sin 2t + E \cos 2t) + e^t(2D \cos 2t - 2E \sin 2t)$$
$$= e^t[(D - 2E) \sin 2t + (2D + E) \cos 2t].$$

See Exercise 4(b) for details; the result will be $x_2 = e^t \sin 2t$. Combining all this information, we get

$$x = \underbrace{(t^2 - t)e^{2t}}_{x_1} + \underbrace{e^t \sin 2t}_{x_2} + C_1 \cos 2t + C_2 \sin 2t$$

as the general solution of the given equation. ∎

SUMMARY OF KEY RESULTS AND TECHNIQUES

Finding one solution of an inhomogeneous equation can be enough (p. 291): If we can find one solution $x_0(t)$ of the inhomogeneous linear equation

$$x^{(n)} + p_{n-1}(t)x^{(n-1)} + \cdots + p_1(t)x' + p_0(t)x = h(t),$$

then we can get all of them by adding the general solution of the corresponding *homogeneous* equation to $x_0(t)$.

Superposition principle (p. 297): To find a solution of an inhomogeneous linear equation $x^{(n)} + p_{n-1}x^{(n-1)} + \cdots + p_1x' + p_0x = h(t)$ where $h(t) = h_1(t) + \cdots + h_k(t)$, it is enough to find a solution $x_i(t)$ to each of the individual equations

$$x^{(n)} + p_{n-1}x^{(n-1)} + \cdots + p_1x' + p_0x = h_i(t) \ (1 \le i \le k);$$

$x(t) = x_1(t) + \cdots + x_k(t)$ is then a solution to the given equation.

Undetermined coefficients (p. 296): To find a solution of the nth-order equation $x^{(n)} + p_{n-1}x^{(n-1)} + \cdots + p_1x' + p_0x = h(t)$, where $p_0, \ldots, p_{n-1}$ are constants and $h(t)$ is any sum of products of polynomials, exponential functions, sines, and cosines, first use $\sin t = \dfrac{1}{2i}(e^{it} - e^{-it})$, $\cos t = \dfrac{1}{2}(e^{it} + e^{-it})$ to write $h(t)$ as the sum of terms of the form $Ct^m e^{\gamma t}$. By the superposition principle, it is then enough to find, for each such term, a solution of $x^{(n)} + p_{n-1}x^{(n-1)} + \cdots + p_1x' + p_0x = Ct^m e^{\gamma t}$. There will be such a solution of the form $x(t) = (A_m t^m + \cdots + A_1 t + A_0)e^{\gamma t}$, unless $e^{\gamma t}$ is a solution of the homogeneous equation, in which case the degree of the polynomial factor in $x(t)$ must be raised (p. 297).

Variation of constants (pp. 291–293): To solve

$$x^{(n)} + p_{n-1}(t)x^{(n-1)} + \cdots + p_1(t)x' + p_0(t)x = h(t),$$

start with a set of basic solutions $x_1(t), \ldots, x_n(t)$ to the *homogeneous* equation, and put $x(t) = \alpha_1(t)x_1(t) + \cdots + \alpha_n(t)x_n(t)$. Requiring

$$\alpha_1' x_1 + \cdots + \alpha_n' x_n = 0, \ldots, \alpha_1' x_1^{(n-2)} + \cdots + \alpha_n' x_n^{(n-2)} = 0$$

SEC. 4.2 HIGHER-ORDER DIFFERENTIAL EQUATIONS (CONTINUED)

simplifies the computation and leads to the simultaneous equations

$$\begin{cases} \alpha'_1 x_1 + \cdots + \alpha'_n x_n = 0 \\ \quad\vdots \\ \alpha'_1 x_1^{(n-2)} + \cdots + \alpha'_n x_n^{(n-2)} = 0 \\ \alpha'_1 x_1^{(n-1)} + \cdots + \alpha'_n x_n^{(n-1)} = h(t) \end{cases}$$

for $\alpha'_1, \ldots, \alpha'_n$. After solving for $\alpha'_1, \ldots, \alpha'_n$, integrate to get $\alpha_1, \ldots, \alpha_n$ and substitute back into $x = \alpha_1 x_1 + \cdots + \alpha_n x_n$ to get the final answer.

EXERCISES

1. Carry out the missing steps (on p. 294) in the solution of the initial value problem in Example 4.2.2.

2. Find the general solution to the differential equation $x^{(3)} + x = 2te^{-t}$ of Example 4.2.4.

3. (a) Check directly that $x = \dfrac{1}{26}\cos 3t - \dfrac{1}{10}\cos t$ is a solution to
   $x'' - 4x = \sin t \sin 2t$ (as claimed in Example 4.2.5).
   (b) Solve the initial value problem from Example 4.2.5:

   $$x'' - 4x = \sin t \sin 2t, \quad x(0) = -\frac{4}{65}, \quad x'(0) = 1.$$

4. (Compare Example 4.2.6.)
   (a) Find a solution of the form $x_1 = (At^2 + Bt + C)e^{2t}$ to the equation
   $x'' + 4x = (8t^2 - 2)e^{2t}$.
   (b) Find a solution of the form $x_2 = e^t(D \sin 2t + E \cos 2t)$ to the equation
   $x'' + 4x = e^t \sin 2t + 4e^t \cos 2t$.
   (c) Solve the initial value problem

   $$x'' + 4x = (8t^2 - 2)e^{2t} + e^t \sin 2t + 4e^t \cos 2t, \quad x(0) = 3, \quad x'(0) = -1.$$

Write the following as sums of terms of the form $Ct^m e^{\gamma t}$.

5. $e^{-t} \sin 3t$
6. $e^{2t} \cos 4t$
7. $t^2 \sin t \cos 3t$
8. $t \sin 4t \cos t$
9. $e^t \sin t \cos t - e^{3t} \sin^2 2t$
10. $e^{2t} \cos^2 t - t^3 \sin 5t$

For each of the following equations, write down *the form* of one solution (using undetermined coefficients). Do not actually solve for the coefficients, but do be sure that a solution of the type you propose will really exist.

11. $\dfrac{d^3 x}{dt^3} - 2 \dfrac{d^2 x}{dt^2} + \dfrac{dx}{dt} - 2x = te^{2t} - 3e^{-2t} + \cos 3t + 2 \sin 3t$.

12. $\dfrac{d^3x}{dt^3} + 2\dfrac{d^2x}{dt^2} + \dfrac{dx}{dt} + 2x = e^{5t} + 4te^{-2t} + \cos 5t.$

13. $\dfrac{d^3y}{dx^3} + 3\dfrac{d^2y}{dx^2} + \dfrac{dy}{dx} + 3y = 2\sin x + \cos x + 2xe^{-x}.$

14. $\dfrac{d^3y}{dx^3} - 3\dfrac{d^2y}{dx^2} + \dfrac{dy}{dx} - 3y = 4\sin x - \cos x - x^2 e^{4x}.$

15. $x^{(3)} + x'' + 3x' - 5x = e^{-t}(\sin 2t + 4\cos 2t) + 2e^t.$

16. $x^{(3)} + x' - 10x = e^{-t}(3\sin 2t - \cos 2t) + te^{2t}.$

Solve the following differential equations and initial value problems.

17. $x^{(3)} - 4x' = e^{3t} - t.$

18. $x^{(3)} - x' = 5t - e^{2t}.$

19. $x^{(4)} - 5x'' + 4x = e^t + \sin t.$

20. $x^{(4)} - 5x'' + 4x = e^{2t} + \cos t.$

21. $x^{(4)} + 5x'' + 4x = e^{2t} + \cos t.$

22. $x^{(4)} + 5x'' + 4x = 3e^{2t} - \sin t.$

23. $x^{(3)} + 2x'' - 3x' = h(t).$

24. $x^{(3)} - 2x'' - 3x' = h(t).$

25. $x^{(3)} - 2x'' + x' - 2x = h(t).$

26. $\dfrac{d^3y}{dx^3} + 2\dfrac{d^2y}{dx^2} + \dfrac{dy}{dx} + 2y = h(x).$

27. $t^3 x^{(3)} + t^2 x'' - 2tx' + 2x = \dfrac{24 - 6t^2}{t^2},\quad x(1) = -1,\quad x'(1) = 5,\quad x''(1) = -12.$

[Hint: $\dfrac{1}{t}, t, t^2$ are all solutions of the homogeneous equation.]

28. $t^3 x^{(3)} + t^2 x'' - 2tx' + 2x = \dfrac{20 + t^3}{t^3},\quad x(1) = 5,\quad x'(1) = -\dfrac{5}{2},\quad x''(1) = 0.$

(See hint for Exercise 27.)

29. $x^2 \dfrac{d^3y}{dx^3} + x\dfrac{d^2y}{dx^2} - 4\dfrac{dy}{dx} = \dfrac{6}{x},\quad y(1) = 3,\ y'(1) = 3,\ y''(1) = 4.$

[Hint: $\dfrac{1}{x}$ and x^3 are both solutions of the homogeneous equation, and another basic solution is easy to find.]

30. $x^2 \dfrac{d^3y}{dx^3} + x\dfrac{d^2y}{dx^2} - 4\dfrac{dy}{dx} = \dfrac{4x-3}{x},\quad y(1) = 0,\ y'(1) = 1,\ y''(1) = -3.$

(See hint for Exercise 29.)

31. Show that if $x_1(t)$ is a solution to

$$x^{(n)} + p_{n-1}x^{(n-1)} + \cdots + p_1 x' + p_0 x = h_1(t) \qquad \text{(with real } p_0, \ldots, p_{n-1}),$$

then $\overline{x_1(t)}$ is a solution to

$$x^{(n)} + p_{n-1}x^{(n-1)} + \cdots + p_1 x' + p_0 x = \overline{h_1(t)}.$$

32. Consider the inhomogeneous equation

$$\frac{d^n x}{dt^n} + p_{n-1}(t)\frac{d^{n-1}x}{dt^{n-1}} + \cdots + p_1(t)\frac{dx}{dt} + p_0(t)x = h(t)$$

and the related homogeneous equation

$$\frac{d^n x}{dt^n} + p_{n-1}(t)\frac{d^{n-1}x}{dt^{n-1}} + \cdots + p_1(t)\frac{dx}{dt} + p_0(t)x = 0.$$

(a) Show that if $x(t)$ is a solution of the inhomogeneous equation and $y(t)$ is a solution of the homogeneous equation, then $x(t) + y(t)$ is a solution of the inhomogeneous equation.

(b) Show that if one can find *one* solution $x_0(t)$ of the inhomogeneous equation, one can then find *all* solutions of that equation by adding each of the solutions of the homogeneous equation to $x_0(t)$.
[*Hint*: Compare Exercise 23 from Section 1.6 and Exercise 41 from Section 2.8.]

33. (a) Use the Wronskian to show that the functions $x_1(t) = t$, $x_2(t) = t \log t$ $(t > 0)$, and $x_3(t) = t^3$ are linearly independent.

*(b) Now show that these three functions are linearly independent *without* using the Wronskian.
[*Hint*: Use the test for linear independence from Section 4.1. That is, assume that a linear combination $\alpha_1 x_1 + \alpha_2 x_2 + \alpha_3 x_3$ is the zero function, and show that then all the coefficients $\alpha_1, \alpha_2, \alpha_3$ must equal zero.]

34. Prove the superposition principle (p. 297).

4.3 HOMOGENEOUS LINEAR SYSTEMS (AUTONOMOUS, DISTINCT REAL EIGENVALUES)

We have seen in Chapter 3 how a second-order linear differential equation can be rewritten as a system of two first-order linear equations (Example 3.1.3), and how, conversely, such a system can sometimes be solved by converting it to a single second-order equation (Section 3.3). These ideas also apply to higher-order linear equations, as well as to systems of these. However, passing from systems to single equations becomes more difficult and laborious as the systems get larger (or the order of the equations in them gets larger). Passing from single equations to systems, on the other hand, doesn't really become more involved, and so these "higher-order cases" are more likely to end up being studied as systems.

EXAMPLE 4.3.1 $x^{(4)} - 5tx^{(3)} + 6x'' - 4x' + t^2 x = 0$.

This fourth-order linear equation can be converted to a system of four first-order equations by introducing new functions $y_1 = x'$, $y_2 = x''$, $y_3 = x^{(3)}$ along with the original unknown function x. Since $x^{(4)} = \dfrac{dx^{(3)}}{dt} = \dfrac{dy_3}{dt}$, the given equation can be rewritten as $\dfrac{dy_3}{dt} = 5ty_3 - 6y_2 + 4y_1 - t^2 x$. The other three first-order equations will simply express that each of the new functions is the derivative of the previous one, and our system will be

$$\begin{cases} \dfrac{dx}{dt} = y_1 \\ \dfrac{dy_1}{dt} = y_2 \\ \dfrac{dy_2}{dt} = y_3 \\ \dfrac{dy_3}{dt} = 5ty_3 - 6y_2 + 4y_1 - t^2 x. \end{cases}$$
■

EXAMPLE 4.3.2 $\dfrac{d^2 x_1}{dt^2} + k_1 x_1 + k_2 x_2 = 0$, $\dfrac{d^2 x_2}{dt^2} + l_1 x_1 + l_2 x_2 = 0$.

A system such as this (where k_1, k_2, l_1, l_2 are constants) can arise when two springs are coupled. (Compare Example 4.6.1, p. 355.) Elimination would not be difficult here: From the first equation we get $x_2 = -\dfrac{1}{k_2} \dfrac{d^2 x_1}{dt^2} - \dfrac{k_1}{k_2} x_1$, and substituting this into the second equation yields a fourth-order equation for x_1. On the other hand, we could introduce new functions $v_1 = \dfrac{dx_1}{dt}$, $v_2 = \dfrac{dx_2}{dt}$ and convert the system to a system of four first-order equations:

SEC. 4.3 HOMOGENEOUS LINEAR SYSTEMS (AUTONOMOUS, DISTINCT REAL EIGENVALUES)

$$\begin{cases} \dfrac{dx_1}{dt} = v_1 \\ \dfrac{dv_1}{dt} = -k_1 x_1 - k_2 x_2 \\ \dfrac{dx_2}{dt} = v_2 \\ \dfrac{dv_2}{dt} = -l_1 x_1 - l_2 x_2 . \end{cases}$$ ∎

EXAMPLE 4.3.3 Although the system

$$\begin{cases} \dfrac{dx_1}{dt} = 3x_1 - x_2 - 6x_3 \\ \dfrac{dx_2}{dt} = 4x_2 + 2x_3 + 6x_4 \\ \dfrac{dx_3}{dt} = 3x_1 - 3x_2 - 7x_3 - 3x_4 \\ \dfrac{dx_4}{dt} = -5x_1 + 3x_2 + 10x_3 + 2x_4 \end{cases}$$

is not particularly unpleasant, it would be quite a chore to convert this system into a single fourth-order equation (try it, and you'll see!) □

In the rest of this chapter we'll assume that we're given a system of n first-order equations in the unknown functions $x_1, \ldots, x_n$. Since we've already studied the case $n = 2$ in Chapter 3, you may want to review corresponding topics from that chapter while reading this one.

Our system will usually be in the form

$$\begin{cases} \dfrac{dx_1}{dt} = f_1(t, x_1, \ldots, x_n) \\ \dfrac{dx_2}{dt} = f_2(t, x_1, \ldots, x_n) \\ \quad \vdots \\ \dfrac{dx_n}{dt} = f_n(t, x_1, \ldots, x_n), \end{cases}$$

where $f_1, f_2, \ldots, f_n$ are given functions. A **solution** of this system consists of n functions $x_1(t), \ldots, x_n(t)$ for which all n equations are satisfied. If we are given initial values $x_1(t_0), \ldots, x_n(t_0)$ along with the system, the solution will be unique provided that the functions $f_1, \ldots, f_n$ and their partial derivatives with respect to $x_1, \ldots, x_n$ are defined and continuous.

If $f_1, \ldots, f_n$ are linear in the unknown functions $x_1, \ldots, x_n$ (but not necessarily in t), the system is called **linear**. For example,

$$\begin{cases} \dfrac{dx_1}{dt} = 3x_1 - t^2 x_2 \\ \dfrac{dx_2}{dt} = tx_2 - x_3 \\ \dfrac{dx_3}{dt} = (\sin t)x_1 \end{cases}$$

is a linear system for which $f_1(t, x_1, x_2, x_3) = 3x_1 - t^2 x_2$, $f_2(t, x_1, x_2, x_3) = tx_2 - x_3$, $f_3(t, x_1, x_2, x_3) = (\sin t)x_1$.

If $f_1, \ldots, f_n$ don't depend explicitly on t, the system is called **autonomous**. The system above and the system from Example 4.3.1 are not autonomous, but the systems from Examples 4.3.2 and 4.3.3 are. If a *linear* system is autonomous (as in Examples 4.3.2 and 4.3.3), it must have constant coefficients; that is, it will be of the form

$$\begin{aligned} \dfrac{dx_1}{dt} &= \ldots x_1 + \ldots x_2 + (\text{etc.}) + \ldots x_n + \cdots \\ &\vdots \\ \dfrac{dx_n}{dt} &= \ldots x_1 + \ldots x_2 + (\text{etc.}) + \ldots x_n + \cdots, \end{aligned}$$

where each $\ldots$ is a constant. As in Section 3.3, such a system will usually have a unique stationary point (where all the right-hand sides are zero); by shifting this point to the origin, we can then assume the system is of the form

$$\begin{cases} \dfrac{dx_1}{dt} = \ldots x_1 + \ldots x_2 + (\text{etc.}) + \ldots x_n \\ \quad\vdots \\ \dfrac{dx_n}{dt} = \ldots x_1 + \ldots x_2 + (\text{etc.}) + \ldots x_n. \end{cases}$$

Note that the systems in Examples 4.3.2 and 4.3.3 are both in this form; until Section 4.6, we'll only study systems of this type. To be less vague about the constants, let's introduce the notation a_{ij} for the coefficient of x_j in the ith equation.[7] For instance, a_{12} will be the coefficient of x_2 on the right-hand side of the first equation. The system will then have the form

$$\begin{cases} \dfrac{dx_1}{dt} = a_{11}x_1 + a_{12}x_2 + \cdots + a_{1n}x_n \\ \dfrac{dx_2}{dt} = a_{21}x_1 + a_{22}x_2 + \cdots + a_{2n}x_n \\ \quad\vdots \\ \dfrac{dx_n}{dt} = a_{n1}x_1 + a_{n2}x_2 + \cdots + a_{nn}x_n. \end{cases}$$

---

[7] Although you might have expected this coefficient to be called a_{ji}, our notation is standard and corresponds to the usual subscript notation for matrices (see Appendix B, Section B.2).

(NOTE: In Chapter 3 we used a, b, c, d for $a_{11}, a_{12}, a_{21}, a_{22}$ and x, y for x_1, x_2, which made the system

$$\begin{cases} \dfrac{dx}{dt} = ax + by \\ \dfrac{dy}{dt} = cx + dy \end{cases}.)$$

As in Chapter 3, we'll think of $(x_1, x_2, \ldots, x_n)$ as an unknown vector depending on t (if you like, an unknown vector function). This vector has derivative $\left(\dfrac{dx_1}{dt}, \dfrac{dx_2}{dt}, \ldots, \dfrac{dx_n}{dt}\right)$, and we can rewrite our system as a vector differential equation

$$\frac{d}{dt}\begin{bmatrix} x_1 \\ x_2 \\ \vdots \\ x_n \end{bmatrix} = \begin{bmatrix} a_{11} & a_{12} & \cdots & a_{1n} \\ a_{21} & a_{22} & \cdots & a_{2n} \\ \vdots & & & \\ a_{n1} & a_{n2} & \cdots & a_{nn} \end{bmatrix} \begin{bmatrix} x_1 \\ x_2 \\ \vdots \\ x_n \end{bmatrix}.$$

The notation can be streamlined even further by denoting the vector $(x_1, x_2, \ldots, x_n)$ by $\mathbf{v}$ and the matrix $\begin{bmatrix} a_{11} & \cdots & a_{1n} \\ \vdots & & \vdots \\ a_{n1} & \cdots & a_{nn} \end{bmatrix}$ by $\mathbf{A}$. If this is done, our vector differential equation becomes $\dfrac{d\mathbf{v}}{dt} = \mathbf{A}\mathbf{v}$.

EXAMPLE 4.3.3 (continued) The system

$$\begin{cases} \dfrac{dx_1}{dt} = 3x_1 - x_2 - 6x_3 \\ \dfrac{dx_2}{dt} = 4x_2 + 2x_3 + 6x_4 \\ \dfrac{dx_3}{dt} = 3x_1 - 3x_2 - 7x_3 - 3x_4 \\ \dfrac{dx_4}{dt} = -5x_1 + 3x_2 + 10x_3 + 2x_4 \end{cases}$$

can be written as $\dfrac{d\mathbf{v}}{dt} = \mathbf{A}\mathbf{v}$ by putting

$$\mathbf{v} = (x_1, x_2, x_3, x_4) \quad \text{and} \quad \mathbf{A} = \begin{bmatrix} 3 & -1 & -6 & 0 \\ 0 & 4 & 2 & 6 \\ 3 & -3 & -7 & -3 \\ -5 & 3 & 10 & 2 \end{bmatrix}. \quad \square$$

As in Section 3.4, our strategy for solving equations (systems) of type $\frac{d\mathbf{v}}{dt} = \mathbf{A}\mathbf{v}$ will be to make a change of coordinates such as to give the matrix an easier form. We'll let $X_1, X_2, \ldots, X_n$ be the new coordinates, so $\mathbf{V} = (X_1, X_2, \ldots, X_n)$ will be the new unknown vector; the two vectors $\mathbf{V}$ and $\mathbf{v}$ are related by some transformation matrix $\mathbf{T}$ such that $\mathbf{v} = \mathbf{T}\mathbf{V}$, $\mathbf{V} = \mathbf{T}^{-1}\mathbf{v}$.

EXAMPLE If $n = 3$ and

$$\mathbf{T} = \begin{bmatrix} 0 & 2 & 6 \\ 1 & -1 & 4 \\ 3 & 4 & 1 \end{bmatrix},$$

then

$$\begin{bmatrix} x_1 \\ x_2 \\ x_3 \end{bmatrix} = \begin{bmatrix} 0 & 2 & 6 \\ 1 & -1 & 4 \\ 3 & 4 & 1 \end{bmatrix} \begin{bmatrix} X_1 \\ X_2 \\ X_3 \end{bmatrix},$$

so $x_1 = 2X_2 + 6X_3$, $x_2 = X_1 - X_2 + 4X_3$, and so on. Meanwhile,

$$\begin{bmatrix} X_1 \\ X_2 \\ X_3 \end{bmatrix} = \begin{bmatrix} 0 & 2 & 6 \\ 1 & -1 & 4 \\ 3 & 4 & 1 \end{bmatrix}^{-1} \begin{bmatrix} x_1 \\ x_2 \\ x_3 \end{bmatrix} = \begin{bmatrix} -\frac{17}{64} & \frac{11}{32} & \frac{7}{32} \\ \frac{11}{64} & -\frac{9}{32} & \frac{3}{32} \\ \frac{7}{64} & \frac{3}{32} & -\frac{1}{32} \end{bmatrix} \begin{bmatrix} x_1 \\ x_2 \\ x_3 \end{bmatrix},$$ [8]

so $X_1 = -\frac{17}{64}x_1 + \frac{11}{32}x_2 + \frac{7}{32}x_3$, and so on. ∎

Now we should investigate how the vector differential equation $\frac{d\mathbf{v}}{dt} = \mathbf{A}\mathbf{v}$ changes when we transform to the new vector $\mathbf{V}$. Since $\mathbf{V} = \mathbf{T}^{-1}\mathbf{v}$ and $\mathbf{v} = \mathbf{T}\mathbf{V}$, we'll get

$$\frac{d\mathbf{V}}{dt} = \mathbf{T}^{-1}\frac{d\mathbf{v}}{dt} = \mathbf{T}^{-1}\mathbf{A}\mathbf{v} = \mathbf{T}^{-1}\mathbf{A}\mathbf{T}\mathbf{V}.$$

Therefore, we would like to choose $\mathbf{T}$ such that the matrix $\mathbf{T}^{-1}\mathbf{A}\mathbf{T}$ is easier to deal with than the matrix $\mathbf{A}$. As in Section 3.4, the optimal situation would be for $\mathbf{T}^{-1}\mathbf{A}\mathbf{T}$ to be a **diagonal** matrix, that is, a matrix of the form

$$\begin{bmatrix} \lambda_1 & & & 0 \\ & \lambda_2 & & \\ & & \ddots & \\ 0 & & & \lambda_n \end{bmatrix}$$

[8] See Appendix B, Example B.3.3, for the computation of the inverse matrix.

(the "large zeros" indicate that all entries off the main diagonal are zero). If we can choose **T** so this happens, our new vector differential equation will be

$$\frac{d\mathbf{V}}{dt} = \begin{bmatrix} \lambda_1 & & 0 \\ & \lambda_2 & \\ 0 & & \ddots \\ & & & \lambda_n \end{bmatrix} \mathbf{V},$$

which is equivalent to the new system

$$\begin{cases} \dfrac{dX_1}{dt} = \lambda_1 X_1 \\ \dfrac{dX_2}{dt} = \lambda_2 X_2 \\ \quad\vdots \\ \dfrac{dX_n}{dt} = \lambda_n X_n. \end{cases}$$

This system is "uncoupled" and can be solved immediately; we get

$$X_1 = C_1 e^{\lambda_1 t},\ X_2 = C_2 e^{\lambda_2 t},\ \ldots,\ X_n = C_n e^{\lambda_n t},$$

where $C_1, C_2, \ldots, C_n$ are arbitrary constants. The general solution of our original system is then given by

$$\begin{bmatrix} x_1 \\ x_2 \\ \vdots \\ x_n \end{bmatrix} = \mathbf{v} = \mathbf{TV} = \mathbf{T} \begin{bmatrix} C_1 e^{\lambda_1 t} \\ C_2 e^{\lambda_2 t} \\ \vdots \\ C_n e^{\lambda_n t} \end{bmatrix}.$$

Now we will see how to look for a suitable **T**. Just as in Section 3.4, in order to have

$$\mathbf{T}^{-1}\mathbf{AT} = \begin{bmatrix} \lambda_1 & & 0 \\ & \lambda_2 & \\ 0 & & \ddots \\ & & & \lambda_n \end{bmatrix},$$

we must have

$$\mathbf{AT} = \mathbf{T} \begin{bmatrix} \lambda_1 & & 0 \\ & \lambda_2 & \\ 0 & & \ddots \\ & & & \lambda_n \end{bmatrix},$$

and this implies that the vectors which form the *columns* of **T** have the special property that the matrix **A** transforms them into multiples of themselves (see Exercise 25). More specifically, **A** transforms the first column of **T** into λ_1 times itself, the second column of **T** into λ_2 times itself, and so on. We see that according to the following definition, each column of **T** is an eigenvector of **A**.

A nonzero vector $\boldsymbol{\xi} = (\xi_1, \xi_2, \ldots, \xi_n)$ is called an **eigenvector** of $\mathbf{A}$ for the **eigenvalue** λ if $\mathbf{A}\boldsymbol{\xi} = \lambda\boldsymbol{\xi}$, that is, $\mathbf{A}\begin{bmatrix}\xi_1\\ \vdots \\ \xi_n\end{bmatrix} = \lambda\begin{bmatrix}\xi_1\\ \vdots \\ \xi_n\end{bmatrix}$.

If we write this out using $\mathbf{A} = \begin{bmatrix} a_{11} & \cdots & a_{1n} \\ \vdots & & \vdots \\ a_{n1} & \cdots & a_{nn}\end{bmatrix}$, we get

$$\begin{cases} a_{11}\xi_1 + a_{12}\xi_2 + \cdots + a_{1n}\xi_n = \lambda\xi_1 \\ a_{21}\xi_1 + a_{22}\xi_2 + \cdots + a_{2n}\xi_n = \lambda\xi_2 \\ \quad \vdots \\ a_{n1}\xi_1 + a_{n2}\xi_2 + \cdots + a_{nn}\xi_n = \lambda\xi_n, \end{cases}$$

which can be rewritten as

$$\begin{cases} (a_{11} - \lambda)\xi_1 + a_{12}\xi_2 + \cdots + a_{1n}\xi_n = 0 \\ a_{21}\xi_1 + (a_{22} - \lambda)\xi_2 + \cdots + a_{2n}\xi_n = 0 \\ \quad \vdots \\ a_{n1}\xi_1 + a_{n2}\xi_2 + \cdots + (a_{nn} - \lambda)\xi_n = 0. \end{cases} \quad (*)$$

These simultaneous equations for $\xi_1, \ldots, \xi_n$ have a nontrivial solution only when

$$\det\begin{bmatrix} a_{11} - \lambda & a_{12} & \cdots & a_{1n} \\ a_{21} & a_{22} - \lambda & \cdots & a_{2n} \\ \vdots & \vdots & \ddots & \vdots \\ a_{n1} & a_{n2} & \cdots & a_{nn} - \lambda \end{bmatrix} = 0 \qquad \text{(see p. 624)}. \quad (**)$$

Conclusion. The eigenvalues of $\mathbf{A}$ are the roots λ of equation $(**)$. In order to find an eigenvector $(\xi_1, \ldots, \xi_n)$ for a specific eigenvalue λ, we solve the simultaneous equations $(*)$. If there are n distinct real eigenvalues $\lambda_1, \ldots, \lambda_n$ for $\mathbf{A}$, then we can use corresponding eigenvectors (in the same order) as columns for a transformation matrix $\mathbf{T}$.[9] $\mathbf{T}$ will then have the property that $\mathbf{T}^{-1}\mathbf{A}\mathbf{T} = \begin{bmatrix} \lambda_1 & & 0 \\ & \ddots & \\ 0 & & \lambda_n \end{bmatrix}$, and the general solution of the differential equation $\dfrac{d\mathbf{v}}{dt} = \mathbf{A}\mathbf{v}$ will be given by $\mathbf{v} = \mathbf{T}\mathbf{V} = \mathbf{T}\begin{bmatrix} C_1 e^{\lambda_1 t} \\ \vdots \\ C_n e^{\lambda_n t} \end{bmatrix}$, with $C_1, \ldots, C_n$ arbitrary constants.

---

[9] It is not too hard to show that $\mathbf{T}$ will have an inverse as long as the eigenvalues $\lambda_1, \ldots, \lambda_n$ are distinct. See textbooks on linear algebra.

EXAMPLE 4.3.4 Find all eigenvalues and eigenvectors for the matrix
$$A = \begin{bmatrix} 2 & 0 & 0 \\ 1 & 4 & 1 \\ -1 & 6 & 3 \end{bmatrix}.$$
Then solve the system $\dfrac{d\mathbf{v}}{dt} = A\mathbf{v}$, where $\mathbf{v} = (x_1, x_2, x_3)$.

The eigenvalues are the roots of the equation
$$\det\begin{bmatrix} 2-\lambda & 0 & 0 \\ 1 & 4-\lambda & 1 \\ -1 & 6 & 3-\lambda \end{bmatrix} = 0,$$
which yields $(2-\lambda)[(4-\lambda)(3-\lambda) - 6] = 0$. We could multiply all this out, but since we want to find the roots, it's better to leave the factor $2 - \lambda$ alone, and we get
$$(2-\lambda)(12 - 7\lambda + \lambda^2 - 6) = 0$$
$$(2-\lambda)(\lambda - 1)(\lambda - 6) = 0.$$
So the eigenvalues are $\lambda_1 = 1$, $\lambda_2 = 2$, and $\lambda_3 = 6$.

(**Reminder:** The order in which the eigenvalues are listed is random. Once the order is chosen, though, it should be maintained consistently throughout the problem, and the eigenvectors—to be found below—should be entered in the columns of the transformation matrix in the same order.)

The eigenvectors for $\lambda = 1$ are the nonzero vectors (ξ_1, ξ_2, ξ_3) for which
$$A\begin{bmatrix} \xi_1 \\ \xi_2 \\ \xi_3 \end{bmatrix} = 1 \cdot \begin{bmatrix} \xi_1 \\ \xi_2 \\ \xi_3 \end{bmatrix},$$
so
$$\begin{cases} 2\xi_1 & = \xi_1 \\ \xi_1 + 4\xi_2 + \xi_3 = \xi_2 \\ -\xi_1 + 6\xi_2 + 3\xi_3 = \xi_3 \end{cases}$$
or, equivalently,
$$\begin{cases} \xi_1 & = 0 \\ \xi_1 + 3\xi_2 + \xi_3 = 0 \\ -\xi_1 + 6\xi_2 + 2\xi_3 = 0 \end{cases}$$
[which we could have gotten directly from (∗) on p. 310].

We soon see that these are the vectors of the form
$$(0, \xi_2, -3\xi_2) = \xi_2(0, 1, -3) \quad \text{with} \quad \xi_2 \neq 0.$$

To get the eigenvectors for $\lambda = 2$, we start over; this time (∗) on p. 310 yields
$$\begin{cases} 0 & = 0 \\ \xi_1 + 2\xi_2 + \xi_3 = 0 \\ -\xi_1 + 6\xi_2 + \xi_3 = 0, \end{cases}$$
from which we find $\xi_1 = 2\xi_2$, $\xi_3 = -4\xi_2$, so the eigenvectors for $\lambda = 2$ look like

$(2\xi_2, \xi_2, -4\xi_2) = \xi_2(2, 1, -4)$ with $\xi_2 \neq 0$. A similar computation [see Exercise 19(a)] shows that the eigenvectors for $\lambda = 6$ are the vectors $\xi_2(0, 1, 2)$ with $\xi_2 \neq 0$.

For the purpose of solving the system, we need only one eigenvector for each eigenvalue, and we can make life easy by choosing $\xi_2 = 1$ in each case. If we put the resulting vectors $(0, 1, -3), (2, 1, -4), (0, 1, 2)$ in the columns of a transformation matrix:

$$\mathbf{T} = \begin{bmatrix} 0 & 2 & 0 \\ 1 & 1 & 1 \\ -3 & -4 & 2 \end{bmatrix},$$

then we are guaranteed that

$$\mathbf{T}^{-1}\mathbf{AT} = \begin{bmatrix} \lambda_1 & 0 & 0 \\ 0 & \lambda_2 & 0 \\ 0 & 0 & \lambda_3 \end{bmatrix} = \begin{bmatrix} 1 & 0 & 0 \\ 0 & 2 & 0 \\ 0 & 0 & 6 \end{bmatrix}.$$

[Of course, you could check this by computing $\mathbf{T}^{-1}\mathbf{AT}$ directly; see Exercise 19(b). For larger matrices this would be a great deal of work, though.] So we'll have

$$\mathbf{V} = \begin{bmatrix} C_1 e^t \\ C_2 e^{2t} \\ C_3 e^{6t} \end{bmatrix}$$

and finally

$$\mathbf{v} = \mathbf{TV} = \begin{bmatrix} 0 & 2 & 0 \\ 1 & 1 & 1 \\ -3 & -4 & 2 \end{bmatrix} \begin{bmatrix} C_1 e^t \\ C_2 e^{2t} \\ C_3 e^{6t} \end{bmatrix}.$$

That is, the general solution of our system is given by

$$\begin{cases} x_1 = 2C_2 e^{2t} \\ x_2 = C_1 e^t + C_2 e^{2t} + C_3 e^{6t} \\ x_3 = -3C_1 e^t - 4C_2 e^{2t} + 2C_3 e^{6t}, \quad C_1, C_2, C_3 \text{ arbitrary.} \end{cases}$$

We can also write the general solution in vector form as

$$(x_1, x_2, x_3) = C_1 e^t \cdot (0, 1, -3) + C_2 e^{2t} \cdot (2, 1, -4) + C_3 e^{6t} \cdot (0, 1, 2).$$

Note that we now have three basic solutions (in vector form):

$$e^t \cdot (0, 1, -3), \; e^{2t} \cdot (2, 1, -4), \; e^{6t} \cdot (0, 1, 2),$$

which are such that *any* solution (in vector form) is a linear combination of these. These basic solutions have an easily recognizable pattern: There is one for each eigenvalue λ, namely $e^{\lambda t}$ times an eigenvector for that eigenvalue. (See Exercises 23 and 24.) ∎

Now that we've seen an example of the method, what problems are there? First of all, we have to find the eigenvalues as the roots of the equation

$$\det \begin{bmatrix} a_{11} - \lambda & a_{12} & \cdots & a_{1n} \\ a_{21} & a_{22} - \lambda & \cdots & a_{2n} \\ \vdots & \vdots & \ddots & \vdots \\ a_{n1} & a_{n2} & \cdots & a_{nn} - \lambda \end{bmatrix} = 0,$$

and this can be a formidable computation. Incidentally, the matrix of which we're taking the determinant here is often rewritten as

$$\begin{bmatrix} a_{11} & a_{12} & \cdots & a_{1n} \\ a_{21} & a_{22} & \cdots & a_{2n} \\ \vdots & \vdots & & \vdots \\ a_{n1} & a_{n2} & \cdots & a_{nn} \end{bmatrix} - \begin{bmatrix} \lambda & & & 0 \\ & \lambda & & \\ & & \ddots & \\ 0 & & & \lambda \end{bmatrix} = \mathbf{A} - \lambda \mathbf{I},$$

where $\mathbf{I}$ is the $n \times n$ identity matrix (see p. 612). The condition for λ to be an eigenvalue of $\mathbf{A}$ then becomes $\det(\mathbf{A} - \lambda \mathbf{I}) = 0$, which is known as the **characteristic equation**[10] of $\mathbf{A}$. The left-hand side $\det(\mathbf{A} - \lambda \mathbf{I})$ is a polynomial of degree n in λ, called the **characteristic polynomial** of $\mathbf{A}$.[11] Just as in the case $n = 2$, there are complications if the characteristic equation has complex or multiple roots, and we'll study these cases in the next section. Meanwhile, we'll assume that the characteristic equation has n distinct real roots $\lambda_1, \ldots, \lambda_n$. Finding these roots may be another matter, especially if n is large. Since it often is a great deal of work to actually expand $\det(\mathbf{A} - \lambda \mathbf{I})$ as a polynomial in λ, and since even after that numerical methods (such as Newton's method) may be needed to find the roots of such a polynomial, methods have been developed to approximate the eigenvalues of $\mathbf{A}$ without using the characteristic equation. (See "Further Reading" following the Exercises.) In this book, the characteristic equations occurring in the exercises should be relatively easy to solve; sometimes one or more of the eigenvalues will be given. Once all eigenvalues are found (assuming, again, that there are n distinct ones), finding the eigenvectors should not be difficult (although for n large it can be very time-consuming). The general solution to the system can then be written down quickly. Here are two more examples.

EXAMPLE 4.3.5 Find the general solution of the system

$$\begin{cases} \dfrac{dx_1}{dt} = 3x_1 - x_2 - 6x_3 \\ \dfrac{dx_2}{dt} = 4x_2 + 2x_3 + 6x_4 \\ \dfrac{dx_3}{dt} = 3x_1 - 3x_2 - 7x_3 - 3x_4 \\ \dfrac{dx_4}{dt} = -5x_1 + 3x_2 + 10x_3 + 2x_4 \end{cases} \quad \text{from Example 4.3.3 (p. 307)}.$$

---

[10] See Exercise 27 for the connection with the characteristic equation for a higher-order differential equation, as defined in Section 4.1.

[11] Some authors refer to $\det(\lambda \mathbf{I} - \mathbf{A})$ as the characteristic polynomial. Since $\det(\lambda \mathbf{I} - \mathbf{A}) = (-1)^n \det(\mathbf{A} - \lambda \mathbf{I})$ (factor out -1 from each row), it doesn't matter much.

In this case, the characteristic equation $\det(\mathbf{A} - \lambda\mathbf{I}) = 0$ is

$$\det\begin{bmatrix} 3-\lambda & -1 & -6 & 0 \\ 0 & 4-\lambda & 2 & 6 \\ 3 & -3 & -7-\lambda & -3 \\ -5 & 3 & 10 & 2-\lambda \end{bmatrix} = 0.$$

Unfortunately, there is no nice computational shortcut from here, unless you have a computer with an algebra program at your disposal. Proceeding by hand, we get (expanding along the first row)

$$(3-\lambda)\det\begin{bmatrix} 4-\lambda & 2 & 6 \\ -3 & -7-\lambda & -3 \\ 3 & 10 & 2-\lambda \end{bmatrix} + \det\begin{bmatrix} 0 & 2 & 6 \\ 3 & -7-\lambda & -3 \\ -5 & 10 & 2-\lambda \end{bmatrix} - 6\det\begin{bmatrix} 0 & 4-\lambda & 6 \\ 3 & -3 & -3 \\ -5 & 3 & 2-\lambda \end{bmatrix} = 0$$

$$(3-\lambda)\{(4-\lambda)[(-7-\lambda)(2-\lambda)+30] - 2[-3(2-\lambda)+9] + 6[-30-3(-7-\lambda)]\}$$
$$+ \{-2[3(2-\lambda)-15] + 6[30+5(-7-\lambda)]\} - 6\{(\lambda-4)[3(2-\lambda)-15] + 6(9-15)\} = 0.$$

Several minutes later, this yields $\lambda^4 - 2\lambda^3 - \lambda^2 + 2\lambda = 0$, which factors as

$$\lambda(\lambda^3 - 2\lambda^2 - \lambda + 2) = 0$$
$$\lambda(\lambda-2)(\lambda^2-1) = 0$$
$$\lambda(\lambda-2)(\lambda-1)(\lambda+1) = 0.$$

Thus the eigenvalues of the matrix $\mathbf{A}$ are $\lambda_1 = 0$, $\lambda_2 = 2$, $\lambda_3 = 1$, $\lambda_4 = -1$. To find an eigenvector for $\lambda = 0$, we have to solve the simultaneous equations

$$\begin{cases} 3\xi_1 - \xi_2 - 6\xi_3 & = 0 \\ 4\xi_2 + 2\xi_3 + 6\xi_4 = 0 \\ 3\xi_1 - 3\xi_2 - 7\xi_3 - 3\xi_4 = 0 \\ -5\xi_1 + 3\xi_2 + 10\xi_3 + 2\xi_4 = 0. \end{cases}$$

This can be done by repeated elimination (for instance, rewrite the first equation as $\xi_2 = 3\xi_1 - 6\xi_3$ and substitute this into the other three equations to get three equations in the three unknowns ξ_1, ξ_3, ξ_4, etc.) or by the streamlined elimination technique known as "row reduction" or "Gauss–Jordan elimination" (see textbooks on linear algebra). The end result will be that the equations are satisfied whenever we have

$$\xi_1 = -\frac{1}{2}\xi_4, \ \xi_2 = -\frac{3}{2}\xi_4, \ \xi_3 = 0, \ \xi_4 \text{ arbitrary.}$$

Since we need only one eigenvector, we can choose $\xi_4 = -2$ (to avoid fractions) and get $(\xi_1, \xi_2, \xi_3, \xi_4) = (1, 3, 0, -2)$ as our eigenvector for $\lambda = 0$. To find an eigenvector for $\lambda = 2$, we have to start all over, this time with

$$\begin{cases} \xi_1 - \xi_2 - 6\xi_3 & = 0 \\ 2\xi_2 + 2\xi_3 + 6\xi_4 = 0 \\ 3\xi_1 - 3\xi_2 - 9\xi_3 - 3\xi_4 = 0 \\ -5\xi_1 + 3\xi_2 + 10\xi_3 & = 0. \end{cases}$$

[These are equations (*) on p. 310 for $\lambda = 2$.] This time, the end result of the elimination will be $\xi_1 = -\dfrac{4}{3}\xi_4$, $\xi_2 = -\dfrac{10}{3}\xi_4$, $\xi_3 = \dfrac{1}{3}\xi_4$, ξ_4 arbitrary; for instance, we can take $\xi_4 = 3$ and get $(-4, -10, 1, 3)$ as our eigenvector for $\lambda = 2$.

$$\text{(Check:)} \quad \underbrace{\begin{bmatrix} 3 & -1 & -6 & 0 \\ 0 & 4 & 2 & 6 \\ 3 & -3 & -7 & -3 \\ -5 & 3 & 10 & 2 \end{bmatrix}}_{\mathbf{A}} \underbrace{\begin{bmatrix} -4 \\ -10 \\ 1 \\ 3 \end{bmatrix}}_{\boldsymbol{\xi}} = \begin{bmatrix} -8 \\ -20 \\ 2 \\ 6 \end{bmatrix} = \underbrace{2}_{\lambda} \cdot \underbrace{\begin{bmatrix} -4 \\ -10 \\ 1 \\ 3 \end{bmatrix}}_{\boldsymbol{\xi}}.)$$

Similar computations show that $(1, 2, 0, -1)$ is an eigenvector for $\lambda = 1$ and $(2, 2, 1, -2)$ is an eigenvector for $\lambda = -1$ (see Exercise 20). Now that we have an eigenvector for each of the four eigenvalues, we can form

$$\mathbf{T} = \begin{bmatrix} 1 & -4 & 1 & 2 \\ 3 & -10 & 2 & 2 \\ 0 & 1 & 0 & 1 \\ -2 & 3 & -1 & -2 \end{bmatrix}.$$

$\uparrow$ eigenvector for $\lambda = 0$ $\qquad$ $\uparrow$ eigenvector for $\lambda = -1$

Then

$$\mathbf{T}^{-1}\mathbf{AT} = \begin{bmatrix} \lambda_1 & & & 0 \\ & \lambda_2 & & \\ & & \lambda_3 & \\ 0 & & & \lambda_4 \end{bmatrix} = \begin{bmatrix} 0 & 0 & 0 & 0 \\ 0 & 2 & 0 & 0 \\ 0 & 0 & 1 & 0 \\ 0 & 0 & 0 & -1 \end{bmatrix}, \quad \mathbf{V} = \begin{bmatrix} C_1 \\ C_2 e^{2t} \\ C_3 e^{t} \\ C_4 e^{-t} \end{bmatrix},$$

and

$$\mathbf{v} = \mathbf{TV} = \begin{bmatrix} 1 & -4 & 1 & 2 \\ 3 & -10 & 2 & 2 \\ 0 & 1 & 0 & 1 \\ -2 & 3 & -1 & -2 \end{bmatrix} \begin{bmatrix} C_1 \\ C_2 e^{2t} \\ C_3 e^{t} \\ C_4 e^{-t} \end{bmatrix}.$$

Therefore, the general solution of our system is given by

$$(x_1, x_2, x_3, x_4) = C_1(1, 3, 0, -2) + C_2 e^{2t}(-4, -10, 1, 3)$$
$$+ C_3 e^{t}(1, 2, 0, -1) + C_4 e^{-t}(2, 2, 1, -2), \quad C_1, C_2, C_3, C_4 \text{ arbitrary.}$$

Once again, we see that whenever we have an eigenvalue λ and a corresponding eigenvector $\boldsymbol{\xi} = (\xi_1, \xi_2, \xi_3, \xi_4)$, we have a basic solution $e^{\lambda t}\boldsymbol{\xi}$. This works in general (see Exercises 23 and 24). ∎

EXAMPLE 4.3.6 Solve the initial value problem

$$\begin{cases} \dfrac{dx_1}{dt} = -7x_1 - 10x_2 + 2x_3, & x_1(0) = 0, \\ \dfrac{dx_2}{dt} = 6x_1 + 9x_2 - 2x_3, & x_2(0) = 4, \\ \dfrac{dx_3}{dt} = -6x_1 - 6x_2 + 5x_3, & x_3(0) = -2. \end{cases}$$

To find the general solution of the system, we first find the eigenvalues:

$$\det \begin{bmatrix} -7-\lambda & -10 & 2 \\ 6 & 9-\lambda & -2 \\ -6 & -6 & 5-\lambda \end{bmatrix} = 0$$

$$(-7-\lambda)[(9-\lambda)(5-\lambda) - 12] + 10[6(5-\lambda) - 12] + 2[-36 + 6(9-\lambda)] = 0.$$

Multiplying out and simplifying, we get

$$-\lambda^3 + 7\lambda^2 - 7\lambda - 15 = 0$$

or

$$\lambda^3 - 7\lambda^2 + 7\lambda + 15 = 0.$$

By inspection, $\lambda = -1$ is a root of this equation, so there is a factor $\lambda + 1$. We get

$$(\lambda + 1)(\lambda^2 - 8\lambda + 15) = 0$$
$$(\lambda + 1)(\lambda - 3)(\lambda - 5) = 0.$$

Thus the eigenvalues are $\lambda_1 = -1$, $\lambda_2 = 3$, $\lambda_3 = 5$. For each eigenvalue λ, the eigenvectors are the (nonzero) solution vectors (ξ_1, ξ_2, ξ_3) of the simultaneous equations

$$\begin{cases} (-7-\lambda)\xi_1 - 10\xi_2 + 2\xi_3 = 0 \\ 6\xi_1 + (9-\lambda)\xi_2 - 2\xi_3 = 0 \\ -6\xi_1 - 6\xi_2 + (5-\lambda)\xi_3 = 0. \end{cases}$$

For $\lambda = -1$, for example, we get

$$\begin{cases} -6\xi_1 - 10\xi_2 + 2\xi_3 = 0 \\ 6\xi_1 + 10\xi_2 - 2\xi_3 = 0 \\ -6\xi_1 - 6\xi_2 + 6\xi_3 = 0, \end{cases}$$

or in the "streamlined" form for row reduction,

$$\begin{bmatrix} -6 & -10 & 2 & | & 0 \\ 6 & 10 & -2 & | & 0 \\ -6 & -6 & 6 & | & 0 \end{bmatrix}.$$

You should find from this that $(2, -1, 1)$ is one eigenvector for $\lambda_1 = -1$. Similarly, $(1, -1, 0)$ is an eigenvector for $\lambda_2 = 3$, and $(1, -1, 1)$ is an eigenvector for $\lambda_3 = 5$.

Therefore, the general solution is given by

$$\begin{bmatrix} x_1 \\ x_2 \\ x_3 \end{bmatrix} = \mathbf{v} = \mathbf{TV} = \begin{bmatrix} 2 & 1 & 1 \\ -1 & -1 & -1 \\ 1 & 0 & 1 \end{bmatrix} \begin{bmatrix} C_1 e^{-t} \\ C_2 e^{3t} \\ C_3 e^{5t} \end{bmatrix}$$

or

$$\begin{cases} x_1 = 2C_1 e^{-t} + C_2 e^{3t} + C_3 e^{5t} \\ x_2 = -C_1 e^{-t} - C_2 e^{3t} - C_3 e^{5t} \\ x_3 = C_1 e^{-t} + C_3 e^{5t}. \end{cases}$$

To finish, we now have to find the values of C_1, C_2, C_3 for which $x_1(0) = 0$, $x_2(0) = 4$, $x_3(0) = -2$. See Exercise 12. ∎

SUMMARY OF KEY CONCEPTS, RESULTS, AND TECHNIQUES

Converting a higher-order equation to a first-order system (p. 304): To convert an nth-order differential equation for the unknown function x to a system of n first-order equations, introduce new unknown functions $y_1 = x'$, $y_2 = x''$, ..., $y_{n-1} = x^{(n-1)}$ along with x.

Homogeneous, autonomous linear system (p. 306): A system of the form

$$\begin{cases} \dfrac{dx_1}{dt} = a_{11} x_1 + \cdots + a_{1n} x_n \\ \quad \vdots \qquad\qquad \vdots \qquad\qquad \vdots \\ \dfrac{dx_n}{dt} = a_{n1} x_1 + \cdots + a_{nn} x_n. \end{cases}$$

This system can be rewritten as the vector differential equation

$$\frac{d\mathbf{v}}{dt} = \mathbf{A}\mathbf{v}, \quad \text{with} \quad \mathbf{v} = (x_1, \ldots, x_n), \quad \mathbf{A} = \begin{bmatrix} a_{11} & \cdots & a_{1n} \\ \vdots & & \vdots \\ a_{n1} & \cdots & a_{nn} \end{bmatrix} \quad \text{(p. 307)}.$$

Change of coordinates (p. 308): Under the change of coordinates $\mathbf{v} = \mathbf{TV}$, $\mathbf{V} = \mathbf{T}^{-1}\mathbf{v}$ the system becomes $\dfrac{d\mathbf{V}}{dt} = \mathbf{T}^{-1}\mathbf{AT}\,\mathbf{V}$.

An **eigenvector** of the matrix $\mathbf{A}$ for the **eigenvalue** λ is a nonzero vector $\boldsymbol{\xi}$ such that $\mathbf{A}\boldsymbol{\xi} = \lambda \boldsymbol{\xi}$ (p. 310). The eigenvalues of $\mathbf{A}$ are the roots of the **characteristic equation** $\det(\mathbf{A} - \lambda \mathbf{I}) = 0$ (p. 313). To find the eigenvectors of $\mathbf{A}$, first find the eigenvalues, then solve $\mathbf{A}\boldsymbol{\xi} = \lambda \boldsymbol{\xi}$, or equivalently

$$\begin{cases} (a_{11} - \lambda)\xi_1 + \cdots + a_{1n}\xi_n = 0 \\ \qquad\qquad\qquad \vdots \\ a_{n1}\xi_1 + \cdots + (a_{nn} - \lambda)\xi_n = 0, \end{cases}$$

for each eigenvalue λ separately.

Solving the system $\dfrac{d\mathbf{v}}{dt} = \mathbf{A}\mathbf{v}$ (pp. 308–310): If $\mathbf{A}$ has n distinct real eigenvalues $\lambda_1, \ldots, \lambda_n$ (see Section 4.4 for the other cases), and we use corresponding eigenvectors (in the same order) as columns of $\mathbf{T}$, then

$$\mathbf{T}^{-1}\mathbf{A}\mathbf{T} = \begin{bmatrix} \lambda_1 & & & 0 \\ & \lambda_2 & & \\ 0 & & \ddots & \\ & & & \lambda_n \end{bmatrix}.$$

Thus the new system will be $\dfrac{dX_1}{dt} = \lambda_1 X_1, \ldots, \dfrac{dX_n}{dt} = \lambda_n X_n$, so we get

$$\mathbf{V} = \begin{bmatrix} C_1 e^{\lambda_1 t} \\ C_2 e^{\lambda_2 t} \\ \vdots \\ C_n e^{\lambda_n t} \end{bmatrix}$$

and, finally,

$$\mathbf{v} = \mathbf{T}\mathbf{V} = \mathbf{T} \begin{bmatrix} C_1 e^{\lambda_1 t} \\ C_2 e^{\lambda_2 t} \\ \vdots \\ C_n e^{\lambda_n t} \end{bmatrix}.$$

The vector functions $e^{\lambda t}\boldsymbol{\xi}$ for each eigenvalue λ and corresponding eigenvector $\boldsymbol{\xi}$ will form a set of basic solutions (pp. 312, 315).

EXERCISES

In each case below, write down a system of first-order differential equations that is equivalent to the given equation or system.

1. $x^{(3)} - 3t^2 x'' + (\sin t)x = 0$.
2. $x^{(3)} + 4tx' - t^2 x = 0$.
3. $\dfrac{d^2 x}{dt^2} - 3\dfrac{dy}{dt} + 4y = t, \quad \dfrac{d^2 y}{dt^2} - 4\dfrac{dx}{dt} + t^2 x = 0$.
4. $\dfrac{d^2 x}{dt^2} + t\dfrac{dy}{dt} - t^2 x = 3, \quad \dfrac{d^2 y}{dt^2} + 5t\dfrac{dy}{dt} - 3ty = 0$.
5. $x^{(3)} - 2tx'' + x = 0, \quad y'' - 4y = 0$.

For each of the following matrices, find all eigenvalues and eigenvectors.

6. $\begin{bmatrix} -3 & 0 & 0 \\ 1 & 2 & 5 \\ 4 & 1 & 6 \end{bmatrix}$

7. $\begin{bmatrix} 2 & 0 & 0 & 0 \\ 0 & 4 & 0 & 0 \\ 0 & -1 & -1 & 2 \\ 0 & 2 & 2 & -4 \end{bmatrix}$

8. $\begin{bmatrix} 5 & 1 & 2 \\ 1 & 5 & 2 \\ 2 & 2 & 4 \end{bmatrix}$

[*Hint*: $\lambda = 2$ is one eigenvalue.]

9. $\begin{bmatrix} 3 & 2 & 2 \\ 2 & 4 & 1 \\ 2 & 1 & 4 \end{bmatrix}$

[*Hint*: $\lambda = 3$ is one eigenvalue.]

10. $\begin{bmatrix} 3 & 0 & 0 & 0 \\ 0 & -1 & 1 & 2 \\ 1 & 0 & 1 & 4 \\ -1 & 2 & -1 & -5 \end{bmatrix}$

11. $\begin{bmatrix} 2 & 0 & 0 & 0 \\ 1 & 3 & 0 & 2 \\ 1 & 1 & 5 & -1 \\ -1 & 2 & 4 & -1 \end{bmatrix}$

12. Finish Example 4.3.6.

Solve the following differential equations and initial value problems.

13. $\dfrac{d\mathbf{v}}{dt} = \mathbf{A}\mathbf{v}$, where $\mathbf{v} = (x_1, x_2, x_3)$, $\mathbf{A} = \begin{bmatrix} 5 & 1 & 2 \\ 1 & 5 & 2 \\ 2 & 2 & 4 \end{bmatrix}$. (Compare Exercise 8.)

14. $\dfrac{d\mathbf{v}}{dt} = \mathbf{A}\mathbf{v}$, where $\mathbf{v} = (x_1, x_2, x_3)$, $\mathbf{A} = \begin{bmatrix} 3 & 2 & 2 \\ 2 & 4 & 1 \\ 2 & 1 & 4 \end{bmatrix}$. (Compare Exercise 9.)

15. $\dfrac{dx_1}{dt} = -9x_1 + 6x_2 - 6x_3$, $\quad \dfrac{dx_2}{dt} = -10x_1 + 7x_2 - 6x_3$, $\quad \dfrac{dx_3}{dt} = 2x_1 - 2x_2 + 3x_3$.

16. $\dfrac{dx_1}{dt} = x_1 + 2x_2 + x_3$, $\quad \dfrac{dx_2}{dt} = 3x_2 + 6x_3$, $\quad \dfrac{dx_3}{dt} = x_2 + 2x_3$.

17. $\dfrac{dx_1}{dt} = 3x_1$, $\quad x_1(0) = 16$, $\quad \dfrac{dx_2}{dt} = -x_2 + x_3 + 2x_4$, $\quad x_2(0) = 4$,

$\dfrac{dx_3}{dt} = x_1 + x_3 + 4x_4$, $\quad x_3(0) = -4$, $\quad \dfrac{dx_4}{dt} = -x_1 + 2x_2 - x_3 - 5x_4$, $\quad x_4(0) = 3$.

(Compare Exercise 10.)

18. $\dfrac{dx_1}{dt} = 2x_1$, $\quad x_1(0) = -5$, $\quad \dfrac{dx_2}{dt} = x_1 + 3x_2 + 2x_4$, $\quad x_2(0) = 2$,

$\dfrac{dx_3}{dt} = x_1 + x_2 + 5x_3 - x_4$, $\quad x_3(0) = 3$, $\quad \dfrac{dx_4}{dt} = -x_1 + 2x_2 + 4x_3 - x_4$,

$x_4(0) = 2$. (Compare Exercise 11.)

19. (a) For Example 4.3.4, find all eigenvectors for $\lambda = 6$. (Be systematic; don't use the fact that the answer is in the text.)

(b) For the same example, check by direct computation that

$$\mathbf{T}^{-1}\mathbf{AT} = \begin{bmatrix} 1 & 0 & 0 \\ 0 & 2 & 0 \\ 0 & 0 & 6 \end{bmatrix}.$$

20. For Example 4.3.5, find all eigenvectors for $\lambda = 1$ and for $\lambda = -1$. (Don't use the fact that the answers are in the text.)

For each of the following systems, indicate what coordinate transformation would be needed to shift the stationary point to the origin.

21. $\dfrac{dx}{dt} = 4x + 3y + 2z$, $\quad \dfrac{dy}{dt} = x + y - 2$, $\quad \dfrac{dz}{dt} = 2x + 2y + z - 1$.

22. $\dfrac{dx}{dt} = 4x - y - z$, $\quad \dfrac{dy}{dt} = 2x + y + 3z - 4$, $\quad \dfrac{dz}{dt} = y + 5z$.

23. Show that if $\boldsymbol{\xi} = (\xi_1, \ldots, \xi_n)$ is an eigenvector of $\mathbf{A}$ for the eigenvalue λ, then $\mathbf{v}(t) = e^{\lambda t}\boldsymbol{\xi}$ is a solution to the system $\dfrac{d\mathbf{v}}{dt} = \mathbf{A}\mathbf{v}$.

*24. (Compare Section 3.4, Exercise 22.)

(a) Show that if $\mathbf{v}_1(t), \ldots, \mathbf{v}_n(t)$ are solutions to the system $\dfrac{d\mathbf{v}}{dt} = \mathbf{A}\mathbf{v}$, then so is any linear combination $C_1\mathbf{v}_1(t) + \cdots + C_n\mathbf{v}_n(t)$, where $C_1, \ldots, C_n$ are constants.

(b) Combine the results of part (a) and of Exercise 23 to show that if $\boldsymbol{\xi}^{(1)}, \ldots, \boldsymbol{\xi}^{(n)}$ are eigenvectors of $\mathbf{A}$ corresponding to the distinct eigenvalues $\lambda_1, \ldots, \lambda_n$, respectively, then $C_1 e^{\lambda_1 t}\boldsymbol{\xi}^{(1)} + \cdots + C_n e^{\lambda_n t}\boldsymbol{\xi}^{(n)}$ will be a solution to the system $\dfrac{d\mathbf{v}}{dt} = \mathbf{A}\mathbf{v}$. Show that this solution (with $C_1, \ldots, C_n$ arbitrary) is the same as the general solution $\mathbf{v} = \mathbf{T}\mathbf{V}$ found in the text.

25. (a) Show that if

$$\mathbf{AT} = \mathbf{T}\begin{bmatrix} \lambda_1 & & & 0 \\ & \lambda_2 & & \\ & & \ddots & \\ 0 & & & \lambda_n \end{bmatrix}$$

and $\boldsymbol{\xi}$ is the vector in the first column of $\mathbf{T}$, then $\mathbf{A}\boldsymbol{\xi} = \lambda_1\boldsymbol{\xi}$.

(b) In the same situation, show that if $\boldsymbol{\xi}$ is the vector in the jth column of $\mathbf{T}$, with $1 \leq j \leq n$, then $\mathbf{A}\boldsymbol{\xi} = \lambda_j\boldsymbol{\xi}$.

26. Show that for any square matrix $\mathbf{A}$ and any invertible matrix $\mathbf{T}$ of the same size, $\det(\mathbf{A} - \lambda\mathbf{I}) = \det(\mathbf{T}^{-1}\mathbf{AT} - \lambda\mathbf{I})$.
[*Hint*: Write $\mathbf{T}^{-1}\mathbf{AT} - \lambda\mathbf{I} = \mathbf{T}^{-1}(\mathbf{A} - \lambda\mathbf{I})\mathbf{T}$. Why is this correct?]
Note that by the result of this exercise, the matrices $\mathbf{A}$ and $\mathbf{T}^{-1}\mathbf{AT}$ have the same characteristic equation.

27. (a) Write down a linear system that is equivalent to the nth-order differential equation

$$\frac{d^n x}{dt^n} + p_{n-1}(t) \frac{d^{n-1} x}{dt^{n-1}} + \cdots + p_1(t) \frac{dx}{dt} + p_0(t) x = h(t).$$

(b) Show that the constant coefficient nth-order equation

$$\frac{d^n x}{dt^n} + p_{n-1} \frac{d^{n-1} x}{dt^{n-1}} + \cdots + p_1 \frac{dx}{dt} + p_0 x = 0$$

is equivalent to an autonomous linear system with matrix

$$\mathbf{A} = \begin{bmatrix} 0 & 1 & 0 & 0 & \cdots & 0 \\ 0 & 0 & 1 & 0 & \cdots & 0 \\ & & & \ddots & \ddots & \vdots \\ & & 0 & & & 0 \\ & & & & & 1 \\ -p_0 & -p_1 & \cdots & & & -p_{n-1} \end{bmatrix}.$$

*(c) Show that the characteristic polynomial $\det(\mathbf{A} - \lambda \mathbf{I})$ of the matrix $\mathbf{A}$ in (b) is $(-1)^n(\lambda^n + p_{n-1}\lambda^{n-1} + \cdots + p_1\lambda + p_0)$. Note that except for the factor $(-1)^n$, this is the characteristic polynomial of the original differential equation. [In linear algebra, the matrix $\mathbf{A}$ from part (b) is called the **companion matrix** of this polynomial.]

FURTHER READING

For an introduction to methods for the approximation of eigenvalues, see Section 8.4 of Burden, Faires, and Reynolds, *Numerical Analysis*, 2nd ed. (Prindle, Weber & Schmidt, 1981).

4.4 HOMOGENEOUS, AUTONOMOUS LINEAR SYSTEMS (CONTINUED)

We continue to consider systems of the form

$$\begin{cases} \dfrac{dx_1}{dt} = a_{11}x_1 + \cdots + a_{1n}x_n \\ \quad\vdots \\ \dfrac{dx_n}{dt} = a_{n1}x_1 + \cdots + a_{nn}x_n, \end{cases}$$

written as $\dfrac{d\mathbf{v}}{dt} = \mathbf{A}\mathbf{v}$ for short. In this section we'll look at what happens if the matrix $\mathbf{A}$ doesn't have n distinct real eigenvalues, or, equivalently, if the characteristic equation $\det(\mathbf{A} - \lambda\mathbf{I}) = 0$ doesn't have n distinct real roots λ. This can happen because some of the roots are complex (nonreal), repeated, or both.

NOTE: We still assume that the matrix $\mathbf{A}$ itself has real entries!

Let's start with the case in which there are complex roots, but no repeated ones. That is, all n roots λ of $\det(\mathbf{A} - \lambda\mathbf{I}) = 0$ are distinct, but not all of them are real. By allowing vectors with complex coordinates (as in Section 3.5), we will still be able to get a (complex) eigenvector for each eigenvalue. These vectors can be used as the columns of a transformation matrix $\mathbf{T}$; $\mathbf{T}^{-1}\mathbf{A}\mathbf{T}$ will then be a diagonal matrix, and we can find the general complex-valued solution to the system by the method of Section 4.3. This general solution will be of the form

$$\mathbf{v} = \mathbf{T}\begin{bmatrix} C_1 e^{\lambda_1 t} \\ C_2 e^{\lambda_2 t} \\ \vdots \\ C_n e^{\lambda_n t} \end{bmatrix},$$

where $\lambda_1, \ldots, \lambda_n$ are the distinct eigenvalues and $C_1, \ldots, C_n$ are arbitrary complex constants. Of course, in most applications we really want to find the real-valued solutions of the system. To see which solutions are real-valued, first note that the complex (nonreal) eigenvalues and eigenvectors come in complex conjugate pairs: If $\boldsymbol{\xi}$ is an eigenvector of $\mathbf{A}$ for the eigenvalue λ, then $\mathbf{A}\boldsymbol{\xi} = \lambda\boldsymbol{\xi}$, so taking conjugates we have $\mathbf{A}\overline{\boldsymbol{\xi}} = \overline{\lambda}\,\overline{\boldsymbol{\xi}}$. (Since the entries of $\mathbf{A}$ are assumed to be real, they won't be affected by complex conjugation.) Therefore, $\overline{\boldsymbol{\xi}}$ is an eigenvector of $\mathbf{A}$ for $\overline{\lambda}$. Now that we know this, let's see how to find the real-valued solutions in a particular example.

EXAMPLE 4.4.1 Find the general (real-valued) solution to the system

$$\begin{cases} \dfrac{dx_1}{dt} = 6x_1 + 6x_2 + x_3 \\ \dfrac{dx_2}{dt} = -4x_1 - 2x_2 - 3x_3 \\ \dfrac{dx_3}{dt} = -3x_1 - 5x_2 + x_3. \end{cases}$$

Here $\mathbf{A} = \begin{bmatrix} 6 & 6 & 1 \\ -4 & -2 & -3 \\ -3 & -5 & 1 \end{bmatrix}$, so $\mathbf{A} - \lambda\mathbf{I} = \begin{bmatrix} 6-\lambda & 6 & 1 \\ -4 & -2-\lambda & -3 \\ -3 & -5 & 1-\lambda \end{bmatrix}$, and the characteristic equation $\det(\mathbf{A} - \lambda\mathbf{I}) = 0$ works out to $\lambda^3 - 5\lambda^2 + 4\lambda + 10 = 0$. By inspection, $\lambda = -1$ is a root of this equation; if we factor out $\lambda + 1$, we find

$$(\lambda + 1)(\lambda^2 - 6\lambda + 10) = 0.$$

$\lambda^2 - 6\lambda + 10 = 0$ has (complex) roots $\lambda = 3 \pm i$ (see Appendix A, Section A.1, Exercise 14). An eigenvector for $\lambda = 3 + i$ is a nonzero vector (ξ_1, ξ_2, ξ_3) for which

$$\begin{cases} 6\xi_1 + 6\xi_2 + \xi_3 = (3+i)\xi_1 \\ -4\xi_1 - 2\xi_2 - 3\xi_3 = (3+i)\xi_2 \\ -3\xi_1 - 5\xi_2 + \xi_3 = (3+i)\xi_3, \end{cases}$$

or, equivalently,

$$\begin{cases} (3-i)\xi_1 + 6\xi_2 + \xi_3 = 0 \\ -4\xi_1 - (5+i)\xi_2 - 3\xi_3 = 0 \\ -3\xi_1 - 5\xi_2 - (2+i)\xi_3 = 0. \end{cases}$$

Elimination yields $\xi_1 = (-2-i)\xi_2$, $\xi_3 = (1+i)\xi_2$, ξ_2 arbitrary, so we can take $\xi_2 = 1$ and get $(-2-i, 1, 1+i)$ as an eigenvector for $\lambda = 3 + i$. Therefore, by conjugation, $(-2+i, 1, 1-i)$ is an eigenvector for $\lambda = 3 - i$. Finally, $(-1, 1, 1)$ is an eigenvector for $\lambda = -1$ (review Section 4.3 if you have difficulty finding this one). So we can take

$$\mathbf{T} = \begin{bmatrix} -2-i & -2+i & -1 \\ 1 & 1 & 1 \\ 1+i & 1-i & 1 \end{bmatrix}; \text{ we then have } \mathbf{T}^{-1}\mathbf{A}\mathbf{T} = \begin{bmatrix} 3+i & 0 & 0 \\ 0 & 3-i & 0 \\ 0 & 0 & -1 \end{bmatrix}, \text{ and}$$

$$\mathbf{v} = \mathbf{T} \begin{bmatrix} C_1 e^{(3+i)t} \\ C_2 e^{(3-i)t} \\ C_3 e^{-t} \end{bmatrix}, \quad C_1, C_2, C_3 \text{ arbitrary complex constants,}$$

will be the general complex-valued solution of our system. If we write out the components of $\mathbf{v}$, we get

$$\begin{cases} x_1 = (-2-i)C_1 e^{(3+i)t} + (-2+i)C_2 e^{(3-i)t} - C_3 e^{-t} \\ x_2 = C_1 e^{(3+i)t} + C_2 e^{(3-i)t} + C_3 e^{-t} \\ x_3 = (1+i)C_1 e^{(3+i)t} + (1-i)C_2 e^{(3-i)t} + C_3 e^{-t}. \end{cases}$$

Now we have to decide which of these solutions are real-valued, that is, for which choices of the complex constants C_1, C_2, C_3 all three functions x_1, x_2, x_3 become real-valued. Since the expression for x_2 above is the least complicated, let's first write x_2 in real and imaginary parts, assuming that the constants have been written that way as $C_1 = A_1 + iB_1$, $C_2 = A_2 + iB_2$, and $C_3 = A_3 + iB_3$. We get

$$\begin{aligned} x_2 &= (A_1 + iB_1)e^{(3+i)t} + (A_2 + iB_2)e^{(3-i)t} + (A_3 + iB_3)e^{-t} \\ &= (A_1 + iB_1)e^{3t}(\cos t + i\sin t) + \cdots \\ &= [(A_1 + A_2)e^{3t}\cos t + (B_2 - B_1)e^{3t}\sin t + A_3 e^{-t}] \\ &\quad + i[(B_1 + B_2)e^{3t}\cos t + (A_1 - A_2)e^{3t}\sin t + B_3 e^{-t}]. \end{aligned}$$

So x_2 will be real provided

$$(B_1 + B_2)e^{3t} \cos t + (A_1 - A_2)e^{3t} \sin t + B_3 e^{-t} = 0.$$

Since the functions $e^{3t} \cos t$, $e^{3t} \sin t$, and e^{-t} are linearly independent (why?), this will happen only when $B_1 + B_2 = 0$, $A_1 - A_2 = 0$, and $B_3 = 0$; in other words, only when $B_2 = -B_1$, $A_2 = A_1$, and $B_3 = 0$. For the three original complex constants, this means: $C_2 = \overline{C}_1$ and C_3 is real. On the other hand, suppose that, indeed, $C_2 = \overline{C}_1$ and C_3 is real. In this case it's easy to see that not only x_2, but also x_1 and x_3 are real-valued functions: In the expressions for x_1, x_2, x_3 above, the first two terms (involving C_1 and C_2) will be complex conjugates of each other, so their sum will be real, while the third term will be real because C_3 is. In fact, if $C_2 = \overline{C}_1$ and C_3 is real, we'll have

$$\begin{aligned}x_1 &= (-2 - i)C_1 e^{(3+i)t} + (-2 + i)\overline{C}_1 e^{(3-i)t} - C_3 e^{-t} \\ &= 2\,\mathrm{Re}[(-2 - i)C_1 e^{(3+i)t}] - C_3 e^{-t},\end{aligned}$$

and similarly

$$x_2 = 2\,\mathrm{Re}(C_1 e^{(3+i)t}) + C_3 e^{-t}, \qquad x_3 = 2\,\mathrm{Re}[(1 + i)C_1 e^{(3+i)t}] + C_3 e^{-t}.$$

As in Section 3.5, we can simplify these expressions slightly by replacing the arbitrary constant C_1 by $\dfrac{1}{2} C_1$. If we then put $C_1 = A_1 + iB_1$ and write out everything, we'll get

$$\begin{cases} x_1 = [(-2A_1 + B_1)\cos t + (A_1 + 2B_1)\sin t]e^{3t} - C_3 e^{-t} \\ x_2 = [A_1 \cos t - B_1 \sin t]e^{3t} + C_3 e^{-t} \\ x_3 = [(A_1 - B_1)\cos t - (A_1 + B_1)\sin t]e^{3t} + C_3 e^{-t}, \end{cases}$$

with A_1, B_1, C_3 arbitrary real constants, as our general solution.

N O T E : This solution in vector form can be written in terms of three basic solutions as

$$\begin{aligned}(x_1, x_2, x_3) = &\,A_1[(-2\cos t + \sin t)e^{3t}, (\cos t)e^{3t}, (\cos t - \sin t)e^{3t}] \\ &+ B_1[(\cos t + 2\sin t)e^{3t}, -(\sin t)e^{3t}, (-\cos t - \sin t)e^{3t}] \\ &+ C_3 e^{-t}(-1, 1, 1).\end{aligned}$$

The third basic solution has the familiar form $e^{\lambda t}\boldsymbol{\xi}$, where $\boldsymbol{\xi} = (-1, 1, 1)$ is an eigenvector for $\lambda = -1$ (compare Section 4.3, Exercise 23). The first two basic solutions look less inviting, but they are actually $\mathrm{Re}(e^{\lambda t}\boldsymbol{\xi})$ and $-\mathrm{Im}(e^{\lambda t}\boldsymbol{\xi})$, where $\boldsymbol{\xi} = (-2 - i, 1, 1 + i)$ is our eigenvector for $\lambda = 3 + i$. This works in general (see Exercise 25 of this section). ∎

Just as in the example above, the general real-valued solution to a system $\dfrac{d\mathbf{v}}{dt} = \mathbf{Av}$ for which all eigenvalues are distinct can always be obtained as follows. First find the eigenvalues and arrange the nonreal ones in conjugate pairs $\lambda, \overline{\lambda}$. Find an eigenvector for each eigenvalue; in each conjugate pair, take the conjugate of the eigenvector for λ as the eigenvector for $\bar{\lambda}$. As usual, form the transformation matrix **T** whose columns are the eigenvectors. Then the real-valued solutions will be exactly those among the complex-valued solutions

$$\mathbf{v} = \mathbf{T} \begin{bmatrix} C_1 e^{\lambda_1 t} \\ C_2 e^{\lambda_2 t} \\ \vdots \\ C_n e^{\lambda_n t} \end{bmatrix}$$

for which the constants C corresponding to real eigenvalues are real and each pair of constants corresponding to a conjugate pair of eigenvalues is conjugate to each other. As in the example, this procedure will lead only to real-valued solutions, because each non-real term will be added to its conjugate. Although it is a bit messy to show that *all* real-valued solutions are obtained this way, the essential idea is the same as in the example: Since all eigenvalues are distinct, the various real exponentials $e^{\lambda t}$ and the real and imaginary parts of the various complex exponentials $e^{\lambda t}$ are linearly independent; therefore, there is no "unexpected" way to combine them that will yield a real-valued solution.[12]

Here is one more example of our method.

EXAMPLE 4.4.2 Solve the initial value problem

$$\frac{d\mathbf{v}}{dt} = \mathbf{A}\mathbf{v}, \quad \mathbf{v}(0) = (20, 6, -6, 26), \quad \text{where} \quad \mathbf{A} = \begin{bmatrix} -6 & 9 & 4 & 4 \\ -2 & 3 & 1 & 1 \\ 2 & -2 & -4 & -2 \\ -8 & 10 & 13 & 7 \end{bmatrix}.$$

N O T E : The given "vector initial condition" is equivalent to the four initial conditions $x_1(0) = 20$, $x_2(0) = 6$, and so on.

First we find the general solution of the system, starting with the characteristic equation $\det(\mathbf{A} - \lambda \mathbf{I}) = 0$, which works out to $\lambda^4 + 5\lambda^2 + 4 = 0$. Since there are no terms with λ^3 or λ (of course, the example is rigged), this equation can be thought of as a quadratic equation in λ^2. It is then easily factored:

$$(\lambda^2 + 1)(\lambda^2 + 4) = 0$$
$$(\lambda - i)(\lambda + i)(\lambda - 2i)(\lambda + 2i) = 0,$$

and we have four distinct complex eigenvalues in conjugate pairs $\pm i$ and $\pm 2i$. For $\lambda = i$ we find $(3 + i, 1 + i, 0, 2)$ as an eigenvector [see Exercise 22(a)], so for $\lambda = -i$ we have the conjugate eigenvector $(3 - i, 1 - i, 0, 2)$. The computation for $\lambda = 2i$ is less pleasant [Exercise 22(b)], but eventually we find $(11 \mp 5i, 4, -3, 13 \mp 2i)$ as a conjugate pair of

---

[12] An alternative procedure starts with the basic solutions $e^{\lambda t}\boldsymbol{\xi}$ for real eigenvalues λ and $\operatorname{Re}(e^{\lambda t}\boldsymbol{\xi})$, $\operatorname{Im}(e^{\lambda t}\boldsymbol{\xi})$ for conjugate pairs λ, $\bar{\lambda}$, then takes all (real) linear combinations of these basic solutions. (See the Note on p. 324, and Exercise 25.)

eigenvectors for $\lambda = \pm 2i$.[13] We now know that the general complex-valued solution to the system is given by

$$\mathbf{v} = \underbrace{\begin{bmatrix} 3+i & 3-i & 11-5i & 11+5i \\ 1+i & 1-i & 4 & 4 \\ 0 & 0 & -3 & -3 \\ 2 & 2 & 13-2i & 13+2i \end{bmatrix}}_{\mathbf{T}} \underbrace{\begin{bmatrix} C_1 e^{it} \\ C_2 e^{-it} \\ C_3 e^{2it} \\ C_4 e^{-2it} \end{bmatrix}}_{\mathbf{V}}.$$

The real-valued solutions are those with $C_2 = \overline{C}_1$, $C_4 = \overline{C}_3$; we can either find the general real-valued solution first and then solve the initial value problem, or we can look right away for the complex constants C_1, C_2, C_3, C_4 for which $\mathbf{v}(0) = (20, 6, -6, 26)$, that is,

$$\begin{bmatrix} 3+i & 3-i & 11-5i & 11+5i \\ 1+i & 1-i & 4 & 4 \\ 0 & 0 & -3 & -3 \\ 2 & 2 & 13-2i & 13+2i \end{bmatrix} \begin{bmatrix} C_1 \\ C_2 \\ C_3 \\ C_4 \end{bmatrix} = \begin{bmatrix} 20 \\ 6 \\ -6 \\ 26 \end{bmatrix}.$$

See Exercise 22(c) and (d) for the rest of the computation. ■

Now we will consider the case in which $\mathbf{A}$ has repeated (real or complex) eigenvalues. For $n = 2$ this case was studied in Section 3.5. As we saw in that section, it is usually impossible in this case to find an invertible matrix $\mathbf{T}$ such that $\mathbf{T}^{-1}\mathbf{A}\mathbf{T}$ is a diagonal matrix, that is, to diagonalize $\mathbf{A}$. The reason is that we usually can't find "enough" eigenvectors to form the columns of such a matrix $\mathbf{T}$. Typically, although not always, there is only one eigenvector (up to nonzero multiples) per eigenvalue, so if there are fewer then n distinct eigenvalues, there will tend to be "too few" eigenvectors also.[14]

EXAMPLE 4.4.3

$$\mathbf{A} = \begin{bmatrix} -9 & -1 & -13 \\ 3 & 2 & 2 \\ 7 & 1 & 11 \end{bmatrix}.$$

The characteristic equation yields $\lambda^3 - 4\lambda^2 - 3\lambda + 18 = 0$ (check this!). By inspection, $\lambda = -2$ is an eigenvalue, and the equation factors as $(\lambda + 2)(\lambda^2 - 6\lambda + 9) = 0$ or $(\lambda + 2)(\lambda - 3)^2 = 0$. Thus there is a repeated eigenvalue $\lambda = 3$ along with $\lambda = -2$. However, the only eigenvector (up to nonzero multiples) for $\lambda = 3$ is $(1, 1, -1)$, while for $\lambda = -2$ we find only $(-2, 1, 1)$ (see Exercise 1). □

---

[13] By convention, if both $\mp$ and $\pm$ signs are used in a statement, the statement should be correct if we consistently choose either the upper or the lower of the indicated signs.

[14] It can be seen easily that if one column of a matrix is a multiple of another, then that matrix is not invertible (see Appendix B, Section B.4, Exercise 20). So using several multiples of the same vector as columns of $\mathbf{T}$ won't work. More generally, it is shown in linear algebra that a square matrix is invertible if and only if the vectors that form its columns are linearly independent (see Section 4.1 for a definition). For an example in which $\mathbf{A}$ is diagonalizable despite having fewer than n eigenvalues, see Section 4.5.

SEC. 4.4 HOMOGENEOUS, AUTONOMOUS LINEAR SYSTEMS (CONTINUED)

In cases like this with "too few" eigenvectors, we can try following the procedure from Section 3.5. That is, we can take the eigenvectors we do have and put them in the first columns of **T**, then complete **T** to a square matrix using an arbitrary final column or columns. Provided that **T** is invertible, we can hope that $\mathbf{T}^{-1}\mathbf{AT}$ will have a simpler form than **A**, making it possible to solve the system $\dfrac{d\mathbf{V}}{dt} = \mathbf{T}^{-1}\mathbf{AT\,V}$ for the new unknown vector $\mathbf{V} = \mathbf{T}^{-1}\mathbf{v}$.

EXAMPLE 4.4.3 (*continued*)

$$\begin{cases} \dfrac{dx_1}{dt} = -9x_1 - x_2 - 13x_3 \\ \dfrac{dx_2}{dt} = 3x_1 + 2x_2 + 2x_3 \\ \dfrac{dx_3}{dt} = 7x_1 + x_2 + 11x_3. \end{cases}$$

We've seen above that the only eigenvectors of $\mathbf{A} = \begin{bmatrix} -9 & -1 & -13 \\ 3 & 2 & 2 \\ 7 & 1 & 11 \end{bmatrix}$ are $(1, 1, -1)$ and $(-2, 1, 1)$. So we can try taking a matrix **T** of the form $\mathbf{T} = \begin{bmatrix} 1 & -2 & \cdots \\ 1 & 1 & \cdots \\ -1 & 1 & \cdots \end{bmatrix}$. For now, the last column of **T** can be chosen arbitrarily, as long as the resulting **T** is invertible; the easiest choice may be $(0, 0, 1)$. We then have

$$\mathbf{T} = \begin{bmatrix} 1 & -2 & 0 \\ 1 & 1 & 0 \\ -1 & 1 & 1 \end{bmatrix}, \quad \mathbf{T}^{-1} = \begin{bmatrix} \dfrac{1}{3} & \dfrac{2}{3} & 0 \\ -\dfrac{1}{3} & \dfrac{1}{3} & 0 \\ \dfrac{2}{3} & \dfrac{1}{3} & 1 \end{bmatrix}^{15}$$

and therefore

$$\mathbf{T}^{-1}\mathbf{AT} = \begin{bmatrix} \dfrac{1}{3} & \dfrac{2}{3} & 0 \\ -\dfrac{1}{3} & \dfrac{1}{3} & 0 \\ \dfrac{2}{3} & \dfrac{1}{3} & 1 \end{bmatrix} \begin{bmatrix} -9 & -1 & -13 \\ 3 & 2 & 2 \\ 7 & 1 & 11 \end{bmatrix} \begin{bmatrix} 1 & -2 & 0 \\ 1 & 1 & 0 \\ -1 & 1 & 1 \end{bmatrix} = \begin{bmatrix} 3 & 0 & -3 \\ 0 & -2 & 5 \\ 0 & 0 & 3 \end{bmatrix}.$$

[15] See Appendix B, Section B.3, for the method of computing the inverse matrix.

So if we make the coordinate change $\mathbf{V} = \mathbf{T}^{-1}\mathbf{v}$, $\mathbf{v} = \mathbf{TV}$, the system becomes

$$\begin{cases} \dfrac{dX_1}{dt} = 3X_1 - 3X_3 \\ \dfrac{dX_2}{dt} = -2X_2 + 5X_3 \\ \dfrac{dX_3}{dt} = 3X_3. \end{cases}$$

The last equation yields $X_3 = Ce^{3t}$, which leaves us with

$$\begin{cases} \dfrac{dX_1}{dt} = 3X_1 - 3Ce^{3t} \\ \dfrac{dX_2}{dt} = -2X_2 + 5Ce^{3t}. \end{cases}$$

Each of these equations can be solved by the method of Section 1.6 (see Exercise 2) and we get $X_1 = (-3Ct + D)e^{3t}$, $X_2 = Ce^{3t} + Ee^{-2t}$, $X_3 = Ce^{3t}$, C, D, E arbitrary. Finally,

$$\underbrace{\begin{bmatrix} x_1 \\ x_2 \\ x_3 \end{bmatrix}}_{\mathbf{v}} = \underbrace{\begin{bmatrix} 1 & -2 & 0 \\ 1 & 1 & 0 \\ -1 & 1 & 1 \end{bmatrix}}_{\mathbf{T}} \underbrace{\begin{bmatrix} (-3Ct + D)e^{3t} \\ Ce^{3t} + Ee^{-2t} \\ Ce^{3t} \end{bmatrix}}_{\mathbf{V}},$$

and we get

$$\begin{cases} x_1 = [-3Ct + (-2C + D)]e^{3t} - 2Ee^{-2t} \\ x_2 = [-3Ct + (C + D)]e^{3t} + Ee^{-2t} \\ x_3 = [3Ct + (2C - D)]e^{3t} + Ee^{-2t} \end{cases}, \quad C, D, E \text{ arbitrary,}$$

as the general solution to our system.

N O T E : The general solution (in vector form) can be rewritten

$$(x_1, x_2, x_3) = Ce^{3t}(-3t - 2, -3t + 1, 3t + 2) + De^{3t}(1, 1, -1) + Ee^{-2t}(-2, 1, 1).$$

Of the three basic solutions $e^{3t}(-3t - 2, -3t + 1, 3t + 2)$, $e^{3t}(1, 1, -1)$, $e^{-2t}(-2, 1, 1)$, the last two are in the form $e^{\lambda t}\boldsymbol{\xi}$ with $\boldsymbol{\xi}$ an eigenvector for λ. (Compare Section 4.3, Exercise 23.) ∎

Unfortunately, we'll now see that there are cases in which this method needs refinement, even for a 3×3 matrix $\mathbf{A}$.

EXAMPLE 4.4.4 The matrix

$$\mathbf{A} = \begin{bmatrix} 2 & 3 & -1 \\ -2 & 3 & 0 \\ 1 & 11 & -2 \end{bmatrix}$$

has characteristic equation $\lambda^3 - 3\lambda^2 + 3\lambda - 1 = 0$ or $(\lambda - 1)^3 = 0$. The one (repeated) eigenvalue $\lambda = 1$ yields only the eigenvector $(1, 1, 4)$ (up to multiples). If we try taking

SEC. 4.4 HOMOGENEOUS, AUTONOMOUS LINEAR SYSTEMS (CONTINUED)

the final two columns of **T** to be (0, 1, 0) and (0, 0, 1), which will make the computation of $\mathbf{T}^{-1}$ particularly fast, we get

$$\mathbf{T} = \begin{bmatrix} 1 & 0 & 0 \\ 1 & 1 & 0 \\ 4 & 0 & 1 \end{bmatrix}, \quad \mathbf{T}^{-1} = \begin{bmatrix} 1 & 0 & 0 \\ -1 & 1 & 0 \\ -4 & 0 & 1 \end{bmatrix}, \quad \mathbf{T}^{-1}\mathbf{A}\mathbf{T} = \begin{bmatrix} 1 & 3 & -1 \\ 0 & 0 & 1 \\ 0 & -1 & 2 \end{bmatrix}.$$

Therefore, if we make the change of coordinates given by $\mathbf{V} = \mathbf{T}^{-1}\mathbf{v}$, the new system will be

$$\begin{cases} \dfrac{dX_1}{dt} = X_1 + 3X_2 - X_3 \\ \dfrac{dX_2}{dt} = X_3 \\ \dfrac{dX_3}{dt} = -X_2 + 2X_3. \end{cases}$$

In this system, none of the equations can be solved separately yet, since each of them is still coupled to at least one of the others. Apparently, then, we have to choose the matrix **T** more carefully. ◾

We were able to solve the system in Example 4.4.3 because the "new" matrix
$$\mathbf{T}^{-1}\mathbf{A}\mathbf{T} = \begin{bmatrix} 3 & 0 & -3 \\ 0 & -2 & 5 \\ 0 & 0 & 3 \end{bmatrix}$$
in that example was upper triangular; that is, all its entries below the main diagonal were zero. This always happens when all columns of **T** except *the last one* are eigenvectors of **A** (as in Example 4.4.3; see Exercise 23), but it can also happen in more general situations—as mentioned below. Whenever the new matrix is upper triangular, the new system can (in principle) be solved by starting with the last equation and working "upward."

EXAMPLE If

$$\mathbf{T}^{-1}\mathbf{A}\mathbf{T} = \begin{bmatrix} 3 & 1 & 3 & -1 \\ 0 & 3 & 4 & 2 \\ 0 & 0 & 3 & 1 \\ 0 & 0 & 0 & 3 \end{bmatrix},$$

then the new system

$$\begin{cases} \dfrac{dX_1}{dt} = 3X_1 + X_2 + 3X_3 - X_4 \\ \dfrac{dX_2}{dt} = 3X_2 + 4X_3 + 2X_4 \\ \dfrac{dX_3}{dt} = 3X_3 + X_4 \\ \dfrac{dX_4}{dt} = 3X_4 \end{cases}$$

can be solved by first finding $X_4 = Ce^{3t}$, then finding X_3 from $\dfrac{dX_3}{dt} = 3X_3 + Ce^{3t}$, then finding X_2, and so on. ∎

It can be shown that one can *always* find an invertible matrix $\mathbf{T}$ such that $\mathbf{T}^{-1}\mathbf{A}\mathbf{T}$ is upper triangular. (If $\mathbf{A}$ has nonreal eigenvalues, $\mathbf{T}$ must be allowed to have nonreal entries.) In fact, there is a systematic procedure to find a $\mathbf{T}$ for which $\mathbf{T}^{-1}\mathbf{A}\mathbf{T}$ has a particularly elegant upper triangular form, known as the **Jordan form**.[16] If $\mathbf{A}$ is diagonalizable with eigenvalues $\lambda_1, \ldots, \lambda_n$, the Jordan form is the usual diagonal form

$$\begin{bmatrix} \lambda_1 & & 0 \\ & \ddots & \\ 0 & & \lambda_n \end{bmatrix}$$

and we have nothing new. See "Further Reading" following the Exercises.

In the next section, we'll present an alternative (although related) approach to the solution of systems of the form $\dfrac{d\mathbf{v}}{dt} = \mathbf{A}\mathbf{v}$. This approach will be better suited to the study of *inhomogeneous* linear systems, to which Section 4.6 gives an introduction.

SUMMARY OF KEY RESULTS AND TECHNIQUES

Solving the system $\dfrac{d\mathbf{v}}{dt} = \mathbf{A}\mathbf{v}$, continued: If $\mathbf{A}$ has n distinct eigenvalues that are not all real, we can get the general complex-valued solution as in Section 4.3, using complex eigenvectors (p. 322). If $\boldsymbol{\xi}$ is an eigenvector of $\mathbf{A}$ for $\lambda = \mu + i\nu$, then $\overline{\boldsymbol{\xi}}$ is an eigenvector for $\overline{\lambda}$ (p. 322). If we take the eigenvectors in (complex) conjugate pairs as columns of the transformation matrix, then the general *real-valued* solution consists of those vector functions

$$\mathbf{v} = \mathbf{T} \begin{bmatrix} C_1 e^{\lambda_1 t} \\ C_2 e^{\lambda_2 t} \\ \vdots \\ C_n e^{\lambda_n t} \end{bmatrix}$$

for which the constants corresponding to real eigenvalues are real and constants corresponding to conjugate eigenvalues are conjugate (p. 325). The vector functions $\text{Re}(e^{\lambda t}\boldsymbol{\xi})$, $\text{Im}(e^{\lambda t}\boldsymbol{\xi})$ for the various nonreal eigenvalues λ and corresponding eigenvectors $\boldsymbol{\xi}$, along with the $e^{\lambda t}\boldsymbol{\xi}$ for real λ, provide basic solutions (p. 324; p. 325, footnote 12).

---

[16] "Jordan" refers not to a country or river in the Middle East, but to the French mathematician Camille Jordan (1838–1922) of "Jordan curve theorem" fame. Accordingly, the name is properly pronounced to scan more like "Taiwan" than like "boredom."

SEC. 4.4 HOMOGENEOUS, AUTONOMOUS LINEAR SYSTEMS (CONTINUED)

If $\mathbf{A}$ has repeated eigenvalues, it is usually impossible to find "enough" eigenvectors for a transformation matrix $\mathbf{T}$ such that $\mathbf{T}^{-1}\mathbf{A}\mathbf{T}$ is diagonal(p. 326). If, by using what eigenvectors we have, we manage to get $\mathbf{T}^{-1}\mathbf{A}\mathbf{T}$ to be upper triangular, then we can still solve the new system $\dfrac{d\mathbf{V}}{dt} = \mathbf{T}^{-1}\mathbf{A}\mathbf{T}\,\mathbf{V}$ by "working up from the bottom" (p. 329).

EXERCISES

1. Let $\mathbf{A} = \begin{bmatrix} -9 & -1 & -13 \\ 3 & 2 & 2 \\ 7 & 1 & 11 \end{bmatrix}$, as in Example 4.4.3 (p. 326).

   (a) Find all eigenvectors for $\lambda = 3$. (Be systematic; don't use the fact that the answer is given in the text.)

   (b) Find all eigenvectors for $\lambda = -2$. (Same instructions.)

2. (Compare p. 328.) Solve the system
   $$\frac{dX_1}{dt} = 3X_1 - 3Ce^{3t}, \qquad \frac{dX_2}{dt} = -2X_2 + 5Ce^{3t},$$
   where C is some complex constant.

For each of the following matrices $\mathbf{A}$:

(a) Find all (real or complex) eigenvalues.

(b) Find a matrix $\mathbf{T}$ (with real or complex entries) such that $\mathbf{T}^{-1}\mathbf{A}\mathbf{T}$ is a diagonal matrix (and give this diagonal matrix), or explain why such a $\mathbf{T}$ does not exist.

3. $\begin{bmatrix} 5 & -1 & 2 \\ 0 & 0 & 1 \\ 0 & -1 & 0 \end{bmatrix}$
4. $\begin{bmatrix} 0 & 0 & -4 \\ 4 & -1 & 4 \\ 1 & 0 & 0 \end{bmatrix}$
5. $\begin{bmatrix} -11 & 3 & 7 \\ -1 & 0 & 1 \\ -13 & 2 & 9 \end{bmatrix}$

6. $\begin{bmatrix} 2 & 1 & -1 \\ 6 & 1 & -2 \\ -1 & 3 & -2 \end{bmatrix}$
7. $\begin{bmatrix} 4 & 1 & 0 \\ 0 & -2 & 0 \\ 0 & 3 & 4 \end{bmatrix}$

8. $\begin{bmatrix} 1 & 2 & 0 & -3 \\ 0 & 0 & 0 & 1 \\ 0 & 4 & 1 & -1 \\ 0 & -1 & 0 & 0 \end{bmatrix}$
9. $\begin{bmatrix} 2 & 3 & 2 & 1 \\ -6 & -4 & 4 & -1 \\ 0 & 0 & 1 & 0 \\ 0 & 0 & 0 & 0 \end{bmatrix}$
10. $\begin{bmatrix} 0 & 1 & 2 & 0 \\ 0 & 5 & -6 & 0 \\ 0 & 3 & -1 & 0 \\ 0 & 4 & 1 & -1 \end{bmatrix}$

11. $\begin{bmatrix} 4 & 0 & 0 & 0 \\ 1 & -10 & -33 & -11 \\ -1 & 2 & 7 & 2 \\ 2 & 5 & 15 & 6 \end{bmatrix}$
12. $\begin{bmatrix} -6 & 2 & -1 & -1 \\ -15 & 5 & -3 & 3 \\ -5 & 2 & -2 & -1 \\ 0 & 0 & 0 & 2 \end{bmatrix}$

Solve the following differential equations and initial value problems.

13. $\begin{cases} \dfrac{dx_1}{dt} = 5x_1 - x_2 + 2x_3, & x_1(0) = 0 \\ \dfrac{dx_2}{dt} = x_3, & x_2(0) = 3 \\ \dfrac{dx_3}{dt} = -x_2, & x_3(0) = -2. \end{cases}$ (Compare Exercise 3.)

14. $\begin{cases} \dfrac{dx_1}{dt} = -4x_3, & x_1(0) = -10 \\ \dfrac{dx_2}{dt} = 4x_1 - x_2 + 4x_3, & x_2(0) = 0 \\ \dfrac{dx_3}{dt} = x_1, & x_3(0) = 1. \end{cases}$ (Compare Exercise 4.)

15. $\dfrac{d\mathbf{v}}{dt} = \mathbf{Av}$, where $\mathbf{v} = (x_1, x_2, x_3)$, $\mathbf{A} = \begin{bmatrix} -11 & 3 & 7 \\ -1 & 0 & 1 \\ -13 & 2 & 9 \end{bmatrix}$. (Compare Exercise 5.)

16. $\dfrac{d\mathbf{v}}{dt} = \mathbf{Av}$, where $\mathbf{v} = (x_1, x_2, x_3)$, $\mathbf{A} = \begin{bmatrix} 2 & 1 & -1 \\ 6 & 1 & -2 \\ -1 & 3 & -2 \end{bmatrix}$. (Compare Exercise 6.)

17. $\dfrac{d\mathbf{v}}{dt} = \mathbf{Av}$, where $\mathbf{A} = \begin{bmatrix} 1 & 2 & 0 & 0 \\ -5 & -5 & 0 & 0 \\ 1 & -2 & 1 & 1 \\ -12 & -10 & -2 & -1 \end{bmatrix}$.

[Hint: $\lambda = i$ is one eigenvalue.]

18. $\dfrac{d\mathbf{v}}{dt} = \mathbf{Av}$, where $\mathbf{A} = \begin{bmatrix} 3 & -5 & 1 & -12 \\ 2 & -3 & -2 & -10 \\ 0 & 0 & 3 & -2 \\ 0 & 0 & 1 & 1 \end{bmatrix}$.

[Hint: $\lambda = i$ is one eigenvalue.]

19. $\begin{cases} \dfrac{dx_1}{dt} = -x_1 + x_2 + x_4, & x_1(0) = -1 \\ \dfrac{dx_2}{dt} = x_1 + 4x_3 - x_4, & x_2(0) = 8 \\ \dfrac{dx_3}{dt} = -2x_1 + x_2 - x_3 + x_4, & x_3(0) = -3 \\ \dfrac{dx_4}{dt} = -7x_1 + 4x_2 - 4x_3 + 5x_4, & x_4(0) = -18. \end{cases}$

SEC. 4.4 HOMOGENEOUS, AUTONOMOUS LINEAR SYSTEMS (CONTINUED)

20. $\dfrac{d\mathbf{v}}{dt} = \mathbf{A}\mathbf{v}$, where $\mathbf{A} = \begin{bmatrix} 1 & -1 & 1 & 0 \\ 0 & 1 & 0 & 0 \\ 0 & 0 & 1 & 0 \\ 0 & 0 & 0 & 2 \end{bmatrix}$.

[*Hint*: Take a good look at $\mathbf{A}$ before you start.]

21. $\dfrac{d\mathbf{v}}{dt} = \mathbf{A}\mathbf{v}$, where $\mathbf{A} = \begin{bmatrix} 2 & 2 & -1 & 0 \\ 0 & 2 & 1 & 0 \\ 0 & 0 & 2 & 0 \\ 0 & 0 & 0 & -3 \end{bmatrix}$. (Same hint as for Exercise 20.)

22. Let $\mathbf{A} = \begin{bmatrix} -6 & 9 & 4 & 4 \\ -2 & 3 & 1 & 1 \\ 2 & -2 & -4 & -2 \\ -8 & 10 & 13 & 7 \end{bmatrix}$, as in Example 4.4.2.

(a) Find an eigenvector for $\lambda = i$. (Be systematic; don't use the fact that the answer is given in the text.)

*(b) Find an eigenvector for $\lambda = 2i$. (Same instructions.)

(c) Find the general real-valued solution to $\dfrac{d\mathbf{v}}{dt} = \mathbf{A}\mathbf{v}$ and use it to find the solution of the initial value problem $\dfrac{d\mathbf{v}}{dt} = \mathbf{A}\mathbf{v}$, $\mathbf{v}(0) = (20, 6, -6, 26)$.

(d) Working directly from the general complex-valued solution of $\dfrac{d\mathbf{v}}{dt} = \mathbf{A}\mathbf{v}$, again find the solution of $\dfrac{d\mathbf{v}}{dt} = \mathbf{A}\mathbf{v}$, $\mathbf{v}(0) = (20, 6, -6, 26)$.

23. (a) Check that

$$\begin{bmatrix} 1 & 2 & 3 \\ 4 & 5 & 6 \\ 7 & 8 & 9 \end{bmatrix} \begin{bmatrix} 1 \\ 0 \\ 0 \end{bmatrix} = \begin{bmatrix} 1 \\ 4 \\ 7 \end{bmatrix}, \begin{bmatrix} 1 & 2 & 3 \\ 4 & 5 & 6 \\ 7 & 8 & 9 \end{bmatrix} \begin{bmatrix} 0 \\ 1 \\ 0 \end{bmatrix} = \begin{bmatrix} 2 \\ 5 \\ 8 \end{bmatrix}, \begin{bmatrix} 1 & 2 & 3 \\ 4 & 5 & 6 \\ 7 & 8 & 9 \end{bmatrix} \begin{bmatrix} 0 \\ 0 \\ 1 \end{bmatrix} = \begin{bmatrix} 3 \\ 6 \\ 9 \end{bmatrix}.$$

(b) More generally, let $\mathbf{A}$ be any $n \times n$ matrix. Let $\mathbf{e}_1, \ldots, \mathbf{e}_n$ be the n vectors in n-space which have one component 1 and all others 0: $\mathbf{e}_1 = (1, 0, \ldots, 0)$, $\mathbf{e}_2 = (0, 1, \ldots, 0), \ldots, \mathbf{e}_n = (0, 0, \ldots, 1)$. [These vectors are called the **standard basis vectors** for n-dimensional space; any vector $\mathbf{v} = (x_1, \ldots, x_n)$ can be written uniquely as a linear combination $\mathbf{v} = x_1\mathbf{e}_1 + x_2\mathbf{e}_2 + \cdots + x_n\mathbf{e}_n$.] Show that $\mathbf{A}\mathbf{e}_1$ equals the first column of $\mathbf{A}$, $\mathbf{A}\mathbf{e}_2$ equals the second column of $\mathbf{A}$, and so on.

*(c) Now suppose that $\mathbf{T}$ is a transformation matrix whose first column is an eigenvector of $\mathbf{A}$ for the eigenvalue λ. Show that the first column of $\mathbf{T}^{-1}\mathbf{AT}$ is $\begin{bmatrix} \lambda \\ 0 \\ \vdots \\ 0 \end{bmatrix}$. (See hint on p. 334.)

[*Hint*: By part (b), the first column of $\mathbf{T}^{-1}\mathbf{AT}$ is $\mathbf{T}^{-1}\mathbf{ATe}_1$. What do you know about $\mathbf{Te}_1$? If you're not that familiar with matrix algebra, Exercise 10(b) from Appendix B, Section B.2 will be helpful.]

(d) Show that if all columns of $\mathbf{T}$ except the last one are eigenvectors of $\mathbf{A}$, then $\mathbf{T}^{-1}\mathbf{AT}$ will be upper triangular.

*24. Show that if $\mathbf{T}^{-1}\mathbf{AT}$ is upper triangular, at least *one* column of $\mathbf{T}$ is an eigenvector of $\mathbf{A}$.
[*Hint*: Which one would it be, do you think?]

25. (a) Show that if $\boldsymbol{\xi}$ is an eigenvector of $\mathbf{A}$ for the complex eigenvalue $\lambda = \mu + i\nu$ (with $\nu \neq 0$), then $\text{Re}(e^{\lambda t}\boldsymbol{\xi})$ and $\text{Im}(e^{\lambda t}\boldsymbol{\xi})$ are solutions to $\dfrac{d\mathbf{v}}{dt} = \mathbf{Av}$.

(b) Suppose that we replace λ in part (a) by $\overline{\lambda}$. What other changes should we then make? How do the solutions we get this way for $\overline{\lambda}$ compare to the ones for λ?

Now suppose that $\mathbf{A}$ has n distinct eigenvalues, of which $\lambda_1, \ldots, \lambda_k$ are real and $\lambda_{k+1} = \sigma_1, \overline{\sigma}_1, \sigma_2, \overline{\sigma}_2, \ldots, \sigma_m, \overline{\sigma}_m$ are in conjugate pairs.

(c) Explain why $k + 2m = n$.

*(d) Suppose that $\boldsymbol{\xi}^{(1)}, \ldots, \boldsymbol{\xi}^{(k)}$ are eigenvectors for $\lambda_1, \ldots, \lambda_k$, respectively, and that $\boldsymbol{\eta}^{(1)}, \ldots, \boldsymbol{\eta}^{(m)}$ are eigenvectors for $\sigma_1, \ldots, \sigma_m$, respectively. Show that any (real) linear combination of $e^{\lambda_1 t} \cdot \boldsymbol{\xi}^{(1)}, \ldots, e^{\lambda_k t} \cdot \boldsymbol{\xi}^{(k)}, \text{Re}[e^{\sigma_1 t} \cdot \boldsymbol{\eta}^{(1)}]$, $\text{Im}[e^{\sigma_1 t} \cdot \boldsymbol{\eta}^{(1)}], \ldots, \text{Re}[e^{\sigma_m t} \cdot \boldsymbol{\eta}^{(m)}], \text{Im}[e^{\sigma_m t} \cdot \boldsymbol{\eta}^{(m)}]$ is a solution of $\dfrac{d\mathbf{v}}{dt} = \mathbf{Av}$.

[*Hint*: Combine the results of Exercises 23 and 24 from Section 4.3 with the result of part (a).]

*(e) Show that the procedure from part (d) leads to the general solution of $\dfrac{d\mathbf{v}}{dt} = \mathbf{Av}$, as described on pp. 324–325.

FURTHER READING

For information on the Jordan form, see Appendix B in Strang, *Linear Algebra and Its Applications*, 2nd ed. (Academic Press, 1980).

4.5 HOMOGENEOUS, AUTONOMOUS LINEAR SYSTEMS (USING MATRIX EXPONENTIALS)

In the last two sections we've studied systems that can be written as vector differential equations $\frac{d\mathbf{v}}{dt} = \mathbf{A}\mathbf{v}$, where $\mathbf{A}$ is an $n \times n$ matrix and $\mathbf{v} = (x_1, \ldots, x_n)$. For $n = 2$ we already studied such systems in Chapter 3. However, the simplest possible case is not $n = 2$, but $n = 1$. In this case the "system" consists of a single equation $\frac{dx}{dt} = ax$, where the constant a is the only entry in the 1×1 matrix $\mathbf{A}$. This equation was already studied in Section 1.3; we saw there that the solutions are given by $x(t) = x(0)e^{at}$. Might there possibly be a version of this formula for general n, found by replacing x by $\mathbf{v}$ and a by $\mathbf{A}$? If so, wouldn't this give us a wonderfully efficient way of solving initial value problems of type $\frac{d\mathbf{v}}{dt} = \mathbf{A}\mathbf{v}$, $\mathbf{v}(0) = \mathbf{v}_0$, where $\mathbf{v}_0$ is a given vector? (We encountered such a problem as Example 4.4.2 in the previous section, and it involved a lot of computation!) In this section we'll see that the answer to the first question is basically "yes," but that the answer to the second question is less clear. However, the formula we'll find for the solutions of $\frac{d\mathbf{v}}{dt} = \mathbf{A}\mathbf{v}$ is important both for theoretical reasons and for use in solving *inhomogeneous* linear systems, which we'll discuss in the next section.

If we literally replace x by $\mathbf{v}$ and a by $\mathbf{A}$ in the formula $x(t) = x(0)e^{at}$, the result is $\mathbf{v}(t) = \mathbf{v}(0)e^{\mathbf{A}t}$, but this new formula appears to be nonsense. For instance, since $\mathbf{A}$ is a matrix, the exponent $\mathbf{A}t$ is undefined: We could multiply $\mathbf{A}$ on the right by another $n \times n$ matrix or by a column vector, say, but not by a number t. However, $t\mathbf{A}$ is defined as the matrix obtained from $\mathbf{A}$ by multiplying all entries by t.[17] So perhaps we should try for $\mathbf{v}(t) = \mathbf{v}(0)e^{t\mathbf{A}}$. What would this mean? Well, $t\mathbf{A}$ is an $n \times n$ matrix, so presumably (this is a bit hazy, for now) $e^{t\mathbf{A}}$ would also be an $n \times n$ matrix. It would be unusual to multiply $\mathbf{v}(0)$ by $e^{t\mathbf{A}}$ in that order [$\mathbf{v}(0)$ would have to be written as a row vector, and usually when we combine vectors with matrices we write them as column vectors]. On the other hand, $\mathbf{v}(t) = e^{t\mathbf{A}}\mathbf{v}(0)$ looks more reasonable: If $e^{t\mathbf{A}}$ is indeed an $n \times n$ matrix, it can be multiplied by the vector $\mathbf{v}(0)$ to yield a new vector that will depend on t. Notice that for $n = 1$ we now get $x(t) = e^{ta}x(0) = x(0)e^{at}$, so our original case hasn't changed. Our program, then, is to make sense out of the formula $\mathbf{v}(t) = e^{t\mathbf{A}}\mathbf{v}(0)$ by defining $e^{t\mathbf{A}}$ as an $n \times n$ matrix in such a way that the formula does indeed give us the solutions to $\frac{d\mathbf{v}}{dt} = \mathbf{A}\mathbf{v}$.

How should we define the exponential of the matrix $t\mathbf{A}$, or of any other $n \times n$ matrix $\mathbf{B}$? Clearly, we can't just raise the number e to a "matrix power." The standard definition, used in many calculus books, of the exponential function as inverse function of the loga-

---

[17] A few authors actually do define $\mathbf{A}t$ to mean the same thing as $t\mathbf{A}$.

rithm won't help, either, unless we first define logarithms for matrices (which seems, and is, at least as hard to do[18]). However, we do have the power series expansion

$$e^x = 1 + x + \frac{x^2}{2!} + \frac{x^3}{3!} + \cdots = \sum_{m=0}^{\infty} \frac{x^m}{m!}$$

for real numbers x, and we could try to replace x by an $n \times n$ matrix **B** in this expansion. Of course, the powers $\mathbf{B}^2$, $\mathbf{B}^3$, and so on, are found by repeated multiplication; for instance, $\mathbf{B}^4 = \mathbf{B} \cdot \mathbf{B} \cdot \mathbf{B} \cdot \mathbf{B} = (\mathbf{B}^2)^2$. Division by $m!$ can be interpreted as multiplication (of all entries) by the constant $\frac{1}{m!}$. This only leaves the first term, 1 ; this term must be replaced by an $n \times n$ matrix, and the "obvious" candidate, because it has similar properties to the number 1, is the $n \times n$ identity matrix **I** . Now we have a tentative definition of the exponential $e^\mathbf{B}$ for any square matrix **B** :

$$e^\mathbf{B} = \mathbf{I} + \mathbf{B} + \frac{1}{2!}\mathbf{B}^2 + \frac{1}{3!}\mathbf{B}^3 + \cdots$$
$$= \sum_{m=0}^{\infty} \frac{1}{m!}\mathbf{B}^m .$$

Note that in the summation, $\mathbf{B}^0$ is interpreted as **I** (and $0! = 1$).

EXAMPLE 4.5.1 Let

$$\mathbf{B} = \begin{bmatrix} 0 & 1 & 0 & 1 \\ 0 & 0 & -1 & 1 \\ 0 & 0 & 0 & 2 \\ 0 & 0 & 0 & 0 \end{bmatrix}.$$

Then

$$\mathbf{B}^2 = \mathbf{BB} = \begin{bmatrix} 0 & 0 & -1 & 1 \\ 0 & 0 & 0 & -2 \\ 0 & 0 & 0 & 0 \\ 0 & 0 & 0 & 0 \end{bmatrix}, \quad \mathbf{B}^3 = \mathbf{BB}^2 = \begin{bmatrix} 0 & 0 & 0 & -2 \\ 0 & 0 & 0 & 0 \\ 0 & 0 & 0 & 0 \\ 0 & 0 & 0 & 0 \end{bmatrix},$$

[18] Unlike in calculus, where $\log x = \int_1^x \frac{1}{t} dt$ $(x > 0)$ provides a direct definition of the logarithm.

while $\mathbf{B}^4 = \mathbf{B}\mathbf{B}^3 = \mathbf{0}$, the "zero matrix" with all entries 0. Therefore, all further powers of $\mathbf{B}$ are $\mathbf{0}$ also, and we get

$$e^{\mathbf{B}} = \mathbf{I} + \mathbf{B} + \frac{1}{2!}\mathbf{B}^2 + \frac{1}{3!}\mathbf{B}^3 + \mathbf{0} + \mathbf{0} + \cdots$$

$$= \begin{bmatrix} 1 & 0 & 0 & 0 \\ 0 & 1 & 0 & 0 \\ 0 & 0 & 1 & 0 \\ 0 & 0 & 0 & 1 \end{bmatrix} + \begin{bmatrix} 0 & 1 & 0 & 1 \\ 0 & 0 & -1 & 1 \\ 0 & 0 & 0 & 2 \\ 0 & 0 & 0 & 0 \end{bmatrix} + \begin{bmatrix} 0 & 0 & -\frac{1}{2} & \frac{1}{2} \\ 0 & 0 & 0 & -1 \\ 0 & 0 & 0 & 0 \\ 0 & 0 & 0 & 0 \end{bmatrix} + \begin{bmatrix} 0 & 0 & 0 & -\frac{1}{3} \\ 0 & 0 & 0 & 0 \\ 0 & 0 & 0 & 0 \\ 0 & 0 & 0 & 0 \end{bmatrix}$$

$$= \begin{bmatrix} 1 & 1 & -\frac{1}{2} & \frac{7}{6} \\ 0 & 1 & -1 & 0 \\ 0 & 0 & 1 & 2 \\ 0 & 0 & 0 & 1 \end{bmatrix}. \quad \blacksquare$$

In this example, it was very helpful that a power of $\mathbf{B}$ was the zero matrix, so that the infinite series for $e^{\mathbf{B}}$ became a finite sum. A matrix $\mathbf{B}$ for which this happens is called **nilpotent**.[19] If $\mathbf{B}$ is not nilpotent, our definition of $e^{\mathbf{B}}$ is a little more questionable, because it involves an infinite series of matrices. Even though we can add matrices entry by entry, we'll get an infinite series for each entry this way, and it's not immediately clear that these series will be convergent. However, using the comparison test for infinite series, it's not too hard to show that the series for each entry will indeed converge (see Exercise 41).

EXAMPLE 4.5.2 Let

$$\mathbf{B} = \begin{bmatrix} 2 & 0 & 0 \\ 0 & 3 & 0 \\ 0 & 0 & -1 \end{bmatrix}.$$

Then

$$\mathbf{B}^2 = \begin{bmatrix} 4 & 0 & 0 \\ 0 & 9 & 0 \\ 0 & 0 & 1 \end{bmatrix}, \quad \mathbf{B}^3 = \begin{bmatrix} 8 & 0 & 0 \\ 0 & 27 & 0 \\ 0 & 0 & -1 \end{bmatrix},$$

and in general

$$\mathbf{B}^m = \begin{bmatrix} 2^m & 0 & 0 \\ 0 & 3^m & 0 \\ 0 & 0 & (-1)^m \end{bmatrix}.$$

[19] From "nil" = zero + "potent" = powerful: "nilpotent" = "zero-powerful" = having a power zero.

Therefore, we have

$$e^{\mathbf{B}} = \sum_{m=0}^{\infty} \frac{1}{m!} \mathbf{B}^m = \mathbf{I} + \mathbf{B} + \frac{1}{2!} \mathbf{B}^2 + \cdots + \frac{1}{m!} \mathbf{B}^m + \cdots$$

$$= \begin{bmatrix} 1 & 0 & 0 \\ 0 & 1 & 0 \\ 0 & 0 & 1 \end{bmatrix} + \begin{bmatrix} 2 & 0 & 0 \\ 0 & 3 & 0 \\ 0 & 0 & -1 \end{bmatrix} + \begin{bmatrix} \frac{4}{2!} & 0 & 0 \\ 0 & \frac{9}{2!} & 0 \\ 0 & 0 & \frac{1}{2!} \end{bmatrix} + \cdots + \begin{bmatrix} \frac{2^m}{m!} & 0 & 0 \\ 0 & \frac{3^m}{m!} & 0 \\ 0 & 0 & \frac{(-1)^m}{m!} \end{bmatrix} + \cdots$$

$$= \begin{bmatrix} 1 + 2 + \frac{4}{2!} + \cdots + \frac{2^m}{m!} + \cdots & 0 & 0 \\ 0 & 1 + 3 + \frac{9}{2!} + \cdots + \frac{3^m}{m!} + \cdots & 0 \\ 0 & 0 & 1 + (-1) + \frac{1}{2!} + \cdots + \frac{(-1)^m}{m!} + \cdots \end{bmatrix}$$

$$= \begin{bmatrix} \sum_{m=0}^{\infty} \frac{2^m}{m!} & 0 & 0 \\ 0 & \sum_{m=0}^{\infty} \frac{3^m}{m!} & 0 \\ 0 & 0 & \sum_{m=0}^{\infty} \frac{(-1)^m}{m!} \end{bmatrix} = \begin{bmatrix} e^2 & 0 & 0 \\ 0 & e^3 & 0 \\ 0 & 0 & e^{-1} \end{bmatrix}.$$

In this computation we've written out the first few terms of the series, but only for illustrative purposes; you're welcome to use summation notation throughout. ∎

In this example, as in Example 4.5.1, we were able to take advantage of a special situation; this time, **B** was a diagonal matrix. In general, it may be hard to find an expression for $\mathbf{B}^m$, let alone to sum the series $\sum_{m=0}^{\infty} \frac{1}{m!} \mathbf{B}^m$ directly.

EXAMPLE 4.5.3 If $\mathbf{B} = \begin{bmatrix} 1 & -3 \\ 1 & 5 \end{bmatrix}$, then we get

$$\mathbf{B}^2 = \begin{bmatrix} -2 & -18 \\ 6 & 22 \end{bmatrix}, \mathbf{B}^3 = \begin{bmatrix} -20 & -84 \\ 28 & 92 \end{bmatrix}, \text{ and } \mathbf{B}^4 = \begin{bmatrix} -104 & -340 \\ 120 & 376 \end{bmatrix},$$

so

$$e^{\mathbf{B}} = \begin{bmatrix} 1 & 0 \\ 0 & 1 \end{bmatrix} + \begin{bmatrix} 1 & -3 \\ 1 & 5 \end{bmatrix} + \frac{1}{2!}\begin{bmatrix} -2 & -18 \\ 6 & 22 \end{bmatrix} + \frac{1}{3!}\begin{bmatrix} -20 & -84 \\ 28 & 92 \end{bmatrix} + \frac{1}{4!}\begin{bmatrix} -104 & -340 \\ 120 & 376 \end{bmatrix} + (?)$$
$$= (??).$$

Not very promising, so far! ☐

Before we try to compute matrix exponentials any further, let's see if they can serve the purpose we intended them for. That is, are the solutions of the system $\frac{d\mathbf{v}}{dt} = \mathbf{A}\mathbf{v}$ really given by $\mathbf{v}(t) = e^{t\mathbf{A}}\mathbf{v}(0)$? Given an "initial value vector" $\mathbf{v}_0$, we can certainly define a vector $\mathbf{v}$ (that depends on t) by $\mathbf{v} = e^{t\mathbf{A}}\mathbf{v}_0$. If this a solution of the system, it will be the unique solution to the initial value problem $\frac{d\mathbf{v}}{dt} = \mathbf{A}\mathbf{v}$, $\mathbf{v}(0) = \mathbf{v}_0$, since for $t = 0$ we get $\mathbf{v}(0) = e^{0 \cdot \mathbf{A}}\mathbf{v}_0 = e^{\mathbf{0}}\mathbf{v}_0 = \mathbf{I}\mathbf{v}_0 = \mathbf{v}_0$ (see Exercise 32). Provided this works for any choice of $\mathbf{v}_0$, the uniqueness shows that all solutions of $\frac{d\mathbf{v}}{dt} = \mathbf{A}\mathbf{v}$ are, indeed, given by the formula $\mathbf{v}(t) = e^{t\mathbf{A}}\mathbf{v}(0)$.

We still need to show that for any $\mathbf{v}_0$, $\mathbf{v} = e^{t\mathbf{A}}\mathbf{v}_0$ is a solution to our system, and this takes a surprising amount of work. It's natural to try to argue as follows: Let $\mathbf{v} = e^{t\mathbf{A}}\mathbf{v}_0$; then we have

$$\frac{d\mathbf{v}}{dt} = \frac{d}{dt}(e^{t\mathbf{A}}\mathbf{v}_0)$$
$$= \frac{d}{dt}(e^{t\mathbf{A}})\mathbf{v}_0 \quad \text{(because } \mathbf{v}_0 \text{ doesn't depend on } t\text{)}$$
$$= \mathbf{A}e^{t\mathbf{A}}\mathbf{v}_0 \quad \text{("chain rule")}$$
$$= \mathbf{A}\mathbf{v}.$$

If we look a little more closely, though, several of the steps above are questionable. $e^{t\mathbf{A}}$ is a matrix; what does $\frac{d}{dt}(e^{t\mathbf{A}})$ mean? Why is $\frac{d}{dt}(e^{t\mathbf{A}}) = \mathbf{A}e^{t\mathbf{A}}$? Exercises 33 and 42 show how to put all this on a firmer footing. Once this has been done, we can use our new method, assuming that we can compute $e^{t\mathbf{A}}$.

EXAMPLE 4.5.4 Solve the initial value problem

$$\begin{cases} \dfrac{dx_1}{dt} = 2x_2 - x_3, & x_1(0) = 5, \\ \dfrac{dx_2}{dt} = 4x_3, & x_2(0) = -1, \\ \dfrac{dx_3}{dt} = 0, & x_3(0) = 6 \end{cases}$$

using a matrix exponential. (This isn't necessarily the fastest method; see Exercise 30.)

We have $\dfrac{d\mathbf{v}}{dt} = \mathbf{Av}$, $\mathbf{v}(0) = (5, -1, 6)$ for the matrix $\mathbf{A} = \begin{bmatrix} 0 & 2 & -1 \\ 0 & 0 & 4 \\ 0 & 0 & 0 \end{bmatrix}$. The solution will be given by $\mathbf{v}(t) = e^{t\mathbf{A}}\mathbf{v}(0)$, so we need to find $e^{t\mathbf{A}}$. We have $t\mathbf{A} = \begin{bmatrix} 0 & 2t & -t \\ 0 & 0 & 4t \\ 0 & 0 & 0 \end{bmatrix}$,

$(t\mathbf{A})^2 = \begin{bmatrix} 0 & 0 & 8t^2 \\ 0 & 0 & 0 \\ 0 & 0 & 0 \end{bmatrix}$, $(t\mathbf{A})^3 = 0$, so

$$e^{t\mathbf{A}} = \mathbf{I} + t\mathbf{A} + \dfrac{1}{2!}(t\mathbf{A})^2 + 0 + 0 + \cdots$$

$$= \begin{bmatrix} 1 & 2t & -t + 4t^2 \\ 0 & 1 & 4t \\ 0 & 0 & 1 \end{bmatrix}$$

and

$$\mathbf{v}(t) = \begin{bmatrix} 1 & 2t & -t + 4t^2 \\ 0 & 1 & 4t \\ 0 & 0 & 1 \end{bmatrix} \begin{bmatrix} 5 \\ -1 \\ 6 \end{bmatrix} = \begin{bmatrix} 5 - 8t + 24t^2 \\ -1 + 24t \\ 6 \end{bmatrix}.$$

In other words, the solution consists of the three functions

$$x_1(t) = 5 - 8t + 24t^2, \qquad x_2(t) = -1 + 24t, \qquad x_3(t) = 6. \qquad \blacksquare$$

We still need a method to compute matrix exponentials in general; so far, we can only compute $e^{\mathbf{B}}$ if $\mathbf{B}$ is either a nilpotent or a diagonal matrix. In the general case the easiest (and often the only) way to compute $e^{\mathbf{B}}$ is to use a change of coordinates, much as in our previous method for solving linear systems. Remember that the effect of changing coordinates using a transformation matrix $\mathbf{T}$ is to change the matrix $\mathbf{B}$ to the new matrix $\mathbf{T}^{-1}\mathbf{BT}$. The exponential of this new matrix is related to the exponential of $\mathbf{B}$, since

$$e^{\mathbf{T}^{-1}\mathbf{BT}} = \sum_{m=0}^{\infty} \dfrac{1}{m!}(\mathbf{T}^{-1}\mathbf{BT})^m$$

$$= \sum_{m=0}^{\infty} \dfrac{1}{m!}\mathbf{T}^{-1}\mathbf{B}^m\mathbf{T} \qquad \text{[see Exercise 43(a)]}$$

$$= \mathbf{T}^{-1}\left(\sum_{m=0}^{\infty} \dfrac{1}{m!}\mathbf{B}^m\right)\mathbf{T} \qquad \text{[see Exercise 43(b)]}$$

$$= \mathbf{T}^{-1}e^{\mathbf{B}}\mathbf{T}.$$

SEC. 4.5 HOMOGENEOUS, AUTONOMOUS LINEAR SYSTEMS (USING MATRIX EXPONENTIALS)

If $T^{-1}BT$ is a matrix (such as a diagonal matrix) whose exponential we can find, then we can use this equation to solve for e^B:

$$e^{T^{-1}BT} = T^{-1}e^BT$$
$$Te^{T^{-1}BT} = e^BT \quad \text{(multiplying by } T \text{ on the left)}$$
$$Te^{T^{-1}BT}T^{-1} = e^B \quad \text{(multiplying by } T^{-1} \text{ on the right)}.$$

EXAMPLE 4.5.5 Compute e^B for $B = \begin{bmatrix} 1 & -3 \\ 1 & 5 \end{bmatrix}$. (In Example 4.5.3 we saw that this is hard to do directly.)

To find a suitable transformation matrix T, we first find the eigenvalues and eigenvectors of B.

$$\det(B - \lambda I) = (1 - \lambda)(5 - \lambda) + 3 = \lambda^2 - 6\lambda + 8 = (\lambda - 2)(\lambda - 4),$$

so we have eigenvalues $\lambda_1 = 2$, $\lambda_2 = 4$; corresponding eigenvectors turn out to be $(3, -1)$ and $(1, -1)$, respectively (check this!). Thus we can take $T = \begin{bmatrix} 3 & 1 \\ -1 & -1 \end{bmatrix}$ and have $T^{-1}BT = \begin{bmatrix} 2 & 0 \\ 0 & 4 \end{bmatrix}$. The exponential of this new matrix is

$$\sum_{m=0}^{\infty} \frac{1}{m!} \begin{bmatrix} 2 & 0 \\ 0 & 4 \end{bmatrix}^m = \sum_{m=0}^{\infty} \begin{bmatrix} \frac{2^m}{m!} & 0 \\ 0 & \frac{4^m}{m!} \end{bmatrix} = \begin{bmatrix} \sum_{m=0}^{\infty} \frac{2^m}{m!} & 0 \\ 0 & \sum_{m=0}^{\infty} \frac{4^m}{m!} \end{bmatrix} = \begin{bmatrix} e^2 & 0 \\ 0 & e^4 \end{bmatrix},$$

so we have

$$e^B = Te^{T^{-1}BT}T^{-1} = \begin{bmatrix} 3 & 1 \\ -1 & -1 \end{bmatrix} \begin{bmatrix} e^2 & 0 \\ 0 & e^4 \end{bmatrix} \begin{bmatrix} \frac{1}{2} & \frac{1}{2} \\ -\frac{1}{2} & -\frac{3}{2} \end{bmatrix}$$

$$= \begin{bmatrix} 3e^2 & e^4 \\ -e^2 & -e^4 \end{bmatrix} \begin{bmatrix} \frac{1}{2} & \frac{1}{2} \\ -\frac{1}{2} & -\frac{3}{2} \end{bmatrix}$$

$$= \begin{bmatrix} \frac{3}{2}e^2 - \frac{1}{2}e^4 & \frac{3}{2}e^2 - \frac{3}{2}e^4 \\ -\frac{1}{2}e^2 + \frac{1}{2}e^4 & -\frac{1}{2}e^2 + \frac{3}{2}e^4 \end{bmatrix}.$$

Note that in order to find e^B, we had to compute the inverse matrix T^{-1} as well as T. ∎

To apply this method to the computation of e^{tA}, we'll need the eigenvalues and eigenvectors of tA. However, all eigenvectors of A are also eigenvectors of tA (see Exercise 34). Thus if A is diagonalizable and T is a transformation matrix such that $T^{-1}AT$ is the diagonal matrix

$$\begin{bmatrix} \lambda_1 & & & 0 \\ & \lambda_2 & & \\ & & \ddots & \\ 0 & & & \lambda_n \end{bmatrix},$$

then we can use the same transformation matrix for tA and get

$$T^{-1}(tA)T = \begin{bmatrix} \lambda_1 t & & & 0 \\ & \lambda_2 t & & \\ & & \ddots & \\ 0 & & & \lambda_n t \end{bmatrix}$$

[see Exercise 35(a)]. We then find

$$e^{T^{-1}(tA)T} = \begin{bmatrix} e^{\lambda_1 t} & & & 0 \\ & e^{\lambda_2 t} & & \\ & & \ddots & \\ 0 & & & e^{\lambda_n t} \end{bmatrix}$$

[Exercise 35(b)], from which

$$e^{tA} = Te^{T^{-1}(tA)T}T^{-1} = T \begin{bmatrix} e^{\lambda_1 t} & & & 0 \\ & e^{\lambda_2 t} & & \\ & & \ddots & \\ 0 & & & e^{\lambda_n t} \end{bmatrix} T^{-1}.$$

Therefore, the solutions of the system $\dfrac{d\mathbf{v}}{dt} = A\mathbf{v}$ will be given by

$$\mathbf{v}(t) = e^{tA}\mathbf{v}(0) = T \begin{bmatrix} e^{\lambda_1 t} & & & 0 \\ & e^{\lambda_2 t} & & \\ & & \ddots & \\ 0 & & & e^{\lambda_n t} \end{bmatrix} T^{-1}\mathbf{v}(0)$$

in this case.

EXAMPLE 4.5.6 Solve the initial value problem

$$\begin{cases} \dfrac{dx_1}{dt} = 6x_1 - 12x_2 - 9x_3, & x_1(0) = -3, \\ \dfrac{dx_2}{dt} = 7x_1 - 14x_2 - 10x_3, & x_2(0) = -10, \\ \dfrac{dx_3}{dt} = -6x_1 + 12x_2 + 9x_3, & x_3(0) = 7. \end{cases}$$

SEC. 4.5 HOMOGENEOUS, AUTONOMOUS LINEAR SYSTEMS (USING MATRIX EXPONENTIALS)

We have $\mathbf{A} = \begin{bmatrix} 6 & -12 & -9 \\ 7 & -14 & -10 \\ -6 & 12 & 9 \end{bmatrix}$; the characteristic equation turns out to be

$$\lambda^3 - \lambda^2 - 6\lambda = 0 \quad \text{or} \quad \lambda(\lambda - 3)(\lambda + 2) = 0.$$

An eigenvector for $\lambda = 0$ is $(2, 1, 0)$; an eigenvector for $\lambda = 3$ is $(1, 1, -1)$; an eigenvector for $\lambda = -2$ is $(12, 17, -12)$. So we can take $\mathbf{T} = \begin{bmatrix} 2 & 1 & 12 \\ 1 & 1 & 17 \\ 0 & -1 & -12 \end{bmatrix}$ and have

$$\mathbf{T}^{-1}(t\mathbf{A})\mathbf{T} = \begin{bmatrix} 0 & 0 & 0 \\ 0 & 3t & 0 \\ 0 & 0 & -2t \end{bmatrix}, \quad e^{\mathbf{T}^{-1}(t\mathbf{A})\mathbf{T}} = \begin{bmatrix} 1 & 0 & 0 \\ 0 & e^{3t} & 0 \\ 0 & 0 & e^{-2t} \end{bmatrix}.$$

Then our solution will be

$$\mathbf{v}(t) = e^{t\mathbf{A}}\mathbf{v}(0) = \mathbf{T}e^{\mathbf{T}^{-1}(t\mathbf{A})\mathbf{T}}\mathbf{T}^{-1}\mathbf{v}(0)$$

$$= \begin{bmatrix} 2 & 1 & 12 \\ 1 & 1 & 17 \\ 0 & -1 & -12 \end{bmatrix} \begin{bmatrix} 1 & 0 & 0 \\ 0 & e^{3t} & 0 \\ 0 & 0 & e^{-2t} \end{bmatrix} \begin{bmatrix} \dfrac{1}{2} & 0 & \dfrac{1}{2} \\ \dfrac{6}{5} & -\dfrac{12}{5} & -\dfrac{11}{5} \\ -\dfrac{1}{10} & \dfrac{1}{5} & \dfrac{1}{10} \end{bmatrix} \begin{bmatrix} -3 \\ -10 \\ 7 \end{bmatrix}$$

$$= \begin{bmatrix} 2 & e^{3t} & 12e^{-2t} \\ 1 & e^{3t} & 17e^{-2t} \\ 0 & -e^{3t} & -12e^{-2t} \end{bmatrix} \begin{bmatrix} 2 \\ 5 \\ -1 \end{bmatrix} = \begin{bmatrix} 4 + 5e^{3t} - 12e^{-2t} \\ 2 + 5e^{3t} - 17e^{-2t} \\ -5e^{3t} + 12e^{-2t} \end{bmatrix}.$$

That is, $x_1(t) = 4 + 5e^{3t} - 12e^{-2t}$, and so on. ∎

As above, suppose that $\mathbf{A}$ is diagonalizable, with

$$\mathbf{T}^{-1}\mathbf{A}\mathbf{T} = \begin{bmatrix} \lambda_1 & & & 0 \\ & \lambda_2 & & \\ & & \ddots & \\ 0 & & & \lambda_n \end{bmatrix}.$$

To see the connection between the formula

$$\mathbf{v}(t) = \mathbf{T}\begin{bmatrix} e^{\lambda_1 t} & & & 0 \\ & e^{\lambda_2 t} & & \\ & & \ddots & \\ 0 & & & e^{\lambda_n t} \end{bmatrix}\mathbf{T}^{-1}\mathbf{v}(0)$$

we found above for the solutions of $\dfrac{d\mathbf{v}}{dt} = \mathbf{A}\mathbf{v}$ and the results of Section 4.3, note that $\mathbf{v}(0)$ can be chosen to be any "initial value vector." As $\mathbf{v}(0)$ runs through all possible vectors, so will $\mathbf{T}^{-1}\mathbf{v}(0)$. If we put $\mathbf{T}^{-1}\mathbf{v}(0) = \begin{bmatrix} C_1 \\ \vdots \\ C_n \end{bmatrix}$, the formula above becomes

$$\mathbf{v}(t) = \mathbf{T} \begin{bmatrix} e^{\lambda_1 t} & & 0 \\ & e^{\lambda_2 t} & \\ & & \ddots \\ 0 & & e^{\lambda_n t} \end{bmatrix} \begin{bmatrix} C_1 \\ C_2 \\ \vdots \\ C_n \end{bmatrix} = \mathbf{T} \begin{bmatrix} C_1 e^{\lambda_1 t} \\ C_2 e^{\lambda_2 t} \\ \vdots \\ C_n e^{\lambda_n t} \end{bmatrix},$$

just as in Section 4.3. Our new approach not only leads to the same result but uses essentially the same computations (of eigenvalues and eigenvectors). It has the advantage of giving an explicit formula for the solution of an initial value problem, but one has to compute $\mathbf{T}^{-1}$ to use this formula. As in Section 4.4, there is no real difficulty with complex eigenvalues (no pun intended), but repeated eigenvalues may cause trouble because $\mathbf{A}$ may no longer be diagonalizable.

EXAMPLE 4.5.7 Solve the initial value problem

$$\begin{cases} \dfrac{dx_1}{dt} = -x_1 + x_2 - x_4, & x_1(0) = 2, \\ \dfrac{dx_2}{dt} = -4x_1 + 4x_2 + 2x_3 - 3x_4, & x_2(0) = 1, \\ \dfrac{dx_3}{dt} = 3x_1 - 2x_2 - x_3 + x_4, & x_3(0) = 0, \\ \dfrac{dx_4}{dt} = -2x_1 + 3x_2 + 2x_3 - 2x_4, & x_4(0) = -1 \end{cases}$$

using a matrix exponential.
Here we have

$$\mathbf{A} = \begin{bmatrix} -1 & 1 & 0 & -1 \\ -4 & 4 & 2 & -3 \\ 3 & -2 & -1 & 1 \\ -2 & 3 & 2 & -2 \end{bmatrix};$$

the characteristic equation works out to $\lambda^4 + 2\lambda^2 + 1 = 0$. This gives us $(\lambda^2 + 1)^2 = 0$ and then $(\lambda - i)^2(\lambda + i)^2 = 0$, so we have a pair of (complex conjugate) double eigenvalues: $\lambda = i$ and $\bar{\lambda} = -i$. Since we usually expect one eigenvector (up to multiples) per eigenvalue, it seems that **A** won't be diagonalizable, but let's find out. If $(\xi_1, \xi_2, \xi_3, \xi_4)$ is an eigenvector for $\lambda = i$, then

$$\begin{cases} -\xi_1 + \xi_2 - \xi_4 = i\xi_1 \\ -4\xi_1 + 4\xi_2 + 2\xi_3 - 3\xi_4 = i\xi_2 \\ 3\xi_1 - 2\xi_2 - \xi_3 + \xi_4 = i\xi_3 \\ -2\xi_1 + 3\xi_2 + 2\xi_3 - 2\xi_4 = i\xi_4 \end{cases} \quad \text{or} \quad \begin{cases} (-1-i)\xi_1 + \xi_2 - \xi_4 = 0 \\ -4\xi_1 + (4-i)\xi_2 + 2\xi_3 - 3\xi_4 = 0 \\ 3\xi_1 - 2\xi_2 + (-1-i)\xi_3 + \xi_4 = 0 \\ -2\xi_1 + 3\xi_2 + 2\xi_3 + (-2-i)\xi_4 = 0. \end{cases}$$

When we solve this system, whether by row reduction or by successive elimination, we find that two of the equations are implied by the other two. For instance, we can use the first equation to write $\xi_2 = (1+i)\xi_1 + \xi_4$; substituting this in the third equation we find $\xi_4 = (1-2i)\xi_1 + (-1-i)\xi_3$, so that $\xi_2 = (1+i)\xi_1 + \xi_4 = (2-i)\xi_1 + (-1-i)\xi_3$. If we now substitute these expressions for ξ_2 and ξ_4 into the two remaining equations, it turns out that they are automatically satisfied. That is, for *any* choices of ξ_1 and ξ_3, we can take $\xi_2 = (2-i)\xi_1 + (-1-i)\xi_3$, $\xi_4 = (1-2i)\xi_1 + (-1-i)\xi_3$ and get a solution $(\xi_1, \xi_2, \xi_3, \xi_4)$ to the simultaneous equations above.[20] Thus we are in the unusual situation here of having several essentially different eigenvectors for one eigenvalue $\lambda = i$; we'll see that **A** is diagonalizable after all. For example, if we take $\xi_1 = 1$, $\xi_3 = 0$ we get the eigenvector $(1, 2-i, 0, 1-2i)$, while we get $(0, -1-i, 1, -1-i)$ for $\xi_1 = 0$, $\xi_3 = 1$. These eigenvectors are not multiples of each other (why?). The complex conjugates of these vectors will be eigenvectors for $\lambda = -i$; we now have four eigenvectors in all, so we can form a square matrix **T** with these columns:

$$\mathbf{T} = \begin{bmatrix} 1 & 1 & 0 & 0 \\ 2-i & 2+i & -1-i & -1+i \\ 0 & 0 & 1 & 1 \\ 1-2i & 1+2i & -1-i & -1+i \end{bmatrix}.$$

We will then have

$$\mathbf{T}^{-1}\mathbf{A}\mathbf{T} = \begin{bmatrix} i & 0 & 0 & 0 \\ 0 & -i & 0 & 0 \\ 0 & 0 & i & 0 \\ 0 & 0 & 0 & -i \end{bmatrix} \quad [21]$$

(make sure you see why the diagonal entries appear in this particular order; see Exercise 44). Therefore (further explanations will follow the computation),

---

[20] We could just as well have expressed, say, ξ_1 and ξ_2 in terms of ξ_3 and ξ_4, but the computation given here avoids fractions.

[21] Of course, this only works because **T** is invertible. See footnote 14 on p. 326 for further discussion of this point.

$$\mathbf{v}(t) = e^{t\mathbf{A}}\mathbf{v}(0) = \mathbf{T}\begin{bmatrix} e^{it} & 0 & 0 & 0 \\ 0 & e^{-it} & 0 & 0 \\ 0 & 0 & e^{it} & 0 \\ 0 & 0 & 0 & e^{-it} \end{bmatrix}\mathbf{T}^{-1}\mathbf{v}(0)$$

$$= \begin{bmatrix} 1 & 1 & 0 & 0 \\ 2-i & 2+i & -1-i & -1+i \\ 0 & 0 & 1 & 1 \\ 1-2i & 1+2i & -1-i & -1+i \end{bmatrix} \begin{bmatrix} e^{it} & 0 & 0 & 0 \\ 0 & e^{-it} & 0 & 0 \\ 0 & 0 & e^{it} & 0 \\ 0 & 0 & 0 & e^{-it} \end{bmatrix} \begin{bmatrix} \frac{1+i}{2} & -\frac{i}{2} & 0 & \frac{i}{2} \\ \frac{1-i}{2} & \frac{i}{2} & 0 & -\frac{i}{2} \\ -\frac{3i}{2} & i & \frac{1+i}{2} & -\frac{i}{2} \\ \frac{3i}{2} & -i & \frac{1-i}{2} & \frac{i}{2} \end{bmatrix} \begin{bmatrix} 2 \\ 1 \\ 0 \\ -1 \end{bmatrix}$$

$$= \begin{bmatrix} e^{it} & e^{-it} & 0 & 0 \\ (2-i)e^{it} & (2+i)e^{-it} & (-1-i)e^{it} & (-1+i)e^{-it} \\ 0 & 0 & e^{it} & e^{-it} \\ (1-2i)e^{it} & (1+2i)e^{-it} & (-1-i)e^{it} & (-1+i)e^{-it} \end{bmatrix} \begin{bmatrix} 1 \\ 1 \\ -\frac{3i}{2} \\ \frac{3i}{2} \end{bmatrix}$$

$$= \begin{bmatrix} e^{it} + e^{-it} \\ \left(\frac{1}{2} + \frac{1}{2}i\right)e^{it} + \left(\frac{1}{2} - \frac{1}{2}i\right)e^{-it} \\ -\frac{3}{2}ie^{it} + \frac{3}{2}ie^{-it} \\ \left(-\frac{1}{2} - \frac{1}{2}i\right)e^{it} + \left(-\frac{1}{2} + \frac{1}{2}i\right)e^{-it} \end{bmatrix} = \begin{bmatrix} 2\,\mathrm{Re}(e^{it}) \\ \mathrm{Re}[(1+i)e^{it}] \\ \mathrm{Re}(-3ie^{it}) \\ \mathrm{Re}[(-1-i)e^{it}] \end{bmatrix} = \begin{bmatrix} 2\cos t \\ \cos t - \sin t \\ 3\sin t \\ -\cos t + \sin t \end{bmatrix}.$$

Here we have used the given initial values, which form the vector $\mathbf{v}(0) = (2, 1, 0, -1)$. Also, the identity $z + \bar{z} = 2\,\mathrm{Re}(z)$ (see Appendix A, Section A.3, Exercise 2) has been used as a shortcut. ∎

EXAMPLE 4.5.8 Solve the initial value problem

$$\begin{cases} \dfrac{dx_1}{dt} = x_1 + 4x_2, & x_1(0) = 1 \\ \dfrac{dx_2}{dt} = -x_1 + 5x_2, & x_2(0) = 1 \end{cases}$$

using a matrix exponential.

The matrix $\mathbf{A} = \begin{bmatrix} 1 & 4 \\ -1 & 5 \end{bmatrix}$ has a double eigenvalue $\lambda = 3$, and this time the only eigenvector (up to multiples) is $(2, 1)$. So $\mathbf{A}$ and $t\mathbf{A}$ are not diagonalizable, which makes computing e^{tA} more difficult. However, we've seen (in Sections 3.5 and 4.4) that we can get $T^{-1}AT$ [or $T^{-1}(tA)T$] to be an upper triangular matrix by using the eigenvector of A as the first column of T. If we take $(0, 1)$, for instance, as the second column, we'll have

$$T = \begin{bmatrix} 2 & 0 \\ 1 & 1 \end{bmatrix}, \quad T^{-1} = \begin{bmatrix} \frac{1}{2} & 0 \\ -\frac{1}{2} & 1 \end{bmatrix},$$

$$T^{-1}(tA)T = \begin{bmatrix} \frac{1}{2} & 0 \\ -\frac{1}{2} & 1 \end{bmatrix} \begin{bmatrix} t & 4t \\ -t & 5t \end{bmatrix} \begin{bmatrix} 2 & 0 \\ 1 & 1 \end{bmatrix} = \begin{bmatrix} 3t & 2t \\ 0 & 3t \end{bmatrix}.$$

There are two ways of computing the exponential of this matrix. It can be done directly (see Exercise 45) or by first rewriting the matrix in a way we'll discuss below. Either way, the result is $e^{T^{-1}AT} = \begin{bmatrix} e^{3t} & 2te^{3t} \\ 0 & e^{3t} \end{bmatrix}$, from which we get

$$e^{tA} = \begin{bmatrix} 2 & 0 \\ 1 & 1 \end{bmatrix} \begin{bmatrix} e^{3t} & 2te^{3t} \\ 0 & e^{3t} \end{bmatrix} \begin{bmatrix} \frac{1}{2} & 0 \\ -\frac{1}{2} & 1 \end{bmatrix} = \begin{bmatrix} (1-2t)e^{3t} & 4te^{3t} \\ -te^{3t} & (1+2t)e^{3t} \end{bmatrix}$$

and finally

$$v(t) = e^{tA}v(0) = \begin{bmatrix} (1-2t)e^{3t} & 4te^{3t} \\ -te^{3t} & (1+2t)e^{3t} \end{bmatrix} \begin{bmatrix} 1 \\ 1 \end{bmatrix} = \begin{bmatrix} (1+2t)e^{3t} \\ (1+t)e^{3t} \end{bmatrix}. \blacksquare$$

As mentioned in Section 4.4, it is always possible to find an invertible matrix T such that $T^{-1}AT$ [and thus $T^{-1}(tA)T$ as well] is upper triangular. However, for most upper triangular matrices it isn't practicable to compute the exponential directly. The method that is used instead is based on a generalization of the formula $e^{x+y} = e^x \cdot e^y$ for real numbers x and y.

Warning: If C and D are arbitrary square matrices of the same size, it is usually *not* true that $e^{C+D} = e^C e^D$. See Exercise 36.

The good news is that the formula $e^{C+D} = e^C \cdot e^D$ does hold if the matrices C and D **commute**, that is, if $CD = DC$. See Exercise 37.

EXAMPLE 4.5.9 Suppose that we want to compute the exponential of $\begin{bmatrix} 3t & 2t \\ 0 & 3t \end{bmatrix}$, as in Example 4.5.8. Then we can write our matrix as $C + D$ with $C = \begin{bmatrix} 3t & 0 \\ 0 & 3t \end{bmatrix}$, $D = \begin{bmatrix} 0 & 2t \\ 0 & 0 \end{bmatrix}$. Since we have $CD = DC = \begin{bmatrix} 0 & 6t^2 \\ 0 & 0 \end{bmatrix}$ (check this!), C and D commute, so we have $e^{C+D} = e^C \cdot e^D$.

However, **C** is diagonal and **D** is nilpotent, so $e^{\mathbf{C}}$ and $e^{\mathbf{D}}$ can be computed quickly and then multiplied together to yield our exponential $e^{\mathbf{C}+\mathbf{D}}$. See Exercise 38. ∎

Although not every upper triangular matrix can be written as the sum of a diagonal and a nilpotent matrix that commute with each other, this *can* be done for every matrix in Jordan form, the special upper triangular form mentioned at the end of Section 4.4. Thus the exponential of any square matrix can, in principle, be computed using transformation to the Jordan form. (For details, see "Further Reading" following the Exercises.) This means that the solutions of any homogeneous, autonomous linear system $\frac{d\mathbf{v}}{dt} = \mathbf{A}\mathbf{v}$ can be found using $\mathbf{v}(t) = e^{t\mathbf{A}}\mathbf{v}(0)$. This way of writing the solutions will be used in the next section when we discuss the method of variation of constants for systems.

SUMMARY OF KEY CONCEPTS, RESULTS, AND TECHNIQUES

Matrix exponential (p. 336): For any square matrix **B**,

$$e^{\mathbf{B}} = \mathbf{I} + \mathbf{B} + \frac{1}{2!}\mathbf{B}^2 + \frac{1}{3!}\mathbf{B}^3 + \cdots = \sum_{m=0}^{\infty} \frac{1}{m!}\mathbf{B}^m.$$

Nilpotent matrix (p. 337): A square matrix, one of whose powers is the zero matrix. If **B** is nilpotent, the series for $e^{\mathbf{B}}$ becomes a finite sum. To compute $e^{\mathbf{B}}$ if **B** is not diagonal or nilpotent, the formula $e^{\mathbf{B}} = \mathbf{T}e^{\mathbf{T}^{-1}\mathbf{B}\mathbf{T}}\mathbf{T}^{-1}$ (p. 341) is helpful. In particular, if **B** is diagonalizable, choose **T** so that $\mathbf{T}^{-1}\mathbf{B}\mathbf{T}$ will be diagonal (using eigenvectors of **B** as the columns of **T**).

Solving $\frac{d\mathbf{v}}{dt} = \mathbf{A}\mathbf{v}$, $\mathbf{v}(0) = \mathbf{v}_0$ using a matrix exponential (p. 339): The solution to this initial value problem is $\mathbf{v}(t) = e^{t\mathbf{A}}\mathbf{v}_0$. Since all eigenvectors of **A** are also eigenvectors of $t\mathbf{A}$, $e^{t\mathbf{A}}$ can be computed by the method above (using **T**), provided that **A** is diagonalizable (p. 342).

Square matrices **C**, **D** of the same size **commute** if $\mathbf{CD} = \mathbf{DC}$ (p. 347). If **C** and **D** commute, then $e^{\mathbf{C}+\mathbf{D}} = e^{\mathbf{C}} \cdot e^{\mathbf{D}}$. This fact can be used, together with $e^{\mathbf{B}} = \mathbf{T}e^{\mathbf{T}^{-1}\mathbf{B}\mathbf{T}}\mathbf{T}^{-1}$, to compute $e^{\mathbf{B}}$ even when **B** is not diagonalizable, by choosing **T** such that $\mathbf{T}^{-1}\mathbf{B}\mathbf{T}$ is the sum of a diagonal matrix and a nilpotent matrix that commute with each other (pp. 347–348).

EXERCISES

Find the exponentials of the following matrices.

1. $\begin{bmatrix} -1 & 0 & 0 \\ 0 & 0 & 0 \\ 0 & 0 & 2 \end{bmatrix}$
2. $\begin{bmatrix} 0 & 0 & 0 \\ 0 & 3 & 0 \\ 0 & 0 & -4 \end{bmatrix}$
3. $\begin{bmatrix} 0 & 1 & 2 \\ 0 & 0 & -1 \\ 0 & 0 & 0 \end{bmatrix}$

SEC. 4.5 HOMOGENEOUS, AUTONOMOUS LINEAR SYSTEMS (USING MATRIX EXPONENTIALS) 349

4. $\begin{bmatrix} 0 & 3 & 5 \\ 0 & 0 & 4 \\ 0 & 0 & 0 \end{bmatrix}$
5. $\begin{bmatrix} t & 2t \\ 3t & 6t \end{bmatrix}$
6. $\begin{bmatrix} 1 & -3 \\ 2 & -6 \end{bmatrix}$

7. $\begin{bmatrix} 0 & 1 \\ -4 & 0 \end{bmatrix}$
8. $\begin{bmatrix} 0 & 9t \\ -t & 0 \end{bmatrix}$
9. $\begin{bmatrix} 0 & 0 & 0 \\ 1 & 0 & 0 \\ -1 & 3 & 0 \end{bmatrix}$

10. $\begin{bmatrix} 1 & 0 & 0 \\ 1 & 2 & 0 \\ -1 & 1 & 3 \end{bmatrix}$
11. $\begin{bmatrix} 3 & 0 & 0 \\ 1 & -1 & 0 \\ 2 & 2 & 1 \end{bmatrix}$
12. $\begin{bmatrix} 2 & 0 & 0 \\ 1 & 4 & 1 \\ -1 & 6 & 3 \end{bmatrix}$

13. $\begin{bmatrix} -3 & 0 & 0 \\ 1 & 2 & 5 \\ 4 & 1 & 6 \end{bmatrix}$

14. $\begin{bmatrix} 6 & 1 & 2 \\ 1 & 6 & 2 \\ 2 & 2 & 5 \end{bmatrix}$
[*Hint*: $\lambda = 3$ is one eigenvalue.]

15. $\begin{bmatrix} -2 & 2 & 2 \\ 2 & -1 & 1 \\ 2 & 1 & -1 \end{bmatrix}$
[*Hint*: $\lambda = -2$ is one eigenvalue.]

16. $t \cdot \begin{bmatrix} -2 & 1 & -2 \\ 0 & 0 & 1 \\ 0 & -1 & 0 \end{bmatrix}$
17. $t \cdot \begin{bmatrix} 0 & 0 & -4 \\ 6 & 3 & 8 \\ 1 & 0 & 0 \end{bmatrix}$

Solve the following initial value problems using matrix exponentials.

18. $\dfrac{dx}{dt} = -5x + 8y, \quad \dfrac{dy}{dt} = 4x - y, \quad x(0) = 0, \quad y(0) = 15.$

19. $\dfrac{dx}{dt} = 4y, \quad \dfrac{dy}{dt} = -\dfrac{1}{2}x + 3y, \quad x(0) = 8, \quad y(0) = -1.$

20. Exercise 17 from Section 4.3 (p. 319).
21. Exercise 18 from Section 4.3 (p. 319).
22. Exercise 13 from Section 4.4 (p. 332).
23. Exercise 14 from Section 4.4 (p. 332).
24. Exercise 19 from Section 4.4 (p. 332).

25. $\begin{cases} \dfrac{dx_1}{dt} = x_2 + 2x_3, \quad x_1(0) = -1, \\ \dfrac{dx_2}{dt} = -x_3, \quad x_2(0) = 4, \\ \dfrac{dx_3}{dt} = 0, \quad x_3(0) = -5. \quad \text{(Compare Exercise 3.)} \end{cases}$

26. $\begin{cases} \dfrac{dx_1}{dt} = x_2 - x_3, & x_1(0) = 4, \\ \dfrac{dx_2}{dt} = 4x_3, & x_2(0) = -1, \\ \dfrac{dx_3}{dt} = 0, & x_3(0) = 10. \end{cases}$

27. $\dfrac{dx_1}{dt} = 5x_1 - x_2, \quad \dfrac{dx_2}{dt} = x_1 + 3x_2, \quad x_1(0) = 5, \quad x_2(0) = -1.$

28. $\dfrac{dx_1}{dt} = -2x_1 + x_2, \quad \dfrac{dx_2}{dt} = -x_1 - 4x_2, \quad x_1(0) = 6, \quad x_2(0) = 0.$

29. $\begin{cases} \dfrac{dx_1}{dt} = 4x_1 + 7x_2 - 6x_3, & x_1(0) = 5, \\ \dfrac{dx_2}{dt} = -12x_1 - 16x_2 + 12x_3, & x_2(0) = -11, \\ \dfrac{dx_3}{dt} = -9x_1 - 10x_2 + 7x_3, & x_3(0) = -11. \end{cases}$

[*Hint*: $\lambda = 1$ is an eigenvalue.]

30. Solve the initial value problem from Example 4.5.4 (pp. 339–340) directly, without using a matrix exponential.

31. Solve the initial value problem from Exercise 26 directly, without using a matrix exponential.

32. Show that for the $n \times n$ zero matrix $\mathbf{0}$, we have $e^{\mathbf{0}} = \mathbf{I}$.

33. (a) Let $\mathbf{A}$ be an $m \times n$ matrix whose entries are functions of t:

$$\mathbf{A}(t) = \begin{bmatrix} a_{11}(t) & \cdots & a_{1n}(t) \\ \vdots & & \vdots \\ a_{m1}(t) & \cdots & a_{mn}(t) \end{bmatrix}.$$

How would you define the derivative $\dfrac{d\mathbf{A}}{dt}$? The integral $\int \mathbf{A}(t)\,dt$?

(b) Show that if $\mathbf{A}$ and $\mathbf{B}$ are matrices of the same size (whose entries are functions of t), then $\dfrac{d(\mathbf{A} + \mathbf{B})}{dt} = \dfrac{d\mathbf{A}}{dt} + \dfrac{d\mathbf{B}}{dt}$ and for any constant number k, $\dfrac{d(k\mathbf{A})}{dt} = k \cdot \dfrac{d\mathbf{A}}{dt}$. ($k\mathbf{A}$ is the matrix obtained from $\mathbf{A}$ by multiplying all entries by k.)

(c) Show that if $\mathbf{A}$ and $\mathbf{B}$ are matrices such that the product $\mathbf{AB}$ is defined, then $\dfrac{d(\mathbf{AB})}{dt} = \dfrac{d\mathbf{A}}{dt}\mathbf{B} + \mathbf{A}\dfrac{d\mathbf{B}}{dt}$.

(d) Show that if $\mathbf{A}$ is an $m \times n$ matrix and $\mathbf{v}$ is a vector in n-space, then $\dfrac{d(\mathbf{Av})}{dt} = \dfrac{d\mathbf{A}}{dt}\mathbf{v} + \mathbf{A}\dfrac{d\mathbf{v}}{dt}$.

(e) Show that in the situation of p. 339 of the text, $\dfrac{d}{dt}(e^{tA}\mathbf{v}_0) = \dfrac{d}{dt}(e^{tA})\mathbf{v}_0$, as claimed.

34. Show that if $\boldsymbol{\xi}$ is an eigenvector of $\mathbf{A}$ for the eigenvalue λ, then $\boldsymbol{\xi}$ is also an eigenvector of $t\mathbf{A}$, for the eigenvalue $t\lambda = \lambda t$.

35. (a) Show that if

$$\mathbf{T}^{-1}\mathbf{A}\mathbf{T} = \begin{bmatrix} \lambda_1 & & & 0 \\ & \lambda_2 & & \\ & & \ddots & \\ 0 & & & \lambda_n \end{bmatrix},$$

then

$$\mathbf{T}^{-1}(t\mathbf{A})\mathbf{T} = \begin{bmatrix} \lambda_1 t & & & 0 \\ & \lambda_2 t & & \\ & & \ddots & \\ 0 & & & \lambda_n t \end{bmatrix}.$$

(b) Show that in this situation,

$$e^{\mathbf{T}^{-1}(t\mathbf{A})\mathbf{T}} = \begin{bmatrix} e^{\lambda_1 t} & & & 0 \\ & e^{\lambda_2 t} & & \\ & & \ddots & \\ 0 & & & e^{\lambda_n t} \end{bmatrix}.$$

36. Let $\mathbf{C} = \begin{bmatrix} 0 & 1 \\ 0 & 0 \end{bmatrix}$, $\mathbf{D} = \begin{bmatrix} 0 & 0 \\ 1 & 0 \end{bmatrix}$.

  (a) Compute $e^{\mathbf{C}}$ and $e^{\mathbf{D}}$.
  (b) Compute $e^{\mathbf{C}+\mathbf{D}}$.
      [*Hint*: $\mathbf{C} + \mathbf{D}$ is diagonalizable.]
  (c) Show that $e^{\mathbf{C}}e^{\mathbf{D}}$, $e^{\mathbf{D}}e^{\mathbf{C}}$, and $e^{\mathbf{C}+\mathbf{D}}$ are all unequal.

*37. Let $\mathbf{C}$ and $\mathbf{D}$ be square matrices of the same size that commute, that is, such that $\mathbf{CD} = \mathbf{DC}$.

  (a) Show that the binomial theorem holds for powers of $\mathbf{C} + \mathbf{D}$:

$$(\mathbf{C} + \mathbf{D})^m = \sum_{k=0}^{m} \binom{m}{k} \mathbf{C}^k \mathbf{D}^{m-k}.$$

  (b) Show that $e^{\mathbf{C}+\mathbf{D}} = e^{\mathbf{C}} e^{\mathbf{D}}$.
      [*Hint*: Follow the steps of Exercise 10 from Appendix A, Section A.3.]

38. Let $\mathbf{C} = \begin{bmatrix} 3t & 0 \\ 0 & 3t \end{bmatrix}$, $\mathbf{D} = \begin{bmatrix} 0 & 2t \\ 0 & 0 \end{bmatrix}$.

  (a) Compute $e^{\mathbf{C}}$ and $e^{\mathbf{D}}$.
  (b) Compute $e^{\mathbf{C}+\mathbf{D}}$. (Note that $\mathbf{C}$ and $\mathbf{D}$ commute.)

*39. Show that any upper triangular matrix which has only zeros along the (main) diagonal is nilpotent.

40. (a) Show that if $\mathbf{C}$ and $\mathbf{D}$ are nilpotent matrices that *commute*, then $\mathbf{C} + \mathbf{D}$ is also nilpotent.
   [*Hint*: Use Exercise 37(a).]
   (b) Give an example of matrices $\mathbf{C}$ and $\mathbf{D}$ (of the same size) such that $\mathbf{C}$ and $\mathbf{D}$ are nilpotent but $\mathbf{C} + \mathbf{D}$ is not.

*41. Let $\mathbf{B}$ be an $n \times n$ matrix and let M be the largest of the absolute values of the entries of $\mathbf{B}$.
   (a) Show that the absolute value of any entry in $\mathbf{B}^2$ is at most nM^2.
   (b) Show that the absolute value of any entry in $\mathbf{B}^m$ (where $m \geq 1$) is at most $n^{m-1}M^m$.
   (c) Show that the infinite series for each entry of $e^{\mathbf{B}}$ will converge provided the series $1 + M + \dfrac{n}{2!}M^2 + \dfrac{n^2}{3!}M^3 + \cdots + \dfrac{n^{m-1}}{m!}M^m + \cdots$ does.
   (d) Prove that the series in part (c) does indeed converge.

42. (a) Show that if the entries of a matrix $\mathbf{B}$ are functions of t and the number k is also a function of t, then $\dfrac{d(k\mathbf{B})}{dt} = \dfrac{dk}{dt}\mathbf{B} + k\dfrac{d\mathbf{B}}{dt}$. (See Exercise 33 for the definition of the derivative of a matrix.)
   (b) Show that $(t\mathbf{A})^m = t^m \mathbf{A}^m$ for any square matrix $\mathbf{A}$, any number t, and any $m \geq 1$.
   (c) Show that if the matrix $\mathbf{A}$ does not depend on t, then
   $$\dfrac{d}{dt}\left(\dfrac{1}{m!}(t\mathbf{A})^m\right) = \mathbf{A} \cdot \dfrac{1}{(m-1)!}(t\mathbf{A})^{m-1}.$$
   (d) Show that if we can differentiate the series for $e^{t\mathbf{A}}$ by differentiating each term separately, then $\dfrac{d}{dt}(e^{t\mathbf{A}}) = \mathbf{A}e^{t\mathbf{A}}$.
   *(e) Show that this is indeed legitimate.

43. (a) Show that for $m \geq 0$, $(\mathbf{T}^{-1}\mathbf{B}\mathbf{T})^m = \mathbf{T}^{-1}\mathbf{B}^m\mathbf{T}$. (Consider the cases $m = 0$ and $m > 0$ separately.)
   (b) Show that $\displaystyle\sum_{m=0}^{\infty} \dfrac{1}{m!} \mathbf{T}^{-1}\mathbf{B}^m \mathbf{T} = \mathbf{T}^{-1}\left(\sum_{m=0}^{\infty} \dfrac{1}{m!} \mathbf{B}^m\right)\mathbf{T}.$

44. (a) In Example 4.5.7, suppose that we had taken $\mathbf{T}$ to be
$$\begin{bmatrix} 1 & 0 & 1 & 0 \\ 2-i & -1-i & 2+i & -1+i \\ 0 & 1 & 0 & 1 \\ 1-2i & -1-i & 1+2i & -1+i \end{bmatrix}.$$
   What would $\mathbf{T}^{-1}\mathbf{A}\mathbf{T}$ be?
   (b) Why do you think $\mathbf{T}$ was chosen as on p. 345, rather than as in part (a)?

*45. Let $\mathbf{B} = \begin{bmatrix} 3t & 2t \\ 0 & 3t \end{bmatrix}.$
   (a) Compute $\mathbf{B}^2$, $\mathbf{B}^3$, and $\mathbf{B}^4$.

(b) Find a formula for $\mathbf{B}^m$.

[*Hint*: First find the diagonal entries, then consider the ratio between the off-diagonal and diagonal entries.]

(c) Show that $e^{\mathbf{B}} = \begin{bmatrix} e^{3t} & \dfrac{2}{3} \sum\limits_{m=0}^{\infty} \dfrac{m \cdot 3^m t^m}{m!} \\ 0 & e^{3t} \end{bmatrix}$.

(d) Show that $\sum\limits_{m=0}^{\infty} \dfrac{m \cdot 3^m t^m}{m!} = 3te^{3t}$, and find $e^{\mathbf{B}}$.

*46. Suppose that we have a *nonautonomous* linear system of the form $\dfrac{d\mathbf{v}}{dt} = \mathbf{A}(t)\mathbf{v}$, for which the entries of the matrix $\mathbf{A}$ depend on t. In the simplest case, $n = 1$, this becomes $\dfrac{dx}{dt} = a(t)x$ and the solutions are given by $x(t) = e^{\int_0^t a(s)ds} x(0)$ (see Section 1.6). One might therefore suspect that in general, the solutions are given by $\mathbf{v}(t) = e^{\int_0^t \mathbf{A}(s)ds} \mathbf{v}(0)$. In this problem we'll show that this *doesn't* work. Let
$$\mathbf{A}(t) = \begin{bmatrix} 1 & 2t \\ 0 & 2 \end{bmatrix}.$$

(a) Show that $\int_0^t \mathbf{A}(s)ds = \begin{bmatrix} t & t^2 \\ 0 & 2t \end{bmatrix}$.

(b) Show that $e^{\int_0^t \mathbf{A}(s)ds} = \begin{bmatrix} e^t & t(e^{2t} - e^t) \\ 0 & e^{2t} \end{bmatrix}$.

(c) Show that $\mathbf{v}(t) = e^{\int_0^t \mathbf{A}(s)ds} \mathbf{v}(0)$ is not a solution to the system $\dfrac{d\mathbf{v}}{dt} = \mathbf{A}(t)\mathbf{v}$, unless $\mathbf{v}(0)$ is a multiple of $(1, 0)$.

(d) Explain what is *wrong* with the following argument: If $\mathbf{v}(t) = e^{\int_0^t \mathbf{A}(s)ds} \mathbf{v}(0)$, then

$$\dfrac{d\mathbf{v}}{dt} = \dfrac{d}{dt}(e^{\int_0^t \mathbf{A}(s)ds})\mathbf{v}(0) \qquad \text{[since } \mathbf{v}(0) \text{ doesn't depend on } t\text{]}$$

$$= \dfrac{d}{dt}\left(\int_0^t \mathbf{A}(s)ds\right) \cdot e^{\int_0^t \mathbf{A}(s)ds} \mathbf{v}(0) \qquad \text{(chain rule)}$$

$$= \mathbf{A}(t)\mathbf{v} \qquad \text{(Fundamental Theorem of Calculus)}.$$

FURTHER READING

For more on computing matrix exponentials in general, see Curtis, *Linear Algebra* (Springer-Verlag, 1984), Sections 24 and 34.

4.6 GENERAL LINEAR SYSTEMS

The most general form of a linear system (of first-order differential equations) is

$$\begin{cases} \dfrac{dx_1}{dt} = a_{11}(t)x_1 + a_{12}(t)x_2 + \cdots + a_{1n}(t)x_n + h_1(t) \\ \dfrac{dx_2}{dt} = a_{21}(t)x_1 + a_{22}(t)x_2 + \cdots + a_{2n}(t)x_n + h_2(t) \\ \vdots \\ \dfrac{dx_n}{dt} = a_{n1}(t)x_1 + a_{n2}(t)x_2 + \cdots + a_{nn}(t)x_n + h_n(t) . \end{cases}$$

Such a system has solutions that are defined whenever all n functions $h_1(t), \ldots, h_n(t)$ as well as all n^2 functions $a_{ij}(t)$ (with $1 \le i \le n$, $1 \le j \le n$) are defined and continuous. We can rewrite the system as a vector differential equation

$$\frac{d\mathbf{v}}{dt} = \mathbf{A}(t)\mathbf{v} + \mathbf{h}(t)$$

for the unknown vector $\mathbf{v} = (x_1, \ldots, x_n)$. Here

$$\mathbf{A}(t) = \begin{bmatrix} a_{11}(t) & \cdots & a_{1n}(t) \\ \vdots & & \vdots \\ a_{n1}(t) & \cdots & a_{nn}(t) \end{bmatrix} \quad \text{and} \quad \mathbf{h}(t) = (h_1(t), \ldots, h_n(t)) .$$

The system is called **homogeneous** if $\mathbf{h}(t) = \mathbf{0}$, that is, if $h_1(t) = \cdots = h_n(t) = 0$.

In the simplest possible case, $n = 1$, a homogeneous linear "system" has the form $\dfrac{dx}{dt} = a(t)x$, and by our results in Section 1.6, the solutions to this equation are given by $x(t) = e^{\int_0^t a(s)ds} x(0)$. At first, one might hope that this could be generalized to $\mathbf{v}(t) = e^{\int_0^t \mathbf{A}(s)ds} \mathbf{v}(0)$ [22] for the solutions of $\dfrac{d\mathbf{v}}{dt} = \mathbf{A}(t)\mathbf{v}$. Unfortunately, this doesn't work in general (see Section 4.5, Exercise 46), although it does work when $\mathbf{A}(t) = \mathbf{A}$ is a constant matrix: In that case, $\int_0^t \mathbf{A}(s)\, ds = t\mathbf{A}$ (see Exercise 14), and, as we saw in Section 4.5, $\mathbf{v}(t) = e^{t\mathbf{A}}\mathbf{v}(0)$ does give the solutions to $\dfrac{d\mathbf{v}}{dt} = \mathbf{A}\mathbf{v}$.

In the general case of a homogeneous system $\dfrac{d\mathbf{v}}{dt} = \mathbf{A}(t)\mathbf{v}$, one can show that there does exist a matrix $\mathbf{\Phi}(t)$, called the **fundamental matrix** of the system, such that the solutions of the system are given by $\mathbf{v}(t) = \mathbf{\Phi}(t)\mathbf{v}(0)$. The n columns of the fundamental matrix are basic solutions, from which all solutions can be found by taking linear combi-

---
[22] The **integral** of a matrix is defined entry by entry [Section 4.5, Exercise 33(a)].

SEC. 4.6 GENERAL LINEAR SYSTEMS

nations. Although these results may seem quite powerful, they are not too hard to prove, using the existence and uniqueness of solutions to initial value problems (see Exercise 19). Unfortunately, *finding* the fundamental matrix is usually very difficult [unless $\mathbf{A}(t) = \mathbf{A}$, in which case $\mathbf{\Phi}(t) = e^{t\mathbf{A}}$].

Now suppose that the homogeneous system $\dfrac{d\mathbf{v}}{dt} = \mathbf{A}(t)\mathbf{v}$ has somehow been solved. It should not come as a complete surprise that to solve the related inhomogeneous system $\dfrac{d\mathbf{v}}{dt} = \mathbf{A}(t)\mathbf{v} + \mathbf{h}(t)$, it is enough to find one solution of that related system (see Exercise 21). If we're able to guess the form of such a solution, we can use the method of undetermined coefficients. This doesn't involve any really new ideas, and we won't give examples here. Often it will be unreasonable to expect to guess the form of a solution, and we'll fall back on the method of variation of constants. Although this method can be carried out in general using the fundamental matrix described above (see Exercise 22), we will only apply it to inhomogeneous systems of the form

$$\frac{d\mathbf{v}}{dt} = \mathbf{A}\mathbf{v} + \mathbf{h}(t)$$

for which the matrix $\mathbf{A}$ has constant entries. You may wonder whether it is really worthwhile to allow $\mathbf{h}$ to be a function of t while insisting that $\mathbf{A}$ be constant, but this situation is actually quite common in applications.

EXAMPLE 4.6.1 In the configuration (setup) shown in Figure 4.6.1, one particle is attached to a spring while a second particle is attached to the first by means of a second spring. (Models involving masses and springs are used in many situations in physics, not only when describing the coupling of train cars and such.) To describe this setup, let x_1 and x_2 denote the respective displacements of the two particles from the position in which both springs are relaxed. Then, as in Example 2.1.3, the first spring will exert a force on the first particle that is (approximately) $-k_1 x_1$ for some spring constant $k_1 > 0$. Meanwhile, the

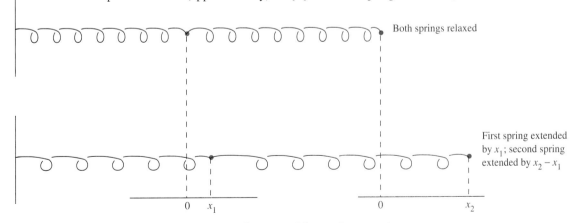

Figure 4.6.1. A particular configuration of two particles and two springs.

second spring is extended a distance $x_2 - x_1$ if $x_2 > x_1$, or compressed a distance $x_1 - x_2$ if $x_1 > x_2$. When the second spring is extended, it pulls the first particle to the right; thus the force of the second spring on the first particle is $k_2(x_2 - x_1)$, where $k_2 > 0$ is another spring constant. Therefore, the total force on the first particle from both springs will be $-k_1 x_1 + k_2(x_2 - x_1)$, while the force on the second particle from the second spring will be $-k_2(x_2 - x_1)$.

We'll assume that the only other force on either particle is an external force $F(t)$ on the second particle. In particular, both particles move without friction. Using Newton's second law, the second-order differential equations for their motion become

$$m_1 x_1'' = -k_1 x_1 + k_2(x_2 - x_1), \qquad m_2 x_2'' = -k_2(x_2 - x_1) + F(t).$$

If we introduce new unknown functions $v_1 = x_1'$, $v_2 = x_2'$ as usual (compare Example 4.3.2, p. 304), we can rewrite this system as

$$\begin{cases} x_1' = v_1 \\ v_1' = -\dfrac{k_1 + k_2}{m_1} x_1 + \dfrac{k_2}{m_1} x_2 \\ x_2' = v_2 \\ v_2' = \dfrac{k_2}{m_2} x_1 - \dfrac{k_2}{m_2} x_2 + \dfrac{F(t)}{m_2}. \end{cases}$$

This is in the form $\dfrac{d\mathbf{v}}{dt} = \mathbf{A}\mathbf{v} + \mathbf{h}(t)$, with unknown vector $\mathbf{v} = (x_1, v_1, x_2, v_2)$, matrix

$$\mathbf{A} = \begin{bmatrix} 0 & 1 & 0 & 0 \\ -\dfrac{k_1 + k_2}{m_1} & 0 & \dfrac{k_2}{m_1} & 0 \\ 0 & 0 & 0 & 1 \\ \dfrac{k_2}{m_2} & 0 & -\dfrac{k_2}{m_2} & 0 \end{bmatrix},$$

and vector (function) $\mathbf{h}(t) = \left(0, 0, 0, \dfrac{F(t)}{m_2}\right)$. ∎

As in Example 4.6.1, the term $\mathbf{h}(t)$ in the vector differential equation $\dfrac{d\mathbf{v}}{dt} = \mathbf{A}\mathbf{v} + \mathbf{h}(t)$ often arises from some external force or influence on an otherwise autonomous system. Accordingly, $\mathbf{h}(t)$ is often called a **forcing term** in the equation.

EXAMPLE 4.6.2 In Example 3.1.5, we had the system

$$\begin{cases} \dfrac{dI}{dt} = -\dfrac{R}{L} I - \dfrac{1}{L} V_C + \dfrac{E(t)}{L} \\ \dfrac{dV_C}{dt} = \dfrac{1}{C} I \end{cases}$$

SEC. 4.6 GENERAL LINEAR SYSTEMS

describing an *RLC* circuit with generator voltage $E(t)$. This system is again of the form $\frac{d\mathbf{v}}{dt} = \mathbf{A}\mathbf{v} + \mathbf{h}(t)$, with unknown vector $\mathbf{v} = (I, V_C)$, matrix

$$\mathbf{A} = \begin{bmatrix} -\frac{R}{L} & -\frac{1}{L} \\ \frac{1}{C} & 0 \end{bmatrix},$$

and forcing term $\mathbf{h}(t) = \left(\frac{E(t)}{L}, 0\right)$. Note that this particular system can probably be solved most easily by elimination (see Example 3.1.5). ∎

To solve the general inhomogeneous system $\frac{d\mathbf{v}}{dt} = \mathbf{A}\mathbf{v} + \mathbf{h}(t)$, we'll imitate the method used in Section 1.6 to solve the inhomogeneous equation $\frac{dx}{dt} + f(t)x = h(t)$. We start with the general solution to the homogeneous system $\frac{d\mathbf{v}}{dt} = \mathbf{A}\mathbf{v}$, which, as we saw in Section 4.5, is given by $\mathbf{v}(t) = e^{t\mathbf{A}}\mathbf{v}(0)$. We then replace the arbitrary constant vector $\mathbf{v}(0)$ by an unknown vector *function* $\mathbf{w}(t)$ and look for solutions to the inhomogeneous system that are of the form $\mathbf{v}(t) = e^{t\mathbf{A}}\mathbf{w}(t)$.

(N O T E : It can be shown [see Exercise 15(b)] that any vector $\mathbf{v}(t)$ can be written in this form, so, just as in Section 1.6, we're only making a substitution; compare p. 52, Note 1.)

By Exercises 33 and 42 of Section 4.5, differentiating our expression $\mathbf{v}(t) = e^{t\mathbf{A}}\mathbf{w}(t)$ yields

$$\frac{d\mathbf{v}}{dt} = \frac{d}{dt}(e^{t\mathbf{A}}) \cdot \mathbf{w}(t) + e^{t\mathbf{A}} \frac{d\mathbf{w}}{dt}$$
$$= \mathbf{A}e^{t\mathbf{A}}\mathbf{w}(t) + e^{t\mathbf{A}} \frac{d\mathbf{w}}{dt}.$$

Therefore, if we substitute $\mathbf{v}(t) = e^{t\mathbf{A}}\mathbf{w}(t)$ into $\frac{d\mathbf{v}}{dt} = \mathbf{A}\mathbf{v} + \mathbf{h}(t)$, we get

$$\mathbf{A}e^{t\mathbf{A}}\mathbf{w}(t) + e^{t\mathbf{A}} \frac{d\mathbf{w}}{dt} = \mathbf{A}e^{t\mathbf{A}}\mathbf{w}(t) + \mathbf{h}(t),$$

from which

$$e^{t\mathbf{A}} \frac{d\mathbf{w}}{dt} = \mathbf{h}(t)$$

and [see Exercise 15(c)]

$$\frac{d\mathbf{w}}{dt} = e^{-t\mathbf{A}}\mathbf{h}(t).$$

Now, assuming that we have found e^{tA}, the right-hand side of this equation can be computed quickly: To get e^{-tA}, just replace t by $-t$ in e^{tA}; then multiply this matrix by the vector $\mathbf{h}(t)$. This right-hand side can then in principle be integrated to get $\mathbf{w}(t)$ (there will be n integration constants, one for each component), and finally $\mathbf{v}(t) = e^{tA}\mathbf{w}(t)$ will give the solutions to our inhomogeneous system. However, even for relatively simple systems, the computations involved can be quite laborious.

EXAMPLE 4.6.3 For the inhomogeneous system

$$\begin{cases} \dfrac{dx_1}{dt} = 6x_1 - 12x_2 - 9x_3 + 3\sin t \\ \dfrac{dx_2}{dt} = 7x_1 - 14x_2 - 10x_3 - 2\cos t \\ \dfrac{dx_3}{dt} = -6x_1 + 12x_2 + 9x_3, \end{cases}$$

we have

$$\mathbf{A} = \begin{bmatrix} 6 & -12 & -9 \\ 7 & -14 & -10 \\ -6 & 12 & 9 \end{bmatrix} \quad \text{and} \quad \mathbf{h}(t) = \begin{bmatrix} 3\sin t \\ -2\cos t \\ 0 \end{bmatrix}.$$

The corresponding homogeneous system occurred in Example 4.5.6 (pp. 342–343), where we saw that

$$e^{tA} = \begin{bmatrix} 2 & 1 & 12 \\ 1 & 1 & 17 \\ 0 & -1 & -12 \end{bmatrix} \begin{bmatrix} 1 & 0 & 0 \\ 0 & e^{3t} & 0 \\ 0 & 0 & e^{-2t} \end{bmatrix} \begin{bmatrix} \frac{1}{2} & 0 & \frac{1}{2} \\ \frac{6}{5} & -\frac{12}{5} & -\frac{11}{5} \\ -\frac{1}{10} & \frac{1}{5} & \frac{1}{10} \end{bmatrix}$$

$$= \begin{bmatrix} 1 + \dfrac{6}{5}e^{3t} - \dfrac{6}{5}e^{-2t} & -\dfrac{12}{5}e^{3t} + \dfrac{12}{5}e^{-2t} & 1 - \dfrac{11}{5}e^{3t} + \dfrac{6}{5}e^{-2t} \\ \dfrac{1}{2} + \dfrac{6}{5}e^{3t} - \dfrac{17}{10}e^{-2t} & -\dfrac{12}{5}e^{3t} + \dfrac{17}{5}e^{-2t} & \dfrac{1}{2} - \dfrac{11}{5}e^{3t} + \dfrac{17}{10}e^{-2t} \\ -\dfrac{6}{5}e^{3t} + \dfrac{6}{5}e^{-2t} & \dfrac{12}{5}e^{3t} - \dfrac{12}{5}e^{-2t} & \dfrac{11}{5}e^{3t} - \dfrac{6}{5}e^{-2t} \end{bmatrix}.$$

Therefore, replacing t by $-t$, we get

$$e^{-tA} = \begin{bmatrix} 1 + \dfrac{6}{5}e^{-3t} - \dfrac{6}{5}e^{2t} & \cdots \\ \vdots & \ddots \end{bmatrix}$$

and

$$e^{-tA}\mathbf{h}(t) = \begin{bmatrix} 1 + \frac{6}{5}e^{-3t} - \frac{6}{5}e^{2t} & -\frac{12}{5}e^{-3t} + \frac{12}{5}e^{2t} & \cdots \\ \vdots & \ddots & \end{bmatrix} \begin{bmatrix} 3\sin t \\ -2\cos t \\ 0 \end{bmatrix}$$

$$= \begin{bmatrix} 3\sin t + \frac{18}{5}e^{-3t}\sin t - \frac{18}{5}e^{2t}\sin t + \frac{24}{5}e^{-3t}\cos t - \frac{24}{5}e^{2t}\cos t \\ \vdots \end{bmatrix}.$$

To finish the computation, we would have to integrate this vector to get $\mathbf{w}(t)$ and then multiply that vector by e^{tA} to get the general solution

$$\mathbf{v}(t) = \begin{bmatrix} x_1(t) \\ x_2(t) \\ x_3(t) \end{bmatrix}$$

to our system.[23] ∎

While such computations can be carried out using good integral tables and/or suitable computer software, all this work may not provide much added insight into the method. Thus in the computational exercises all systems have just two unknown functions, although it could be argued that such systems could actually be solved more quickly using the elimination method.

Here is one more example, in which an inhomogeneous system describing an electrical network is first set up and then solved. [The actual computation of the solutions starts at (∗) below.]

EXAMPLE 4.6.4 In the electrical network shown in Figure 4.6.2 (p. 360), the same four elements occur as in Example 3.1.5. However, this time the resistor and inductor are connected in parallel. If we let I denote the current through the left branch of the circuit while I_R and I_L denote the currents through the resistor and inductor respectively, we see that $I = I_R + I_L$. Also, if V is the voltage drop across the resistor (which equals the voltage drop across the inductor, since the two are in parallel), then $V = RI_R = L\frac{dI_L}{dt}$. Finally, if V_C is the voltage drop across the capacitor, then $E(t) = V_C + V$ (by Kirchhoff's second law; see p. 147) while

---

[23] One might also consider the method of undetermined coefficients, which would involve looking for a solution of the form

$$\begin{bmatrix} x_1 \\ x_2 \\ x_3 \end{bmatrix} = \begin{bmatrix} A\sin t + B\cos t \\ C\sin t + D\cos t \\ E\sin t + F\cos t \end{bmatrix}.$$

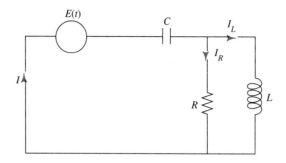

Figure 4.6.2. The electrical network of Example 4.6.4.

$I = C\dfrac{dV_C}{dt}$. So we have five equations in all, for the five unknown functions I, I_R, I_L, V_C, V:

$$\begin{cases} I = I_R + I_L & (1) \\ V = RI_R & (2) \\ V = L\dfrac{dI_L}{dt} & (3) \\ E(t) = V_C + V & (4) \\ I = C\dfrac{dV_C}{dt} & (5) \end{cases}$$

Since only the derivatives of I_L and V_C occur in these equations, it seems natural to try to get a system of differential equations involving only these two unknown functions. In fact, the others can be eliminated:

$$V = L\frac{dI_L}{dt} \Rightarrow \frac{dI_L}{dt} = \frac{1}{L}V = \frac{1}{L}[E(t) - V_C] \quad \text{[from (4)]}$$
$$= -\frac{1}{L}V_C + \frac{E(t)}{L};$$

$$I = C\frac{dV_C}{dt} \Rightarrow \frac{dV_C}{dt} = \frac{1}{C}I = \frac{1}{C}(I_R + I_L) \quad \text{[from (1)]}$$
$$= \frac{1}{C}\left(\frac{V}{R} + I_L\right) \quad \text{[from (2)]}$$
$$= \frac{1}{C}\left(\frac{E(t) - V_C}{R} + I_L\right) \quad \text{[from (4)]}$$
$$= \frac{1}{C}I_L - \frac{1}{RC}V_C + \frac{E(t)}{RC}.$$

SEC. 4.6 GENERAL LINEAR SYSTEMS

The system is now in the form

$$\frac{d}{dt}\begin{bmatrix} I_L \\ V_C \end{bmatrix} = \mathbf{A} \cdot \begin{bmatrix} I_L \\ V_C \end{bmatrix} + \mathbf{h}(t) \quad \text{with} \quad \mathbf{A} = \begin{bmatrix} 0 & -\frac{1}{L} \\ \frac{1}{C} & -\frac{1}{RC} \end{bmatrix}, \quad \mathbf{h}(t) = \begin{bmatrix} \frac{E(t)}{L} \\ \frac{E(t)}{RC} \end{bmatrix}.$$

To illustrate the computation, we'll choose $R = \frac{1}{3}$, $C = \frac{1}{2}$, $L = \frac{1}{4}$ [24] (arbitrary choices, except that they give pleasant numerical results) and also change the notation I_L, V_C for the unknown functions to x_1, x_2. We then have the system

(*)
$$\begin{cases} \dfrac{dx_1}{dt} = -4x_2 + 4E(t) \\ \dfrac{dx_2}{dt} = 2x_1 - 6x_2 + 6E(t) \end{cases}$$

to solve. As an answer we can expect to find formulas for $x_1(t)$ and $x_2(t)$ containing integrals involving the generator voltage $E(t)$.

First we find $e^{t\mathbf{A}}$ for $\mathbf{A} = \begin{bmatrix} 0 & -4 \\ 2 & -6 \end{bmatrix}$ by the method of Section 4.5. $\mathbf{A}$ is found to have eigenvectors $(1, 1)$ and $(2, 1)$ for the eigenvalues $\lambda = -4$ and $\lambda = -2$, respectively (see Exercise 16), so we have $\mathbf{T}^{-1}\mathbf{A}\mathbf{T} = \begin{bmatrix} -4 & 0 \\ 0 & -2 \end{bmatrix}$ for $\mathbf{T} = \begin{bmatrix} 1 & 2 \\ 1 & 1 \end{bmatrix}$. Therefore,

$$e^{t\mathbf{A}} = \mathbf{T} e^{\mathbf{T}^{-1}(t\mathbf{A})\mathbf{T}} \mathbf{T}^{-1} = \begin{bmatrix} 1 & 2 \\ 1 & 1 \end{bmatrix} \begin{bmatrix} e^{-4t} & 0 \\ 0 & e^{-2t} \end{bmatrix} \begin{bmatrix} -1 & 2 \\ 1 & -1 \end{bmatrix}$$

$$= \begin{bmatrix} -e^{-4t} + 2e^{-2t} & 2e^{-4t} - 2e^{-2t} \\ -e^{-4t} + e^{-2t} & 2e^{-4t} - e^{-2t} \end{bmatrix}.$$

Now, since our system has the form $\dfrac{d\mathbf{v}}{dt} = \mathbf{A}\mathbf{v} + \mathbf{h}(t)$ with $\mathbf{h}(t) = \begin{bmatrix} 4E(t) \\ 6E(t) \end{bmatrix}$, making the "change of variable" $\mathbf{v}(t) = e^{t\mathbf{A}}\mathbf{w}(t)$ will yield

$$\frac{d\mathbf{w}}{dt} = e^{-t\mathbf{A}}\mathbf{h}(t) = \begin{bmatrix} -e^{4t} + 2e^{2t} & 2e^{4t} - 2e^{2t} \\ -e^{4t} + e^{2t} & 2e^{4t} - e^{2t} \end{bmatrix} \begin{bmatrix} 4E(t) \\ 6E(t) \end{bmatrix}$$

$$= \begin{bmatrix} (8e^{4t} - 4e^{2t})E(t) \\ (8e^{4t} - 2e^{2t})E(t) \end{bmatrix}.$$

Thus we have

$$\mathbf{w}(t) = \begin{bmatrix} \int (8e^{4t} - 4e^{2t})E(t)\,dt \\ \int (8e^{4t} - 2e^{2t})E(t)\,dt \end{bmatrix}$$

---

[24] Of course, R, C, and L are assumed to be expressed in units compatible with those for V and I. For instance, in the mksA system, V is expressed in volts, I in amperes, R in ohms, C in farads, and L in henrys.

and finally

$$\begin{bmatrix} x_1(t) \\ x_2(t) \end{bmatrix} = e^{t\mathbf{A}}\mathbf{w}(t)$$

$$= \begin{bmatrix} (-e^{-4t} + 2e^{-2t})[\int (8e^{4t} - 4e^{2t})E(t)\,dt] + (2e^{-4t} - 2e^{-2t})[\int (8e^{4t} - 2e^{2t})E(t)\,dt] \\ (-e^{-4t} + e^{-2t})[\int (8e^{4t} - 4e^{2t})E(t)\,dt] + (2e^{-4t} - e^{-2t})[\int (8e^{4t} - 2e^{2t})E(t)\,dt] \end{bmatrix}.$$

Note that each indefinite integral is determined only up to a constant of integration, but that the constant must be chosen the same way each time that integral occurs. By adding constants C_1 and C_2, respectively, to

$$\int (8e^{4t} - 4e^{2t})E(t)\,dt \quad \text{and} \quad \int (8e^{4t} - 2e^{2t})E(t)\,dt,$$

we add the vector

$$\begin{bmatrix} (-e^{-4t} + 2e^{-2t})C_1 + (2e^{-4t} - 2e^{-2t})C_2 \\ (-e^{-4t} + e^{-2t})C_1 + (2e^{-4t} - e^{-2t})C_2 \end{bmatrix} = e^{t\mathbf{A}} \begin{bmatrix} C_1 \\ C_2 \end{bmatrix}$$

to the solution—no big surprise, since this is the general solution of the homogeneous equation. ∎

SUMMARY OF KEY CONCEPTS, RESULTS, AND TECHNIQUES

Fundamental matrix (p. 354): For a homogeneous linear system $\dfrac{d\mathbf{v}}{dt} = \mathbf{A}(t)\mathbf{v}$, the fundamental matrix is the matrix $\mathbf{\Phi}(t)$ such that the solutions of the system are given by $\mathbf{v}(t) = \mathbf{\Phi}(t)\mathbf{v}(0)$. The fundamental matrix is usually hard to find; however, if $\mathbf{A}(t) = \mathbf{A}$ is constant, then $\mathbf{\Phi}(t) = e^{t\mathbf{A}}$ (pp. 354–355 and Section 4.5).

Forcing term (p. 356): The term (vector function) $\mathbf{h}(t)$ in an inhomogeneous linear system $\dfrac{d\mathbf{v}}{dt} = \mathbf{A}(t)\mathbf{v} + \mathbf{h}(t)$.

Variation of constants (p. 357): To solve $\dfrac{d\mathbf{v}}{dt} = \mathbf{A}\mathbf{v} + \mathbf{h}(t)$ by variation of constants, put $\mathbf{v}(t) = e^{t\mathbf{A}}\mathbf{w}(t)$. This yields $\dfrac{d\mathbf{w}}{dt} = e^{-t\mathbf{A}}\mathbf{h}(t)$. To compute the right-hand side, find $e^{t\mathbf{A}}$ as in Section 4.5 and then get $e^{-t\mathbf{A}}$ by replacing t by $-t$. Finally, integrate $e^{-t\mathbf{A}}\mathbf{h}(t)$ (each component separately) to get $\mathbf{w}(t)$, then get the general solution from $\mathbf{v}(t) = e^{t\mathbf{A}}\mathbf{w}(t)$.

EXERCISES

Solve the following inhomogeneous linear systems using the method of variation of constants.

1. $\dfrac{dx_1}{dt} = x_1 - 3x_2, \quad \dfrac{dx_2}{dt} = 2x_1 - 6x_2 + 13\sin t.$

SEC. 4.6 GENERAL LINEAR SYSTEMS

2. $\dfrac{dx_1}{dt} = x_1 - x_2 + 8, \quad \dfrac{dx_2}{dt} = -x_1 + x_2 + 10 \cos t.$

3. $\dfrac{dx_1}{dt} = x_2 + 3e^t, \quad \dfrac{dx_2}{dt} = -4x_1 + 2e^t.$

4. $\dfrac{dx_1}{dt} = x_2 - e^{-2t}, \quad \dfrac{dx_2}{dt} = -x_1 + 3e^{-2t}.$

5. $\dfrac{dx_1}{dt} = x_2 + 3\cos^2 t, \quad \dfrac{dx_2}{dt} = -x_1 + 2.$

6. $\dfrac{dx_1}{dt} = -16x_2 - 5, \quad \dfrac{dx_2}{dt} = x_1 + \sin^2 4t.$

7. $\dfrac{dx_1}{dt} = -x_1 + 2x_2 + 4e^t, \quad \dfrac{dx_2}{dt} = 3x_1 + 4x_2 - 2e^{3t}.$

8. $\dfrac{dx_1}{dt} = 5x_1 + x_2 + e^{-t} - 4, \quad \dfrac{dx_2}{dt} = -6x_1 + e^{8t}.$

9. $\dfrac{dx_1}{dt} = -3x_1 + x_2 + t^2, \quad \dfrac{dx_2}{dt} = -9x_1 + 3x_2 + 10.$

10. $\dfrac{dx_1}{dt} = 4x_1 - x_2 - 3t, \quad \dfrac{dx_2}{dt} = 16x_1 - 4x_2 - \dfrac{6}{t}.$

11. Suppose that the particles in Example 4.6.1 were suspended from the ceiling (by the springs) and gravity was taken into account. Set up a system of the form $\dfrac{d\mathbf{v}}{dt} = \mathbf{A}\mathbf{v} + \mathbf{h}(t)$ that would describe this situation.

12. Set up a system of the form $\dfrac{d\mathbf{v}}{dt} = \mathbf{A}\mathbf{v} + \mathbf{h}(t)$ which describes the configuration of Figure 4.6.3, given an external force $F(t)$ on the third particle.

Mass m_1 Mass m_2 Mass m_3

Figure 4.6.3. A configuration of three particles and three springs.

13. Suppose that the particles in Exercise 12 were suspended from the ceiling (by the springs) and gravity was taken into account. Set up a system of the form $\dfrac{d\mathbf{v}}{dt} = \mathbf{A}\mathbf{v} + \mathbf{h}(t)$ describing this situation.

14. Show that if $\mathbf{A}(t) = \mathbf{A}$ is a constant matrix, then $\displaystyle\int_0^t \mathbf{A}(s)\,ds = t\mathbf{A}$.

15. (a) Show that for any $n \times n$ matrix $\mathbf{A}$, $e^{t\mathbf{A}}e^{-t\mathbf{A}} = \mathbf{I} = e^{-t\mathbf{A}}e^{t\mathbf{A}}$.
    (b) Show that any vector $\mathbf{v}(t)$ can be written in the form $\mathbf{v}(t) = e^{t\mathbf{A}}\mathbf{w}(t)$ for a suitable vector $\mathbf{w}(t)$.
    [*Hint:* What would $\mathbf{w}(t)$ be?]
    (c) Show that $e^{t\mathbf{A}} \dfrac{d\mathbf{w}}{dt} = \mathbf{h}(t)$ implies $\dfrac{d\mathbf{w}}{dt} = e^{-t\mathbf{A}}\mathbf{h}(t)$.

16. Find the eigenvalues and eigenvectors of $\mathbf{A} = \begin{bmatrix} 0 & -4 \\ 2 & -6 \end{bmatrix}$ and check them against those given in Example 4.6.4.

17. (a) Find a system of the form $\dfrac{d\mathbf{v}}{dt} = \mathbf{A}\mathbf{v} + \mathbf{h}(t)$ that describes the electrical network shown in Figure 4.6.4.
    (b) Solve the system for $R = \dfrac{3}{8}$, $C = \dfrac{1}{3}$, $L = \dfrac{1}{4}$. [Your answer will include integrals involving the unknown generator voltage $E(t)$.]

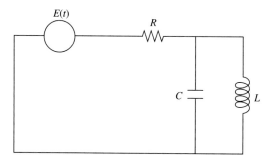

Figure 4.6.4. The electrical network of Exercise 17.

18. (a) Find a system of the form $\dfrac{d\mathbf{v}}{dt} = \mathbf{A}\mathbf{v} + \mathbf{h}(t)$ that describes the electrical network shown in Figure 4.6.5 (p. 365).
    (b) Solve the system for $R = 0.7$, $C = 0.1$, $L = 1$. [Your answer will include integrals involving the unknown generator voltage $E(t)$.]

19. In this exercise we consider solutions of the vector differential equation
    $$\dfrac{d\mathbf{v}}{dt} = \mathbf{A}(t)\mathbf{v}.$$
    (a) Show that if $\mathbf{v}_1(t)$ and $\mathbf{v}_2(t)$ are solutions, then so is $\mathbf{v}_1(t) + \mathbf{v}_2(t)$.
    (b) Show that if $\mathbf{v}(t)$ is a solution, then so is $\alpha \mathbf{v}(t)$, where α is a constant (number).
    *(c) Let $\mathbf{e}_1, \ldots, \mathbf{e}_n$ be the n vectors in n-space that have one component 1 and all others 0:
    $$\mathbf{e}_1 = (1, 0, 0, \ldots, 0), \; \mathbf{e}_2 = (0, 1, 0, \ldots, 0), \ldots, \mathbf{e}_n = (0, 0, 0, \ldots, 1).$$

SEC. 4.6 GENERAL LINEAR SYSTEMS

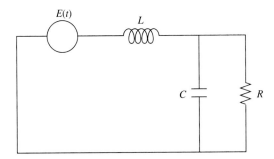

Figure 4.6.5. The electrical network of Exercise 18.

[These **standard basis vectors** also appeared in Exercise 23(b) from Section 4.4.] Let $w_1(t)$ be defined as the unique solution of the initial value problem $\frac{d\mathbf{v}}{dt} = \mathbf{A}(t)\mathbf{v}$, $\mathbf{v}(0) = \mathbf{e}_1$, let $\mathbf{w}_2(t)$ be the unique solution of $\frac{d\mathbf{v}}{dt} = \mathbf{A}(t)\mathbf{v}$, $\mathbf{v}(0) = \mathbf{e}_2$, and so on. Let $\mathbf{v}(t)$ be any solution of $\frac{d\mathbf{v}}{dt} = \mathbf{A}(t)\mathbf{v}$; suppose that $\mathbf{v}(0) = (c_1, c_2, \ldots, c_n) = c_1\mathbf{e}_1 + c_2\mathbf{e}_2 + \cdots + c_n\mathbf{e}_n$. Show that
$$\mathbf{v}(t) = c_1\mathbf{w}_1(t) + c_2\mathbf{w}_2(t) + \cdots + c_n\mathbf{w}_n(t).$$
[*Hint:* Show that both sides are solutions of the same initial value problem.]

*(d) Finally, let $\mathbf{\Phi}(t)$ be the square matrix whose columns are $\mathbf{w}_1(t), \mathbf{w}_2(t), \ldots, \mathbf{w}_n(t)$. Show that $\mathbf{\Phi}(t)$ is a fundamental matrix, that is, that the solutions of $\frac{d\mathbf{v}}{dt} = \mathbf{A}(t)\mathbf{v}$ are given by $\mathbf{v}(t) = \mathbf{\Phi}(t)\mathbf{v}(0)$.
[*Hint:* Let $\mathbf{v}(0) = (c_1, \ldots, c_n)$. Use part (c) along with Exercise 23(b) from Section 4.4.]

*20. Let $\mathbf{\Phi}(t)$ be a fundamental matrix for the system $\frac{d\mathbf{v}}{dt} = \mathbf{A}(t)\mathbf{v}$.
   (a) Show that for any vector $\mathbf{v}_0$, $\mathbf{\Phi}'(t)\mathbf{v}_0 = \mathbf{A}(t)\mathbf{\Phi}(t)\mathbf{v}_0$.
   (b) Show that $\mathbf{\Phi}'(t) = \mathbf{A}(t)\mathbf{\Phi}(t)$. [Use Exercise 23(b) from Section 4.4.]

21. Consider the inhomogeneous system $\frac{d\mathbf{v}}{dt} = \mathbf{A}(t)\mathbf{v} + \mathbf{h}(t)$ and the related homogeneous system $\frac{d\mathbf{v}}{dt} = \mathbf{A}(t)\mathbf{v}$.
   (a) Show that if $\mathbf{v}(t)$ is a solution of the inhomogeneous system and $\mathbf{w}(t)$ is a solution of the homogeneous system, then $\mathbf{v}(t) + \mathbf{w}(t)$ is a solution of the inhomogeneous system.
   (b) Show that if one can find *one* solution $\mathbf{v}_1(t)$ of the inhomogeneous system, one can then find *all* solutions of that system by adding each of the solutions of the homogeneous system to $\mathbf{v}_1(t)$.

22. Suppose $\Phi(t)$ is a fundamental matrix for the system $\dfrac{d\mathbf{v}}{dt} = \mathbf{A}(t)\mathbf{v}$. Show that $\mathbf{v}(t) = \Phi(t)\mathbf{w}(t)$ is a solution to the inhomogeneous system $\dfrac{d\mathbf{v}}{dt} = \mathbf{A}(t)\mathbf{v} + \mathbf{h}(t)$ if and only if $\Phi(t)\mathbf{w}'(t) = \mathbf{h}(t)$. [Use Exercise 20(b).] Explain how any inhomogeneous system of the form $\dfrac{d\mathbf{v}}{dt} = \mathbf{A}(t)\mathbf{v} + \mathbf{h}(t)$ can be solved once the fundamental matrix $\Phi(t)$ for the homogeneous system $\dfrac{d\mathbf{v}}{dt} = \mathbf{A}(t)\mathbf{v}$ is found.

FURTHER READING

For more about fundamental matrices, especially in the case that $\mathbf{A}(t)$ is periodic, see Chapter 3 in Coddington and Levinson, *Theory of Ordinary Differential Equations* (McGraw-Hill, 1955), especially Section 5.

CHAPTER 5

Power Series Methods

5.1 INTRODUCTION

We now resume the study of linear second-order differential equations. In Chapter 2 we dealt mainly with the constant-coefficient case; in this chapter the emphasis will be on the case in which the coefficients are polynomials or rational functions.[1] As you might expect, this case is harder, and often we won't be able to get a "neat" form for the solutions. Still, although the techniques of this chapter are a bit more difficult, they are quite important; in Section 7.4 we'll see how they can eventually be used to construct solutions to a variety of partial differential equations that arise in applications, particularly in physics.

Throughout this chapter, we'll assume that we have a homogeneous linear second-order equation

$$\frac{d^2y}{dx^2} + p(x)\frac{dy}{dx} + q(x)y = 0$$

for the unknown function y of the variable x.[2] [We saw in Section 2.8 that if this equation can be solved, then so can any inhomogeneous equation $\frac{d^2y}{dx^2} + p(x)\frac{dy}{dx} + q(x)y = h(x)$ with the same coefficients.]

---

[1] Quotients of polynomials are called rational functions, just as quotients of integers are called rational numbers.

[2] Of course, we could also let $x(t)$ be the unknown function, and the equation would then have the form $x'' + p(t)x' + q(t)x = 0$. However, in the applications mentioned, the variable usually *isn't* the time, and in Section 5.5 it will be more convenient if the variable isn't called t.

Now suppose that we are given the initial value problem

$$\frac{d^2y}{dx^2} + p(x)\frac{dy}{dx} + q(x)y = 0, \qquad y(0) = y_0, \quad y'(0) = z_0,$$

where y_0 and z_0 are given numbers. We know from Theorem 2.2.2 that this initial value problem has a unique solution provided p and q are continuous on an interval containing $x = 0$ (in its interior), and that the solution $y(x)$ will be defined on this same interval. What else can we say about this solution $y(x)$? We can use the equation $y'' + p(x)y' + q(x)y = 0$ by substituting $x = 0$ to get

$$y''(0) = -p(0)y'(0) - q(0)y(0) = -p(0)z_0 - q(0)y_0.$$

That is, not only do we know $y(0)$ and $y'(0)$ (which were given in the initial conditions), but we know the *second* derivative of y at $x = 0$, as well. We can also compute all the higher derivatives of the function y at $x = 0$. For instance, first rewriting the equation as $y'' = -p(x)y' - q(x)y$ and then differentiating yields

$$y^{(3)} = -p(x)y'' - p'(x)y' - q(x)y' - q'(x)y; \quad \text{therefore,}$$

$$y^{(3)}(0) = -p(0)y''(0) - [p'(0) + q(0)]y'(0) - q'(0)y(0).$$

By the above we have $y''(0) = -p(0)z_0 - q(0)y_0$ along with $y'(0) = z_0$, $y(0) = y_0$; substituting all these in, we get

$$y^{(3)}(0) = [(p(0))^2 - p'(0) - q(0)]z_0 + [p(0)q(0) - q'(0)]y_0.$$

EXAMPLE 5.1.1 $(x^2 - 1)y'' + 4y' - 2y = 0, \quad y(0) = 1, \quad y'(0) = 1.$

First we divide through by $x^2 - 1$ to obtain

$$y'' = \frac{-4}{x^2 - 1}y' + \frac{2}{x^2 - 1}y. \qquad (*)$$

In particular, $y''(0) = 4y'(0) - 2y(0) = 2$. Also, differentiating $(*)$, we have

$$y^{(3)} = \frac{8x}{(x^2 - 1)^2}y' - \frac{4}{x^2 - 1}y'' - \frac{4x}{(x^2 - 1)^2}y + \frac{2}{x^2 - 1}y',$$

so $y^{(3)}(0) = 4y''(0) - 2y'(0) = 6$. In this way, we can compute as many of the higher derivatives of y at $x = 0$ as we want, *without ever having found the actual solution y.* Note, however, that with each successive differentiation the computation will become more involved. ∎

Recall from calculus that the **Taylor polynomials** of the function $y(x)$ near $x = 0$ are those polynomials that give the best possible approximation of y near $x = 0$ by a polynomial of a certain degree.[3] For instance, the Taylor polynomial "of degree 0" is the

---

[3] That is, the only approximation by such a polynomial for which the error satisfies the inequality below.

constant $y(0)$, while the Taylor polynomial "of degree 1" is nothing but the equation $y(0) + y'(0)x$ of the tangent line at $x = 0$ (compare Fact 1 on p. 2). The Taylor polynomial "of degree n," or **nth-order** Taylor polynomial, is given by

$$y(0) + y'(0)x + \frac{y''(0)}{2!}x^2 + \cdots + \frac{y^{(n)}(0)}{n!}x^n .\;^4$$

Note that the coefficients of this polynomial can be found right away (by dividing by the successive factorials) from the successive derivatives of the function y at $x = 0$. This may suggest to you a way of finding an approximate solution to the differential equation: Given the initial value problem $\frac{d^2y}{dx^2} + p(x)\frac{dy}{dx} + q(x)y = 0$, $y(0) = y_0$, $y'(0) = z_0$, we have seen how to compute the higher derivatives of y at $x = 0$; compute an appropriate number of them and use these to find the Taylor polynomial of appropriate degree for y. This Taylor polynomial will then be an approximate solution to the initial value problem.

The trouble with this "method" is that it is usually impossible to tell what is an "appropriate" number of higher derivatives to compute. It is true that there are estimates from calculus for the error we make if we replace $y(x)$ by its nth-order Taylor polynomial. That is, if the **remainder term** $R_n(x)$ [5] is defined by

$$y(x) = y(0) + y'(0)x + \cdots + \frac{y^{(n)}(0)}{n!}x^n + R_n(x),$$

then there are estimates for $|R_n(x)|$. For instance, $|R_n(x)| \leq M_n \frac{|x|^{n+1}}{(n+1)!}$, where M_n is a constant such that $|y^{(n+1)}(\xi)| \leq M_n$ for all ξ in the interval between 0 and x. The problem is that all these estimates, like the one just given, involve the behavior of the next higher derivative $y^{(n+1)}$, *not just at the point* 0 but on an interval containing 0. We will have no way of saying anything about $y^{(n+1)}$ on such an interval if we only know the nth-order Taylor polynomial of $y(x)$, and so we will have no way of telling how good our approximation might be! [Note that the $(n+1)$st derivative of that Taylor polynomial will be identically zero—and thus not give any information—since the polynomial has degree (at most) n.]

For instance, for the initial value problem from Example 5.1.1,

$$(x^2 - 1)y'' + 4y' - 2y = 0, \quad y(0) = 1, \quad y'(0) = 1,$$

we know that the third-order Taylor polynomial of the solution will be $1 + x + x^2 + x^3$, since we have $y(0) = 1$, $y'(0) = 1$, $y''(0) = 2$, $y^{(3)}(0) = 6 = 3!$. However, it is not too useful to conclude that $1 + x + x^2 + x^3$ is an approximate solution to the differential equation, since it is unclear how good the approximation is.

---

[4] The degree of this polynomial really is n, unless $y^{(n)}(0) = 0$, in which case it is less than n.
[5] Notations for this remainder term vary widely; $R_{n+1}(x)$, $R_n(y, x)$, and other variations occur. We won't use the notation much.

Still, we can salvage our idea of "solution by repeated differentiation" in some cases, as follows. Suppose that we could compute *all* the higher derivatives of y at $x = 0$. Then we would know all the Taylor polynomials of y, as well as the **Taylor series**

$$\sum_{n=0}^{\infty} \frac{y^{(n)}(0)}{n!} x^n = y(0) + y'(0)x + \frac{y^{(2)}(0)}{2!} x^2 + \cdots + \frac{y^{(n)}(0)}{n!} x^n + \cdots,$$

whose partial sums are the Taylor polynomials. This infinite series may not be convergent; even if it is convergent, the sum may not be $y(x)$ (see Exercise 16). However, there are large classes of functions, including the exponential, trigonometric, and inverse trigonometric functions, which are given by their Taylor series, at least on an interval around $x = 0$. We could hope that $y(x)$ was such a function, and if so we would have an expression for $y(x)$ as a convergent power series.[6]

EXAMPLE 5.1.2

$y'' - y = 0$, $y(0) = 1$, $y'(0) = 1$.

We easily see that if $y(x)$ is the solution to this initial value problem, then *all* the higher derivatives of y at $x = 0$ equal 1:

$$y'' = y \implies y''(0) = y(0) = 1;$$

$$y^{(3)} = y' \implies y^{(3)}(0) = y'(0) = 1;$$

$$y^{(4)} = y'' \implies y^{(4)}(0) = y''(0) = 1; \text{ and so on.}$$

Assuming that y is given by its Taylor series, we have

$$y = y(0) + y'(0)x + \frac{y''(0)}{2!} x^2 + \cdots + \frac{y^{(n)}(0)}{n!} x^n + \cdots$$

$$= 1 + x + \frac{x^2}{2!} + \cdots + \frac{x^n}{n!} + \cdots = \sum_{n=0}^{\infty} \frac{x^n}{n!}.$$

You may recognize this as the Taylor series expansion for e^x (see Section 5.2). Of course, we could also have found the solution $y = e^x$ to this particular initial value problem by the methods of Chapter 2. ∎

Usually, it is not practical to compute the higher derivatives of y at $x = 0$ one at a time; as in Example 5.1.1, the computations tend to expand as more derivatives are taken. To get a first look at the method that is actually used, let's work through the initial value problem from that example.

---

[6] Basic information on power series will be reviewed in Section 5.2.

EXAMPLE 5.1.3 $(x^2 - 1)y'' + 4y' - 2y = 0$, $y(0) = 1$, $y'(0) = 1$.

To begin with, we will assume that the solution $y(x)$ is given by its Taylor series $\sum_{n=0}^{\infty} \frac{y^{(n)}(0)}{n!} x^n$. We'll abbreviate the coefficients by $a_n = \frac{y^{(n)}(0)}{n!}$ [in particular, they include $a_0 = y(0) = 1$, $a_1 = y'(0) = 1$], so we have

$$y = \sum_{n=0}^{\infty} a_n x^n = a_0 + a_1 x + a_2 x^2 + \cdots + a_n x^n + \cdots.$$

If we differentiate both sides, we get

$$y' = a_1 + 2a_2 x + 3a_3 x^2 + \cdots + na_n x^{n-1} + (n+1)a_{n+1} x^n + \cdots$$

and then

$$y'' = 2a_2 + 6a_3 x + 12a_4 x^2 + \cdots + n(n-1)a_n x^{n-2} + (n+1)na_{n+1} x^{n-1}$$
$$+ (n+2)(n+1)a_{n+2} x^n + \cdots.$$

The idea is to substitute these power series for y, y', and y'' into the differential equation $(x^2 - 1)y'' + 4y' - 2y = 0$. To do so, we first compute

$$x^2 y'' = 2a_2 x^2 + \cdots + n(n-1)a_n x^n + \cdots$$

and subtract the series above for y'' to get

$$(x^2 - 1)y'' = -2a_2 - 6a_3 x + [2a_2 - 12a_4]x^2 + \cdots$$
$$+ [n(n-1)a_n - (n+2)(n+1)a_{n+2}]x^n + \cdots.$$

(NOTE: In the long run we won't write out all the terms; for instance, we could just have written $(x^2 - 1)y'' = \sum_{n=0}^{\infty} [n(n-1)a_n - (n+2)(n+1)a_{n+2}]x^n$. More on how to streamline computations of this sort in Sections 5.2 and 5.3; for now, it might be a good idea to put in more detail than strictly necessary.)

Now we combine our power series for $(x^2 - 1)y''$, y', and y, and we get

$(x^2 - 1)y'' + 4y' - 2y$
$= -2a_2 - 6a_3 x + [2a_2 - 12a_4]x^2 + \cdots + [n(n-1)a_n - (n+2)(n+1)a_{n+2}]x^n + \cdots$
$ + 4a_1 + 8a_2 x + 12a_3 x^2 + \cdots + 4(n+1)a_{n+1} x^n + \cdots$
$ - 2a_0 + 2a_1 x - 2a_2 x^2 - \cdots - 2a_n x^n - \cdots$
$= [-2a_0 + 4a_1 - 2a_2] + [-2a_1 + 8a_2 - 6a_3]x + [12a_3 - 12a_4]x^2 + \cdots$
$ + [(n(n-1) - 2)a_n + 4(n+1)a_{n+1} - (n+2)(n+1)a_{n+2}]x^n + \cdots.$

Here we have combined all terms with the same power of x. This power series must be zero, and it will be zero if all its coefficients are zero, that is, if

$$-2a_0 + 4a_1 - 2a_2 = 0$$
$$-2a_1 + 8a_2 - 6a_3 = 0$$
$$12a_3 - 12a_4 = 0$$
$$\vdots$$

$$[n(n-1) - 2]a_n + 4(n+1)a_{n+1} - (n+2)(n+1)a_{n+2} = 0. \qquad (*)$$

Once again, the first three equations are just the special cases $n = 0$, $n = 1$, $n = 2$ of the general equation $(*)$.

Since we have $a_0 = y(0) = 1$, $a_1 = y'(0) = 1$ from the given initial conditions, we can use the equation $-2a_0 + 4a_1 - 2a_2 = 0$ to find $a_2 = 1$. Once we have a_2, we can use $-2a_1 + 8a_2 - 6a_3 = 0$ to find a_3, then $12a_3 - 12a_4 = 0$ to find a_4, and so on. In general, once we have a_n and a_{n+1}, we can use $(*)$ to find

$$a_{n+2} = \frac{4}{n+2} a_{n+1} + \frac{n(n-1) - 2}{(n+1)(n+2)} a_n.$$

A relation of this kind between the coefficients of a power series (or the members of any other sequence of numbers), in which each coefficient is given in terms of one or more of the preceding coefficients, is called a **recurrence relation**.

In our example, once we find that $a_2 = 1$, $a_3 = 1$, $a_4 = 1$ (check this!), it is natural to suspect a pattern: It looks like $a_n = 1$ for all n. Once this pattern is suspected, we can use the recurrence relation to check whether it is correct, as follows. Suppose that for some n, $a_n = a_{n+1} = 1$. Then the recurrence relation yields

$$a_{n+2} = \frac{4}{n+2} + \frac{n(n-1) - 2}{(n+1)(n+2)} = 1.$$

In other words, whenever two coefficients are equal to 1, so is the next coefficient. This means that, indeed, all the coefficients will be 1, and our solution will be

$$y = 1 + x + x^2 + \cdots + x^n + \cdots.$$

This is a **geometric series** with ratio x, convergent for $|x| < 1$; its sum is $y = \dfrac{1}{1-x}$. We can now check directly that $y = \dfrac{1}{1-x}$ is indeed the solution of our initial value problem (see Exercise 13). Note that this solution is defined not only for $|x| < 1$, but for all x with $x \neq 1$; however, the Taylor series $y = 1 + x + x^2 + \cdots$ converges only for $|x| < 1$. ∎

In the next section we'll review some facts, which were used at various points in this example, about power series; we'll continue to use this material as we look more generally at power series solutions to linear second-order differential equations.[7] [Note that in Example 5.1.3, "the end justified the means": We found a solution $y = \dfrac{1}{1-x}$ that could be checked directly. In general, though, it will not be so easy to find a pattern for the coefficients a_n, and it may not be possible to write the solution in closed form (in our case, as $\dfrac{1}{1-x}$) rather than as a power series.]

SUMMARY OF KEY CONCEPTS, RESULTS, AND TECHNIQUES

Given an initial value problem $\dfrac{d^2y}{dx^2} + p(x)\dfrac{dy}{dx} + q(x)y = 0$, $y(0) = y_0$, $y'(0) = z_0$, we can find $y''(0)$ by substituting $x = 0$, and $y^{(3)}(0)$, $y^{(4)}(0)$, and so on, by first differentiating and then substituting $x = 0$ (p. 368).

Taylor polynomials (p. 369): The nth-order Taylor polynomial of $y(x)$ near $x = 0$ is
$$y(0) + y'(0)x + \frac{y''(0)}{2!}x^2 + \cdots + \frac{y^{(n)}(0)}{n!}x^n.$$

Taylor series (p. 370): The Taylor series for $y(x)$ near $x = 0$ is the infinite series whose partial sums are the Taylor polynomials, that is, $\sum_{n=0}^{\infty} \dfrac{y^{(n)}(0)}{n!}x^n$. Many important functions are given by their Taylor series.

Geometric series with ratio x (p. 372):
$$\sum_{n=0}^{\infty} x^n = 1 + x + x^2 + \cdots + x^n + \cdots = \frac{1}{1-x} \qquad (|x| < 1).$$

Recurrence relation for a sequence (a_n) (p. 372): A relation between the various a_n, which gives each one in terms of one or more of the previous ones.

To solve an initial value problem of the form
$$\frac{d^2y}{dx^2} + p(x)\frac{dy}{dx} + q(x)y = 0, \qquad y(0) = y_0, \quad y'(0) = z_0,$$
we can try substituting
$$y = \sum_{n=0}^{\infty} a_n x^n = a_0 + a_1 x + a_2 x^2 + \cdots + a_n x^n + \cdots$$
to find a recurrence relation for the coefficients a_n. From the initial conditions we get $a_0 = y_0$, $a_1 = z_0$, and we can then use these values to find a_2, a_3, and so on. (See Example 5.1.3.)

---

[7] There is no theoretical reason not to develop such a theory for linear higher-order equations as well. Since most applications involve second-order equations and the computations for these are already quite extensive, we won't pursue this further. On the other hand, linear first-order equations can be solved by the methods of Section 1.6.

EXERCISES

In each of the following cases, find $y''(0)$ and $y^{(3)}(0)$.

1. $y'' + (\sin x)y' + (\cos x)y = 0$, $y(0) = 1$, $y'(0) = -1$.
2. $y'' - (\cos x)y' + (\sin x)y = 0$, $y(0) = 1$, $y'(0) = 2$.
3. $e^x y'' + (1-x)y' - 3xy = 0$, $y(0) = 1$, $y'(0) = 0$.
4. $e^x y'' - (x+3)y' + 2xy = 0$, $y(0) = -1$, $y'(0) = 1$.

5. Consider the initial value problem $y'' - 4y = 0$, $y(0) = 1$, $y'(0) = 2$.
   (a) Without solving the differential equation, find $y''(0)$, $y^{(3)}(0)$, and $y^{(4)}(0)$.
   (b) Find an expression for $y^{(n)}(0)$.
   (c) Assuming that y is given by its Taylor series, write down a power series for y. Do you recognize this series?
   (d) Check your results from part (c) by solving the initial value problem by the methods of Chapter 2.

6. Consider the initial value problem $y'' - y = 0$, $y(0) = 1$, $y'(0) = -1$.
   (a) Without solving the differential equation, find $y''(0)$, $y^{(3)}(0)$, and $y^{(4)}(0)$.
   (b) Find an expression for $y^{(n)}(0)$.
   (c) Assuming that y is given by its Taylor series, write down a power series for y. Do you recognize this series?
   (d) Check your results from part (c) by solving the initial value problem by the methods of Chapter 2.

In each of the following cases:

(a) Compute a_2, a_3, and a_4, then guess a formula for a_n.
(b) Use the recurrence relation to check whether your formula is correct.

7. $a_{n+2} = 4a_{n+1} - 3a_n$, $a_0 = 1$, $a_1 = 3$.
8. $a_{n+2} = 2a_{n+1} - a_n$, $a_0 = 1$, $a_1 = 2$.
9. $a_{n+2} = 4a_{n+1} - 4a_n$, $a_0 = 1$, $a_1 = 2$.
*10. $a_{n+2} = 4a_{n+1} - 4a_n$, $a_0 = 0$, $a_1 = 2$.
11. $a_{n+2} = na_{n+1} + (2n+2)a_n$, $a_0 = 1$, $a_1 = 1$.
12. $a_{n+2} = (n+3)a_{n+1} - (n+1)a_n$, $a_0 = 1$, $a_1 = 1$.

13. (See p. 372.) Check directly that $y = \dfrac{1}{1-x}$ is the solution of the initial value problem $(x^2 - 1)y'' + 4y' - 2y = 0$, $y(0) = 1$, $y'(0) = 1$.

14. Consider the initial value problem
$$(x^2 - 1)y'' - 4y' - 2y = 0, \quad y(0) = 1, \quad y'(0) = -1.$$
(a) Assuming that $y = \sum_{n=0}^{\infty} a_n x^n$, find a recurrence relation for the coefficients a_n.

(b) Use the given initial conditions to find a_2, a_3, a_4, and a_5, and guess a formula for a_n.

(c) Use the recurrence relation to check whether your formula is correct.

(d) Write the solution $y = \sum_{n=0}^{\infty} a_n x^n$ in closed form. For what x is the function you get defined? For what x does the power series actually converge?

(e) Check directly that your function from part (d) is the solution of the given problem.

15. Consider the initial value problem
$$(1 + x)y'' + (1 - x)y' - y = 0, \qquad y(0) = -1, \quad y'(0) = 1.$$

(a) Show that substitution of $y = \sum_{n=0}^{\infty} a_n x^n$ leads to the recurrence relation
$$(n + 2)a_{n+2} + (n + 1)a_{n+1} - a_n = 0.$$

(b) Compute a_2, a_3, a_4, a_5 from the given initial conditions, and guess a formula for a_n.

(c) Use the recurrence relation to show that your formula is correct.

(d) Write the solution $y = \sum_{n=0}^{\infty} a_n x^n$ in closed form. For what x is the function you get defined? For what x does the power series actually converge?

(e) Check directly that your function from part (d) is the solution of the given problem.

*16. Consider the function y given by $y(x) = \begin{cases} e^{-1/x^2} & (x \neq 0) \\ 0 & (x = 0). \end{cases}$

(a) Show that y is differentiable at $x = 0$. [*Hint*: Use the definition of the derivative (p. 2); to compute $\lim_{h \to 0} \dfrac{1}{h} e^{-1/h^2}$, put $k = \dfrac{1}{h}$. Note that k approaches ∞ or $-\infty$ according to whether h approaches 0 from above or from below.]

(b) Show that y is differentiable everywhere and that its derivative is given by
$$y'(x) = \begin{cases} \dfrac{2}{x^3} e^{-1/x^2} & (x \neq 0) \\ 0 & (x = 0). \end{cases}$$

(c) Show that y' is differentiable everywhere and that
$$y''(x) = \begin{cases} \left(\dfrac{4}{x^6} - \dfrac{6}{x^4}\right) e^{-1/x^2} & (x \neq 0) \\ 0 & (x = 0). \end{cases}$$

It can be shown that *all* the higher derivatives of y exist and are zero for $x = 0$.

Therefore, although the Taylor series

$$y(0) + y'(0)x + \frac{y''(0)}{2!}x^2 + \cdots + \frac{y^{(n)}(0)}{n!}x^n + \cdots$$

converges for all x, it is identically zero. In particular,

$$y(x) \neq \sum_{n=0}^{\infty} \frac{y^{(n)}(0)}{n!} x^n \quad \text{for } x \neq 0.$$

(d) Show that $y(x)$ satisfies the differential equation $x^6 y'' + 3x^5 y' - 4y = 0$.

17. Show that if we are "inspired" by the constant-coefficient case from Chapter 2 to try to solve $\dfrac{d^2 y}{dx^2} + p(x)\dfrac{dy}{dx} + q(x)y = 0$ by looking for a solution of the form $y = e^{f(x)}$, we get the nonlinear second-order equation (which is usually hopeless) $f''(x) + [f'(x)]^2 + p(x)f'(x) + q(x) = 0$ for $f(x)$.

5.2 POWER SERIES

While much of the material in this section should be familiar from calculus, some of it goes beyond calculus material. The computational technique called "shifting the index of summation" (pp. 381–382) will be especially useful in the rest of this chapter. The material on multiplying and dividing power series on pages 384 and 389 will only come up in some of the more difficult computations (in exercises) toward the end of the chapter, and you might decide to skip it on first reading. However, the technique for finding the radius of convergence in the case of a rational function (p. 390) and the concept of an analytic function (p. 385) will already be used in Section 5.3. We start by reviewing the basic ideas.

An infinite series of the form $a_0 + a_1 x + a_2 x^2 + \cdots = \sum_{n=0}^{\infty} a_n x^n$, in which x is a variable, is called a **power series** in x. The numbers $a_0, a_1, a_2, \ldots$ are the **coefficients** of this power series.

Given a power series $\sum_{n=0}^{\infty} a_n x^n$, it is usually important to know for which values of x the series converges. Using the ratio test, it can be shown that the series is absolutely convergent[8] for $|x| < R$ and divergent for $|x| > R$, where R, the **radius of convergence**[9] of

---

[8] That is, $\Sigma |a_n x^n|$ is convergent. This implies that $\Sigma a_n x^n$ is also.
[9] If you wonder why this is called a *radius*, see Appendix A, p. 600.

SEC. 5.2 POWER SERIES

the series, can be 0, a finite positive number, or ∞. (If $R = 0$, the series converges for $x = 0$ only; if $R = \infty$, the series is absolutely convergent for all x.) Again using the ratio test, one finds that $R = \lim_{n \to \infty} \left| \dfrac{a_n}{a_{n+1}} \right|$, provided this limit exists; fortunately, in practice this is very often the case (see below for an important exception). Note that the ratio test gives no information for $x = \pm R$; other tests[10] are used to decide whether the series converges for those particular ("endpoint") values of x.

EXAMPLE 5.2.1 Consider the power series

$$\sum_{n=1}^{\infty} (-1)^{n-1} \frac{x^n}{n} = x - \frac{x^2}{2} + \frac{x^3}{3} - \frac{x^4}{4} + \cdots.$$

Here $a_n = \dfrac{(-1)^{n-1}}{n}$ (except for the constant term $a_0 = 0$), so

$$\left| \frac{a_n}{a_{n+1}} \right| = \left| \frac{n+1}{n} \right| = \left| 1 + \frac{1}{n} \right| \quad \text{and} \quad R = \lim_{n \to \infty} \left| 1 + \frac{1}{n} \right| = 1.$$

Hence the power series is absolutely convergent for $|x| < 1$ and divergent for $|x| > 1$. For $x = 1$, $\sum_{n=1}^{\infty} (-1)^{n-1} \dfrac{x^n}{n} = \sum_{n=1}^{\infty} \dfrac{(-1)^{n-1}}{n}$ converges by the alternating series test, but is not absolutely convergent. For $x = -1$, $\sum_{n=1}^{\infty} (-1)^{n-1} \dfrac{x^n}{n} = \sum_{n=1}^{\infty} \dfrac{-1}{n}$ is divergent. In summary, our power series converges for x in the interval $(-1, 1]$ and diverges for all other x. ∎

The most common case in which the formula $R = \lim_{n \to \infty} \left| \dfrac{a_n}{a_{n+1}} \right|$ breaks down is that of a **gapped** power series, for which $a_{n+1} = 0$ for infinitely many n, so that infinitely many quotients $\dfrac{a_n}{a_{n+1}}$ are undefined. In this case the radius of convergence is usually found by applying the ratio test directly to the series (and thus avoiding the "missing" terms $a_n x^n$ for which $a_n = 0$).

EXAMPLE 5.2.2 $x - \dfrac{x^3}{2 \cdot 3} + \dfrac{x^5}{2^2 \cdot 5} - \dfrac{x^7}{2^3 \cdot 7} + \cdots.$

To write this series using summation notation, note that only odd powers of x occur; since odd numbers n can be written in the form $n = 2k + 1$, the series can be written as

$$\sum_{k=0}^{\infty} c_k x^{2k+1}, \quad \text{where} \quad c_0 = 1, \quad c_1 = \frac{-1}{2 \cdot 3}, \quad c_2 = \frac{1}{2^2 \cdot 5}, \quad \text{and so on}.$$

---

[10] Such as the integral test, the alternating series test, and the (limit) comparison test. See textbooks on calculus.

Note that the (nonzero) coefficients are $c_k = \dfrac{(-1)^k}{2^k \cdot (2k+1)}$; therefore, our series is $\displaystyle\sum_{k=0}^{\infty} \dfrac{(-1)^k x^{2k+1}}{2^k \cdot (2k+1)}$. By the ratio test, this series converges if $L < 1$ and diverges if $L > 1$, where

$$L = \lim_{k \to \infty} \left| \dfrac{(-1)^{k+1} x^{2k+3}/[2^{k+1} \cdot (2k+3)]}{(-1)^k x^{2k+1}/[2^k \cdot (2k+1)]} \right| = \lim_{k \to \infty} \dfrac{x^2}{2} \cdot \dfrac{2k+1}{2k+3} = \dfrac{1}{2} x^2.$$

So the series converges for $\frac{1}{2} x^2 < 1$, or $|x| < \sqrt{2}$, and diverges for $|x| > \sqrt{2}$; $R = \sqrt{2}$ is its radius of convergence. ∎

If a power series $\displaystyle\sum_{n=0}^{\infty} a_n x^n$ is absolutely convergent for all x in some interval, then its sum is a function of x on that interval. In some cases, this will be a "known" function.

EXAMPLE 5.2.3 $\displaystyle\sum_{n=0}^{\infty} 2^n x^n$ is absolutely convergent for $|x| < \dfrac{1}{2}$. (Here $a_n = 2^n$ for all n, so we have $R = \lim_{n \to \infty} \dfrac{2^n}{2^{n+1}} = \dfrac{1}{2}$.) This is a geometric series $1 + 2x + (2x)^2 + \cdots$, so its sum is $\dfrac{1}{1-2x}$ (still for $|x| < \dfrac{1}{2}$). Note that while this *function* $\dfrac{1}{1-2x}$ is also defined outside the interval $(-\dfrac{1}{2}, \dfrac{1}{2})$, although not at $x = \dfrac{1}{2}$, its *power series* expansion

$$\dfrac{1}{1-2x} = 1 + 2x + 4x^2 + \cdots$$

is valid only for $|x| < \dfrac{1}{2}$. ∎

In other cases, a "new" function will be defined by the power series.

EXAMPLE 5.2.4 Consider the power series

$$\sum_{n=0}^{\infty} \dfrac{x^n}{n! \log(n+2)} = \dfrac{1}{\log 2} + \dfrac{x}{\log 3} + \dfrac{x^2}{2 \log 4} + \cdots.$$

SEC. 5.2 POWER SERIES

This time, we have

$$R = \lim_{n \to \infty} \frac{(n+1)!\,\log(n+3)}{n!\,\log(n+2)}$$

$$= \lim_{n \to \infty} (n+1)\frac{\log(n+3)}{\log(n+2)} = \infty$$

(see Exercise 63), so the power series converges (absolutely) for all x. Therefore, if we put

$$f(x) = \sum_{n=0}^{\infty} \frac{x^n}{n!\,\log(n+2)},$$ this function will be defined for all x. ∎

When we find solutions of differential equations in this chapter, they will usually be "new" functions given by power series expansions. At first sight, these expressions for the "new" functions may not seem particularly useful. However, they are often very well suited for numerical computation (using a computer, or even a calculator). After all, the partial sums of a power series are just polynomials, and provided the power series converges, the values of these polynomials will approximate the sum of the series to any desired degree of accuracy.

EXAMPLE 5.2.5 In Section 5.6 we'll find the solution

$$y(x) = \sum_{m=0}^{\infty} \frac{(-1)^m}{2^{2m} m!\,(m+1)!} x^{2m+1} = x - \frac{x^3}{8} + \frac{x^5}{192} - \frac{x^7}{9216} + \cdots$$

to the differential equation $x^2 y'' + xy' + (x^2 - 1)y = 0$ [see Section 5.6, Exercise 16(a)]. The power series converges for all x [Exercise 62(a) of this section]; suppose that we want to find its value for $x = 2$ to within 0.001. Since the series

$$\sum_{m=0}^{\infty} \frac{(-1)^m}{2^{2m} m!\,(m+1)!} 2^{2m+1} = \sum_{m=0}^{\infty} \frac{(-1)^m \cdot 2}{m!\,(m+1)!}$$

for $y(2)$ is an alternating series with terms of decreasing size, the error we make if we cut off the series after a certain term is less than the size (absolute value) of the next term. So if we cut off after the term for $m = M$, the error will be less than $\dfrac{2}{(M+1)!(M+2)!}$. Therefore, we want to choose M such that

$$\frac{2}{(M+1)!\,(M+2)!} \le 0.001.$$

The first such M is $M = 4$ [see Exercise 62(b)], so the desired value will be given (to within 0.001) by

$$\sum_{m=0}^{4} \frac{(-1)^m \cdot 2}{m!\,(m+1)!} = 2 - 1 + \frac{1}{6} - \frac{1}{72} + \frac{1}{1440}.$$ ∎

It is by no means always easy to tell how many terms in a power series should be taken to get a particular accuracy. If many terms are needed, the situation may be further complicated by round-off error (so that "more" isn't necessarily "better"); see textbooks on numerical analysis for more on this.

Meanwhile, since we will be working with functions that are given by power series expansions, it will be useful to be able to make various computations with such functions. Fortunately, one can add, subtract, differentiate, integrate, and even multiply power series as if they were just finite sums (i.e., polynomials $a_0 + a_1 x + \cdots + a_N x^N$ with a finite degree N), *provided that the power series are absolutely convergent*.

In fact, if $f(x) = \sum_{n=0}^{\infty} a_n x^n$ and $g(x) = \sum_{n=0}^{\infty} b_n x^n$ are absolutely convergent power series on some interval, then (more explanations will follow)

$$f(x) + g(x) = \sum_{n=0}^{\infty} (a_n + b_n) x^n,$$

$$f(x) - g(x) = \sum_{n=0}^{\infty} (a_n - b_n) x^n,$$

$$f'(x) = \sum_{n=1}^{\infty} n a_n x^{n-1} = \sum_{n=0}^{\infty} (n+1) a_{n+1} x^n,$$

$$\int f(x)\, dx = \sum_{n=0}^{\infty} \frac{a_n}{n+1} x^{n+1} + C = C + \sum_{n=1}^{\infty} \frac{a_{n-1}}{n} x^n,$$

$$f(x) \cdot g(x) = \sum_{n=0}^{\infty} \left(\sum_{k=0}^{n} a_k b_{n-k} \right) x^n,$$

and these are all absolutely convergent power series on that same interval. The first two of the five formulas above should be self-explanatory, and we'll now consider the others individually.

1. **Differentiating** a power series. If we differentiate

$$f(x) = a_0 + a_1 x + a_2 x^2 + \cdots + a_n x^n + \cdots$$

term by term, we get

$$f'(x) = 0 + a_1 + 2 a_2 x + \cdots + n a_n x^{n-1} + \cdots,$$

that is,

$$f'(x) = \sum_{n=1}^{\infty} n a_n x^{n-1}.\text{ [11]}$$

---

[11] This is the same as $\sum_{n=0}^{\infty} n a_n x^{n-1}$, because the term with $n = 0$ is zero anyway.

Obviously, this is again a power series, but it isn't quite in the standard form. This can be awkward, for example if we want to add this series to another power series $\sum_{n=0}^{\infty} b_n x^n$. To rewrite our series $\sum_{n=1}^{\infty} n a_n x^{n-1}$ in a more standard form, we proceed as follows. Instead of numbering the terms

$$a_1 + 2a_2 x + 3a_3 x^2 + \cdots + n a_n x^{n-1} + \cdots$$
$$n = 1 \quad n = 2 \quad n = 3 \quad \cdots$$

as we did above, we now number them

$$a_1 \quad + \quad 2a_2 x \quad + \quad 3a_3 x^2 + \cdots + (k+1)a_{k+1} x^k + \cdots$$
$$k = 0 \quad\quad k = 1 \quad\quad k = 2$$
$$(k+1 = 1) \quad (k+1 = 2) \quad (k+1 = 3) \cdots.$$

That is, we introduce a new index of summation, k, which starts at 0; note that $k + 1 = n$, so $k = n - 1$. Using k, we can write our series as $\sum_{k=0}^{\infty} (k+1) a_{k+1} x^k$. Now note that the *name* of the index doesn't matter; we could have used m or q instead of k, which would have given us $\sum_{m=0}^{\infty} (m+1) a_{m+1} x^m$ or $\sum_{q=0}^{\infty} (q+1) a_{q+1} x^q$. In particular, we can choose to return to the letter n (which is our usual letter for an index), and write our series for the derivative as

$$f'(x) = \sum_{n=0}^{\infty} (n+1) a_{n+1} x^n.$$

Note that the meaning of the letter n has changed from when we wrote

$$f'(x) = \sum_{n=1}^{\infty} n a_n x^{n-1};$$

what used to be called n is now called $n + 1$. Accordingly, instead of $n a_n x^{n-1}$ we have $(n+1) a_{n+1} x^n$, and instead of starting with $n = 1$ we start with $n + 1 = 1$, that is, with $n = 0$.

The technique of rewriting a series using "a different n," as in $\sum_{n=0}^{\infty} (n+1) a_{n+1} x^n$ instead of $\sum_{n=1}^{\infty} n a_n x^{n-1}$, is called **shifting the index of summation**.

EXAMPLE 5.2.6   Starting with the geometric series expansion

$$\frac{1}{1-x} = 1 + x + x^2 + x^3 + \cdots = \sum_{n=0}^{\infty} x^n \qquad (|x| < 1) \qquad \text{(see p. 372)}$$

and taking derivatives on both sides, we get

$$\frac{1}{(1-x)^2} = 1 + 2x + 3x^2 + \cdots = \sum_{n=1}^{\infty} nx^{n-1} = \sum_{n=0}^{\infty} (n+1)x^n.$$

Once again, although $\sum_{n=1}^{\infty} nx^{n-1} = \sum_{n=0}^{\infty} (n+1)x^n$, the letter n on the left side does not correspond to n on the right [obviously, we don't have $nx^{n-1} = (n+1)x^n$]; instead, n on the left corresponds to $(n+1)$ on the right. To make this very explicit, we can put in an intermediate step, in which we replace n by $(n+1)$ throughout the formula:

$$\sum_{n=1}^{\infty} nx^{n-1} = \sum_{(n+1)=1}^{\infty} (n+1)x^{(n+1)-1} = \sum_{n=0}^{\infty} (n+1)x^n.$$

Or, if you prefer, you can use a new letter k, with $k = n - 1$, as a "temporary index" (later to be renamed n):

$$\sum_{n=1}^{\infty} nx^{n-1} = \sum_{k=0}^{\infty} (k+1)x^k = \sum_{n=0}^{\infty} (n+1)x^n. \qquad \blacksquare$$

2. **Integrating** a power series. If we integrate

$$f(x) = a_0 + a_1 x + a_2 x^2 + \cdots + a_n x^n + \cdots$$

term by term, we will get

$$\int f(x)\,dx = C + a_0 x + a_1 \frac{x^2}{2} + \cdots + a_n \frac{x^{n+1}}{n+1} + \cdots$$

$$= C + \sum_{n=0}^{\infty} a_n \frac{x^{n+1}}{n+1},$$

where C is a constant of integration. If we wish to rewrite this in the standard form, displaying the term with x^n instead of the term with x^{n+1}, we should again shift the index of summation. This time, we can replace n by $n - 1$ throughout the formula to get

$$\sum_{n=0}^{\infty} a_n \frac{x^{n+1}}{n+1} = \sum_{(n-1)=0}^{\infty} a_{n-1} \frac{x^{(n-1)+1}}{(n-1)+1} = \sum_{n=1}^{\infty} \frac{a_{n-1}}{n} x^n.$$

Or, if you prefer, you can introduce $k = n + 1$ as a "temporary index" to get the same result:

$$\sum_{n=0}^{\infty} a_n \frac{x^{n+1}}{n+1} = \sum_{k=1}^{\infty} a_{k-1} \frac{x^k}{k} = \sum_{n=1}^{\infty} a_{n-1} \frac{x^n}{n}.$$

Either way, we end up with

$$\int f(x)\,dx = C + \sum_{n=1}^{\infty} \frac{a_{n-1}}{n} x^n.$$

C, the constant term in the new power series, is the constant of integration.

EXAMPLE 5.2.7 Find a power series expansion for $\log(1 + x)$ ($|x| < 1$).

First we find an expansion for the derivative $\dfrac{1}{1+x}$. We know that

$$\frac{1}{1-x} = 1 + x + x^2 + x^3 + \cdots = \sum_{n=0}^{\infty} x^n \qquad (|x| < 1),$$

and replacing x by $-x$ throughout, we obtain

$$\frac{1}{1+x} = 1 - x + x^2 - x^3 + \cdots = \sum_{n=0}^{\infty} (-x)^n = \sum_{n=0}^{\infty} (-1)^n x^n \qquad (|-x| = |x| < 1).$$

Now we can integrate both sides to get

$$\log(1 + x) = C + x - \frac{x^2}{2} + \frac{x^3}{3} - \frac{x^4}{4} + \cdots = C + \sum_{n=0}^{\infty} (-1)^n \frac{x^{n+1}}{n+1}$$

$$= C + \sum_{n=1}^{\infty} (-1)^{n-1} \frac{x^n}{n} \qquad (|x| < 1).$$

To determine the constant of integration C, we can substitute $x = 0$, since we can compute $\log(1 + 0) = 0$ and $\sum_{n=1}^{\infty} (-1)^{n-1} \dfrac{0^n}{n} = 0$ independently. We find that $0 = C + 0$, so $C = 0$ and

$$\log(1 + x) = \sum_{n=1}^{\infty} (-1)^{n-1} \frac{x^n}{n} = x - \frac{x^2}{2} + \frac{x^3}{3} - \frac{x^4}{4} + \cdots \qquad (|x| < 1).$$

Again, the *function* $\log(1 + x)$ is defined for all $x > -1$, but the *power series* expansion is not valid for $x > 1$. (It turns out that for $x = 1$, the formula

$$\log 2 = 1 - \frac{1}{2} + \frac{1}{3} - \frac{1}{4} + \cdots$$

is still correct.) ∎

3. **Multiplying** two power series. If the two series $f(x) = a_0 + a_1 x + a_2 x^2 + \cdots$ and $g(x) = b_0 + b_1 x + b_2 x^2 + \cdots$ are multiplied as if they were polynomials, that is, using the distributive law and collecting terms of the same degree in x, the result is

$$f(x) \cdot g(x) = a_0 b_0 + (a_0 b_1 + a_1 b_0) x + (a_0 b_2 + a_1 b_1 + a_2 b_0) x^2 + \cdots.$$

If we want to write this using the summation convention, what should the coefficient of x^n be? Well, to get a term with x^n, we have to multiply a term in $f(x)$ of degree at most n, say $a_k x^k$ with $k \leq n$, with the term in $g(x)$ whose degree is $n - k$, that is, $b_{n-k} x^{n-k}$. This yields the term $a_k x^k \cdot b_{n-k} x^{n-k} = a_k b_{n-k} x^n$ in the product $f(x) \cdot g(x)$. To find the coefficient of x^n in the product, we have to collect *all* such terms with $k \leq n$. That is, the coefficient of x^n is $\sum_{k=0}^{n} a_k b_{n-k} = a_0 b_n + a_1 b_{n-1} + \cdots + a_n b_0$. Therefore, we have

$$f(x) \cdot g(x) = \sum_{n=0}^{\infty} \left(\sum_{k=0}^{n} a_k b_{n-k} \right) x^n.$$ This formula is of limited practical use, since the coefficients $\sum_{k=0}^{n} a_k b_{n-k}$ are unwieldy. Still, it is important that the multiplication may be carried out "naively" provided the series are absolutely convergent.

EXAMPLE 5.2.8 Find the power series expansion of $\dfrac{x^2 + 1}{1 - x}$ ($|x| < 1$).

We multiply the polynomial $1 + x^2$ by the power series $\dfrac{1}{1 - x} = 1 + x + x^2 + \cdots$ to obtain

$$\frac{1 + x^2}{1 - x} = (1 + x^2) + (x + x^3) + (x^2 + x^4) + \cdots$$

$$= 1 + x + 2x^2 + 2x^3 + 2x^4 + \cdots$$

$$= 1 + x + 2 \sum_{n=2}^{\infty} x^n.$$

Another way to set up the computation would be

$$\frac{1}{1 - x} = 1 + x + x^2 + x^3 + \cdots$$

$$\frac{x^2}{1 - x} = \phantom{1 + x +{}} x^2 + x^3 + x^4 + x^5 + \cdots$$

$$+ \overline{}$$

$$\frac{1 + x^2}{1 - x} = 1 + x + 2x^2 + 2x^3 + 2x^4 + \cdots.$$

Of course, this series expansion isn't always more helpful than the original expression $\dfrac{x^2 + 1}{1 - x}$, but it is in some situations—for instance, if you need to add another power series in x to it. ∎

So far, we haven't discussed division of power series, and we'll do so at the end of this section; otherwise, we have the computational tools we'll need to deal with power series. However, we'll also need the following general ideas.

Suppose that on some interval $(-R, R)$, a function is given by an absolutely convergent power series in x, say $f(x) = \sum_{n=0}^{\infty} a_n x^n$. Then we've seen that

$$f'(x) = a_1 + 2a_2 x + 3a_3 x^2 + \cdots = \sum_{n=0}^{\infty} (n + 1)a_{n+1} x^n;$$

since this is again an absolutely convergent power series, we can keep differentiating, and we get

$$f''(x) = 2a_2 + 6a_3 x + \cdots = \sum_{n=0}^{\infty} (n + 1)(n + 2)a_{n+2} x^n,$$

$$f^{(3)}(x) = \sum_{n=0}^{\infty} (n + 1)(n + 2)(n + 3)a_{n+3} x^n, \quad \text{and so on.}$$

In particular, for $x = 0$ we have $f(0) = a_0, f'(0) = a_1, f''(0) = 2a_2, f^{(3)}(0) = 6a_3$, and so on. Therefore, the coefficients of the power series are given by

$$a_0 = f(0), \quad a_1 = f'(0), \quad a_2 = \frac{f''(0)}{2}, \quad a_3 = \frac{f^{(3)}(0)}{3!}, \quad \text{and so on.}$$

That is, the power series $\sum_{n=0}^{\infty} a_n x^n$ is nothing but the Taylor series $\sum_{n=0}^{\infty} \dfrac{f^{(n)}(0)}{n!} x^n$ of the function. In other words, **if a function can be written as the sum of an absolutely convergent power series** in x on an interval $(-R, R)$, **then that series must be the Taylor series of that function**. In particular, *two* power series in x can only have *equal* sums on such an interval if they have the same coefficients. A key special case of this is that if $\sum_{n=0}^{\infty} a_n x^n = 0$ for all x (or for all x with $|x| < R$, R a positive number), then *all* the coefficients $a_0, a_1, a_2, \ldots$ must be zero.

A function that has an absolutely convergent power series expansion [on some interval $(-R, R)$ around $x = 0$] is called **analytic** (at $x = 0$). One way of showing that a function is analytic is to use the estimates for the remainder term $R_n(x)$ (referred to on p. 369) to show that the Taylor series converges and has the function itself as its sum. (For an example where the Taylor series converges but does not have the function itself as its sum, see Section 5.1, Exercise 16.) This can be done, for instance, for the functions $\sin x$, $\cos x$, and e^x; see textbooks on calculus for details. Another way is to derive the function from known analytic functions by differentiation, integration, multiplication, composition, and so forth, as indicated above for the examples $\dfrac{1}{(1-x)^2}$, $\log(1+x)$, and $\dfrac{1+x^2}{1-x}$. Of course, if a "new" function, such as $\sum_{n=0}^{\infty} \dfrac{x^n}{n!\log(n+2)}$ in Example 5.2.4, is defined by an absolutely convergent power series, it is automatically analytic.

The following list shows some common analytic functions and their expansions as power series in x:

$$e^x = 1 + x + \frac{x^2}{2!} + \cdots = \sum_{n=0}^{\infty} \frac{x^n}{n!} \qquad (\text{all } x)$$

$$\cos x = 1 - \frac{x^2}{2!} + \frac{x^4}{4!} - \frac{x^6}{6!} + \cdots = \sum_{n=0}^{\infty} (-1)^n \frac{x^{2n}}{(2n)!} \qquad (\text{all } x)$$

$$\sin x = x - \frac{x^3}{3!} + \frac{x^5}{5!} - \frac{x^7}{7!} + \cdots = \sum_{n=0}^{\infty} (-1)^n \frac{x^{2n+1}}{(2n+1)!} \qquad (\text{all } x)$$

$$\arctan x = x - \frac{x^3}{3} + \frac{x^5}{5} - \frac{x^7}{7} + \cdots = \sum_{n=0}^{\infty} (-1)^n \frac{x^{2n+1}}{2n+1} \qquad (|x| < 1)$$

$$\log(1+x) = x - \frac{x^2}{2} + \frac{x^3}{3} - \frac{x^4}{4} + \cdots = \sum_{n=0}^{\infty} (-1)^n \frac{x^{n+1}}{n+1} \qquad (|x| < 1)$$

$$\frac{1}{1-x} = 1 + x + x^2 + \cdots = \sum_{n=0}^{\infty} x^n \qquad (|x| < 1).$$

So far, we have considered only power series expansions in terms of x. As we have seen, any function that has such an expansion must be infinitely differentiable (that is, its first, second, third, ... derivatives must exist) at $x = 0$. In particular, functions such as $\sqrt{x}$, $x^{5/3}$, $\log x$, and $\dfrac{1}{x}$ cannot be written as power series in x. However, each of these functions can still be studied by power series methods, provided we are willing to shift our focus from $x = 0$ to some other value of x where the function is better behaved.

If we want to study the function $f(x)$ near $x = c$, it's a good idea to use $x - c$ as a new variable (note that for $x = c$, $x - c = 0$) and thus to look for an expansion of $f(x)$ in terms of $x - c$. To do so, we can first write $f(x) = f(c + (x - c))$ to show how the function depends on the new variable, and then apply our earlier methods.

EXAMPLE 5.2.9 Consider the function $\log x$ near $x = 1$. To find a power series expansion in $x - 1$, we write $\log x = \log(1 + (x-1))$. Since we have the formula

$$\log(1 + x) = x - \frac{x^2}{2} + \frac{x^3}{3} - \frac{x^4}{4} + \cdots = \sum_{n=0}^{\infty} (-1)^n \frac{x^{n+1}}{n+1} \qquad (|x| < 1)$$

from the list above, we can simply replace x by $x - 1$ to obtain

$$\log x = (x-1) - \frac{(x-1)^2}{2} + \cdots = \sum_{n=0}^{\infty} (-1)^n \frac{(x-1)^{n+1}}{n+1} \qquad (|x-1| < 1).$$

Note that $|x - 1| < 1$ means $0 < x < 2$; as you might expect, the expansion does not give a value for "$\log 0$." □

Let's see what happens in general. To study $f(x)$ near $x = c$, let $x_1 = x - c$ be the new variable. In terms of x_1, we have the function $g(x_1) = f(c + x_1) = f(x)$. [In Example 5.2.9, $x_1 = x - 1$; $g(x_1) = \log(1 + x_1) = \log x$.] If $g(x_1)$ is analytic at $x_1 = 0$, then $g(x_1) = \sum_{n=0}^{\infty} \frac{g^{(n)}(0)}{n!} x_1^n$. Now the higher derivatives of g at $x_1 = 0$ are nothing but the higher derivatives of f at $x = c$:

$g(x_1) = f(c + x_1) \qquad \Rightarrow \qquad g(0) = f(c)$

$g'(x_1) = f'(c + x_1)$ (by the chain rule!) $\Rightarrow g'(0) = f'(c)$

$g''(x_1) = f''(c + x_1) \qquad \Rightarrow \qquad g''(0) = f''(c)$, and so on.

So our power series expansion will have the form

$$f(x) = g(x_1) = \sum_{n=0}^{\infty} \frac{f^{(n)}(c)}{n!} x_1^n,$$

that is,

$$f(x) = \sum_{n=0}^{\infty} \frac{f^{(n)}(c)}{n!} (x - c)^n.$$

This is called the **Taylor expansion** of $f(x)$ near (or about) $x = c$. $f(x)$ is called **analytic** at $x = c$ if the expansion is valid, that is, if $g(x_1)$ is analytic at $x_1 = 0$.

EXAMPLE 5.2.9 (continued) $\log x$ is not analytic at $x = 0$, but we have seen that $\log x = \sum_{n=0}^{\infty} (-1)^n \frac{(x-1)^{n+1}}{n+1}$ ($|x - 1| < 1$), so $\log x$ is analytic at $x = 1$. Equivalently, $\log(1 + x)$ is analytic at $x = 0$. ■

EXAMPLE 5.2.10 $f(x) = x^{5/3}$ is not analytic at $x = 0$ [$f''(0)$ doesn't even exist]. Consider $f(x)$ near $x = 1$. The derivatives are

$$f'(x) = \frac{5}{3}x^{2/3} \quad \Rightarrow \quad f'(1) = \frac{5}{3}$$

$$f''(x) = \frac{10}{9}x^{-1/3} \quad \Rightarrow \quad f''(1) = \frac{10}{9}$$

$$f^{(3)}(x) = -\frac{10}{27}x^{-4/3} \quad \Rightarrow \quad f^{(3)}(1) = -\frac{10}{27}$$

$$f^{(4)}(x) = \frac{40}{81}x^{-7/3} \quad \Rightarrow \quad f^{(4)}(1) = \frac{40}{81}, \quad \text{and so on.}$$

So provided $f(x)$ is analytic at $x = 1$, the expansion in powers of $x - 1$ will look like

$$x^{5/3} = [1 + (x - 1)]^{5/3}$$

$$= 1 + \frac{5}{3}(x - 1) + \frac{10}{9 \cdot 2!}(x - 1)^2 - \frac{10}{27 \cdot 3!}(x - 1)^3 + \frac{40}{81 \cdot 4!}(x - 1)^4 + \cdots.$$

It can be shown that this is indeed the case. ∎

EXAMPLE 5.2.11 Consider $f(x) = e^x$ near $x = 1$. All the higher derivatives $f^{(n)}(1)$ equal e, so the Taylor expansion will be

$$e^x = e + e(x - 1) + e\frac{(x - 1)^2}{2!} + e\frac{(x - 1)^3}{3!} + \cdots,$$

provided e^x is analytic at $x = 1$. This can be shown directly, or we can use the fact that

$$e^{x-1} = 1 + (x - 1) + \frac{(x - 1)^2}{2!} + \frac{(x - 1)^3}{3!} + \cdots$$

(this is just the usual formula with x replaced by $x - 1$) and multiply both sides by e. It is an interesting exercise (see Exercise 64) to show directly that the two expressions for e^x,

$$e + e(x - 1) + e\frac{(x - 1)^2}{2!} + \cdots \quad \text{and} \quad 1 + x + \frac{x^2}{2!} + \cdots,$$

are equal! ∎

In the rest of this chapter we will concentrate on power series expansions in x, that is, on the behavior of solutions near $x = 0$. This is in order not to have to write $x - c$ in all the formulas. If the behavior of solutions near $x = c$, $c \neq 0$, comes up in some application, you can change to the new variable $x - c$ as illustrated above.

SEC. 5.2 POWER SERIES

To conclude this section, we consider **division** of power series, which is possible provided the denominator is not zero. More precisely, if $f(x) = \sum_{n=0}^{\infty} a_n x^n$ and $g(x) = \sum_{n=0}^{\infty} b_n x^n$ are analytic at $x = 0$ and if $g(0) = b_0 \neq 0$, then a power series expansion $\dfrac{f(x)}{g(x)} = \sum_{n=0}^{\infty} c_n x^n$ exists, so the quotient $\dfrac{f(x)}{g(x)}$ is again analytic at $x = 0$. In particular, any rational function (that is, any quotient of polynomials) is analytic at $x = 0$, provided the denominator is not zero there. For example, $\dfrac{x+1}{x^2 - x + 1}$ has an expansion as a power series in x.

If the coefficients c_n are needed, often the easiest way to find them is to note that since $f(x) = \dfrac{f(x)}{g(x)} g(x)$, we must have

$$\sum_{n=0}^{\infty} a_n x^n = \sum_{n=0}^{\infty} b_n x^n \cdot \sum_{n=0}^{\infty} c_n x^n.$$

Then multiply out the right-hand side as explained on p. 384; finally, solve for $c_0, c_1, c_2,$ and so on, in turn.

EXAMPLE 5.2.12 To write $\dfrac{x+1}{x^2 - x + 1} = c_0 + c_1 x + c_2 x^2 + \cdots$, multiply through by $x^2 - x + 1$:

$$x + 1 = (x^2 - x + 1)(c_0 + c_1 x + c_2 x^2 + \cdots)$$
$$= c_0 + (c_1 - c_0)x + (c_2 - c_1 + c_0)x^2 + (c_3 - c_2 + c_1)x^3 + \cdots.$$

Comparing coefficients, we have

$$c_0 = 1$$
$$c_1 - c_0 = 1 \implies c_1 = 2$$
$$c_2 - c_1 + c_0 = 0 \implies c_2 = c_1 - c_0 = 1$$
$$c_3 - c_2 + c_1 = 0 \implies c_3 = c_2 - c_1 = -1$$
$$c_4 - c_3 + c_2 = 0 \implies c_4 = c_3 - c_2 = -2$$
$$c_5 - c_4 + c_3 = 0 \implies c_5 = c_4 - c_3 = -1, \text{ and so on.}$$

Each of the c_n after c_1 is the difference of the two preceding ones, so the sequence $c_0, c_1, c_2, \ldots$ will be the periodic sequence

$$1, 2, 1, -1, -2, -1, 1, 2, 1, \ldots.$$

The power series expansion becomes

$$\frac{x+1}{x^2 - x + 1} = 1 + 2x + x^2 - x^3 - 2x^4 - x^5 + x^6 + 2x^7 + \cdots. \quad \blacksquare$$

Warning: The expansion for $\dfrac{f(x)}{g(x)}$ may have a smaller radius of convergence than either of the ones for $f(x)$ and $g(x)$. For instance, recall that the expansion $\dfrac{1}{1-x} = \sum_{n=0}^{\infty} x^n$ has radius of convergence 1, although the power series 1 and $1 - x$ are, of course, convergent everywhere.

It turns out that for a rational function, it is often enough for our purposes (in Section 5.3) to find the *radius of convergence* of its power series expansion in x, and that this can be done *without* finding the expansion itself. We now show how this is done,[12] using the specific example $\dfrac{(x-2)(x-1)(x^2+1)}{(x-2)^2(x-1)(x^2+2)}$ to illustrate each step.

1. Cancel all common factors from numerator and denominator. In the example, this leaves us with $\dfrac{x^2+1}{(x-2)(x^2+2)}$.

2. Find all *complex* roots of the denominator (including the real ones). In the example, the denominator has roots $x = 2$ (of the factor $x - 2$) and $x = \pm i\sqrt{2}$ (of the quadratic factor $x^2 + 2$).

3. The radius of convergence is the smallest of these absolute values. In the example, since $|i\sqrt{2}| = |-i\sqrt{2}| = \sqrt{2}$ is less than $|2| = 2$, the radius of convergence is $\sqrt{2}$.

SUMMARY OF KEY CONCEPTS, RESULTS, AND TECHNIQUES

Radius of convergence of a power series $\sum_{n=0}^{\infty} a_n x^n$ (p. 376): The "number" R (which can be ∞) such that the series converges absolutely for $|x| < R$ and diverges for $|x| > R$. Provided the limit exists, we have $R = \lim_{n \to \infty} \left| \dfrac{a_n}{a_{n+1}} \right|$; for a **gapped** power series (with infinitely many coefficients 0) R can usually be found using the ratio test.

Computations with power series (p. 380): Absolutely convergent power series can be added, subtracted, differentiated, integrated, and multiplied as if they were polynomials. For division, see below.

Equality of power series (p. 385): If two power series in x have equal sums on some interval $(-R, R)$, then their coefficients must be the same. In particular, if $\sum_{n=0}^{\infty} a_n x^n = 0$ for all x in $(-R, R)$, then $a_n = 0$ for all n.

---

[12] The justification of this method requires the theory of functions of a complex variable and is beyond the scope of this book.

SEC. 5.2 POWER SERIES

Shifting the index of summation (pp. 381–382): To write power series such as $\sum_{n=1}^{\infty} c_n x^{n-1}$ and $\sum_{n=0}^{\infty} a_n x^{n+1}$ in standard form, shift the index of summation. In these particular cases n is replaced by $n+1$ and $n-1$, respectively, which yields $\sum_{n=0}^{\infty} c_{n+1} x^n$ and $\sum_{n=1}^{\infty} a_{n-1} x^n$, respectively.

Analytic function (p. 385): A function is called analytic at $x = 0$ if it is given by a power series $\sum_{n=0}^{\infty} a_n x^n$ with a positive radius of convergence. If so, that power series is unique; it is the Taylor series of the function. To show that a function is analytic, one can either show that it is the sum of its Taylor series, or derive it by addition, differentiation, and so on, from known analytic functions. (See p. 386 for a list of such functions.) A function $f(x)$ is called **analytic** at $x = c$ if it is analytic as a function of $x - c$ at $x - c = 0$ (p. 387). If so, $f(x)$ is given [on some interval $(c - R, c + R)$] by its Taylor series about $x = c$: $f(x) = \sum_{n=0}^{\infty} \frac{f^{(n)}(c)}{n!} (x - c)^n$.

Division of power series (p. 389): If $f(x)$ and $g(x)$ are analytic at $x = 0$ and $g(0) \neq 0$, then $\frac{f(x)}{g(x)}$ is also analytic at $x = 0$. To find the power series expansion $\frac{f(x)}{g(x)} = \sum_{n=0}^{\infty} c_n x^n$ from $f(x) = \sum_{n=0}^{\infty} a_n x^n$ and $g(x) = \sum_{n=0}^{\infty} b_n x^n$, write $\sum_{n=0}^{\infty} a_n x^n = \sum_{n=0}^{\infty} b_n x^n \cdot \sum_{n=0}^{\infty} c_n x^n$, multiply out the right-hand side, and solve for the c_n.

Radius of convergence of the power series in x for a rational function (p. 390): After all common factors have been canceled, the radius of convergence is the smallest of the absolute values of the (real and complex) roots of the denominator.

EXERCISES

Write out the first three (nonzero) terms, and find the radius of convergence, for each of the following power series.

1. $\sum_{n=1}^{\infty} (-1)^n \frac{x^n}{n^2}$.

2. $\sum_{n=0}^{\infty} (-1)^n \frac{x^n}{n+2}$.

3. $\sum_{n=0}^{\infty} \frac{10^n x^n}{n!}$.

4. $\sum_{n=0}^{\infty} \frac{(-1)^n x^n}{10^{n+1} n!}$.

5. $\sum_{k=1}^{\infty} \dfrac{(-1)^{k+1} x^{2k}}{k^2 \cdot 4^k}$.

6. $\sum_{k=0}^{\infty} \dfrac{k x^{2k+1}}{5^k}$.

For each of the following power series, write the series in summation notation, then find the radius of convergence.

7. $\dfrac{x^2}{4 \cdot 2} - \dfrac{x^4}{4 \cdot 4} + \dfrac{x^6}{4 \cdot 6} - \dfrac{x^8}{4 \cdot 8} + \cdots$.

8. $\dfrac{3x}{2} + \dfrac{5x^3}{2^2} + \dfrac{7x^5}{2^3} + \dfrac{9x^7}{2^4} + \cdots$.

9. $\dfrac{2x^2}{3} - \dfrac{4x^3}{3^2} + \dfrac{6x^4}{3^3} - \dfrac{8x^5}{3^4} + \cdots$.

10. $\dfrac{x}{2!} + \dfrac{x^4}{5!} + \dfrac{x^7}{8!} + \dfrac{x^{10}}{11!} + \cdots$.

Write each of the following power series in "standard" form (that is, displaying the term with x^n).

11. $\sum_{n=1}^{\infty} n(n+5) x^{n-1}$.

12. $\sum_{n=1}^{\infty} \dfrac{n}{n^2 + 3} x^{n-1}$.

13. $\sum_{n=0}^{\infty} n^2 x^{n+1}$.

14. $\sum_{n=0}^{\infty} \dfrac{x^{n+1}}{n+5}$.

15. $\sum_{n=0}^{\infty} \dfrac{n(n-1)}{3^n} x^{n-2}$.

16. $\sum_{n=0}^{\infty} \dfrac{n(n-1)(n-2)}{(n+1)^2} x^{n-3}$.

17. $\sum_{n=0}^{\infty} \dfrac{2^n}{n+3} x^{n+2}$.

18. $\sum_{n=0}^{\infty} \dfrac{4}{3n-1} x^{n+1}$.

19. $\sum_{n=0}^{\infty} a_n x^n + \sum_{n=1}^{\infty} b_n x^{n-1} - 3 \sum_{n=0}^{\infty} c_n x^{n+1}$.

[*Hint*: You will have to show the constant term separately.]

20. $\sum_{n=1}^{\infty} a_n x^{n-1} - \sum_{n=0}^{\infty} b_n x^n + 5 \sum_{n=0}^{\infty} c_n x^{n+1}$. (Same hint as for Exercise 19.)

21. $\dfrac{d}{dx} \left[\sum_{n=1}^{\infty} \dfrac{(-1)^n x^n}{n^2} \right]$.

22. $\dfrac{d^2}{dx^2} \left[\sum_{n=1}^{\infty} \dfrac{4x^n}{2^n \cdot n^2} \right]$.

For each of the following functions, state whether or not the function is analytic at $x = 0$, and justify your statement.

23. $f(x) = x \sin x$.

24. $f(x) = \dfrac{\cos x}{x - 2}$.

25. $f(x) = \dfrac{x - 2}{\sin x}$.

26. $f(x) = x \cos x$.

27. $f(x) = 2x^4 + x^{10/3}$.

28. $f(x) = x^{11/3} - 3x^2 + 5$.

29. $f(x) = \sum_{n=0}^{\infty} \dfrac{x^n}{2^n + 1}$.

30. $f(x) = \sum_{n=0}^{\infty} \dfrac{3^n x^n}{n!}$.

31. $f(x) = \sum_{n=0}^{\infty} \dfrac{n! \, x^n}{3^n}$.

32. $f(x) = \sum_{n=0}^{\infty} (2^n + 1)x^n$.

For each of the following functions, find its power series expansion in x, and state for which x the series converges. (The list on p. 386 may be helpful.)

33. e^{3x}

34. e^{-2x}

35. $\dfrac{x^3 - 5}{x - 1}$

36. $\dfrac{x^2 - 5x}{x + 1}$

37. $\arctan(2x)$

38. $\log\left(1 + \dfrac{x}{3}\right)$

39. $\log(3 + x)$ [*Hint*: Use Exercise 38.]

40. $\sin(x^2)$

For each of the following functions, find the radius of convergence of its power series expansion in x (without finding the actual expansion).

41. $\dfrac{(x^2 + 2)(x^2 + 5)}{(x^2 + 3)(x^2 - 9)}$

42. $\dfrac{(x^2 - 1)(x + 3)}{(x^2 - 3)(x^2 + 9)}$

43. $\dfrac{(x - 2)^3(x + 5)}{(x^2 - 5x + 6)(x + 3)}$

44. $\dfrac{(x - 1)(x - 2)^2}{(x^2 - 3x + 2)(x^2 + 2)}$

45. (a) Find the power series expansion in x of $\dfrac{1}{1 + x^2}$. For what x does this series converge?
   (b) Use part (a) to derive the expansion of $\arctan x$.

46. (a) Find the power series expansion in x of $y = xe^x$.
   (b) Use part (a) to find the expansion of $\dfrac{dy}{dx}$.
   (c) Use parts (a) and (b) to show that $y = xe^x$ is a solution of the differential equation $x\dfrac{dy}{dx} = xy + y$.

47. (a) Find the power series expansion in x of $y = \sin 2x$.
   (b) Use part (a) to find the expansions of $\dfrac{dy}{dx}$ and $\dfrac{d^2y}{dx^2}$.
   (c) Use parts (a) and (b) to show that $y = \sin 2x$ is a solution of the differential equation $\dfrac{d^2y}{dx^2} + 4y = 0$.

For each of the following functions, find the first four nonzero terms in its power series expansion in x.

48. $\dfrac{x+1}{x^2 - 2x + 3}$

49. $\dfrac{2x - 7}{x^2 + 4x + 1}$

50. $f(x) = \begin{cases} \dfrac{\sin x}{x} & (x \neq 0) \\ 1 & (x = 0) \end{cases}$

*51. $\tan x$ [*Hint:* $\tan x = \dfrac{\sin x}{\cos x}$.]

For each of the following functions, find its power series expansion *near the indicated point*, and indicate for what values of x this expansion is valid.

52. e^x; $x = -2$.

53. $\sin x$; $x = \dfrac{\pi}{2}$.

54. $\sin x$; $x = \dfrac{\pi}{4}$.

55. $\dfrac{1}{x}$; $x = 5$.

56. $\log x$; $x = 3$. [*Hint:* Use Exercise 39.]

*57. $\log x$; $x = e$.
[*Hint:* $\log(ea) = \log e + \log a = 1 + \log a$.]

For each of the following power series, show whether the series converges for the given value of x. If so, find the sum of the series for that x, to within 0.01.

58. $\displaystyle\sum_{n=1}^{\infty} \dfrac{x^n}{n}$; $x = 1$.

59. $\displaystyle\sum_{n=1}^{\infty} \dfrac{x^n}{n}$; $x = -\dfrac{1}{2}$.

60. $\displaystyle\sum_{k=0}^{\infty} \dfrac{(-1)^k k}{(2k+1)!} x^{2k}$; $x = 1$.

61. $\displaystyle\sum_{k=0}^{\infty} \dfrac{(-1)^k k!}{2k+1} x^{2k}$; $x = 0.01$.

62. (a) Show that the power series from Example 5.2.5 converges for all x.

(b) Show that the smallest value of M such that $\dfrac{2}{(M+1)!\,(M+2)!} \leq 0.001$ is $M = 4$.

63. Show that $\displaystyle\lim_{n \to \infty} (n+1) \dfrac{\log(n+3)}{\log(n+2)} = \infty$.

*64. Show by rearranging terms (you may assume absolute convergence) that

$$e + e(x-1) + e\dfrac{(x-1)^2}{2!} + \cdots = 1 + x + \dfrac{x^2}{2!} + \cdots.$$

5.3 FIRST CASE: $x = 0$ AN ORDINARY POINT

We will now return to the differential equation $\dfrac{d^2y}{dx^2} + p(x)\dfrac{dy}{dx} + q(x)y = 0$ studied in Section 5.1. As indicated, our main interest will be in the case when $p(x)$ and $q(x)$ are polynomials or rational functions. If they are rational functions, the denominators may be zero at certain points; this will cause problems, which we will discuss in the following sections.

EXAMPLE 5.3.1 $x^2(x-1)y'' - xy' - (x+3)(x-1)y = 0$.

In our "standard" form the equation becomes $y'' - \dfrac{1}{x(x-1)}y' - \dfrac{x+3}{x^2}y = 0$, so we have $p(x) = \dfrac{-1}{x(x-1)}$, $q(x) = -\dfrac{x+3}{x^2}$. For $x = 1$, $p(x)$ is undefined; for $x = 0$, both $p(x)$ and $q(x)$ are undefined. As a result, there may not be solutions to the equation that satisfy initial conditions such as $y(0) = y_0$, $y'(0) = z_0$ or $y(1) = y_1$, $y'(1) = z_1$ (see Exercise 16). We'll return to the behavior near $x = 0$ of solutions to the equation in Section 5.7. ∎

EXAMPLE 5.3.2 $y'' - 2xy' + \lambda y = 0$. This equation, in which λ is an unspecified constant, is called the **Hermite equation**[13]; we'll see in Section 7.4 how it arises from a basic partial differential equation in quantum mechanics, the Schrödinger equation for the harmonic oscillator. For the Hermite equation, $p(x) = -2x$ and $q(x) = \lambda$ are polynomials, and of course they are defined for all x. ☐

In this section we will assume that $p(x)$ and $q(x)$ are defined and continuous in an interval around $x = 0$, so that any initial value problem $\dfrac{d^2y}{dx^2} + p(x)\dfrac{dy}{dx} + q(x)y = 0$, $y(0) = y_0$, $y'(0) = z_0$ has a solution (by Theorem 2.2.2). Furthermore, we will assume that $p(x)$ and $q(x)$ are *analytic* (at $x = 0$); that is, there are absolutely convergent power series expansions $p(x) = \sum_{n=0}^{\infty} p_n x^n$, $q(x) = \sum_{n=0}^{\infty} q_n x^n$ valid on some interval around $x = 0$. In this case, $x = 0$ is said to be an **ordinary point** for the differential equation. We have seen in Section 5.2 that this includes the case in which $p(x)$ and $q(x)$ are rational functions, provided the denominators are not zero at $x = 0$.

---

[13] This equation, and certain polynomials associated with it (see p. 400, footnote 16) are named after the French mathematician Hermite (1822–1901), although he was not the first to encounter the polynomials—which appear in an 1859 paper by the Russian mathematician Chebyshev (1821–1894).

We will attack the initial value problem $\dfrac{d^2y}{dx^2} + p(x)\dfrac{dy}{dx} + q(x)y = 0$, $y(0) = y_0$, $y'(0) = z_0$ using the method of Section 5.1; that is, we'll look for an analytic solution of the form $y(x) = \sum_{n=0}^{\infty} a_n x^n$. The following theorem (which we won't prove) assures us that the search will be successful.

Theorem 5.3.1 If $p(x)$ and $q(x)$ are analytic (at $x = 0$), then the solution of the initial value problem

$$\dfrac{d^2y}{dx^2} + p(x)\dfrac{dy}{dx} + q(x)y = 0, \qquad y(0) = y_0, \quad y'(0) = z_0$$

is an analytic function $y = \sum_{n=0}^{\infty} a_n x^n$. Note that $a_0 = y(0) = y_0$, $a_1 = y'(0) = z_0$ are given by the initial conditions. The series $\sum_{n=0}^{\infty} a_n x^n$ is absolutely convergent at least on the interval $(-R, R)$ on which the series $\sum_{n=0}^{\infty} p_n x^n$ and $\sum_{n=0}^{\infty} q_n x^n$ for $p(x)$ and $q(x)$ both converge.

EXAMPLE 5.3.2 (continued) For the Hermite equation $y'' - 2xy' + \lambda y = 0$ all solutions can be written as power series that are absolutely convergent for all x, since

$$p(x) = -2x = -2x + 0 \cdot x^2 + 0 \cdot x^3 + \cdots$$

and

$$q(x) = \lambda = \lambda + 0 \cdot x + 0 \cdot x^2 + \cdots$$

are analytic and their series converge for all x. ☐

EXAMPLE 5.3.3 Consider the **Legendre equation**[14]

$$(1 - x^2)y'' - 2xy' + \nu(\nu + 1)y = 0,$$

in which ν is an unspecified constant. The reason that the coefficient of y is called $\nu(\nu + 1)$ rather than just ν [15] has to do with the way the equation, like the Hermite

---

[14] This equation, and certain polynomials associated with it (see p. 402, footnote 17), are named after the French mathematician Legendre (1752–1833), who first encountered the polynomials in a study of the attraction of spheroids (i.e., ellipsoids of revolution) and planetary motion.

[15] Note that this imposes a restriction on the coefficient: For any real number ν, we have

$$\nu(\nu + 1) = (\nu + \dfrac{1}{2})^2 - \dfrac{1}{4} \geq -\dfrac{1}{4}.$$

SEC. 5.3 FIRST CASE: $x = 0$ AN ORDINARY POINT

equation, arises from a basic partial differential equation. In fact, the Legendre equation comes up when one studies the partial differential equation for electrical potential (the Laplace equation) in spherical coordinates; see Section 7.4, Exercise 16.

To put the equation in "standard" form, as needed for Theorem 5.3.1, we divide by $1 - x^2$ to get

$$y'' - \frac{2x}{1-x^2} y' + \frac{\nu(\nu+1)}{1-x^2} y = 0.$$

Thus $p(x) = \dfrac{-2x}{1-x^2}$, $q(x) = \dfrac{\nu(\nu+1)}{1-x^2}$, and we want to find the interval on which the series for $p(x)$ and $q(x)$ converge. In this case, these series can be found explicitly: Start with

$$\frac{1}{1-x} = 1 + x + x^2 + x^3 + \cdots \quad (|x| < 1) \quad (\text{see p. 386}),$$

and replace x by x^2 to get

$$\frac{1}{1-x^2} = 1 + x^2 + (x^2)^2 + (x^2)^3 + \cdots \quad (|x^2| < 1)$$

$$= 1 + x^2 + x^4 + x^6 + \cdots \quad (|x| < 1),$$

since $|x^2| < 1$ is equivalent to $|x| < 1$. Therefore, the series for $p(x)$ and $q(x)$ are

$$p(x) = \frac{-2x}{1-x^2} = -2x - 2x^3 - 2x^5 \cdots$$

and

$$q(x) = \frac{\nu(\nu+1)}{1-x^2} = \nu(\nu+1) + \nu(\nu+1)x^2 + \nu(\nu+1)x^4 + \cdots,$$

and both series converge on the interval $(-1, 1)$. Theorem 5.3.1 now guarantees us that all the solutions to the Legendre equation will be analytic (at $x = 0$), with power series that are absolutely convergent *at least* for $|x| < 1$. (As we'll see on p. 402, in certain cases such a series might converge on a larger interval, perhaps even for all x.)

We could have reached the same conclusion, *without* actually finding the series for $p(x)$ and $q(x)$, by finding the radii of convergence of these series, as described on p. 390. Since $p(x)$ and $q(x)$ have the same denominator $1 - x^2 = (1 - x)(1 + x)$, whose roots 1 and -1 both have absolute value 1, the radius of convergence of each of their series *had* to be 1. □

Now that Theorem 5.3.1 has reassured us that we will have convergent power series solutions, the method of finding the solutions is just as in Section 5.1: Simply substitute $y = \sum\limits_{n=0}^{\infty} a_n x^n$ into the equation and solve for the coefficients a_n. A recurrence relation for the a_n will be found. From this relation one can find all the a_n (in principle), or at least as many of the a_n as desired.

EXAMPLE 5.3.4   We return to the Hermite equation $y'' - 2xy' + \lambda y = 0$ from Example 5.3.2 and look for solutions of the form

$$y = \sum_{n=0}^{\infty} a_n x^n = a_0 + a_1 x + \cdots.$$

As we saw in Section 5.2, we may differentiate term by term, and thus $y' = \sum_{n=0}^{\infty} n a_n x^{n-1}$, so $-2xy' = -2 \sum_{n=0}^{\infty} n a_n x^n$. Also, $y'' = \sum_{n=0}^{\infty} n(n-1) a_n x^{n-2} = \sum_{n=2}^{\infty} n(n-1) a_n x^{n-2}$ (the terms with $n = 0$ and $n = 1$, corresponding to the second derivatives of a_0 and $a_1 x$, are zero anyway), so we can shift the index of summation to get

$$y'' = \sum_{n=2}^{\infty} n(n-1) a_n x^{n-2} = \sum_{(n+2)=2}^{\infty} (n+2)((n+2)-1) a_{n+2} x^{(n+2)-2}$$

$$= \sum_{n=0}^{\infty} (n+2)(n+1) a_{n+2} x^n.$$

Now that we have seen how to write all the power series we need, y'', $-2xy'$, and λy, in the usual form, we can combine them:

$$y'' - 2xy' + \lambda y = \sum_{n=0}^{\infty} (n+2)(n+1) a_{n+2} x^n - \sum_{n=0}^{\infty} 2n a_n x^n + \sum_{n=0}^{\infty} \lambda a_n x^n$$

$$= \sum_{n=0}^{\infty} [(n+2)(n+1) a_{n+2} - (2n - \lambda) a_n] x^n.$$

For y to be a solution, this power series must be zero, so

$$(n+2)(n+1) a_{n+2} - (2n - \lambda) a_n = 0$$

$$a_{n+2} = \frac{2n - \lambda}{(n+2)(n+1)} a_n \qquad \text{for all } n.$$

This is the recurrence relation for our equation. For the first few values of $n = 0, 1, 2, 3, \ldots$ the relation yields

$$a_2 = -\frac{\lambda}{2} a_0, \quad a_3 = \frac{2 - \lambda}{3 \cdot 2} a_1, \quad a_4 = \frac{4 - \lambda}{4 \cdot 3} a_2, \quad a_5 = \frac{6 - \lambda}{5 \cdot 4} a_3, \quad \ldots.$$

SEC. 5.3 FIRST CASE: $x = 0$ AN ORDINARY POINT

Notice that once a_0 is given, all the even-numbered a's, that is, a_2, a_4, and so on, are determined by the recurrence relation:

$$a_2 = -\frac{\lambda}{2} a_0$$

$$a_4 = \frac{4 - \lambda}{4 \cdot 3} a_2 = \frac{(4 - \lambda)(-\lambda)}{4!} a_0$$

$$a_6 = \frac{8 - \lambda}{6 \cdot 5} a_4 = \frac{(8 - \lambda)(4 - \lambda)(-\lambda)}{6!} a_0, \quad \text{and so on.}$$

Similarly, a_1 determines all the odd-numbered coefficients. a_0 and a_1 are arbitrary; they are the initial values for y and its derivative, since $y(0) = a_0$, $y'(0) = a_1$. Let us look at two particular cases.

Case 1 $y(0) = 1$, $y'(0) = 0$. For this initial value problem, $a_1 = 0$, so the power series will have only terms of even degree, starting with $a_0 = 1$. If we call the particular solution we get $y_1(x)$, then we have

$$y_1(x) = 1 - \frac{\lambda}{2} x^2 + \frac{(4 - \lambda)(-\lambda)}{4!} x^4 + \frac{(8 - \lambda)(4 - \lambda)(-\lambda)}{6!} x^6 + \cdots.$$

Case 2 $y(0) = 0$, $y'(0) = 1$. This time, the power series has only terms of odd degree. Let's call it $y_2(x)$:

$$y_2(x) = x + \frac{2 - \lambda}{3!} x^3 + \frac{(6 - \lambda)(2 - \lambda)}{5!} x^5 + \cdots.$$

If we now return to the general case $y(0) = a_0$, $y'(0) = a_1$, we see that we can write the (general) solution in terms of the two particular solutions above:

$$y(x) = a_0 + a_1 x - \frac{\lambda}{2} a_0 x^2 + \frac{2 - \lambda}{3!} a_1 x^3 + \frac{(4 - \lambda)(-\lambda)}{4!} a_0 x^4 + \frac{(6 - \lambda)(2 - \lambda)}{5!} a_1 x^5 + \cdots$$

$$= a_0 \left[1 - \frac{\lambda}{2} x^2 + \frac{(4 - \lambda)(-\lambda)}{4!} x^4 + \cdots \right] + a_1 \left[x + \frac{2 - \lambda}{3!} x^3 + \frac{(6 - \lambda)(2 - \lambda)}{5!} x^5 + \cdots \right]$$

$$= a_0 y_1(x) + a_1 y_2(x).$$

Note once again that by Theorem 5.3.1, all these power series are absolutely convergent for all x.

Using our terminology from Section 2.2, we see that $y_1(x)$ and $y_2(x)$ form a pair of basic solutions of the linear differential equation $y'' - 2xy' + \lambda y = 0$. (Why is neither of y_1, y_2 a constant multiple of the other?)

An interesting special case occurs when λ is a nonnegative even integer. For instance, when $\lambda = 4$, the series for $y_1(x)$ breaks off after the term with x^2, since $4 - \lambda = 0$. So for $\lambda = 4$, $y_1(x) = 1 - 2x^2$. Similarly, whenever $\lambda \geq 0$ is an even integer, one of the two power series (y_1 or y_2 according to whether λ is one of 0, 4, 8, . . . or one of 2, 6, 10, . . .) breaks off, so it is actually a polynomial. The *other* power series does not break off, so the differential equation has just one polynomial solution, up to multiplication by a constant; its degree is the integer $m = \frac{1}{2}\lambda$. For instance, for $m = 0, 1, 2, 3$ we have polynomial solutions $1, x, 1 - 2x^2, x - \frac{2}{3}x^3$. [16] ∎

EXAMPLE 5.3.5

We now return to the Legendre equation $(1 - x^2)y'' - 2xy' + \nu(\nu + 1)y = 0$ from Example 5.3.3. On p. 397, we divided through by $1 - x^2$ to get $y'' + p(x)y' + q(x)y = 0$ with

$$p(x) = \frac{-2x}{1 - x^2} = -2x - 2x^3 - 2x^5 \cdots \quad (|x| < 1),$$

$$q(x) = \frac{\nu(\nu + 1)}{1 - x^2} = \nu(\nu + 1)(1 + x^2 + x^4 + \cdots) \quad (|x| < 1),$$

and concluded that all the solutions will have absolutely convergent power series expansions for $|x| < 1$. To find the solutions, we could substitute $y = \sum_{n=0}^{\infty} a_n x^n$ in the "standard" form of the equation and start multiplying power series. However, it is much more convenient to substitute $y = \sum_{n=0}^{\infty} a_n x^n$ in the original equation. Differentiating term by term, we have

$$y' = \sum_{n=0}^{\infty} n a_n x^{n-1}, \quad -2xy' = -\sum_{n=0}^{\infty} 2n a_n x^n,$$

$$y'' = \sum_{n=0}^{\infty} n(n-1) a_n x^{n-2}, \quad -x^2 y'' = -\sum_{n=0}^{\infty} n(n-1) a_n x^n,$$

[16] In some applications it is convenient to multiply these solutions by constants in such a way that the coefficient of x^m is 2^m, as in 1, 2x, $-2(1 - 2x^2) = 4x^2 - 2$, $-12(x - \frac{2}{3}x^3) = 8x^3 - 12x$. Once this is done, the polynomials are called the **Hermite polynomials** and the polynomial of degree m (or **order** m) corresponding to $\lambda = 2m$ is denoted by $H_m(x)$. (See also Exercise 26.)

SEC. 5.3 FIRST CASE: $x = 0$ AN ORDINARY POINT

and also $y'' = \sum_{n=0}^{\infty} (n + 2)(n + 1)a_{n+2}x^n$. (If you're confused, go back and re-read Example 5.3.4.) Combining these series, we get

$$(1 - x^2)y'' - 2xy' + \nu(\nu + 1)y$$

$$= \sum_{n=0}^{\infty} (n + 2)(n + 1)a_{n+2}x^n - \sum_{n=0}^{\infty} n(n - 1)a_n x^n - \sum_{n=0}^{\infty} 2na_n x^n + \sum_{n=0}^{\infty} \nu(\nu + 1)a_n x^n$$

$$= \sum_{n=0}^{\infty} [(n + 2)(n + 1)a_{n+2} + (\nu(\nu + 1) - n(n + 1))a_n]x^n.$$

Therefore, the recurrence relation will be

$$(n + 2)(n + 1)a_{n+2} + [\nu(\nu + 1) - n(n + 1)]a_n = 0$$

or

$$a_{n+2} = \frac{n(n + 1) - \nu(\nu + 1)}{(n + 1)(n + 2)} a_n.$$

Just as for the Hermite equation, a_0 determines all the even-degree coefficients and a_1 all the odd-degree ones. Once again, if we denote by $y_1(x)$ the special solution with $a_0 = 1$, $a_1 = 0$:

$$y_1(x) = 1 - \frac{\nu(\nu + 1)}{2} x^2 + \frac{[-\nu(\nu + 1)][6 - \nu(\nu + 1)]}{4!} x^4 + \cdots$$

and by $y_2(x)$ the special solution with $a_0 = 0$, $a_1 = 1$:

$$y_2(x) = x + \frac{2 - \nu(\nu + 1)}{3!} x^3 + \frac{[2 - \nu(\nu + 1)][12 - \nu(\nu + 1)]}{5!} x^5 + \cdots,$$

then the general solution is $y(x) = a_0 y_1(x) + a_1 y_2(x)$.

Note that if ν is a positive integer or zero, one of the two power series will break off [$y_1(x)$ if ν is even, $y_2(x)$ if ν is odd]. For instance, for $\nu = 0, 2, 4$, respectively, we have

$$y_1(x) = 1, \ 1 - 3x^2, \ 1 - 10x^2 + \frac{35}{3} x^4,$$

while for $\nu = 1, 3, 5$ we have

$$y_2(x) = x, \ x - \frac{5}{3} x^3, \ x - \frac{14}{3} x^3 + \frac{21}{5} x^5.$$

So whenever ν is an integer $m \geq 0$, there is a unique polynomial solution of degree m (up to constant multiples).[17] In particular, that (finite) power series will converge for all x, not just for $|x| < 1$. ∎

EXAMPLE 5.3.6 We next consider the **Airy equation**,[18] $y'' - xy = 0$, whose solutions are used in the theory of diffraction of radio waves around the earth's surface. [This and much other information about specific applications of solutions to "named" differential equations in this chapter can be found in Lebedev, *Special Functions and Their Applications* (Dover, 1972).]

Since we have $p(x) = 0$, $q(x) = -x$, Theorem 5.3.1 (p. 396) assures us that the solutions will be power series $y = \sum_{n=0}^{\infty} a_n x^n$ that converge absolutely for all x. Writing the terms in the equation as series, we get

$$xy = x \sum_{n=0}^{\infty} a_n x^n = \sum_{n=0}^{\infty} a_n x^{n+1} = \sum_{n-1=0}^{\infty} a_{n-1} x^{(n-1)+1} = \sum_{n=1}^{\infty} a_{n-1} x^n$$

and

$$y'' = \sum_{n=0}^{\infty} (n+2)(n+1) a_{n+2} x^n.$$

Note that the constant term is missing in the series for xy, so to combine the two series it is a good idea to take the constant term separately in y'', as well:

$$y'' = 2a_2 + \sum_{n=1}^{\infty} (n+2)(n+1) a_{n+2} x^n$$

$$y'' - xy = 2a_2 + \sum_{n=1}^{\infty} [(n+2)(n+1) a_{n+2} - a_{n-1}] x^n.$$

---

[17] In some applications it is useful to multiply these solutions by constants in such a way that they have the value 1 for $x = 1$, as in 1, x, $-\frac{1}{2}(1 - 3x^2) = \frac{3}{2}x^2 - \frac{1}{2}$, $-\frac{3}{2}(x - \frac{5}{3}x^3) = \frac{5}{2}x^3 - \frac{3}{2}x$. Once this is done, the polynomials are called the **Legendre polynomials** and the polynomial of degree (or **order**) m is denoted by $P_m(x)$.

[18] Although this equation and its solutions, the **Airy functions**, are named after the English astronomer Sir George Airy (1801–1892), it appears [see Watson, *Bessel Functions*, 2nd ed. (Macmillan, 1945), pp. 188–189 for a fuller account] that while Airy encountered a particular solution in the form of an integral, it was the well-known mathematical physicist Sir George Stokes (1819–1903) who made the connection to the differential equation.

So the recurrence relation will be

$$(n+1)(n+2)a_{n+2} - a_{n-1} = 0 \implies a_{n+2} = \frac{a_{n-1}}{(n+1)(n+2)} \quad (n \geq 1);$$

we have the additional equation $2a_2 = 0$ from the constant term. In particular, this time, a_0 does not determine all the even-degree coefficients; instead, a_0 determines $a_3 = \frac{a_0}{2 \cdot 3}$, $a_6 = \frac{a_3}{5 \cdot 6} = \frac{a_0}{2 \cdot 3 \cdot 5 \cdot 6}$, and so on. Similarly, a_1 determines the coefficients $a_4 = \frac{a_1}{3 \cdot 4}$, $a_7 = \frac{a_4}{6 \cdot 7} = \frac{a_1}{3 \cdot 4 \cdot 6 \cdot 7}$, and so on. What about the other coefficients? Well, $a_2 = 0$, so $a_5 = \frac{a_2}{4 \cdot 5} = 0$, $a_8 = \frac{a_5}{7 \cdot 8} = 0$, and so on.

If we denote the special solution with $a_0 = 1$, $a_1 = 0$ by $y_1(x)$, then

$$y_1(x) = 1 + \frac{1}{2 \cdot 3} x^3 + \frac{1}{2 \cdot 3 \cdot 5 \cdot 6} x^6 + \frac{1}{2 \cdot 3 \cdot 5 \cdot 6 \cdot 8 \cdot 9} x^9 + \cdots.$$

Similarly, for $a_0 = 0$, $a_1 = 1$ we have

$$y_2(x) = x + \frac{1}{3 \cdot 4} x^4 + \frac{1}{3 \cdot 4 \cdot 6 \cdot 7} x^7 + \frac{1}{3 \cdot 4 \cdot 6 \cdot 7 \cdot 9 \cdot 10} x^{10} + \cdots.$$

The general solution is $y(x) = a_0 y_1(x) + a_1 y_2(x)$. As noted, all these series converge absolutely for all x. ∎

In each of the examples above we found two basic solutions $y_1(x)$ and $y_2(x)$, corresponding to $a_0 = 1$, $a_1 = 0$ and $a_0 = 0$, $a_1 = 1$, respectively, and the general solution then came out to be $y(x) = a_0 y_1(x) + a_1 y_2(x)$. This will always work, and it really has nothing to do with power series; it is simply a result of the theory of Section 2.2. The conditions $a_0 = 1$, $a_1 = 0$ and $a_0 = 0$, $a_1 = 1$ are just the initial conditions $y(0) = 1$, $y'(0) = 0$ and $y(0) = 0$, $y'(0) = 1$, respectively. If $y_1(x)$ and $y_2(x)$ are the solutions corresponding to these initial conditions, they are not constant multiples of each other (why?), and thus they form a pair of basic solutions. The solution $y(x)$ with $y(0) = a_0$, $y'(0) = a_1$ is a solution to the same initial value problem as $a_0 y_1(x) + a_1 y_2(x)$, and so $y(x) = a_0 y_1(x) + a_1 y_2(x)$, as claimed.

In each of our examples so far, the recurrence relation was relatively uncomplicated, and we were able to find a pattern to the coefficients of the power series solutions. As our final example illustrates, this is not always the case.

EXAMPLE 5.3.7 $(3x + 1)y'' + (x - 1)y' - 2y = 0$.

We have $p(x) = \frac{x - 1}{3x + 1}$, $q(x) = \frac{-2}{3x + 1}$, so the radius of convergence of both the series for $p(x)$ and $q(x)$ is $\frac{1}{3}$ (the absolute value of the root $-\frac{1}{3}$ of $3x + 1$). Thus by

Theorem 5.3.1, there will be absolutely convergent power series solutions for $|x| < \dfrac{1}{3}$.

When we try to find them by substituting $y = \sum\limits_{n=0}^{\infty} a_n x^n$ into the differential equation, we first get

$$xy' = \sum_{n=0}^{\infty} n a_n x^n, \qquad y' = \sum_{n=0}^{\infty} (n+1) a_{n+1} x^n,$$

$$xy'' = \sum_{n=0}^{\infty} n(n-1) a_n x^{n-1} = \sum_{n=0}^{\infty} (n+1) n a_{n+1} x^n,$$

$$y'' = \sum_{n=0}^{\infty} (n+2)(n+1) a_{n+2} x^n.$$

Combining all these yields the recurrence relation

$$3(n+1) n a_{n+1} + (n+2)(n+1) a_{n+2} + n a_n - (n+1) a_{n+1} - 2 a_n = 0,$$

which simplifies to

$$a_{n+2} = -\frac{3n+2}{n+2} a_{n+1} - \frac{n-2}{(n+1)(n+2)} a_n.$$

If we look for the solution $y_1(x)$ with $a_0 = 1$, $a_1 = 0$, the recurrence relation gives us

$$a_2 = -a_1 + a_0 = 1 \quad \text{(for } n=0\text{)}, \qquad a_3 = -\frac{5}{3} a_2 + \frac{1}{6} a_1 = -\frac{5}{3},$$

$$a_4 = -2 a_3 = \frac{10}{3}, \qquad a_5 = -\frac{11}{5} a_4 - \frac{1}{20} a_3 = -\frac{29}{4},$$

but the more coefficients we compute, the less of a pattern there seems to be. We now know that

$$y_1(x) = 1 + x^2 - \frac{5}{3} x^3 + \frac{10}{3} x^4 - \frac{29}{4} x^5 + \cdots,$$

but without a clear idea of what might come next. As for the solution $y_2(x)$ with $a_0 = 0$, $a_1 = 1$, its power series expansion starts off

$$y_2(x) = x - x^2 + \frac{11}{6} x^3 - \frac{11}{3} x^4 + \cdots$$

(see Exercise 15), and again no pattern appears. Even such "messy" series, however, can be helpful for numerical computations, especially if $|x|$ is small enough for the series to converge rapidly. ∎

SEC. 5.3 FIRST CASE: $x = 0$ AN ORDINARY POINT

SUMMARY OF KEY CONCEPTS, RESULTS, AND TECHNIQUES

Ordinary point (p. 395): $x = 0$ is called an ordinary point for the differential equation $\frac{d^2y}{dx^2} + p(x)\frac{dy}{dx} + q(x)y = 0$ if $p(x)$ and $q(x)$ are analytic at $x = 0$. In this case, the solutions of the differential equation are also analytic. Specifically, the solution of any initial value problem

$$\frac{d^2y}{dx^2} + p(x)\frac{dy}{dx} + q(x)y = 0, \quad y(0) = y_0, \quad y'(0) = z_0$$

is given by a power series $y = \sum_{n=0}^{\infty} a_n x^n$ (with $a_0 = y_0$, $a_1 = z_0$) which is absolutely convergent at least on the interval $(-R, R)$ where the series expansions for $p(x)$ and $q(x)$ both converge (Theorem 5.3.1, p. 396). To actually find these "power series solutions," use the technique from Section 5.1: substitute $y = \sum_{n=0}^{\infty} a_n x^n$ into the differential equation to get a recurrence relation for the a_n. The general solution will have the form $y = a_0 y_1 + a_1 y_2$, where the basic solutions $y_1(x)$ and $y_2(x)$ are the power series solutions satisfying $y_1(0) = 1$, $y_1'(0) = 0$ and $y_2(0) = 0$, $y_2'(0) = 1$, respectively (p. 403). For some important differential equations, there are special cases in which polynomial solutions exist. Specifically, the Hermite equation $y'' - 2xy' + \lambda y = 0$ has polynomial solutions if λ is a nonnegative even integer (p. 400), while the Legendre equation $(1 - x^2)y'' - 2xy' + \nu(\nu + 1)y = 0$ has polynomial solutions if ν is a nonnegative integer (pp. 401–402).

EXERCISES

For each of the following differential equations:

(a) State for which value of R the power series expansions of the solutions are guaranteed to converge for $|x| < R$.
(b) Find the recurrence relation.
(c) Write down the first three nonzero terms for each of two basic (power series) solutions.

1. $(x^2 + x - 2)y'' - y' + y = 0$.
2. $(x^2 - 4x + 4)y'' + 2y' - y = 0$.
3. $(x + 3)y'' - y' + 2y = 0$.
4. $(x - 4)y'' + 3y' + y = 0$.
5. $y'' - xy' + 3xy = 0$.
6. $y'' + xy' - 6xy = 0$.
7. $(x - 2)y'' + (2x + 1)y' - y = 0$.
8. $(x + 2)y'' + (x - 4)y' + 3y = 0$.

For each of the following differential equations:

(a) State for which value of R the power series expansions of the solutions are guaranteed to converge for $|x| < R$.
(b) Find the recurrence relation.
(c) Find two basic (power series) solutions.

9. $y'' - 5xy' + 3y = 0$.
10. $y'' + 4xy' - 2y = 0$.
11. $(x^2 - 3)y'' + xy' - y = 0$.
12. $(x^2 + 4)y'' + xy' - y = 0$.
13. $y'' + 3xy = 0$.
14. $y'' + 8xy = 0$.

15. Show that in Example 5.3.7, the coefficients given for the power series expansion of $y_2(x)$ are indeed correct, and find the coefficient of x^5 in this expansion.

16. (Compare Example 5.3.1.)
   (a) Show that if the initial value problem
   $$x^2(x - 1)y'' - xy' - (x + 3)(x - 1)y = 0, \qquad y(0) = y_0, \quad y'(0) = z_0$$
   has a solution, then $y_0 = 0$.
   (b) Show that if the initial value problem
   $$x^2(x - 1)y'' - xy' - (x + 3)(x - 1)y = 0, \qquad y(1) = y_1, \quad y'(1) = z_1$$
   has a solution, then $z_1 = 0$.

For each of the following differential equations, find the values of λ for which there is a *polynomial* solution.

17. $y'' + 4xy' - \lambda y = 0$.
18. $y'' - 3xy' - \lambda y = 0$.
19. $(x^2 - 4)y'' + xy' - \lambda y = 0$.
20. $(x^2 + 2)y'' + xy' + \lambda y = 0$.

21. Although the Hermite equation and polynomials were not named after Chebyshev (see p. 395, footnote 13), the **Chebyshev equation** $(1 - x^2)y'' - xy' + \nu^2 y = 0$, where ν is an unspecified constant, was.
   (a) Find the recurrence relation for the Chebyshev equation.
   (b) Exhibit a pair of basic (power series) solutions. What do you know about their radius of convergence?
   (c) For what values of ν will there be polynomial solutions? Exhibit polynomial solutions of degree m for $m = 0, 1, 2, 3, 4$. [If these polynomials are multiplied by constants in such a way that the coefficient of x^m is 2^{m-1} (for $m > 0$), they are then called the **Chebyshev polynomials**. These polynomials crop up in many areas of mathematics, including numerical analysis. See "Further Reading" below.]

22. (a) Show that the recurrence relation for the equation $(x - 1)y'' - xy' + y = 0$ is
$$a_{n+2} = \frac{n}{n + 2} a_{n+1} + \frac{1 - n}{(n + 1)(n + 2)} a_n.$$

SEC. 5.3 FIRST CASE: $x = 0$ AN ORDINARY POINT

(b) Solve the initial value problem
$$(x - 1)y'' - xy' + y = 0, \qquad y(0) = 0, \quad y'(0) = 1.$$

(c) Solve the initial value problem
$$(x - 1)y'' - xy' + y = 0, \qquad y(0) = 1, \quad y'(0) = 1.$$

Write your answer in closed form (after you find the power series).

(d) Find the general solution, in closed form, of $(x - 1)y'' - xy' + y = 0$.

23. (a) Show that the recurrence relation for the equation
$$(1 - 2x)y'' - 8xy' - 8y = 0 \text{ is}$$

$$a_{n+2} = \frac{2n}{n+2} a_{n+1} + \frac{8}{n+2} a_n.$$

(b) Solve the initial value problem
$$(1 - 2x)y'' - 8xy' - 8y = 0, \qquad y(0) = 1, \quad y'(0) = 2.$$

Write your answer in closed form (after you find the power series).

*(c) Use the method of reduction of order to find the general solution of
$$(1 - 2x)y'' - 8xy' - 8y = 0.$$

24. (a) Show that the recurrence relation for the equation
$$-(x + 2)y'' + (2x + 2)y' + 2y = 0 \text{ is}$$

$$a_{n+2} = \frac{1}{n+2} a_n + \frac{2-n}{2(n+2)} a_{n+1}.$$

(b) Solve the initial value problem
$$-(x + 2)y'' + (2x + 2)y' + 2y = 0, \qquad y(0) = \frac{1}{2}, \quad y'(0) = -\frac{1}{4},$$

and write your answer in closed form.

*(c) Use the method of reduction of order to find the general solution of
$$-(x + 2)y'' + (2x + 2)y' + 2y = 0.$$

25. (a) Solve the initial value problem
$$(1 - x)(1 - 3x)y'' + (12x - 8)y' + 6y = 0, \qquad y(0) = 1, \quad y'(0) = 1,$$

and write your answer in closed form.

(b) As part (a), but with initial conditions $y(0) = 1$, $y'(0) = 3$.

(c) Find the general solution of
$$(1 - x)(1 - 3x)y'' + (12x - 8)y' + 6y = 0.$$

*26. In this exercise we show that the Hermite polynomials, as defined in footnote 16 on p. 400, are given by $H_m(x) = (-1)^m e^{x^2} \dfrac{d^m(e^{-x^2})}{dx^m}$, that is, by multiplying the mth derivative of e^{-x^2} by $(-1)^m e^{x^2}$.

(a) Find $(-1)^m e^{x^2} \dfrac{d^m(e^{-x^2})}{dx^m}$ for $m = 0, 1, 2$, and check that you have indeed found the appropriate Hermite polynomials.

(b) Show that for m even, $m \geq 2$,

$$H_m(x) = \frac{2^m m!}{(-4)(-8)\cdots(4-2m)(-2m)} y_1(x),$$

where $y_1(x)$ is the first basic solution from p. 399, with $\lambda = 2m$.
[*Hint*: You know that $H_m(x)$ is a multiple of $y_1(x)$; you just have to check that this is the correct multiple.]

(c) Show that for m odd,

$$H_m(x) = \frac{2^m m!}{(-4)(-8)\cdots(6-2m)(2-2m)} y_2(x),$$

where $y_2(x)$ is the second basic solution from p. 399, with $\lambda = 2m$.

(d) Show that the derivative $y_2'(x)$ of the second basic solution is equal to the first basic solution $y_1(x)$, but with λ replaced by $\lambda - 2$.

(e) Show that for m odd, $H_m'(x) = 2m H_{m-1}(x)$.

(f) Show that for m even, $m \geq 2$, we still have $H_m'(x) = 2m H_{m-1}(x)$.
[*Hint*: Part (f) will be similar to part (e) once you have something like part (d) for $y_1'(x)$.]

(g) Show that $H_m''(x) - 2x H_m'(x) + 2m H_m(x) = 0$.
[*Hint*: Look at the Hermite equation.]

(h) Combine the results of parts (e), (f), and (g) to show that for $m \geq 1$, $H_m(x) = 2x H_{m-1}(x) - H_{m-1}'(x)$.

(i) Show that for $m \geq 1$, $\dfrac{d}{dx}[H_{m-1}(x)e^{-x^2}] = -H_m(x)e^{-x^2}$.

(j) Show, by induction on m, that $H_m(x) = (-1)^m e^{x^2} \dfrac{d^m(e^{-x^2})}{dx^m}$.

FURTHER READING

For more on Chebyshev polynomials, see Rivlin, *The Chebyshev Polynomials* (Wiley, 1974).

5.4 SINGULAR POINTS: INTRODUCTION

We continue our study of the differential equation $\dfrac{d^2y}{dx^2} + p(x)\dfrac{dy}{dx} + q(x)y = 0$, where $p(x)$ and $q(x)$ are rational functions. However, we will no longer assume that $p(x)$ and $q(x)$ are defined at $x = 0$, and so our method from Section 5.3 may no longer work.

EXAMPLE 5.4.1 $xy'' + (1 - x)y' + \nu y = 0$, where ν is an unspecified constant. This is the **Laguerre equation**[19]; there are sometimes (as we'll see) polynomial solutions, and these come up repeatedly in mathematical physics—for instance, in the theory of the hydrogen atom. Note that unless $\nu = 0$, neither $p(x) = \dfrac{1}{x} - 1$ nor $q(x) = \dfrac{\nu}{x}$ is defined at $x = 0$. □

EXAMPLE 5.4.2 $x(x - 2)\dfrac{d^2y}{dx^2} + x^2 \dfrac{dy}{dx} - (x - 3)y = 0$.

Here $p(x) = \dfrac{x}{x - 2}$ is defined at $x = 0$, but $q(x)$ is not. ■

EXAMPLE 5.4.3 $xy'' + y' + xy = 0$.

This equation, a special case of the Bessel equation (which we'll study in Section 5.6), arises from the partial differential equation describing the oscillation of a heavy chain, or rope, suspended from the ceiling[20]—a problem already considered, and largely solved, by Daniel Bernoulli in the 1730s. Strangely enough, $x = 0$ corresponds to the *free* end of the chain, yet $p(x) = \dfrac{1}{x}$ is undefined at $x = 0$. ■

A point where $p(x)$, $q(x)$, or both are undefined is called a **singular point** for the differential equation[21]; all other points are called **ordinary points**. Thus $x = 0$ is a singular point in all three examples above; in Example 5.4.2, $x = 2$ is also a singular point.

Your first inclination may well be to ignore singular points; after all, there are only finitely many of them (a polynomial has only a finite number of roots), so they can easily be avoided. In each of the examples above, we could take $x - \dfrac{1}{2}$ as the variable instead of

---

[19] This equation, and certain polynomials associated with it (see p. 412, footnote 23) are named after the French mathematician Laguerre (1834–1886), although, as in the case of Hermite (p. 395, footnote 13), "his" polynomials had previously appeared in Chebyshev's 1859 paper.

[20] See Section 7.4, Exercise 17.

[21] If it's not known that $p(x)$ and $q(x)$ are rational functions, then a singular point is defined as one where $p(x)$ and $q(x)$ are not both analytic. See Section 5.7.

x; since $\frac{1}{2}$ is an ordinary point, we will find absolutely convergent power series expansions in $x - \frac{1}{2}$ for all the solutions of the differential equation. Of course, these expansions will probably not converge for all x; Theorem 5.3.1 will only guarantee that they converge for $\left| x - \frac{1}{2} \right| < \frac{1}{2}$, that is, for $0 < x < 1$. Still, since $p(x)$ is not defined at $x = 0$, can one expect any better?

In a sense, no. If $x = 0$ is a singular point, then solutions will have a tendency to behave "strangely" there: They may become undefined, or they may cease to be differentiable, or they may be defined on one side of $x = 0$ and not on the other, and so forth. However, that is just what makes singular points so important and why they must be studied carefully. It is clear that it will be important in applications not only to know that "some drastic change might occur in the solution" at a certain value of x, but also just what that change can be expected to be like! Conversely, we may know in advance that a certain type of change *cannot* occur and then use this to rule out certain solutions. This will happen, for instance, in the case of the heavy chain mentioned in Example 5.4.3; see Section 7.4, Exercise 17.

Unfortunately, changing the variable and considering a power series expansion about some other point is unlikely to give any insight into the problem. Power series expansions $\sum_{n=0}^{\infty} a_n(x - c)^n$ are designed to give information on what happens near $x = c$: what the higher derivatives at c are, how to compute the function approximately near $x = c$, and so on. Sometimes such expansions are useful for all x; after all, "near" is a relative concept. However, when a power series expansion no longer converges, it is practically useless[22] for our purposes.

EXAMPLES

$$\log x = \log[1 + (x - 1)] = (x - 1) - \frac{(x - 1)^2}{2} + \frac{(x - 1)^3}{3} - \frac{(x - 1)^4}{4} + \cdots \quad \text{(see p. 387)};$$

$$\frac{1}{x} = \frac{1}{1 - -(x - 1)} = 1 - (x - 1) + (x - 1)^2 - (x - 1)^3 + \cdots$$

(see p. 386, or by differentiation from the expansion above);

$$x^{5/3} = 1 + \frac{5}{3}(x - 1) + \frac{5}{9}(x - 1)^2 - \frac{5}{81}(x - 1)^3 + \cdots \quad \text{(see p. 388)}.$$

---

[22] And potentially dangerous, because one might think that the partial sums would provide reasonable approximations to the function when, in fact, they do nothing of the sort.

All these three expansions are absolutely convergent for $0 < x < 2$ (that is, for $|x-1| < 1$), and they are useful—to establish the behavior of $\log x$, $\frac{1}{x}$, and $x^{5/3}$ near $x = 1$. However, they do not help much if we want to understand how these functions behave near $x = 0$. ■

The rest of this chapter will be devoted to the study of the case when $x = 0$ is a singular point of the equation $\frac{d^2y}{dx^2} + p(x)\frac{dy}{dx} + q(x)y = 0$. We will assume for now that $p(x)$ and $q(x)$ are rational functions; we can then multiply through by the denominators, after which the equation will look like $f_2(x)\frac{d^2y}{dx^2} + f_1(x)\frac{dy}{dx} + f_0(x)y = 0$ for suitable polynomials $f_2(x)$, $f_1(x)$, and $f_0(x)$. We have $f_2(0) = 0$, since $x = 0$ is a singular point. Note that the examples on p. 409 are already in this form; for instance, for the Laguerre equation, $f_2(x) = x$, $f_1(x) = 1 - x$, $f_0(x) = \nu$.

How are we to solve such an equation $f_2(x)y'' + f_1(x)y' + f_0(x)y = 0$, where f_0, f_1, and f_2 are polynomials and $f_2(0) = 0$? One idea would be to still look for solutions of the form $y = \sum_{n=0}^{\infty} a_n x^n$, even though we can no longer use Theorem 5.3.1 to conclude that *all* solutions must be of this form. Let's see what happens in a few examples if this is done.

EXAMPLE 5.4.4 $x^2 y'' + (3x - 1)y' + y = 0$.

If we substitute $y = \sum_{n=0}^{\infty} a_n x^n$ and differentiate term by term, we have

$$x^2 y'' = \sum_{n=0}^{\infty} n(n-1) a_n x^n, \quad 3xy' = \sum_{n=0}^{\infty} 3n a_n x^n, \quad -y' = -\sum_{n=0}^{\infty} (n+1) a_{n+1} x^n,$$

so

$$x^2 y'' + (3x - 1)y' + y = \sum_{n=0}^{\infty} [n(n-1)a_n + 3n a_n - (n+1)a_{n+1} + a_n] x^n.$$

Thus the recurrence relation will be

$$n(n-1)a_n + 3n a_n - (n+1)a_{n+1} + a_n = 0$$
$$(n^2 + 2n + 1)a_n = (n+1)a_{n+1}$$
$$a_{n+1} = (n+1)a_n.$$

Note that the first coefficient a_0 determines a_1, a_2, and so on, in turn. This already shows that not *all* the solutions are found in this way, since we can only impose one initial condition $y(0) = a_0$. For instance, taking $a_0 = 1$, we find $a_1 = 1$, $a_2 = 2$, $a_3 = 6 = 3!$, $a_4 = 24 = 4!$; since $(n + 1)! = (n + 1)n!$, this pattern will persist and we get $a_n = n!$, $y = \sum_{n=0}^{\infty} n! \, x^n$. Now let's investigate when this power series converges. Its radius of convergence is given (see pp. 376–377) by

$$\lim_{n \to \infty} \left| \frac{a_n}{a_{n+1}} \right| = \lim_{n \to \infty} \frac{n!}{(n+1)!} = \lim_{n \to \infty} \frac{1}{n+1} = 0 \quad \text{(ouch!)}.$$

So the power series converges for $x = 0$ only. It is a "formal" solution; that is, when it is differentiated term by term, it seems to satisfy the equation, but it is not an actual solution: $y = \sum_{n=0}^{\infty} n! \, x^n$ does not have a value except for $x = 0$. ∎

EXAMPLE 5.4.5 $xy'' + (1 - x)y' + \nu y = 0$, the Laguerre equation from Example 5.4.1. If we substitute $y = \sum_{n=0}^{\infty} a_n x^n$ and differentiate termwise, we have

$$xy'' = \sum_{n=0}^{\infty} n(n+1)a_{n+1}x^n, \quad y' = \sum_{n=0}^{\infty} (n+1)a_{n+1}x^n, \quad -xy' = \sum_{n=0}^{\infty} -na_n x^n,$$

and thus

$$xy'' + (1 - x)y' + \nu y = \sum_{n=0}^{\infty} [n(n+1)a_{n+1} + (n+1)a_{n+1} - na_n + \nu a_n] x^n.$$

The recurrence relation will be

$$(n+1)^2 a_{n+1} + (\nu - n)a_n = 0$$

$$a_{n+1} = \frac{n - \nu}{(n+1)^2} a_n.$$

As in Example 5.4.4, a_0 determines all the other coefficients. For instance, if we take $a_0 = 1$, we get

$$y = 1 - \nu x + \frac{(1-\nu)(-\nu)}{2^2} x^2 + \frac{(2-\nu)(1-\nu)(-\nu)}{3^2 \cdot 2^2} x^3 + \cdots.$$

Note that if ν is a positive integer or zero, the power series will break off, and we'll have a polynomial of degree ν.[23] Otherwise, the series will continue indefinitely, and its radius

---

[23] These polynomials (with $a_0 = 1$) are called the **Laguerre polynomials**; the Laguerre polynomial of degree m is denoted by $L_m(x)$. Thus the first few Laguerre polynomials are

$$L_0(x) = 1, \; L_1(x) = 1 - x, \; L_2(x) = 1 - 2x + \frac{1}{2}x^2, \; L_3(x) = 1 - 3x + \frac{3}{2}x^2 - \frac{1}{6}x^3.$$

of convergence will be

$$\lim_{n \to \infty} \left| \frac{a_n}{a_{n+1}} \right| = \lim_{n \to \infty} \frac{(n+1)^2}{|n - \nu|} = \infty.$$

That is, our power series solution, which is unique up to constant multiples, will converge for all x (whether or not ν is a nonnegative integer).

Starting with the solution $y_1(x) = 1 - \nu x + \frac{(1-\nu)(-\nu)}{2^2} x^2 + \cdots$ found above, one can now find a second basic solution to the Laguerre equation using the method of reduction of order (from Section 2.4). Although this second solution is first found in the form of an integral [see Exercises 17(a), 18, and 19], it can be rewritten in the form

$$y_2(x) = \begin{cases} y_1(x) \log x + \sum_{n=0}^{\infty} b_n x^n & (x > 0) \\ y_1(x) \log(-x) + \sum_{n=0}^{\infty} b_n x^n & (x < 0) \end{cases} \quad \text{[see Exercises 17(b) and 20]}.$$

Here $\sum_{n=0}^{\infty} b_n x^n$ is analytic at $x = 0$, that is, absolutely convergent on some interval $(-R, R)$. Note, however, that because of the "logarithmic part," y_2 is undefined for $x = 0$; in fact, $\lim_{x \to 0} y_2(x) = -\infty$. In Section 5.7 we'll return to the material of this paragraph in a more general context. ∎

EXAMPLE 5.4.6

$x^2 y'' - 2xy' + 2y = 0$.

If we again substitute $y = \sum_{n=0}^{\infty} a_n x^n$ and differentiate "formally," that is, without considering convergence for the moment, our equation becomes

$$\sum_{n=0}^{\infty} [n(n-1)a_n - 2na_n + 2a_n] x^n = 0.$$

That is, we must have $n(n-1)a_n - 2na_n + 2a_n = 0$ for all n. Note that this is *not* a recurrence relation, since only one coefficient occurs! If we factor out a_n and simplify, we get $(n-1)(n-2)a_n = 0$. Therefore, we must have $a_n = 0$ *unless* $n = 1$ or $n = 2$: All coefficients $a_3, a_4, \ldots$ of degree > 2, as well as the constant term a_0, must be zero. If $n = 1$ or $n = 2$, on the other hand, there is no condition on a_n at all, because the equation $(n-1)(n-2)a_n = 0$ will then automatically be true. That is, *any* function $y = a_1 x + a_2 x^2$, a_1, a_2 arbitrary, will be a solution. In particular, x and x^2 form a pair of basic solutions, so (by Theorem 2.2.3) we've found *all* solutions of our second-order equation, at least on any interval where $x \neq 0$. (It turns out that this restriction can be removed for our particular equation; see Exercise 16.) Surprisingly, all these solutions are actually analytic at the singular point $x = 0$! ∎

EXAMPLE 5.4.7 $x^2 y'' - 2xy' + \frac{5}{4} y = 0$.

The computation is very similar to the one in the previous example; substituting $y = \sum_{n=0}^{\infty} a_n x^n$ yields

$$n(n-1)a_n - 2na_n + \frac{5}{4} a_n = 0$$

$$\left(n - \frac{1}{2}\right)\left(n - \frac{5}{2}\right) a_n = 0 \quad \text{for all } n.$$

Since n is never $\frac{1}{2}$ or $\frac{5}{2}$, this shows that all the a_n are zero, and so we only get the trivial solution 0 as a power series. However, this computation does show how a solution may be found: Just as in the previous example getting $(n-1)(n-2)a_n = 0$ meant that a_1 and a_2 could be chosen arbitrarily and that $a_1 x + a_2 x^2$ was then always a solution, in this example getting $(n - \frac{1}{2})(n - \frac{5}{2}) a_n = 0$ means that it should be possible to choose $a_{1/2}$ and $a_{5/2}$ arbitrarily and have $a_{1/2} x^{1/2} + a_{5/2} x^{5/2}$ as a solution. Sure enough, $x^{1/2}$ and $x^{5/2}$ form a pair of basic solutions to the equation (check this by substitution), and so the general solution is $\alpha \cdot x^{1/2} + \beta \cdot x^{5/2}$, α and β arbitrary constants. (Note that these solutions are defined only for $x \geq 0$; we'll see later what happens for $x < 0$.) ∎

At this point you may be a bit confused by the variety of things that can happen at a singular point, from no (nonzero) analytic solution at all (as in Examples 5.4.4 and 5.4.7) through one analytic solution (up to constant multiples; as in Example 5.4.5) to all analytic solutions (as in Example 5.4.6). Actually, all this is often quite predictable, as we'll see in Section 5.7. First, though, we'll look carefully at two particularly important cases of differential equations with a singular point, starting in the next section with the simplest case: that of Euler equations. Knowing the behavior of the solutions to these equations, which include the equations in Examples 5.4.6 and 5.4.7, will be helpful later when dealing with more complicated cases.

SUMMARY OF KEY CONCEPTS, RESULTS, AND TECHNIQUES

Singular point (p. 409): For a differential equation $\frac{d^2 y}{dx^2} + p(x) \frac{dy}{dx} + q(x) y = 0$ with rational functions $p(x)$ and $q(x)$, a singular point is a point where $p(x)$, $q(x)$, or both are undefined; all other points are **ordinary points**. If $x = 0$ is an ordinary point, we can use the method of Section 5.3 to find power series solutions in x. If $x = 0$ is a singular point, we can still try to find nonzero solutions of the form $y = \sum_{n=0}^{\infty} a_n x^n$ (p. 411).

SEC. 5.4 SINGULAR POINTS: INTRODUCTION

However, caution is needed; no such solutions may exist, and worse, there may be "formal" solutions that appear to work but are actually divergent series (see Example 5.4.4, p. 411). On the other hand, the Laguerre equation $xy'' + (1 - x)y' + \nu y = 0$ has a convergent power series solution, which is a polynomial if ν is a nonnegative integer (pp. 412–413). There even are equations for which all solutions are analytic at $x = 0$, although $x = 0$ is a singular point (see Example 5.4.6, p. 413).

EXERCISES

Find the singular points for each of the following differential equations.

1. $x(x^2 - 3)y'' - 4x^2 y' + 2xy = 0.$
2. $x(x^2 - 4)y'' + 3y' - 6xy = 0.$
3. $x(x + 1)y'' + (x + 1)^2 y' - 4y = 0.$
4. $x(x + 1)y'' + (1 - x^2)y' + (2x + 2)y = 0.$

For each of the following differential equations:

(a) Find all "formal" power series solutions (in x). (If there is no clear pattern to the coefficients, list the first three nonzero terms.)
(b) Determine which of your solutions from part (a) actually converge on some interval $(-R, R)$, and find their radius of convergence.

5. $xy'' - 3xy' - y = 0.$
6. $xy'' + 5xy' - 2y = 0.$
7. $x^2 y'' - 6y = 0.$
8. $x^2 y'' - 27y = 0.$
9. $(x^2 - 2x)y'' + 3y = 0.$
10. $(x^2 + 4x)y'' - 5y = 0.$
11. $x^2 y'' - y' + y = 0.$
12. $x^2 y'' + 5y' - y = 0.$
13. $x^3 y'' + xy' - y = 0.$
14. $x^3 y'' - 2xy' - y = 0.$

15. (a) Show that $x(x - 1)y'' + (x + 3)y' + xy = 0$ has a nonzero "formal" power series solution.
    [*Hint*: The first several terms are zero.]
    *(b) Find the radius of convergence for this power series.

16. (Compare Example 5.4.6.) Let $y(x)$ be a solution of $x^2 y'' - 2xy' + 2y = 0$ on an interval $(-R, R)$ about $x = 0$.
    (a) Show that there are constants α and β such that for x in the interval $(0, R)$, $y(x) = \alpha x + \beta x^2$.
    (b) Show that there are constants γ and δ such that for x in the interval $(-R, 0)$, $y(x) = \gamma x + \delta x^2$.
    (c) Note that since $y''(0)$ is defined, $y'(x)$ must be continuous at $x = 0$. Use this to show that $\alpha = \gamma$.
    *(d) Show that $\beta = \delta$, as well, and that $y(x) = \alpha x + \beta x^2$ for all x in $(-R, R)$.

17. (a) Find a second basic solution (besides $y_1 = 1$) to the Laguerre equation of order zero $xy'' + (1 - x)y' = 0$. Your answer will be in the form of an integral.

(b) Show that for $x > 0$, your answer can be rewritten as

$$\log x + x + \frac{x^2}{4} + \frac{x^3}{3 \cdot 3!} + \frac{x^4}{4 \cdot 4!} + \cdots,$$

while for $x < 0$ you get

$$\log(-x) + x + \frac{x^2}{4} + \frac{x^3}{3 \cdot 3!} + \frac{x^4}{4 \cdot 4!} + \cdots.$$

18. Using the method of reduction of order, show that a second basic solution [besides the Laguerre polynomial $L_1(x) = 1 - x$] to the Laguerre equation of order one $xy'' + (1 - x)y' + y = 0$ is given by $(1 - x) \int \dfrac{e^x}{x(1-x)^2} dx$. What is the role of the arbitrary integration constant in this formula?

*19. Use the method of reduction of order to show that if $y_1(x)$ is a nonzero solution to the Laguerre equation $xy'' + (1 - x)y' + \nu y = 0$, then so is

$$y_2(x) = y_1(x) \int \frac{e^x}{x[y_1(x)]^2} dx.$$

What is the role of the arbitrary integration constant in this formula?

*20. Let $y_1(x) = 1 - \nu x + \dfrac{(1-\nu)(-\nu)}{2^2} x^2 + \cdots$ be the power series solution to the Laguerre equation, found on p. 412.

(a) Show that $\dfrac{e^x}{[y_1(x)]^2}$ has a power series expansion with constant term 1, which converges on some interval $(-R, R)$.
[*Hint*: See Section 5.2 for division of power series, but don't actually try to *compute* the expansion!]

(b) Show that $\dfrac{e^x}{x[y_1(x)]^2}$ can be written as $\dfrac{1}{x} + \sum_{n=0}^{\infty} d_n x^n$, where $\sum_{n=0}^{\infty} d_n x^n$ is analytic at $x = 0$.

(c) Show that the second solution $y_2(x)$ from Exercise 19 is of the form

$$y_2(x) = \begin{cases} y_1(x) \log x + \sum_{n=0}^{\infty} b_n x^n & (x > 0) \\ y_1(x) \log(-x) + \sum_{n=0}^{\infty} b_n x^n & (x < 0), \end{cases}$$

where $\sum_{n=0}^{\infty} b_n x^n$ is analytic at $x = 0$.

(d) Show that the constant of integration can be chosen so as to get $b_0 = 0$.

(e) By substituting the expression from part (c) for $y_2(x)$ into the Laguerre equation, show that the b_n satisfy the recurrence relation

$$(n+1)^2 b_{n+1} + (\nu - n)b_n + 2(n+1)a_{n+1} - a_n = 0,$$

where a_n denotes the coefficient of x^n in the power series expansion for $y_1(x)$.

FURTHER READING

For a more general form of the Laguerre polynomials, see Lebedev, *Special Functions and Their Applications* (Dover, 1972), Sections 4.17 and 4.18.

5.5 EULER EQUATIONS

A differential equation of the form $x^2 \dfrac{d^2y}{dx^2} + ax \dfrac{dy}{dx} + by = 0$, where a and b are constants, is often called an **Euler equation**. Other names, which we won't use, include Cauchy–Euler equation, Cauchy equation, and equidimensional equation.[24]

Having seen two examples (for $a = -2, b = 2$ and $a = -2, b = \dfrac{5}{4}$) of the solution of such equations as Examples 5.4.6 and 5.4.7, we know that it is reasonable to look for solutions of the form $y = x^\lambda$, where the exponent λ is not necessarily an integer. However, if λ is not an integer, then $x^\lambda = e^{\lambda \log x}$ is usually defined for $x > 0$ only. For the moment, then, we'll assume that $x > 0$. (Since we're interested in the behavior of the solutions near $x = 0$, eventually we'll have to look at the case $x < 0$, as well.)

When we substitute $y = x^\lambda$ into the Euler equation $x^2 \dfrac{d^2y}{dx^2} + ax \dfrac{dy}{dx} + by = 0$, we get $\lambda(\lambda - 1)x^\lambda + a\lambda x^\lambda + bx^\lambda = 0$, which will be true provided $\lambda(\lambda - 1) + a\lambda + b = 0$, that is,

$$\lambda^2 + (a-1)\lambda + b = 0.$$

---

[24] Although, to be sure, these equations were studied by the great mathematicians Euler (1707–1783) and Cauchy (1789–1857), Euler's teacher Jean Bernoulli (1667–1748), one of a celebrated family of Swiss mathematicians, already knew a way to solve them. See "Further Reading" following the Exercises.

So if λ satisfies this quadratic equation, then x^λ will be a solution of the Euler equation (and vice versa). In particular, if $(a-1)^2 > 4b$, then there will be two distinct roots $\lambda_{1,2} = \dfrac{-(a-1) \pm \sqrt{(a-1)^2 - 4b}}{2}$ of the quadratic equation, and the functions $x^{\lambda_1}, x^{\lambda_2}$ will form a pair of basic solutions for the Euler equation. The general solution (still for $x > 0$) will therefore be given by $\alpha x^{\lambda_1} + \beta x^{\lambda_2} = \alpha e^{\lambda_1 \log x} + \beta e^{\lambda_2 \log x}$, α and β arbitrary constants.

This result may remind you of the general solution of a constant-coefficient second-order equation in the easiest case, that is, the case that the characteristic equation has distinct roots. We found in Section 2.3 that if those roots were λ_1, λ_2, then that general solution was $\alpha e^{\lambda_1 t} + \beta e^{\lambda_2 t}$, with α and β arbitrary constants. Comparing this to our result $\alpha e^{\lambda_1 \log x} + \beta e^{\lambda_2 \log x}$, it looks as if $\log x$ is playing the role of t, and also as if our equation $\lambda^2 + (a-1)\lambda + b = 0$ might be the characteristic equation.

Let's see whether there really is such a connection! That is, let's introduce a new variable $t = \log x$ in the Euler equation $x^2 \dfrac{d^2 y}{dx^2} + ax \dfrac{dy}{dx} + by = 0$ and see whether we get a constant-coefficient equation. We're now going to think of y as (an unknown) function of t, so we should express everything in the differential equation in terms of t and the derivatives $\dfrac{dy}{dt}, \dfrac{d^2 y}{dt^2}$. Here goes:

First of all, $x = e^t$, so $\dfrac{dx}{dt} = e^t$ and (by the chain rule)

$$\frac{dy}{dx} = \frac{dy}{dt} \bigg/ \frac{dx}{dt} = \frac{dy}{dt} \bigg/ e^t$$

$$= e^{-t} \frac{dy}{dt}.$$

Now we can use this to work on the second derivative:

$$\frac{d^2 y}{dx^2} = \frac{d}{dx}\left(\frac{dy}{dx}\right) = \frac{d}{dt}\left(\frac{dy}{dx}\right) \bigg/ \frac{dx}{dt}$$

$$= \frac{d}{dt}\left(e^{-t} \frac{dy}{dt}\right) \bigg/ e^t = e^{-t} \frac{d}{dt}\left(e^{-t} \frac{dy}{dt}\right)$$

$$= -e^{-2t} \frac{dy}{dt} + e^{-2t} \frac{d^2 y}{dt^2}$$

$$= e^{-2t} \left(\frac{d^2 y}{dt^2} - \frac{dy}{dt}\right).$$

Finally, the Euler equation becomes

$$(e^t)^2 \cdot e^{-2t} \left(\frac{d^2 y}{dt^2} - \frac{dy}{dt}\right) + ae^t \cdot e^{-t} \frac{dy}{dt} + by = 0,$$

SEC. 5.5 EULER EQUATIONS

which simplifies to

$$\frac{d^2y}{dt^2} + (a-1)\frac{dy}{dt} + by = 0.$$

Sure enough, this is a constant-coefficient equation, and its characteristic equation is exactly the equation $\lambda^2 + (a-1)\lambda + b = 0$ that we found earlier!

Notice that in making the connection between the Euler equation and the constant-coefficient equation, we didn't use the condition $(a-1)^2 > 4b$ at all. (We did use $x > 0$, though, when we put $t = \log x$.) Therefore, we can use this connection to find the solutions of the Euler equation (for $x > 0$) in the remaining two cases $(a-1)^2 = 4b$ and $(a-1)^2 < 4b$, since we know the solutions of the constant-coefficient equation in those cases from Sections 2.4 and 2.5, respectively.

EXAMPLE 5.5.1

$$x^2 \frac{d^2y}{dx^2} - x\frac{dy}{dx} + y = 0 \quad (x > 0).$$

This Euler equation (for which $a = -1$, $b = 1$) is equivalent to

$$\frac{d^2y}{dt^2} - 2\frac{dy}{dt} + y = 0, \quad \text{where} \quad t = \log x.$$

The characteristic equation $\lambda^2 - 2\lambda + 1 = 0$ has a double root $\lambda = 1$, so by Section 2.4 a pair of basic solutions is given by

$$y_1 = e^t = x, \quad y_2 = te^t = (\log x)x = x \log x,$$

and the general solution to our Euler equation is

$$y = \alpha x + \beta x \log x, \quad \alpha, \beta \text{ arbitrary}.$$

[For graphs of the basic solutions, see Figure 5.5.1(v), p. 423.] ∎

EXAMPLE 5.5.2

$$x^2 \frac{d^2y}{dx^2} + 3x\frac{dy}{dx} + 5y = 0 \quad (x > 0).$$

This time we have $a = 3$, $b = 5$, and the equivalent constant-coefficient equation is

$$\frac{d^2y}{dt^2} + 2\frac{dy}{dt} + 5y = 0 \quad (t = \log x).$$

The characteristic equation has complex conjugate roots $\lambda = -1 \pm 2i$. By Section 2.5, $\text{Re}[e^{(-1+2i)t}] = e^{-t}\cos 2t$ and $\text{Im}[e^{(-1+2i)t}] = e^{-t}\sin 2t$ form a pair of basic real-valued solutions. In terms of x, these are

$$e^{-t}\cos 2t = x^{-1}\cos(2 \log x) \quad \text{and} \quad e^{-t}\sin 2t = x^{-1}\sin(2 \log x),$$

so the general real-valued solution to our original equation is

$$y = x^{-1}[\alpha \cos(2 \log x) + \beta \sin(2 \log x)], \quad \alpha, \beta \text{ arbitrary}.$$

[For graphs of the basic solutions, see Figure 5.5.1(ix), p. 423.] ∎

In hindsight, we could have made the change of variables $x = e^t$, $t = \log x$ right away when we first encountered Euler equations in Section 5.4, but how were we to know? The route we followed—first looking for power series solutions, then recognizing that solutions sometimes had the form x^λ, finally changing the variable—is much more typical of what happens in practice (if you're very lucky, at that) when you're confronted with a new type of equation.

How about the case $x < 0$? A similar change of variables is used, and here's one way of finding the trick. For $x > 0$ we took $t = \log x$, which is no longer possible if $x < 0$. But $\log x = \int \dfrac{dx}{x}$ (suppressing the integration constant), so we could also have written $t = \int \dfrac{dx}{x}$, and *this* we can do again if x is negative. That is, in the case $x < 0$, we can try $t = \int \dfrac{dx}{x} = \log(-x)$, which corresponds to $x = -e^t$. Indeed, if we make the change of variables $x = -e^t$, $t = \log(-x)$ in the Euler equation $x^2 \dfrac{d^2y}{dx^2} + ax \dfrac{dy}{dx} + by = 0$ ($x < 0$), it turns out (see Exercise 13) that we get *the same* constant-coefficient equation $\dfrac{d^2y}{dt^2} + (a - 1) \dfrac{dy}{dt} + by = 0$ that we found earlier for the case $x > 0$. Therefore, in terms of t the solutions will "look" the same; in terms of x, they will be the old solutions with x replaced by $-x$.

EXAMPLE 5.5.3 $2x^2 y'' - 7xy' + 10y = 0$.

This is an Euler equation with $a = -\dfrac{7}{2}$, $b = 5$. The equivalent constant-coefficient equation, either for $x > 0$ or for $x < 0$, is $\dfrac{d^2y}{dt^2} - \dfrac{9}{2} \dfrac{dy}{dt} + 5y = 0$, where $t = \log x$ if $x > 0$ and $t = \log(-x)$ if $x < 0$. We find roots $\lambda_1 = 2$ and $\lambda_2 = \dfrac{5}{2}$ for the characteristic equation $\lambda^2 - \dfrac{9}{2}\lambda + 5 = 0$, so e^{2t} and $e^{5t/2}$ are basic solutions. That is, for $x > 0$ our equation has basic solutions $e^{2\log x} = x^2$ and $e^{5/2 \log x} = x^{5/2}$, while for $x < 0$ we have basic solutions $e^{2\log(-x)} = (-x)^2 = x^2$ and $(-x)^{5/2}$. ∎

We can now write down the solutions to the Euler equation $x^2 \dfrac{d^2y}{dx^2} + ax \dfrac{dy}{dx} + by = 0$, both for $x > 0$ and for $x < 0$, in all cases, using the general formulas from Sections 2.3 to 2.5. Rather than memorizing all the new formulas below, however, it might be a good idea simply to remember that the change of variables $\begin{cases} t = \log x & (x > 0) \\ t = \log(-x) & (x < 0) \end{cases}$ converts the Euler equation into a constant-coefficient one, whose characteristic equation is $\lambda^2 + (a - 1)\lambda + b = 0$.

SEC. 5.5 EULER EQUATIONS

Solutions to the Euler equation $\quad x^2 \dfrac{d^2y}{dx^2} + ax\dfrac{dy}{dx} + by = 0$.

Case 1 $(a-1)^2 > 4b$. If λ_1 and λ_2 are the (distinct, real) roots of $\lambda^2 + (a-1)\lambda + b = 0$, then $x^{\lambda_1}(= e^{\lambda_1 t})$ and x^{λ_2} form a pair of basic solutions for $x > 0$, while $(-x)^{\lambda_1}$ and $(-x)^{\lambda_2}$ are basic solutions for $x < 0$. The general solution is

$$\begin{cases} \alpha x^{\lambda_1} + \beta x^{\lambda_2} & (x > 0) \\ \alpha (-x)^{\lambda_1} + \beta (-x)^{\lambda_2} & (x < 0) \end{cases}.^{25}$$

Case 2 $(a-1)^2 = 4b$. If λ is the double root of $\lambda^2 + (a-1)\lambda + b = 0$, then x^λ and $x^\lambda \log x$ $(= te^{\lambda t})$ form a pair of basic solutions for $x > 0$, while $(-x)^\lambda$ and $(-x)^\lambda \log(-x)$ are basic solutions for $x < 0$. The general solution is

$$\begin{cases} x^\lambda [\alpha + \beta \log x] & (x > 0) \\ (-x)^\lambda [\alpha + \beta \log(-x)] & (x < 0) \end{cases}.$$

Case 3 $(a-1)^2 < 4b$. If $\lambda = \mu + i\nu$, $\bar\lambda = \mu - i\nu$ are the complex conjugate roots of $\lambda^2 + (a-1)\lambda + b = 0$, then

$$x^\mu \cos(\nu \log x) \qquad [= e^{\mu t} \cos \nu t = \operatorname{Re}(e^{\lambda t})]$$

and

$$x^\mu \sin(\nu \log x) \qquad [= e^{\mu t} \sin \nu t = \operatorname{Im}(e^{\lambda t})]$$

form a pair of basic solutions for $x > 0$, while

$$(-x)^\mu \cos[\nu \log(-x)] \quad \text{and} \quad (-x)^\mu \sin[\nu \log(-x)]$$

are basic solutions for $x < 0$. The general solution is

$$\begin{cases} x^\mu [\alpha \cos(\nu \log x) + \beta \sin(\nu \log x)] & (x > 0) \\ (-x)^\mu [\alpha \cos(\nu \log(-x)) + \beta \sin(\nu \log(-x))] & (x < 0) \end{cases}.$$

In each case above, α and β are arbitrary constants.

Another way to think about (and remember?) the above is to go back to our approach on p. 417, where we showed that if λ is a solution to $\lambda^2 + (a-1)\lambda + b = 0$, then x^λ is a solution to the Euler equation (for $x > 0$). *This is still true for complex λ,* provided that we define $x^\lambda = e^{\lambda \log x}$ (but how else would we define a power with a complex exponent?); see Exercise 16. Of course, if $\lambda = \mu + i\nu$ is a complex number, x^λ will be a complex-valued solution; as usual, to get real-valued solutions we take real and imaginary parts:

$$\operatorname{Re}(x^\lambda) = \operatorname{Re}[e^{(\mu + i\nu)\log x}] = e^{\mu \log x} \cos(\nu \log x) = x^\mu \cos(\nu \log x),$$

$$\operatorname{Im}(x^\lambda) = x^\mu \sin(\nu \log x).$$

Again, for $x < 0$, replace x by $-x$ in all these formulas.

---

[25] Many authors would write this as $\alpha |x|^{\lambda_1} + \beta |x|^{\lambda_2}$, using $t = \log|x|$ as a convenient shorthand to cover both cases ($x > 0$ and $x < 0$). For some of the dangers of this notation, see Exercises 19 and 20 in this section, as well as Exercise 22 from Section 1.3.

Now that we have the solutions, how do they behave near the singular point $x = 0$? There are really two separate questions here, a "one-sided" and a "two-sided" one. The "one-sided" question, which we'll consider first, is how solutions defined for $x > 0$ behave as x approaches 0 from the right (by symmetry, the behavior of solutions defined for $x < 0$, as x approaches 0 from the left, will be similar). The "two-sided" question is whether there are any solutions that are defined on *both sides* of $x = 0$ *as well as* at $x = 0$ itself; that is, functions y such that y, $\frac{dy}{dx}$, and $\frac{d^2y}{dx^2}$ are defined at $x = 0$ and on both sides, and satisfy $x^2 \frac{d^2y}{dx^2} + ax \frac{dy}{dx} + by = 0$.

Back to the "one-sided" question. If λ is real, the behavior of x^λ is known from calculus; for $\lambda > 0$, $\lim_{x \to 0^+} x^\lambda = 0$, while for $\lambda < 0$, $\lim_{x \to 0^+} x^\lambda = \infty$. For the "transitional" value $\lambda = 0$, x^λ is the constant function 1 (for $x \neq 0$). So if we are in Case 1 above (λ_1, λ_2 distinct, real), then much of the behavior of the solutions will depend only on the signs of λ_1, λ_2; for instance, if $\lambda_1 > 0$, $\lambda_2 < 0$, there will be just one solution (up to constant multiples) that doesn't "blow up" as $x \to 0^+$, whereas if $\lambda_1 > 0$, $\lambda_2 > 0$, all solutions will approach zero as $x \to 0^+$. Parts (i), (ii), and (iii) of Figure 5.5.1 show specific examples of the behavior of basic solutions in Case 1 (for various values of a and b). Note that when $0 < \lambda < 1$, although the function x^λ approaches 0 as $x \to 0^+$, its *derivative* $\lambda x^{\lambda-1}$ approaches ∞: The graph is "vertical" as it leaves the origin.

In Case 2 (double root λ), we have to consider the second basic solution $x^\lambda \log x$ along with x^λ. As $x \to 0^+$, $\log x$ approaches $-\infty$, but very "slowly," in the sense that if $\lambda > 0$, $x^\lambda \log x$ will approach zero. For example, for $\lambda = 1$,

$$\lim_{x \to 0^+} x \log x = \lim_{\frac{1}{x} \to \infty} \frac{-\log\left(\frac{1}{x}\right)}{\frac{1}{x}} = \lim_{q \to \infty} \frac{-\log q}{q},$$

which is zero by l'Hôpital's rule. (See Exercise 15 for general $\lambda > 0$.) On the other hand, for $\lambda \leq 0$ we have $\lim_{x \to 0^+} x^\lambda \log x = -\infty$. Specific examples of the behavior of basic solutions in Case 2 are shown in Figure 5.5.1(iv), (v), and (vi). (See also Exercise 17.)

Finally, for Case 3 the behavior of the functions $\cos(\nu \log x)$ and $\sin(\nu \log x)$ (which are multiplied by x^μ) as $x \to 0^+$ is important. Assuming that $\nu > 0$ (why can we do so?), $\nu \log x$ will approach $-\infty$ as $x \to 0^+$, so $\cos(\nu \log x)$ and $\sin(\nu \log x)$ will fluctuate more and more quickly between 1 and -1, while maintaining their "phase difference." Therefore, $x^\mu \cos(\nu \log x)$ and $x^\mu \sin(\nu \log x)$ will show similar fluctuations between x^μ and $-x^\mu$. For specific examples, see Figure 5.5.1(vii), (viii), and (ix). (See also Exercise 18.)

One reason we've gone into so much detail is that, as we'll see in Section 5.7, the behavior as $x \to 0^+$ of the solutions to Euler equations provides "models" for the behavior of solutions to equations of a much more general form (still with a singular point at $x = 0$). In fact, a specific example of this approach will already show up in Section 5.6.

Meanwhile, we'll conclude this section with a quick look at the pitfalls of the "two-sided" question above: What about solutions defined at $x = 0$ *and* on *both* sides?

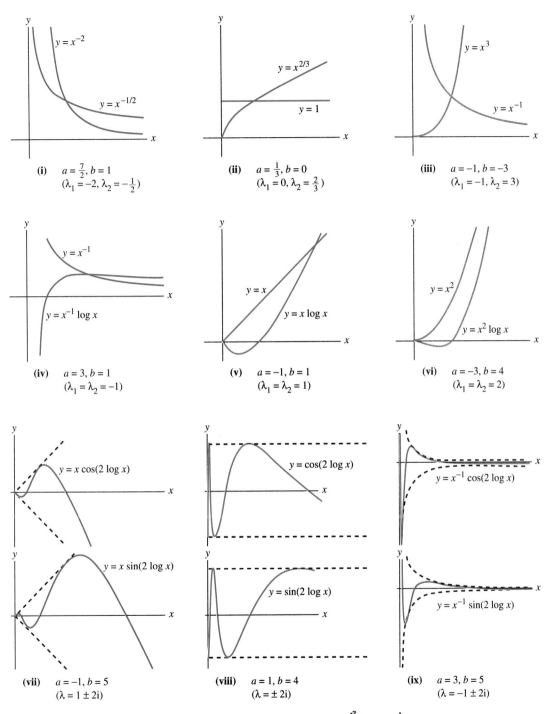

Figure 5.5.1. Pairs of basic solutions to the Euler equation $x^2 \dfrac{d^2y}{dx^2} + ax\dfrac{dy}{dx} + by = 0$ $(x > 0)$ for various choices of a and b.

In a few cases, functions we listed on p. 421 as basic solutions for $x > 0$ are already defined for all x; specifically, x^λ for nonnegative integers λ [26] and for positive rational λ with odd denominator, such as $\lambda = \dfrac{5}{3}$. Careful, though! If by a solution on an interval we mean (as we should) a function y such that y, $\dfrac{dy}{dx}$, and $\dfrac{d^2y}{dx^2}$ are defined *everywhere* on that interval and satisfy the equation, then a function such as $x^{5/3}$ does *not* qualify as a "two-sided" solution about $x = 0$, since for $y = x^{5/3}$, $\dfrac{d^2y}{dx^2} = \dfrac{10}{9} x^{-1/3}$ is undefined at $x = 0$. However, $y = x^\lambda$ *does* qualify if $\lambda > 2$ and λ is rational with odd denominator, or if λ is any nonnegative integer.

On the other hand, there are also cases in which we can come up with solutions by "patching together" solutions for $x > 0$ and for $x < 0$. Sometimes, in fact, there are more solutions than we might like!

EXAMPLE 5.5.4

$x^2 \dfrac{d^2y}{dx^2} - 8x \dfrac{dy}{dx} + 20y = 0$.

The characteristic equation factors as $(\lambda - 4)(\lambda - 5) = 0$, so for $x > 0$, x^4, x^5 is a pair of basic solutions, and $\alpha x^4 + \beta x^5$ is the general solution. This is also true for $x < 0$, since $(-x)^4 = x^4$ and $(-x)^5 = -x^5$ have the same linear combinations as x^4, x^5. However, there are solutions that are defined *for all x* and that are *not* of the form $\alpha x^4 + \beta x^5$; one example is

$$y = \begin{cases} 2x^4 - x^5 & (x \leq 0) \\ 3x^4 + 10x^5 & (x \geq 0) \end{cases}.$$

Note that for this function, y, $\dfrac{dy}{dx}$, and $\dfrac{d^2y}{dx^2}$ really are defined for all x and satisfy our Euler equation. (See also Exercise 28.) Of course, y is not analytic at $x = 0$; the *analytic* (power series) solutions have the form $\alpha x^4 + \beta x^5$. ∎

This example illustrates a theme from science fiction, but also from serious physics: If you go through a singular point, you may not look the same when you reappear on the other side!

SUMMARY OF KEY CONCEPTS, RESULTS, AND TECHNIQUES

Euler equation (p. 417): A differential equation of the form $x^2 \dfrac{d^2y}{dx^2} + ax \dfrac{dy}{dx} + by = 0$, with a, b constants. If λ is a root of the quadratic equation $\lambda^2 + (a - 1)\lambda + b = 0$, then $y = x^\lambda = e^{\lambda \log x}$ ($x > 0$) is a solution to the Euler equation (p. 417); this is still true,

---

[26] Strictly speaking, x^0 is not defined for $x = 0$. If we define $0^0 = 1$, though, all will be well.

SEC. 5.5 EULER EQUATIONS

but x^λ will be complex-valued, for complex λ (p. 421). For $x > 0$, the substitution $x = e^t$, $t = \log x$ transforms the Euler equation into the constant-coefficient equation $\dfrac{d^2y}{dt^2} + (a-1)\dfrac{dy}{dt} + by = 0$, whose characteristic equation is the quadratic equation for λ shown above (pp. 418–419). For $x < 0$, putting $x = -e^t$, $t = \log(-x)$ has the same result (p. 420).

Since constant-coefficient equations can be solved by the methods of Sections 2.3 to 2.5, we can get the general solution of any Euler equation this way, both for $x > 0$ and for $x < 0$; the results are listed on p. 421.

The various possibilities for the behavior of solutions to the Euler equation as $x \to 0^+$ are considered on p. 422 and illustrated in Figure 5.5.1 (p. 423).

Studying solutions that are valid for all x rather than just for $x > 0$ or for $x < 0$ is possible but tricky; in some cases, there are few if any such solutions, while in other cases, there are uncomfortably many (p. 424).

EXERCISES

Solve the following differential equations and initial value problems.

1. $x^2 y'' - 3xy' + 3y = 0$, $y(1) = -4$, $y'(1) = 2$.
2. $x^2 y'' - 4xy' + 6y = 0$, $y(1) = 4$, $y'(1) = 1$.
3. $x^2 y'' - 3xy' + 4y = 0$, $y(-1) = 2$, $y'(-1) = 0$.
4. $x^2 y'' + 5xy' + 4y = 0$ $(x < 0)$.
5. $x^2 y'' - 3xy' + 5y = 0$ $(x < 0)$.
6. $x^2 y'' + 5xy' + 5y = 0$ $(x > 0)$.
7. $xy'' + 5y' = 0$, $y(2) = 3$, $y'(2) = 1$.
8. $xy'' - 3y' = 0$, $y(-\tfrac{1}{2}) = 1$, $y'(-\tfrac{1}{2}) = 3$.
9. $x^3 \dfrac{d^3 y}{dx^3} + x^2 \dfrac{d^2 y}{dx^2} - 2x \dfrac{dy}{dx} + 2y = 0$ $(x < 0)$.
10. $x^3 \dfrac{d^3 y}{dx^3} - 2x^2 \dfrac{d^2 y}{dx^2} + 3x \dfrac{dy}{dx} - 3y = 0$ $(x > 0)$. (To check your answer, see Example 4.2.2.)
11. $x^2 y'' - 4xy' + 6y = x^3$ $(x > 0)$. (Use variation of constants.)
12. $x^2 y'' - 3xy' + 3y = x$ $(x < 0)$. (Same suggestion.)
13. Show that the change of variables $x = -e^t$, $t = \log(-x)$ transforms the Euler equation $x^2 \dfrac{d^2 y}{dx^2} + ax \dfrac{dy}{dx} + by = 0$ $(x < 0)$ into $\dfrac{d^2 y}{dt^2} + (a-1)\dfrac{dy}{dt} + by = 0$ (compare pp. 418–420).

14. Suppose that λ is a double root of the equation $\lambda^2 + (a-1)\lambda + b = 0$.
    (a) Show that $\lambda = \dfrac{1-a}{2}$.
    (b) Starting with the solution x^λ of the Euler equation $x^2 \dfrac{d^2y}{dx^2} + ax \dfrac{dy}{dx} + by = 0$ ($x > 0$), find the general solution *using* the method of *reduction of order*.

15. Show that for any $\lambda > 0$, $\lim_{x \to 0^+} x^\lambda \log x = 0$ (see p. 422).

    [*Hint*: $\log x = \dfrac{1}{\lambda} \log x^\lambda$; as $x \to 0^+$, $x^\lambda \to 0^+$.]

16. (a) Show that for $\lambda = \mu + i\nu$, the derivative of the complex-valued function $y = x^\lambda = e^{\lambda \log x}$ ($x > 0$) is $y' = \lambda x^{\lambda - 1}$.
    [*Hint*: Write y in real and imaginary parts, then use the definition from p. 108 to find y'.]
    (b) Show that if $\lambda = \mu + i\nu$ is a solution to $\lambda^2 + (a-1)\lambda + b = 0$, then $y = x^\lambda$ is a solution to $x^2 y'' + axy' + by = 0$ (as claimed on p. 421).

17. (a) Show that when $0 < \lambda \le 1$, although the function $x^\lambda \log x$ approaches 0 as $x \to 0^+$, its derivative approaches $-\infty$. Explain what this has to do with Figure 5.5.1(v).
    (b) Show that when $\lambda > 1$, the derivative of $x^\lambda \log x$ approaches 0 as $x \to 0^+$. Explain the connection with Figure 5.5.1(vi).

18. Investigate what happens to the derivative of $x^\mu \cos(\nu \log x)$, $\nu > 0$, as $x \to 0^+$, in each of the following cases: (a) $0 < \mu < 1$; (b) $\mu = 1$; (c) $\mu > 1$.

19. Show that although $y = |x|$ is a solution of $2x^2 y'' - 3xy' + 3y = 0$ for $x > 0$ and also for $x < 0$, $y = |x|$ is *not* a solution for all x. However, show that $y = x$ is a solution for all x.

20. (a) Show that $y_1 = |x|^2$ and $y_2 = |x|^3$ are both solutions of $x^2 y'' - 4xy' + 6y = 0$ for all x, and that y_1, y_2 form a pair of basic solutions for $x > 0$ and also for $x < 0$.
    (b) However, show that $y = 2x^2 + x^3$ is an analytic solution of $x^2 y'' - 4xy' + 6y = 0$ which is *not* a linear combination of y_1 and y_2. Show explicitly how y is a linear combination of y_1 and y_2 if we restrict to the interval $(0, \infty)$, and if we restrict to the interval $(-\infty, 0)$.

Sketch two basic solutions to the Euler equation $x^2 y'' + axy' + by = 0$ for $x < 0$, in each of the following cases.

21. $a = 0$, $b = -6$.   22. $a = -1$, $b = 1$.
23. $a = 5$, $b = 4$.   24. $a = -5$, $b = 9$.
25. $a = 1$, $b = 1$.   26. $a = -1$, $b = 5$.

*27. (a) In a case such as that of Figure 5.5.1(i) (p. 423) where both basic solutions x^{-2} and $x^{-1/2}$ "blow up" as $x \to 0^+$, one could imagine that some combination of them might *not* blow up. Show, however, that no such "nice" linear combination exists. That is, show that

$$\lim_{x \to 0^+} (\alpha x^{-2} + \beta x^{-1/2})$$

is only finite if $\alpha = \beta = 0$.

(b) Show that the analogous statement is true for the case of Figure 5.5.1(ix).

*28. (See Example 5.5.4, p. 424.)

(a) Show that we can specify $y(1) = y_1$, $y'(1) = z_1$, $y(-1) = y_{-1}$, $y'(-1) = z_{-1}$, *all at once*, arbitrarily, and get a solution of $x^2 \dfrac{d^2 y}{dx^2} - 8x \dfrac{dy}{dx} + 20y = 0$, defined for all x, which meets the specifications. Note that this shows not only that the initial conditions $y(-1) = y_{-1}$, $y'(-1) = z_{-1}$ don't determine the function y uniquely, but also that they have *no influence at all* on what y will be for $x > 0$ (beyond the singular point).

(b) Show that the phenomenon described in part (a) occurs for the Euler equation $x^2 y'' + axy' + by = 0$ whenever $\lambda^2 + (a-1)\lambda + b = 0$ has two distinct real roots λ_1, λ_2 with $\lambda_1 > 2$, $\lambda_2 > 2$.

*29. Investigate whether the phenomenon described in Exercise 28 ever occurs when $\lambda^2 + (a-1)\lambda + b = 0$ has a double root. If so, for what values of the double root?

*30. Investigate whether the phenomenon described in Exercise 28 ever occurs when $\lambda^2 + (a-1)\lambda + b = 0$ has complex roots $\lambda, \overline{\lambda} = \mu \pm i\nu$ ($\nu > 0$). If so, for what values of μ and ν?

FURTHER READING

Much information about the history of solution methods for differential equations can be found in Appendix A of Ince, *Ordinary Differential Equations* (Dover, 1944); specifically, Jean Bernoulli's solution method for Euler equations is at the beginning of Section A·4.

5.6 THE BESSEL EQUATION

In this section we'll study the **Bessel equation** (of **order** ν)

$$x^2 y'' + xy' + (x^2 - \nu^2)y = 0,$$

where ν is an unspecified constant.[27] This is a very important example of a differential equation with a singular point at $x = 0$. Its solutions have a wide variety of applications to topics in mathematical physics and engineering, such as heat conduction and the propagation and diffraction of electromagnetic waves, especially in cases involving cylinders (for instance, wires and cables).

As explained on p. 410, we're particularly interested in the behavior of the solutions near $x = 0$. (Near any other point $x = c$, we can find power series expansions for the solutions in terms of $x - c$, by the method of Section 5.3.) Just as for Euler equations, it's easiest to consider the cases $x > 0$ and $x < 0$ separately, and if $y(x)$ is a solution for $x > 0$, then $y(-x)$ is a solution for $x < 0$ (see Exercise 5). We will therefore concentrate on the case $x > 0$.

What can we say about the behavior of the *coefficients* as x approaches 0 ? Well, x^2 and x certainly approach 0, and these coefficients are in as simple a form as one could hope. On the other hand, $x^2 - \nu^2$ approaches the constant $-\nu^2$; provided $\nu \neq 0$, we can say that $x^2 - \nu^2$ behaves "like $-\nu^2$" as $x \to 0$. In terms of Taylor expansions, the first *nonzero* Taylor polynomial of x^2 is x^2, of x is x, and of $x^2 - \nu^2$ is $-\nu^2$. So we might suspect (or at least hope) that the solutions to the Bessel equation will behave similarly to the solutions of $x^2 y'' + xy' - \nu^2 y = 0$, as far as their behavior as $x \to 0$ is concerned. Admittedly, this idea is quite vague, but it will provide helpful motivation.

First, then, let's solve the simpler equation $x^2 y'' + xy' - \nu^2 y = 0$ $(x > 0)$. This is an Euler equation with $a = 1$, $b = -\nu^2$, so it is equivalent (under the substitution $t = \log x$) to $\dfrac{d^2 y}{dt^2} - \nu^2 y = 0$. The characteristic equation $\lambda^2 - \nu^2 = 0$ has distinct roots $\lambda = \pm \nu$ except for $\nu = 0$, when it has a double root $\lambda = 0$. Thus, if $\nu \neq 0$ we have basic solutions $e^{\nu t}$, $e^{-\nu t}$, that is, x^ν, $x^{-\nu}$, while if $\nu = 0$, $\log x$ and 1 form a pair of basic solutions.

We now see that if our suspicion is correct, the Bessel equation

$$x^2 y'' + xy' + (x^2 - \nu^2)y = 0 \quad (x > 0)$$

will have a solution that "behaves like x^ν" as x approaches 0. If we write this solution as $y = x^\nu \cdot z$, then the function $z = y x^{-\nu}$ should approach 1 as $x \to 0^+$. We can get a differential equation for z by substituting $y = x^\nu z$ into the Bessel equation (see Exercise 2); the result is

$$x^{\nu+2} z'' + (2\nu + 1) x^{\nu+1} z' + x^{\nu+2} z = 0$$

$$xz'' + (2\nu + 1) z' + xz = 0.$$

---

[27] This equation is named after the German astronomer and mathematician Bessel (1784–1846), who encountered it, for integers ν, while studying planetary motion. Euler had also studied the equation for integers ν, in a 1764 investigation of the vibrations of a stretched circular membrane. Apparently, E. Lommel (1837–1899) was the first to consider the case of general ν systematically. See "Further Reading" following the Exercises.

SEC. 5.6 THE BESSEL EQUATION

Now what? This equation still has a singular point at $x = 0$ (unless $\nu = -\frac{1}{2}$), so at first sight it seems as if our substitution $y = x^\nu z$ may not have helped. However, we think that our new equation will have a solution $z(x)$ that approaches 1 as $x \to 0^+$, so why not look for a solution that has a power series expansion $z(x) = \sum_{n=0}^{\infty} b_n x^n$ with $b_0 = 1$?[28] One reason this might work is that our new equation is similar to the Laguerre equation (from Section 5.4) in the sense that the coefficient of the second derivative is just x (rather than x^2); we've seen (in Example 5.4.5) that the Laguerre equation has a convergent power series solution.

All right, let's substitute $z = \sum_{n=0}^{\infty} b_n x^n$ in the equation $xz'' + (2\nu + 1)z' + xz = 0$, assume absolute convergence, and hope for the best. We have

$$xz'' = \sum_{n=0}^{\infty} n(n+1)b_{n+1}x^n, \quad (2\nu + 1)z' = \sum_{n=0}^{\infty} (2\nu + 1)(n+1)b_{n+1}x^n,$$

and

$$xz = \sum_{n=1}^{\infty} b_{n-1}x^n.$$

Combining these and taking the constant term separately, we get

$$(2\nu + 1)b_1 + \sum_{n=1}^{\infty} [(n + 2\nu + 1)(n+1)b_{n+1} + b_{n-1}]x^n = 0.$$

This yields

$$(2\nu + 1)b_1 = 0; \quad (n + 2\nu + 1)b_{n+1} = \frac{-b_{n-1}}{n+1} \quad (n \geq 1). \quad (*)$$

Let's assume for the moment that 2ν is not a negative integer. Then $2\nu + 1 \neq 0$ and $n + 2\nu + 1 \neq 0$ for all $n \geq 1$, so we get $b_1 = 0$, $b_{n+1} = \frac{-b_{n-1}}{(n+1)(n+2\nu+1)}$ $(n \geq 1)$.

Hence b_0 determines $b_2 = \frac{-b_0}{2(2\nu + 2)}$, $b_4 = \frac{-b_2}{4(2\nu + 4)}$, and so on, in turn, while $b_1 = 0$ implies $b_3 = 0$, $b_5 = 0$, and so on, in turn. In particular, we have a solution with $b_0 = 1$, as we hoped, provided, of course, that the series converges:

$$z(x) = 1 - \frac{1}{2(2\nu + 2)}x^2 + \frac{1}{2 \cdot 4(2\nu + 2)(2\nu + 4)}x^4$$
$$- \frac{1}{2 \cdot 4 \cdot 6(2\nu + 2)(2\nu + 4)(2\nu + 6)}x^6 + \cdots.$$

---

[28] Note that if y indeed "behaves like x^ν" as $x \to 0^+$, it would be impossible to write y, rather than z, as a power series, unless ν is an integer. See also Exercise 1.

Using the ratio test, it is not hard to see that this series actually converges absolutely for all x (see Exercise 3).

So far, we've seen that if 2ν is not a negative integer, there is a solution $y(x) = x^\nu z(x)$ to the Bessel equation $x^2 y'' + xy' + (x^2 - \nu^2)y = 0$ $(x > 0)$, where $z(x)$ is the power series above. Now note that if we *replace ν by $-\nu$ throughout*, the Bessel equation doesn't change, and so if -2ν isn't a negative integer either, we'll have a second solution to that same equation. This second solution will be $x^{-\nu}$ times the result of replacing ν by $-\nu$ in $z(x)$ above. Of course, if $\nu = 0$, the second solution will be the same as the first, but otherwise they'll form a pair of basic solutions (see Exercise 8). That is, unless 2ν is an integer, we have found the general solution to the Bessel equation:

If 2ν is not an integer,[29] **then**

$$y_1(x) = x^\nu \left[1 - \frac{x^2}{2(2\nu + 2)} + \frac{x^4}{2 \cdot 4(2\nu + 2)(2\nu + 4)} - \frac{x^6}{2 \cdot 4 \cdot 6(2\nu + 2)(2\nu + 4)(2\nu + 6)} + \cdots \right]$$

and

$$y_2(x) = x^{-\nu} \left[1 - \frac{x^2}{2(-2\nu + 2)} + \frac{x^4}{2 \cdot 4(-2\nu + 2)(-2\nu + 4)} - \frac{x^6}{2 \cdot 4 \cdot 6(-2\nu + 2)(-2\nu + 4)(-2\nu + 6)} + \cdots \right]$$

form a pair of basic solutions to the Bessel equation[30] **(for $x > 0$); the general solution is $\alpha y_1(x) + \beta y_2(x)$, α and β arbitrary.**

Note that as $x \to 0^+$, $y_1(x)$ will behave like x^ν in the sense that $\lim_{x \to 0^+} \frac{y_1(x)}{x^\nu} = 1$; similarly, $y_2(x)$ will behave like $x^{-\nu}$.

If 2ν is an even integer $\neq 0$, then one of the two functions $y_1(x)$, $y_2(x)$ displayed above will be undefined, because from some point on the denominators in its series will be zero. If $2\nu = 0$, both functions are defined, but they are equal. Thus if 2ν is any even integer, more work is definitely needed; we'll return to this case below.

On the other hand, if 2ν is an odd integer, then the difficulty with equations (*) on p. 429 (that either for ν or for $-\nu$, we won't be able to find b_1 or some b_{n+1} because we can't divide by 0) is not that serious; see Exercise 13. In fact, if 2ν is odd, the two functions $y_1(x)$, $y_2(x)$ displayed above are defined, their series are absolutely convergent for all x (Exercise 3), and $y_1(x)$, $y_2(x)$ still form a pair of basic solutions to the Bessel equation (Exercise 8).

---

[29] In fact, as we'll soon see, the statement is true as long as ν is not an integer.

[30] It is customary in the literature to multiply y_1 and y_2 by certain constants. Specifically, if for $\nu > 0$ we define $\Gamma(\nu) = \int_0^\infty t^{\nu-1} e^{-t} \, dt$, then $J_\nu(x) = \frac{1}{\nu \Gamma(\nu) 2^\nu} y_1(x)$ is the **Bessel function of the first kind** of order ν, while $J_{-\nu}(x) = \frac{\nu \Gamma(\nu) 2^\nu \sin \pi \nu}{\pi} y_2(x)$ is the Bessel function of the first kind of order $-\nu$. See Exercise 10.

SEC. 5.6 THE BESSEL EQUATION

EXAMPLE 5.6.1   Let $\nu = \dfrac{1}{2}$; note that $2\nu = 1$ is an odd integer. We have

$$y_1(x) = x^{1/2}\left(1 - \frac{x^2}{2\cdot 3} + \frac{x^4}{2\cdot 4 \cdot 3 \cdot 5} - \frac{x^6}{2\cdot 4\cdot 6 \cdot 3 \cdot 5 \cdot 7} + \cdots\right)$$

$$= x^{1/2}\left(1 - \frac{x^2}{3!} + \frac{x^4}{5!} - \frac{x^6}{7!} + \cdots\right).$$

The series in parentheses may look vaguely familiar; checking the list on p. 386, we see that $\sin x = x - \dfrac{x^3}{3!} + \dfrac{x^5}{5!} - \dfrac{x^7}{7!} + \cdots$, so we can rewrite our first basic solution as

$$y_1(x) = x^{-1/2}\left(x - \frac{x^3}{3!} + \frac{x^5}{5!} - \frac{x^7}{7!} + \cdots\right) = x^{-1/2}\sin x.$$

The other basic solution is

$$y_2(x) = x^{-1/2}\left(1 - \frac{x^2}{2\cdot 1} + \frac{x^4}{2\cdot 4\cdot 1\cdot 3} - \frac{x^6}{2\cdot 4\cdot 6 \cdot 1 \cdot 3 \cdot 5} + \cdots\right)$$

$$= x^{-1/2}\left(1 - \frac{x^2}{2!} + \frac{x^4}{4!} - \frac{x^6}{6!} + \cdots\right) = x^{-1/2}\cos x,$$

and the general solution is

$$y = x^{-1/2}(\alpha \sin x + \beta \cos x).$$

Note that despite the factor $x^{-1/2}$, not all nonzero solutions "blow up" as $x \to 0^+$; instead, $y_1(x)$ behaves like $x^{1/2}$. See Figure 5.6.1.

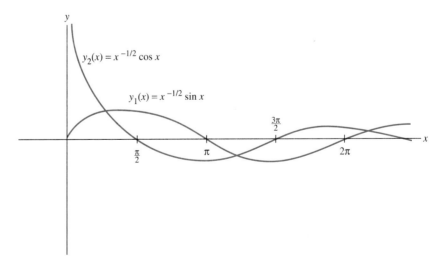

Figure 5.6.1. A pair of basic solutions (for $x > 0$) to the Bessel equation of order $\nu = \frac{1}{2}$.

You may also have noticed that in this example (that is, for $\nu = \frac{1}{2}$) we could have found the general solution more quickly by switching to $\nu = -\frac{1}{2}$ and using the equation $xz'' + (2\nu + 1)z' + xz = 0$ from p. 428 (see Exercise 4). ∎

Just as in the example above, *if 2ν is any positive odd integer*, we can rewrite our first basic solution

$$y_1(x) = x^\nu \left[1 - \frac{x^2}{2(2\nu + 2)} + \cdots \right] \quad \text{as} \quad x^{-\nu} \left[x^{2\nu} - \frac{x^{2\nu+2}}{2(2\nu + 2)} + \cdots \right].$$

The new expression in brackets is then a power series, since 2ν is a positive integer. Thus in this case there are *two* "basic" power series $z(x)$, namely

$$x^{2\nu} - \frac{x^{2\nu+2}}{2(2\nu + 2)} + \cdots \quad \text{and} \quad 1 - \frac{x^2}{2(-2\nu + 2)} + \cdots,$$

such that $x^{-\nu}z(x)$ is a solution of the Bessel equation. (In the example, these are the series expansions of $\sin x$ and $\cos x$, respectively.) Again, these series could have been found directly from equations (∗) on p. 429; see Exercise 13.

We now return to the only remaining case—the one in which 2ν is an even integer; that is, *ν is an integer*. As it turns out, in applications this is the most commonly used case! We'll assume that $\nu \geq 0$ (if not, just switch ν and $-\nu$) and we then have one solution, which is actually a power series, from our earlier work:

$$y_1(x) = x^\nu \left[1 - \frac{x^2}{2(2\nu + 2)} + \frac{x^4}{2 \cdot 4(2\nu + 2)(2\nu + 4)} \right.$$
$$\left. - \frac{x^6}{2 \cdot 4 \cdot 6(2\nu + 2)(2\nu + 4)(2\nu + 6)} + \cdots \right]$$
$$= x^\nu - \frac{x^{\nu+2}}{2(2\nu + 2)} + \cdots.$$

In fact, this series is absolutely convergent for all x (Exercise 3) and we don't even need to restrict the solution to $x > 0$. On the other hand, as noted on p. 430, our usual "second solution" $x^{-\nu} [1 - \frac{x^2}{2(-2\nu + 2)} + \cdots]$ is now useless: either undefined (if $\nu > 0$) or equal to $y_1(x)$ (if $\nu = 0$).

One approach to finding a second basic solution, whether or not ν is an integer, is by the method of reduction of order (from Section 2.4). In fact, the computation is similar to the one for the Laguerre equation (Section 5.4, Exercises 19 and 20). To begin with, the method shows that since $y_1(x)$ is a (nonzero) solution to the Bessel equation, so is

$$y_2(x) = y_1(x) \int \frac{dx}{x[y_1(x)]^2},$$

SEC. 5.6 THE BESSEL EQUATION

which is determined up to constant multiples of $y_1(x)$ (see Exercise 7). Obviously, this answer is not in very satisfactory form. Not only is the expression awkward, but if ν is not an integer, it's quite unclear how the new solution might relate to the series solution $x^{-\nu}[1 - \dfrac{x^2}{2(-2\nu + 2)} + \cdots]$ we found earlier.

To improve on the integral form for $y_2(x)$ given above, we can use the series expansion for $y_1(x)$ and the fact that power series can be multiplied, divided, and integrated (see Section 5.2). Actually attempting to carry out these operations in complete detail would be heroic, foolhardy, or both; however, it is feasible to find the *form* of the result without finding the coefficients yet. When this is done, it turns out (see Exercise 17) that after a suitable choice of integration constant, $y_2(x)$ can be expanded as follows (for $x > 0$):

$$y_2(x) = x^{-\nu} \cdot \sum_{n=0}^{\infty} b_n x^n \qquad \text{if } \nu \text{ is not an integer;}$$

$$y_2(x) = By_1(x) \log x + x^{-\nu} \cdot \sum_{n=0}^{\infty} b_n x^n \qquad \text{if } \nu \text{ is an integer.}$$

Here B is a constant, and $\sum_{n=0}^{\infty} b_n x^n$ is analytic at $x = 0$ (in each case). The "strange" factor $\log x$ first shows up as $\int \dfrac{dx}{x}$.

If ν is not an integer, this is nothing new, and we already know what the coefficients b_n are (see p. 430). But in the case we were stuck on, when ν is an integer, we now know what type of solution to look for! To actually find a second basic solution when ν is an integer, we can substitute

$$y_2(x) = By_1(x) \log x + x^{-\nu} \cdot \sum_{n=0}^{\infty} b_n x^n$$

into the Bessel equation and solve for the coefficients $B, b_0, b_1, b_2, \ldots$ from the resulting recurrence relation.

EXAMPLE 5.6.2 Find the general solution, for $x > 0$, of the Bessel equation of order 0 (that is, for $\nu = 0$). The equation is $x^2 y'' + xy' + x^2 y = 0$, which we might as well rewrite as

$$xy'' + y' + xy = 0.$$

As mentioned in Example 5.4.3, this equation comes up in the study of oscillations of a heavy chain suspended from the ceiling.

We have the solution

$$y_1(x) = 1 - \dfrac{x^2}{2 \cdot 2} + \dfrac{x^4}{2 \cdot 4 \cdot 2 \cdot 4} - \dfrac{x^6}{2 \cdot 4 \cdot 6 \cdot 2 \cdot 4 \cdot 6} + \cdots.$$

Note that $2 \cdot 4 \cdot 6 \cdot 2 \cdot 4 \cdot 6 = 2^6 \cdot 1 \cdot 2 \cdot 3 \cdot 1 \cdot 2 \cdot 3 = 2^6 \, (3!)^2$, and so on, so using summation notation we can write

$$y_1(x) = \sum_{m=0}^{\infty} \frac{(-1)^m}{2^{2m}(m!)^2} x^{2m}.$$

By the above, we can expect a solution of the form $y_2(x) = By_1(x) \log x + \sum_{n=0}^{\infty} b_n x^n$, so we should find y_2' and xy_2'' for this function and substitute them, along with xy_2, into our differential equation. Here goes:

$$y_2 = By_1 \log x + \sum_{n=0}^{\infty} b_n x^n$$

$$y_2' = By_1' \log x + \frac{By_1}{x} + \sum_{n=0}^{\infty} (n+1) b_{n+1} x^n$$

$$y_2'' = By_1'' \log x + \frac{2By_1'}{x} - \frac{By_1}{x^2} + \sum_{n=0}^{\infty} (n+1) n b_{n+1} x^{n-1},$$

so

$$xy_2'' = Bxy_1'' \log x + 2By_1' - \frac{By_1}{x} + \sum_{n=0}^{\infty} (n+1) n b_{n+1} x^n.$$

Therefore,

$$xy_2'' + y_2' + xy_2 = Bxy_1'' \log x + 2By_1' - \frac{By_1}{x} + \sum_{n=0}^{\infty} (n+1) n b_{n+1} x^n$$

$$+ By_1' \log x + \frac{By_1}{x} + \sum_{n=0}^{\infty} (n+1) b_{n+1} x^n$$

$$+ Bxy_1 \log x + \sum_{n=1}^{\infty} b_{n-1} x^n$$

$$= B(xy_1'' + y_1' + xy_1) \log x + 2By_1' + \sum_{n=0}^{\infty} (n+1)^2 b_{n+1} x^n + \sum_{n=1}^{\infty} b_{n-1} x^n.$$

Fortunately, the first term on the right disappears entirely, since we know that y_1 is a solution and thus $xy_1'' + y_1' + xy_1 = 0$. This leaves us with

$$xy_2'' + y_2' + xy_2 = 2By_1' + \sum_{n=0}^{\infty} (n+1)^2 b_{n+1} x^n + \sum_{n=1}^{\infty} b_{n-1} x^n$$

$$= 2By_1' + b_1 + \sum_{n=1}^{\infty} [(n+1)^2 b_{n+1} + b_{n-1}] x^n;$$

for y_2 to be a solution, this expression must be zero. Now remember that

$$y_1 = \sum_{m=0}^{\infty} \frac{(-1)^m}{2^{2m}(m!)^2} x^{2m}, \quad \text{so} \quad y_1' = \sum_{m=0}^{\infty} \frac{(-1)^m \cdot 2m}{2^{2m}(m!)^2} x^{2m-1}$$

and we get

$$\sum_{m=0}^{\infty} \frac{(-1)^m \cdot 4Bm}{2^{2m}(m!)^2} x^{2m-1} + b_1 + \sum_{n=1}^{\infty} [(n+1)^2 b_{n+1} + b_{n-1}] x^n = 0. \quad (*)$$

Don't panic! The only constant term on the left is b_1, so we have $b_1 = 0$. For *even* n, say $n = 2m$, there is no term with $x^n = x^{2m}$ in the first summation, so we get

$$(n+1)^2 b_{n+1} + b_{n-1} = 0,$$

that is,

$$(2m+1)^2 b_{2m+1} + b_{2m-1} = 0 \quad \text{or} \quad b_{2m+1} = \frac{-b_{2m-1}}{(2m+1)^2}.$$

Since we already know $b_1 = 0$, this gives us $b_3 = -\frac{b_1}{9} = 0$ (for $m = 1$), $b_5 = \frac{-b_3}{25} = 0$, and so on.

For *odd* n, say $n = 2m - 1$, on the other hand, both summations in $(*)$ have a term with $x^n = x^{2m-1}$, and we get

$$\frac{(-1)^m \cdot 4Bm}{2^{2m}(m!)^2} + (2m)^2 b_{2m} + b_{2m-2} = 0.$$

This yields

$$b_{2m} = -\frac{1}{(2m)^2} b_{2m-2} + \frac{(-1)^{m+1}}{2^{2m}(m!)^2 m} B \quad (m \geq 1).$$

Once we choose b_0 and B, this recurrence relation will give us $b_2, b_4, \ldots$ in turn. The easiest choice might be $b_0 = 1$, $B = 0$, but then we get

$$b_2 = -\frac{1}{2^2} \quad \text{(for } m = 1\text{)},$$

$$b_4 = -\frac{1}{4^2} \cdot -\frac{1}{2^2} = \frac{1}{(2 \cdot 4)^2} \quad \text{(for } m = 2\text{)},$$

$$b_6 = -\frac{1}{6^2} \cdot \frac{1}{(2 \cdot 4)^2} = \frac{-1}{(2 \cdot 4 \cdot 6)^2}, \quad \text{and so on,}$$

and our "new" solution $y_2 = By_1 \log x + \sum_{n=0}^{\infty} b_n x^n$ turns out to be nothing but

$$1 - \frac{x^2}{2^2} + \frac{x^4}{2^2 \cdot 4^2} - \frac{x^6}{2^2 \cdot 4^2 \cdot 6^2} + \cdots,$$ which is our original solution y_1 back again.

So let's choose $b_0 = 0$, $B = 1$ instead. The recurrence then yields

$$b_2 = \frac{1}{2^2} = \frac{1}{4} \quad \text{(for } m = 1\text{),}$$

$$b_4 = \frac{-\frac{1}{4}}{4^2} - \frac{1}{2^4 \cdot (2!)^2 \cdot 2} = -\frac{3}{128} \quad \text{(for } m = 2\text{),}$$

and in general

$$b_{2m} = \frac{(-1)^{m+1}(1 + \frac{1}{2} + \cdots + \frac{1}{m})}{2^{2m}(m!)^2} \quad \text{(see Exercise 12).}$$

If we use the (standard) abbreviation $H_m = 1 + \frac{1}{2} + \cdots + \frac{1}{m}$ for the mth partial sum of the harmonic series, we get

$$y_2 = By_1 \log x + \sum_{n=0}^{\infty} b_n x^n$$

$$= y_1 \log x + \sum_{m=1}^{\infty} b_{2m} x^{2m} \quad \text{(since we have } B = 1 \text{ and } b_0 = b_1 = b_3 = b_5 = \cdots = 0\text{)}$$

$$y_2(x) = y_1(x) \log x + \sum_{m=1}^{\infty} \frac{(-1)^{m+1} H_m}{2^{2m}(m!)^2} x^{2m}.$$

This is a second basic solution of the Bessel equation of order 0; it behaves like $\log x$ as $x \to 0^+$ (see Exercise 6).

The general solution is given by

$$\alpha y_1 + \beta y_2 = (\alpha + \beta \log x) y_1 + \beta \sum_{m=1}^{\infty} \frac{(-1)^{m+1} H_m}{2^{2m}(m!)^2} x^{2m}$$

$$= (\alpha + \beta \log x) \sum_{m=0}^{\infty} \frac{(-1)^m}{2^{2m}(m!)^2} x^{2m} + \beta \sum_{m=1}^{\infty} \frac{(-1)^{m+1} H_m}{2^{2m}(m!)^2} x^{2m}.$$

This expression looks formidable, but the series converge very rapidly for any given x close to 0, so they are quite suitable for numerical work. Note that as $x \to 0^+$, all solutions with $\beta \neq 0$ "blow up" because of the factor $\log x$. Thus if we know (perhaps on physical grounds) that a solution remains bounded as $x \to 0^+$, then that solution must have the form αy_1 (see Section 7.4, Exercise 17). ∎

In the next section we'll consider to what extent the methods we used for the Bessel equation can be applied to other differential equations.

SEC. 5.6 THE BESSEL EQUATION

SUMMARY OF KEY CONCEPTS, RESULTS, AND TECHNIQUES

The **Bessel equation of order** ν is $x^2 y'' + xy' + (x^2 - \nu^2)y = 0$ (p. 428). If $y(x)$ is a solution for $x > 0$, then $y(-x)$ is a solution for $x < 0$ (see Exercise 5); from now on we consider $x > 0$ only. Since the coefficients of the Bessel equation behave (as $x \to 0^+$) like those of the Euler equation $x^2 y'' + xy' - \nu^2 y = 0$, which has x^ν as a solution, we may expect to have a solution of the Bessel equation that "behaves like x^ν" (p. 428). In fact, unless ν is a negative integer,

$$y_1(x) = x^\nu \left[1 - \frac{x^2}{2(2\nu + 2)} + \frac{x^4}{2 \cdot 4 (2\nu + 2)(2\nu + 4)} - \frac{x^6}{2 \cdot 4 \cdot 6 (2\nu + 2)(2\nu + 4)(2\nu + 6)} + \cdots \right]$$

is a solution of the Bessel equation of order ν; the series converges absolutely for all x. If ν is not an integer, this solution and the one obtained from it by replacing ν by $-\nu$ form a pair of basic solutions (p. 430). If ν is half a positive odd integer, both these solutions can be written as $x^{-\nu}$ times a power series in x (p. 432). For example, for $\nu = \frac{1}{2}$ they are $x^{-1/2} \sin x$ and $x^{-1/2} \cos x$ (p. 431). If $\nu \geq 0$ is an integer, the Bessel equation has a second basic solution of the form $y_2(x) = By_1(x) \log x + x^{-\nu} \cdot \sum_{n=0}^{\infty} b_n x^n$ (p. 433). To find such a solution, substitute y_2 into the Bessel equation to get a recurrence relation for B and the b_n (for the explicit computation for $\nu = 0$, see Example 5.6.2).

EXERCISES

1. Show directly that unless ν is an integer, there can be no nonzero power series solution $y(x) = \sum_{n=0}^{\infty} a_n x^n$ to the Bessel equation of order ν.

2. Show that if $y = x^\nu z$ $(x > 0)$ is a solution to the Bessel equation of order ν, then z satisfies the differential equation $xz'' + (2\nu + 1)z' + xz = 0$ (as claimed on p. 428).

3. Show that if ν is not a negative integer, the series

$$1 - \frac{1}{2(2\nu + 2)} x^2 + \frac{1}{2 \cdot 4 (2\nu + 2)(2\nu + 4)} x^4 - \frac{1}{2 \cdot 4 \cdot 6 (2\nu + 2)(2\nu + 4)(2\nu + 6)} x^6 + \cdots$$

$$= \sum_{m=0}^{\infty} \frac{(-1)^m}{2 \cdot 4 \cdot \ldots \cdot 2m \cdot (2\nu + 2)(2\nu + 4) \cdots (2\nu + 2m)} x^{2m}$$

converges absolutely for all x. (Compare pp. 429–430, 432.)

4. Consider the Bessel equation of order $\frac{1}{2}$: $x^2y'' + xy' + (x^2 - \frac{1}{4})y = 0$.

   (a) Why is this the same as the Bessel equation of order $-\frac{1}{2}$?

   (b) Solve the equation without using power series, by using the result of Exercise 2 for $\nu = -\frac{1}{2}$.

   (c) Solve the initial value problem $x^2y'' + xy' + (x^2 - \frac{1}{4})y = 0$, $y(\pi) = 2$, $y'(\pi) = -1$.

5. Show that if $y(x)$ is a solution to the Bessel equation of order ν, then so is $w(x) = y(-x)$.
   [*Hint*: By the chain rule, $w'(x) = -y'(-x)$.]
   Note that if $y(x)$ is defined for $x > 0$, then $w(x)$ is defined for $x < 0$.

6. Show that for the second basic solution $y_2(x)$ (which we found in Example 5.6.2) to the Bessel equation of order 0, $\lim_{x \to 0^+} \frac{y_2(x)}{\log x} = 1$.

*7. Use the method of reduction of order to show that if $y_1(x)$ is a nonzero solution to the Bessel equation of order ν, then so is $y_2(x) = y_1(x) \int \frac{dx}{x[y_1(x)]^2}$. What is the role of the arbitrary integration constant in this formula?

8. Show that unless $\nu = 0$, the two functions $y_1(x)$, $y_2(x)$ displayed on p. 430 are not constant multiples of each other. (Assume that ν is not an integer, so that both functions are defined.)

9. Show that if $\nu \geq 0$ is an integer, then the power series solution $y_1(x)$ (see p. 432) to the Bessel equation of order ν can be written as

$$y_1(x) = \nu!\, 2^\nu \sum_{m=0}^{\infty} \frac{(-1)^m}{m!\,(\nu+m)!} \left(\frac{x}{2}\right)^{2m+\nu}.$$

10. (Compare p. 430, footnote 30.) Let $\Gamma(\nu) = \int_0^\infty t^{\nu-1} e^{-t}\, dt$.

   *(a) Show that this improper integral converges for any $\nu > 0$.

   [*Hint*: Use a comparison with $\int_0^1 t^{\nu-1}\, dt$ to show convergence at the lower bound.]

   (b) Show, using integration by parts, that $\Gamma(\nu + 1) = \nu\Gamma(\nu)$ $(\nu > 0)$.

   (c) Use part (b) to show that $\Gamma(\nu) = (\nu - 1)!$ for all positive *integers* ν.

   (d) Let ν be a positive integer. Use part (c) and Exercise 9 to show that the Bessel function $J_\nu(x)$ of the first kind, which can be defined as $\frac{1}{\nu\Gamma(\nu)2^\nu} y_1(x)$, can also be written as

$$J_\nu(x) = \sum_{m=0}^{\infty} \frac{(-1)^m}{m!\,(\nu+m)!} \left(\frac{x}{2}\right)^{2m+\nu}.$$

SEC. 5.6 THE BESSEL EQUATION

*(e) Show that whether or not ν is an integer, for $\nu > 0$ we have

$$J_\nu(x) = \sum_{m=0}^{\infty} \frac{(-1)^m}{m!\,\Gamma(\nu + m + 1)} \left(\frac{x}{2}\right)^{2m+\nu}.$$

[*Note*: This formula can actually be used to define $J_\nu(x)$ for *any* ν, unless ν is a negative integer. However, to do so in general, we first have to define $\Gamma(\nu)$ for negative ν, which cannot be done by the improper integral above. [For more on the **gamma function** Γ, see a book on advanced calculus or Artin, *The Gamma Function* (Holt, Rinehart and Winston, 1964).] The way in which $J_{-\nu}(x)$ is defined in footnote 30 on p. 430 avoids this difficulty.]

11. (a) Show that the *derivative* of the power series solution

$$y_1(x) = \sum_{m=0}^{\infty} \frac{(-1)^m}{2^{2m}\,(m!)^2} x^{2m}$$

(see Example 5.6.2) to the Bessel equation of order 0 is equal to $-\dfrac{1}{2}$ times our "standard" power series solution to the Bessel equation *of order 1*.

*(b) Without using power series, show that, more generally, if y is any solution to the Bessel equation of order 0, then its derivative y' is a solution to the Bessel equation of order 1.

12. (See pp. 435–436.) Given $b_0 = 0$, $B = 1$, and the recurrence

$$b_{2m} = \frac{-1}{(2m)^2} b_{2m-2} + \frac{(-1)^{m+1}}{2^{2m}(m!)^2 m} B \qquad (m \geq 1):$$

(a) Check that $b_2 = \dfrac{1}{4}$ and $b_4 = -\dfrac{3}{128}$; find b_6 and b_8.

(b) Show that

$$b_{2m} = \frac{(-1)^{m+1}\left(1 + \dfrac{1}{2} + \cdots + \dfrac{1}{m}\right)}{2^{2m}(m!)^2} \qquad \text{for all } m \geq 1.$$

13. Let 2ν be an odd positive integer.

(a) Explain, using (∗) on p. 429 and without doing any computation, why the function $y = x^{-\nu} \cdot \sum_{n=0}^{\infty} b_n x^n$ will be a solution to the Bessel equation of order ν when the recurrence relation $(-2\nu + 1)b_1 = 0$; $(n - 2\nu + 1)b_{n+1} = \dfrac{-b_{n-1}}{n+1}$ ($n \geq 1$) is satisfied.

(b) Show that by taking $b_0 = 1$, $b_1 = 0$ this yields the solution

$$y_2(x) = x^{-\nu}\left[1 - \frac{x^2}{2(-2\nu + 2)} + \cdots\right]$$

displayed on p. 430.

(c) Let $\nu = \dfrac{1}{2}$. Show, directly from the recurrence relation, that there is also a nonzero solution with only odd degree terms in the power series, of the form $y = x^{-1/2}(x + b_3 x^3 + b_5 x^5 + \cdots)$. Find b_3 and b_5.

(d) Similarly, show that for $\nu = \dfrac{3}{2}$ there is a solution of the form
$y = x^{-3/2}(x^3 + b_5 x^5 + \cdots)$. Find b_5.

(e) Show that in general (for 2ν odd, positive) there is a solution of the form $y = x^{-\nu}(x^{2\nu} + b_{2\nu+2} x^{2\nu+2} + \cdots)$. Find $b_{2\nu+2}$. Compare your solution to the solution $y_1(x)$ displayed on p. 430.

*14. In Bessel's study of planetary motion (see p. 428, footnote 27) he encountered the integral $y(x) = \displaystyle\int_0^{2\pi} \cos(n\theta - x\sin\theta)\,d\theta$, where n is a positive integer. [See McLachlan, *Bessel Functions for Engineers*, 2nd ed. (Oxford University Press, 1955), Section 1.20 for the context.] Show that $y(x)$ is a solution to the Bessel equation of order n.

[*Hint*: It is legitimate here to find y' and y'' by differentiating inside the integral— with respect to x, of course. Later on, integration by parts will be helpful.]

*15. Show that for any $\nu \geq 0$, the function

$$y(x) = x^\nu \int_0^\pi \cos(x\cos\theta)\sin^{2\nu}\theta\,d\theta$$

is a solution to the Bessel equation of order ν.

[*Hint*: To find y' and y'', use the product rule; as in Exercise 14, you may differentiate "under the integral sign."]

*16. In this exercise we'll "find" a second basic solution, for $x > 0$, of the Bessel equation of order 1: $x^2 y'' + xy' + (x^2 - 1)y = 0$.

(a) Show that the solution $y_1(x)$ from p. 430 can be written as

$$y_1(x) = \sum_{m=0}^\infty \frac{(-1)^m}{2^{2m} m!\,(m+1)!} x^{2m+1}.$$

(b) Assume that $y_2(x) = By_1(x)\log x + x^{-1} \cdot \displaystyle\sum_{n=0}^\infty b_n x^n$ is a solution (see p. 433). Show that

$$2Bxy_1'(x) + x^{-1}\left[\sum_{n=0}^\infty (n^2 - 2n)b_n x^n + \sum_{n=2}^\infty b_{n-2} x^n\right] = 0.$$

(c) Show that

$$2B\sum_{m=0}^\infty \frac{(-1)^m(2m+1)}{2^{2m} m!\,(m+1)!} x^{2m+1} - b_1 + \sum_{n=2}^\infty [(n^2 - 2n)b_n + b_{n-2}]x^{n-1} = 0.$$

(d) Show that $b_1 = 0$ and that for *odd* n, $n \geq 3$, $b_n = \dfrac{b_{n-2}}{2n - n^2}$; then show that $b_n = 0$ for all odd n.

(e) Show that for *even* n, say $n = 2m$, we have
$$\dfrac{(-1)^{m-1}(2m - 1) \cdot 2}{2^{2m-2}(m - 1)!\, m!} B + (4m^2 - 4m)b_{2m} + b_{2m-2} = 0 \quad (m \geq 1).$$

(f) Show that for $m = 1$, we get $2B + b_0 = 0$ from part (e), while for $m > 1$, we get
$$b_{2m} = \dfrac{-b_{2m-2}}{2m(2m - 2)} + (-1)^m \dfrac{(2m - 1) \cdot 2}{2^{2m-2} \cdot 2m(2m - 2)(m - 1)!\, m!} B.$$

(g) Show that if we take $B = 0$ and $b_2 = 1$, we get our original solution back.

(h) Show that by taking $B = 1$ and $b_2 = 0$, we get a second basic solution of the form
$$y_2(x) = y_1(x) \log x + x^{-1}\left(-2 + \dfrac{3}{32} x^4 - \dfrac{7}{1152} x^6 + \cdots\right).$$

*17. We've seen that for the Bessel equation of order ν ($\nu \geq 0$, $x > 0$) there is a solution of the form $y_1(x) = x^\nu(1 + \ldots x^2 + \ldots x^4 + \cdots)$, where dots denote coefficients that we won't keep exact track of, and that $y_2(x) = y_1(x) \displaystyle\int \dfrac{dx}{x[y_1(x)]^2}$ is also a solution. In this exercise we show that $y_2(x)$ can be written in the (more useful) form shown on p. 433.

(a) Show that $[y_1(x)]^2 = x^{2\nu}(1 + \ldots x^2 + \ldots x^4 + \cdots)$.

(b) Show that
$$\int \dfrac{dx}{x[y_1(x)]^2} = \int \dfrac{1 + \ldots x^2 + \ldots x^4 + \cdots}{x^{2\nu+1}}\, dx.$$

(c) Show that if ν is not an integer, we have
$$\int \dfrac{dx}{x[y_1(x)]^2} = x^{-2\nu}(\ldots + \ldots x^2 + \ldots x^4 + \cdots).$$

(d) Still assuming that ν is not an integer, show that
$$y_2(x) = x^{-\nu}(\ldots + \ldots x^2 + \ldots x^4 + \cdots) = x^{-\nu} \cdot \sum_{n=0}^{\infty} b_n x^n,$$

as claimed on p. 433.

(e) Now assume that ν is an integer. Use the fact that
$$\int \dfrac{x^{2\nu}}{x^{2\nu+1}}\, dx = \int \dfrac{dx}{x} = \log x + C \quad (x > 0)$$

to show that $y_2(x)$ has the form claimed on p. 433.

FURTHER READING

For more about the history of the Bessel equation and about the properties of its solutions, see Watson, [*A Treatise on the Theory of*] *Bessel Functions*, 2nd ed. (Macmillan, 1945).

5.7 SINGULAR POINTS: CONCLUSION

We now return to the general linear homogeneous equation $\dfrac{d^2y}{dx^2} + p(x)\dfrac{dy}{dx} + q(x)y = 0$ and the behavior of its solutions near $x = 0$. (To find their behavior near $x = c$, we can change the variable to $x - c$; compare Section 5.2, pp. 386–388.) If $p(x)$ and $q(x)$ are analytic at $x = 0$, our differential equation has an ordinary point at $x = 0$, and all its solutions are also analytic there; their power series expansions can, in principle, be found by the method of Section 5.3. The remaining case, then, is that of a *singular point* at $x = 0$, for which $p(x)$ and $q(x)$ are not both analytic there. [If $p(x)$ and $q(x)$ are rational functions, this just means that $p(0)$ and $q(0)$ are not both defined.]

In Section 5.6 we saw that for the Bessel equation $x^2y'' + xy' + (x^2 - \nu^2)y = 0$, that is, when $p(x) = \dfrac{1}{x}$, $q(x) = 1 - \dfrac{\nu^2}{x^2}$, we could get an idea of the behavior of the solutions by first solving the "approximating" Euler equation $x^2y'' + xy' - \nu^2 y = 0$. We might hope that this works more generally. Along these lines, a natural first question would be: When *is* there an "approximating" Euler equation?

If we multiply through our general equation by x^2 to get the "right" first term x^2y'', the result is $x^2y'' + x^2p(x)y' + x^2q(x)y = 0$, and we want to know when this can be "approximated" by an Euler equation $x^2y'' + axy' + by = 0$. For the Bessel equation, we did this "approximating" by taking the first (nonzero) term in the Taylor expansion of each coefficient. For this to work in the general case, we need Taylor expansions of the form

$$x^2 p(x) = ax + \ldots x^2 + \ldots x^3 + \cdots \quad \text{and} \quad x^2 q(x) = b + \ldots x + \ldots x^2 + \cdots,$$

SEC. 5.7 SINGULAR POINTS: CONCLUSION

or, equivalently,

$$xp(x) = a + \ldots x + \ldots x^2 + \cdots \quad \text{and} \quad x^2q(x) = b + \ldots x + \ldots x^2 + \cdots.$$

That is, both $xp(x)$ and $x^2q(x)$ should have power series expansions in x.

EXAMPLE 5.7.1 $x^2y'' + (\sin x)y' - (2x + 3)y = 0$.

For this equation we have $x^2p(x) = \sin x$, $x^2q(x) = -2x - 3$, so

$$xp(x) = \frac{\sin x}{x} = \frac{x - \dfrac{x^3}{3!} + \cdots}{x} = 1 - \frac{x^2}{3!} + \cdots \quad \text{and} \quad x^2q(x) = -3 - 2x.$$

Thus we have $a = 1$, $b = -3$, and the "approximating" Euler equation is therefore $x^2y'' + xy' - 3y = 0$.

Note that $xp(x) = \dfrac{\sin x}{x}$ is, strictly speaking, undefined at $x = 0$. However, $xp(x)$ is given for $x \neq 0$ by the power series $1 - \dfrac{x^2}{3!} + \cdots$, and so it is natural to consider $1 = \lim_{x \to 0} xp(x)$ as the "value" of $xp(x)$ at $x = 0$. Once this is done, $xp(x)$ will be analytic at $x = 0$. ∎

Just as in this example, an "approximating" Euler equation will exist in general *provided $xp(x)$ and $x^2q(x)$ are both analytic at $x = 0$* [once they are given the values

$$a = \lim_{x \to 0} xp(x) \quad \text{and} \quad b = \lim_{x \to 0} x^2q(x),$$

respectively, there]. When this is true, the singular point at $x = 0$ is said to be a **regular** singular point; otherwise, the singular point is **irregular**.[31]

EXAMPLE 5.7.2 $x^2y'' + (3x - 1)y' + y = 0$.

This is the equation from Example 5.4.4, where we saw that there are no convergent power series solutions. We have $x^2p(x) = 3x - 1$, $x^2q(x) = 1$, so $xp(x) = 3 - \dfrac{1}{x}$, $x^2q(x) = 1$. Since $3 - \dfrac{1}{x}$ is *not* analytic at $x = 0$, $x = 0$ is an irregular singular point of our equation. ∎

---

[31] Similarly, a singular point at $x = c$ of the differential equation $y'' + p(x)y' + q(x)y = 0$ is called **regular** if $(x - c)p(x)$ and $(x - c)^2q(x)$ are both analytic at $x = c$ (when given their limits there as values), and **irregular** otherwise.

EXAMPLE 5.7.3

$$y'' - \frac{4x-2}{x}y' + \frac{3x^2 - 5x + 1}{x^2(x+2)}y = 0.$$

Here $xp(x) = 2 - 4x$, $x^2 q(x) = \dfrac{3x^2 - 5x + 1}{x + 2}$, both of which are analytic at $x = 0$, so $x = 0$ is a regular singular point. Since $a = \lim_{x \to 0} xp(x) = 2$ and $b = \lim_{x \to 0} x^2 q(x) = \dfrac{1}{2}$, the "approximating" Euler equation is $x^2 y'' + 2xy' + \dfrac{1}{2} y = 0$. ∎

In the important special case that $p(x)$ and $q(x)$ are both rational functions, the situation is as follows. If $p(0)$ and $q(0)$ are both defined, $x = 0$ is an ordinary point. Otherwise, if $p(x)$ has at most one factor x in the denominator and $q(x)$ has at most two factors x there (so that $xp(x)$ and $x^2 q(x)$ won't have any factors x in the denominator), $x = 0$ is a regular singular point. This is illustrated by Example 5.7.3, where we have $p(x) = \dfrac{2 - 4x}{x}$, $q(x) = \dfrac{3x^2 - 5x + 1}{x^2(x+2)}$. Finally, if $p(x)$ has more than one factor x in the denominator [as in Example 5.7.2, where $p(x) = \dfrac{3x - 1}{x^2}$] or $q(x)$ has more than two factors x in the denominator, $x = 0$ is an irregular singular point.

For the rest of this section, we'll look at differential equations with a regular singular point at $x = 0$; the theory for an irregular singular point is more difficult. So we will assume that we have the following situation:

$y'' + p(x)y' + q(x)y = 0$, $p(x)$ or $q(x)$ (perhaps both) not analytic at $x = 0$, $xp(x)$ and $x^2 q(x)$ analytic at $x = 0$ with $\lim_{x \to 0} xp(x) = a$ and $\lim_{x \to 0} x^2 q(x) = b$. We then have the "approximating" Euler equation $x^2 y'' + axy' + by = 0$.

Our approach is essentially the same as in the special case of the Bessel equation (from Section 5.6); it is called the **method of Frobenius**.[32] We will consider the cases $x > 0$ and $x < 0$ separately, and we start with the case $x > 0$.

Recall (from Section 5.5) that if λ is a root of $\lambda^2 + (a - 1)\lambda + b = 0$, then x^λ will be a solution to the Euler equation. We can then look for a solution to our original equation $y'' + p(x)y' + q(x)y = 0$ which "behaves like" x^λ. More precisely, we look for such a solution of the form $y = x^\lambda \cdot \sum_{n=0}^{\infty} a_n x^n$, where $\sum_{n=0}^{\infty} a_n x^n$ is an absolutely convergent power series with $a_0 = 1$. To do so, we can substitute $y = \sum_{n=0}^{\infty} a_n x^{n+\lambda}$ into the "original" equation $x^2 y'' + x^2 p(x) y' + x^2 q(x) y = 0$. Since $x^2 p(x)$ and $x^2 q(x)$ are given by power

---

[32] This approach was developed by the German mathematician Frobenius (1849–1917) in an 1873 paper, which built on the work of Fuchs (1833–1902) and others.

SEC. 5.7 SINGULAR POINTS: CONCLUSION

series (in general), this substitution may require multiplying power series. However, if $p(x)$ and $q(x)$ are rational functions, we can avoid this by clearing all denominators before substituting $y = \sum_{n=0}^{\infty} a_n x^{n+\lambda}$. In any case, in principle the substitution can be carried out, and the result will be a recurrence relation for the coefficients a_n. Starting with $a_0 = 1$, usually all the coefficients can then be found, and we will have one solution $y = x^\lambda \cdot \sum_{n=0}^{\infty} a_n x^n$. If we can do this for both roots of $\lambda^2 + (a-1)\lambda + b = 0$, we will get a pair of basic solutions to our equation.

EXAMPLE 5.7.4 $2x^2 y'' + (x - 2x^2)y' + (x - 3)y = 0$.

Here we have $xp(x) = \frac{1}{2} - x$, $x^2 q(x) = \frac{1}{2}x - \frac{3}{2}$, so $x = 0$ is a regular singular point with $a = \frac{1}{2}$, $b = -\frac{3}{2}$. x^λ will be a solution (for $x > 0$) to the "approximating" Euler equation $x^2 y'' + \frac{1}{2} xy' - \frac{3}{2} y = 0$ when $\lambda^2 + (\frac{1}{2} - 1)\lambda - \frac{3}{2} = 0$, that is, for $\lambda = -1$ and for $\lambda = \frac{3}{2}$.

Let's start with $\lambda = -1$ and look for a solution of the form

$$y = x^{-1} \cdot \sum_{n=0}^{\infty} a_n x^n = \sum_{n=0}^{\infty} a_n x^{n-1}$$

to our differential equation. We get

$$y' = \sum_{n=0}^{\infty} (n-1) a_n x^{n-2}, \quad y'' = \sum_{n=0}^{\infty} (n-1)(n-2) a_n x^{n-3},$$

and so
$2x^2 y'' + (x - 2x^2)y' + (x - 3)y$

$$= \sum_{n=0}^{\infty} 2(n-1)(n-2) a_n x^{n-1} + \sum_{n=0}^{\infty} (n-1) a_n x^{n-1} - \sum_{n=0}^{\infty} 2(n-1) a_n x^n$$

$$+ \sum_{n=0}^{\infty} a_n x^n - \sum_{n=0}^{\infty} 3 a_n x^{n-1}$$

$$= \sum_{n=0}^{\infty} [2(n-1)(n-2) + (n-1) - 3] a_n x^{n-1} + \sum_{n=0}^{\infty} [-2(n-1) + 1] a_n x^n$$

$$= \sum_{n=0}^{\infty} [2n^2 - 5n] a_n x^{n-1} + \sum_{n=0}^{\infty} [3 - 2n] a_n x^n.$$

Since the first term of the first sum is zero (which, as we'll see, is no accident), we can leave it off and then shift the index of summation:

$$\cdots = \sum_{n=1}^{\infty} [2n^2 - 5n]a_n x^{n-1} + \sum_{n=0}^{\infty} [3 - 2n]a_n x^n$$

$$= \sum_{n=0}^{\infty} [2(n+1)^2 - 5(n+1)]a_{n+1} x^n + \sum_{n=0}^{\infty} [3 - 2n]a_n x^n$$

$$= \sum_{n=0}^{\infty} [(2n^2 - n - 3)a_{n+1} + (3 - 2n)a_n]x^n.$$

We can now read off the recurrence relation $(2n^2 - n - 3)a_{n+1} + (3 - 2n)a_n = 0$, which simplifies to

$$a_{n+1} = \frac{a_n}{n+1}.$$

Note that a_0 determines all the other coefficients, so up to constant multiples we'll get one solution (at least a "formal" one, since we haven't considered convergence yet!). For instance, if we take $a_0 = 1$, we'll get $a_1 = 1$, $a_2 = \frac{1}{2}$, $a_3 = \frac{1}{6}$, ..., $a_n = \frac{1}{n!}$, so the power series turns out to be a familiar one, which converges for all x:

$$\sum_{n=0}^{\infty} a_n x^n = \sum_{n=0}^{\infty} \frac{x^n}{n!} = e^x \quad \text{(see p. 386)}.$$

Of course, the solution

$$y_1 = x^{-1} e^x$$

is undefined for $x = 0$ and "blows up" as $x \to 0$.

For $\lambda = \frac{3}{2}$ we have a similar computation, starting with

$$y = x^{3/2} \cdot \sum_{n=0}^{\infty} a_n x^n = \sum_{n=0}^{\infty} a_n x^{n+3/2}.$$

(For details, see Exercise 15.) We then come up with

$$2x^2 y'' + (x - 2x^2)y' + (x - 3)y = \sum_{n=0}^{\infty} [2n^2 + 5n]a_n x^{n+3/2} + \sum_{n=0}^{\infty} [-2n - 2]a_n x^{n+5/2}.$$

Once again, the first term of the first sum is zero, and we can shift the index of summation and combine the sums. This leads to the recurrence relation

$$a_{n+1} = \frac{2}{2n+7} a_n,$$

and once again we get one "formal" solution up to constant multiples, namely

$$y_2 = x^{3/2} \cdot \left(1 + \frac{2}{7}x + \frac{4}{7 \cdot 9}x^2 + \frac{8}{7 \cdot 9 \cdot 11}x^3 + \cdots\right)$$

$$= x^{3/2} \cdot \sum_{n=0}^{\infty} \frac{2^n}{7 \cdot 9 \cdot \ldots \cdot (2n+5)} x^n.$$

One can use the ratio test to show that the power series converges for all x, and this time the solution approaches 0 as $x \to 0^+$.

Since we now have a pair of basic solutions to our differential equation, the general solution (still for $x > 0$) is

$$c_1 y_1 + c_2 y_2 = c_1 \frac{e^x}{x} + c_2 x^{3/2} \cdot \sum_{n=0}^{\infty} \frac{2^n}{7 \cdot 9 \cdot \ldots \cdot (2n+5)} x^n, \qquad c_1, c_2 \text{ arbitrary.} \blacksquare$$

As you might expect, computations using the method of Frobenius are usually considerably more involved than the ones in the example above. The recurrence relations tend to be much more complicated, and it is often not realistic to expect to find a pattern for the coefficients of the power series. Fortunately, the following theorem (which we won't prove; see "Further Reading" following the Exercises) guarantees that the method outlined so far will work in most cases—at least in principle!

Theorem 5.7.1 If $xp(x)$ and $x^2 q(x)$ are analytic (at $x = 0$) with $\lim_{x \to 0} xp(x) = a$ and $\lim_{x \to 0} x^2 q(x) = b$, and if the quadratic equation $\lambda^2 + (a-1)\lambda + b = 0$ has two roots λ_1, λ_2 **whose difference $\lambda_1 - \lambda_2$ is not an integer**, then the differential equation $y'' + p(x)y' + q(x)y = 0$ has a pair of basic solutions (for $x > 0$) that are of the form $y_1(x) = x^{\lambda_1} \sum_{n=0}^{\infty} a_n x^n$,

$y_2(x) = x^{\lambda_2} \sum_{n=0}^{\infty} b_n x^n$ with $a_0 = 1, b_0 = 1$. (Note that because $a_0 = b_0 = 1$, the solutions y_1 and y_2 "behave like" x^{λ_1} and x^{λ_2}, respectively, as $x \to 0^+$.) The power series $\sum_{n=0}^{\infty} a_n x^n$ and

$\sum_{n=0}^{\infty} b_n x^n$ are absolutely convergent at least on the interval $(-R, R)$ on which the series for $xp(x)$ and $x^2 q(x)$ both converge.[33]

---

[33] By contrast, for an *irregular* singular point one can have a formal solution that doesn't converge on any interval; see Example 5.4.4.

The quadratic equation $\lambda^2 + (a - 1)\lambda + b = 0$, which is associated with the "approximating" Euler equation $x^2 y'' + axy' + by = 0$, is often called the **indicial equation** of our differential equation $y'' + p(x)y' + q(x)y = 0$ at the singular point $x = 0$. Now why should it matter whether or not the difference of the roots of the indicial equation is an integer? To get an idea, let's go back to the Bessel equation $x^2 y'' + xy' + (x^2 - \nu^2)y = 0$ from Section 5.6. There we had $\lambda^2 - \nu^2 = 0$ as the indicial equation, so $\lambda_1 = \nu$, $\lambda_2 = -\nu$, $\lambda_1 - \lambda_2 = 2\nu$; on the other hand, the recurrence relation (*) on p. 429 was in danger of breaking down exactly *when 2ν was an integer*. (As it turned out, the second basic solution had a truly different form only when ν was an integer.)

For now, let's assume that $\lambda_1 - \lambda_2$ is not an integer, so we do have a pair of basic solutions of the form $y_1 = x^{\lambda_1} \sum_{n=0}^{\infty} a_n x^n$, $y_2 = x^{\lambda_2} \sum_{n=0}^{\infty} b_n x^n$. To find these, it's usually faster *not* to do two entirely separate computations for λ_1 and λ_2, but instead to put the series $y = x^\lambda \sum_{n=0}^{\infty} a_n x^n = \sum_{n=0}^{\infty} a_n x^{n+\lambda}$ into the differential equation *without* specifying λ yet.

EXAMPLE 5.7.5

$y'' - \dfrac{1}{x(x-1)} y' - \dfrac{x+3}{x^2} y = 0$.

This is the equation from Example 5.3.1 (p. 395). Since

$$xp(x) = \frac{-1}{x-1} = 1 + x + x^2 + \cdots \; (|x| < 1) \quad \text{and} \quad x^2 q(x) = -3 - x$$

are analytic at $x = 0$, we have a regular singular point there, with

$$a = \lim_{x \to 0} \frac{-1}{x-1} = 1 \quad \text{and} \quad b = -3.$$

The indicial equation is $\lambda^2 - 3 = 0$. Its roots $\lambda_1 = \sqrt{3}$, $\lambda_2 = -\sqrt{3}$ don't differ by an integer, so the theorem assures us of a pair of basic solutions $x^{\sqrt{3}} \sum_{n=0}^{\infty} a_n x^n$, $x^{-\sqrt{3}} \sum_{n=0}^{\infty} b_n x^n$ which "behave like" $x^{\sqrt{3}}$, $x^{-\sqrt{3}}$ respectively as $x \to 0^+$ (and so that the series converge at least for all x with $|x| < 1$).

To find these solutions, we'll substitute $y = x^\lambda \sum_{n=0}^{\infty} a_n x^n = \sum_{n=0}^{\infty} a_n x^{n+\lambda}$. First, though, we clear all denominators from the equation:

$$(x^3 - x^2)y'' - xy' - (x^2 + 2x - 3)y = 0.$$

SEC. 5.7 SINGULAR POINTS: CONCLUSION

Substitution then yields

$$\sum_{n=0}^{\infty}(n+\lambda)(n+\lambda-1)a_n x^{n+\lambda+1} - \sum_{n=0}^{\infty}(n+\lambda)(n+\lambda-1)a_n x^{n+\lambda}$$

$$-\sum_{n=0}^{\infty}(n+\lambda)a_n x^{n+\lambda} - \sum_{n=0}^{\infty}a_n x^{n+\lambda+2} - \sum_{n=0}^{\infty}2a_n x^{n+\lambda+1} + \sum_{n=0}^{\infty}3a_n x^{n+\lambda} = 0.$$

Note that the beginnings of these series don't match up very well; for three of the series, the first term (for $n=0$) features x^λ, for two of them, we have $x^{\lambda+1}$, and in one case we have $x^{\lambda+2}$. Let's write out the terms for x^λ and $x^{\lambda+1}$ separately, then shift indices of summation and combine the remaining series:

$$\lambda(\lambda-1)a_0 x^{\lambda+1} + \sum_{n=1}^{\infty}(n+\lambda)(n+\lambda-1)a_n x^{n+\lambda+1} - \lambda(\lambda-1)a_0 x^\lambda - (\lambda+1)\lambda a_1 x^{\lambda+1}$$

$$-\sum_{n=2}^{\infty}(n+\lambda)(n+\lambda-1)a_n x^{n+\lambda} - \lambda a_0 x^\lambda - (\lambda+1)a_1 x^{\lambda+1} - \sum_{n=2}^{\infty}(n+\lambda)a_n x^{n+\lambda}$$

$$-\sum_{n=0}^{\infty}a_n x^{n+\lambda+2} - 2a_0 x^{\lambda+1} - \sum_{n=1}^{\infty}2a_n x^{n+\lambda+1} + 3a_0 x^\lambda + 3a_1 x^{\lambda+1} + \sum_{n=2}^{\infty}3a_n x^{n+\lambda} = 0.$$

$$[-\lambda(\lambda-1)a_0 - \lambda a_0 + 3a_0]x^\lambda$$
$$+ [\lambda(\lambda-1)a_0 - (\lambda+1)\lambda a_1 - (\lambda+1)a_1 - 2a_0 + 3a_1]x^{\lambda+1}$$
$$+ \sum_{n=0}^{\infty}[(n+1+\lambda)(n+\lambda)a_{n+1} - (n+2+\lambda)(n+1+\lambda)a_{n+2}$$
$$- (n+2+\lambda)a_{n+2} - a_n - 2a_{n+1} + 3a_{n+2}]x^{n+\lambda+2} = 0.$$

From the term with x^λ, we see that $-\lambda(\lambda-1)a_0 - \lambda a_0 + 3a_0 = 0$ or $(3-\lambda^2)a_0 = 0$, but we knew this already: It is guaranteed by the indicial equation $\lambda^2 - 3 = 0$. As in Example 5.7.4, this is no accident. In fact, Exercise 37a shows that the term with x^λ will always have the form $F(\lambda)a_0 x^\lambda$, where $F(\lambda) = \lambda^2 + (a-1)\lambda + b$. [We have an extra factor (-1) because to avoid denominators, we didn't start with the "standard" form $x^2 y'' + x^2 p(x) y' + x^2 q(x) = 0$ of our equation.]

Moving on to the rest of the series, we get

$$\lambda(\lambda-1)a_0 - (\lambda+1)\lambda a_1 - (\lambda+1)a_1 - 2a_0 + 3a_1 = 0,$$

which simplifies to

$$a_1 = \frac{\lambda^2 - \lambda - 2}{(\lambda+1)^2 - 3} a_0,$$

and also

$$a_{n+2} = \frac{[(n+1+\lambda)(n+\lambda) - 2]a_{n+1} - a_n}{(n+2+\lambda)^2 - 3}.$$

Note that the denominator for a_1 is $F(\lambda + 1)$ and the one for a_{n+2} is $F(n + 2 + \lambda)$, where $F(\lambda) = \lambda^2 - 3$ is the quadratic polynomial from the indicial equation. This pattern will always occur [see Exercise 37(b)]; more on this shortly.

We can now choose either $\lambda = \sqrt{3}$ or $\lambda = -\sqrt{3}$ and find as many of the coefficients as we like of the corresponding series. For example, taking $\lambda = \sqrt{3}$ and $a_0 = 1$, we get

$$a_1 = \frac{(\sqrt{3})^2 - \sqrt{3} - 2}{(\sqrt{3} + 1)^2 - 3} = -\frac{7}{11} + \frac{3}{11}\sqrt{3},$$

$$a_2 = \frac{[(1 + \sqrt{3})\sqrt{3} - 2]\left(-\frac{7}{11} + \frac{3}{11}\sqrt{3}\right) - 1}{(\sqrt{3} + 2)^2 - 3} = -\frac{3}{88} - \frac{5}{88}\sqrt{3},$$

and so on (?), so the solution has the form

$$y(x) = x^{\sqrt{3}}\left(1 + \frac{-7 + 3\sqrt{3}}{11}x - \frac{3 + 5\sqrt{3}}{88}x^2 + \cdots\right).$$

This is the only solution (up to constant multiples) which doesn't "blow up" as $x \to 0^+$. (Why?) ∎

Up to now, we've looked only at the case $x > 0$, and you might expect another round of long computations for $x < 0$. Fortunately, it turns out that in the situation of Theorem 5.7.1, all we have to do to get a pair of basic solutions for $x < 0$ from our basic solutions for $x > 0$ is to replace x^{λ_1}, x^{λ_2} by $(-x)^{\lambda_1}$, $(-x)^{\lambda_2}$ respectively, *while leaving the power series alone* (see Exercise 36). For instance, in the example above, one solution for $x < 0$ will look like

$$y(x) = (-x)^{\sqrt{3}}\left(1 + \frac{-7 + 3\sqrt{3}}{11}x - \frac{3 + 5\sqrt{3}}{88}x^2 + \cdots\right).$$

Note that this is consistent with what we did for the Bessel equation, even though there we replaced x by $-x$ throughout (p. 428), because for the Bessel equation the power series contained only even powers of x, which do indeed remain unchanged when x is replaced by $-x$.

So what can go wrong with our method? Perhaps the most obvious problem, but not a very serious one, is that the indicial equation $\lambda^2 + (a - 1)\lambda + b = 0$ may not have any real roots. However, we've seen that in this case the Euler equation has complex-valued solutions x^λ, $x^{\bar\lambda}$ ($x > 0$), where $\lambda = \mu + i\nu$, $\bar\lambda = \mu - i\nu$ are the two conjugate roots of the indicial equation. Not surprisingly, we can then find a complex-valued solution of the form

$$x^\lambda \cdot \sum_{n=0}^{\infty} a_n x^n = \sum_{n=0}^{\infty} a_n x^{n+\lambda}, \quad a_0 = 1,$$

to our differential equation. Once the (complex) coefficients a_n have been found, we take the real and imaginary parts of this solution; they will form a pair of basic solutions that will behave like $x^\mu \cos(\nu \log x)$ and $x^\mu \sin(\nu \log x)$, respectively, as $x \to 0^+$. (Recall that $x^\lambda = x^\mu [\cos(\nu \log x) + i \sin(\nu \log x)]$.) It is possible to avoid the use of complex powers and coefficients by looking right away for solutions of the form

$$x^{\mu}\left[\cos(\nu \log x) \cdot \sum_{n=0}^{\infty} b_n x^n + \sin(\nu \log x) \cdot \sum_{n=0}^{\infty} c_n x^n\right] \quad \text{with real } b_n, c_n,$$

but this will usually make the computations even more painful.

A more serious problem will occur if the indicial equation has a double root, because our method will then only give us one basic solution. (This happened, for example, for the Bessel equation of order 0; see Section 5.6.) It is also possible that the recurrence relation for the coefficients a_n of the power series "breaks down" because of division by 0. As we saw in Example 5.7.5 (see also Exercise 37), the denominator for a_m always turns out to be $F(\lambda + m)$, where $F(\lambda) = \lambda^2 + (a - 1)\lambda + b$ is the quadratic polynomial from the indicial equation. So for the recurrence to break down, $F(\lambda + m)$ must be zero, that is, $\lambda + m$ must be a root of the indicial equation, where m is a positive integer and λ *is also a root of the indicial equation*. That is, the two roots of the indicial equation must differ by a positive integer $m = (\lambda + m) - \lambda$. In this case, it may be impossible to find a solution

$$x^{\lambda} \sum_{n=0}^{\infty} a_n x^n, \ a_0 = 1,$$ corresponding to the *smaller* of the two roots.[34]

In discussing these troublesome cases, let's assume that the roots of the indicial equation are λ_1 and λ_2 with $\lambda_1 \geq \lambda_2$ and $\lambda_1 - \lambda_2$ an integer. We can still find a solution

$$y_1(x) = x^{\lambda_1} \sum_{n=0}^{\infty} a_n x^n, \ a_0 = 1,$$ corresponding to the larger of the roots (or the double root,

if $\lambda_1 = \lambda_2$). This solution can then be used to find a second basic solution by the method of reduction of order. Just as for the Bessel equation (see p. 432), this second solution is first found in integral form (Exercise 34). It can then be shown (Exercise 35) that the solution can be rewritten as

$$y_2(x) = B y_1(x) \log x + x^{\lambda_2} \sum_{n=0}^{\infty} b_n x^n,$$

where B is a constant and $\sum_{n=0}^{\infty} b_n x^n$ is an absolutely convergent power series.[35] If $\lambda_1 > \lambda_2$,

B may be zero; if so, the second basic solution has the form $x^{\lambda_2} \sum_{n=0}^{\infty} b_n x^n$, after all. (This happens, for example, for the Bessel equation of order ν when 2ν is an odd integer.) If

---

[34] Note that in this case, the roots must be real (why?).

[35] For $x < 0$, we get $y_1(x) = (-x)^{\lambda_1} \sum_{n=0}^{\infty} a_n x^n$ and $y_2(x) = B y_1(x) \log(-x) + (-x)^{\lambda_2} \sum_{n=0}^{\infty} b_n x^n$, with the same coefficients a_n, b_n, and B.

$\lambda_1 = \lambda_2$, B is *not* zero, and there definitely is a second solution (found by dividing by B) of the form $y_2(x) = y_1(x) \log x + x^{\lambda_1} \sum_{n=0}^{\infty} b_n x^n$, which includes the "logarithmic part."
This was to be expected, since in this case the "approximating" Euler equation has basic solutions x^{λ_1}, $x^{\lambda_1} \log x$. [Examples include the Bessel equation of order 0 (Example 5.6.2, pp. 433–436) and the Laguerre equation (Example 5.4.5, pp. 412–413).]

In general, if $\lambda_1 - \lambda_2$ is an integer, the best way to actually find the coefficients B, b_0, b_1, ... in the expression

$$y_2(x) = By_1(x) \log x + x^{\lambda_2} \sum_{n=0}^{\infty} b_n x^n$$

is to substitute it in and solve the resulting recurrence relation, as shown in Example 5.6.2 for the Bessel equation of order 0. However, if there is a chance that $B = 0$, you might not want to rush into this computation! Instead, you could first find all solutions of the form $x^\lambda \sum_{n=0}^{\infty} a_n x^n = \sum_{n=0}^{\infty} a_n x^{n+\lambda}$; if you find such a solution with $\lambda = \lambda_2$, $a_0 = 1$ as well as one with $\lambda = \lambda_1$, you won't need the logarithmic part. Here is an example.

EXAMPLE 5.7.6 $(x^2 - x^3)y'' + (x^3 - 6x^2 + 6x)y' + (2x^2 - 6x + 6)y = 0$.
Here

$$xp(x) = \frac{x(x^3 - 6x^2 + 6x)}{x^2 - x^3} = \frac{x^2 - 6x + 6}{1 - x}, \quad x^2 q(x) = \frac{2x^2 - 6x + 6}{1 - x}.$$

Both of these are analytic at $x = 0$, so we have a regular singular point there. The indicial equation is $\lambda^2 + 5\lambda + 6 = 0$ [see Exercise 32(a)], and we find $\lambda_1 = -2$, $\lambda_2 = -3$ (with $\lambda_1 \geq \lambda_2$); unfortunately, $\lambda_1 - \lambda_2$ is an integer. Thus we can expect basic solutions of the form

$$y_1(x) = x^{-2} \sum_{n=0}^{\infty} a_n x^n, \quad a_0 = 1,$$

$$y_2(x) = By_1(x) \log x + x^{-3} \sum_{n=0}^{\infty} b_n x^n;$$

incidentally, note that *all* nonzero solutions will "blow up" as $x \to 0^+$. Let's first find all solutions of the form $y = x^\lambda \sum_{n=0}^{\infty} a_n x^n = \sum_{n=0}^{\infty} a_n x^{n+\lambda}$, $a_0 = 1$. [The worst, and most likely, thing that can happen is that we'll only find $y_1(x)$.] After we substitute this expression for

SEC. 5.7 SINGULAR POINTS: CONCLUSION

y into our differential equation and carry out a rather long computation [see Exercise 32(b)] similar to the one in Example 5.7.5, we end up with

$$(\lambda^2 + 5\lambda + 6)a_0 x^\lambda + [(\lambda^2 + 7\lambda + 12)a_1 - (\lambda^2 + 5\lambda + 6)a_0]x^{\lambda+1}$$

(*)

$$+ \sum_{n=0}^{\infty} [((n + 2 + \lambda)(n + 7 + \lambda) + 6)a_{n+2} - ((n + 1 + \lambda)(n + 6 + \lambda) + 6)a_{n+1} + (n + 2 + \lambda)a_n]x^{n+\lambda+2} = 0.$$

From the term with x^λ we see, as expected, that $\lambda = -2$ or $\lambda = -3$. If we take $\lambda = -2$ first, the equation simplifies to

$$2a_1 x^{-1} + \sum_{n=0}^{\infty} [(n^2 + 5n + 6)a_{n+2} - (n^2 + 3n + 2)a_{n+1} + na_n]x^n = 0,$$

so we get

$$a_1 = 0, \quad a_{n+2} = \frac{n^2 + 3n + 2}{n^2 + 5n + 6} a_{n+1} + \frac{n}{n^2 + 5n + 6} a_n$$

$$a_{n+2} = \frac{n+1}{n+3} a_{n+1} + \frac{n}{n^2 + 5n + 6} a_n.$$

The first few coefficients are

$$a_0 = 1, \ a_1 = 0, \ a_2 = \frac{1}{3} a_1 + 0 \cdot a_0 = 0, \ a_3 = \frac{1}{2} a_2 + \frac{1}{12} a_1 = 0, \ldots.$$

But wait! Once two coefficients in a row are zero, the recurrence relation, which expresses a_{n+2} in terms of a_{n+1} and a_n, tells us that *all* further coefficients are zero. In other words, we've found

$$y = x^{-2}$$

as one basic solution.

Now let's look at $\lambda = -3$. Remarkably, in (*) the terms with x^λ and $x^{\lambda+1}$ *both* disappear for $\lambda = -3$, and we get

$$\sum_{n=0}^{\infty} [(n^2 + 3n + 2)a_{n+2} - (n^2 + n)a_{n+1} + (n - 1)a_n]x^{n-1} = 0,$$

which yields

$$a_{n+2} = \frac{n}{n+2} a_{n+1} - \frac{n-1}{n^2 + 3n + 2} a_n.$$

Note that although we want $a_0 = 1$, a_1 is completely arbitrary! [See Exercise 32(c).] If we take $a_1 = 0$, for instance, we get

$$a_2 = 0 \cdot a_1 + \frac{1}{2} a_0 = \frac{1}{2},$$

$$a_3 = \frac{1}{3} a_2 - 0 \cdot a_1 = \frac{1}{6},$$

$$a_4 = \frac{1}{2} a_3 - \frac{1}{12} a_2 = \frac{1}{24},$$

which is enough to suspect a pattern. Certainly we'll get a second basic solution of the form

$$y = x^{-3}\left(1 + \frac{1}{2} x^2 + \frac{1}{6} x^3 + \frac{1}{24} x^4 + \cdots\right);$$

see Exercise 32 for the rest.

So why didn't the recurrence relation break down for $\lambda = -3$, as you might have expected from the discussion on p. 451? To see the reason, look closely at the term with $x^{\lambda+1}$ in (*), namely,

$$[(\lambda^2 + 7\lambda + 12)a_1 - (\lambda^2 + 5\lambda + 6)a_0]x^{\lambda+1}.$$

For $\lambda = -3$ the coefficient of a_1 within this term is zero. Usually, this would spell disaster, because it would imply that the only way the term could be zero would be to have $a_0 = 0$. However, by a lucky coincidence (that is, because the example was rigged) the coefficient of a_0 within the term is zero *as well*, and there is nothing impossible about the situation.

Don't count on such luck; most computations for $\lambda_1 - \lambda_2$ an integer are at least as difficult as Exercise 16 from Section 5.6 (which deals with the case of the Bessel equation of order 1). ∎

SUMMARY OF KEY CONCEPTS, RESULTS, AND TECHNIQUES

Regular singular point (p. 443): Assume that $y'' + p(x)y' + q(x)y = 0$ has a singular point at $x = 0$. The equation has a regular singular point there if $xp(x)$ and $x^2q(x)$ are both analytic there, once they are given the values $a = \lim_{x \to 0} xp(x)$ and $b = \lim_{x \to 0} x^2 q(x)$, respectively. In this case, $x^2y'' + axy' + by = 0$ serves as an "approximating" Euler equation.

Irregular singular point (p. 443): A singular point that is not regular.

Indicial equation (p. 448): The quadratic equation $\lambda^2 + (a - 1)\lambda + b = 0$ associated with the "approximating" Euler equation (at a regular singular point).

SEC. 5.7 SINGULAR POINTS: CONCLUSION

Method of Frobenius (p. 444): Given a differential equation $y'' + p(x)y' + q(x)y = 0$ with a regular singular point at $x = 0$, look for solutions of the form

$$y = x^\lambda \sum_{n=0}^{\infty} a_n x^n = \sum_{n=0}^{\infty} a_n x^{n+\lambda}, \quad a_0 = 1 \quad (x > 0),$$

where λ is a (real or complex) root of the indicial equation; note that as $x \to 0^+$, such a solution "behaves like" the solution x^λ of the "approximating" Euler equation. If the indicial equation has distinct roots which don't differ by an integer, two basic solutions of the form shown exist, and the power series involved will be absolutely convergent at least on the same interval $(-R, R)$ as those for $xp(x)$, $x^2 q(x)$ (Theorem 5.7.1, p. 447).

To find these solutions, substitute $y = \sum_{n=0}^{\infty} a_n x^{n+\lambda}$ into the differential equation, without specifying λ yet (p. 448). The resulting term with x^λ will confirm the indicial equation $\lambda^2 + (a - 1)\lambda + b = 0$ (p. 449), while the rest of the resulting series yields a recurrence relation for the a_n. If the two roots of the indicial equation differ by an integer, this recurrence relation may break down when λ is the smaller of the two roots (p. 451). In this case, as in the case of equal roots, there will be a second basic solution of the form $y_2(x) = By_1(x) \log x + x^{\lambda_2} \sum_{n=0}^{\infty} b_n x^n$ along with the first basic solution

$y_1(x) = x^{\lambda_1} \sum_{n=0}^{\infty} a_n x^n$ (where λ_1, λ_2, with $\lambda_1 \geq \lambda_2$, are the roots of the indicial equation; pp. 451–452).

EXERCISES

For each of the following differential equations:

(a) Determine whether $x = 0$ is a regular singular point, an irregular singular point, or an ordinary point.
(b) If $x = 0$ is a regular singular point, find the "approximating" Euler equation.
(c) State what you know about the form of a pair of basic solutions for $x > 0$, and about their behavior as $x \to 0^+$.

1. $x^2 y'' + (5x - x^2)y' + (2x + 3)y = 0$.
2. $x^2 y'' + (6x + 2x^2)y' + (4 - x)y = 0$.
3. $2x^2 y'' + (5x - x^2)y' + (3x - 2)y = 0$.
4. $2x^2 y'' + (3x^2 - x)y' + (3x - 2)y = 0$.
5. $x^3 y'' - 2xy' + y = 0$.
6. $xy'' - x^2 y' + (x^3 - x)y = 0$.

7. $x^2y'' + (5\sin x - x^2)y' + (7x + 4)y = 0$.
8. $x^2y'' + (x^2 - 3\sin x)y' + (4 - 2x)y = 0$.
9. $x^2y'' + x^2y' + (x^4 - \sin^2 x)y = 0$.
10. $x^3y'' + x^2y' + (3x - 4)y = 0$.
11. $xy'' + 2y' - (x + 3)y = 0$.
12. $2xy'' - 3y' + (4x - 1)y = 0$.
13. $x^2y'' + (x\cos x)y' + 2e^x y = 0$.
14. $x^2y'' + xe^x y' + (5\cos x)y = 0$.

15. (See p. 446.)

   (a) Show that for $y = \sum_{n=0}^{\infty} a_n x^{n+3/2}$, we have

   $$2x^2 y'' + (x - 2x^2)y' + (x - 3)y$$
   $$= \sum_{n=0}^{\infty} [2n^2 + 5n]a_n x^{n+3/2} + \sum_{n=0}^{\infty} [-2n - 2]a_n x^{n+5/2}.$$

   (b) Show that if $y = \sum_{n=0}^{\infty} a_n x^{n+3/2}$ is a solution to

   $$2x^2 y'' + (x - 2x^2)y' + (x - 3)y = 0, \text{ then } a_{n+1} = \frac{2}{2n+7} a_n.$$

   (c) Check that this recurrence relation yields the solution y_2 given in Example 5.7.4, and that the power series converges for all x.

16. Consider the differential equation $x^2(1 - 2x)y'' + x(x + 3)y' + (x - 1)y = 0$.

   (a) Show that $x = 0$ is a regular singular point, and find the "approximating" Euler equation.

   (b) Show that if $y = \sum_{n=0}^{\infty} a_n x^{n+\lambda}$ is a solution, then $(\lambda^2 + 2\lambda - 1)a_0 = 0$ and for all $n \geq 0$,

   $$[(n + \lambda + 1)(n + \lambda + 3) - 1]a_{n+1} = [(n + \lambda)(2n + 2\lambda - 3) - 1]a_n.$$

   (c) Find the first three nonzero terms of each of two basic solutions to the given equation.

   *(d) Show that in each case, the radius of convergence of $\sum_{n=0}^{\infty} a_n x^n$ is $\frac{1}{2}$.

   [Hint: Compute $\lim_{n\to\infty} \left|\frac{a_n}{a_{n+1}}\right|$.] Explain, by considering the differential equation itself, why this is reasonable.

Find the general solution of each of the following differential equations (for $x > 0$).

17. $2x^2 y'' + (8x^2 + 5x)y' + (8x^2 + 10x - 2)y = 0$.
18. $2x^2 y'' - (8x^2 + 3x)y' + (8x^2 + 6x - 3)y = 0$.
19. $4x^2 y'' - (8x^2 + 4x)y' + (4x + 3)y = 0$.
20. $9x^2 y'' + (6x - 9x^2)y' - (3x + 2)y = 0$.
*21. $9x^2 y'' + (18x^2 + 3x)y' + (9x^2 + 3x + 1)y = 0$.

For each of the following differential equations, find two basic solutions for $x > 0$ (if no pattern is apparent, give at least three nonzero terms of each solution).

22. $2x^2 y'' + (3x - 2x^2)y' - (x + 1)y = 0$.
23. $2x^2 y'' - (3x + 2x^2)y' + (5x - 7)y = 0$.
*24. $x^2 y'' + 3xy' + (x + 5)y = 0$.
*25. $x^2 y'' - xy' + (2x + 9)y = 0$.

For each of the following differential equations, find at least one nonzero solution (give at least three nonzero terms) for $x > 0$.

26. $x^2 y'' + 2xy' + xy = 0$.
27. $x^2 y'' - 3xy' + 2xy = 0$.
28. $x^2 y'' + 2xy' + (x + \frac{1}{4})y = 0$.
29. $x^2 y'' - xy' + (3x + 1)y = 0$.
30. $x^2 y'' + 2xy' + (x - 6)y = 0$.
31. $x^2 y'' + (x - 6)y = 0$.

32. Consider the differential equation
$$(x^2 - x^3)y'' + (x^3 - 6x^2 + 6x)y' + (2x^2 - 6x + 6)y = 0$$
from Example 5.7.6.
(a) Find the indicial equation.
(b) Show that if $y = \sum_{n=0}^{\infty} a_n x^{n+\lambda}$ is a solution, equation (*) on p. 453 must hold.
(c) Once we find $y = x^{-2}$ as one basic solution, we should expect that in the recurrence for $\lambda = -3$, a_1 will be completely arbitrary (if there is a solution for the recurrence at all). Explain why.
(d) As shown on pp. 453–454, the recurrence relation
$$a_{n+2} = \frac{n}{n+2} a_{n+1} - \frac{n-1}{n^2 + 3n + 2} a_n$$
together with the "starting values" $a_0 = 1$, $a_1 = 0$ yields $a_2 = \frac{1}{2}$, $a_3 = \frac{1}{6}$, $a_4 = \frac{1}{24}$. Guess a formula for a_n ($n \geq 2$) and use the recurrence relation to check that your formula is correct for all n.

(e) How would the results of part (d) change if we chose $a_1 = \alpha$ rather than $a_1 = 0$?

(f) Find the general solution of the differential equation. Write your answer *in closed form*.

33. Let $g(x)$ and $h(x)$ be analytic functions, and consider the differential equation
$$xy'' + g(x)y' + h(x)y = 0.$$
(a) Show that $x = 0$ is a regular singular point of the equation.

*(b) Show (using the results from the text) that there is at least one nonzero power series solution $y = \sum_{n=0}^{\infty} a_n x^n$. Is there necessarily one with $a_0 = 1$?

34. Let y_1 be a nonzero solution of $y'' + p(x)y' + q(x)y = 0$. Using the method of reduction of order, show that
$$y_2(x) = y_1(x) \cdot \int \frac{dx}{[y_1(x)]^2 e^{\int p(x)\,dx}}$$
is also a solution.

*35. Let $x = 0$ be a regular singular point of $y'' + p(x)y' + q(x)y = 0$. As in the text, let $x^2 p(x) = ax + \ldots x^2 + \cdots$ and $x^2 q(x) = b + \ldots x + \cdots$, where dots denote coefficients that we won't keep exact track of. In this exercise we'll see how to rewrite the "second solution" from Exercise 34 in the form shown on p. 451.

(a) Show that for $x > 0$, and with a suitable choice of the integration constant, we have
$$\int p(x)\,dx = a \log x + \ldots x + \ldots x^2 + \cdots.$$

(b) Show that $e^{\int p(x)\,dx} = x^a(1 + \ldots x + \ldots x^2 + \cdots)$.

[*Hint:* You may use that the composition of analytic functions is analytic.]

(c) Let λ_1 and λ_2 be the roots of the indicial equation, and assume that a solution $y_1(x) = x^{\lambda_1}(1 + \ldots x + \ldots x^2 + \cdots)$ to the differential equation has been found. Show that
$$[y_1(x)]^2 \cdot e^{\int p(x)\,dx} = x^{2\lambda_1 + a}(1 + \ldots x + \ldots x^2 + \cdots).$$

(d) Show that if $2\lambda_1 + a$ is not an integer, then the second solution
$$y_2(x) = y_1(x) \cdot \int \frac{dx}{[y_1(x)]^2 e^{\int p(x)\,dx}}$$
found in Exercise 34 has the form
$$y_2(x) = x^{\lambda_2}(\ldots + \ldots x + \ldots x^2 + \cdots).$$

[*Hint:* First show that since $\lambda^2 + (a-1)\lambda + b = (\lambda - \lambda_1)(\lambda - \lambda_2)$, we have $1 - a = \lambda_1 + \lambda_2$.]

Note that this is, up to a constant multiple, the "usual" form of the second basic solution.

(e) Show that if $\lambda_1 - \lambda_2$ is an integer and $\lambda_1 \geq \lambda_2$, then $y_2(x)$ has the form shown on p. 451.
[*Hint*: If $\lambda_1 - \lambda_2 \geq 0$ is an integer, then $2\lambda_1 + a$ is a positive integer.]

36. (a) Show that $y(x) = x^\lambda z(x)$ is a solution of $y'' + p(x)y' + q(x)y = 0$ for $x > 0$ if and only if $z(x)$ satisfies the differential equation

$$x^2 z'' + [2\lambda x + p(x)x^2]z' + [\lambda(\lambda - 1) + \lambda p(x)x + q(x)x^2]z = 0.$$

(b) Show that $y(x) = (-x)^\lambda z(x)$ is a solution of $y'' + p(x)y' + q(x)y = 0$ for $x < 0$ if and only if $z(x)$ satisfies *the same* differential equation as in part (a).

*(c) Explain why the method explained on p. 450 for getting (basic) solutions for $x < 0$ from (basic) solutions for $x > 0$ is legitimate.

*37. As in the text and in Exercise 35, let $x = 0$ be a regular singular point of $x^2 y'' + x^2 p(x) y' + x^2 q(x) y = 0$ and let

$$x^2 p(x) = ax + \ldots x^2 + \cdots, \quad x^2 q(x) = b + \ldots x + \cdots.$$

Let $F(\lambda) = \lambda^2 + (a - 1)\lambda + b$ be the quadratic polynomial from the indicial equation.

(a) Show that if $y = x^\lambda \sum_{n=0}^{\infty} a_n x^n = \sum_{n=0}^{\infty} a_n x^{n+\lambda}$ is substituted into the differential equation (and terms with equal powers of x are combined), the term with x^λ in the result will be $F(\lambda) a_0 x^\lambda$.

(b) Show that from the term with $x^{n+\lambda}$, we get a recurrence relation of the form

$$F(n + \lambda) a_n = \text{some expression in } a_{n-1}, a_{n-2}, \ldots, a_0.$$

FURTHER READING

For proofs of Theorem 5.3.1 and Theorem 5.7.1, see Appendix A to Chapter 5 of Simmons, *Differential Equations with Applications and Historical Notes* (McGraw-Hill, 1972).

CHAPTER 6

The Laplace Transform

6.1 INTRODUCTION

In this chapter we return to the consideration of initial value problems such as

$$x'' + px' + qx = h(t), \qquad x(0) = x_0, \quad x'(0) = v_0,$$

where $x = x(t)$ is the unknown function, p and q are constants, and $h(t)$ is a given function. By Theorem 2.2.2 (p. 87), such an initial value problem has a unique solution provided the function $h(t)$ is continuous. However, this assumption may not be satisfied, for instance in the situation of Example 3.1.5. There we found the equation

$$LC \frac{d^2 V_C}{dt^2} + RC \frac{dV_C}{dt} + V_C = E(t)$$

for the capacitor voltage $V_C(t)$ in an RLC circuit. If we rewrite this equation as

$$\frac{d^2 V_C}{dt} + \frac{R}{L} \frac{dV_C}{dt} + \frac{1}{LC} V_C = \frac{E(t)}{LC},$$

it will be in the form above, with constant coefficients $p = \dfrac{R}{L}$ and $q = \dfrac{1}{LC}$, but the function $h(t) = \dfrac{E(t)}{LC}$ may not be continuous. Specifically, although while the generator is running its voltage $E(t)$ is likely to be either (approximately) constant (for a DC[1] genera-

---

[1] DC, direct current; AC, alternating current.

tor) or given by a "sine wave" (for AC), $E(t)$ will jump abruptly from or to zero when the generator is switched on or off, respectively.

For example, if the power source is "standard" 110 V, 50 Hz AC which is switched on at $t = 1$, the function $E(t)$ will have the form

$$E(t) = \begin{cases} 0 & (t < 1) \\ 110 \cos(100\pi t - \phi) & (t > 1), \end{cases}$$

where the constant ϕ is determined by the phase of the generator (see Exercise 25). It is not clear whether $E(t)$ is even defined for $t = 1$; certainly $E(t)$ will not be continuous there (except if $\cos \phi = 0$, which is unlikely). On the other hand, on the time intervals $(-\infty, 1)$ and $(1, \infty)$ the function $E(t)$ is well defined and continuous. See Figure 6.1.1.

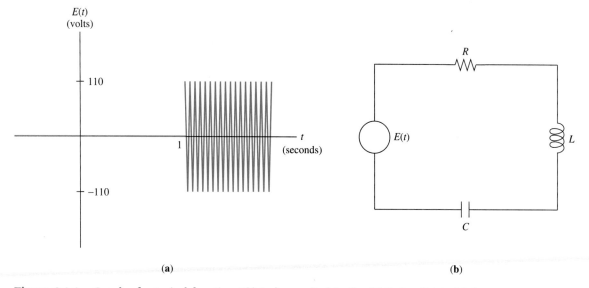

Figure 6.1.1. Graph of a typical function $E(t)$ to be applied to the RLC circuit (at right).

One could argue that $E(t)$ doesn't actually "jump" at $t = 1$ but, instead, makes a very fast, continuous transition from zero. This idea is not too helpful in practice, because it is hard to determine what form such a transition function might take, and because computations using such a function would be long and complicated. On the other hand, this point of view does reinforce the intuitive idea that given $V_C(0)$ and $\left.\dfrac{dV_C}{dt}\right|_{t=0}$, an equation such as the above ought to have a unique solution, even though $E(t)$ is not continuous. This solution should, after all, indicate what actually happens in the circuit if the experiment is carried out. (Since $C\dfrac{dV_C}{dt}$ is the current, the initial conditions indicate the voltage across the capacitor and the current through the circuit at the start of the experiment.)

SEC. 6.1 INTRODUCTION

How can the solution actually be found? One method starts by considering the intervals $(-\infty, 1)$ and $(1, \infty)$, on each of which $E(t)$ is continuous, separately. On the interval $(-\infty, 1)$ there is no particular difficulty; we have an initial value problem

$$\frac{d^2 V_C}{dt^2} + \frac{R}{L}\frac{dV_C}{dt} + \frac{1}{LC}V_C = 0 \quad (t < 1), \qquad V_C(0) = A_0, \qquad \left.\frac{dV_C}{dt}\right|_{t=0} = B_0$$

for some A_0 and B_0, and we can solve for V_C using the methods of Chapter 2. Assume that this has been done. On the other interval, $(1, \infty)$, we have the equation

$$\frac{d^2 V_C}{dt^2} + \frac{R}{L}\frac{dV_C}{dt} + \frac{1}{LC}V_C = \frac{110}{LC}\cos(100\pi t - \phi) \quad (t > 1),$$

and we can certainly find its *general* solution. However, we're not given a value for V_C or its derivative for any t in the interval $(1, \infty)$, so at first sight we may be stuck with the general solution.

Let's look at what happens for $t = 1$, though. The generator voltage $E(t)$ jumps, but we don't really expect $V_C(t)$ to *jump* in response: It will take time for the charges on the capacitor to rearrange themselves. Thus it is reasonable to expect $V_C(t)$ to be continuous at $t = 1$. Similarly, the current $C\frac{dV_C}{dt}$ in the circuit may be assumed to be continuous. Therefore, in the equation

$$\frac{d^2 V_C}{dt^2} + \frac{R}{L}\frac{dV_C}{dt} + \frac{1}{LC}V_C = \frac{E(t)}{LC}$$

the last two terms on the left-hand side *will be continuous, even at $t = 1$*. As a result, $\frac{d^2 V_C}{dt^2}$, just like $\frac{E(t)}{LC}$ on the right-hand side, will be undefined at $t = 1$.

Once we assume that V_C and $\frac{dV_C}{dt}$ are continuous at $t = 1$, *this will give us initial values* $V_C(1)$ and $\left.\frac{dV_C}{dt}\right|_{t=1}$ for our equation

$$\frac{d^2 V_C}{dt^2} + \frac{R}{L}\frac{dV_C}{dt} + \frac{1}{LC}V_C = \frac{110}{LC}\cos(100\pi t - \phi) \quad (t > 1),$$

because we have already found $V_C(t)$ on the interval $(-\infty, 1)$. We can then use these "new" initial values, together with the general solution, to find the specific solution on the interval $(1, \infty)$.

More generally, if $E(t)$ is any function with finitely many jumps, an initial value problem of the form

$$\frac{d^2 V_C}{dt^2} + \frac{R}{L}\frac{dV_C}{dt} + \frac{1}{LC}V_C = \frac{E(t)}{LC}, \qquad V_C(0) = A_0, \qquad \left.\frac{dV_C}{dt}\right|_{t=0} = B_0$$

can be solved as follows. First find the general solution of the differential equation on each of the separate intervals on which $E(t)$ is continuous. Then get the solution of the initial value problem by starting on the interval containing $t = 0$ [where $V_C(0)$ and $\dfrac{dV_C}{dt}\bigg|_{t=0}$ are given] and extending the solution from there to adjoining intervals, using the fact that $V_C(t)$ and $\dfrac{dV_C}{dt}$ are continuous *even where $E(t)$ is not*.

In fact, the same method can be used for any initial value problem

$$x'' + p(t)x' + q(t)x = h(t), \qquad x(0) = x_0, \quad x'(0) = v_0,$$

where $p(t)$ and $q(t)$ are continuous and $h(t)$ is continuous except for finitely many jumps, provided that the associated homogeneous equation can be solved (see Exercise 31). The resulting solution $x(t)$, which is unique, will have a continuous derivative $x'(t)$ everywhere, but its second derivative will be undefined where $h(t)$ has jumps.

To be precise, a function $h(t)$ has a **jump** at t_0 if h is not continuous there, but the one-sided limits $h(t_0^-) = \lim\limits_{t \to t_0^-} h(t)$ and $h(t_0^+) = \lim\limits_{t \to t_0^+} h(t)$ do exist (and are finite). A function that is continuous except for finitely many jumps is called **piecewise continuous**.

The main drawback of using the "interval by interval" method above to deal with piecewise continuous functions is the length of the computations. This should be clear from the following example, which is similar to the case $R = 0$ of the circuit considered earlier (although the constants have been changed to streamline the computation!).

EXAMPLE 6.1.1 Solve the initial value problem

$$x'' + 9x = h(t), \quad x(0) = 1, \; x'(0) = 1, \qquad \text{where} \qquad h(t) = \begin{cases} 0 & \left(t < \dfrac{\pi}{2}\right) \\ \sin t & \left(t > \dfrac{\pi}{2}\right). \end{cases}$$

First we consider the interval $\left(-\infty, \dfrac{\pi}{2}\right)$. On this interval the equation is $x'' + 9x = 0$, with characteristic equation $\lambda^2 + 9 = 0$, roots $\lambda = \pm 3i$, and general real-valued solution $x(t) = \alpha \cos 3t + \beta \sin 3t$. From the initial conditions $x(0) = 1$, $x'(0) = 1$ we get $\alpha = 1$, $\beta = \dfrac{1}{3}$, so we have

$$x(t) = \cos 3t + \frac{1}{3} \sin 3t \qquad \textit{for } t < \frac{\pi}{2}.$$

Since the solution and its derivative

$$x'(t) = -3 \sin 3t + \cos 3t \qquad \left(t < \frac{\pi}{2}\right)$$

SEC. 6.1 INTRODUCTION

are continuous *even* at $t = \dfrac{\pi}{2}$, we can actually write

$$x(t) = \cos 3t + \frac{1}{3}\sin 3t \quad \text{for } t \le \frac{\pi}{2};$$

in particular, $x(\dfrac{\pi}{2}) = \cos\dfrac{3\pi}{2} + \dfrac{1}{3}\sin\dfrac{3\pi}{2} = -\dfrac{1}{3}$. Similarly, $x'(\dfrac{\pi}{2}) = 3$.

On the interval $(\dfrac{\pi}{2}, \infty)$, we have the equation $x'' + 9x = \sin t$. Using either the method of undetermined coefficients (from Section 2.6) or that of variation of constants (from Section 2.8), we find the general solution

$$x(t) = \frac{1}{8}\sin t + \alpha \cos 3t + \beta \sin 3t \quad ^2 \quad \left(t > \frac{\pi}{2}\right),$$

whose derivative is

$$x'(t) = \frac{1}{8}\cos t - 3\alpha \sin 3t + 3\beta \cos 3t.$$

We now use the "initial values" $x(\dfrac{\pi}{2}) = -\dfrac{1}{3}$ and $x'(\dfrac{\pi}{2}) = 3$ that we found above from the previous interval; since $x(t)$ and $x'(t)$ are continuous at $t = \dfrac{\pi}{2}$, we can substitute $t = \dfrac{\pi}{2}$ in the equations for the new interval to find α and β:

$$\begin{cases} -\dfrac{1}{3} = x(\dfrac{\pi}{2}) = \dfrac{1}{8}\sin\dfrac{\pi}{2} + \alpha \cos\dfrac{3\pi}{2} + \beta \sin\dfrac{3\pi}{2} \\ 3 = x'(\dfrac{\pi}{2}) = \dfrac{1}{8}\cos\dfrac{\pi}{2} - 3\alpha \sin\dfrac{3\pi}{2} + 3\beta \cos\dfrac{3\pi}{2}. \end{cases}$$

This yields $\beta = \dfrac{1}{3} + \dfrac{1}{8} = \dfrac{11}{24}$, $\alpha = 1$.

Therefore, the solution to our initial value problem is

$$x(t) = \begin{cases} \cos 3t + \dfrac{1}{3}\sin 3t & \left(t \le \dfrac{\pi}{2}\right) \\ \dfrac{1}{8}\sin t + \cos 3t + \dfrac{11}{24}\sin 3t & \left(t \ge \dfrac{\pi}{2}\right). \end{cases}$$

To repeat, even though $x(t)$ is given by two different formulas, $x(t)$ and $x'(t)$ are continuous *everywhere*. ∎

---

[2] Of course, these are not the same constants α and β as on the interval $(-\infty, \dfrac{\pi}{2})$.

What makes the computation above a bit tedious is not that the homogeneous equation $x'' + 9x = 0$ is difficult, but that initial conditions have to be used for each separate interval in order to solve for the constants in the general solution. Later in this chapter we'll present a method that avoids this; in fact, it doesn't yield a general solution at all unless this is specifically needed. Accordingly, this new method, the use of the so-called Laplace transform, is often very useful in applications when particular constant-coefficient initial value problems arise, notably in electrical engineering. However, the new method has its own drawbacks; for instance, it often requires the use of tables. It also may seem highly artificial at first, since the proper mathematical context of the Laplace transform is beyond the scope of this book. What's more, we'll need some preliminary work before the method can be presented at all (in Section 6.2). Please bear with it! In the rest of this section we will define the Laplace transform and derive a few of its basic properties.

First we recall the concept of improper integral from calculus. If $f(t)$ is a continuous function, or a function that is continuous except for finitely many jumps, then we can consider $\int_0^T f(t)\,dt$ for various upper bounds T. In particular, we may ask whether $\lim_{T\to\infty} \int_0^T f(t)\,dt$ exists. If this limit does exist, it is called the **(improper) integral** of $f(t)$ from 0 to ∞ and denoted by $\int_0^\infty f(t)\,dt$. [In practice, we often write $\int_0^\infty f(t)\,dt$ before we know whether the improper integral actually exists.] Since $\int_0^T f(t)\,dt$ is equal to the area between $t = 0$, $t = T$, the graph of f, and the t-axis (those parts of the area that are below the axis to be counted negatively), $\int_0^\infty f(t)\,dt$ can be thought of as the "unbounded" area, to the right of $t = 0$, between the graph of f and the t-axis (with the same sign convention); see Figure 6.1.2. Of course, it is not essential that the lower bound be zero; we can define $\int_a^\infty f(t)\,dt = \lim_{T\to\infty} \int_a^T f(t)\,dt$ for any number a, provided this limit exists.

EXAMPLE 6.1.2

$$\int_0^\infty e^{-t}\,dt = \lim_{T\to\infty} \int_0^T e^{-t}\,dt = \lim_{T\to\infty} (1 - e^{-T}) = 1.\quad\blacksquare$$

EXAMPLE 6.1.3

$$\int_0^\infty \frac{dt}{t^2 + 1} = \lim_{T\to\infty} \int_0^T \frac{dt}{t^2 + 1} = \lim_{T\to\infty} \arctan T = \frac{\pi}{4}.\quad\blacksquare$$

SEC. 6.1 INTRODUCTION

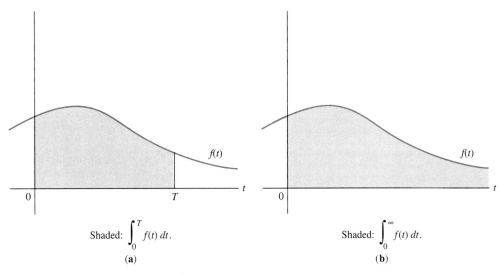

Figure 6.1.2. Improper integral.

EXAMPLE 6.1.4 $\int_1^\infty \frac{1}{t}\,dt$ [3] $= \lim_{T\to\infty} \log T$. Since $\log T$ grows without bound as $T\to\infty$, this limit doesn't exist and neither does $\int_1^\infty \frac{1}{t}\,dt$. ∎

EXAMPLE 6.1.5 $\int_1^\infty \frac{1}{t^p}\,dt = \lim_{T\to\infty} \int_1^T \frac{1}{t^p}\,dt = \lim_{T\to\infty} \frac{1}{1-p}(T^{1-p}-1)$ $(p\neq 1)$. This limit exists if $p>1$ (the exponent of T is negative then, so $T^{1-p}\to 0$ as $T\to\infty$), but not if $p<1$. Using the result of the previous example for the case $p=1$, we get

$$\int_1^\infty \frac{dt}{t^p} = \begin{cases} \dfrac{1}{p-1} & (p>1) \\ \text{does not exist} & (p\leq 1). \end{cases}$$ ∎

There is a strong analogy between the existence or nonexistence of the improper integral $\int_0^\infty f(t)\,dt$ and the convergence or divergence of the infinite series $\sum_{n=0}^\infty f(n)$. (In fact, if f is a *monotonic* function the two are equivalent—this is the "integral test" for infinite

---

[3] Since $\dfrac{1}{t}$ is not defined at $t=0$, considering $\int_0^\infty \frac{1}{t}\,dt$ would involve looking at another type of improper integral, which we won't need.

series.) Accordingly, the words "convergence" (instead of existence) and "divergence" are used for improper integrals, as well. Just as for infinite series, one can often find out whether or not an integral converges without actually computing it, by comparing it with a known improper integral. The "comparison test" for improper integrals is as follows:

Theorem 6.1.1 If $|f(t)| \leq g(t)$ for all t (or for all sufficiently large t) and $\int_a^\infty g(t)\,dt$ converges, then $\int_a^\infty f(t)\,dt$ converges as well.

If both functions are positive, this simply means: If the integral of the larger function converges (finite area below the higher graph), then the integral of the smaller function converges (finite area below the lower graph). See Figure 6.1.3.

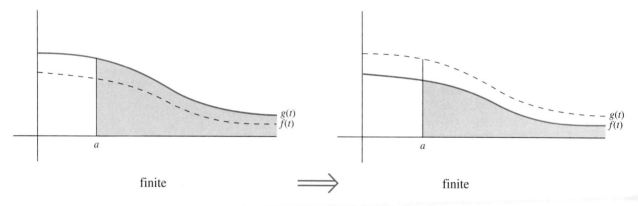

Figure 6.1.3. Comparison test for improper integrals.

EXAMPLE 6.1.6 $\int_1^\infty \dfrac{dt}{t^3 + t^2 + 2t + 7}$ converges, since $0 \leq \dfrac{1}{t^3 + t^2 + 2t + 7} \leq \dfrac{1}{t^3}$ and $\int_1^\infty \dfrac{dt}{t^3}$ converges (by Example 6.1.5). ■

EXAMPLE 6.1.7 $\int_1^\infty \dfrac{\sin t}{t^2}\,dt$ converges, since $\left|\dfrac{\sin t}{t^2}\right| \leq \dfrac{1}{t^2}$. ■

EXAMPLE 6.1.8 $\int_0^\infty t^3 e^{-t}\,dt$ exists, since for t large enough we have $t^3 \leq e^{t/2}$ (see Exercise 7), hence $0 \leq t^3 e^{-t} \leq e^{-t/2}$, and since $\int_0^\infty e^{-t/2}\,dt$ exists (why?). As it happens, $\int_0^\infty t^3 e^{-t}\,dt$ can be computed without too much trouble. ■

SEC. 6.1 INTRODUCTION

Now we'll actually define the Laplace transform. Given a function $f(t)$, we first introduce a *new variable s*, and we consider the function of t obtained by multiplying $f(t)$ by e^{-st}. Then we take the improper integral of this new function: $\int_0^\infty f(t)e^{-st}\,dt$. Of course, this integral will depend on s, and in many cases it will exist for some values of s and not for others. This improper integral, as a function of s, is called the **Laplace transform** of the original function f; it will usually be denoted by $(\mathcal{L}f)(s)$. Thus, the formula defining the Laplace transform is

$$(\mathcal{L}f)(s) = \int_0^\infty f(t)e^{-st}\,dt.$$

NOTATIONAL NOTE: We write $(\mathcal{L}f)(s)$ to emphasize that the Laplace transform of the function f is a new function $\mathcal{L}f$ of a new variable s. Many other notations, such as $\mathcal{L}f(s)$, $\mathcal{L}[f](s)$, $\mathcal{L}\{f(t)\}$, and so forth, occur in the literature. We will use the notation $\mathcal{L}\{f(t)\}$ only if $f(t)$ is a complicated expression in t that we don't want to abbreviate; for instance, $\mathcal{L}\{\sin t \cos t - t\}$ would mean "$(\mathcal{L}f)(s)$, where $f(t) = \sin t \cos t - t$ is the original function." Many authors also write $F(s)$ to indicate the Laplace transform of $f(t)$. Although this is a convenient shorthand notation, it isn't always used in the same way—some authors use capital letters for functions of t and the corresponding lower-case letters for their Laplace transforms! Regardless of which notation is chosen, the Laplace transform of a function of t is always a function of the new variable s, as shown in the examples below.

[Unfortunately, it is still too early to see what all this has to do with initial value problems of the type considered earlier. When Sieur (= Lord) Pierre Simon de Laplace, famed French astronomer and mathematician (1749–1827), first considered the transform, he had other applications, to probability theory, in mind. It was Oliver Heaviside (1850–1925), an electrical engineer, who first noticed that the Laplace transform could simplify calculations concerning electrical circuits.]

EXAMPLE 6.1.9 Let $f(t) = 1$ be constant. Then

$$(\mathcal{L}f)(s) = \int_0^\infty e^{-st}\,dt = \lim_{T\to\infty} \left[\frac{e^{-st}}{-s}\right]_{t=0}^{t=T} \qquad (s \neq 0)$$

$$= \lim_{T\to\infty} \frac{e^{-sT} - 1}{-s}.$$

If $s > 0$, then $e^{-sT} \to 0$ as $T \to \infty$, so the improper integral exists and equals $\dfrac{1}{s}$. On the other hand, if s is negative, then $e^{-sT} \to \infty$ as $T \to \infty$ and the improper integral diverges. For $s = 0$ the computation above does not apply, but we see directly that $(\mathcal{L}f)(0) = \int_0^\infty 1\,dt$ does not exist.

To summarize, we have found that the Laplace transform of $f(t) = 1$ is defined for $s > 0$ and equals $(\mathcal{L}f)(s) = \dfrac{1}{s}$. [4] ∎

EXAMPLE 6.1.10 Let $f(t) = t$. Then

$$(\mathcal{L}f)(s) = \int_0^\infty t e^{-st}\, dt = \lim_{T\to\infty} \int_0^T t e^{-st}\, dt.$$

We can integrate te^{-st} by parts:

$$\int t e^{-st}\, dt = t \cdot \frac{e^{-st}}{-s} + \int \frac{e^{-st}}{s}\, dt \qquad (s \ne 0)$$

$$= -\frac{t}{s} e^{-st} - \frac{e^{-st}}{s^2} + C.$$

Therefore,

$$\int_0^\infty t e^{-st}\, dt = \lim_{T\to\infty} \left(-\frac{T}{s} e^{-sT} - \frac{e^{-sT}}{s^2} + \frac{1}{s^2} \right).$$

For $s > 0$, Te^{-st} and e^{-sT} approach zero as $T \to \infty$, so the Laplace transform is defined and equal to $\dfrac{1}{s^2}$. For $s \le 0$, the Laplace transform is undefined (what happens for $s = 0$?). ∎

EXAMPLE 6.1.11 If $f(t) = t^n$, n a positive integer, we can compute $(\mathcal{L}f)(s)$ by repeatedly integrating by parts (as in Example 6.1.10). See Exercises 16 to 18. ∎

EXAMPLE 6.1.12 Let $f(t) = e^{2t}$. Then

$$(\mathcal{L}f)(s) = \int_0^\infty e^{2t} e^{-st}\, dt = \int_0^\infty e^{-(s-2)t}\, dt.$$

Note that this is the same integral as in Example 6.1.9, but with s replaced by $s - 2$. Therefore, this Laplace transform exists for $s - 2 > 0$ (i.e., for $s > 2$) and equals $\dfrac{1}{s-2}$. Similarly, if a is any constant and $f(t) = e^{at}$, then

$$(\mathcal{L}f)(s) = \frac{1}{s-a} \qquad (s > a). \quad \blacksquare$$

---

[4] Of course, the function $\dfrac{1}{s}$ is defined for any $s \ne 0$, but for $s < 0$ this is no longer the value of the improper integral.

EXAMPLE 6.1.13 Let $f(t) = \begin{cases} -1 & (t < 7) \\ 1 & (t > 7) \end{cases}$.

Then $(\mathcal{L}f)(s) = \int_0^\infty f(t)e^{-st}dt$ consists of two parts:

$$(\mathcal{L}f)(s) = -\int_0^7 e^{-st}\,dt + \int_7^\infty e^{-st}\,dt.$$

The first part is an elementary integral; the second part is the same improper integral as in Example 6.1.9, but with the lower bound 0 replaced by 7. Thus the second part is

$$\int_7^\infty e^{-st}\,dt = \int_0^\infty e^{-st}\,dt - \int_0^7 e^{-st}\,dt = \frac{1}{s} - \int_0^7 e^{-st}\,dt \qquad (s > 0)$$

and hence

$$(\mathcal{L}f)(s) = -2\int_0^7 e^{-st}\,dt + \frac{1}{s} = \frac{1}{s}(2e^{-7s} - 1) \qquad (s > 0).$$

$(\mathcal{L}f)(s)$ is not defined for $s \leq 0$. (Why?) ∎

The examples above are all similar in one respect. All the Laplace transforms we computed are defined for s large enough (specifically $s > 0$, $s > 2$, or $s > a$), but none of them is defined for all s. This happens because the exponential function e^t is the most rapidly growing type of common function (as $t \to \infty$). As a result, for most functions $f(t)$ that we will encounter, $f(t)e^{-st}$ will approach zero rapidly as $t \to \infty$ provided s is large enough; the improper integral $\int_0^\infty f(t)e^{-st}\,dt$ will then converge. On the other hand, $f(t)e^{-st}$ will *grow* rapidly as $t \to \infty$ if s is "negative enough," so the improper integral will not exist for such s. [Exercise 35 provides examples of functions $f(t)$ that don't follow this pattern.]

In practice we are usually not too interested in knowing exactly for which s the Laplace transform of f is defined (provided that it is defined for s large enough), since we won't be substituting actual values for s. We are interested in knowing *what function of s* $\mathcal{L}f$ is. The reason for our interest is that one can sometimes find $(\mathcal{L}f)(s)$ without knowing the original function $f(t)$, knowing only that $f(t)$ is the solution to a certain initial value problem. One can then "backtrack" from the function $(\mathcal{L}f)(s)$ and find out what the original function $f(t)$ must be! We'll return to this in the next section. Meanwhile, we'll derive some basic properties of the Laplace transform.

If $f(t)$ is the sum of two functions $f_1(t)$ and $f_2(t)$, then $(\mathcal{L}f)(s)$ will be the sum of their Laplace transforms. For if $f(t) = f_1(t) + f_2(t)$, then we have

$$\begin{aligned}(\mathcal{L}f)(s) &= \int_0^\infty f(t)e^{-st}\,dt = \int_0^\infty [f_1(t) + f_2(t)]e^{-st}\,dt \\ &= \int_0^\infty f_1(t)e^{-st}\,dt + \int_0^\infty f_2(t)e^{-st}\,dt \qquad \text{(provided both these integrals exist)} \\ &= (\mathcal{L}f_1)(s) + (\mathcal{L}f_2)(s).\end{aligned}$$

Similarly, the Laplace transform of a constant times a function is that same constant times the Laplace transform of the function: $(\mathcal{L}(\alpha f))(s) = \alpha \cdot (\mathcal{L}f)(s)$ for any constant α. Combining this with the previous fact, we see that the Laplace transform of a linear combination of functions is the corresponding linear combination of the Laplace transforms:

$$(\mathcal{L}(\alpha f_1 + \beta f_2))(s) = \alpha \cdot (\mathcal{L}f_1)(s) + \beta \cdot (\mathcal{L}f_2)(s).$$

(See p. 89 for the definition of linear combination.)

EXAMPLE 6.1.14 Find the Laplace transform of $f(t) = 3 - 7t + 5e^{2t}$.

Since this function is a linear combination of the functions 1, t, and e^{2t}, whose Laplace transforms we found earlier, we can combine the results of Examples 6.1.9, 6.1.10, and 6.1.12 to see that

$$(\mathcal{L}f)(s) = 3 \cdot \frac{1}{s} - 7 \cdot \frac{1}{s^2} + 5 \cdot \frac{1}{s-2} = \frac{8s^2 - 13s + 14}{s^2(s-2)}.$$

Another way to write this would be

$$\mathcal{L}\{3 - 7t + 5e^{2t}\} = \frac{8s^2 - 13s + 14}{s^2(s-2)}$$

(see the "notational note" on p. 469). ∎

A transform with the properties mentioned above, that is, such that the transform of a linear combination of functions is the corresponding linear combination of their transforms, is called a **linear transform** or a **linear operator**. Many linear operators besides the Laplace transform occur in mathematics[5]; we will encounter another one, the inverse Laplace transform, in the next section. See also Exercises 32 to 34.

The definition of the Laplace transform can be applied to *complex-valued* functions $f(t)$, as well. Just as complex-valued functions can be differentiated (see p. 108), they can be integrated, either by taking the real and imaginary parts separately or by using any of the standard rules of integration. (Since integration is an inverse operation to differentiation, this should not be surprising.) Using complex-valued functions is sometimes helpful even in problems that do not involve complex numbers at the outset.

EXAMPLE 6.1.15 Find the Laplace transforms of $f_1(t) = \cos t$ and $f_2(t) = \sin t$.

It is possible to find the required integrals $\int_0^\infty \cos t \, e^{-st} \, dt$ and $\int_0^\infty \sin t \, e^{-st} \, dt$ by integration by parts (see Exercise 19). Another method uses the fact that $\cos t$ and $\sin t$ are

---

[5] In fact, linear operators are of great importance in the general context of abstract vector spaces.

SEC. 6.1 INTRODUCTION

the real and imaginary parts of the complex-valued function $f(t) = e^{it} = \cos t + i \sin t$ (see Appendix A, p. 601). Taking the Laplace transform of $f(t)$, we get

$$(\mathcal{L}f)(s) = \int_0^\infty e^{(i-s)t}\, dt.$$

Since $\int_0^T e^{(i-s)t}\, dt = \dfrac{1}{i-s}[e^{(i-s)T} - 1]$, the improper integral exists for $s > 0$ [6] and equals $\dfrac{-1}{i-s} = \dfrac{s+i}{s^2+1}$. Therefore, the Laplace transform of $\cos t$ is the real part $\operatorname{Re}(\dfrac{s+i}{s^2+1}) = \dfrac{s}{s^2+1}$, while the Laplace transform of $\sin t$ is $\operatorname{Im}(\dfrac{s+i}{s^2+1}) = \dfrac{1}{s^2+1}$. Both these Laplace transforms are defined for $s > 0$.

NOTE: It may seem surprising that the functions $\dfrac{s}{s^2+1}$ and $\dfrac{1}{s^2+1}$ are defined for all s, while the improper integrals $\int_0^\infty \cos t\, e^{-st}\, dt$ and $\int_0^\infty \sin t\, e^{-st}\, dt$ converge only for $s > 0$. One can get a better understanding of this and other phenomena involving the Laplace transform by allowing s to take on *complex* values. The Laplace transform will then (usually) converge if the *real part* of s is large enough. In this particular case, the integrals will converge for $\operatorname{Re}(s) > 0$. For $\operatorname{Re}(s) = 0$, problems arise; for $s = \pm i$, in particular, $\dfrac{s}{s^2+1}$ and $\dfrac{1}{s^2+1}$ are undefined. We will not pursue this further; as mentioned above, it is not that important for us to know just for which s a particular Laplace transform is defined. ■

Table 6.1.1 (on p. 474), which will be expanded later in the chapter, lists some frequently occurring Laplace transforms.

SUMMARY OF KEY CONCEPTS, RESULTS, AND TECHNIQUES

Jump (p. 464): A function $h(t)$ is said to have a jump at t_0 if h is not continuous there, but the one-sided limits $h(t_0^-) = \lim_{t \to t_0^-} h(t)$ and $h(t_0^+) = \lim_{t \to t_0^+} h(t)$ do exist (and are finite).

Piecewise continuous function (p. 464): A function that is continuous except for finitely many jumps. To solve an initial value problem $x'' + p(t)x' + q(t)x = h(t)$, $x(0) = x_0$, $x'(0) = v_0$ when $h(t)$ is piecewise continuous, one can first find the general solution on each interval on which $h(t)$ is continuous, then extend the solution to the initial value problem from one such interval to the next by using the continuity of $x(t)$ and $x'(t)$. (See

---

[6] $e^{(i-s)T} = e^{-sT} e^{iT}$ is a complex number whose absolute value is e^{-sT} (see Appendix A, p. 601); hence it approaches 0 as $T \to \infty$, provided $s > 0$.

Table 6.1.1 Some frequently used Laplace transforms

$f(t)$	$(\mathscr{L}f)(s) = F(s)$	Proof
1	$\dfrac{1}{s}$	Example 6.1.9, p. 469
e^{at}	$\dfrac{1}{s-a}$	Example 6.1.12, p. 470
$\sin bt$	$\dfrac{b}{s^2 + b^2}$	Exercises 20, 21, p. 476
$\cos bt$	$\dfrac{s}{s^2 + b^2}$	Exercises 20, 21, p. 476
t^n	$\dfrac{n!}{s^{n+1}}$	Exercise 18, p. 475
$e^{at} \sin bt$	$\dfrac{b}{(s-a)^2 + b^2}$	Exercise 22, p. 476
$e^{at} \cos bt$	$\dfrac{s-a}{(s-a)^2 + b^2}$	Exercise 22, p. 476
$t^n e^{at}$	$\dfrac{n!}{(s-a)^{n+1}}$	Exercise 24, p. 476

In this table, a and b are arbitrary constants.

Example 6.1.1.) Alternatively, the Laplace transform can be used, as we'll see later in this chapter.

Comparison test for improper integrals (p. 468): If f and g are piecewise continuous functions, $|f(t)| \leq g(t)$ for all sufficiently large t, and $\int_a^\infty g(t)\,dt$ converges, then $\int_a^\infty f(t)\,dt$ converges as well.

Laplace transform (p. 469): The Laplace transform of $f(t)$ is defined to be the function $(\mathscr{L}f)(s) = \int_0^\infty f(t)e^{-st}\,dt$. Typically, $(\mathscr{L}f)(s)$ will be defined only for s large enough (p. 471). Laplace transforms of some common functions can be found in Table 6.1.1 above. The Laplace transform is a **linear operator** (p. 472); that is, the Laplace transform of a linear combination of functions equals the corresponding linear combination of their transforms:

$$[\mathscr{L}(\alpha f_1 + \beta f_2)](s) = \alpha \cdot (\mathscr{L}f_1)(s) + \beta \cdot (\mathscr{L}f_2)(s).$$

The Laplace transform can be applied to complex-valued functions by transforming their real and imaginary parts separately (p. 472).

EXERCISES

Using Table 6.1.1 (p. 474), find the Laplace transforms of each of the following functions.

1. $4t^3 - 2t^2 + 5t + 10$
2. $5t^4 - t + 6$
3. $4 \cos t - 3 \sin 2t$
4. $e^{3t}(\cos 2t + 3 \sin 2t - 5)$
5. $e^{-2t}(\sin 2t + t^2 - 3)$
6. $5e^t + 2 \cos 4t - 10t^2$

7. Show that for t large enough, $t^3 \leq e^{t/2}$.
   [*Hint*: Remember l'Hôpital's rule?]

For each of the following improper integrals, determine whether the integral converges. (Use the comparison test, or compute the integral explicitly.)

8. $\displaystyle\int_1^\infty \frac{dt}{t^2 + 5}$.
9. $\displaystyle\int_1^\infty \frac{dt}{t\sqrt{t}}$.
10. $\displaystyle\int_1^\infty \frac{dt}{10 + \sqrt{t}}$.
11. $\displaystyle\int_1^\infty \frac{\cos t}{t^2}\, dt$.
12. $\displaystyle\int_0^\infty t^{15} e^{-t}\, dt$.
13. $\displaystyle\int_1^\infty \frac{e^t}{t^{15}}\, dt$.
14. $\displaystyle\int_0^\infty (\sin t)e^{-t/10}\, dt$.
15. $\displaystyle\int_2^\infty \frac{dt}{t^2 - 1}$.

16. Compute the Laplace transform of $f(t) = t^2$, and show for which values of s it is defined.

17. Repeat Exercise 16 for $f(t) = t^3$.

18. (a) Use the results of Example 6.1.10 (p. 470) and Exercises 16 and 17 above to guess a formula for the Laplace transform of $f(t) = t^n$ (with n a positive integer).
    *(b) Prove your formula by induction on n.

19. (a) Show that $\displaystyle\int_0^T \cos t\, e^{-st}\, dt = \frac{1}{s}\left(1 - \frac{\cos T}{e^{sT}} - \int_0^T \sin t\, e^{-st}\, dt\right)$.

    (b) Show that for $s > 0$,

    $$\int_0^\infty \cos t\, e^{-st}\, dt = \frac{1}{s}\left(1 - \int_0^\infty \sin t\, e^{-st}\, dt\right)$$

    and

    $$\int_0^\infty \sin t\, e^{-st}\, dt = \frac{1}{s}\int_0^\infty \cos t\, e^{-st}\, dt.$$

    (c) Use the results of part (b) to show that

    $$\int_0^\infty \cos t\, e^{-st}\, dt = \frac{s}{1 + s^2} \quad (s > 0)$$

and

$$\int_0^\infty \sin t \, e^{-st} \, dt = \frac{1}{1+s^2} \quad (s > 0).$$

20. Compute the Laplace transforms of the functions $\cos bt$ and $\sin bt$, where b is an arbitrary constant, using a complex exponential (as in Example 6.1.15).

21. Repeat Exercise 20, this time using integration by parts (as in Exercise 19).

22. Using the results of Exercise 20 or Exercise 21, compute the Laplace transforms of the functions $e^{at} \sin bt$ and $e^{at} \cos bt$. (Compare Example 6.1.12, p. 470).

23. Prove the formula $\mathcal{L}\{f(t)e^{at}\} = (\mathcal{L}f)(s - a)$, in which a is an arbitrary constant.

24. Use the results of Exercises 18 and 23 to compute the Laplace transform of $t^n e^{at}$.

25. Explain why for a 110 V, 50 Hz power source the generator voltage has the form $E(t) = 110 \cos(100\pi t - \phi)$.

Solve the following initial value problems using the "interval by interval" method from pp. 463–464, as in Example 6.1.1.

26. $x'' + 4x = h(t)$, $x(0) = 1$, $x'(0) = 2$, where

$$h(t) = \begin{cases} 0 & (t < \pi) \\ \cos t & (t > \pi). \end{cases}$$

27. $x'' + 9x = h(t)$, $x(0) = 1$, $x'(0) = 1$, where

$$h(t) = \begin{cases} 0 & \left(t < \dfrac{\pi}{2}\right) \\ \sin t & \left(\dfrac{\pi}{2} < t < \dfrac{2\pi}{3}\right) \\ 0 & \left(t > \dfrac{2\pi}{3}\right). \end{cases}$$

[*Hint*: The result of Example 6.1.1 can help.]

28. $x'' + 4x = h(t)$, $x(0) = 1$, $x'(0) = 2$, where

$$h(t) = \begin{cases} 0 & (t < \pi) \\ \cos t & (\pi < t < 2\pi) \\ 2 & (2\pi < t < 4\pi) \\ 0 & (t > 4\pi). \end{cases}$$

[*Hint*: The result of Exercise 26 can help.]

29. $x'' - 4x' + 3x = h(t)$, $x(0) = 2$, $x'(0) = 4$, where

$$h(t) = \begin{cases} 0 & (t < 3) \\ 2 & (3 < t < 5) \\ 0 & (t > 5). \end{cases}$$

30. An *RLC* circuit, with $R = 1$, $L = 1$, and $C = 4$, has as its generator a 6 Volt (DC) battery, which is switched on at $t = 1$ (second) and left on thereafter. At $t = 0$, the circuit was "dead," with no charge on the capacitor and no current through the circuit.
    (a) Find the capacitor voltage $V_C(t)$.
    (b) Find the current $I_C(t)$ through the circuit.
    (c) What happens to V_C and I_C as $t \to \infty$? Explain why your results are reasonable.

31. (See p. 464). Explain how once one can solve the homogeneous equation $x'' + p(t)x' + q(t)x = 0$, one can also (in principle) solve the initial value problem $x'' + p(t)x' + q(t)x = h(t)$, $x(0) = x_0$, $x'(0) = v_0$, where $p(t)$, $q(t)$ are continuous and $h(t)$ is continuous except for finitely many jumps.

32. Define the **differentiation operator** $\mathcal{D}$ by $(\mathcal{D}f)(s) = f'(s)$; in other words, $\mathcal{D}f$ is nothing but the derivative of the function f. Show that $\mathcal{D}$ is a linear operator.

33. Suppose that $p_0(s), p_1(s), \ldots, p_{n-1}(s)$ are arbitrary functions. Define a differential operator $\mathcal{M}$ by

$$(\mathcal{M}f)(s) = f^{(n)}(s) + p_{n-1}(s)f^{(n-1)}(s) + \cdots + p_1(s)f'(s) + p_0(s)f(s).$$

Show that $\mathcal{M}$ is a linear operator. Note that for a function f, the condition $\mathcal{M}f = 0$ is equivalent to f being a solution of the nth-order linear differential equation $f^{(n)} + p_{n-1}f^{(n-1)} + \cdots + p_0 f = 0$.

34. Define an integration operator $\mathcal{I}$ by $(\mathcal{I}f)(s) = \int_0^s f(t)\,dt$. Show that $\mathcal{I}$ is a linear operator.

*35. (a) Let $f(t) = e^{-t^2}$. Show that the Laplace transform $(\mathcal{L}f)(s)$ is defined for *all* values of s.
    (b) Show that the Laplace transform of the function e^{t^2} is not defined for *any* value of s.

*36. Show that if $f(t) \geq 0$ for all t, and if $(\mathcal{L}f)(s)$ is defined for $s = a$, then $(\mathcal{L}f)(s)$ will in fact be defined for all $s \geq a$.

*37. In this problem we consider the Laplace transform of $f(t) = t^n$ when n is *not* an integer.
    (a) Show that if $n > 0$, then $(\mathcal{L}f)(s)$ converges for all $s > 0$.
    [*Hint*: If N is an integer with $n < N$, then $t^n e^{-st} < t^N e^{-st}$ for $t > 1$. Use l'Hôpital's rule N times to show that $t^N e^{-st/2} \to 0$ as $t \to \infty$; conclude that $t^n e^{-st} < e^{-st/2}$ for t large enough.]
    (b) Show that $(\mathcal{L}f)(s) = \dfrac{(\mathcal{L}f)(1)}{s^{n+1}}$.
    [*Hint*: Use a change of variable.]
    (c) By comparing the result of part (b) to the entry for t^n in Table 6.1.1, explain why it would be reasonable to *define* $n!$ as $n! = \int_0^\infty t^n e^{-t}\,dt$, even when n is not an integer.

(d) Using the definition from part (c), show that $(\frac{1}{2})! = \int_0^\infty e^{-u^2}\,du$. This integral is known to equal $\frac{\sqrt{\pi}}{2}$ (many books on multivariable calculus show why). Conclude that $\mathcal{L}\{\sqrt{t}\} = \frac{\sqrt{\pi}}{2s\sqrt{s}}$.

(e) Using the definition from part (c), show that $(n+1)! = (n+1)n!$, even when n is not an integer.

(f) Find $\mathcal{L}\{t\sqrt{t}\}$ and $\mathcal{L}\{t^2\sqrt{t}\}$.

FURTHER READING

For more about "generalized factorials" (as in Exercise 37), see Artin, *The Gamma Function* (Holt, Rinehart and Winston, 1964; you may want to start with Section 2), or a book on advanced calculus. Parts (a) to (c) of Exercise 10 in Section 5.6 are also relevant.

6.2 TRANSFORMING INITIAL VALUE PROBLEMS

In Section 6.1, we found the Laplace transform of the function $f(t) = t$ by integrating by parts (Example 6.1.10, p. 470). Let's see what happens if we try to carry out the same integration by parts for a general Laplace transform.

We start with

$$(\mathcal{L}f)(s) = \int_0^\infty f(t)e^{-st}\,dt = \lim_{T\to\infty}\int_0^T f(t)e^{-st}\,dt,$$

and we then consider the integral

$$\int_0^T f(t)e^{-st}\,dt = f(t)\cdot\frac{e^{-st}}{-s}\bigg]_{t=0}^{t=T} + \int_0^T f'(t)\frac{e^{-st}}{s}\,dt \qquad (s\neq 0)$$

$$= \frac{f(T)e^{-sT}}{-s} + \frac{f(0)}{s} + \int_0^T f'(t)\frac{e^{-st}}{s}\,dt.$$

To find the Laplace transform, we must take the limit of this as $T\to\infty$. However, if the Laplace transform exists at all, then $f(T)e^{-sT}$ will approach 0 for $T\to\infty$, and in the limit we'll get

$$\int_0^\infty f(t)e^{-st}\,dt = \frac{f(0)}{s} + \int_0^\infty f'(t)\frac{e^{-st}}{s}\,dt$$

$$= \frac{1}{s}[f(0) + \int_0^\infty f'(t)e^{-st}\,dt].$$

Note that we have taken the factor $\dfrac{1}{s}$ outside the integral on the right-hand side; this is allowed because $\dfrac{1}{s}$ is a constant for integration over t. Note, also, that the integral $\int_0^\infty f'(t)e^{-st}\,dt$ which now appears on the right-hand side is nothing but the *Laplace transform of the derivative* $f'(t)$. We therefore have the formula

$$(\mathcal{L}f)(s) = \frac{1}{s}[f(0) + (\mathcal{L}f')(s)],$$

which expresses the Laplace transform of f in terms of the transform of f'. We can turn this around to get a formula for $\mathcal{L}f'$ in terms of $\mathcal{L}f$:

$$(\mathcal{L}f)(s) = \frac{1}{s}[f(0) + (\mathcal{L}f')(s)]$$

$$s(\mathcal{L}f)(s) = f(0) + (\mathcal{L}f')(s)$$

$$(\mathcal{L}f')(s) = s \cdot (\mathcal{L}f)(s) - f(0). \qquad (*)$$

Even though this argument requires that $s \neq 0$, this is not really necessary; as long as s is large enough so that the Laplace transform $(\mathcal{L}f)(s)$ exists, $(\mathcal{L}f')(s)$ will also exist and be given by $(*)$ above (see Exercise 44).

EXAMPLE Let $f(t) = \cos t$; then $(\mathcal{L}f)(s) = \dfrac{s}{s^2+1}$ (see p. 473). Therefore, the Laplace transform of $f'(t) = -\sin t$ will be $\dfrac{s^2}{s^2+1} - \cos 0 = \dfrac{s^2}{s^2+1} - 1 = \dfrac{-1}{s^2+1}$. This checks with our result from p. 473. ∎

The process above can be repeated to get the Laplace transforms of the higher derivatives of f. For instance,

$$(\mathcal{L}f'')(s) = s \cdot (\mathcal{L}f')(s) - f'(0)$$
$$= s \cdot [s \cdot (\mathcal{L}f)(s) - f(0)] - f'(0),$$

so

$$(\mathcal{L}f'')(s) = s^2(\mathcal{L}f)(s) - sf(0) - f'(0)$$

gives the Laplace transform of the second derivative. The general result is

$$(\mathcal{L}f^{(n)})(s) = s^n(\mathcal{L}f)(s) - s^{n-1}f(0) - s^{n-2}f'(0) \cdots - f^{(n-1)}(0);$$

this is true provided the Laplace transform of f exists (and, of course, the nth derivative of f exists). See Exercise 45.

We are now in a position to start applying the Laplace transform method to solve initial value problems. Given an initial value problem, the procedure is to apply the Laplace transform to both sides of the equation. Since we know how to express the transforms of the derivatives of the unknown function in terms of the Laplace transform of that function itself, this will give us an equation for that Laplace transform.

EXAMPLE 6.2.1 $x'' + 9x = 1$, $x(0) = 2$, $x'(0) = -1$.

We know by Theorem 2.2.2 that this initial value problem has a unique solution $x(t)$; we can consider its Laplace transform $(\mathcal{L}x)(s)$ (assuming that it exists[7]), and also

$$(\mathcal{L}x'')(s) = s^2(\mathcal{L}x)(s) - sx(0) - x'(0)$$
$$= s^2(\mathcal{L}x)(s) - 2s + 1,$$

where we have substituted the given initial values. Therefore, we can take the Laplace transform of both sides of the original equation $x'' + 9x = 1$ to obtain

$$s^2(\mathcal{L}x)(s) - 2s + 1 + 9(\mathcal{L}x)(s) = (\mathcal{L}1)(s) = \frac{1}{s} \quad \text{(see Example 6.1.9)}$$

$$(s^2 + 9) \cdot (\mathcal{L}x)(s) = \frac{1}{s} + 2s - 1$$

$$(\mathcal{L}x)(s) = \frac{\frac{1}{s} + 2s - 1}{s^2 + 9} = \frac{2s^2 - s + 1}{s(s^2 + 9)}.$$

At this point we don't know what the unknown function $x(t)$ is yet, but we have found its Laplace transform. [We'll find $x(t)$ on p. 483 below.] □

To solve our initial value problem, we now need a method to find $x(t)$ from $(\mathcal{L}x)(s)$. However, it is not clear that such a method should exist: There might be several functions $x(t)$ with the same Laplace transform.

EXAMPLE The functions f_1, f_2, f_3 (see Figure 6.2.1) given by

$f_1(t) = \sin t$

$f_2(t) = \begin{cases} 0 & (t \leq 0) \\ \sin t & (t \geq 0) \end{cases}$

$f_3(t) = \begin{cases} \sin t & (t \neq 2\pi) \\ 1 & (t = 2\pi) \end{cases}$

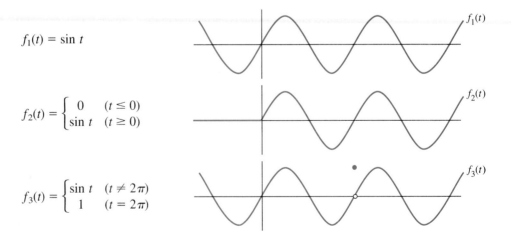

Figure 6.2.1. Three functions with the same Laplace transform.

[7] The basis for this assumption will be discussed on pp. 483–484.

SEC. 6.2 TRANSFORMING INITIAL VALUE PROBLEMS

all have the *same* Laplace transform (specifically, $\frac{1}{s^2+1}$). The reason f_1 and f_2 have the same transform is that $(\mathcal{L}f)(s) = \int_0^\infty f(t)e^{-st}\,dt$ depends only on the values of $f(t)$ for $t \geq 0$, since the integral is taken from $t = 0$ on. As for f_1 and f_3, although these functions are unequal for $t = 2\pi$, this does not affect the integral:

$$\int_0^\infty f_3(t)e^{-st}\,dt = \int_0^{2\pi} \sin t\, e^{-st}\,dt + \int_{2\pi}^\infty \sin t\, e^{-st}\,dt$$

$$= \int_0^\infty \sin t\, e^{-st}\,dt = \int_0^\infty f_1(t)e^{-st}\,dt. \quad \blacksquare$$

As illustrated by this example, we can only hope to find $x(t)$ *for* $t \geq 0$ from the Laplace transform $(\mathcal{L}x)(s)$ and, even then, two functions $x_1(t)$ and $x_2(t)$ which are equal except for finitely many values of t will have the same Laplace transform. However, this last problem will not really arise, since $x(t)$ has to be *continuous* to be an acceptable solution to the initial value problem.[8] As for the restriction to $t \geq 0$, in applications where t denotes the time, it is often enough to know $x(t)$ for $t \geq 0$, since this describes how the experiment will develop from the given initial values at $t = 0$. From a mathematical point of view, if one can find a formula giving $x(t)$ for $t \geq 0$, this formula will often hold for $t < 0$ as well [provided that $h(t)$ has no jumps for $t \leq 0$], because of the uniqueness of the solution.

EXAMPLE 6.2.2 $x'' + x = 0$, $x(0) = 0$, $x'(0) = 1$.

Taking the Laplace transform of both sides, we get

$$(\mathcal{L}x'')(s) + (\mathcal{L}x)(s) = 0$$

$$s^2(\mathcal{L}x)(s) - 1 + (\mathcal{L}x)(s) = 0 \quad \text{[using } x(0) = 0,\ x'(0) = 1\text{]}$$

$$(\mathcal{L}x)(s) = \frac{1}{s^2+1}.$$

If we can conclude from this that $x(t) = \sin t$ for $t \geq 0$, we will have solved our initial value problem: $\sin t$ will be the solution not just for $t \geq 0$, but for all t. $\quad\blacksquare$

What we are looking for, then, is a way to find a continuous function $x(t)$, $t \geq 0$, with a given Laplace transform. Fortunately, if such a function can be found, it has to be the right one, because of

Lerch's Theorem[9]   There can be only one continuous function $f(t)$ ($t \geq 0$) that has a given Laplace transform $F(s) = (\mathcal{L}f)(s)$.

---

[8] Even for an equation $x'' + px' + qx = h(t)$ in which $h(t)$ has jumps (see Section 6.1).

[9] In 1903, the Czech mathematician M. Lerch (1860–1922) published a proof of this result (actually, of a more general one). Such a proof is well beyond the scope of this book.

Note that Lerch's theorem does *not* say that any function of s can occur as a Laplace transform. In fact, there are many common functions of s, such as $F(s) = 1$ or $F(s) = s$, which can never be Laplace transforms; it can be shown that a function which is a Laplace transform must approach zero as $s \to \infty$. (Notice that this is so for all functions in the second column of Table 6.1.1, p. 474.) What the theorem *does* say is that *if* a function $F(s)$ is the Laplace transform of a continuous function $f(t)$, then $f(t)$, $t \geq 0$, is uniquely determined. In this situation, $f(t)$ is called the **inverse Laplace transform** of $F(s)$; we will use the notation $f(t) = \mathcal{L}^{-1}\{F(s)\}$ or $f(t) = (\mathcal{L}^{-1}F)(t)$ (compare the "notational note" on p. 469).

EXAMPLE 6.2.3 If $F(s) = \dfrac{1}{s^2 + 1}$, then $(\mathcal{L}^{-1}F)(t) = \sin t$, since this is the *continuous* function of t whose Laplace transform is $F(s)$. ■

EXAMPLE 6.2.4 $\mathcal{L}^{-1}\left\{\dfrac{1}{s}\right\} = 1$, since we have seen (Example 6.1.9) that $(\mathcal{L}1)(s) = \dfrac{1}{s}$. More generally, in each row of Table 6.1.1, p. 474, the function in the first column is the inverse Laplace transform of the function in the second column. ■

The inverse Laplace transform is a *linear* transform, because the Laplace transform itself is one. For instance, to find the inverse Laplace transform of the sum of two functions $F(s)$ and $G(s)$, one can take the inverse transforms of $F(s)$ and $G(s)$ separately (assuming that these exist) and add them. In other words,

$$\mathcal{L}^{-1}\{F(s) + G(s)\} = \mathcal{L}^{-1}\{F(s)\} + \mathcal{L}^{-1}\{G(s)\}.$$

More generally, for any constants α and β, we have

$$\mathcal{L}^{-1}\{\alpha \cdot F(s) + \beta \cdot G(s)\} = \alpha \cdot \mathcal{L}^{-1}\{F(s)\} + \beta \cdot \mathcal{L}^{-1}\{G(s)\},$$

or, equivalently,

$$(\mathcal{L}^{-1}(\alpha \cdot F + \beta \cdot G))(t) = \alpha \cdot (\mathcal{L}^{-1}F)(t) + \beta \cdot (\mathcal{L}^{-1}G)(t),$$

provided the inverse transforms on the right exist.

EXAMPLE 6.2.5 Find the inverse Laplace transform of $F(s) = \dfrac{3}{s} + \dfrac{4}{s^2 + 1}$.

Since $\mathcal{L}^{-1}\left\{\dfrac{1}{s}\right\} = 1$ and $\mathcal{L}^{-1}\left\{\dfrac{1}{s^2 + 1}\right\} = \sin t$, we have $\mathcal{L}^{-1}\{F(s)\} = 3 + 4 \sin t$. ■

As in the example above, we can find the inverse Laplace transform of any function that is a linear combination of functions in the second column of Table 6.1.1. Often, a

EXAMPLE 6.2.6

We now complete the solution of the initial value problem $x'' + 9x = 1$, $x(0) = 2$, $x'(0) = -1$. In Example 6.2.1 (p. 480) we saw that $(\mathcal{L}x)(s) = \dfrac{2s^2 - s + 1}{s(s^2 + 9)}$. To write this rational function as a linear combination of simpler fractions, put

$$\frac{2s^2 - s + 1}{s(s^2 + 9)} = \frac{A}{s} + \frac{Bs + C}{s^2 + 9}$$

and solve for the coefficients A, B, and C, obtaining $A = \dfrac{1}{9}$, $B = \dfrac{17}{9}$, $C = -1$. (See the section on methods of integration in a calculus book for details.) We then have

$$(\mathcal{L}x)(s) = \frac{\frac{1}{9}}{s} + \frac{\frac{17}{9}s - 1}{s^2 + 9}.$$

Turning to Table 6.1.1 (p. 474), we see that $\dfrac{1}{s-a}$, $\dfrac{b}{s^2 + b^2}$, and $\dfrac{s}{s^2 + b^2}$ all appear as Laplace transforms. Specifically, taking $a = 0$ and $b = 3$ (to get $b^2 = 9$), we have

$$\frac{1}{s} = \mathcal{L}\{1\}, \quad \frac{3}{s^2 + 9} = \mathcal{L}\{\sin 3t\}, \quad \frac{s}{s^2 + 9} = \mathcal{L}\{\cos 3t\}$$

or

$$\mathcal{L}^{-1}\left\{\frac{1}{s}\right\} = 1, \quad \mathcal{L}^{-1}\left\{\frac{1}{s^2 + 9}\right\} = \frac{1}{3}\sin 3t, \quad \mathcal{L}^{-1}\left\{\frac{s}{s^2 + 9}\right\} = \cos 3t.$$

Therefore, starting with $(\mathcal{L}x)(s) = \dfrac{\frac{1}{9}}{s} + \dfrac{\frac{17}{9}s - 1}{s^2 + 9}$ and taking inverse Laplace transforms, we get

$$x(t) = \frac{1}{9}\mathcal{L}^{-1}\left\{\frac{1}{s}\right\} + \frac{17}{9}\mathcal{L}^{-1}\left\{\frac{s}{s^2 + 9}\right\} - \mathcal{L}^{-1}\left\{\frac{1}{s^2 + 9}\right\}$$

$$= \frac{1}{9} + \frac{17}{9}\cos 3t - \frac{1}{3}\sin 3t$$

as the solution of our initial value problem.

Wait a minute, though. At the beginning of our computation (back on p. 480) we assumed that $(\mathcal{L}x)(s)$ exists for the solution $x(t)$ of our initial value problem $x'' + 9x = 1$, $x(0) = 2$, $x'(0) = -1$. How do we know that this is true?

One way to resolve this problem is to *check directly* that the function we found, $x(t) = \frac{1}{9} + \frac{17}{9}\cos 3t - \frac{1}{3}\sin 3t$, really does satisfy $x'' + 9x = 1$ along with the initial conditions $x(0) = 2$, $x'(0) = -1$ (see Exercise 1). This check also is a safeguard against mechanical errors in the computations! ∎

It can be shown (see Exercise 48) that $(\mathcal{L}x)(s)$ will always exist, at least for large enough s, when $x(t)$ is a solution of a constant-coefficient equation $x'' + px' + qx = h(t)$ for which the "forcing term" $h(t)$ is a sum of products of polynomials, exponential functions, sines, and cosines. However, if you are not in one of these "standard" special cases, it will be easier to check directly (as suggested above) that the final answer is correct than to show in advance that $(\mathcal{L}x)(s)$ must exist.

Here are several more examples of the use of the Laplace transform to solve initial value problems.

EXAMPLE 6.2.7 Solve $x'' - 3x' + 2x = 10e^{-3t}$, $x(0) = 0$, $x'(0) = 0$ (Section 2.6, Exercise 27) using the Laplace transform.

Since $x(0) = 0$ and $x'(0) = 0$, we have $(\mathcal{L}x')(s) = s(\mathcal{L}x)(s)$, $(\mathcal{L}x'')(s) = s^2(\mathcal{L}x)(s)$, so we get the "transformed equation"

$$(s^2 - 3s + 2)(\mathcal{L}x)(s) = \mathcal{L}(10e^{-3t})$$

$$= \frac{10}{s+3} \quad \text{(from Table 6.1.1).}$$

Therefore,

$$(\mathcal{L}x)(s) = \frac{10}{(s^2 - 3s + 2)(s+3)} = \frac{10}{(s-1)(s-2)(s+3)}.$$

The partial fraction decomposition has the form $\frac{A}{s-1} + \frac{B}{s-2} + \frac{C}{s+3}$, and we find $A = -\frac{5}{2}$, $B = 2$, $C = \frac{1}{2}$. Taking the inverse Laplace transform yields

$$x(t) = -\frac{5}{2}\mathcal{L}^{-1}\left\{\frac{1}{s-1}\right\} + 2\mathcal{L}^{-1}\left\{\frac{1}{s-2}\right\} + \frac{1}{2}\mathcal{L}^{-1}\left\{\frac{1}{s+3}\right\}$$

$$= -\frac{5}{2}e^t + 2e^{2t} + \frac{1}{2}e^{-3t}.$$

Again, it's a good idea to check this answer directly (see Exercise 2)! ∎

EXAMPLE 6.2.8 $x'' - 4x' + 3x = -9t$, $x(0) = 1$, $x'(0) = 0$.

Taking Laplace transforms, we get

$$s^2(\mathcal{L}x)(s) - s - 4[s(\mathcal{L}x)(s) - 1] + 3(\mathcal{L}x)(s) = \frac{-9}{s^2},$$

SEC. 6.2 TRANSFORMING INITIAL VALUE PROBLEMS

which simplifies to

$$(\mathcal{L}x)(s) = \frac{s - 4 - \frac{9}{s^2}}{s^2 - 4s + 3} = \frac{s^2(s-4) - 9}{s^2(s^2 - 4s + 3)} = \frac{s^3 - 4s^2 - 9}{s^2(s-1)(s-3)}.$$

We see that $(\mathcal{L}x)(s)$ can be written as

$$\frac{As + B}{s^2} + \frac{C}{s - 1} + \frac{D}{s - 3} = \frac{A}{s} + \frac{B}{s^2} + \frac{C}{s - 1} + \frac{D}{s - 3},$$

at which point we can use Table 6.1.1 to find the inverse Laplace transform of this expression. See Exercise 24 for the rest. ∎

The Laplace transform can also be used to find particular solutions to *systems* of differential equations.

EXAMPLE 6.2.9 Solve the initial value problem

$$\frac{dx}{dt} = -5x + 8y, \quad \frac{dy}{dt} = 4x - y, \quad x(0) = 0, \quad y(0) = 15$$

(Section 3.4, Exercise 15) using the Laplace transform.

Transforming the equations using the given initial values, we get

$$s(\mathcal{L}x)(s) = -5(\mathcal{L}x)(s) + 8(\mathcal{L}y)(s)$$

and

$$s(\mathcal{L}y)(s) - 15 = 4(\mathcal{L}x)(s) - (\mathcal{L}y)(s).^{10}$$

This is a pair of simultaneous equations for $(\mathcal{L}x)(s)$ and $(\mathcal{L}y)(s)$; if we write $X = (\mathcal{L}x)(s)$, $Y = (\mathcal{L}y)(s)$ for short, the equations simplify to

$$\begin{cases} (s + 5)X - 8Y = 0 \\ -4X + (s + 1)Y = 15. \end{cases}$$

From the first of these equations we get $Y = \frac{s + 5}{8} X$; substituting this into the second equation, after some computation we find

$$X = \frac{120}{s^2 + 6s - 27}, \quad Y = \frac{15s + 75}{s^2 + 6s - 27}.$$

To get back to $x(t) = \mathcal{L}^{-1}\{X\}$ and $y(t) = \mathcal{L}^{-1}\{Y\}$, we first factor the denominator as $s^2 + 6s - 27 = (s + 9)(s - 3)$ and write

$$X = \frac{-10}{s + 9} + \frac{10}{s - 3}, \quad Y = \frac{5}{s + 9} + \frac{10}{s - 3}.$$

[10] Once again, we are assuming that the Laplace transforms exist; comments similar to those on p. 484 apply.

Finally,

$$x(t) = -10\mathcal{L}^{-1}\left\{\frac{1}{s+9}\right\} + 10\mathcal{L}^{-1}\left\{\frac{1}{s-3}\right\}$$
$$= -10e^{-9t} + 10e^{3t}$$

and

$$y(t) = 5e^{-9t} + 10e^{3t}. \quad \blacksquare$$

You may wonder whether there is a general method for computing inverse Laplace transforms, rather than looking them up in a table. Unfortunately, the only completely general method involves integration along a path in the complex plane, which is beyond the scope of this book. However, in the rest of this chapter we'll see several ways (besides partial fraction decomposition) to find particular inverse Laplace transforms that don't occur in Table 6.1.1.

You may also wonder whether using the Laplace transform to solve an initial value problem is actually preferable to our earlier methods. This is often a matter of taste; for instance, in Examples 6.2.7 and 6.2.9, the methods of Sections 2.6 and 3.4, respectively, would work equally well. (See also Exercise 47.) However, when the function $h(t)$ has jumps, the Laplace transform method is usually the method of choice. We'll study this case in the next section.

SUMMARY OF KEY CONCEPTS, RESULTS, AND TECHNIQUES

Lerch's theorem (p. 481): Given $F(s)$, there can be only one continuous function $f(t)$ ($t \geq 0$) whose Laplace transform is $F(s) = (\mathcal{L}f)(s)$. If such an $f(t)$ exists, it is called the **inverse Laplace transform** of $F(s)$ and is denoted by $\mathcal{L}^{-1}\{F(s)\}$ or $(\mathcal{L}^{-1}F)(t)$ (p. 482).

The inverse Laplace transform is linear:

$$\mathcal{L}^{-1}\{\alpha \cdot F(s) + \beta \cdot G(s)\} = \alpha \cdot \mathcal{L}^{-1}\{F(s)\} + \beta \cdot \mathcal{L}^{-1}\{G(s)\},$$

provided both inverse transforms on the right exist (p. 482).

The Laplace transforms of the derivatives of f are given by

$$(\mathcal{L}f')(s) = s(\mathcal{L}f)(s) - f(0), \quad (\mathcal{L}f'')(s) = s^2(\mathcal{L}f)(s) - sf(0) - f'(0), \ldots,$$
$$(\mathcal{L}f^{(n)})(s) = s^n(\mathcal{L}f)(s) - s^{n-1}f(0) - s^{n-2}f'(0) \cdots - f^{(n-1)}(0). \quad (\text{p. 479})$$

Given an initial value problem $x'' + px' + qx = h(t)$, $x(0) = x_0$, $x'(0) = v_0$, we can apply the Laplace transform to both sides of the equation. Using the expressions for $(\mathcal{L}x'')(s)$ and $(\mathcal{L}x')(s)$ given by the initial values and the formulas above, we can then solve for $(\mathcal{L}x)(s)$, as in Example 6.2.1, p. 480. If we can find the inverse Laplace transform of $(\mathcal{L}x)(s)$, we'll get the desired solution $x(t)$. To reduce $(\mathcal{L}x)(s)$ to a form in which this

EXERCISES

(Table 6.1.1, p. 474, will often be helpful in solving these exercises.)

1. (See p. 484.) Check directly that $x(t) = \dfrac{1}{9} + \dfrac{17}{9}\cos 3t - \dfrac{1}{3}\sin 3t$ satisfies
$x'' + 9x = 1$, $x(0) = 2$, $x'(0) = -1$.

2. (See p. 484.) Check directly that $x(t) = -\dfrac{5}{2}e^t + 2e^{2t} + \dfrac{1}{2}e^{-3t}$ satisfies
$x'' - 3x' + 2x = 10e^{-3t}$, $x(0) = 0$, $x'(0) = 0$.

For each of the following initial value problems, find the Laplace transform $(\mathcal{L}x)(s)$ (assuming it exists) of the unknown function $x(t)$.

3. $x'' - 4x = 3$, $x(0) = 1$, $x'(0) = 5$.
4. $x'' + 2x' + x = \sin t$, $x(0) = 2$, $x'(0) = 0$.
5. $x'' + 2x' + x = e^{-t}$, $x(0) = 1$, $x'(0) = -1$.
6. $x'' - 2x' + x = 3e^t$, $x(0) = 0$, $x'(0) = 2$.
7. $x'' - 2x' - 3x = \cos t + e^{-t}$, $x(0) = 1$, $x'(0) = -1$.
8. $x'' + 2x' + 2x = t^2 - 1$, $x(0) = 3$, $x'(0) = 1$.

Find the inverse Laplace transform of each of the following functions.

9. $\dfrac{1}{s} - \dfrac{3}{s+1}$

10. $\dfrac{1}{(s-3)^2} + \dfrac{1}{s^2+4}$

11. $\dfrac{1}{(s-2)^3} - \dfrac{5}{s^2+4} + \dfrac{2}{s-3}$

12. $\dfrac{1}{s(s+1)}$

13. $\dfrac{1}{s(s-4)}$

14. $\dfrac{2s^2+s+1}{s(s-1)(s-2)}$

15. $\dfrac{s^2-s+3}{s(s^2-4)}$

16. $\dfrac{1}{s^2+8s+20}$
[*Hint:* Complete the square.]

17. $\dfrac{1}{s^2-8s+20}$

18. $\dfrac{3s^2-22s+60}{s(s^2-8s+20)}$

19. $\dfrac{s^2+4s+20}{s(s^2+8s+20)}$

20. $\dfrac{2s+1}{(s^2+1)(s^2+4)}$

21. $\dfrac{s}{(s^2+1)(s^2+9)}$ **22.** $\dfrac{s^2+s-18}{s(s^2+9)}$ **23.** $\dfrac{4s^2+3s+20}{s(s^2+4)}$

24. Finish Example 6.2.8, and check directly that your answer is correct.

Use the Laplace transform to solve each of the following initial value problems.

25. $x' - 4x = 0$, $x(0) = 3$.

26. $x' + 2x = 0$, $x(0) = -5$.

27. $x'' - 5x' + 6x = 0$, $x(0) = 2$, $x'(0) = -1$. (Example 2.3.3.)

28. $5x'' - 10x' = 0$, $x(0) = 1$, $x'(0) = -1$. (p. 101, Exercise 8.)

29. $x'' - 6x' + 9x = 0$, $x(0) = 2$, $x'(0) = 0$. (p. 106, Exercise 13.)

30. $x'' + 10x' + 25x = 0$, $x(0) = -1$, $x'(0) = -1$. (p. 106, Exercise 12.)

31. $x'' + 6x' + 10x = 0$, $x(0) = 1$, $x'(0) = 3$. (p. 111, Exercise 19.)

32. $x'' - 2x' + 2x = 0$, $x(0) = -1$, $x'(0) = 5$. (p. 111, Exercise 23.)

33. $x'' + 4x' + 3x = 1 - t$, $x(0) = 0$, $x'(0) = 0$. (p. 117, Exercise 21.)

34. $x'' - 4x' + 3x = t + 1$, $x(0) = 0$, $x'(0) = 0$. (p. 117, Exercise 20.)

35. $x'' + 4x' + 4x = e^{-2t}$, $x(0) = 3$, $x'(0) = -1$. (p. 117, Exercise 29.)

36. $x'' + 2x' + x = 2e^{-t}$, $x(0) = 0$, $x'(0) = 0$. (p. 117, Exercise 28.)

37. $x'' + 3x' + 2x = 5\cos t - e^t$, $x(0) = 0$, $x'(0) = 1$. (p. 118, Exercise 33.)

38. $x'' - 3x' + 2x = 5\sin t + e^{-t}$, $x(0) = 0$, $x'(0) = 1$. (p. 117, Exercise 32.)

39. $x^{(3)} - 4x' = 3e^t - 2$, $x(0) = \dfrac{3}{2}$, $x'(0) = \dfrac{15}{2}$, $x''(0) = 9$.

40. $x^{(3)} - x' = 3e^{2t} + 1$, $x(0) = 2$, $x'(0) = -\dfrac{5}{2}$, $x''(0) = \dfrac{11}{2}$.

41. $\begin{cases} \dfrac{dx}{dt} = x + y + 3, & x(0) = 5, \\ \dfrac{dy}{dt} = y - 1, & y(0) = 2. \end{cases}$ (p. 183, Exercise 7.)

42. $\begin{cases} \dfrac{dx}{dt} = 4y, & x(0) = 8, \\ \dfrac{dy}{dt} = -\dfrac{1}{2}x + 3y, & y(0) = -1. \end{cases}$ (p. 198, Exercise 14.)

43. $\begin{cases} \dfrac{dx}{dt} = x - y, & x(0) = 1, \\ \dfrac{dy}{dt} = 5x - y, & y(0) = 0. \end{cases}$ (Example 3.3.3.)

44. Show, by integrating by parts, that $(\mathcal{L}f')(s) = s \cdot (\mathcal{L}f)(s) - f(0)$ whenever $(\mathcal{L}f)(s)$ is defined.

*45. Show, by induction on n, that
$$(\mathcal{L}f^{(n)})(s) = s^n(\mathcal{L}f)(s) - s^{n-1}f(0) - s^{n-2}f'(0) \cdots - f^{(n-1)}(0).$$

*46. Find the inverse Laplace transform of $\dfrac{1}{(s^2 + 1)^2}$.

[*Hint*: Use $s^2 + 1 = (s + i)(s - i)$ to get a preliminary answer in terms of complex exponentials; these should no longer appear in your final answer.]

*47. (a) Show that the characteristic equation of the matrix $\begin{bmatrix} -5 & 8 \\ 4 & -1 \end{bmatrix}$ associated to the system of Example 6.2.9 (p. 485) is related (how?) to the denominator $s^2 + 6s - 27$ which occurred in the computation of Example 6.2.9.

(b) Show that this same phenomenon occurs for any initial value problem
$$\frac{dx}{dt} = ax + by, \quad \frac{dy}{dt} = cx + dy, \quad x(0) = x_0, \quad y(0) = y_0.$$

*48. (a) Show that the Laplace transform of $f(t) = t^n e^{\gamma t}$, where $\gamma = a + bi$ is a complex number, is defined for $s > a$.

(b) Show that if $x(t)$ is a solution of a constant-coefficient differential equation $x'' + px' + qx = h(t)$ for which $h(t)$ is a sum of products of polynomials, exponential functions, sines, and cosines, then $(\mathcal{L}x)(s)$ exists for s large enough.

[*Hint*: Use the results of Section 4.2 to show that $x(t)$ is a linear combination of functions of the form $t^n e^{\gamma t}$.]

*49. (a) Show that if $F(s)$ is the Laplace transform of $f(t)$, then $F'(s)$ is the Laplace transform of $-t \cdot f(t)$. (Assume that you may differentiate under the integral sign.)

(b) Find the Laplace transform of $t^2 \sin t$.

6.3 STEP FUNCTIONS

In order to apply the Laplace transform method to initial value problems

$$x'' + px' + qx = h(t), \quad x(0) = x_0, \quad x'(0) = v_0$$

for which the function $h(t)$ has jumps, we have to be able to compute the Laplace transform of such a function $h(t)$. This will be a sum of integrals over the intervals on which $h(t)$ is continuous.

EXAMPLE

Let $h(t) = \begin{cases} 1 & (t < \frac{\pi}{2}) \\ \cos t & (\frac{\pi}{2} < t < \pi) \\ 1 & (t > \pi) \end{cases}$. Then

$$(\mathscr{L}h)(s) = \int_0^{\pi/2} e^{-st}\, dt + \int_{\pi/2}^{\pi} \cos t \, e^{-st}\, dt + \int_{\pi}^{\infty} e^{-st}\, dt. \quad \blacksquare$$

Although one could try to compute these integrals separately for each individual case that comes up, it is usually easier to express $h(t)$ in terms of certain "basic step functions" and to use the properties of these step functions, which we will now define.

Let $c \geq 0$ be a constant. We define the **basic step function** at c, $u_c(t)$, by

$$u_c(t) = \begin{cases} 0 & (t < c) \\ 1 & (t > c) \end{cases} \quad \text{(see Figure 6.3.1)}.$$

NOTE: There is really nothing to prevent us from defining $u_c(t)$ for negative c, as well. However, we'll be using these step functions to compute Laplace transforms, which only involve $t \geq 0$, and so for our purposes any basic step function $u_c(t)$ with $c < 0$ might as well be the constant 1.

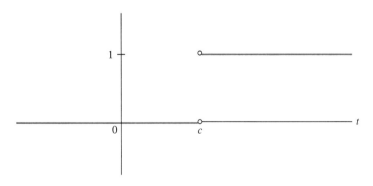

Figure 6.3.1. The basic step function $u_c(t)$.

SEC. 6.3 STEP FUNCTIONS

The Laplace transform of $u_c(t)$ is not hard to find:

$$(\mathcal{L}u_c)(s) = \int_0^\infty u_c(t)e^{-st}\,dt$$

$$= \int_0^c 0 \cdot e^{-st}\,dt + \int_c^\infty 1 \cdot e^{-st}\,dt$$

$$= \int_c^\infty e^{-st}\,dt = \frac{e^{-cs}}{s} \quad \text{(for } s > 0\text{)}. \tag{$*$}$$

Any function which has finitely many jumps and which is constant in between jumps can be written as a linear combination of basic step functions, and so its Laplace transform can be found also.

EXAMPLE 6.3.1 Find the Laplace transform of

$$h(t) = \begin{cases} 0 & (t < 1) \\ 2 & (1 < t < 2) \\ -1 & (2 < t < 4) \\ 3 & (t > 4). \end{cases}$$

Since $h(t)$ has jumps at $t = 1$, $t = 2$, and $t = 4$, we'll need the step functions $u_1(t)$, $u_2(t)$, and $u_4(t)$. Of these three, $u_2(t)$ and $u_4(t)$ are zero for $t < 2$, so we need $2u_1(t)$ to get the right behavior at $t = 1$:

$$2u_1(t) = \begin{cases} 0 & (t < 1) \\ 2 & (t > 1). \end{cases}$$

If we now subtract $3u_2(t)$, we'll also get the right jump at $t = 2$:

$$2u_1(t) - 3u_2(t) = \begin{cases} 0 & (t < 1) \\ 2 & (1 < t < 2) \\ -1 & (t > 2). \end{cases}$$

Finally, we add $4u_4(t)$ to get the jump at $t = 4$, and we find that

$$h(t) = 2u_1(t) - 3u_2(t) + 4u_4(t).$$

Hence, by $(*)$ and the linearity of the Laplace transform, we have

$$(\mathcal{L}h)(s) = 2(\mathcal{L}u_1)(s) - 3(\mathcal{L}u_2)(s) + 4(\mathcal{L}u_4)(s)$$

$$= \frac{1}{s}(2e^{-s} - 3e^{-2s} + 4e^{-4s}). \quad \blacksquare$$

A function that is not constant between jumps can still be written as a sum of multiples of basic step functions, but their coefficients will no longer be constants.

EXAMPLE 6.3.2

$$h(t) = \begin{cases} 1 & (t < \frac{\pi}{2}) \\ \cos t & (\frac{\pi}{2} < t < \pi) \\ 2 & (t > \pi) \end{cases} \quad \text{can be written as}$$

$$h(t) = 1 + (\cos t - 1)u_{\pi/2}(t) + (2 - \cos t)u_\pi(t) \,. \quad \square$$

We see that in order to find the Laplace transform of such a function, it would help if we could find the transform of a term of the form $g(t)u_c(t)$. [In the example above, one such term has $g(t) = \cos t - 1$, $c = \frac{\pi}{2}$, while another has $g(t) = 2 - \cos t$, $c = \pi$.] So let's write down the transform of such a term, in general:

$$\mathcal{L}\{g(t)u_c(t)\} = \int_0^\infty g(t)u_c(t)e^{-st}\, dt$$

$$= \int_c^\infty g(t)e^{-st}\, dt \,.$$

We're now rid of the awkward $u_c(t)$, but the integral is no longer from 0 to ∞. This can be remedied, though, by changing the variable to $t_1 = t - c$, and we then get

$$\mathcal{L}\{g(t)u_c(t)\} = \int_0^\infty g(t_1 + c)e^{-s(t_1 + c)}\, dt_1$$

$$= e^{-cs} \cdot \int_0^\infty g(t_1 + c)e^{-st_1}\, dt_1 \,.$$

In the last step we used that $e^{-s(t_1 + c)} = e^{-cs} \cdot e^{-st_1}$ and that since e^{-cs} doesn't depend on t_1, it can be taken out of the integral. Now note that the new integral on the right is itself a Laplace transform; to see this more clearly, let's change the dummy variable from t_1 to t (which doesn't affect the value of the integral). This yields

$$\mathcal{L}\{g(t)u_c(t)\} = e^{-cs} \cdot \int_0^\infty g(t + c)e^{-st}\, dt \,.$$

That is, the Laplace transform of $g(t)u_c(t)$ is e^{-cs} times the Laplace transform of $g(t + c)$, or in a formula:

$$\mathcal{L}\{g(t)u_c(t)\} = e^{-cs} \cdot \mathcal{L}\{g(t + c)\} \tag{1}$$

(Remember that both sides of this formula represent functions *of s* !)

EXAMPLE 6.3.3

$$\mathcal{L}\{(\cos t - 1)u_{\pi/2}(t)\} = e^{-\pi s/2} \cdot \mathcal{L}\left\{\cos\left(t + \frac{\pi}{2}\right) - 1\right\}$$

$$= e^{-\pi s/2} \cdot \mathcal{L}\{-\sin t - 1\}$$

$$= e^{-\pi s/2} \cdot \left(\frac{-1}{s^2 + 1} - \frac{1}{s}\right) \quad \text{(by Table 6.1.1, p. 474).}$$

SEC. 6.3 STEP FUNCTIONS

Similarly, $\mathcal{L}\{(2 - \cos t)u_\pi(t)\} = e^{-\pi s}\left(\dfrac{2}{s} + \dfrac{s}{s^2 + 1}\right)$. We can now find the Laplace transform of the function

$$h(t) = 1 + (\cos t - 1)u_{\pi/2}(t) + (2 - \cos t)u_\pi(t)$$

from Example 6.3.2. Specifically, we get

$$(\mathcal{L}h)(s) = (\mathcal{L}1)(s) + \mathcal{L}\{(\cos t - 1)u_{\pi/2}(t)\} + \mathcal{L}\{(2 - \cos t)u_\pi(t)\}$$

$$= \dfrac{1}{s} - e^{-\pi s/2}\left(\dfrac{1}{s^2 + 1} + \dfrac{1}{s}\right) + e^{-\pi s}\left(\dfrac{2}{s} + \dfrac{s}{s^2 + 1}\right)$$

$$= \dfrac{1}{s}(1 - e^{-\pi s/2} + 2e^{-\pi s}) + \dfrac{se^{-\pi s} - e^{-\pi s/2}}{s^2 + 1}. \quad \blacksquare$$

Now that we're in a position to find the "transformed equation" of an initial value problem for which $h(t)$ has jumps, let's see what actually happens for the initial value problem from Example 6.1.1.

EXAMPLE 6.3.4 $x'' + 9x = h(t)$, $x(0) = 1$, $x'(0) = 1$, where

$$h(t) = \begin{cases} 0 & \left(t < \dfrac{\pi}{2}\right) \\ \sin t & \left(t > \dfrac{\pi}{2}\right). \end{cases}$$

We have $h(t) = (\sin t)u_{\pi/2}(t)$, and so we get

$$(\mathcal{L}x'')(s) + 9(\mathcal{L}x)(s) = \mathcal{L}\{(\sin t)u_{\pi/2}(t)\}$$

$$s^2(\mathcal{L}x)(s) - s - 1 + 9(\mathcal{L}x)(s) = e^{-\pi s/2} \cdot \mathcal{L}\left\{\sin\left(t + \dfrac{\pi}{2}\right)\right\}$$

[using the given initial conditions on the left and formula (1) (p. 492) on the right]

$$(s^2 + 9)(\mathcal{L}x)(s) - s - 1 = e^{-\pi s/2} \cdot \mathcal{L}\{\cos t\}$$

$$= e^{-\pi s/2} \cdot \dfrac{s}{s^2 + 1} \quad \text{(from Table 6.1.1)}.$$

Solving for the Laplace transform of x yields

$$(\mathcal{L}x)(s) = \dfrac{s + 1}{s^2 + 9} + e^{-\pi s/2}\dfrac{s}{(s^2 + 1)(s^2 + 9)}.$$

As you can see, it isn't obvious how to find the inverse transform of the second term above; we'll clear up this problem, and finish the example later. $\square$

In general, we can expect a term of the form $e^{-cs}G(s)$ to occur in $(\mathcal{L}x)(s)$ each time $h(t)$ has a jump at c. [In the example above, $c = \dfrac{\pi}{2}$ and $G(s) = \dfrac{s}{(s^2+1)(s^2+9)}$.] So it would help to have a formula for $\mathcal{L}^{-1}\{e^{-cs}G(s)\}$, in other words, to be able to find a continuous function whose Laplace transform is $e^{-cs}G(s)$.

To find such a function, note that in formula (1) (p. 492) the right-hand side does have the factor e^{-cs} we want:

$$\mathcal{L}\{g(t)u_c(t)\} = e^{-cs} \cdot \mathcal{L}\{g(t+c)\}.$$

In order to simplify the right-hand side, we can replace the function $g(t)$ by the "shifted" function $g_1(t) = g(t-c)$, which has the effect of replacing $g(t+c)$ by $g_1(t+c) = g(t)$. We then have

$$\mathcal{L}\{g(t-c)u_c(t)\} = e^{-cs} \cdot \mathcal{L}\{g(t)\},$$

and now we can arrange to get $\mathcal{L}\{g(t)\} = G(s)$ by taking $g(t)$ to be the inverse Laplace transform $\mathcal{L}^{-1}\{G(s)\}$ (provided it exists). That is, if we let $g(t) = \mathcal{L}^{-1}\{G(s)\}$, we have $\mathcal{L}\{g(t-c)u_c(t)\} = e^{-cs}G(s)$, **and so $g(t-c)u_c(t)$ is a candidate for the inverse transform of $e^{-cs}G(s)$.**

NOTE: It is tempting to go ahead and write

$$\mathcal{L}^{-1}\{e^{-cs}G(s)\} = g(t-c)u_c(t),$$

but keep in mind that $g(t-c)u_c(t)$ might not be continuous—it might have a jump discontinuity at c. For instance,

$$\cos\left(t - \frac{\pi}{2}\right)u_{\pi/2}(t) = (\sin t)u_{\pi/2}(t) = \begin{cases} 0 & \left(t < \dfrac{\pi}{2}\right) \\ \sin t & \left(t > \dfrac{\pi}{2}\right) \end{cases}$$

has a jump at $t = \dfrac{\pi}{2}$. On the other hand,

$$\sin\left(t - \frac{\pi}{2}\right)u_{\pi/2}(t) = \begin{cases} 0 & \left(t < \dfrac{\pi}{2}\right) \\ -\cos t & \left(t > \dfrac{\pi}{2}\right) \end{cases}$$

does not have a jump; this function can be made continuous everywhere by defining it to be 0 for $t = \dfrac{\pi}{2}$.

EXAMPLE 6.3.4 (continued) Let's see what $g(t-c)u_c(t)$ actually looks like in the case from our example, that is, for $c = \dfrac{\pi}{2}$, $G(s) = \dfrac{s}{(s^2+1)(s^2+9)}$. First we need to find $g(t) = \mathcal{L}^{-1}\{G(s)\}$, which can be

SEC. 6.3 STEP FUNCTIONS

done using the method of partial fractions. The result is $g(t) = \frac{1}{8}(\cos t - \cos 3t)$ (see Section 6.2, Exercise 21), so

$$g(t-c)u_c(t) = \frac{1}{8}\left[\cos\left(t-\frac{\pi}{2}\right) - \cos 3\left(t-\frac{\pi}{2}\right)\right]u_{\pi/2}(t)$$

$$= \frac{1}{8}\left(\sin t + \sin 3t\right)u_{\pi/2}(t)$$

$$= \begin{cases} 0 & \left(t < \frac{\pi}{2}\right) \\ \frac{1}{8}(\sin t + \sin 3t) & \left(t > \frac{\pi}{2}\right). \end{cases}$$

Since $\sin\frac{\pi}{2} + \sin\frac{3\pi}{2} = 0$, this function does *not* have a jump at $t = \frac{\pi}{2}$; if we define it to be 0 for $t = \frac{\pi}{2}$, it will be continuous. So $g(t-c)u_c(t)$ really is the inverse transform of $e^{-cs}G(s)$ in this case; that is,

$$\mathcal{L}^{-1}\left\{e^{-\pi s/2}\frac{s}{(s^2+1)(s^2+9)}\right\} = \begin{cases} 0 & \left(t \le \frac{\pi}{2}\right) \\ \frac{1}{8}(\sin t + \sin 3t) & \left(t \ge \frac{\pi}{2}\right). \end{cases}$$

Now we can finish the computation. Back on p. 493 we had

$$(\mathcal{L}x)(s) = \frac{s+1}{s^2+9} + e^{-\pi s/2}\frac{s}{(s^2+1)(s^2+9)},$$

and applying the inverse Laplace transform to both sides, we get

$$x(t) = \mathcal{L}^{-1}\left\{\frac{s}{s^2+9}\right\} + \mathcal{L}^{-1}\left\{\frac{1}{s^2+9}\right\} + \mathcal{L}^{-1}\left\{e^{-\pi s/2}\frac{s}{(s^2+1)(s^2+9)}\right\}$$

$$= \cos 3t + \frac{1}{3}\sin 3t + \begin{cases} 0 & \left(t \le \frac{\pi}{2}\right) \\ \frac{1}{8}(\sin t + \sin 3t) & \left(t \ge \frac{\pi}{2}\right) \end{cases}$$

$$= \begin{cases} \cos 3t + \frac{1}{3}\sin 3t & \left(t \le \frac{\pi}{2}\right) \\ \cos 3t + \frac{11}{24}\sin 3t + \frac{1}{8}\sin t & \left(t \ge \frac{\pi}{2}\right), \end{cases}$$

the answer we originally found on p. 465. ■

It should not be too surprising that in the example above, $g(t-c)u_c(t)$ turned out to be continuous[11]; after all, we had already seen in Section 6.1 that the initial value problem would have a continuous solution despite the jump in $h(t)$. This will work in general, and so for the terms $e^{-cs}G(s)$ that will arise in our computations, we do have

$$\mathcal{L}^{-1}\{e^{-cs}G(s)\} = g(t-c)u_c(t), \quad \text{where} \quad g(t) = \mathcal{L}^{-1}\{G(s)\}.\text{[12]} \qquad (2)$$

Here are two more examples of using the Laplace transform to solve an initial value problem in which $h(t)$ has jumps.

EXAMPLE 6.3.5

$$x'' - 4x' + 4x = \begin{cases} 0 & (t<1) \\ 6 & (t>1) \end{cases}, \quad x(0)=2, \quad x'(0)=0.$$

We can rewrite the equation as $x'' - 4x' + 4x = 6u_1(t)$, and taking Laplace transforms of both sides gives us

$$(\mathcal{L}x'')(s) - 4(\mathcal{L}x')(s) + 4(\mathcal{L}x)(s) = 6\mathcal{L}\{u_1(t)\}$$

$$s^2(\mathcal{L}x)(s) - 2s - 4[s(\mathcal{L}x)(s) - 2] + 4(\mathcal{L}x)(s) = 6\frac{e^{-s}}{s}$$

[using (*) from p. 491, or formula (1) from p. 492]. Solving for $(\mathcal{L}x)(s)$ yields

$$(\mathcal{L}x)(s) = \frac{2s-8}{s^2-4s+4} + \frac{6e^{-s}}{s(s^2-4s+4)}.$$

Since $s^2 - 4s + 4 = (s-2)^2$, we have partial fraction decompositions of the form

$$\frac{2s-8}{s^2-4s+4} = \frac{A}{s-2} + \frac{B}{(s-2)^2}$$

and

$$\frac{6}{s(s^2-4s+4)} = \frac{C}{s} + \frac{D}{s-2} + \frac{E}{(s-2)^2}.$$

The coefficients turn out to be $A=2$, $B=-4$, $C=\frac{3}{2}$, $D=-\frac{3}{2}$, $E=3$ [see Exercise 21(a)], so we get

$$(\mathcal{L}x)(s) = \frac{2}{s-2} - \frac{4}{(s-2)^2} + e^{-s}\left[\frac{3/2}{s} - \frac{3/2}{s-2} + \frac{3}{(s-2)^2}\right]$$

$$= \frac{2}{s-2} - \frac{4}{(s-2)^2} + e^{-s}G(s),$$

---

[11] That is, after defining it to be zero at $t=c$.

[12] One can actually extend the definition of the inverse Laplace transform to allow for piecewise continuous results, although some care is needed at the jumps to make sure that the results are uniquely defined. Formula (2) remains correct in this broader context.

SEC. 6.3 STEP FUNCTIONS

where $G(s) = \dfrac{3/2}{s} - \dfrac{3/2}{s-2} + \dfrac{3}{(s-2)^2}$. Therefore,

$$x(t) = 2\mathcal{L}^{-1}\left\{\dfrac{1}{s-2}\right\} - 4\mathcal{L}^{-1}\left\{\dfrac{1}{(s-2)^2}\right\} + \mathcal{L}^{-1}\{e^{-s}G(s)\}.$$

Now note that in Table 6.1.1 (p. 474), $\dfrac{n!}{(s-a)^{n+1}}$ appears as a Laplace transform; in particular, for $n = 1$, $a = 2$ we get $\mathcal{L}^{-1}\left\{\dfrac{1}{(s-2)^2}\right\} = te^{2t}$. Also,

$$\mathcal{L}^{-1}\{e^{-s}G(s)\} = g(t-1)u_1(t) \quad \text{[by formula (2), p. 496]},$$

where

$$g(t) = \mathcal{L}^{-1}\{G(s)\}$$

$$= \dfrac{3}{2}\mathcal{L}^{-1}\left\{\dfrac{1}{s}\right\} - \dfrac{3}{2}\mathcal{L}^{-1}\left\{\dfrac{1}{s-2}\right\} + 3\mathcal{L}^{-1}\left\{\dfrac{1}{(s-2)^2}\right\}$$

$$= \dfrac{3}{2} - \dfrac{3}{2}e^{2t} + 3te^{2t}.$$

That is,

$$\mathcal{L}^{-1}\{e^{-s}G(s)\} = \left[\dfrac{3}{2} - \dfrac{3}{2}e^{2(t-1)} + 3(t-1)e^{2(t-1)}\right]u_1(t).$$

Substituting all this into our expression for $x(t)$ above, we end up with

$$x(t) = 2e^{2t} - 4te^{2t} + \left[\dfrac{3}{2} - \dfrac{3}{2}e^{2(t-1)} + 3(t-1)e^{2(t-1)}\right]u_1(t)$$

$$= \begin{cases} 2e^{2t} - 4te^{2t} & (t < 1) \\ 2e^{2t} - 4te^{2t} + \dfrac{3}{2} - \dfrac{3}{2}e^{2(t-1)} + 3(t-1)e^{2(t-1)} & (t > 1) \end{cases}$$

$$= \begin{cases} 2e^{2t} - 4te^{2t} & (t < 1) \\ \dfrac{3}{2} + \left(2 - \dfrac{9}{2e^2}\right)e^{2t} + \left(\dfrac{3}{e^2} - 4\right)te^{2t} & (t > 1). \end{cases}$$

(See Exercise 21(b) for the last step.) ∎

EXAMPLE 6.3.6 $x'' - 4x' + 3x = h(t)$, $x(0) = 3$, $x'(0) = 1$, where

$$h(t) = \begin{cases} 0 & (t < 2) \\ t & (2 < t < 4) \\ 6 & (t > 4). \end{cases}$$

Using the given initial conditions, we find the "transformed equation"
$$s^2(\mathcal{L}x)(s) - 3s - 1 - 4[s(\mathcal{L}x)(s) - 3] + 3(\mathcal{L}x)(s) = \mathcal{L}\{h(t)\}$$
$$(s^2 - 4s + 3)(\mathcal{L}x)(s) - 3s + 11 = \mathcal{L}\{h(t)\}.$$

To work out $\mathcal{L}\{h(t)\}$, we first write $h(t)$ in terms of basic step functions:
$$h(t) = tu_2(t) + (6 - t)u_4(t)$$
$$\mathcal{L}\{h(t)\} = \mathcal{L}\{tu_2(t)\} + \mathcal{L}\{(6 - t)u_4(t)\}$$
$$= e^{-2s} \cdot \mathcal{L}\{t + 2\} + e^{-4s} \cdot \mathcal{L}\{6 - (t + 4)\} \quad \text{[by formula (1), p. 492]}$$
$$= e^{-2s}(\mathcal{L}\{t\} + 2\mathcal{L}\{1\}) + e^{-4s}(-\mathcal{L}\{t\} + 2\mathcal{L}\{1\})$$
$$= e^{-2s}\left(\frac{1}{s^2} + \frac{2}{s}\right) + e^{-4s}\left(-\frac{1}{s^2} + \frac{2}{s}\right)$$
$$= e^{-2s} \cdot \frac{2s + 1}{s^2} + e^{-4s} \cdot \frac{2s - 1}{s^2}.$$

Substituting this into the "transformed equation" above and solving for $(\mathcal{L}x)(s)$, we find
$$(\mathcal{L}x)(s) = \frac{3s - 11}{s^2 - 4s + 3} + e^{-2s}\frac{2s + 1}{s^2(s^2 - 4s + 3)} + e^{-4s}\frac{2s - 1}{s^2(s^2 - 4s + 3)}. \quad (*)$$

We now need to find the inverse Laplace transform of each of the three terms on the right. In each case, we'll use partial fractions; for the last two terms, we'll also need formula (2) from p. 496. Specifically, the partial fraction decompositions turn out to be

$$\frac{3s - 11}{s^2 - 4s + 3} = \frac{4}{s - 1} - \frac{1}{s - 3},$$

$$\frac{2s + 1}{s^2(s^2 - 4s + 3)} = \frac{\frac{10}{9}s + \frac{1}{3}}{s^2} - \frac{\frac{3}{2}}{s - 1} + \frac{\frac{7}{18}}{s - 3},$$

$$\frac{2s - 1}{s^2(s^2 - 4s + 3)} = \frac{\frac{2}{9}s - \frac{1}{3}}{s^2} - \frac{\frac{1}{2}}{s - 1} + \frac{\frac{5}{18}}{s - 3}$$

[see Exercise 22(a)]. The inverse Laplace transforms of these rational functions are, therefore,

$$\mathcal{L}^{-1}\left\{\frac{3s - 11}{s^2 - 4s + 3}\right\} = 4e^t - e^{3t},$$

$$\mathcal{L}^{-1}\left\{\frac{2s + 1}{s^2(s^2 - 4s + 3)}\right\} = \frac{10}{9} + \frac{1}{3}t - \frac{3}{2}e^t + \frac{7}{18}e^{3t},$$

$$\mathcal{L}^{-1}\left\{\frac{2s - 1}{s^2(s^2 - 4s + 3)}\right\} = \frac{2}{9} - \frac{1}{3}t - \frac{1}{2}e^t + \frac{5}{18}e^{3t}.$$

Now we use (2) to find the inverse transforms for the last two terms in (*):

$$\mathcal{L}^{-1}\left\{e^{-2s}\frac{2s+1}{s^2(s^2-4s+3)}\right\} = \left[\frac{10}{9} + \frac{1}{3}(t-2) - \frac{3}{2}e^{(t-2)} + \frac{7}{18}e^{3(t-2)}\right]u_2(t),$$

$$\mathcal{L}^{-1}\left\{e^{-4s}\frac{2s-1}{s^2(s^2-4s+3)}\right\} = \left[\frac{2}{9} + \frac{1}{3}(t-4) - \frac{1}{2}e^{(t-4)} + \frac{5}{18}e^{3(t-4)}\right]u_4(t).$$

Assembling the three inverse transforms for the right-hand side of (*), we finally get

$$x(t) = 4e^t - e^{3t} + \left[\frac{10}{9} + \frac{1}{3}(t-2) - \frac{3}{2}e^{(t-2)} + \frac{7}{18}e^{3(t-2)}\right]u_2(t)$$

$$+ \left[\frac{2}{9} - \frac{1}{3}(t-4) - \frac{1}{2}e^{(t-4)} + \frac{5}{18}e^{3(t-4)}\right]u_4(t)$$

$$= \begin{cases} 4e^t - e^{3t} & (t \le 2) \\ \left(4 - \frac{3}{2e^2}\right)e^t + \left(\frac{7}{18e^6} - 1\right)e^{3t} + \frac{1}{3}t + \frac{4}{9} & (2 \le t \le 4) \\ \left(4 - \frac{3}{2e^2} - \frac{1}{2e^4}\right)e^t + \left(\frac{7}{18e^6} + \frac{5}{18e^{12}} - 1\right)e^{3t} + 2 & (t \ge 4). \end{cases}$$

[See Exercise 22(b) and (c) for the last step.]

N O T E : If we were only interested in the solution for $t > 4$, say, we could have predicted immediately that since $x(t) = 2$ is a constant solution of $x'' - 4x' + 3x = 6$ and since e^t, e^{3t} are solutions of the homogeneous equation, our solution would have the form $\alpha e^t + \beta e^{3t} + 2$ $(t > 4)$ for suitable constants α and β. However, actually finding the constants α and β is quite another story! ∎

SUMMARY OF KEY CONCEPTS, RESULTS, AND TECHNIQUES

The **basic step function** at c is defined by $u_c(t) = \begin{cases} 0 & (t < c) \\ 1 & (t > c) \end{cases}$ and has Laplace transform $(\mathcal{L}u_c)(s) = \dfrac{e^{-cs}}{s}$ (pp. 490–491). Any piecewise continuous function can be written in terms of basic step functions, one for each jump (pp. 491–492). Its Laplace transform can then be found using the formula $\mathcal{L}\{g(t)u_c(t)\} = e^{-cs} \cdot \mathcal{L}\{g(t+c)\}$ (p. 492).

When using the Laplace transform to solve an initial value problem $x'' + px' + qx = h(t)$, $x(0) = x_0$, $x'(0) = v_0$ for which $h(t)$ is piecewise continuous, one gets an expression for $(\mathcal{L}x)(s)$ which contains a term, of the form $e^{-cs}G(s)$, corresponding to each value $t = c$ where $h(t)$ has a jump (p. 494). The inverse Laplace transform of such a term is given by $\mathcal{L}^{-1}\{e^{-cs}G(s)\} = g(t-c)u_c(t)$, where $g(t) = \mathcal{L}^{-1}\{G(s)\}$ [and it is understood that $g(t-c)u_c(t)$ is made continuous by defining it to be zero at $t = c$] (p. 496). [For a general function g, $g(t-c)u_c(t)$ may have a jump at c, but this will not occur in the situation described here.]

EXERCISES

(Table 6.1.1, p. 474, will often be helpful in solving these exercises.)

For each of the following functions $h(t)$:

(a) Write $h(t)$ in terms of basic step functions.
(b) Compute the Laplace transform of $h(t)$.

1. $h(t) = \begin{cases} 0 & (t < 1) \\ 1 & (1 < t < 2) \\ 0 & (t > 2). \end{cases}$

2. $h(t) = \begin{cases} 2 & (t < 1) \\ -2 & (1 < t < 3) \\ 0 & (3 < t < 5) \\ 1 & (t > 5). \end{cases}$

3. $h(t) = \begin{cases} \sin t & (t < 0) \\ \cos t & (0 < t < \pi) \\ 1 & (t > \pi). \end{cases}$

4. $h(t) = \begin{cases} 1 & (t < 0) \\ \cos t & (0 < t < \pi) \\ \sin t & (t > \pi). \end{cases}$

5. $h(t) = \begin{cases} 0 & (t < 2) \\ 2t & (2 < t < 3) \\ 2e^t & (t > 3). \end{cases}$

6. $h(t) = \begin{cases} e^t & (t < 1) \\ 1 & (1 < t < 5) \\ t & (t > 5). \end{cases}$

7. $h(t) = \begin{cases} 1 - e^{2t} & (t < \pi) \\ 0 & (\pi < t < 2\pi) \\ \cos t & (t > 2\pi). \end{cases}$

8. $h(t) = \begin{cases} e^{2t} - 1 & (t < \pi) \\ \sin t & (\pi < t < 2\pi) \\ 1 & (t > 2\pi). \end{cases}$

For each of the following initial value problems, find $(\mathcal{L}x)(s)$.

9. $x'' - 4x' + 3x = h(t)$, $x(0) = 2$, $x'(0) = -1$, where $h(t)$ is the function from Exercise 1.

10. $x'' + 5x' + 4x = h(t)$, $x(0) = 1$, $x'(0) = 3$, where $h(t)$ is the function from Exercise 2.

11. $x'' - 4x' = h(t)$, $x(0) = 3$, $x'(0) = 0$, where $h(t)$ is the function from Exercise 7.

12. $x'' + 4x' + 4x = h(t)$, $x(0) = 0$, $x'(0) = -2$, where $h(t)$ is the function from Exercise 8.

Solve each of the following initial value problems using the Laplace transform.

13. $x' + 3x = \begin{cases} 0 & (t < 2) \\ -1 & (t > 2) \end{cases}$, $x(0) = 4$.

14. $x' - 2x = \begin{cases} 0 & (t < 1) \\ 3 & (t > 1) \end{cases}$, $x(0) = 5$.

15. $x'' - 4x' + 3x = h(t)$, $x(0) = 3$, $x'(0) = 7$, where $h(t) = \begin{cases} 0 & (t < 2) \\ -1 & (2 < t < 4) \\ 0 & (t > 4). \end{cases}$

16. $x'' + 3x' + 2x = h(t)$, $x(0) = 5$, $x'(0) = 7$, where $h(t) = \begin{cases} 0 & (t < 1) \\ 1 & (1 < t < 3) \\ 0 & (t > 3). \end{cases}$

SEC. 6.3 STEP FUNCTIONS

17. $x'' - 4x = \begin{cases} 0 & (t < 2) \\ t & (t > 2) \end{cases}$, $x(0) = 1$, $x'(0) = 6$.

18. $x'' + 4x = \begin{cases} 0 & (t < \pi) \\ \sin 3t & (t > \pi) \end{cases}$, $x(0) = 3$, $x'(0) = -4$.

19. $x'' + 4x' + 3x = h(t)$, $x(0) = -1$, $x'(0) = 3$, where $h(t) = \begin{cases} 0 & (t < 2) \\ -t & (2 < t < 4) \\ 0 & (t > 4) \end{cases}$.

20. $x'' - 3x' + 2x = h(t)$, $x(0) = 5$, $x'(0) = 2$, where $h(t) = \begin{cases} 0 & (t < 1) \\ t & (1 < t < 3) \\ 3 & (t > 3) \end{cases}$.

21. (a) Find the partial fraction decompositions for $\dfrac{2s - 8}{s^2 - 4s + 4}$ and $\dfrac{6}{s(s^2 - 4s + 4)}$.
    (See Example 6.3.5, but don't use the fact that the answers appear there.)
    (b) Justify the last step in Example 6.3.5.

22. (a) Find the partial fraction decompositions for $\dfrac{3s - 11}{s^2 - 4s + 3}$, $\dfrac{2s + 1}{s^2(s^2 - 4s + 3)}$, and $\dfrac{2s - 1}{s^2(s^2 - 4s + 3)}$. (See Example 6.3.6, but don't use the fact that the answers appear there.)
    (b) Show that for $t > 2$,

    $$4e^t - e^{3t} + \left[\frac{10}{9} + \frac{1}{3}(t - 2) - \frac{3}{2}e^{(t-2)} + \frac{7}{18}e^{3(t-2)}\right]u_2(t)$$
    $$= \left(4 - \frac{3}{2e^2}\right)e^t + \left(\frac{7}{18e^6} - 1\right)e^{3t} + \frac{1}{3}t + \frac{4}{9}.$$

    (c) Justify the last step in Example 6.3.6 (just before the Note).

23. Redo Exercise 26 from Section 6.1 (p. 476) using the Laplace transform.

24. Redo Exercise 27 from Section 6.1 using the Laplace transform.

25. Redo Exercise 28 from Section 6.1 using the Laplace transform.

26. Show that if the function g is continuous, then $g(t - c)u_c(t)$ has a jump discontinuity at c if and only if $g(0) \neq 0$. (Compare the Note on p. 494.)

27. (a) Find a function of t whose Laplace transform is $\dfrac{e^{-\pi s}}{s^2 + 1}$. Is this function the inverse Laplace transform of $\dfrac{e^{-\pi s}}{s^2 + 1}$, according to our definition? Explain!

    (b) Find a function of t whose Laplace transform is $\dfrac{se^{-\pi s}}{s^2 + 1}$. Is this function the inverse Laplace transform of $\dfrac{se^{-\pi s}}{s^2 + 1}$, according to our definition? Explain!

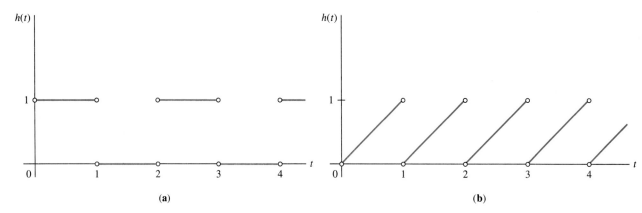

Figure 6.3.2. Two periodic functions with jumps.

*28. If the function $h(t)$ has *infinitely many* jumps, it *may* still be possible to find $(\mathcal{L}h)(s)$. An especially common and important case occurs when $h(t)$ is *periodic* and has only finitely many jumps in each period; Figure 6.3.2 shows examples of such functions. Suppose that $h(t)$ is *any* periodic function (which may or may not have jumps) and that T is the period of $h(t)$, so $h(t) = h(t + T)$ for all t.
(a) Write $(\mathcal{L}h)(s)$ as the sum of an infinite series of definite integrals over the intervals $[0, T]$, $[T, 2T]$, and so on.
(b) By making a change of variable in each of these definite integrals, show that
$$(\mathcal{L}h)(s) = (1 + e^{-sT} + e^{-2sT} + \cdots) \int_0^T h(t) e^{-st}\, dt.$$
(c) Show that $(\mathcal{L}h)(s) = \dfrac{\int_0^T h(t) e^{-st}\, dt}{1 - e^{-sT}}$ $(s > 0)$.

Use the result of Exercise 28 to find the Laplace transforms of the following functions.

29. $h(t) = \begin{cases} 1 & (0 < t < 1,\ 2 < t < 3,\ \ldots,\ 2k < t < 2k+1,\ \ldots) \\ 0 & (1 < t < 2,\ 3 < t < 4,\ \ldots,\ 2k+1 < t < 2k+2,\ \ldots) \end{cases}$

[This is the **square wave** shown in Figure 6.3.2(a).]

[*Hint*: To find $\int_0^2 h(t) e^{-st}\, dt$, split up the interval $[0, 2]$.]

30. $h(t) = \begin{cases} 1 & (0 < t < 1,\ 2 < t < 3,\ \ldots,\ 2k < t < 2k+1,\ \ldots) \\ -1 & (1 < t < 2,\ 3 < t < 4,\ \ldots,\ 2k+1 < t < 2k+2,\ \ldots) \end{cases}$

(Another square wave.)

31. $h(t) = \begin{cases} t & (0 < t < 1) \\ t - 1 & (1 < t < 2) \\ \vdots & \\ t - k & (k < t < k+1) \\ \vdots & \end{cases}$

[This is the **sawtooth wave** shown in Figure 6.3.2(b).]

32. $h(t) = \begin{cases} \dfrac{1}{2}t & (0 < t < 2) \\ \dfrac{1}{2}t - 1 & (2 < t < 4) \\ \vdots & \\ \dfrac{1}{2}t - k & (2k < t < 2k+2) \\ \vdots & \end{cases}$ (Another sawtooth wave.)

33. $h(t) = \cos t$. (The answer should be no surprise.)

34. $h(t) = \sin t$. (Same comment.)

6.4 THE CONVOLUTION INTEGRAL

In this section we'll solve the following problem: Find the inverse Laplace transform of a *product* of functions, given the inverse transforms of the functions individually. First, though, let's look at an example, to see how a concrete case of this problem might come up.

EXAMPLE 6.4.1 Solve the initial value problem $x'' + x = \cos t$, $x(0) = 2$, $x'(0) = -1$ using the Laplace transform.

The "transformed equation" is $s^2(\mathcal{L}x)(s) - 2s + 1 + (\mathcal{L}x)(s) = \dfrac{s}{s^2 + 1}$, and when we solve for the Laplace transform of x we get

$$(\mathcal{L}x)(s) = \frac{2s - 1}{s^2 + 1} + \frac{s}{(s^2 + 1)^2} = \frac{2s}{s^2 + 1} - \frac{1}{s^2 + 1} + \frac{s}{(s^2 + 1)^2}.$$

Finding the inverse Laplace transforms of the first two terms on the right is no problem; we have $\mathcal{L}^{-1}\left\{\dfrac{2s}{s^2 + 1}\right\} = 2\cos t$, $\mathcal{L}^{-1}\left\{\dfrac{1}{s^2 + 1}\right\} = \sin t$. This leaves us with the issue of how to find $\mathcal{L}^{-1}\left\{\dfrac{s}{(s^2+1)^2}\right\}$. Since both factors in the denominator are the same, the method of partial fractions won't help.[13] However, we do have

$$\frac{s}{(s^2+1)^2} = \frac{s}{s^2+1} \cdot \frac{1}{s^2+1},$$

and we know the inverse transform of each of the factors $\dfrac{s}{s^2+1}$, $\dfrac{1}{s^2+1}$ on the right. $\square$

Returning to the general problem, let $F(s) = (\mathcal{L}f)(s)$ and $G(s) = (\mathcal{L}g)(s)$ be the functions whose inverse Laplace transforms $f(t)$ and $g(t)$ are known; we want to find the inverse transform of the product $F(s) \cdot G(s)$. So we have

$$F(s) = \int_0^\infty f(t)e^{-st}\,dt, \qquad G(s) = \int_0^\infty g(t)e^{-st}\,dt,$$

and we want to write $F(s) \cdot G(s)$ as a Laplace transform. To do so, we'll first rewrite the product as an iterated (double) integral[14]; let's change the names of the ''dummy'' (integration) variables for F and G to t_1 and t_2, respectively, so we don't get them mixed up. Here goes:

$$F(s) \cdot G(s) = \int_0^\infty f(t_1)e^{-st_1}\,dt_1 \cdot \int_0^\infty g(t_2)e^{-st_2}\,dt_2$$

$$= \int_0^\infty \left[\int_0^\infty f(t_1)e^{-st_1}\,dt_1\right] g(t_2)e^{-st_2}\,dt_2$$

[since the integral in parentheses is a constant factor for integration over t_2]

$$= \int_0^\infty \int_0^\infty f(t_1)e^{-st_1} g(t_2)e^{-st_2}\,dt_1\,dt_2$$

[since $g(t_2)e^{-st_2}$ is a constant factor for integration over t_1]

$$= \int_0^\infty \int_0^\infty f(t_1)g(t_2)e^{-s(t_1+t_2)}\,dt_1\,dt_2.$$

Note that the exponential inside this iterated integral still looks like e^{-st}, but with $t_1 + t_2$ instead of t. To emphasize this, we'll use $t = t_1 + t_2$ as a new variable for the inner integral. The change of variable in the inner integral will yield

$$\int_0^\infty f(t_1)g(t_2)e^{-s(t_1+t_2)}\,dt_1 = \int_{t_2}^\infty f(t-t_2)g(t_2)e^{-st}\,dt$$

---

[13] Unless you use the complex factorization $(s^2+1)^2 = (s-i)^2(s+i)^2$; see Exercise 30.

[14] For background information on double integrals, see any textbook covering multivariable calculus.

($dt = dt_1$, $t_1 = t - t_2$. Note the new lower bound!), so we get

$$F(s) \cdot G(s) = \int_0^\infty \int_{t_2}^\infty f(t - t_2)g(t_2)e^{-st}\, dt\, dt_2.$$

Since we want to interpret this as the Laplace transform of some function, it would be better if the *outer* integral were of the form $\int_0^\infty \cdots e^{-st}\, dt$. Thus it seems natural to reverse the order of integration in the iterated integral above. Even though the integral is improper, this can be done[15] (although we won't prove it) by the usual method. That is, first we use the bounds on the iterated integral to see that we're integrating over the region in the t, t_2-plane described by the inequalities $0 \leq t_2 < \infty$, $t_2 \leq t < \infty$ (or $t_2 \geq 0$, $t \geq t_2$). Then we sketch this region (Figure 6.4.1). Finally, we describe the region again, this time using constant bounds on t: $0 \leq t < \infty$, $0 \leq t_2 < t$.

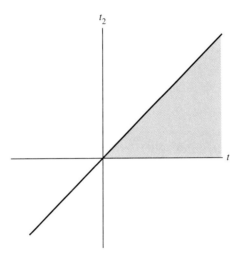

Figure 6.4.1. The region of integration in the t, t_2-plane.

Using these new bounds, we have

$$F(s) \cdot G(s) = \int_0^\infty \int_0^t f(t - t_2)g(t_2)e^{-st}\, dt_2\, dt.$$

Since the factor e^{-st} is constant for integration over t_2, we can take it outside the inner integral. Therefore,

$$F(s) \cdot G(s) = \int_0^\infty \left(\int_0^t f(t - t_2)g(t_2)\, dt_2 \right) e^{-st}\, dt.$$

---

[15] Under the assumption that $(\mathcal{L}f)(s) = F(s)$ and $(\mathcal{L}g)(s) = G(s)$ for all large enough s; compare the discussion on p. 471.

But now this is recognizable as a Laplace transform; in fact, we've written $F(s) \cdot G(s)$ as the Laplace transform of the expression $\int_0^t f(t - t_2)g(t_2) \, dt_2$ in parentheses. Note that since t_2 is just a "dummy" variable, this expression really is a function of the upper bound t only; also, we can replace t_2 by the more usual letter τ, and we then have

$$F(s) \cdot G(s) = \mathcal{L}\left\{\int_0^t f(t - \tau)g(\tau) \, d\tau\right\}.$$

That is, $F(s) \cdot G(s) = (\mathcal{L}h)(s)$, where h is the function defined by

$$h(t) = \int_0^t f(t - \tau)g(\tau) \, d\tau$$

for all $t \geq 0$. This function is called the **convolution** of f and g, and we write

$$h = f * g;$$

we've seen that

$$(\mathcal{L}f)(s) \cdot (\mathcal{L}g)(s) = F(s) \cdot G(s) = (\mathcal{L}(f * g))(s).$$

In words, reading from right to left, the Laplace transform of the convolution is the product of the Laplace transforms. Now provided $f * g$ is continuous, we have solved our problem, because then

$$\mathcal{L}^{-1}\{F(s) \cdot G(s)\} = (f * g)(t), \text{ where } f(t) = \mathcal{L}^{-1}\{F(s)\}, \, g(t) = \mathcal{L}^{-1}\{G(s)\}.$$

That is, the inverse Laplace transform of a product is the convolution of the inverse Laplace transforms.

EXAMPLE 6.4.1 (continued) Remember that we were looking for $\mathcal{L}^{-1}\left\{\dfrac{s}{(s^2 + 1)^2}\right\}$, which we can express as $\mathcal{L}^{-1}\{F(s) \cdot G(s)\}$ with $F(s) = \dfrac{s}{s^2 + 1}$, $G(s) = \dfrac{1}{s^2 + 1}$. We know that

$$f(t) = \mathcal{L}^{-1}\left\{\frac{s}{s^2 + 1}\right\} = \cos t, \quad g(t) = \mathcal{L}^{-1}\left\{\frac{1}{s^2 + 1}\right\} = \sin t,$$

so the inverse Laplace transform we want is the convolution

$$(f * g)(t) = \int_0^t \cos(t - \tau) \sin \tau \, d\tau,$$

provided this works out (as it will) to be a continuous function of t. To compute the integral, we can integrate by parts twice [see Exercise 28(a)], use complex exponentials [Exercise 28(b)], or—this is probably the slickest way—use the trigonometric identity $\sin p \cos q = \dfrac{1}{2}[\sin(p - q) + \sin(p + q)]$ [16] to rewrite the integrand:

---

[16] This can be shown by applying the addition formula for sin to both terms on the right. Two other useful identities of the same type are

$$\sin p \sin q = \frac{1}{2}[\cos(p - q) - \cos(p + q)], \quad \cos p \cos q = \frac{1}{2}[\cos(p - q) + \cos(p + q)].$$

SEC. 6.4 THE CONVOLUTION INTEGRAL

$$\cos(t-\tau)\sin\tau = \sin\tau\cos(t-\tau) = \frac{1}{2}[\sin(2\tau-t) + \sin t],$$

so

$$\int_0^t \cos(t-\tau)\sin\tau\, d\tau = \frac{1}{2}\int_0^t [\sin(2\tau-t) + \sin t]\, d\tau$$

$$= \frac{1}{2}\left[-\frac{1}{2}\cos(2\tau-t) + \tau\sin t\right]_{\tau=0}^{\tau=t}$$

$$= \frac{1}{2}\left[-\frac{1}{2}\cos(t) + t\sin t + \frac{1}{2}\cos(-t) - 0\right]$$

$$= \frac{1}{2}t\sin t.$$

Now that we have $\mathcal{L}^{-1}\left\{\dfrac{s}{(s^2+1)^2}\right\} = \dfrac{1}{2}t\sin t$, we can combine this with our earlier work on p. 503 to get

$$x(t) = 2\cos t - \sin t + \frac{1}{2}t\sin t$$

as the solution of the original initial value problem $x'' + x = \cos t$, $x(0) = 2$, $x'(0) = -1$. ∎

EXAMPLE 6.4.2   Find the inverse Laplace transform of $\dfrac{1}{s(s-4)}$.

This can be done using partial fractions (see Section 6.2, Exercise 13), but let's see what happens if we use convolution instead. We have $\mathcal{L}^{-1}\left\{\dfrac{1}{s}\right\} = 1$, $\mathcal{L}^{-1}\left\{\dfrac{1}{s-4}\right\} = e^{4t}$, so provided that the convolution of $f(t) = 1$ and $g(t) = e^{4t}$ is continuous, it will be our answer. The convolution is given by

$$(f*g)(t) = \int_0^t 1\cdot e^{4\tau}\, d\tau = \frac{1}{4}e^{4\tau}\Big]_{\tau=0}^{t} = \frac{1}{4}(e^{4t} - 1),$$

so we have

$$\mathcal{L}^{-1}\left\{\frac{1}{s(s-4)}\right\} = \frac{1}{4}(e^{4t} - 1). \quad\blacksquare$$

As it turns out, $f*g$ will always be continuous, provided that f and g are. In fact, even if f and g are arbitrary functions that are only *piecewise* continuous, we can still define their convolution $f*g$ by

$$(f*g)(t) = \int_0^t f(t-\tau)g(\tau)\, d\tau,$$

and $f * g$ will be continuous (not just piecewise). We won't prove this (see "Further Reading" following the Exercises), but here is an example.

EXAMPLE 6.4.3 Find $1 * u_c$, where the number 1 is short for the constant function with that value, and u_c is the basic step function (see p. 490).

We have $(1 * u_c)(t) = \int_0^t u_c(\tau)\, d\tau$. If $t < c$, then $u_c(\tau)$ will be zero on the entire interval $[0, t]$, so the integral will be zero, as well. On the other hand, if $t \geq c$ we have $\int_0^t u_c(\tau)\, d\tau = \int_c^t 1\, d\tau = t - c$. Therefore,

$$(1 * u_c)(t) = \begin{cases} 0 & (t < c) \\ t - c & (t \geq c). \end{cases}$$

Note that this function is, indeed, continuous for all t. ■

Since convolution of functions corresponds to multiplication of their Laplace transforms, it should not be surprising that convolution and multiplication have many properties in common. For instance, convolution is commutative, that is, for any two functions f and g we have $g * f = f * g$. This is not hard to show:

$$(g * f)(t) = \int_0^t g(t - \tau)f(\tau)\, d\tau$$

$$= \int_t^0 g(\tau_1)f(t - \tau_1) \cdot -d\tau_1 \quad \text{(changing the integration variable to } \tau_1 = t - \tau\text{)}$$

$$= \int_0^t f(t - \tau_1)g(\tau_1)\, d\tau_1$$

$$= (f * g)(t).$$

Convolution is also associative [Exercise 32(b)] and distributive over addition [Exercise 32(a)]. On the other hand, 1 is *not* an identity for convolution; that is, $1 * f$ is usually not equal to f, as we've seen in Examples 6.4.2 and 6.4.3 [see also Exercise 29(a)]. In fact, because 1 is the identity for multiplication, any function that was an identity for convolution would have 1 *as its Laplace transform*, and as mentioned on p. 482, this is impossible. However, in the next section we'll encounter a sort of "pseudo-function" which *does* have Laplace transform 1 (and acts as an identity for convolution), and which is useful in some computations.

While convolution can be used in many situations to find inverse Laplace transforms, other methods are often more efficient. For instance, in Example 6.2.7 we needed to find $\mathcal{L}^{-1}\left\{\dfrac{10}{(s - 1)(s - 2)(s + 3)}\right\}$. We did so by partial fractions, which is faster than taking

SEC. 6.4 THE CONVOLUTION INTEGRAL

the convolution of the three functions $\mathcal{L}^{-1}\left\{\dfrac{10}{s-1}\right\} = 10e^t$, $\mathcal{L}^{-1}\left\{\dfrac{1}{s-2}\right\} = e^{2t}$, $\mathcal{L}^{-1}\left\{\dfrac{1}{s+3}\right\} = e^{-3t}$ (see Exercise 27). As it turns out, however, convolution is more than just a computational tool for finding inverse Laplace transforms; the convolution integral, especially in the alternative form of Exercise 33, is quite important in several areas of mathematics.

For a final example, let's see how convolution could have been used to handle an example from Section 6.3 involving jumps.

EXAMPLE 6.4.4 Find $\mathcal{L}^{-1}\left\{e^{-\pi s/2}\dfrac{s}{(s^2+1)(s^2+9)}\right\}$ (this is the hard part of Example 6.3.4).

We want a (continuous) function whose Laplace transform is $e^{-\pi s/2}\dfrac{s}{(s^2+1)(s^2+9)}$; from p. 491 (or Table 6.4.1, p. 511) we know that $(\mathcal{L} u_{\pi/2})(s) = \dfrac{e^{-\pi s/2}}{s}$, and so it might be a good idea to write

$$e^{-\pi s/2}\dfrac{s}{(s^2+1)(s^2+9)} = \dfrac{e^{-\pi s/2}}{s} \cdot \dfrac{s^2}{(s^2+1)(s^2+9)}$$

$$= \dfrac{e^{-\pi s/2}}{s} \cdot \dfrac{s}{s^2+1} \cdot \dfrac{s}{s^2+9}.$$

For each of the three factors on the right, we know a function with that Laplace transform; the functions are $f(t) = u_{\pi/2}(t)$, $g(t) = \cos t$, and $h(t) = \cos 3t$, respectively. Therefore, the convolution $f * (g * h)$ of these three functions will have the Laplace transform we want. First, let's compute $g * h$:

$$(g * h)(t) = \int_0^t \cos(t - \tau) \cos 3\tau \, d\tau$$

$$= \int_0^t \dfrac{1}{2}[\cos(t - \tau - 3\tau) + \cos(t - \tau + 3\tau)] \, d\tau \quad \text{(see p. 506, footnote 16)}$$

$$= \dfrac{1}{2}\left[-\dfrac{1}{4}\sin(t - 4\tau) + \dfrac{1}{2}\sin(t + 2\tau)\right]_{\tau=0}^{t}$$

$$= -\dfrac{1}{8}\sin(-3t) + \dfrac{1}{4}\sin 3t - \dfrac{1}{8}\sin t$$

$$= \dfrac{3}{8}\sin 3t - \dfrac{1}{8}\sin t.^{17}$$

---

[17] Again, it might have been faster to get this as $\mathcal{L}^{-1}\left\{\dfrac{s^2}{(s^2+1)(s^2+9)}\right\}$, using partial fractions.

Now we want to find $f * (g * h)$. To avoid getting the awkward expression $u_{\pi/2}(t - \tau)$ in the integral, though, it might be better to switch the order to $(g * h) * f$:

$$(f * (g * h))(t) = ((g * h) * f)(t)$$

$$= \int_0^t (g * h)(t - \tau) f(\tau) \, d\tau$$

$$= \int_0^t \left(\frac{3}{8} \sin(3t - 3\tau) - \frac{1}{8} \sin(t - \tau) \right) u_{\pi/2}(\tau) \, d\tau.$$

Just as in Example 6.4.3 (p. 508), this integral is zero for $t < \frac{\pi}{2}$, while for $t \geq \frac{\pi}{2}$ we get

$$(f * (g * h))(t) = \int_{\pi/2}^t \left(\frac{3}{8} \sin(3t - 3\tau) - \frac{1}{8} \sin(t - \tau) \right) d\tau$$

$$= \left[\frac{1}{8} \cos(3t - 3\tau) - \frac{1}{8} \cos(t - \tau) \right]_{\tau = \pi/2}^t$$

$$= \frac{1}{8} - \frac{1}{8} - \frac{1}{8} \cos\left(3t - \frac{3\pi}{2} \right) + \frac{1}{8} \cos\left(t - \frac{\pi}{2} \right)$$

$$= \frac{1}{8} \sin 3t + \frac{1}{8} \sin t.$$

So we have

$$\mathcal{L}^{-1} \left\{ e^{-\pi s/2} \frac{s}{(s^2 + 1)(s^2 + 9)} \right\} = \begin{cases} 0 & \left(t < \frac{\pi}{2} \right) \\ \frac{1}{8} \sin 3t + \frac{1}{8} \sin t & \left(t \geq \frac{\pi}{2} \right), \end{cases}$$

which agrees with our result from p. 495. ∎

Table 6.4.1 (p. 511) is an expanded version of Table 6.1.1; along with specific Laplace transforms, some useful general formulas are listed. The unexplained notation in the last two lines of the table will be introduced in the next section. (Far more extensive tables are available; see "Further Reading" following the Exercises.)

SUMMARY OF KEY CONCEPTS, RESULTS, AND TECHNIQUES

Convolution (p. 506): The convolution $f * g$ of two functions f and g is the function defined by $(f * g)(t) = \int_0^t f(t - \tau) g(\tau) \, d\tau$ ($t \geq 0$). Convolution is commutative, associative, and distributive over addition (p. 508). If f and g are piecewise continuous, then $f * g$ is continuous and

$$(\mathcal{L}(f * g))(s) = (\mathcal{L}f)(s) \cdot (\mathcal{L}g)(s) \quad \text{(pp. 506–507)}.$$

SEC. 6.4 THE CONVOLUTION INTEGRAL

Table 6.4.1 Laplace transforms

$f(t) = \mathcal{L}^{-1}\{F(s)\}$	$(\mathcal{L}f)(s) = F(s)$	Proof
$\alpha g_1(t) + \beta g_2(t)$	$\alpha G_1(s) + \beta G_2(s)$	pp. 471–472
$g'(t)$	$sG(s) - g(0)$	Section 6.2, Exercise 44
$g''(t)$	$s^2 G(s) - sg(0) - g'(0)$	(See next line)
$g^{(n)}(t)$	$s^n G(s) - s^{n-1}g(0) \cdots - g^{(n-1)}(0)$	Section 6.2, Exercise 45
$(g_1 * g_2)(t) = \int_0^t g_1(t-\tau)g_2(\tau)\,d\tau$	$G_1(s)G_2(s)$	pp. 504–506
$\int_0^t g(\tau)\,d\tau$	$\dfrac{G(s)}{s}$	Section 6.4, Exercise 29
$-t \cdot g(t)$	$G'(s)$	Section 6.2, Exercise 49a
$u_c(t)$	$\dfrac{e^{-cs}}{s}$	p. 491
$g(t)u_c(t)$	$e^{-cs}\mathcal{L}\{g(t+c)\}$	p. 492
$g(t-c)u_c(t)$	$e^{-cs}G(s)$	pp. 494, 496
1	$\dfrac{1}{s}$	Example 6.1.9
$\sin bt$	$\dfrac{b}{s^2+b^2}$	Section 6.1, Exercises 20, 21
$\cos bt$	$\dfrac{s}{s^2+b^2}$	Section 6.1, Exercises 20, 21
t^n	$\dfrac{n!}{s^{n+1}}$	Section 6.1, Exercise 18
e^{at}	$\dfrac{1}{s-a}$	Example 6.1.12
$e^{at}\sin bt$	$\dfrac{b}{(s-a)^2+b^2}$	Section 6.1, Exercise 22
$e^{at}\cos bt$	$\dfrac{s-a}{(s-a)^2+b^2}$	Section 6.1, Exercise 22
$t^n e^{at}$	$\dfrac{n!}{(s-a)^{n+1}}$	Section 6.1, Exercise 24
$g(t)e^{at}$	$G(s-a)$	Section 6.1, Exercise 23
$t\sin bt$	$\dfrac{2bs}{(s^2+b^2)^2}$	Section 6.4, Exercise 8
$t\cos bt$	$\dfrac{s^2-b^2}{(s^2+b^2)^2}$	Section 6.4, Exercise 13
$\delta(t)$	1	p. 518
$\delta(t-t_0)$	e^{-st_0}	p. 520

Therefore, the inverse Laplace transform of a product is the convolution of the inverse Laplace transforms:

$$\mathcal{L}^{-1}\{F(s) \cdot G(s)\} = (f * g)(t), \quad \text{where} \quad f(t) = \mathcal{L}^{-1}\{F(s)\}, \ g(t) = \mathcal{L}^{-1}\{G(s)\} \ (\text{p. 506}).$$

This formula can often be used as an alternative to, or together with, the methods of Sections 6.2 and 6.3.

Table 6.4.1 (p. 511) summarizes many of the results of the current chapter.

EXERCISES

In each of the following cases, find $(f * g)(t)$.

1. $f(t) = t^2, \quad g(t) = t + 5$.
2. $f(t) = t, \quad g(t) = t^2 - 1$.
3. $f(t) = \sin 4t, \quad g(t) = \cos 2t$.
4. $f(t) = \cos 4t, \quad g(t) = \sin 3t$.
5. $f(t) = e^{3t}, \quad g(t) = e^{-t}$.
6. $f(t) = e^{-2t}, \quad g(t) = e^{t}$.

Use convolution to find the inverse Laplace transform of each of the following.

7. $\dfrac{1}{(s^2+1)^2}$

8. $\dfrac{s}{(s^2+9)^2}$

9. $\dfrac{s^2}{(s^2+b^2)^2}$ ($b \neq 0$ an arbitrary constant)

10. $\dfrac{1}{(s^2+b^2)^2}$ ($b \neq 0$)

11. $\dfrac{s^2 - b^2}{(s^2+b^2)^2}$

[*Hint*: Combine the results of Exercises 9 and 10.]

12. $\dfrac{s}{(s^2+b^2)^2}$

13. $\dfrac{1}{(s+4)(s-5)}$

14. $\dfrac{1}{(s-2)(s+7)}$

15. $\dfrac{1}{s^2(s^2+16)}$

16. $\dfrac{1}{s^2(s^2+b^2)}$ ($b \neq 0$)

17. $\dfrac{1}{(s-a)^2}$

18. $\dfrac{1}{(s-a)(s-b)}$ ($a \neq b$)

19. $\dfrac{e^{-\pi s}}{s^2+1}$ [Compare Section 6.3, Exercise 27(a).]

20. $\dfrac{e^{-2s}}{s(s^2-4s+3)}$

21. $\dfrac{e^{-4s}}{s(s^2-3s+2)}$

Solve the following initial value problems using the Laplace transform (and Table 6.4.1, p. 511).

22. $x'' + 4x = \sin 2t, \quad x(0) = 1, \quad x'(0) = -5$.
23. $x'' + 9x = -2\cos 3t, \quad x(0) = 4, \quad x'(0) = 0$.

24. $x'' + x = t \sin t$, $x(0) = 3$, $x'(0) = -2$.

25. $x'' + 4x = t \sin 2t$, $x(0) = 0$, $x'(0) = 0$.

*26. $x'' + x = t \cos t + \sin t$, $x(0) = 0$, $x'(0) = 0$.

27. Find $\mathcal{L}^{-1}\left\{\dfrac{10}{(s-1)(s-2)(s+3)}\right\}$ by taking the convolution of $10e^t$, e^{2t}, and e^{-3t} (see pp. 508–509), and check your answer with Example 6.2.7.

28. (See p. 506.) Find $\displaystyle\int_0^t \cos(t-\tau)\sin\tau\, d\tau$ as follows:

(a) By integrating by parts twice.

(b) By using complex exponentials.

29. (a) Find $(1 * f)(t)$ for an arbitrary function f.

(b) Show that $\mathcal{L}\left\{\displaystyle\int_0^t f(\tau)\, d\tau\right\} = \dfrac{(\mathcal{L}f)(s)}{s}$, provided $(\mathcal{L}f)(s)$ exists (for s large enough).

*30. Use $(s^2+1)^2 = (s-i)^2(s+i)^2$ and partial fractions to find $\mathcal{L}^{-1}\left\{\dfrac{s}{(s^2+1)^2}\right\}$ (see Example 6.4.1).

31. (a) Show that for any function f,

$$(f * u_c)(t) = \begin{cases} 0 & (t<c) \\ \displaystyle\int_c^t f(t-\tau)\, d\tau & (t \geq c). \end{cases}$$

*(b) Show that

$$(u_{c_1} * u_{c_2})(t) = \begin{cases} 0 & (t < c_1 + c_2) \\ t - c_1 - c_2 & (t \geq c_1 + c_2). \end{cases}$$

32. (a) Show that for any functions f, g, and h, $f * (g + h) = f * g + f * h$.

*(b) Show that for any functions f, g, and h, $f * (g * h) = (f * g) * h$.

33. Note that $(f * g)(t)$ has been defined only for $t \geq 0$, and that $(\mathcal{L}f)(s)$, $(\mathcal{L}g)(s)$ only depend on the values of $f(t)$, $g(t)$ for $t \geq 0$. Thus our formulas for the convolution will remain true if we *redefine* $f(t)$ and $g(t)$ to be *zero* for $t < 0$. We can also *define* $(f * g)(t) = 0$ for $t < 0$. Show that if all this is done, then

$$(f * g)(t) = \int_{-\infty}^{\infty} f(t-\tau)g(\tau)\, d\tau \quad \text{for all real numbers } t.$$

*34. Solve the initial value problem $x'' + 4x = h(t)$, $x(0) = 3$, $x'(0) = -1$, where

$$h(t) = \begin{cases} 0 & (t < \pi) \\ \cos 2t & (\pi < t < 2\pi) \\ 0 & (t > 2\pi). \end{cases}$$

*35. Solve the initial value problem $x'' + x = h(t)$, $x(0) = -2$, $x'(0) = 5$, where

$$h(t) = \begin{cases} 0 & (t < \pi) \\ \sin t & (\pi < t < 2\pi) \\ 0 & (t > 2\pi). \end{cases}$$

[*Hint*: Use the result of Exercise 7.]

FURTHER READING

For a more extensive table of Laplace transforms, and for a sketch of a proof that if f and g are piecewise continuous, then $f*g$ is continuous, see Churchill, *Operational Mathematics*, 3rd ed. (McGraw-Hill, 1972).

6.5 THE DELTA "FUNCTION"

In certain applications, the "forcing term" $h(t)$ in an initial value problem

$$x'' + px' + qx = h(t), \quad x(0) = x_0, \; x'(0) = v_0$$

is a function that is hard to determine, but that is known to feature a brief, sudden upward surge or "spike," followed immediately by a return to "normal." For instance, $h(t)$ may be zero for all t except for a very short time interval during which $h(t)$ is very large. This phenomenon will occur when a "passive" electrical circuit (without a generator, or with the generator turned off) is struck by lightning. It can also occur as a result of a mechanical collision, as in the following example.

EXAMPLE 6.5.1 In the experiment shown in Figure 6.5.1, two particles move along the *x*-axis. The first particle has mass m and is attached to a spring with spring constant k; therefore, assuming that it moves without friction, we'll have $mx'' + kx = h(t)$, where $h(t)$ is the external force on the particle. This external force is provided by a collision with the second particle, which has mass $\frac{1}{4}m$ and which has been approaching the first particle along the *x*-axis with constant speed v_0. The particles collide at $t = 0$; until then, the first particle has been at equilibrium. If the collision is elastic (that is, no kinetic energy is lost), then what can be said about the force $h(t)$?

SEC. 6.5 THE DELTA "FUNCTION"

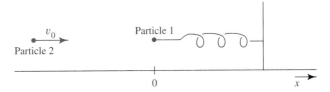

Figure 6.5.1. What is the force $h(t)$ on particle 1 from particle 2?

It seems clear that $h(t)$ is zero for $t < 0$ (before the second particle hits) and again almost immediately after $t = 0$ (once the collision is over). To find out more about $h(t)$, let's consider the velocities v_1, v_2 of the first and second particles, respectively, *right after* the collision. We can find v_1 and v_2 using the basic conservation laws of mechanics, which yield

$$\frac{1}{4} m v_0 = m v_1 + \frac{1}{4} m v_2 \qquad \text{(conservation of momentum)}$$

and

$$\frac{1}{2}\left(\frac{1}{4}m\right)v_0^2 = \frac{1}{2} m v_1^2 + \frac{1}{2}\left(\frac{1}{4}m\right)v_2^2 \qquad \text{(conservation of energy[18])}.$$

These equations simplify to $v_0 = 4v_1 + v_2$, $v_0^2 = 4v_1^2 + v_2^2$, and we then find $v_1 = \frac{2}{5}v_0$, $v_2 = -\frac{3}{5}v_0$ (see Exercise 13). So right after the collision, the second particle will be "retreating" with speed $\frac{3}{5}v_0$, while the first particle "moves forward" with speed $\frac{2}{5}v_0$.

Now suppose that the collision lasts from $t = 0$ until $t = \epsilon$. During this time interval, the acceleration of the first particle is $\frac{1}{m} h(t)$, by Newton's second law.[19] Due to this acceleration, the velocity of the first particle changes from 0 (for $t = 0$) to $\frac{2}{5}v_0$ (for $t = \epsilon$). Since the velocity is obtained by integrating the acceleration, it follows that

$$\frac{2}{5}v_0 - 0 = \int_0^\epsilon \frac{1}{m} h(t)\, dt, \qquad \text{that is,} \qquad \int_0^\epsilon h(t)\, dt = \frac{2}{5} m v_0.$$

---

[18] Here we're assuming that the first particle moves so little during the actual collision that its potential energy (due to compression of the spring) right afterward is negligible. This assumption is realistic because the collision is very brief.

[19] This assumes that the spring force is negligible during the collision. Again, this is realistic because during the brief time interval $[0, \epsilon]$, the particle will move very little from its equilibrium position.

Since $h(t)$ is zero outside the interval $[0, \epsilon]$, we might as well write this as

$$\int_{-\infty}^{\infty} h(t)\,dt = \frac{2}{5} mv_0\,.$$

In other words, what we know about the function $h(t)$ are the value of $\int_{-\infty}^{\infty} h(t)\,dt$, along with the fact that $h(t)$ is zero outside some interval $[0, \epsilon]$, where ϵ is "very small."

N O T E : You may wonder whether we shouldn't just assume $\epsilon = 0$: Shouldn't the collision be instantaneous, for $t = 0$ only? However, if we do have $h(t) = 0$ for all $t \neq 0$ and if $h(0) = h_0$ is any finite number, then

$$\int_{-\infty}^{\infty} h(t)\,dt = \int_{-\infty}^{0} h(t)\,dt + \int_{0}^{\infty} h(t)\,dt = 0 + 0 = 0\,,$$

which contradicts the above. We'll return to this issue soon. □

If we want to use the Laplace transform method to solve an initial value problem $x'' + px' + qx = h(t)$, $x(0) = x_0$, $x'(0) = v_0$ in which $h(t)$ has a "spike," we'll need to find $(\mathcal{L}h)(s)$, so let's see what we can say about such a Laplace transform. As in the example, we'll assume that $h(t)$ is zero outside an interval $[0, \epsilon]$ and that $\int_{-\infty}^{\infty} h(t)\,dt$ is a given positive number. For simplicity, we'll take this number to be 1 [this can be arranged by multiplying $h(t)$ by a constant factor], so

$$\int_{-\infty}^{\infty} h(t)\,dt = 1\,.$$

Of course, there are many possibilities for such a "spike" function; Figure 6.5.2 shows a few, of which the "flat spike" shown in Figure 6.5.2(a) is the easiest to work with. To compute its Laplace transform, we use the results of Section 6.3: We have

$$h_1(t) = \begin{cases} 0 & (t < 0) \\ \dfrac{1}{\epsilon} & (0 \leq t \leq \epsilon) \\ 0 & (t > \epsilon) \end{cases} = \frac{1}{\epsilon}[u_0(t) - u_\epsilon(t)]\,,$$

so

$$(\mathcal{L}h_1)(s) = \frac{1}{\epsilon}[(\mathcal{L}u_0)(s) - (\mathcal{L}u_\epsilon)(s)]$$

$$= \frac{1}{\epsilon}\left(\frac{1}{s} - \frac{e^{-\epsilon s}}{s}\right) = \frac{1 - e^{-\epsilon s}}{\epsilon s}\,.$$

For the other "spike" functions shown in Figure 6.5.2, the computations are longer (see Exercises 15 to 17), but eventually we find

SEC. 6.5 THE DELTA "FUNCTION"

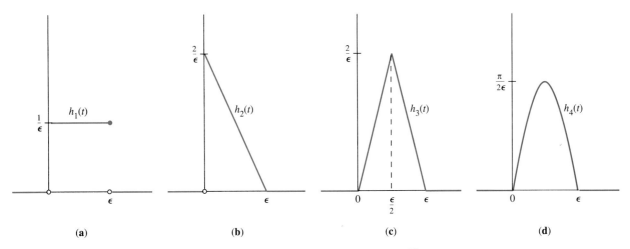

Figure 6.5.2. Some "spike" functions $h(t)$, zero outside $[0, \epsilon]$, for which $\int_{-\infty}^{\infty} h(t)\, dt = 1$.

$$(\mathscr{L}h_2)(s) = \frac{2}{\epsilon^2 s^2} (e^{-\epsilon s} - 1 + \epsilon s),$$

$$(\mathscr{L}h_3)(s) = \frac{4}{\epsilon^2 s^2} (1 - 2e^{-\epsilon s/2} + e^{-\epsilon s}),$$

$$(\mathscr{L}h_4)(s) = \frac{\pi^2}{2} \cdot \frac{1 + e^{-\epsilon s}}{\pi^2 + \epsilon^2 s^2}.$$

At first, these results may seem discouraging. The four Laplace transforms above look quite different; since the actual "spike" function h probably won't be any of h_1, h_2, h_3, h_4, how can we possibly say anything about $(\mathscr{L}h)(s)$? What we haven't used yet, though, is the idea that ϵ is "very small." In fact, as noted above, in our example it would have been nice to be able to take $\epsilon = 0$ (to get an instantaneous collision). Although we can't actually take $\epsilon = 0$, we can look at what happens to the Laplace transforms in the limit as ϵ approaches zero. For example, we can consider $\lim_{\epsilon \to 0^+} (\mathscr{L}h_1)(s)$, and we get an indeterminate form to which l'Hôpital's rule applies:

$$\lim_{\epsilon \to 0^+} (\mathscr{L}h_1)(s) = \lim_{\epsilon \to 0^+} \frac{1 - e^{\epsilon s}}{\epsilon s} \stackrel{*}{=} \lim_{\epsilon \to 0^+} \frac{se^{-\epsilon s}}{s} = \lim_{\epsilon \to 0^+} e^{-\epsilon s} = 1.^{[20]}$$

Thus the smaller ϵ gets, the more $(\mathscr{L}h_1)(s)$ will "approach the constant function 1" (for fixed s, at least; see Exercise 23). Similar computations for $(\mathscr{L}h_2)(s)$ and $(\mathscr{L}h_3)(s)$ give the same result [see Exercises 15(c) and 16(c)], and for $(\mathscr{L}h_4)(s)$ we get

---

[20] The symbol $\stackrel{*}{=}$ indicates that the limits to either side of it are equal provided the one on the right exists (which is one of the conditions for l'Hôpital's rule).

$$\lim_{\epsilon \to 0^+} \frac{\pi^2}{2} \cdot \frac{1 + e^{-\epsilon s}}{\pi^2 + \epsilon^2 s^2} = \frac{\pi^2}{2} \cdot \frac{1 + 1}{\pi^2} = 1, \quad \text{as well!}$$

It should now seem plausible, at least, that regardless of the exact form of the "spike" function $h(t)$ with $\int_{-\infty}^{\infty} h(t)\,dt = 1$, its Laplace transform approaches the constant function 1 as the interval $[0, \epsilon]$ on which $h(t)$ is nonzero shrinks down to the origin. Therefore, if we want to make a mathematical model of a lightning strike, a collision between particles, or a hammer blow, all of which are instantaneous (or nearly so) and occur at $t = 0$, we may expect reasonable results if we don't specify the function $h(t)$, but instead compute with its Laplace transform as if that were a *constant* function of s. In particular, if $\int_{-\infty}^{\infty} h(t)\,dt = 1$, then we've seen that in such a model, the Laplace transform $(\mathscr{L}h)(s)$ should be the constant function 1.

Unfortunately, as mentioned on p. 482, the constant function 1 cannot actually occur as the Laplace transform of *any* function $f(t)$. Still, you might wonder: Since we got 1 as the limit of $(\mathscr{L}h_1)(s)$ as $\epsilon \to 0^+$, couldn't we take the limit of

$$h_1(t) = \begin{cases} 0 & (t < 0) \\ \dfrac{1}{\epsilon} & (0 \leq t \leq \epsilon) \\ 0 & (t > \epsilon) \end{cases}$$

as $\epsilon \to 0^+$ and get a function whose Laplace transform is 1? Well, $\lim\limits_{\epsilon \to 0^+} h_1(t) = 0$ for any $t \neq 0$, since for ϵ small enough, t won't be in the interval $[0, \epsilon]$. On the other hand, $\lim\limits_{\epsilon \to 0^+} h_1(0) = \lim\limits_{\epsilon \to 0^+} \dfrac{1}{\epsilon} = \infty$, so we see that the "function" $\lim\limits_{\epsilon \to 0^+} h_1(t)$ "blows up" for $t = 0$.

It is a convenient notational fiction to pretend that the limit $\lim\limits_{\epsilon \to 0^+} h_1(t)$ *is a function* whose Laplace transform is 1. This "pseudo-function" is called the **(Dirac) delta "function"**[21] or **unit impulse function**; it is denoted by $\delta(t)$. By the above, $\delta(t) = 0$ for $t \neq 0$, while $\delta(0)$ is undefined, and $(\mathscr{L}\delta)(s) = 1$.

The mathematical ground we're on is definitely shaky. In fact, if $\delta(t)$ were an ordinary function, the fact that $\delta(t) = 0$ for all $t > 0$ would show that the Laplace transform $(\mathscr{L}\delta)(s) = \int_0^\infty \delta(t)e^{-st}\,dt$ would be zero, not 1. Apparently, then, integrals involving the delta "function" cannot be treated like integrals involving ordinary functions; more on this below. (Intuitively, the delta "function" is concentrated at one point, $t = 0$, but since it is infinite there, this one point can make a nonzero contribution to the integral.) But then what else will happen if we allow this "function"? What other "functions" are we going

---

[21] The delta "function" was first defined by Kirchhoff and Heaviside, but it is named after the British physicist P. A. M. Dirac (1902–), who used it in his fundamental work in quantum mechanics. (See "Further Reading" following the Exercises.)

SEC. 6.5 THE DELTA "FUNCTION"

to accept, and how will we work with them? (See Exercise 22 for another issue.) Satisfactory answers to these and other, similar, questions were not available until the late 1940s, when the French mathematician Laurent Schwartz (1915–) developed his theory of **distributions** or "generalized functions," which is well beyond the scope of this book. (See Exercise 29 for a hint of the ideas involved.) Meanwhile, however, electrical engineers and physicists had been making successful ad hoc computations using the delta "function" for dozens of years. In that spirit, we'll largely ignore the questions raised above, and from now on (as is customary), we'll no longer use quotation marks when referring to the delta function.

So how should we treat integrals involving the delta function, such as $(\mathcal{L}\delta)(s) = \int_0^\infty \delta(t)e^{-st}\,dt$? Remember that the answer 1 that we want for this integral originally came up as $\lim_{\epsilon \to 0^+} (\mathcal{L}h_1)(s) = \lim_{\epsilon \to 0^+} \int_0^\infty h_1(t)e^{-st}\,dt$. We can now *define* any integral involving the delta function in this same way. That is, **any integral involving $\delta(t)$ is to be interpreted as the limit as $\epsilon \to 0^+$ of the corresponding integral involving**

$$h_1(t) = \begin{cases} 0 & (t < 0) \\ \dfrac{1}{\epsilon} & (0 \leq t \leq \epsilon) \\ 0 & (t > \epsilon). \end{cases}$$

For instance, we have now defined $\int_{-\infty}^\infty \delta(t)\,dt$ as $\lim_{\epsilon \to 0^+} \int_{-\infty}^\infty h_1(t)\,dt = \lim_{\epsilon \to 0^+} 1 = 1$. In fact, the delta function is often defined as the "function" for which $\delta(t) = 0$ for $t \neq 0$ while $\int_{-\infty}^\infty \delta(t)\,dt = 1$. Intuitively, we can think of the graph of $\delta(t)$ as having an "infinitely narrow infinite spike" (at $t = 0$) whose area is 1.

It is shown in Exercise 24 that for any continuous function $f(t)$, the definition above will give us

$$\int_{-\infty}^\infty \delta(t)f(t)\,dt = \int_0^\infty \delta(t)f(t)\,dt = f(0) \quad [22]. \tag{*}$$

For example, taking $f(t) = e^{-st}$, we see again that the Laplace transform of $\delta(t)$ is $\int_0^\infty \delta(t)e^{-st}\,dt = e^{-s\cdot 0} = 1$.

NOTE: We can now also see that, as anticipated on p. 508, the delta function acts as an identity for convolution, at least with continuous functions. For if f is any continuous function, then for any $t > 0$,

---

[22] In the literature, integrals involving the delta function are often taken only over intervals, such as $(-\infty, \infty)$, with 0 in their interior, and $\delta(t)$ is often defined as a limit of "flat spikes" that are symmetrical about $t = 0$, rather than extending only to the right. However, the "one-sided" approach is more convenient for the Laplace transform.

$$(f * \delta)(t) = \int_0^t f(t - \tau)\delta(\tau)\, d\tau$$

$$= \int_0^\infty f(t - \tau)\delta(\tau)\, d\tau \quad \text{[see Exercise 27(a)]}$$

$$= f(t - 0) \quad \text{[by (*) from p. 519]}$$

$$= f(t).$$

Admittedly, for $t = 0$ the situation is less clear; see Exercise 27(b).

So far, we have only considered "spikes" at $t = 0$. To study similar phenomena at $t = t_0$, we can simply replace t by $t - t_0$ in the above. In particular, the function $\delta(t - t_0)$ describes an "infinite spike of unit area" at $t = t_0$; we have $\delta(t - t_0) = 0$ for $t \neq t_0$. Also, for any continuous function $f(t)$, we have

$$\int_{-\infty}^\infty \delta(t - t_0) f(t)\, dt = \int_{-\infty}^\infty \delta(t_1) f(t_1 + t_0)\, dt_1 \quad \text{(where } t_1 = t - t_0\text{; see Exercise 26)}$$

$$= \int_{-\infty}^\infty \delta(t) f(t + t_0)\, dt \quad \text{(renaming the "dummy" variable)}$$

$$= f(0 + t_0) \quad \text{[by (*) from p. 519]}$$

$$= f(t_0).$$

Provided $t_0 \geq 0$, the integral can also be taken from 0 to ∞ [since $\delta(t - t_0)$ is zero anyway for negative t [23]] and we have

$$\int_{-\infty}^\infty \delta(t - t_0) f(t)\, dt = \int_0^\infty \delta(t - t_0) f(t)\, dt = f(t_0) \quad (t_0 \geq 0).$$

In particular, the Laplace transform of $\delta(t - t_0)$ is

$$\mathcal{L}\{\delta(t - t_0)\} = \int_0^\infty \delta(t - t_0) e^{-st}\, dt = e^{-st_0}.$$

Now let's look at a specific example, in which we have an initial value problem involving a delta function.

EXAMPLE 6.5.2 Suppose that a "free" particle of mass 5 moves along the x-axis with velocity -2, starting at position 3, until $t = 2$, when a very brief, violent force causes it to reverse direction. For instance, the particle might be a puck sliding down the ice (without friction) toward a hockey player, who takes a slap shot at $t = 2$. If the force is given by $100\delta(t - 2)$, what will the position function $x(t)$ of the particle be?

---

[23] More precisely, since the functions $h_1(t - t_0)$ of which $\delta(t - t_0)$ is the limit are zero anyway for negative t.

SEC. 6.5 THE DELTA "FUNCTION"

By Newton's second law, the initial value problem is $5x'' = 100\delta(t-2)$, $x(0) = 3$, $x'(0) = -2$. Taking Laplace transforms, we get

$$(\mathcal{L}x'')(s) = 20\mathcal{L}\{\delta(t-2)\}$$

$$s^2(\mathcal{L}x)(s) - 3s + 2 = 20e^{-2s}$$

$$(\mathcal{L}x)(s) = \frac{3s - 2 + 20e^{-2s}}{s^2} = \frac{3}{s} - \frac{2}{s^2} + e^{-2s} \cdot \frac{20}{s^2}.$$

The inverse Laplace transforms of the first two terms on the right are $\mathcal{L}^{-1}\left\{\frac{3}{s}\right\} = 3$ and $\mathcal{L}^{-1}\left\{-\frac{2}{s^2}\right\} = -2t$. As for the third term, it is of the form $e^{-cs}G(s)$, so by formula (2) from p. 496 we have

$$\mathcal{L}^{-1}\left\{e^{-2s} \cdot \frac{20}{s^2}\right\} = g(t-2)u_2(t), \quad \text{where } g(t) = \mathcal{L}^{-1}\left\{\frac{20}{s^2}\right\} = 20t$$

$$= 20(t-2)u_2(t).$$

Combining the three inverse Laplace transforms yields

$$x(t) = 3 - 2t + 20(t-2)u_2(t)$$

as the solution to the initial value problem.

Let's look at this solution more carefully. If we write it out rather than using a step function, we get $x(t) = \begin{cases} 3 - 2t & (t < 2) \\ 18t - 37 & (t \geq 2) \end{cases}$. This function is continuous, *but not differentiable*, at $t = 2$; in fact, its derivative, the velocity, has a jump at $t = 2$:

$$x'(t) = \begin{cases} -2 & (t < 2) \\ 18 & (t > 2) \end{cases} = -2 + 20u_2(t).$$

Note that according to our differential equation, the derivative $x''(t)$ of $x'(t)$ should be $20\delta(t-2)$, that is, $\frac{d}{dt}[-2 + 20u_2(t)] = 20\delta(t-2)$. What it looks like, then, is that the "function" $\delta(t-2)$ is the "derivative" of the step function $u_2(t)$. ∎

Could we say, more generally, that the delta function $\delta(t-t_0)$ is the "derivative" of the unit step function $u_{t_0}(t)$? Certainly, the derivative of $u_{t_0}(t)$ is zero for all $t \neq t_0$. On the other hand, this derivative is undefined at $t = t_0$ (where the step function jumps from 0 to 1), and we are back on dangerous ground. However, the delta function $\delta(t-t_0)$, which is not defined either for $t = t_0$, has several properties that you would expect for a "derivative" of $u_{t_0}(t)$ (see Exercise 28). In the theory of distributions the concept of "derivative" is successfully extended so that $\delta(t-t_0)$ is indeed the derivative of $u_{t_0}(t)$. In particular, $\delta(t) = u'_0(t)$, and the delta function is sometimes *defined* this way.

Just as in Example 6.5.2, when we have a differential equation of the form

$$x'' + px' + qx = r\delta(t - t_0),$$

in which p, q, and r are constants, we can expect to find *continuous* solutions $x(t)$, but we cannot expect them to be *differentiable* for $t = t_0$. Instead, $x'(t)$ will have a jump of height r at $t = t_0$, which will make *its* derivative $x''(t)$ behave more or less like $r\delta(t - t_0)$ (and the two sides of the equation will then match). As a result, we have to be careful not to include a value for $x'(t_0)$ among our initial conditions, since $x'(t_0)$ will turn out to be undefined. Strictly speaking, for example, the "initial value problem"

$$x'' + 3x' + 2x = \delta(t), \quad x(0) = 0, \; x'(0) = 0$$

will not have any solutions, since $x'(0)$ will be undefined for every solution of the differential equation. Nevertheless, such "initial value problems" are found (and solved) in the literature, with the understanding that $x'(t_0)$ really stands for $x'(t_0^-) = \lim_{t \to t_0} x'(t)$, the value of the derivative "before the jump."

EXAMPLE 6.5.3

Let's return to the collision experiment from Example 6.5.1. We've seen that the position function $x(t)$ of the first particle satisfies $mx'' + kx = h(t)$, where $h(t)$ is an unspecified "spike" function with $\int_{-\infty}^{\infty} h(t)\, dt = \frac{2}{5} mv_0$. Also, the first particle was at rest until the moment $t = 0$ of impact. We can therefore set up the initial value problem

$$mx'' + kx = \frac{2}{5} mv_0\, \delta(t), \quad x(0) = 0, \; x'(0^-) = 0.$$

It turns out that in solving such a problem using the Laplace transform, all will be well if we use $x'(0^-)$ in the formulas instead of $x'(0)$.[24] For instance, in our case, we have

$$(\mathcal{L}x'')(s) = s^2(\mathcal{L}x)(s) - sx(0) - x'(0^-) = s^2(\mathcal{L}x)(s)$$

and the "transformed equation" becomes

$$(ms^2 + k)(\mathcal{L}x)(s) = \frac{2}{5} mv_0\, \mathcal{L}\{\delta(t)\} = \frac{2}{5} mv_0$$

$$(\mathcal{L}x)(s) = \frac{\frac{2}{5} mv_0}{ms^2 + k}.$$

Using the formula $\mathcal{L}^{-1}\left\{\dfrac{b}{s^2 + b^2}\right\} = \sin bt$ (see Table 6.4.1, p. 511), we get the answer

$$x(t) = \frac{2v_0}{5\sqrt{k/m}} \sin\left(\sqrt{\frac{k}{m}}\, t\right).$$

Note that this is only valid for $t \geq 0$ (the inverse Laplace

---

[24] This is certainly in the spirit of ad hoc computation, as mentioned earlier! It also explains why $x'(0^-)$ is often written as $x'(0)$ in the literature.

SEC. 6.5 THE DELTA "FUNCTION"

transform is only determined for $t \geq 0$, anyway). Since the particle was at rest until $t = 0$, we have

$$x(t) = \begin{cases} 0 & (t \leq 0) \\ \dfrac{2v_0}{5\sqrt{k/m}} \sin\left(\sqrt{\dfrac{k}{m}}\, t\right) & (t \geq 0) \end{cases}.$$

As expected, the velocity jumps from 0 to $\dfrac{2}{5} v_0$ at $t = 0$ [see Exercise 11(a)]. In fact, one can find $x(t)$ directly from this jump in velocity, without using the delta function or the Laplace transform [see Exercise 11(b)]. Exercise 12 explores, in a specific numerical case, what happens if friction is introduced into our model. ∎

We conclude this section with two computational examples.

EXAMPLE 6.5.4 Solve the initial value problem

$$x'' - 2x' + x = 3 + 4\delta(t), \qquad x(0) = 0, \quad x'(0^-) = 4.$$

Taking the Laplace transform of both sides and using the initial conditions, we get

$$s^2(\mathcal{L}x)(s) - 4 - 2s(\mathcal{L}x)(s) + (\mathcal{L}x)(s) = \frac{3}{s} + 4$$

$$(s^2 - 2s + 1)(\mathcal{L}x)(s) = \frac{3}{s} + 8$$

$$(\mathcal{L}x)(s) = \frac{3 + 8s}{s(s^2 - 2s + 1)}.$$

Since the denominator factors as $s(s-1)^2$, we have a partial fraction decomposition of the form $\dfrac{3 + 8s}{s(s^2 - 2s + 1)} = \dfrac{A}{s} + \dfrac{B}{s-1} + \dfrac{C}{(s-1)^2}$. Specifically, we get [see Exercise 18(a)]

$$(\mathcal{L}x)(s) = \frac{3}{s} - \frac{3}{s-1} + \frac{11}{(s-1)^2},$$

so

$$x(t) = 3\,\mathcal{L}^{-1}\left\{\frac{1}{s}\right\} - 3\,\mathcal{L}^{-1}\left\{\frac{1}{s-1}\right\} + 11\,\mathcal{L}^{-1}\left\{\frac{1}{(s-1)^2}\right\}$$

$$x(t) = 3 - 3e^t + 11te^t \qquad \text{(from Table 6.4.1, p. 511).}$$

NOTE: Once again, this answer is only valid for $t \geq 0$. See Exercise 18(b) and (c). ∎

EXAMPLE 6.5.5 Solve the initial value problem

$$x'' - 4x' + 3x = 8\delta(t-1) + 12u_2(t), \qquad x(0) = 1, \quad x'(0) = 5.$$

Taking the Laplace transform of both sides, we get

$$s^2(\mathcal{L}x)(s) - s - 5 - 4[s(\mathcal{L}x)(s) - 1] + 3(\mathcal{L}x)(s) = 8e^{-s} + \frac{12e^{-2s}}{s}.$$

Solving for $(\mathcal{L}x)(s)$ then yields

$$(\mathcal{L}x)(s) = \frac{s+1}{s^2 - 4s + 3} + e^{-s} \cdot \frac{8}{s^2 - 4s + 3} + e^{-2s} \cdot \frac{12}{s(s^2 - 4s + 3)}.$$

Using the partial fraction decompositions

$$\frac{s+1}{s^2 - 4s + 3} = \frac{2}{s-3} - \frac{1}{s-1},$$

$$\frac{8}{s^2 - 4s + 3} = \frac{4}{s-3} - \frac{4}{s-1},$$

and

$$\frac{12}{s(s^2 - 4s + 3)} = \frac{2}{s-3} - \frac{6}{s-1} + \frac{4}{s},$$

we get the inverse Laplace transforms

$$\mathcal{L}^{-1}\left\{\frac{s+1}{s^2 - 4s + 3}\right\} = 2e^{3t} - e^t,$$

$$\mathcal{L}^{-1}\left\{e^{-s} \cdot \frac{8}{s^2 - 4s + 3}\right\} = g(t-1)u_1(t)$$

[where $g(t) = \mathcal{L}^{-1}\left\{\dfrac{8}{s^2 - 4s + 3}\right\} = 4e^{3t} - 4e^t$]

$$= (4e^{3(t-1)} - 4e^{t-1})u_1(t),$$

and

$$\mathcal{L}^{-1}\left\{e^{-2s} \cdot \frac{12}{s(s^2 - 4s + 3)}\right\} = f(t-2)u_2(t)$$

[where $f(t) = \mathcal{L}^{-1}\left\{\dfrac{12}{s(s^2 - 4s + 3)}\right\} = 2e^{3t} - 6e^t + 4$]

$$= [2e^{3(t-2)} - 6e^{t-2} + 4]u_2(t).$$

SEC. 6.5 THE DELTA "FUNCTION"

Therefore,

$$x(t) = 2e^{3t} - e^t + [4e^{3(t-1)} - 4e^{t-1}]u_1(t) + [2e^{3(t-2)} - 6e^{t-2} + 4]u_2(t)$$

$$= \begin{cases} 2e^{3t} - e^t & (t \le 1) \\ \left(2 + \dfrac{4}{e^3}\right)e^{3t} - \left(1 + \dfrac{4}{e}\right)e^t & (1 \le t \le 2) \\ \left(2 + \dfrac{4}{e^3} + \dfrac{2}{e^6}\right)e^{3t} - \left(1 + \dfrac{4}{e} + \dfrac{6}{e^2}\right)e^t + 4 & (t \ge 2) \end{cases}$$

[see Exercise 19(a)]. As expected, $x(t)$ is continuous for all t, but $x'(t)$ has a jump of height 8 at $t = 1$ [see Exercise 19(b)]. ∎

SUMMARY OF KEY CONCEPTS, RESULTS, AND TECHNIQUES

In some applications, the "forcing term" $h(t)$ in an equation $x'' + px' + qx = h(t)$ is a "spike function" which is nonzero outside a small interval, and for which $\int_{-\infty}^{\infty} h(t)\, dt$ is known (p. 516). For example,

$$h_1(t) = \begin{cases} 0 & (t < 0) \\ \dfrac{1}{\epsilon} & (0 \le t \le \epsilon) \\ 0 & (t > \epsilon) \end{cases}$$

is such a function, for which $\int_{-\infty}^{\infty} h_1(t)\, dt = 1$; as $\epsilon \to 0^+$, the Laplace transform $(\mathscr{L}h_1)(s)$ approaches 1 (p. 517).

The **delta function** $\delta(t)$, which is not really a function, can be defined as $\lim_{\epsilon \to 0^+} h_1(t)$ (p. 518). Any integral involving $\delta(t)$ is defined as the limit of the corresponding integral involving $h_1(t)$; in particular, $\int_{-\infty}^{\infty} \delta(t)\, dt = 1$ (p. 519) and $(\mathscr{L}\delta)(s) = 1$. While $\delta(t) = 0$ for $t \ne 0$, the graph of $\delta(t)$ has an "infinitely narrow spike of area 1" at $t = 0$. For any continuous function $f(t)$, $\int_{-\infty}^{\infty} \delta(t)f(t)\, dt = \int_{0}^{\infty} \delta(t)f(t)\, dt = f(0)$ (p. 519; Exercise 24). Also, $f * \delta = f$ (p. 520). For any $t_0 \ge 0$, $\delta(t - t_0)$ has similar properties to $\delta(t)$, but with respect to $t = t_0$ rather than $t = 0$. For instance, for any continuous $f(t)$,

$$\int_{-\infty}^{\infty} \delta(t - t_0) f(t)\, dt = \int_{0}^{\infty} \delta(t - t_0) f(t)\, dt = f(t_0) \quad \text{(p. 520).}$$

We have $\mathscr{L}\{\delta(t - t_0)\} = e^{-st_0}$ (p. 520).

The delta function $\delta(t)$ can be considered to be the derivative of the unit step function $u_0(t)$; more generally, $\delta(t - t_0) = u'_{t_0}(t)$ (p. 521).

For a differential equation of the form $x'' + px' + qx = r\delta(t - t_0)$, the solutions will be continuous at $t = t_0$, but their derivatives will have jumps of height r there (p. 522). In particular, the initial value problem

$$x'' + px' + qx = r\delta(t), \qquad x(0) = x_0, \quad x'(0) = v_0$$

will have no solutions (if $r \neq 0$) unless $x'(0)$ is interpreted to mean $x'(0^-) = \lim_{t \to 0^-} x(t)$.

If this is done, the usual formulas for $(\mathcal{L}x')(s)$ and $(\mathcal{L}x'')(s)$ can be used in solving the initial value problem (p. 522).

EXERCISES

Solve the following initial value problems. (Table 6.4.1, p. 511, will help.)

1. $x'' + 4x = 3\delta(t - 2\pi)$, $x(0) = 1$, $x'(0) = -4$.
2. $x'' + 9x = 2\delta(t - \pi)$, $x(0) = 4$, $x'(0) = 0$.
3. $x'' + 4x' + 3x = 6t - 2 + \delta(t - 1)$, $x(0) = 0$, $x'(0) = 0$.
4. $x'' - 4x' + 3x = 12t + 1 + \delta(t - 2)$, $x(0) = 0$, $x'(0) = 0$.
5. $x'' - 4x' + 3x = e^{2t} + \delta(t)$, $x(0) = 0$, $x'(0^-) = 0$.
6. $x'' + 4x' + 3x = e^{-2t} + 5\delta(t)$, $x(0) = 0$, $x'(0^-) = 0$.
7. $x'' - 4x' + 4x = e^{2t} + \delta(t - 1)$, $x(0) = 2$, $x'(0) = 0$.
8. $x'' + 4x' + 4x = e^{-2t} + \delta(t - 3)$, $x(0) = -4$, $x'(0) = 0$.
9. $x'' - x = 2\delta(t) + 10t - 5$, $x(0) = 3$, $x'(0^-) = -5$.
10. $x'' - 4x = -3\delta(t) + 8t + 2$, $x(0) = -1$, $x'(0^-) = 0$.

11. (See Example 6.5.3.)
    (a) Show that if
    $$x(t) = \begin{cases} 0 & (t \leq 0) \\ \dfrac{2v_0}{5\sqrt{k/m}} \sin\left(\sqrt{\dfrac{k}{m}}\, t\right) & (t \geq 0), \end{cases}$$
    the velocity $x'(t)$ jumps from 0 to $\dfrac{2}{5}v_0$ at $t = 0$.
    (b) Conversely, suppose you know that right after the collision, the first particle leaves the origin with velocity $\dfrac{2}{5}v_0$, subject only to the spring force. Find $x(t)$ for $t \geq 0$ (from scratch), without using the Laplace transform.

SEC. 6.5 THE DELTA "FUNCTION"

12. Suppose that in Example 6.5.3, we have $m = 1$, $k = 4$, $v_0 = \dfrac{5}{2}$, so the initial value problem becomes $x'' + 4x = \delta(t)$, $x(0) = 0$, $x'(0^-) = 0$.
    (a) Find the solution (from scratch, using the Laplace transform) and graph it.
    (b) Suppose that we introduce a frictional force $-\gamma x'$ with $\gamma = 2$ on the first particle, so that we get $x'' + 2x' + 4x = \delta(t)$, $x(0) = 0$, $x'(0^-) = 0$. Solve this initial value problem using the Laplace transform and Table 6.4.1.
    [*Hint* for part (b) only: At some point, complete the square.] Graph the solution.
    (c) Repeat part (b) for $\gamma = 4$.
    (d) Repeat part (b) for $\gamma = 5$.
    (e) Compare the velocity immediately after the collision in parts (a), (b), (c), and (d). How does it vary with γ?
    (f) What do your results from parts (a), (b), (c), and (d) have to do with the different cases of mechanical vibration described in Section 2.7?

13. (a) Show that the equations $v_0 = 4v_1 + v_2$, $v_0^2 = 4v_1^2 + v_2^2$ have *two* solutions for v_1 and v_2, and find these solutions.
    (b) Explain how we know which solution to choose in the situation of Example 6.5.1.

14. Suppose that in the collision experiment from Example 6.5.1, the second particle had an arbitrary mass M (rather than $\dfrac{1}{4}m$).
    (a) Find the velocities v_1 and v_2 after the collision.
    (b) Is it possible that after the collision, the first particle will still be at rest? If so, for what M?
    (c) Is it possible that after the collision, the second particle will be at rest? If so, for what M?

15. (a) Express the function $h_2(t)$ from Figure 6.5.2(b) (p. 517) in terms of step functions.
    (b) Show that $(\mathcal{L}h_2)(s) = \dfrac{2}{\epsilon^2 s^2}(e^{-\epsilon s} - 1 + \epsilon s)$.
    (c) Show that $\lim\limits_{\epsilon \to 0^+}(\mathcal{L}h_2)(s) = 1$.

16. (a) Express the function $h_3(t)$ from Figure 6.5.2(c) in terms of step functions.
    (b) Show that $(\mathcal{L}h_3)(s) = \dfrac{4}{\epsilon^2 s^2}(1 - 2e^{-\epsilon s/2} + e^{-\epsilon s})$.
    (c) Show that $\lim\limits_{\epsilon \to 0^+}(\mathcal{L}h_3)(s) = 1$.
    [*Hint*: Write $(\mathcal{L}h_3)(s)$ as a square.]

17. (a) Given that the function $h_4(t)$ from Figure 6.5.2(d) describes one arch of a sine wave on the interval $[0, \epsilon]$, express $h_4(t)$ in terms of step functions.

(b) Show that $(\mathcal{L}h_4)(s) = \dfrac{\pi^2}{2} \cdot \dfrac{1 + e^{-\epsilon s}}{\pi^2 + \epsilon^2 s^2}$.

18. (See Example 6.5.4.)

    (a) Find the partial fraction decomposition for $\dfrac{3 + 8s}{s(s^2 - 2s + 1)}$. (Don't use the fact that the answer is given in the text.)

    (b) Consider the answer $x(t) = 3 - 3e^t + 11te^t$ to the initial value problem of Example 6.5.4. Compute $x'(0)$ for this answer. Explain the apparent contradiction between your result and the initial condition $x'(0^-) = 4$.

    (c) Find a formula for the solution $x(t)$ to $x'' - 2x' + x = 3 + 4\delta(t)$, $x(0) = 0$, $x'(0^-) = 4$ which is valid *for all* t (not just for $t \geq 0$).

19. (a) Show that the last step in Example 6.5.5 [in which $x(t)$ is written out without step functions] is correct.

    (b) Show that $x(t)$ is continuous for all t, but that $x'(t)$ has a jump of height 8 at $t = 1$.

    (c) Show that $x'(t)$ is continuous at $t = 2$, but that $x''(t)$ has a jump of height 12 at $t = 2$, and explain why this should be.

20. (a) Find a "function" whose Laplace transform is $\dfrac{s}{s - 1}$.

    [*Hint*: $\dfrac{s}{s-1} = 1 + \dfrac{1}{s-1}$.]

    *(b) Use your result from part (a) and convolution to find the inverse Laplace transform of $\dfrac{s}{s-1} \cdot \dfrac{1}{s-2}$.

    [*Hint*: The fact that $\delta(t)$ is an identity for convolution will help.]

    (c) Check your result from part (b) by using partial fractions.

21. (a) Find a "function" whose Laplace transform is $\dfrac{2s^2 - s + 1}{s^2 + 9}$.

    [*Hint*: First use "long" division.]

    *(b) Use your result from part (a) and convolution to find the inverse Laplace transform of $\dfrac{2s^2 - s + 1}{s(s^2 + 9)}$. Check your answer against the one on p. 483.

22. Show that for the functions $h_3(t)$ and $h_4(t)$ from Figure 6.5.2,
$$\lim_{\epsilon \to 0^+} h_3(t) = \lim_{\epsilon \to 0^+} h_4(t) = 0$$

for all t, *including* $t = 0$. Thus we cannot justify writing
$$\delta(t) = \lim_{\epsilon \to 0^+} h_3(t) \quad \text{or} \quad \delta(t) = \lim_{\epsilon \to 0^+} h_4(t)$$

even to the limited extent that we can justify writing $\delta(t) = \lim_{\epsilon \to 0^+} h_1(t)$.

23. (See p. 517.) Show that although for any particular value of s, we have $\lim_{\epsilon \to 0^+} (\mathscr{L}h_1)(s) = 1$, it is also true that for *any* $\epsilon > 0$, no matter how small, $\lim_{s \to \infty} (\mathscr{L}h_1)(s) = 0$. Thus for any given ϵ, we can make $(\mathscr{L}h_1)(s)$ arbitrarily small (and very unlike 1) by taking s large enough!

24. Let $f(t)$ be any continuous function.

  (a) Show that if
  $$h_1(t) = \begin{cases} 0 & (t < 0) \\ \dfrac{1}{\epsilon} & (0 \leq t \leq \epsilon) \\ 0 & (t \geq \epsilon), \end{cases}$$
  then
  $$\int_{-\infty}^{\infty} h_1(t)f(t)\,dt = \int_0^{\infty} h_1(t)f(t)\,dt = \frac{1}{\epsilon}\int_0^{\epsilon} f(t)\,dt.$$

  (b) Show that if m_ϵ, M_ϵ are the minimum and maximum values, respectively, of $f(t)$ on the closed interval $[0, \epsilon]$, then $m_\epsilon \leq \dfrac{1}{\epsilon}\int_0^{\epsilon} f(t)\,dt \leq M_\epsilon$.

  (c) Show that there is some number t_ϵ in the interval $[0, \epsilon]$ such that
  $$f(t_\epsilon) = \frac{1}{\epsilon}\int_0^{\epsilon} f(t)\,dt.$$

  (d) Show that $\lim_{\epsilon \to 0^+} \dfrac{1}{\epsilon}\int_0^{\epsilon} f(t)\,dt = f(0).$

  (e) Show that if we define integrals involving the delta function as on p. 519, then statement (∗) on p. 519 is true.

*25. (a) Show that in the collision experiment from Example 6.5.1, the particles will not collide a second time, even though the first particle will eventually go to the left of its equilibrium position.
[*Hint:* Use the solution for $x(t)$ from p. 523. You can show graphically that no second collision can occur.]

  (b) Show that if the experiment is modified so the second particle has mass $\dfrac{5}{6}m$ (rather than $\dfrac{1}{4}m$), then the particles *will* collide again.
[*Hint:* Use the results of Exercise 14(a). You will not be able to get an exact answer for the time of the second collision, but you can show graphically that it must occur.]

*26. Explain how to define $\displaystyle\int_{-\infty}^{\infty} \delta(t - t_0)f(t)\,dt$ as a limit as $\epsilon \to 0^+$ of an integral, and show why we then have

$$\int_{-\infty}^{\infty} \delta(t - t_0)f(t)\, dt = \int_{-\infty}^{\infty} \delta(t_1)f(t_1 + t_0)\, dt_1.$$

*27. (See p. 520.) **(a)** Explain why for any $t > 0$,
$$\int_0^t f(t - \tau)\delta(\tau)\, d\tau = \int_0^{\infty} f(t - \tau)\delta(\tau)\, d\tau.$$
[*Hint*: Write the integrals as limits and use that for ϵ small enough, $\epsilon < t$.]

(b) For $t = 0$, this argument breaks down. To remedy this and get the convolution exactly "right" for the delta function, the convolution can be defined as in Exercise 33 of Section 6.4, that is, by
$$(f * g)(t) = \int_{-\infty}^{\infty} f(t - \tau)g(\tau)\, d\tau.$$
Show that if this is done, then $(f * \delta)(t) = f(t)$ for all continuous functions $f(t)$.

*28. (See p. 521.) **(a)** Show that $\int_a^b \delta(t - t_0)\, dt = u_{t_0}(b) - u_{t_0}(a)$, in each of the following cases: (i) $b > a > t_0$; (ii) $b > t_0 > a$; (iii) $t_0 > b > a$. Note that this is a property you would expect if $\delta(t - t_0)$ is to be the derivative of $u_{t_0}(t)$.

(b) Show that if $f(t)$ is a continuous function such that $L = \lim_{t \to \infty} f(t)$ exists, then
$$\int_{-\infty}^{\infty} u_{t_0}(t)f'(t)\, dt = L - f(t_0).$$
[*Hint*: On what interval is the integrand not zero?]

(c) In the situation of part (b), show that
$$\int_{-\infty}^{\infty} \delta(t - t_0)f(t)\, dt = u_{t_0}(t)f(t)\Big]_{-\infty}^{\infty} - \int_{-\infty}^{\infty} u_{t_0}(t)f'(t)\, dt.$$
[*Hint*: Compute both sides separately.] Note that if $\delta(t - t_0)$ is the derivative of $u_{t_0}(t)$, this is an integration by parts.

*29. **(a)** Let $f(t)$ be any continuous function. Define a **functional** (a function defined on functions) F_f, which takes continuous functions $g(t)$ to real numbers, by $F_f(g(t)) = \int_{-\infty}^{\infty} f(t)g(t)\, dt$. Unfortunately, this improper integral will only converge for certain functions $g(t)$, and this creates a lot of trouble (which we'll ignore). Show that F_f is a **linear** functional in the sense that
$$F_f(\alpha \cdot g_1(t) + \beta \cdot g_2(t)) = \alpha F_f(g_1(t)) + \beta F_f(g_2(t))$$
provided the improper integrals on the right converge.

(b) Define another functional Δ by $\Delta(g(t)) = g(0)$. Show that Δ is also a linear functional.

(c) Show that in some sense, $\Delta = F_\delta$.

[*Hint*: See formula (∗) on p. 519.] Thus we can put actual functions and the delta function on an equal footing by thinking of them all as defining linear functionals.

FURTHER READING

For the early history of the delta function, see Chapter 4, Part 2 in Lützen, *The Prehistory of the Theory of Distributions* (Springer-Verlag, 1982).

CHAPTER 7

Appetizers

INTRODUCTION

This chapter has a different format, and to some extent a different purpose, from the other chapters. What you'll find here, rather than a systematic exposition of basic material, is a series of glimpses ahead into areas that you may want to study in more detail. As a result, it seems especially likely that as you read this chapter, many questions will occur to you. Some of the answers to these questions may turn out to be in the exercises, which often explore additional new material. For others, you may want to turn to the books suggested for further reading at the end of each section.

7.1 DIFFERENCE EQUATIONS (RECURRENCE RELATIONS)

In the applications considered so far, we've studied functions of a continuous real variable t; that is, our functions have been defined for arbitrary real t (at least within some interval). In practice, however, there are many cases in which a function is defined only for discrete values of t, for instance only when t is an integer, or even only when t is a positive (or a nonnegative) integer. For example, $x(t)$ might be the Dow Jones closing average t days after January 1, 1993, or the number of bushels of wheat harvested in Kansas in the year $1992 + t$, or the number of vehicles passing across the Golden Gate Bridge during the month t months after the bridge was first opened to traffic. In other cases, a function may be defined for all t, but it may only be possible to get information about the function for certain discrete values of t; for example, $x(t)$ might be the size of a certain population, known only in the years in which a periodic census is taken.

When we study a function only for integer values of t, especially if these are limited to positive t or to nonnegative t, it is customary to refer to the function as a **sequence**[1] and to use subscripts for the values of t. That is, we write

$$x_1 = x(1), \ x_2 = x(2), \ \ldots, \ x_t = x(t), \ \ldots \ .$$

The notation $(x_t)_{t\geq 1}$, or (x_t) for short, is used for this particular sequence. The individual values $x_1, x_2, \ldots$ are known as the **terms** of the sequence. For example, the first few terms of the sequence $(x_t) = (5^t - 3t - 10)$ are $x_1 = 5^1 - 3 \cdot 1 - 10 = -8$, $x_2 = 9$, and so on. Similarly, the sequence $(2^t)_{t\geq 0}$ starts off 1, 2, 4, 8, 16, $\ldots$.

Now suppose that we have a function x of this kind. How can we measure how fast x is changing? Clearly, the derivative $x'(t) = \lim\limits_{h \to 0} \dfrac{x(t+h) - x(t)}{h}$ is undefined, since when t and $t + h$ are restricted to be integers, we can't let h approach 0 in any sensible way. The best we can do is look at how much x changes "from one time to the next," that is, consider the *difference* between the "new" value $x(t + 1)$ and the "old" value $x(t)$. [Note that if we take $h = 1$ in the difference quotient $\dfrac{x(t+h) - x(t)}{h}$ without taking the limit, we get exactly this difference $x(t + 1) - x(t)$.] Of course, this difference will usually depend on t, so we get a new function, which is called the **difference function** or **forward difference function**[2] of x and denoted by Δx: $\Delta x(t) = x(t+1) - x(t)$. In "sequence notation" this reads $(\Delta x)_t = x_{t+1} - x_t$.

Just as one can repeat the process of taking the derivative of a function of a continuous variable and get higher-order derivatives, one can get higher-order difference functions for a function of an integer variable t:

$$(\Delta^2 x)_t = (\Delta x)_{t+1} - (\Delta x)_t = x_{t+2} - x_{t+1} - (x_{t+1} - x_t) = x_{t+2} - 2x_{t+1} + x_t,$$
$$(\Delta^3 x)_t = (\Delta^2 x)_{t+1} - (\Delta^2 x)_t,$$

and so on. (See Exercise 7 for the general pattern.) A **difference equation** for the unknown sequence x is an equation in which one or more of the difference functions of x occur. Difference equations can be classified much like differential equations; for instance, the **order** of a difference equation is the order of the highest difference function to occur. One can even define partial difference equations involving functions of several discrete variables (see Exercise 1), but for now we'll only look at difference equations involving one unknown function x of one integer variable t.

EXAMPLE 7.1.1 A family has its main investment in yearly certificates of deposit, which mature and give 10% interest at the end of each year. The family then "rolls over" the money plus interest, along with $2000 that they are able to save during the year, into new certificates of

---

[1] Some authors, especially in statistics, call it a **time series** instead. This can be confusing, because it blurs the important distinction between (infinite) sequences and series that is maintained elsewhere in mathematics.

[2] In some applications the **backward difference function** given by $\delta x(t) = x(t) - x(t-1)$ is used instead.

SEC. 7.1 DIFFERENCE EQUATIONS (RECURRENCE RELATIONS)

deposit. If they start with $10,000, how much will they have after t years (assuming that the interest rate and yearly family savings both remain constant)?

Let x_t be the amount, in thousands of dollars, that the family has after t years. Since they start with $10,000, we know that $x_0 = 10$. At the end of year $t + 1$ the family's capital will have increased from the end of year t by 10% of x_t (in interest) plus 2 (in savings) thousands of dollars, so we also know that $(\Delta x)_t = \frac{1}{10} x_t + 2$. Since this is true for all integers $t \geq 0$, we have the difference equation $\Delta x = \frac{1}{10} x + 2$; in fact, we have the initial value problem $\Delta x = \frac{1}{10} x + 2$, $x_0 = 10$, which will have a unique solution (x_t) (can you see why?).

Note that the equation $(\Delta x)_t = \frac{1}{10} x_t + 2$ can be rewritten as $x_{t+1} - x_t = \frac{1}{10} x_t + 2$ or $x_{t+1} = \frac{11}{10} x_t + 2$. In this last form, we have a **recurrence relation** giving the "next" value x_{t+1} in terms of the "present value" x_t; of course, we still have the initial condition $x_0 = 10$. Later in this section, we'll see how to solve the initial value problem. □

EXAMPLE 7.1.2 Suppose that you are running an experiment (or doing a computation) which gives you the sequence of numbers $x_1 = 3$, $x_2 = 4$, $x_3 = 9$, $x_4 = 20$, $x_5 = 39$, $x_6 = 68$, $x_7 = 109$, ???, and you wonder whether there is any pattern to this sequence. One way to try to find out is to make a **table of differences** (that is, a table of values of difference functions):

				$t \longrightarrow$			
	1	2	3	4	5	6	7
x	3	4	9	20	39	68	109
Δx	1	5	11	19	29	41	
$\Delta^2 x$	4	6	8	10	12		
$\Delta^3 x$	2	2	2	2			

Note that each row in this table can be found from the one directly above it. Looking at the last row, it seems likely that the function x is a solution to the third-order difference equation $\Delta^3 x = 2$. If you believe that this is so, you can predict the values of $x_8, x_9, \ldots$ by continuing the table of differences to the right, this time working from the bottom up (see Exercise 2). ■

Just as in Example 7.1.1, a difference equation can always be rewritten as a recurrence relation. This process can be reversed, as well (see Exercise 5), so there is really no need to have a separate theory of difference equations, provided one can solve recurrence relations. On the other hand, we don't have any general methods for solving recurrence relations yet; in Chapter 5, when recurrence relations for the coefficients a_n came up when we were looking for power series solutions $\sum a_n x^n$ to differential equations, we only computed a few of the a_n and hoped to recognize a pattern.

The easiest recurrence relations to solve are probably those of the form $x_{t+1} = rx_t$ [equivalent to $\Delta x = (r-1)x$; see Exercise 6(a)] with constant r. If we assume that the initial term of the sequence is $x_0 = a$, then we have $x_1 = ar$, $x_2 = ar^2$, and so on, and the sequence is a **geometric sequence** (ar^t). On the other hand, the easiest difference equations to solve may be those of the form $\Delta x = d$ with constant d. If we again let $x_0 = a$ be the initial term, we find $x_1 = a + d$, $x_2 = a + 2d$, and so on, and we have an **arithmetic sequence** $(a + td)$.

The methods used to solve more complicated recurrence relations and difference equations are often strikingly similar to the methods used for analogous differential equations.

EXAMPLE 7.1.3 Let's reconsider the first-order difference equation $\Delta x = \dfrac{1}{10} x + 2$ from Example 7.1.1.

An analogous first-order differential equation would be $\dfrac{dx}{dt} = \dfrac{1}{10} x + 2$. Recall (from Section 1.6) that one way to solve this linear equation, which would work more generally for any equation of the form $\dfrac{dx}{dt} = \dfrac{1}{10} x + h(t)$, would be by variation of constants: Start with the general solution $x = Ae^{t/10}$ of the homogeneous equation $\dfrac{dx}{dt} = \dfrac{1}{10} x$, then replace A by $A(t)$ and look for solutions of the form $x = A(t)e^{t/10}$ of the inhomogeneous equation. Now let's try this method for our difference equation, or the equivalent recurrence relation $x_{t+1} = \dfrac{11}{10} x_t + 2$. We've seen that the solutions to the homogeneous relation $x_{t+1} = \dfrac{11}{10} x_t$ are geometric sequences (x_t) with $x_t = a\left(\dfrac{11}{10}\right)^t$, a a constant, so let's look for solutions of the form $x_t = a_t \left(\dfrac{11}{10}\right)^t$ of the given (inhomogeneous) relation. We get

$$a_{t+1}\left(\frac{11}{10}\right)^{t+1} = \frac{11}{10} \cdot a_t \left(\frac{11}{10}\right)^t + 2, \text{ that is,}$$

$$(a_{t+1} - a_t)\left(\frac{11}{10}\right)^{t+1} = 2$$

$$a_{t+1} - a_t = 2 \cdot \left(\frac{10}{11}\right)^{t+1}.$$

What we have here are the values of the difference function Δa; we can choose a_0 arbitrarily and then get

$$a_1 = a_0 + (a_1 - a_0) = a_0 + 2 \cdot \frac{10}{11},$$

$$a_2 = a_0 + (a_1 - a_0) + (a_2 - a_1) = a_0 + 2 \cdot \frac{10}{11} + 2 \cdot \left(\frac{10}{11}\right)^2, \ldots,$$

$$a_t = a_0 + 2 \cdot \left[\frac{10}{11} + \left(\frac{10}{11}\right)^2 + \cdots + \left(\frac{10}{11}\right)^t\right]$$

$$= a_0 + 2 \cdot \frac{10}{11} \cdot \frac{1 - \left(\frac{10}{11}\right)^t}{1 - \frac{10}{11}} = a_0 + 20 \cdot \left[1 - \left(\frac{10}{11}\right)^t\right].$$

Finally, we have $x_t = a_t \cdot \left(\frac{11}{10}\right)^t$, so

$$x_t = a_0 \cdot \left(\frac{11}{10}\right)^t + 20 \cdot \left[\left(\frac{11}{10}\right)^t - 1\right], \quad a_0 \text{ arbitrary.}$$

Note that for $t = 0$ we have $x_0 = a_0$. Thus the solution to the initial value problem $\Delta x = \frac{1}{10} x + 2$, $x_0 = 10$ from Example 7.1.1 will be

$$x_t = 10 \cdot \left(\frac{11}{10}\right)^t + 20 \cdot \left[\left(\frac{11}{10}\right)^t - 1\right] = 30 \cdot \left(\frac{11}{10}\right)^t - 20. \quad \blacksquare$$

EXAMPLE 7.1.4 A particularly famous sequence of integers is the **Fibonacci sequence**[3] 1, 1, 2, 3, 5, 8, 13, ..., in which each term (or **Fibonacci number**) is the sum of the two preceding terms. In other words, the sequence is given by the recurrence relation $x_{t+2} = x_t + x_{t+1}$ together with the initial conditions $x_1 = x_2 = 1$.

If we convert the recurrence relation to a difference equation, we get [Exercise 5(c)] $\Delta^2 x + \Delta x + x = 0$, which is analogous to the differential equation $\frac{d^2 x}{dt^2} + \frac{dx}{dt} + x = 0$.

Since this is a second-order linear homogeneous equation, it can be solved by finding two basic solutions; we also know that we can look for these solutions to be of the form $x = e^{\lambda t}$. Returning to our difference equation, it can be shown [Exercise 10(b)] that once again, all solutions can be found from two basic solutions. What form should we expect

---

[3] "Fibonacci" was a nickname for Leonardo of Pisa (1170–1250), the outstanding European mathematician of his day. See Exercise 11 for the connection between Fibonacci and our particular sequence.

these basic solutions to have? In the differential equation case, $e^{\lambda t}$ was a solution of the first-order homogeneous equation $\dfrac{dx}{dt} = \lambda x$. The analogous first-order difference equation $\Delta x = \lambda x$ [or $x_{t+1} = (\lambda + 1)x_t$] has, as we've seen, solutions $(a(\lambda + 1)^t)$. So we may want to look for basic solutions of the form (r^t) (with r constant) this time.[4]

If we do substitute $x_t = r^t$ into the recurrence relation $x_{t+2} = x_t + x_{t+1}$, we get

$$r^{t+2} = r^t + r^{t+1} \quad \text{or} \quad r^t(r^2 - r - 1) = 0.$$

Since we want this to be true for all t (and since we want $r \neq 0$ —why?) we need to have $r^2 - r - 1 = 0$, or by the quadratic formula, $r = \dfrac{1 \pm \sqrt{5}}{2}$. Thus we have two basic solutions to our recurrence relation, one with $x_t = \left(\dfrac{1+\sqrt{5}}{2}\right)^t$ and the other with $x_t = \left(\dfrac{1-\sqrt{5}}{2}\right)^t$. Any solution is therefore of the form

$$(x_t) = \left(\alpha\left(\dfrac{1+\sqrt{5}}{2}\right)^t + \beta\left(\dfrac{1-\sqrt{5}}{2}\right)^t\right)$$

with constants α and β. In particular, from the initial conditions for the Fibonacci sequence we get [Exercise 10(d)] $\alpha = 1/\sqrt{5}$, $\beta = -1/\sqrt{5}$, and thus we have the formula[5]

$$x_t = \dfrac{1}{\sqrt{5}}\left[\left(\dfrac{1+\sqrt{5}}{2}\right)^t - \left(\dfrac{1-\sqrt{5}}{2}\right)^t\right]$$

for the tth Fibonacci number. ∎

EXAMPLE 7.1.5 Consider the sequence of integers determined by the initial value problem

$$x_{t+2} = 4x_{t+1} - 5x_t, \quad x_0 = 1, \quad x_1 = 0.$$

To get an idea of what the sequence looks like, you might be tempted to compute a few terms: $1, 0, -5, -20, -55, -120, \ldots$ (check these!). At this point you may well feel that you have a pretty good idea of how the sequence will behave in the long run.

Here are the next few terms (after $x_5 = -120$): $-205, -220, 145, 1680, 5995, \ldots$ Surprise! What do you think will happen next?

If we use the same approach as in Example 7.1.4 and look for basic solutions of the form $x_t = r^t$ to our recurrence relation $x_{t+2} = 4x_{t+1} - 5x_t$, we get $r^2 - 4r + 5 = 0$, which has complex roots $r = 2 \pm i$. Thus the general complex-valued solution to the recurrence relation will be $x_t = C_1(2+i)^t + C_2(2-i)^t$, where C_1 and C_2 are arbitrary

---

[4] You may notice that for $r > 0$, we can write $r^t = e^{t \log r}$, so we have exponential functions again; however, in this context (where t is restricted to integers) it is more natural to write r^t.

[5] This formula is known as **Binet's formula** after Jacques Binet (1786–1856).

complex constants. Our particular initial value problem will have

$$x_t = \left(\frac{1}{2} + i\right)(2 + i)^t + \left(\frac{1}{2} - i\right)(2 - i)^t$$

as its solution [see Exercise 12(a)]. This can be rewritten [see Exercise 12(b)] as

$$x_t = (\sqrt{5})^t(\cos t\theta - 2 \sin t\theta) \quad \text{with} \quad \theta = \arctan \frac{1}{2}.$$

We see that, while the factor $(\sqrt{5})^t$ will make the numbers x_t tend to grow, the factor $\cos t\theta - 2 \sin t\theta$ will make them oscillate; this accounts for the "unexpected" behavior we found. ∎

As we have seen in Chapter 3, differential equations often occur in systems; this is equally true of difference equations and recurrence relations. Since this section is intended as a quick introduction, let's avoid notational problems by only considering the case of two unknown sequences x, y.[6]

The easiest type of system of two differential equations for two unknown functions x, y would be a homogeneous linear autonomous system, given by $\begin{bmatrix} dx/dt \\ dy/dt \end{bmatrix} = \mathbf{A} \begin{bmatrix} x \\ y \end{bmatrix}$ for some 2×2 matrix $\mathbf{A}$. Similarly, the easiest type of system of difference equations looks like $\begin{bmatrix} \Delta x \\ \Delta y \end{bmatrix} = \mathbf{A} \begin{bmatrix} x \\ y \end{bmatrix}$ for some 2×2 matrix $\mathbf{A} = \begin{bmatrix} a & b \\ c & d \end{bmatrix}$. Such a system can be rewritten as

$$\begin{cases} x_{t+1} - x_t = ax_t + by_t \\ y_{t+1} - y_t = cx_t + dy_t \end{cases}$$

or as the equivalent system of recurrence relations

$$\begin{cases} x_{t+1} = (a + 1)x_t + by_t \\ y_{t+1} = cx_t + (d + 1)y_t. \end{cases}$$

Note that according to these recurrence relations, we have

$$\begin{bmatrix} x_{t+1} \\ y_{t+1} \end{bmatrix} = \begin{bmatrix} a+1 & b \\ c & d+1 \end{bmatrix} \begin{bmatrix} x_t \\ y_t \end{bmatrix} = (\mathbf{A} + \mathbf{I}) \begin{bmatrix} x_t \\ y_t \end{bmatrix},$$

where, as usual, $\mathbf{I} = \begin{bmatrix} 1 & 0 \\ 0 & 1 \end{bmatrix}$ denotes the identity matrix. Therefore, starting with initial values x_0, y_0, we get

$$\begin{bmatrix} x_1 \\ y_1 \end{bmatrix} = (\mathbf{A} + \mathbf{I}) \begin{bmatrix} x_0 \\ y_0 \end{bmatrix}, \quad \begin{bmatrix} x_2 \\ y_2 \end{bmatrix} = (\mathbf{A} + \mathbf{I}) \begin{bmatrix} x_1 \\ y_1 \end{bmatrix} = (\mathbf{A} + \mathbf{I})^2 \begin{bmatrix} x_0 \\ y_0 \end{bmatrix}, \quad \text{and so on,}$$

---

[6] If we were to introduce n unknown sequences $x_1, \ldots, x_n$, we could no longer use the subscript notation x_t for the tth term of the sequence x.

and in general

$$\begin{bmatrix} x_t \\ y_t \end{bmatrix} = (\mathbf{A} + \mathbf{I})^t \begin{bmatrix} x_0 \\ y_0 \end{bmatrix}.$$

Thus the system can be solved if we can find the powers of the matrix $\mathbf{A} + \mathbf{I}$. This can easily be done, provided $\mathbf{A}$ is diagonalizable (see Exercise 16).

EXAMPLE 7.1.6 In a certain suburban neighborhood, it has been observed that each week of the summer, 40% of the lawns which are mowed that week will not be mowed again the following week. On the other hand, 80% of the lawns that are not mowed one week will be mowed the following week. If all lawns are mowed the first week of summer, what percentage of lawns will be mowed 9 weeks later?

Let x_t be the percentage of lawns mowed t weeks after the first week of summer, and let y_t be the percentage unmowed. Since all lawns are mowed to begin with, we have

$$x_0 = 100, \; y_0 = 0.$$

Subsequently, of the total percentage x_t mowed t weeks later, $0.4x_t$ will not be mowed the $(t+1)$st week and thus will contribute to y_{t+1} rather than x_{t+1}. On the other hand, $0.8y_t$ will contribute to x_{t+1} rather than y_{t+1}. Thus we have

$$\begin{cases} (\Delta x)_t = -0.4x_t + 0.8y_t \\ (\Delta y)_t = 0.4x_t - 0.8y_t, \end{cases}$$

or

$$\begin{bmatrix} \Delta x \\ \Delta y \end{bmatrix} = \mathbf{A} \begin{bmatrix} x \\ y \end{bmatrix} \quad \text{with} \quad \mathbf{A} = \begin{bmatrix} -0.4 & 0.8 \\ 0.4 & -0.8 \end{bmatrix}.$$

Note that $(\Delta x)_t + (\Delta y)_t = 0$; this should not be surprising, since

$$x_{t+1} + y_{t+1} = x_t + y_t = 100 \quad \text{(why?)}.$$

We can rewrite our system of difference equations as

$$\begin{cases} x_{t+1} = 0.6x_t + 0.8y_t \\ y_{t+1} = 0.4x_t + 0.2y_t, \end{cases}$$

or

$$\begin{bmatrix} x_{t+1} \\ y_{t+1} \end{bmatrix} = \begin{bmatrix} 0.6 & 0.8 \\ 0.4 & 0.2 \end{bmatrix} \begin{bmatrix} x_t \\ y_t \end{bmatrix}.$$

Therefore, in the tenth week of summer (when $t = 9$) we'll have

$$\begin{bmatrix} x_9 \\ y_9 \end{bmatrix} = \begin{bmatrix} 0.6 & 0.8 \\ 0.4 & 0.2 \end{bmatrix}^9 \begin{bmatrix} x_0 \\ y_0 \end{bmatrix} = \begin{bmatrix} 0.6 & 0.8 \\ 0.4 & 0.2 \end{bmatrix}^9 \begin{bmatrix} 100 \\ 0 \end{bmatrix}.$$

See Exercise 17 for the rest of the computation.

SEC. 7.1 DIFFERENCE EQUATIONS (RECURRENCE RELATIONS)

NOTE: One can think of $p_t = \dfrac{x_t}{100}$ and $q_t = \dfrac{y_t}{100}$ as the probabilities that a lawn chosen at random will or won't be mowed. Note that $p_t + q_t = 1$: p_t and q_t are "**complementary**" probabilities. Loosely speaking, a process in which complementary probabilities change from one time to the next in a way independent of the previous history of the process (in our example, p_{t+1} and q_{t+1} only depend on p_t and q_t) is called a **Markov process**. If the new probabilities are obtained from the old ones by multiplication by a constant matrix $\mathbf{B}$ with nonnegative entries, we have a **Markov chain**. In our example, $\mathbf{B} = \mathbf{A} + \mathbf{I} = \begin{bmatrix} 0.6 & 0.8 \\ 0.4 & 0.2 \end{bmatrix}$. Although you may feel the example is a bit simplistic, Markov chains, often with more than two complementary probabilities, are of considerable importance. Besides their occurrence in probability theory, they have extensive applications in the social sciences. ■

EXERCISES

1. Let x be a function of two integer variables t and u. Define the **partial differences** $\Delta_t x$ and $\Delta_u x$ of x with respect to t and u, respectively. Give a definition of a **partial difference equation** for the unknown function x.
   [NOTE: Partial difference equations are widely used in numerical approximations to the solutions of partial differential equations.]

2. In Example 7.1.2 (p. 535), assume that we have $(\Delta^3 x)_t = 2$ for all t. By continuing the table of differences to the right, find x_8, x_9, and x_{10}.

For each of the following sequences, find some pattern, then predict the next two terms of the sequence.

3. $-2, 0, 11, 32, 67, 123, 210, \ldots$.

4. $2\dfrac{1}{2}, 6, 10\dfrac{1}{2}, 15\dfrac{1}{2}, 20\dfrac{5}{6}, 26\dfrac{5}{12}, \ldots$.

5. Rewrite the following recurrence relations as difference equations.
   (a) $x_{t+1} = 5x_t + 2$.
   (b) $x_{t+2} = 3x_{t+1} - x_t + 4$.
   (c) $x_{t+2} = x_t + x_{t+1}$.

6. (a) Show that the recurrence relation $x_{t+1} = rx_t$ is equivalent to the difference equation $\Delta x = (r-1)x$.
   (b) Find the recurrence relation that is equivalent to the difference equation $\Delta x = d$.

7. Show (by induction on n) that
$$(\Delta^n x)_t = x_{t+n} - nx_{t+n-1} + \dfrac{n(n-1)}{2} x_{t+n-2} \cdots + (-1)^n x_t = \sum_{k=0}^{n} (-1)^k \binom{n}{k} x_{t+n-k}.$$

Note the similarity with the binomial expansion of $(x - 1)^n$:

$$(x - 1)^n = \sum_{k=0}^{n} (-1)^k \binom{n}{k} x^{n-k}.$$

Can you see why the two formulas should be similar?

8. Define what it means for a difference equation to be **linear**; also, what it means for a recurrence relation to be linear. Using Exercise 7, show that if a difference equation is linear, the corresponding recurrence relation is linear also.

9. Define what it means for a linear difference equation to be **homogeneous**; also, what it means for a linear recurrence relation to be homogeneous. Show that if a linear difference equation is homogeneous, the corresponding recurrence relation is homogeneous as well.

10. (a) Show that an initial value problem of the form $x_{t+2} = x_t + x_{t+1}$, $x_1 = a$, $x_2 = b$ always has a unique solution. (This solution is called the **Lucas sequence**[7] with initial terms a, b.)

    *(b) Show that all solutions of the recurrence relation $x_{t+2} = x_t + x_{t+1}$ (or of the equivalent difference equation $\Delta^2 x + \Delta x + x = 0$) can be found from two basic solutions.

    (c) Use the basic solutions $x_t = \left(\dfrac{1 \pm \sqrt{5}}{2}\right)^t$ found in the text to get a formula for the Lucas sequence with initial terms a, b.

    (d) Show that for the Fibonacci sequence (with $a = b = 1$) the formula from part (c) reduces to the formula given in the text.

11. Nearly 800 years ago, Leonardo of Pisa posed the following problem (for $t = 12$) in his mathematical work "Liber Abaci." Assume that a newborn pair of rabbits will produce a pair of offspring each month, starting when they are 2 months old. Each pair of offspring will produce its own offspring starting 2 months after its own birth, and so on. If we start with one newborn pair of rabbits, how many pairs of rabbits will we have after t months, assuming that no rabbits ever die? Show that this problem leads to the initial value problem

    $$x_{t+2} = x_{t+1} + x_t, \quad x_0 = 1, \quad x_1 = 1.$$

12. (a) Show that the solution to $x_{t+2} = 4x_{t+1} - 5x_t$, $x_0 = 1$, $x_1 = 0$ is given by

    $$x_t = \left(\frac{1}{2} + i\right)(2 + i)^t + \left(\frac{1}{2} - i\right)(2 - i)^t.$$

    (b) Show that this solution can be rewritten as $x_t = (\sqrt{5})^t (\cos t\theta - 2 \sin t\theta)$ with $\theta = \arctan \dfrac{1}{2}$.

    [*Hint:* Write $2 + i$ in polar form.]

---

[7] After Edouard Lucas (1842–1891), who studied such "generalized Fibonacci sequences."

(c) Show that the general real-valued solution to $x_{t+2} = 4x_{t+1} - 5x_t$ is given by $x_t = \text{Re}[(A + Bi)(2 + i)^t]$, where A and B are arbitrary real constants.
(d) Show that this solution can be rewritten as $x_t = (\sqrt{5})^t (A \cos t\theta - B \sin t\theta)$ with $\theta = \arctan \dfrac{1}{2}$.

13. (a) Show that the recurrence relation $x_{t+2} = -9x_t + 6x_{t+1}$ has a unique solution of the form $x_t = r^t$. (Find r.)
    (b) By analogy with the method of reduction of order for differential equations, we can try finding solutions of the form $x_t = a_t r^t$ to the recurrence relation in part (a). Show that this leads to a recurrence relation for (a_t) that is equivalent to the difference equation $\Delta^2 a = 0$.
    (c) Use part (b) to find the general solution of the recurrence relation $x_{t+2} = -9x_t + 6x_{t+1}$.

14. Find conditions on the real numbers b and c for the recurrence relation $x_{t+2} = bx_t + cx_{t+1}$ to have:
    (a) Two distinct "basic" solutions of the form $x_t = r^t$, r real.
    (b) Exactly one solution of the form $x_t = r^t$, r real.
    (c) Two complex conjugate solutions of the form $x_t = \rho^t$, $x_t = (\bar\rho)^t$, ρ a nonreal complex number.

15. Solve the inhomogeneous first-order linear recurrence relation $x_{t+1} = ax_t + b$, where a and b are constants. (Start with the homogeneous case and use variation of constants.)

16. Recall that the matrix $\mathbf{A}$ is *diagonalizable* if there exists an invertible matrix $\mathbf{T}$ such that $\mathbf{T}^{-1}\mathbf{AT}$ is a diagonal matrix.
    (a) Show that if $\mathbf{A}$ is diagonalizable, then $\mathbf{A} + \mathbf{I}$ is also diagonalizable.
    (b) Suppose that $\mathbf{T}^{-1}(\mathbf{A} + \mathbf{I})\mathbf{T} = \mathbf{D}$. Show that $(\mathbf{A} + \mathbf{I})^t = \mathbf{T}\mathbf{D}^t\mathbf{T}^{-1}$ for all positive integers t.
    [*Hint*: First do the case $t = 1$.]

17. Let $\mathbf{A} = \begin{bmatrix} -0.4 & 0.8 \\ 0.4 & -0.8 \end{bmatrix}$.
    (a) Find all eigenvalues and eigenvectors for $\mathbf{A} + \mathbf{I} = \begin{bmatrix} 0.6 & 0.8 \\ 0.4 & 0.2 \end{bmatrix}$.
    (b) Find an invertible matrix $\mathbf{T}$ such that $\mathbf{T}^{-1}(\mathbf{A} + \mathbf{I})\mathbf{T}$ is a diagonal matrix $\mathbf{D}$, and find $\mathbf{D}$.
    (c) Using Exercise 16(b), find $(\mathbf{A} + \mathbf{I})^9$.
    (d) In Example 7.1.6, find the percentage of lawns that will be mowed in the tenth week of summer.

18. On a certain tropical island, the weather is usually "good," but there are occasional spells of "bad" weather. If the weather is good one day, there is a 5% chance that it will be bad the next day. On the other hand, if the weather is bad one day, there is a 40% chance that it will be good the next day.
    (a) A tourist arrives on a day with good weather. What is the probability that the weather will also be good when she leaves 2 weeks later?

*(b) Suppose that you arrive on the island on a random day. What is the probability that the weather will be good?

   [*Hint*: Suppose that the weather is good with probability p and therefore bad with probability $1 - p$. What will these probabilities be for the next day? Since your arrival day is random, how should the probabilities for the next day compare with p and $1 - p$?]

19. Let $\mathbf{B} = \begin{bmatrix} \alpha & \beta \\ \gamma & \delta \end{bmatrix}$ be the matrix for a Markov chain, that is, $\begin{bmatrix} p_{t+1} \\ q_{t+1} \end{bmatrix} = \mathbf{B} \begin{bmatrix} p_t \\ q_t \end{bmatrix}$, where (p_t), (q_t) are sequences of complementary probabilities.

   *(a) Show that unless the probabilities are constant (i.e., $p_t = p_{t+1} = \cdots$), we have $\alpha + \gamma = \beta + \delta = 1$.

   [*Hint*: Use $p_t + q_t = p_{t+1} + q_{t+1} = \cdots = 1$.]

   (b) Show that unless the probabilities are constant, we have $\mathbf{B} = \mathbf{A} + \mathbf{I}$ with
   $$\mathbf{A} = \begin{bmatrix} -\gamma & \beta \\ \gamma & -\beta \end{bmatrix}.$$

   (c) Show that, whether or not the probabilities are constant, 1 is an eigenvalue of $\mathbf{B}$.

   [*Hint*: If the probabilities are constant, it's easy to find an eigenvector; otherwise, use part (a) or (b).]

FURTHER READING

Two recent textbooks on difference equations are Mickens, *Difference Equations* (Van Nostrand Reinhold, 1987) and Kelley and Peterson, *Difference Equations* (Academic Press, 1991). The latter includes a chapter on systems of difference equations.

7.2 NUMERICAL METHODS

As we have seen, there are important cases in which differential equations can be solved analytically, that is, in which one can derive explicit or implicit formulas for the unknown functions. We have also seen that even when this is not possible, one can sometimes get qualitative information about the behavior of the solutions. However, for many applications this is clearly not enough, and even implicit formulas for the solutions may not be enough; often explicit numerical information about the solutions is needed instead. For example, although it might not be possible or necessary to find a formula for $x(t)$, one might need to know the values of $x(10)$, $x(20)$, and $x(30)$ to within 1% accuracy.

The branch of mathematics that analyzes methods of numerical computation (usually to be carried out on a computer) to make sure that they can deliver results within a desired range of accuracy is called **numerical analysis**. This difficult subject also deals with issues such as the efficiency of computations and the effects of inaccuracies in input data on the results of computations. Naturally, such issues arise not only in connection with differential equations but also when solving algebraic equations, finding eigenvalues and eigenvectors of large matrices, and so forth. For more information on them, you should refer to textbooks or courses on numerical analysis. Meanwhile, to illustrate some of the ideas that are used in numerical computation of solutions to differential equations, in this section we will discuss a few, relatively straightforward, approximation methods.

The problem we'll actually consider is the following. Given an initial value problem of the form

$$\frac{dx}{dt} = f(t, x), \qquad x(t_0) = x_0,$$

how can we, for some given time T, find an approximate value for $x(T)$? We'll assume that f and $\frac{\partial f}{\partial x}$ are defined and continuous, so that the initial value problem has a unique solution defined on some interval around t_0 (see p. 40); we'll also assume that T is in this interval, so that there really is a unique value for $x(T)$. We want to approximate that value without actually solving the differential equation.

To get an idea of how this might be done, consider the direction field (see p. 42) of the differential equation. Recall that the line segment of the direction field through any point (t, x) has slope $f(t, x)$ and is tangent to the solution curve through that point. By starting at (t_0, x_0) and following a succession of such line segments, we should be able to get an approximate value for $x(T)$ (see Figure 7.2.1, p. 546).

Of course, this is still a bit vague; we haven't decided yet which, or how many, line segments to use. In practice, the time interval from t_0 to T is usually divided into n equal pieces, where n is some number chosen in advance, and n line segments corresponding to these n subintervals are then used. The choice of n may be difficult; it will depend on considerations such as the amount of time available for the computation and the precision required in the answer.

Once n is chosen, we have a partition of the time interval from t_0 to $t_n = T$, with intermediate times $t_1 = t_0 + \Delta t$, $t_2 = t_0 + 2\Delta t$, ..., $t_{n-1} = t_0 + (n-1)\Delta t$, where

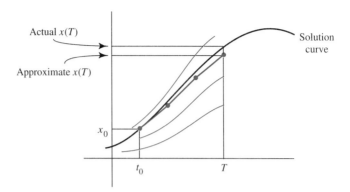

Figure 7.2.1. Approximating $x(T)$ using the direction field.

$\Delta t = \dfrac{T - t_0}{n}$. We can then successively find the endpoints of the line segments shown in Figure 7.2.2, as follows. The first segment goes through (t_0, x_0) and has slope $f(t_0, x_0)$, so its endpoint is $(t_0 + \Delta t, x_0 + f(t_0, x_0) \Delta t)$. In other words, starting with the point (t_0, x_0) which is on the solution curve, we have found a point (t_1, x_1) which is approximately on the solution curve, with $x_1 = x_0 + f(t_0, x_0) \Delta t$. Continuing in this way, we get a point (t_2, x_2) which is approximately on the solution curve, with $x_2 = x_1 + f(t_1, x_1) \Delta t$, and so on. Eventually, we compute $x_n = x_{n-1} + f(t_{n-1}, x_{n-1}) \Delta t$, and this is our approximate value for $x(t_n) = x(T)$.

The method presented here is called **Euler's method** after its inventor. Although it is simple, it has some disadvantages, and you might enjoy trying to think of them and of improvements that could be made before a few of these are discussed below. First, however, here is an example.

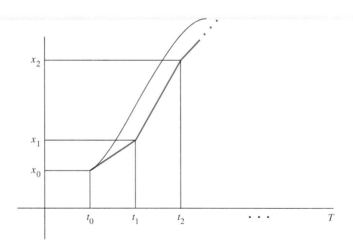

Figure 7.2.2. Euler's method.

SEC. 7.2 NUMERICAL METHODS

EXAMPLE 7.2.1 Use Euler's method with $n = 4$ to approximate $x(2)$, where $x(t)$ is the solution to the initial value problem $\dfrac{dx}{dt} = \dfrac{3t^2}{2x}$, $x(0) = 1$.

Actually, this initial value problem can be solved exactly (see Exercise 1), and it turns out that $x(2) = 3$. Now for Euler's method: Since we are given $t_0 = 0$, $T = 2$, and $n = 4$, we have $\Delta t = \dfrac{2}{4} = \dfrac{1}{2}$, and the intermediate times are $t_1 = \dfrac{1}{2}$, $t_2 = 1$, $t_3 = \dfrac{3}{2}$. Since $f(t, x) = \dfrac{3t^2}{2x}$, we then find

$$x_1 = x_0 + f(t_0, x_0)\,\Delta t = 1 + \frac{3 \cdot 0^2}{2 \cdot 1} \cdot \frac{1}{2} = 1;$$

$$x_2 = x_1 + f(t_1, x_1)\,\Delta t = 1 + \frac{3 \cdot (1/2)^2}{2 \cdot 1} \cdot \frac{1}{2} = 1\frac{3}{16} = \frac{19}{16};$$

$$x_3 = x_2 + f(t_2, x_2)\,\Delta t = \frac{19}{16} + \frac{3 \cdot 1^2}{2 \cdot 19/16} \cdot \frac{1}{2} = \frac{19}{16} + \frac{12}{19} = \frac{553}{304};$$

finally,

$$x_4 = x_3 + f(t_3, x_3)\,\Delta t = \frac{553}{304} + \frac{3 \cdot (3/2)^2}{2 \cdot 553/304} \cdot \frac{1}{2} = \frac{553}{304} + \frac{513}{553} \approx 2.746746.$$

So the approximate value given by Euler's method with $n = 4$ is between 2.7 and 2.8, clearly not very impressive as an approximation of 3. On the other hand, $\Delta t = \dfrac{1}{2}$ is still quite large, and we can hope for better approximations if we divide the time interval into more pieces. Table 7.2.1 on p. 548 shows the effect of repeatedly doubling the number of subintervals (that is, halving Δt). Of course, these calculations were not done by hand; in Exercise 2 you are asked to write a computer program "shell" for use in carrying out computations such as the above. ∎

The results in the table suggest that for this particular example, the error in the approximation is roughly halved each time n is doubled; for instance, the error is $3 - 2.5 = 0.5$ for $n = 2$ and $3 - 2.746746 \approx 0.253$ for $n = 4$. That is, the error is roughly proportional to Δt. It turns out that this is true in general when Euler's method is applied to a "reasonable" initial value problem. Although this is not easy to prove precisely (see "Further Reading" following the Exercises), an intuitive idea of why it might work is not too hard to come up with. First of all, when the function $x(t)$ is approximated on the first subinterval $(t_0, t_0 + \Delta t)$ by its tangent line at t_0, the error is bounded by $M_1 \dfrac{(\Delta t)^2}{2}$, where M_1 is a bound for the (absolute value of the) second derivative on that subinterval. (Compare p. 369.) Now in Euler's method we use n such approximations in all, although admittedly

Table 7.2.1 Results of approximating $x(2)$ [a] by Euler's method, given $\dfrac{dx}{dt} = \dfrac{3t^2}{2x}$, $x(0) = 1$

n	Approximate value of $x(2)$
1	1
2	2.5
4	2.746746
8	2.870815
16	2.935068
32	2.967468
64	2.983719
128	2.991856
256	2.995928
512	2.997964

[a] Actual value: $x(2) = 3$.

after the first step we can no longer claim to be using the tangent line to the actual solution $x(t)$. So it seems reasonable that the (total) error from all these approximations will be bounded by something like $nM \dfrac{(\Delta t)^2}{2}$, where M is an overall bound for the second derivative of $x(t)$ on the whole interval (t_0, T). Since $n \Delta t = T - t_0$, we have

$$nM \frac{(\Delta t)^2}{2} = (T - t_0) \frac{M}{2} \Delta t,$$

which is indeed proportional to Δt.

Although the results in the table are shown to six decimal places, you may notice that even the approximation using $n = 512$ is correct only to two decimal places. Remember that in practice, the exact answer will *not* be known, and numerical analysis will be needed to get an estimate of how close a given approximation will be.

One reason that the results from Euler's method are often not as accurate as one might wish is that if all the solution curves have the same concavity, the approximation will be "systematically" too high or too low. For instance, suppose that all the solution curves are concave up. Then the tangent line at (t_0, x_0) will lie below the solution curve, and so (t_1, x_1) will be on a "lower" solution curve. The line segment from (t_1, x_1) to (t_2, x_2), which is tangent to this lower curve, will lie below *it*, and so forth. (See Figure 7.2.3.)

SEC. 7.2 NUMERICAL METHODS

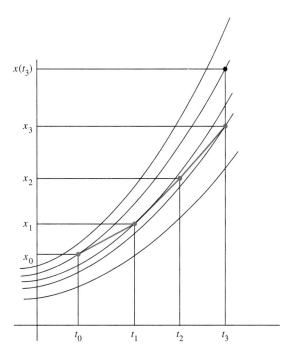

Figure 7.2.3. When all solution curves are concave up, the approximate value given by Euler's method will always be below the actual value.

Perhaps the most obvious way to try to deal with this problem is to take the concavity into account by using the second derivative $x'' = \dfrac{d^2x}{dt^2}$ along with $x' = \dfrac{dx}{dt}$. For example, instead of using the tangent line to approximate the solution curve from t_0 to t_1:

$$x(t) \approx x(t_0) + x'(t_0) \cdot (t - t_0),$$

we can use the second-order Taylor approximation

$$x(t) \approx x(t_0) + x'(t_0) \cdot (t - t_0) + \frac{x''(t_0)}{2} \cdot (t - t_0)^2.$$

Of course, this idea requires us to find the second derivative, but this can be done as follows. Starting with the given equation $x' = f(t, x)$, differentiate both sides with respect to t. On the right-hand side we need the chain rule (see p. 9), and the result will be $x'' = \dfrac{\partial f}{\partial t} + \dfrac{\partial f}{\partial x}\dfrac{dx}{dt}$. However, since $\dfrac{dx}{dt} = f(t, x)$, this yields $x'' = \dfrac{\partial f}{\partial t} + \dfrac{\partial f}{\partial x}f$.

EXAMPLE 7.2.2 If, as in Example 7.2.1, we have $x' = \dfrac{dx}{dt} = \dfrac{3t^2}{2x}$, then

$$x'' = \frac{d^2x}{dt^2} = \frac{6t}{2x} - \frac{3t^2}{2x^2}\frac{dx}{dt} = 3\frac{t}{x} - \frac{9t^4}{4x^3}. \qquad \square$$

Once we have found an expression for x'', let's say $x'' = g(t, x)$, we proceed in a way that is similar to Euler's method. That is, we first subdivide the time interval using intermediate times

$$t_1 = t_0 + \Delta t, \ t_2 = t_0 + 2\Delta t, \ \ldots, \ t_{n-1} = t_0 + (n-1)\Delta t.$$

Next, we successively compute the following:

$$x_1 = x_0 + f(t_0, x_0)\Delta t + \frac{g(t_0, x_0)}{2}(\Delta t)^2,$$

$$x_2 = x_1 + f(t_1, x_1)\Delta t + \frac{g(t_1, x_1)}{2}(\Delta t)^2, \quad \text{and so on.}$$

Note that at each step, we're using the second-order Taylor polynomial to approximate a solution curve on a subinterval. However, this will only be the "actual" solution curve for our initial value problem at the first step, because the points (t_1, x_1), (t_2, x_2), and so on, won't lie exactly on the "actual" curve. Eventually, we find

$$x_n = x_{n-1} + f(t_{n-1}, x_{n-1})\Delta t + \frac{g(t_{n-1}, x_{n-1})}{2}(\Delta t)^2$$

as our approximate value for $x(T)$.

This method is sometimes called the *three-term Taylor series method* (since the second-order Taylor polynomials used have three terms); we'll call it the **three-term method** for short.

EXAMPLE 7.2.2 (continued) Use the three-term method with $n = 2$ to approximate $x(2)$, where $x(t)$ is once again the solution to $\dfrac{dx}{dt} = \dfrac{3t^2}{2x}$, $x(0) = 1$.

This time we have $\Delta t = 1$, $t_1 = 1$, and as we saw earlier, we have $g(t, x) = 3\dfrac{t}{x} - \dfrac{9t^4}{4x^3}$ along with $f(t, x) = \dfrac{3t^2}{2x}$. Thus we get

$$x_1 = 1 + f(0, 1) + \frac{g(0, 1)}{2} = 1$$

and therefore

$$x_2 = 1 + f(1, 1) + \frac{g(1, 1)}{2} = 1 + \frac{3}{2} + \frac{1}{2}\left(3 - \frac{9}{4}\right) = 2.875.$$

Note that this approximation to 3 is (slightly) better than the one obtained by Euler's method for $n = 8$! (See Table 7.2.1.)[8] Table 7.2.2 shows the effect of repeatedly doubling n; see also Exercises 8 and 9. ∎

Table 7.2.2 Results of approximating $x(2)$ [a] by the three-term method, given $\dfrac{dx}{dt} = \dfrac{3t^2}{2x}$, $x(0) = 1$

n	Approximate value of $x(2)$
1	1
2	2.875
4	2.999085
8	2.999782
16	2.999969
32	2.999996
64	3.000000
128	3.000000
256	3.000000
512	2.999999

[a] Actual value: $x(2) = 3$.

The last entry in Table 7.2.2 is rather disturbing. Even though the error is small, it is disconcerting that an error appears at all for $n = 512$ after the approximation was correct (to six decimal places) for $n = 64$, $n = 128$, and $n = 256$. This phenomenon, which is likely to get worse if n is increased further, is due to accumulated **round-off error**. In brief, since numbers are stored in a computer with limited (although very high) precision, at each arithmetic step in a computation a rounding off may occur. Usually the computer carries through calculations using several more digits than are printed out, so after computations with relatively few steps the printed digits will tend to be reliable. However, when computations get longer, round-off errors may accumulate to the point of affecting the

---

[8] Keep in mind, though, that for any particular value of n, the three-term method requires more computation than Euler's method.

answer given by the computer. (For an illustration, see Exercise 10.) Thus it is not only for efficiency, but also for accuracy, that numerical analysis is used to estimate how large n needs to be for the approximation to be precise enough. Simply taking some "very large" value of n would run the risk of substantial round-off error.

Although the three-term (Taylor series) method is certainly an improvement over Euler's method, it does require the computation of the values of $g = \frac{\partial f}{\partial t} + \frac{\partial f}{\partial x} f$ at each of the points (t_0, x_0), (t_1, x_1), and so on. For some functions this takes much more time than just computing values of f, and so it is desirable to have an alternative method which only involves computing $f(t, x)$ for various t and x. One such improvement on Euler's original method is **Heun's method**.

Heun's method, like the three-term method, aims at avoiding the systematic error (which was illustrated in Figure 7.2.3) due to the concavity of the solution curves. The method is based on the idea (see Figure 7.2.4) that if (t_0, x_0) and (t_1, x_1) are points on a solution curve, then the slope of the line connecting them can be expected to be closer to $\frac{1}{2}[f(t_0, x_0) + f(t_1, x_1)]$, the average of the slopes of the curve at x_0 and at x_1, than to $f(t_0, x_0)$ [or, for that matter, to $f(t_1, x_1)$]. Therefore, if we start at (t_0, x_0) and use a line segment to get an *approximate* point (t_1, x_1), it would be better if this segment had slope $\frac{1}{2}[f(t_0, x_0) + f(t_1, x_1)]$ rather than $f(t_0, x_0)$.

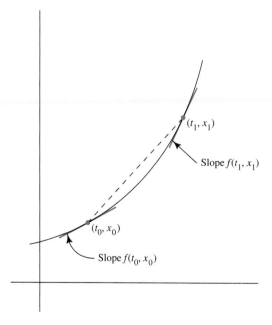

Figure 7.2.4. The slope of the dashed line is close to the average of $f(t_0, x_0)$ and $f(t_1, x_1)$.

At first sight, this idea leads to a dead end, because there is no way for us to compute $\frac{1}{2}[f(t_0, x_0) + f(t_1, x_1)]$ if we don't know x_1 yet.[9] The trick is first to find a "crude" approximation, say $\bar{x}_1$, using Euler's method, and then to use $\frac{1}{2}[f(t_0, x_0) + f(t_1, \bar{x}_1)]$ as the slope of the first line segment. In other words, we successively compute:

$$\bar{x}_1 = x_0 + f(t_0, x_0)\,\Delta t$$

$$x_1 = x_0 + \frac{1}{2}[f(t_0, x_0) + f(t_1, \bar{x}_1)]\,\Delta t$$

$$\bar{x}_2 = x_1 + f(t_1, x_1)\,\Delta t$$

$$x_2 = x_1 + \frac{1}{2}[f(t_1, x_1) + f(t_2, \bar{x}_2)]\,\Delta t, \quad \text{and so on,}$$

until we eventually reach

$$x_n = x_{n-1} + \frac{1}{2}[f(t_{n-1}, x_{n-1}) + f(t_n, \bar{x}_n)]\,\Delta t,$$

which will be our approximation to $x(T)$.

EXAMPLE 7.2.3 Use Heun's method with $n = 2$ to approximate (once more) $x(2)$, where $x(t)$ is the solution to $\dfrac{dx}{dt} = \dfrac{3t^2}{2x}$, $x(0) = 1$.

We have $\Delta t = 1$, $t_1 = 1$, and we get

$$\bar{x}_1 = 1 + \frac{3 \cdot 0^2}{2 \cdot 1} \cdot 1 = 1 \quad \text{(which would be } x_1 \text{ if we were using Euler's method),}$$

$$x_1 = 1 + \frac{1}{2}\left(\frac{3 \cdot 0^2}{2 \cdot 1} + \frac{3 \cdot 1^2}{2 \cdot 1}\right) \cdot 1 = \frac{7}{4},$$

$$\bar{x}_2 = \frac{7}{4} + \frac{3 \cdot 1^2}{2 \cdot 7/4} \cdot 1 = \frac{73}{28},$$

$$x_2 = \frac{7}{4} + \frac{1}{2}\left(\frac{3 \cdot 1^2}{2 \cdot 7/4} + \frac{3 \cdot 2^2}{2 \cdot 73/28}\right) \cdot 1 = \frac{6805}{2044} \approx 3.329256.$$

This is our approximation to the exact value 3.

---

[9] Also, the equation $x_1 - x_0 = \frac{1}{2}[f(t_0, x_0) + f(t_1, x_1)](t_1 - t_0)$, expressing that (t_1, x_1) is on the line segment with the desired slope, is not likely to be easily solved for x_1; note that x_1 occurs on both sides of the equation.

Note that since at each step the function f has to be evaluated twice [for instance, to get x_2, both $f(t_1, x_1)$ and, later, $f(t_2, \bar{x}_2)$ have to be found], Heun's method will take at least twice as long as Euler's method for the same n. Thus the answer above is not impressive, because it is actually a worse approximation than 2.746746, which was found in Example 7.2.1 with an equal number of evaluations of $\dfrac{3t^2}{2x}$. However, if the number of subintervals is repeatedly doubled, Heun's method soon outperforms Euler's method in our example, as Table 7.2.3 shows; see also Exercises 14 and 15. In general, it turns out that both for Heun's method and for the three-term method, the error tends to be roughly proportional to $(\Delta t)^2$ (see Exercise 22). ■

Table 7.2.3 Results of approximating $x(2)$ [a] by Heun's method, given $\dfrac{dx}{dt} = \dfrac{3t^2}{2x}$, $x(0) = 1$

n	Approximate value of $x(2)$
1	7
2	3.329256
4	3.057229
8	3.012099
16	3.002800
32	3.000675
64	3.000166
128	3.000041
256	3.000010
512	3.000002

[a] Actual value: $x(2) = 3$.

Perhaps the ultimate refinement of Euler's method along the lines of Heun's method (that is, a stepwise computation, starting with x_0, of $x_1, x_2, \ldots, x_n$, each from its predecessor, which does not involve any partial derivatives of f) is the **Runge–Kutta** method. This method, which tends to be quite fast and accurate, with error roughly proportional to $(\Delta t)^4$, is presented in Exercises 17 to 21. However, the motivation for this method is well beyond the scope of this book. Other methods we haven't considered include **multistep** methods, in which the computation of x_2 may involve not only x_1 but also x_0, the computation of x_3 may involve x_0, x_1, and x_2, and so on.

Finally, all these methods can be generalized to systems of first-order differential equations. In particular, they also can be used for higher-order equations, by first rewriting those as systems of first-order equations.

SEC. 7.2 NUMERICAL METHODS

EXERCISES

1. Solve the initial value problem $\dfrac{dx}{dt} = \dfrac{3t^2}{2x}$, $x(0) = 1$, and show that for the solution, $x(2) = 3$.

2. Using whatever computer language is convenient and comfortable for you, write a computer program "shell" (or "driver") for Euler's method. The shell will take x_0, t_0, n, and T as given and will assume that $f(t, x)$ can be found for any known t and x. It should then compute the approximate value x_n for $x(T)$, where $x(t)$ is the solution of the initial value problem $\dfrac{dx}{dt} = f(t, x)$, $x(t_0) = x_0$, using Euler's method with the given value of n.

3. Use your "shell" from Exercise 2, or a calculator, to verify the results in Table 7.2.1 (p. 548). [Since you now know that $f(t, x) = \dfrac{3t^2}{2x}$, $t_0 = 0$, and so on, you can make your shell into an actual program that only needs a value of n in order to run.]

4. (a) Let $x(t)$ be the solution to the initial value problem $\dfrac{dx}{dt} = e^{-t^2}$, $x(0) = 0$.

    Using your "shell" from Exercise 2, or a calculator, approximate $x(1)$ by Euler's method, for various values of n.
   (b) For what definite integral did you just find approximate values?
   (c) Are these approximate values too high or too low? Why?

5. Let $x(t)$ be the solution to the initial value problem $\dfrac{dx}{dt} = 1 + t^2 x^2$, $x(0) = 0$.

    Using your "shell" from Exercise 2, or a calculator, approximate $x(1)$ by Euler's method, for various values of n.

6. Write a computer program "shell" as in Exercise 2, but for the three-term method. Take x_0, t_0, n, and T as given and assume that for any known t and x, $f(t, x)$ and $g(t, x)$ can be found.

7. Use your "shell" from Exercise 6, or a calculator, to verify the results in Table 7.2.2 (p. 551). (Depending on the precision of your computer or calculator, you may or may not encounter round-off error near the end of the table.)

8. (a) Let $x(t)$ be the solution to the initial value problem $\dfrac{dx}{dt} = e^{-t^2}$, $x(0) = 0$.

    Using your "shell" from Exercise 6, or a calculator, approximate $x(1)$ by the three-term method, for various values of n.
   (b) Compare your results from part (a) with those of Exercise 4.

9. (a) Let $x(t)$ be the solution to the initial value problem $\dfrac{dx}{dt} = 1 + t^2 x^2$, $x(0) = 0$.

    Using your "shell" from Exercise 6, or a calculator, approximate $x(1)$ by the three-term method, for various values of n.
   (b) Compare your results from part (a) with those of Exercise 5.

10. Assume that a computer is programmed to find a sequence of numbers $x_0, x_1, x_2, \ldots$ by starting with x_0 and successively evaluating

$$x_1 = 2x_0 - 1, \; x_2 = 2x_1 - 1, \ldots, x_n = 2x_{n-1} - 1, \ldots.$$

Suppose also that $x_0 = 1.0000001 = 1 + 10^{-7}$, but that the computer keeps only six significant digits and therefore starts with $x_0 = 1$.
(a) What will the computer find for x_{10}?
(b) What is the actual value of x_{10}? How many of the digits given by the computer are correct?

11. (a) Recall that if $x' = f(t, x)$, then $x'' = g(t, x)$ with $g = \dfrac{\partial f}{\partial t} + \dfrac{\partial f}{\partial x} f$. Find a similar expression for the third derivative $x^{(3)} = h(t, x)$.
(b) Devise a four-term (Taylor series) method.
(c) In practice, the four-term method is seldom used. Why would this be?

12. Write a computer program "shell" as in Exercise 2, but for Heun's method.

13. Use your "shell" from Exercise 12, or a calculator, to verify the results in Table 7.2.3 (p. 554).

14. (a) Let $x(t)$ be the solution to the initial value problem $\dfrac{dx}{dt} = e^{-t^2}$, $x(0) = 0$.
Using your "shell" from Exercise 12, or a calculator, approximate $x(1)$ by Heun's method, for various values of n.
(b) Compare your results from part (a) with those of Exercises 4 and 8.

15. (a) Let $x(t)$ be the solution to the initial value problem $\dfrac{dx}{dt} = 1 + t^2 x^2$, $x(0) = 0$.
Using your "shell" from Exercise 12, or a calculator, approximate $x(1)$ by Heun's method, for various values of n.
(b) Compare your results from part (a) with those of Exercises 5 and 9.

16. Suppose that in the initial value problem $\dfrac{dx}{dt} = f(t, x)$, $x(t_0) = x_0$, the function $f(t, x)$ actually depends only on t, say $f(t, x) = F(t)$.
(a) Show that approximating $x(T)$ is equivalent to approximating the definite integral $\displaystyle\int_{t_0}^{T} F(t)\, dt$.
(b) Show that in this situation, Heun's method gives the same result for the definite integral as the trapezoidal rule does.

17. The Runge–Kutta method (developed around 1900 by the applied mathematicians Runge and Kutta) for approximating $x(T)$, where $x(t)$ is the solution to the initial value problem $\dfrac{dx}{dt} = f(t, x)$, $x(t_0) = x_0$, is as follows. As usual, first divide the

time interval into n equal pieces using the intermediate times $t_1, \ldots, t_{n-1}$. To get x_1 from x_0, first compute the following:

$$a_1 = f(t_0, x_0);$$

$$b_1 = f\left(t_0 + \frac{1}{2}\Delta t, x_0 + \frac{1}{2}a_1 \Delta t\right);$$

$$c_1 = f\left(t_0 + \frac{1}{2}\Delta t, x_0 + \frac{1}{2}b_1 \Delta t\right);$$

$$d_1 = f(t_1, x_0 + c_1 \Delta t).$$

Then $x_1 = x_0 + \frac{1}{6}\Delta t(a_1 + 2b_1 + 2c_1 + d_1)$. Using similar abbreviations, $x_2 = x_1 + \frac{1}{6}\Delta t(a_2 + 2b_2 + 2c_2 + d_2)$, and so on; as usual, x_n is our approximate value for $x(T)$. Write a computer program "shell" as in Exercise 2 for the Runge–Kutta method.

18. Use your "shell" from Exercise 17, or a calculator, to verify the results in Table 7.2.4.

Table 7.2.4 Results of approximating $x(2)$ [a] by the Runge–Kutta method, given $\dfrac{dx}{dt} = \dfrac{3t^2}{2x}$, $x(0) = 1$

n	Approximate value of $x(2)$
1	3.309091
2	3.011003
4	3.000726
8	3.000041
16	3.000002
32	3.000000
64	3.000000
128	3.000000
256	3.000000
512	3.000000

[a] Actual value: $x(2) = 3$.

19. (a) Let $x(t)$ be the solution to the initial value problem $\dfrac{dx}{dt} = e^{-t^2}$, $x(0) = 0$.

   Using your "shell" from Exercise 17, or a calculator, approximate $x(1)$ by the Runge–Kutta method, for various values of n.

   (b) Compare your results from part (a) with those of Exercises 4, 8, and 14.

20. (a) Let $x(t)$ be the solution to the initial value problem $\dfrac{dx}{dt} = 1 + t^2 x^2$, $x(0) = 0$.

   Using your "shell" from Exercise 17, or a calculator, approximate $x(1)$ by the Runge–Kutta method, for various values of n.

   (b) Compare your results from part (a) with those of Exercises 5, 9, and 15.

21. Suppose that in the initial value problem $\dfrac{dx}{dt} = f(t, x)$, $x(t_0) = x_0$, the function $f(t, x)$ actually depends only on t, say $f(t, x) = F(t)$. (Compare Exercise 16.) Show that in this situation, the Runge–Kutta method is equivalent to approximating
$$\int_{t_0}^{T} F(t)\, dt$$
using Simpson's rule.

22. Give a (rough) intuitive argument suggesting why for the three-term method, the error tends to be proportional to $(\Delta t)^2$.

FURTHER READING

For an example of a multistep method, see Section 8.6 in Boyce and DiPrima, *Elementary Differential Equations and Boundary Value Problems,* 5th ed. (Wiley, 1992). Careful numerical analysis of Euler's method is provided in the opening chapters of Gear, *Numerical Initial Value Problems in Ordinary Differential Equations* (Prentice-Hall, 1971) and Henrici, *Discrete Variable Methods in Ordinary Differential Equations* (Wiley, 1962).

7.3 SUCCESSIVE APPROXIMATION

In this section we return to the Existence and Uniqueness Theorem for first-order differential equations; we consider the following form of this theorem.

Theorem (Existence and Uniqueness). If f and $\dfrac{\partial f}{\partial x}$ are defined and continuous in some rectangle containing (t_0, x_0) in its interior, then the initial value problem $\dfrac{dx}{dt} = f(t, x)$, $x(t_0) = x_0$ has a unique solution on some interval containing t_0 in its interior.

Although no indication of the proof was given at the time, this fact was already stated in Section 1.5, and re-reading that section at this point may be worthwhile. By considering (in Example 1.5.1) the initial value problem $\dfrac{dx}{dt} = x^{1/3}$, $x(0) = 0$, which does *not* have a unique solution on any interval around 0, we saw that some condition on $\dfrac{\partial f}{\partial x}$ really is needed to get the uniqueness of the solution.

Meanwhile, you may well wonder how one can ever prove that a solution to the initial value problem exists at all. The idea will be to construct such a solution by successive approximation. That is, we'll define a sequence of functions, none of which is the solution, but which get closer and closer to being solutions; the limit of this sequence of functions will be the desired solution.

Since this project will involve various techniques that may well be unfamiliar to you (related to questions such as: How can one tell whether a sequence of functions has a limit? How can one prove anything about the limit?), we'll first look at an example of a less complicated proof based on the same idea of successive approximation. Specifically, we'll use successive approximation to show that there exists a positive number x with $x^2 = 5$.

This may seem obvious; after all, for $x = \sqrt{5}$ we have $x^2 = 5$. But how do we know that such a real number $\sqrt{5}$ exists? In high school, when $\sqrt{5}$ is first defined, its existence is usually taken for granted: "$\sqrt{5}$ is 'the' positive real number whose square is 5." Probably the easiest way to show that such a number exists is to use the Intermediate Value Theorem. For example, one can define a continuous function g by $g(x) = x^2 - 5$. Since $g(2) = -1$ and $g(3) = 4$ and 0 is between -1 and 4, there must be a value of x between 2 and 3 for which $g(x) = 0$, that is, $x^2 = 5$. However, this proof that $\sqrt{5}$ exists does not say anything about how to find the number $\sqrt{5}$, except that it is between 2 and 3.

An alternative proof that $\sqrt{5}$ exists starts with the observation that if x is a positive number with $x^2 < 5$, then $(\dfrac{5}{x})^2 > 5$ [see Exercise 1(a)]. So if x is a "guess" for $\sqrt{5}$—a number that might be $\sqrt{5}$—and x turns out to be too small, then $\dfrac{5}{x}$ will be too large. It then seems reasonable that the average $\frac{1}{2}(x + \frac{5}{x})$ of x and $\frac{5}{x}$ might be a better guess. Similarly, if x is a positive number with $x^2 > 5$, then $(\frac{5}{x})^2 < 5$ [see Exercise 1(b)], so $\frac{5}{x}$ is too small while x is too large, and again $\frac{1}{2}(x + \frac{5}{x})$ might be a better guess. In fact, it's not hard to show (see Exercise 2) that if $x^2 \ne 5$ but $4 \le x^2 \le 6$, then $\frac{1}{2}(x + \frac{5}{x})$ will indeed be a better guess, in the sense that $[\frac{1}{2}(x + \frac{5}{x})]^2$ will be closer to 5 than x^2.

In particular, if we start with $x = 2$ and keep replacing x by $\frac{1}{2}(x + \frac{5}{x})$, we'll get a sequence of better and better approximations, that is, positive numbers whose squares are closer and closer to 5. This sequence (x_n) is determined by the recurrence relation $x_{n+1} = \frac{1}{2}(x_n + \frac{5}{x_n})$ along with the initial condition $x_1 = 2$. (If you've read Section 7.1, note that the methods from that section don't apply, because our recurrence relation is not linear.) To six decimal places, the first few terms of the sequence are $x_1 = 2$, $x_2 = 2.25$, $x_3 = 2.236111$, $x_4 = 2.236068$, ... After the first step, the sequence is decreasing (see Exercise 3). On the other hand, the sequence is certainly bounded below; for one thing, all the terms are positive (why?). Since a basic property of the real numbers is that every bounded monotonic sequence has a limit,[10] our sequence certainly has a limit; let that limit be $L = \lim_{n \to \infty} x_n$. Now consider the recurrence relation $x_{n+1} = \frac{1}{2}(x_n + \frac{5}{x_n})$ and take the limit on both sides. Since as $n \to \infty$, we also have $n + 1 \to \infty$, on the left-hand side we get $\lim_{n \to \infty} x_{n+1} = L$. As for the right-hand side, since the expression $\frac{1}{2}(x + \frac{5}{x})$ varies continuously with the positive number x, we have

$$\lim_{n \to \infty} \frac{1}{2}\left(x_n + \frac{5}{x_n}\right) = \frac{1}{2}\left(\lim_{n \to \infty} x_n + \frac{5}{\lim_{n \to \infty} x_n}\right) = \frac{1}{2}\left(L + \frac{5}{L}\right).$$

Therefore, since the limits of both sides of the recurrence relation must be equal, we get $L = \frac{1}{2}(L + \frac{5}{L})$, which simplifies to $L = \frac{5}{L}$ or $L^2 = 5$. Thus we have constructed a positive number L (the limit of our sequence 2, 2.25, 2.236111, . . .) whose square is 5.

The idea of the procedure of successive approximation for square roots seems to have been known to the Babylonians, who apparently computed $\sqrt{2}$ to six decimal places in this way (see "Further Reading" following the Exercises). Much later, Newton (1642–1727) devised a far more general method, which is often taught in calculus courses, for approximating a zero of a differentiable function. When applied to the particular function

---

[10] That is, every bounded increasing sequence has a limit, and every bounded decreasing sequence has a limit.

SEC. 7.3 SUCCESSIVE APPROXIMATION

g for which $g(x) = x^2 - 5$, Newton's method leads to the same recurrence relation $x_{n+1} = \frac{1}{2}(x_n + \frac{5}{x_n})$ as above (see Exercise 4).

An even more general principle for finding a real number x by successive approximation is as follows. First find a suitable[11] equation of the form $x = F(x)$ to which the desired number is a solution. [In the case of $x = \sqrt{5}$ above, let $F(x) = \frac{1}{2}(x + \frac{5}{x})$.] Then, starting with a suitable[11] initial "guess" x_1, form a sequence (x_n) using the recurrence relation $x_{n+1} = F(x_n)$. If we can show that the sequence has a limit, say $L = \lim_{n \to \infty} x_n$, then taking limits on both sides of the recurrence relation yields $L = \lim_{n \to \infty} F(x_n)$. Provided F is continuous, we also have $\lim_{n \to \infty} F(x_n) = F(\lim_{n \to \infty} x_n)$, and so $L = F(\lim_{n \to \infty} x_n) = F(L)$. In other words, L is a solution of the equation $x = F(x)$, and it only remains to be shown that L is the desired solution. [In our example above, $\pm\sqrt{5}$ are the two solutions of $x = F(x)$, but since all the terms in the sequence 2, 2.25, ... are positive, its limit is $\sqrt{5}$ rather than $-\sqrt{5}$.]

Now (finally!) back to the Existence and Uniqueness Theorem. First of all, we would like to construct a solution to an initial value problem such as $\frac{dx}{dt} = f(t, x), x(t_0) = x_0$. By changing coordinates in the t,x-plane to $t - t_0, x - x_0$, we can assume that $t_0 = x_0 = 0$ (see Exercise 5). So given a "reasonable" f, we would like to be able to construct a function $x(t)$ for which $\frac{dx}{dt} = f(t, x), x(0) = 0$. This will be done using the same general principle as in the previous paragraph, only this time for a function $x = x(t)$ rather than a number x. To begin with, we should find a suitable equation of the form $x(t) = F(x)(t)$, or $x = F(x)$ for short, to which the desired function is a solution. Starting with $\frac{dx}{dt} = f(t, x)$, it seems natural to integrate both sides, which gives us

$$x = \int f(t, x)\, dt, \quad \text{or in "longhand"} \quad x(t) = \int f(t, x(t))\, dt.$$

The integral on the right is only defined up to a constant, but since we also want $x(0) = 0$, we must take $x(t) = \int_0^t f(s, x(s))\, ds$ (see p. 4). The **integral equation**

$$x(t) = \int_0^t f(s, x(s))\, ds$$

is in the right form $x = F(x)$, where $F(x)(t) = \int_0^t f(s, x(s))\, ds$, and Exercise 6 asks you to show that the new equation $x = F(x)$ is equivalent to the original initial value problem

---

[11] The word *suitable* here is certainly—and deliberately—a bit vague!

$$\frac{dx}{dt} = f(t, x), \ x(0) = 0.$$ That is, any continuous function x that is a solution to the integral equation is also a solution to the initial value problem, and vice versa.

EXAMPLE The initial value problem $\frac{dx}{dt} = 3t^2(x + 2)$, $x(0) = 0$ is equivalent to the integral equation $x(t) = \int_0^t 3s^2 [x(s) + 2] \, ds$. ∎

Having rewritten our equation as $x = F(x)$, the next step in our program is to define a sequence of functions (x_n) by the recurrence relation $x_{n+1} = F(x_n)$, starting with a suitable initial guess x_1. In general, we don't really have anything to go by in making this initial guess, except for the condition $x(0) = 0$ which we want the eventual solution to satisfy. So it seems reasonable (and certainly easy) to take our initial guess to be the "zero function" defined by $x_1(t) = 0$ for all t. Starting with this, we then define the sequence (x_n) of functions using the recurrence relation $x_{n+1}(t) = \int_0^t f(s, x_n(s)) \, ds$.

EXAMPLE 7.3.1 If we again start with the initial value problem $\frac{dx}{dt} = 3t^2(x + 2)$, $x(0) = 0$, the first four functions in our sequence will be x_1, x_2, x_3, x_4, defined by

$x_1(t) = 0$ for all t;

$x_2(t) = \int_0^t 3s^2 [x_1(s) + 2] \, ds = \int_0^t 6s^2 \, ds = 2t^3$;

$x_3(t) = \int_0^t 3s^2 [x_2(s) + 2] \, ds = \int_0^t 3s^2 (2s^3 + 2) \, ds = t^6 + 2t^3$;

$x_4(t) = \int_0^t 3s^2 (s^6 + 2s^3 + 2) \, ds = \frac{1}{3} t^9 + t^6 + 2t^3$. ∎

EXAMPLE 7.3.2 On the other hand, for the initial value problem $\frac{dx}{dt} = e^{-x^2 - t^2}$, $x(0) = 0$, we get

$x_1(t) = 0$ for all t;

$x_2(t) = \int_0^t e^{-s^2} \, ds$,

and already no explicit formula for $x_2(t)$ is available. (This particular function x_2 is important in statistics and is sometimes referred to as the error function, erfc.) ∎

As you can imagine, the second example above is much more typical than the first; usually, one cannot expect to find explicit formulas for the functions $x_2, x_3, \ldots$. How-

SEC. 7.3 SUCCESSIVE APPROXIMATION

ever, by the Fundamental Theorem of Calculus, they are all differentiable functions with the property that $\dfrac{d(x_{n+1}(t))}{dt} = f(t, x_n(t))$. Or are they?

There is one potential problem. We started off by assuming that f (and $\dfrac{\partial f}{\partial x}$) was defined and continuous on some rectangle containing the origin. If f and x_n are continuous, then the composite function $f(s, x_n(s))$ will be continuous and the Fundamental Theorem will apply, but this is known *only as long as all points $(s, x_n(s))$ stay within the rectangle*. Exercise 7 shows how this can be guaranteed by, if necessary, restricting the interval for t, and we won't pursue this technical point further.

To continue our program of finding a solution to $\dfrac{dx}{dt} = f(t, x)$ by successive approximation, the next step should be to show that a limit function $L = \lim\limits_{n \to \infty} x_n$ exists. Here, however, several other technical questions arise. The first, and easiest, question is: What is meant by the statement $L = \lim\limits_{n \to \infty} x_n$ when the x_n are functions? As you might expect, it means that for any value of t (in the interval being considered), $L(t) = \lim\limits_{n \to \infty} x_n(t)$. Now, though, it is not clear how we can expect to prove that such a function L exists. And even assuming *that* for the moment, it is not clear that L will be continuous, let alone differentiable. In fact, it's not hard to find examples of sequences of continuous functions whose limits are not continuous. Figure 7.3.1 suggests how this can happen; see also Exercises 9 and 10.

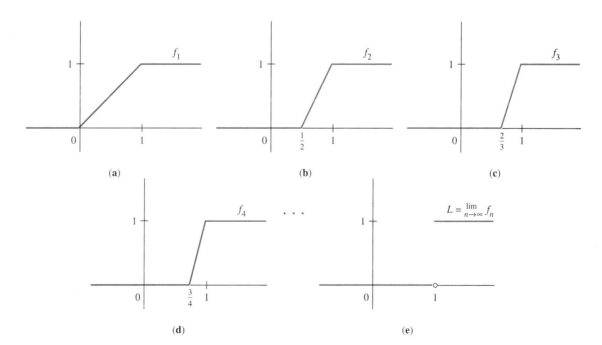

Figure 7.3.1. A sequence of continuous functions whose limit is not continuous.

To see how we might prove that $L(t) = \lim_{n \to \infty} x_n(t)$ is defined for all t, let's return to Example 7.3.1 (p. 562). There we had the initial value problem

$$\frac{dx}{dt} = 3t^2(x+2), \quad x(0) = 0,$$

and we found

$$x_2(t) = 2t^3, \quad x_3(t) = t^6 + 2t^3, \quad x_4(t) = \frac{1}{3}t^9 + t^6 + 2t^3;$$

note how each function seems to be obtained from the previous one by adding one more term. In fact, one can be much more specific about this; if you enjoy finding patterns, you may want to try Exercise 11 before reading the next paragraph.

It turns out (see Exercise 11) that for any $n \geq 2$,

$$x_n(t) = \sum_{k=1}^{n-1} 2 \frac{t^{3k}}{k!} = 2t^3 + t^6 + \frac{t^9}{3} + \cdots + \frac{2t^{3n-3}}{(n-1)!}.$$

In other words, the various $x_n(t)$ (for different n) are the partial sums of the series $\sum_{k=1}^{\infty} 2 \frac{t^{3k}}{k!}$. Since this series converges for all t [see Exercise 12(a)], the sequence $(x_n(t))$ of partial sums also converges for all t, and we'll have $L(t) = \lim_{n \to \infty} x_n(t) = \sum_{k=1}^{\infty} 2 \frac{t^{3k}}{k!}$. In fact, in this particular example one can find an expression for $L(t)$ in closed form (without a summation) and use that expression to show that the function L is a solution to the initial value problem [see Exercise 12(b) and (c)].

As mentioned earlier, the situation of Example 7.3.1 is not typical. In fact, it is exceptional both in that the $x_n(t)$ turned out to be partial sums of a power series and in that the sum of that series could be written in closed form. However, the $x_n(t)$ are *always* the partial sums of *some* series. Specifically, the series

$$\sum_{n=1}^{\infty} [x_{n+1}(t) - x_n(t)] = [x_2(t) - x_1(t)] + [x_3(t) - x_2(t)] + \cdots + [x_{n+1}(t) - x_n(t)] + \cdots$$

$$= x_2(t) + [x_3(t) - x_2(t)] + \cdots + [x_{n+1}(t) - x_n(t)] + \cdots \quad [12]$$

has $x_2(t), x_3(t), \ldots, x_{n+1}(t), \ldots$ as its partial sums. This may seem like an artificial point of view at first, but it has the advantage that one can show that the sequence $(x_n(t))$ converges by showing that the series $\sum [x_{n+1}(t) - x_n(t)]$ converges; to do so, we have the convergence tests from calculus at our disposal. Since we don't expect explicit formulas for the $x_n(t)$ in general, a test such as the ratio test, which would involve

---

[12] Recall that $x_1(t) = 0$ for all t.

$\lim_{n\to\infty} \left| \dfrac{x_{n+1}(t) - x_n(t)}{x_n(t) - x_{n-1}(t)} \right|$, seems hopeless. On the other hand, if we could get a reasonable estimate for $|x_{n+1}(t) - x_n(t)|$, we might be able to show using the comparison test that $\sum [x_{n+1}(t) - x_n(t)]$ converges. So let's look at the difference $x_{n+1}(t) - x_n(t)$.

Since (for $n \geq 2$) we have $x_n(t) = \int_0^t f(s, x_{n-1}(s))\,ds$ as well as the similar expression for $x_{n+1}(t)$, we get

$$x_{n+1}(t) - x_n(t) = \int_0^t [f(s, x_n(s)) - f(s, x_{n-1}(s))]\,ds,$$

and so

$$|x_{n+1}(t) - x_n(t)| \leq \int_0^t |f(s, x_n(s)) - f(s, x_{n-1}(s))|\,ds.\ ^{13}$$

For any particular value of s, we can think of the expression $f(s, x_n(s)) - f(s, x_{n-1}(s))$ whose absolute value occurs in the integral as $G(x_n(s)) - G(x_{n-1}(s))$, where $G(x) = f(s, x)$. By the mean value theorem from calculus, this can be rewritten as $G'(\xi) \cdot (x_n(s) - x_{n-1}(s))$ for some ξ between $x_{n-1}(s)$ and $x_n(s)$.[14] (See Exercise 8 for details on this and the following statements.) On the other hand, $G'(x) = \dfrac{\partial f}{\partial x}(s, x)$, and since $\dfrac{\partial f}{\partial x}$ is defined and continuous on the rectangle, there exists a number K such that $|G'(\xi)| \leq K$ for all s. We then have the estimate

$$|x_{n+1}(t) - x_n(t)| \leq \int_0^t |f(s, x_n(s)) - f(s, x_{n-1}(s))|\,ds \qquad \text{(from above)}$$

$$= \int_0^t |G'(\xi)| \cdot |x_n(s) - x_{n-1}(s)|\,ds$$

$$|x_{n+1}(t) - x_n(t)| \leq K \int_0^t |x_n(s) - x_{n-1}(s)|\,ds. \qquad (*)$$

Using inequality $(*)$, we can estimate $|x_{n+1}(t) - x_n(t)|$ as a function of t if we already have an estimate for $|x_n(s) - x_{n-1}(s)|$ as a function of s. In other words, in the infinite series of functions $\sum (x_{n+1} - x_n)$, we can (loosely speaking) get information about the size of each term from the previous one. We'll soon see how this works in practice.

To get started, note that $x_2(t) - x_1(t) = x_2(t) = \int_0^t f(s, 0)\,ds$, so

$$|x_2(t) - x_1(t)| \leq \int_0^t |f(s, 0)|\,ds.\ ^{15}$$

---

[13] Actually, we're assuming here not only that t is in the interval being considered, but also that $t \geq 0$. If $t < 0$, the integral on the right is negative and has to be replaced by its absolute value; see Exercise 13.

[14] Since ξ may depend on s, it would be more accurate to write $\xi(s)$ for ξ.

[15] Again, assuming $t \geq 0$; see Exercise 13 for the case $t < 0$.

Since f is defined and continuous on the rectangle, so is $|f|$, and so $|f|$ has a maximum value M there. In particular, $|f(s, 0)| \leq M$ for all s between 0 and t, so we have

$$|x_2(t) - x_1(t)| \leq \int_0^t M\, ds = Mt. \quad (**)$$

Now we can start using inequality (*) from p. 565. For $n = 2$, (*) becomes

$$|x_3(t) - x_2(t)| \leq K \int_0^t |x_2(s) - x_1(s)|\, ds.$$

Replacing t by s in (**), we have $|x_2(s) - x_1(s)| \leq Ms$, and therefore

$$|x_3(t) - x_2(t)| \leq K \int_0^t Ms\, ds = \frac{1}{2} KMt^2.$$

Once we have this, (*) for $n = 3$ yields

$$|x_4(t) - x_3(t)| \leq K \int_0^t |x_3(s) - x_2(s)|\, ds \leq K \int_0^t \frac{1}{2} KMs^2\, ds = \frac{K^2 M}{3 \cdot 2} t^3.$$

Proceeding in this way, we get

$$|x_{n+1}(t) - x_n(t)| \leq \frac{K^{n-1} M}{n!} t^n \quad \text{[see Exercise 14(a)]}.$$

This estimate is good enough to show by the comparison test that $\sum [x_{n+1}(t) - x_n(t)]$ converges [Exercise 14(c)]. Thus the sequence $(x_n(t))$ converges for all t in the interval being considered [that is, all t for which there are points (t, x) in the rectangle]. In fact, the estimate is even good enough to show that the limit function L given by $L(t) = \lim_{n \to \infty} x_n(t)$ is continuous. However, this is probably done most easily by using a concept, known as "uniform convergence," from advanced calculus, which is a bit beyond the scope of this book.

Once it is known that L is continuous, we can ask whether L is a solution of the integral equation $x = F(x)$, that is, whether $L(t) = \int_0^t f(s, L(s))\, ds$ for all t. (Note that if L was not continuous, the integral on the right might be undefined.) It's natural to start with the recurrence relation

$$x_{n+1}(t) = \int_0^t f(s, x_n(s))\, ds$$

and take limits (as $n \to \infty$) on both sides; this yields

$$L(t) = \lim_{n \to \infty} \int_0^t f(s, x_n(s))\, ds.$$

If we assume for the moment that we can interchange the limit and integral signs, we get

$$L(t) = \int_0^t \lim_{n \to \infty} f(s, x_n(s))\, ds.$$

SEC. 7.3 SUCCESSIVE APPROXIMATION

Since f is continuous, we know that

$$\lim_{n\to\infty} f(s, x_n(s)) = f(s, \lim_{n\to\infty} x_n(s)) = f(s, L(s)),$$

and we can conclude that

$$L(t) = \int_0^t f(s, L(s))\, ds.$$

As we had seen earlier, L being a solution of this integral equation is equivalent to L being a solution of the original initial value problem, so we have finally shown that the initial value problem $\dfrac{dx}{dt} = f(t, x)$, $x(0) = 0$ has a solution.

Hold on, though. How do we know that we can interchange the limit and integral signs? Exercise 15 gives an example of a similar-looking situation in which such an interchange is not allowed. Once again, uniform convergence is helpful in showing that the interchange is valid in our case, and we will not do so here.

Now that we have considered the "existence" part of the Existence and Uniqueness Theorem, how about "uniqueness"? We would like to show that if L_1 and L_2 are both solutions of the integral equation $x = F(x)$, then $L_1 = L_2$. First of all, an argument similar to our proof of (*) (see Exercise 16) shows that if $L_1 = F(L_1)$ and $L_2 = F(L_2)$, then

$$|L_1(t) - L_2(t)| \le K \int_0^t |L_1(s) - L_2(s)|\, ds \quad \text{``for all } t\text{''}$$

(actually, for all $t \ge 0$ in the interval being considered). If we abbreviate the absolute value of the difference between the functions L_1 and L_2 by Δ, so that

$$\Delta(t) = |L_1(t) - L_2(t)|,$$

then we have

$$0 \le \Delta(t) \le K \int_0^t \Delta(s)\, ds,$$

and we would like to conclude from this that $\Delta(t) = 0$ for all t. An argument similar to the one on p. 566 [see Exercise 17(a)] shows that for some constant N, $\Delta(t) \le \dfrac{K^n N}{n!} t^n$ for all n. Since $\dfrac{K^n N}{n!} t^n \to 0$ as $n \to \infty$ [see Exercise 17(c)], $\Delta(t) = 0$, and we are done.

EXERCISES

1. Let x be a positive real number.
   (a) Show that if $x^2 < 5$, then $\left(\dfrac{5}{x}\right)^2 > 5$.
   (b) Show that if $x^2 > 5$, then $\left(\dfrac{5}{x}\right)^2 < 5$.

(c) What happens if $x^2 = 5$?
(d) What happens if x is allowed to be negative?

2. (a) Show that $[\frac{1}{2}(x + \frac{5}{x})]^2 - 5 = \frac{1}{4}\frac{(x^2-5)^2}{x^2}$.

(b) Show that if $x^2 \neq 5$ and $4 \leq x^2 \leq 6$, then $[\frac{1}{2}(x + \frac{5}{x})]^2$ is closer to 5 than x^2. What happens if $x^2 = 5$?

3. (a) Show that if x is a positive real number and $x^2 > 5$, then $x > \frac{1}{2}(x + \frac{5}{x})$.

*(b) Show that if a is any positive real number with $a^2 \neq 5$ and we define a sequence (x_n) by $x_1 = a$, $x_{n+1} = \frac{1}{2}(x_n + \frac{5}{x_n})$, then the sequence will be decreasing after the first step. [Use Exercises 2(a) and 3(a).]

*4. If you know Newton's method (from calculus) for approximating a zero of a function, show that applying this method with starting point $x_1 = 2$ to the function g defined by $g(x) = x^2 - 5$ yields the same sequence (x_n) as was found on p. 560.

5. Show that $\frac{dx}{dt} = f(t, x)$, $x(t_0) = x_0$ is equivalent to $\frac{dX}{dt} = g(T, X)$, $X(0) = 0$, where $X = x - x_0$, $T = t - t_0$, $g(T, X) = f(T + t_0, X + x_0)$. Also show that if f and $\frac{\partial f}{\partial x}$ are defined and continuous on a rectangle containing (t_0, x_0) in its interior, then g and $\frac{\partial g}{\partial X}$ are defined and continuous on a rectangle containing $(0, 0)$ in its interior.

6. Show that the integral equation $x = F(x)$, where $F(x)(t) = \int_0^t f(s, x(s))\, ds$, is equivalent to the initial value problem $\frac{dx}{dt} = f(t, x)$, $x(0) = 0$, provided f is continuous.
[*Hint*: If x is a continuous function that is a solution to the integral equation, then since f is continuous, $f(s, x(s))$ is a continuous function of s. $\frac{dF(x(t))}{dt}$ can then be found using the Fundamental Theorem of Calculus.]

*7. Suppose that f and $\frac{\partial f}{\partial x}$ are defined and continuous on a rectangle containing $(0, 0)$ in its interior and that we define a sequence (x_n) of functions as on p. 562.
(a) Show that there is a rectangle of the form $|t| \leq a$, $|x| \leq b$ (with $a, b > 0$) where f and $\frac{\partial f}{\partial x}$ are defined and continuous.
(b) Show that $|f(t, x)|$ has a maximum value, say M, on this rectangle.

(c) Show that if, along with f, the function x_n is continuous on this rectangle, then $|x_{n+1}(t)| \leq M|t|$ for all t with $|t| \leq a$. (Use the recurrence relation.)

(d) Show that if $|s| \leq \dfrac{b}{M}$ and $|s| \leq a$, then $(s, x_{n+1}(s))$ is in the rectangle from part (a). What happens if $M = 0$?

(e) Let a_1 be the minimum (the smaller) of $\dfrac{b}{M}$ and a. Show that on the interval $|t| \leq a_1$, all the x_n are differentiable functions with the property that
$$\frac{dx_{n+1}(t)}{dt} = f(t, x_n(t)).$$

*8. Suppose that f and $\dfrac{\partial f}{\partial x}$ are defined and continuous on a rectangle containing $(0, 0)$ in its interior. Suppose also that a sequence of functions has been defined as on p. 562, and that the rectangle has been adjusted if necessary (see p. 563 and Exercise 7) so that "all" points $(s, x_n(s))$ are within the rectangle.

(a) Show that for any "reasonable" value of s, the function G of one variable defined by $G(x) = f(s, x)$ is differentiable on a "reasonable" interval for x, with derivative $\dfrac{\partial f}{\partial x}(s, x)$. Make this statement precise by defining what the word "reasonable" means in each of the two places where it occurs.

(b) Show that for any "reasonable" value of s, there exists a number ξ (which will depend on s) between $x_{n-1}(s)$ and $x_n(s)$ for which
$$f(s, x_n(s)) - f(s, x_{n-1}(s)) = G'(\xi) \cdot (x_n(s) - x_{n-1}(s)).$$

(c) Show that $\left|\dfrac{\partial f}{\partial x}\right|$ has a maximum value, say K, on the rectangle.

(d) Show that for all "reasonable" s,
$$|f(s, x_n(s)) - f(s, x_{n-1}(s))| \leq K |x_n(s) - x_{n-1}(s)|.$$

9. Figure 7.3.1 illustrates a sequence of continuous functions whose limit is not continuous. For example, the function f_1 can be defined explicitly by
$$f_1(x) = \begin{cases} 0 & \text{if } x < 0 \\ x & \text{if } 0 \leq x \leq 1 \\ 1 & \text{if } x > 1. \end{cases}$$

(a) Give similar explicit expressions for $f_2, f_3,$ and f_4.
(b) Give an explicit expression for f_n.
(c) Using part (b), find an explicit expression for $L = \lim_{n \to \infty} f_n$, and show that although L is defined for all x, L is not continuous everywhere.

10. Define a sequence (f_n) of functions by $f_n(x) = \arctan(nx)$. Note that the f_n are differentiable for all x.
(a) Sketch the graphs of $f_1, f_2,$ and f_3.

(b) Use your graphs to predict what the graph of $L = \lim_{n \to \infty} f_n$ will look like.

(c) Find an explicit expression for $L = \lim_{n \to \infty} f_n$ and prove that your expression is correct.

(*Note*: As in Exercise 9, the expression will consist of separate formulas, valid on different intervals.)

(d) Show that although the f_n are differentiable for all x and L is defined for all x, L is not continuous, let alone differentiable, everywhere.

11. (a) In Example 7.3.1, compute $x_5(t)$ and $x_6(t)$.
   *(b) Conjecture a general formula for $x_n(t)$ on the basis of your results from part (a) and the expressions for $x_2(t)$, $x_3(t)$, $x_4(t)$ found on p. 562.
   [*Hint*: Which powers of t occur? What are their coefficients? It may be easier to find the pattern if you divide all coefficients by 2.]
   (c) Prove your conjecture from part (b) by mathematical induction (on n).

12. (a) Show that the series $\sum_{k=1}^{\infty} 2\dfrac{t^{3k}}{k!}$ converges for all t.

   *(b) Find an expression for the sum $\sum_{k=1}^{\infty} 2\dfrac{t^{3k}}{k!}$ in closed form.

   [*Hint*: Recall that $e^t = \sum_{k=0}^{\infty} \dfrac{t^k}{k!}$. Notice that this summation, unlike the given one, starts at $k = 0$.]

   (c) Use your result from part (b) to check that the function L which is given by

   $$L(t) = \sum_{k=1}^{\infty} 2\frac{t^{3k}}{k!}$$

   is the solution to $\dfrac{dx}{dt} = 3t^2(x + 2)$, $x(0) = 0$.

13. Adapt the arguments on pages 565 and 566 to the case of negative t.

14. (a) Show by induction on n that in the situation on p. 566,

    $$|x_{n+1}(t) - x_n(t)| \leq \frac{K^{n-1}M}{n!}t^n \qquad (t \geq 0).$$

    (b) Show that for $t < 0$ we still have $|x_{n+1}(t) - x_n(t)| \leq \dfrac{K^{n-1}M}{n!}|t|^n$.

    (c) Show that $\sum [x_{n+1}(t) - x_n(t)]$ converges.

    [*Hint*: $\dfrac{K^{n-1}M}{n!}|t|^n = \dfrac{M}{K}\dfrac{(K|t|)^n}{n!}$; $\sum_{n=0}^{\infty} \dfrac{x^n}{n!}$, the Taylor series for e^x, converges for all real x.]

15. (a) Show that for any x with $0 \leq x \leq 1$, $\lim_{n \to \infty} \dfrac{2nx}{(1 + nx^2)^2} = 0$.

(b) Evaluate $\displaystyle\int_0^1 \dfrac{2nx}{(1 + nx^2)^2} \, dx$.

(c) Show that $\displaystyle\lim_{n \to \infty} \int_0^1 \dfrac{2nx}{(1 + nx^2)^2} \, dx \neq \int_0^1 \left[\lim_{n \to \infty} \dfrac{2nx}{(1 + nx^2)^2} \right] dx$.

16. In the situation of p. 567, show that if $L_1 = F(L_1)$ and $L_2 = F(L_2)$, then for all $t \geq 0$ in the interval being considered we have

$$|L_1(t) - L_2(t)| \leq K \int_0^t |L_1(s) - L_2(s)| \, ds \, .$$

17. (Compare Exercise 14.)

(a) In the situation of p. 567, show that there is a constant N such that

$$\Delta(t) \leq \dfrac{K^n N}{n!} t^n \text{ for all } n \ (t \geq 0).$$

[*Hint*: The argument is very similar to the one on p. 566. Try to get an idea first of what N might be, then use induction on n.]

(b) Show that for $t < 0$ we still have $\Delta(t) \leq \dfrac{K^n N}{n!} |t|^n$.

(c) Show that $\Delta(t) = 0$ for all t in the interval being considered.

[*Hint*: For all real x, since $\displaystyle\sum_{n=0}^{\infty} \dfrac{x^n}{n!}$ converges, $\dfrac{x^n}{n!} \to 0$ as $n \to \infty$.]

FURTHER READING

For more on Babylonian estimates for $\sqrt{2}$, see Section 3.3 in Resnikoff and Wells, *Mathematics in Civilization* (Holt, Rinehart and Winston, 1973).

7.4 PARTIAL DIFFERENTIAL EQUATIONS

In this section we will consider some partial differential equations, in particular the wave and Schrödinger equations, and see how they give rise to ordinary differential equations such as the Hermite equation. Recall that a partial differential equation involves one or more partial derivatives of an unknown function u of several variables; we'll consider only the case of a function $u(x, t)$ of two variables. In applications, the variable x is often interpreted as a position, while t represents the time.

EXAMPLE 7.4.1 If a flexible string (such as a violin string) that stretches along the x-axis is vibrating in a vertical plane, then the vertical displacement of the spring at position x and time t is given by a function $u(x, t)$ (see Figure 7.4.1). It can be shown that if there are no external forces on the string (that is, gravity, air resistance, etc., are neglected), the displacement will satisfy the differential equation $\dfrac{\partial^2 u}{\partial x^2} - k\dfrac{\partial^2 u}{\partial t^2} = 0$, where k is a positive constant. This equation is known as the **wave equation**.

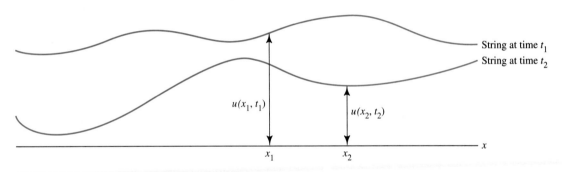

Figure 7.4.1. Vibrating string.

On the other hand, if there is a vertical force $F(x, t)$ per unit length on the string, the differential equation becomes

$$\frac{\partial^2 u}{\partial x^2} - k\frac{\partial^2 u}{\partial t^2} = \frac{1}{\rho} F(x, t).$$

Here ρ is a constant (the mass density of the string). To confuse matters, this equation is also referred to as the **wave equation**. Both versions (homogeneous and nonhomogeneous) of the wave equation occur frequently in mathematical physics. ∎

Just like an ordinary differential equation (see Section 1.4), a partial differential equation is called **linear** if it is linear in the unknown function and its derivatives (although not

SEC. 7.4 PARTIAL DIFFERENTIAL EQUATIONS

necessarily in the variables). For instance, each wave equation is linear, but the equation $u\dfrac{\partial^2 u}{\partial x^2} - k\dfrac{\partial^2 u}{\partial t^2} = 0$ is not. We will consider only *homogeneous linear* PDE; for these, any linear combination of solutions is again a solution.

As we have seen, for an nth-order homogeneous linear *ordinary* differential equation there are n basic solutions such that any solution is a linear combination of these. For partial differential equations, even very simple ones, there are many more solutions.

EXAMPLE 7.4.2 $\dfrac{\partial u}{\partial x} = 0$.

Remember that u is assumed to be a function of x and t; we are looking for such functions whose derivative with respect to x is zero, that is, which don't "really" depend on x. Any function of t alone will be a solution: If $u = f(t)$, then $\dfrac{\partial u}{\partial x} = 0$. ∎

EXAMPLE 7.4.3 $\dfrac{\partial^2 u}{\partial x^2} - \dfrac{\partial^2 u}{\partial t^2} = 0$.

This is the case $k = 1$ of the homogeneous wave equation from Example 7.4.1. Here are a few solutions:

$$u = a + bx + ct + dxt, \quad \text{where } a, b, c, \text{ and } d \text{ are arbitrary constants;}$$

$$u = \sin(ax)\cos(at), \quad \text{where } a \text{ is an arbitrary constant;}$$

$$u = \frac{1}{x+t}.$$

See also Exercise 1. ∎

Because partial differential equations have so many solutions, it is very important in practice to have additional information about the value of the unknown function or its derivatives for specific x, t. If such information is available, one has a **boundary value problem** if the side conditions are for specific x, an **initial value problem** if they are for one specific time t_0, and a mixed initial value/boundary value problem otherwise.

EXAMPLE 7.4.4 If the ends of a vibrating string are fixed on the x-axis at $x = 0$ and at $x = 2$ and there is no external force on the string, we have the boundary value problem

$$\frac{\partial^2 u}{\partial x^2} - k\frac{\partial^2 u}{\partial t^2} = 0, \quad u(0, t) = 0, \quad u(2, t) = 0.$$

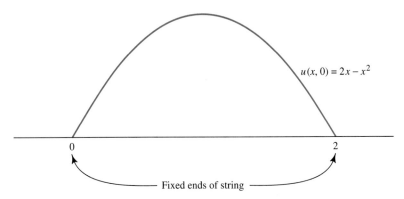

Figure 7.4.2. Initial position of vibrating string.

If, in addition, we know that the string was released at $t = 0$ and that its displacement at that time was given by $2x - x^2$ (see Figure 7.4.2), we have

$$\frac{\partial^2 u}{\partial x^2} - k\frac{\partial^2 u}{\partial t^2} = 0, \quad u(0, t) = 0, \quad u(2, t) = 0, \quad u(x, 0) = 2x - x^2.$$

Finally, if we know that the string was released with velocity zero, then we also have $\frac{\partial u}{\partial t}(x, 0) = 0$. With this much information, it is physically reasonable that one should be able to predict the future motion of the string, and this can be shown to be true: The initial value/boundary value problem

$$\frac{\partial^2 u}{\partial x^2} - k\frac{\partial^2 u}{\partial t^2} = 0, \quad u(0, t) = 0, \quad u(2, t) = 0, \quad u(x, 0) = 2x - x^2, \quad \frac{\partial u}{\partial t}(x, 0) = 0$$

has a unique solution. ☐

If we were solving an initial value problem for an ordinary differential equation, we would usually find the most general solution to the equation first and then choose the integration constants so that the initial conditions would be satisfied. This method usually does *not* work well for initial value/boundary value problems involving partial differential equations, because there are so many solutions that it is often hard to find them all. Instead, one tries to find a limited set of "basic" solutions whose properties make it likely that they can be combined to give a solution to the particular given problem. In particular, if the boundary conditions or initial conditions are changed, the type of basic solutions used may need to be changed. Even so, there will usually be infinitely many basic solutions, which are combined into "infinite linear combinations" using infinite series or even integrals.

EXAMPLE 7.4.5 Suppose that we had found the solutions $u = \sin(ax)\cos(at)$, where a is an arbitrary constant, to the equation $\dfrac{\partial^2 u}{\partial x^2} - \dfrac{\partial^2 u}{\partial t^2} = 0$. We could then combine a finite number of them to get a solution such as

$$u = 2\sin(x)\cos(t) + 4\sin(3x)\cos(3t) - \frac{1}{2}\sin(5x)\cos(5t).$$

We could also combine infinitely many of them to get, for example,

$$u = \sin(x)\cos(t) + \frac{1}{2}\sin(2x)\cos(2t) + \frac{1}{4}\sin(3x)\cos(3t) + \cdots$$

$$= \sum_{n=1}^{\infty} \frac{1}{2^{n-1}} \sin(nx)\cos(nt).$$

It is easy to show that this particular series converges for all x, t and not too hard to show that the result is a function $u(x, t)$ that satisfies our differential equation (in particular, the second partial derivatives of u exist).

Finally, we could combine *all* the solutions with $0 \leq a \leq 1$, say, by taking an integral over variable a. For example, we could take

$$u = \int_0^1 \sin(ax)\cos(at)\,da = \frac{1}{t^2 - x^2}\left[t\sin(ax)\sin(at) + x\cos(ax)\cos(at)\right]_{a=0}^{a=1}$$

$$= \frac{t\sin x \sin t + x\cos x \cos t - x}{t^2 - x^2};$$

however, we might just as well take a different "continuous linear combination" such as
$u = \displaystyle\int_0^1 \frac{\sin(ax)\cos(at)}{a + 2}\,da$, in which case we would have to leave $u(x, t)$ in integral form. ∎

Suppose now that we had the initial value/boundary value problem

$$\frac{\partial^2 u}{\partial x^2} - \frac{\partial^2 u}{\partial t^2} = 0, \quad u(0, t) = 0, \quad u(2, t) = 0, \quad u(x, 0) = 2x - x^2, \quad \frac{\partial u}{\partial t}(x, 0) = 0$$

from Example 7.4.4. Among the solutions $\sin(ax)\cos(at)$ that we are using, there are some that satisfy at least the boundary conditions $u(0, t) = 0$, $u(2, t) = 0$. To find out which, note that for $u = \sin(ax)\cos(at)$, we have $u(0, t) = \sin(0)\cos(at) = 0$ for any a, along with $u(2, t) = \sin(2a)\cos(at)$. We see that to get $u(2, t) = 0$ (for all t) we need

$\sin(2a) = 0$; that is, $2a$ should be a multiple of π. This leads us to take

$$a = \frac{n\pi}{2}, \quad n = 1, 2, \ldots,{}^{16}$$

and we find the solutions

$$u = \sin\frac{n\pi x}{2}\cos\frac{n\pi t}{2} \quad \text{with} \quad u(0, t) = u(2, t) = 0.$$

Any linear combination (finite or infinite) of these will still satisfy $u(0, t) = u(2, t) = 0$, so it is reasonable to try to find such a combination which also satisfies $u(x, 0) = 2x - x^2$, $\frac{\partial u}{\partial t}(x, 0) = 0$. Specifically, let's try to find coefficients c_n such that the series

$$u(x, t) = \sum_{n=1}^{\infty} c_n \sin\frac{n\pi x}{2}\cos\frac{n\pi t}{2}$$

yields a function which satisfies these initial conditions. This will work (if we ignore matters of convergence, differentiability of the series, etc.) provided

$$\sum_{n=1}^{\infty} c_n \sin\frac{n\pi x}{2} = 2x - x^2$$

(see Exercise 3). The problem finally becomes, then, to find coefficients c_n such that $\sum c_n \sin\frac{n\pi x}{2} = 2x - x^2$. More generally, if the initial position of the string was given by a function $u(x, 0) = f(x)$ [with $f(0) = f(2) = 0$, because the ends are fixed], the problem would be to find coefficients c_n such that $f(x) = \sum_{n=1}^{\infty} c_n \sin\frac{n\pi x}{2}$.

Series such as $\sum_{n=1}^{\infty} c_n \sin\frac{n\pi x}{2}$, $\sum_{n=0}^{\infty} d_n \cos\frac{n\pi x}{2} = d_0 + \sum_{n=1}^{\infty} d_n \cos\frac{n\pi x}{2}$, and more generally $d_0 + \sum_{n=1}^{\infty} (c_n \sin\frac{n\pi x}{L} + d_n \cos\frac{n\pi x}{L})$ with constant L, are called **trigonometric series** or **Fourier series**.[17] They are basic tools in much of mathematics and physics; a surprising range of functions can be represented by them. For instance, it turns out that *any* continuous function $f(x)$ defined on the interval $0 \le x \le 2$ and for which $f(0) = f(2) = 0$ can be written on that interval as the sum of a "sine series":

$$f(x) = \sum_{n=1}^{\infty} c_n \sin\frac{n\pi x}{2} \quad (0 \le x \le 2).$$

[16] For negative n we'll get essentially the same solutions. (Why?)

[17] After the French mathematician Fourier (1768–1830), who used such series extensively in his investigations of heat conduction.

Given $f(x)$, the coefficients c_n can (in principle) be found by computing certain integrals. See Exercise 4 for more on this. In particular, this can be done for $f(x) = 2x - x^2$, and in Exercise 5 you are asked to show that the solution of the initial value/boundary value problem posed above will be

$$u(x, t) = \sum_{k=0}^{\infty} \frac{32}{(2k+1)^3 \pi^3} \sin \frac{(2k+1)\pi x}{2} \cos \frac{(2k+1)\pi t}{2}.$$

Now that we have seen that the special solutions $u = \sin(ax)\cos(at)$ to the wave equation can be very helpful, you may wonder how these solutions were found. The trick is to look for solutions of the special form $u(x, t) = F(x)G(t)$, that is, solutions that can be written as the product of two functions, one of which depends only on x while the other depends only on t. If we substitute the product $u(x, t) = F(x)G(t)$ into the wave equation $\frac{\partial^2 u}{\partial x^2} - \frac{\partial^2 u}{\partial t^2} = 0$, we get

$$F''(x)G(t) - F(x)G''(t) = 0.$$

Dividing by $F(x)G(t)$ yields

$$\frac{F''(x)}{F(x)} - \frac{G''(t)}{G(t)} = 0,$$

that is,

$$\frac{F''(x)}{F(x)} = \frac{G''(t)}{G(t)}.$$

Note that the left-hand side of this last equation depends only on x (and not on t), while the right-hand side depends on t (and not on x). Therefore, the only way the two sides can actually be equal is for them to be constant. If we call this constant γ, we get

$$\frac{F''(x)}{F(x)} = \gamma \quad \text{and} \quad \frac{G''(t)}{G(t)} = \gamma.$$

At this point we have two ordinary differential equations, one for an unknown function in each of the variables. We can now solve these equations separately (of course, the solutions will depend on γ), and then the solutions $F(x), G(t)$ for any particular value of γ can be multiplied together to yield solutions $u(x, t) = F(x)G(t)$ of our original partial differential equation. Again, we emphasize that only a few of the solutions of the PDE are obtained in this way, but by combining them many others can be found. See Exercise 8 for more details in the case of our particular equation $\frac{\partial^2 u}{\partial x^2} - \frac{\partial^2 u}{\partial t^2} = 0$.

This method of finding solutions to partial differential equations and initial value/boundary value problems is called the method of **separation of variables** (not to be confused with the method with the same name for solving separable first-order ODE; see Section 1.2). It is very commonly used in mathematical physics, and most of the special linear second-order equations (with nonconstant coefficients) studied in Chapter 5 arise in this way from important partial differential equations. As a specific example, we'll now show how an important case of the Schrödinger equation from quantum mechanics gives rise to the Hermite equation studied in Section 5.3.[18]

Consider a particle of mass μ [19] that moves in one dimension. In this situation, the unknown function in the Schrödinger equation is a complex-valued function $u(x, t)$ called the **wave function** of the particle.[20] Roughly speaking, the physical interpretation of this function is that $|u(x, t)|^2$ measures the probability of the particle being at position x at time t.

N O T E : The term *wave function* is not directly related to the wave equation discussed earlier in this section. To make things even more confusing, some authors refer to the Schrödinger equation as the "quantum-mechanical wave equation."

Now suppose that the potential energy of the particle at x is given by the function $V(x)$. Then the partial differential equation satisfied by the wave function is the **Schrödinger equation**

$$i\hbar \frac{\partial u}{\partial t} = -\frac{\hbar^2}{2\mu} \frac{\partial^2 u}{\partial x^2} + V(x)u.$$

Here i is, as usual, the imaginary unit (with $i^2 = -1$), while $\hbar$ is a physical constant ($\hbar = \frac{h}{2\pi}$, where h is Planck's constant).

If we rewrite the equation as

$$i\hbar \frac{\partial u}{\partial t} + \frac{\hbar^2}{2\mu} \frac{\partial^2 u}{\partial x^2} - V(x)u = 0,$$

we see that it is a homogeneous linear equation. Let's use the method of separation of variables, but with the notation traditional in quantum mechanics, and look for solutions of the form $u(x, t) = \psi(x)F(t)$. This yields

$$i\hbar \psi(x) F'(t) + \frac{\hbar^2}{2\mu} \psi''(x) F(t) - V(x) \psi(x) F(t) = 0,$$

and after dividing through by $\psi(x) F(t)$,

$$i\hbar \frac{F'(t)}{F(t)} + \frac{\hbar^2}{2\mu} \frac{\psi''(x)}{\psi(x)} - V(x) = 0.$$

[18] As will be seen from the dates of Hermite (1822–1901) and Schrödinger (1887–1961), this is not how, historically, the Hermite equation first arose! In fact, this is one of several famous examples in which mathematical theory predated applications that in retrospect might be thought of as "motivating" the development of the theory.

[19] In quantum mechanics, the letter m is usually reserved for the magnetic quantum number, so μ is used for the mass.

[20] The usual notation for this function in quantum mechanics is $\psi(x, t)$. However, this tends to create confusion (at first) with the function $\psi(x)$ that will be introduced soon.

SEC. 7.4 PARTIAL DIFFERENTIAL EQUATIONS

The first term depends only on t while the others depend only on x, so there is a constant E for which

$$i\hbar \frac{F'(t)}{F(t)} = E, \quad \frac{\hbar^2}{2\mu} \frac{\psi''(x)}{\psi(x)} - V(x) = -E.$$

The first of these equations is easily solved and yields

$$F(t) = Ce^{-iEt/\hbar}.$$

From the physical interpretation of $|u(x, t)|^2$ one can now see that E must be real unless the particle is "escaping" to ∞ or $-\infty$ (see Exercise 11). However, you may notice that this argument is a bit fishy, because so far we are only looking for special solutions of the form $u(x, t) = \psi(x)F(t)$ of the Schrödinger equation, and it isn't clear that these special solutions will have the same "physical properties" as the solutions one can eventually get by combining them. Rather than pursuing this issue further, we now return to the second equation, which is $\frac{\hbar^2}{2\mu} \frac{\psi''(x)}{\psi(x)} - V(x) = -E$, or, equivalently,

$$\psi''(x) + \frac{2\mu}{\hbar^2}[E - V(x)]\psi(x) = 0,$$

with the understanding that E is a real constant.

Of course, this second-order differential equation (which is sometimes called the **time-independent Schrödinger equation**) depends on the potential energy function $V(x)$. Recall (see p. 169) that if the particle moves under the influence of a force $F(x)$, the potential energy is given by $V(x) = -\int_0^x F(s)\, ds$. We now consider two specific examples.

EXAMPLES   A. For a *free* particle, $F(x) = 0$, so $V(x) = 0$ and we get $\psi''(x) + \frac{2\mu E}{\hbar^2} \psi(x) = 0$. Since $\frac{2\mu E}{\hbar^2}$ is a constant, this equation is easily solved. See Exercise 12.

B. Now suppose that $F(x) = -kx$; then $V(x) = \int_0^x ks\, ds = \frac{1}{2}kx^2$. This situation is familiar from classical mechanics, where it occurs for a particle on a spring (see Sections 2.1 and 2.7). Recall that for this (undamped) case we found that the particle oscillates with period $2\pi\sqrt{\frac{\mu}{k}}$; that is, the position function has the form $x = C_1 \sin(\omega t + C_2)$ with $\omega = \sqrt{\frac{k}{\mu}}$. It is traditional to maintain the notation $\omega = \sqrt{\frac{k}{\mu}}$ in the quantum-mechanical case. Accordingly, we'll rewrite the constant k as $k = \mu\omega^2$, which makes the potential energy $V(x) = \frac{1}{2}\mu\omega^2 x^2$ and our differential equation

$$\psi''(x) + \frac{2\mu}{\hbar^2}(E - \frac{1}{2}\mu\omega^2 x^2)\psi(x) = 0. \quad \blacksquare$$

Example B is quite important; the potential function $V(x) = \frac{1}{2}kx^2 = \frac{1}{2}\mu\omega^2 x^2$ is the easiest possible type of function which has a minimum (at $x = 0$), and it can be used to approximate other, more complicated, such potential functions. A particle for which $V(x) = \frac{1}{2}\mu\omega^2 x^2$ is called a **harmonic oscillator**. We'll now show that the equation

$$\psi''(x) + \frac{2\mu}{\hbar^2}\left(E - \frac{1}{2}\mu\omega^2 x^2\right)\psi(x) = 0$$

for the quantum-mechanical harmonic oscillator is really the Hermite equation in disguise. First we change the variable from x to $\xi = \sqrt{\frac{\mu\omega}{\hbar}}\, x$; this changes the equation to

$$\frac{d^2\psi}{d\xi^2} + \left(\frac{2E}{\hbar\omega} - \xi^2\right)\psi = 0 \quad \text{(see Exercise 13)}.$$

This can be rewritten

$$\frac{d^2\psi}{d\xi^2} = \left(\xi^2 - \frac{2E}{\hbar\omega}\right)\psi,$$

so we're looking for a function whose second derivative is some (nonconstant) multiple of itself. This may suggest trying for some sort of exponential solution. If we try substituting $\psi(\xi) = e^{f(\xi)}$, we get $f''(\xi) + [f'(\xi)]^2 + (\frac{2E}{\hbar\omega} - \xi^2) = 0$ [see Exercise 14(a)]. This nonlinear equation does not look promising, except if we can arrange things such that both $[f'(\xi)]^2 = \xi^2$ and $f''(\xi) = \frac{-2E}{\hbar\omega}$. $[f'(\xi)]^2 = \xi^2$ implies $f(\xi) = \pm\frac{1}{2}\xi^2 + C$ [Exercise 14(b)]; a physical argument similar to the "fishy" one on p. 579 above shows that we need to take the minus sign if we ever want the particle to be around. If we do try $f(\xi) = -\frac{1}{2}\xi^2 + C$, we find that $f''(\xi) = -1$. So if we are very lucky and $\frac{2E}{\hbar\omega} = 1$, then we do have a solution $\psi(\xi) = e^{-\xi^2/2}$ to our equation $\frac{d^2\psi}{d\xi^2} + (\frac{2E}{\hbar\omega} - \xi^2)\psi = 0$. But what do we do if $\frac{2E}{\hbar\omega} \neq 1$?

The trick used is similar to the method of reduction of order. Since we know that $\psi(\xi) = e^{-\xi^2/2}$ is a solution in one case, let's put

$$\psi(\xi) = e^{-\xi^2/2}\, v(\xi)$$

in general, where $v(\xi)$ is a new unknown function. In the particular case $\frac{2E}{\hbar\omega} = 1$ we know that the simple function $v(\xi) = 1$ will give us a solution, so maybe in general the equation for $v(\xi)$ will turn out to be easier than the one for $\psi(\xi)$. In Exercise 15 you're asked to

SEC. 7.4 PARTIAL DIFFERENTIAL EQUATIONS

show that the result of the substitution $\psi(\xi) = e^{-\xi^2/2} v(\xi)$ is the new equation

$$\frac{d^2v}{d\xi^2} - 2\xi \frac{dv}{d\xi} + \left(\frac{2E}{\hbar\omega} - 1\right)v = 0.$$

This is exactly the Hermite equation for $\lambda = \frac{2E}{\hbar\omega} - 1$.

It turns out that the solutions to the Hermite equation give rise to functions $\psi(\xi) = e^{-\xi^2/2} v(\xi)$ that are physically unacceptable (because they grow as $\xi \to \infty$) *unless* λ is a nonnegative even integer, in which case there is a polynomial solution $v(\xi)$ (see p. 400). Thus the only acceptable values for E are those for which $\frac{2E}{\hbar\omega} - 1 = 0, 2, 4, \ldots$, in other words, for which

$$E = \frac{1}{2}\hbar\omega, \frac{3}{2}\hbar\omega, \frac{5}{2}\hbar\omega, \ldots.$$

It then turns out that these values of E can be interpreted as **energy levels** for the particle; thus the quantum-mechanical model explains the existence of discrete energy levels for, say, the hydrogen atom.

EXERCISES

1. (a) Show that for any number $\alpha \neq 0$, $u = \dfrac{1}{x + \alpha t}$ is a solution of the wave equation $\dfrac{\partial^2 u}{\partial x^2} - k \dfrac{\partial^2 u}{\partial t^2} = 0$ for suitable k. How does k depend on α?

    (b) When is $u = \sin(\alpha x)\cos(\beta t)$ a solution to the (homogeneous) wave equation? For what k?

2. A string is tied to the top of a 3-foot-high post and stretched straight to a point on the ground 10 feet from the bottom of the post, as shown in Figure 7.4.3. In this initial position, the string has velocity 0 everywhere. The free end of the string then starts moving up and down according to the equation $y = 3 - 3 \cos t$. (Note that for $t = 0$, $y = 0$ and $\dfrac{dy}{dt} = 0$, as required by the earlier givens.) Set up an initial value/boundary value problem describing the motion of the string under the assumption that there are no external forces on the string.

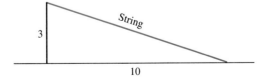

Figure 7.4.3. Initial position of string.

3. Show formally, that is, without worrying about convergence and such, that
$$u(x, t) = \sum_{n=1}^{\infty} c_n \sin \frac{n\pi x}{2} \cos \frac{n\pi t}{2}$$
satisfies the conditions $u(x, 0) = 2x - x^2$,
$$\frac{\partial u}{\partial t}(x, 0) = 0 \text{ provided } \sum_{n=1}^{\infty} c_n \sin \frac{n\pi x}{2} = 2x - x^2.$$

4. (a) Show that for positive integers m, n,
$$\int_0^2 \sin \frac{m\pi x}{2} \sin \frac{n\pi x}{2} \, dx = \begin{cases} 1 & (m = n) \\ 0 & (m \ne n). \end{cases}$$

[*Hint*: Use the identity $\sin p \sin q = \frac{1}{2}[\cos(p - q) - \cos(p + q)]$.]

(b) Show formally that if $f(x) = \sum_{n=1}^{\infty} c_n \sin \frac{n\pi x}{2}$, then for any positive integer m,
$$c_m = \int_0^2 \sin \left(\frac{m\pi x}{2}\right) f(x) \, dx.$$

[*Hint*: Start with the right-hand side.]

5. Assume that, as stated on pp. 576–577, one can write
$$2x - x^2 = \sum_{n=1}^{\infty} c_n \sin \frac{n\pi x}{2} \qquad (0 \le x \le 2).$$

(a) Using Exercise 4(b), show that $c_m = \dfrac{32}{m^3 \pi^3}$ for m odd, $c_m = 0$ for m even.

(Use integration by parts.)

(b) Show that
$$2x - x^2 = \sum_{k=0}^{\infty} \frac{32}{(2k + 1)^3 \pi^3} \sin \frac{(2k + 1)\pi x}{2} \qquad (0 \le x \le 2).$$

(c) Show that the solution to the initial value/boundary value problem from p. 575 (and Example 7.4.4) is
$$u(x, t) = \sum_{k=0}^{\infty} \frac{32}{(2k + 1)^3 \pi^3} \sin \frac{(2k + 1)\pi x}{2} \cos \frac{(2k + 1)\pi t}{2}.$$

(d) By taking $t = 0$ and $x = 1$, show that
$$1 - \frac{1}{3^3} + \frac{1}{5^3} - \frac{1}{7^3} + \cdots = \sum_{k=0}^{\infty} \frac{(-1)^k}{(2k + 1)^3} = \frac{\pi^3}{32}. \quad {}^{21}$$

[21] If you can find a similar result for $1 + \dfrac{1}{3^3} + \dfrac{1}{5^3} + \dfrac{1}{7^3} + \cdots$, mathematical fame (and probably fortune) will be yours.

6. (a) Show that if $f(x) = \sum_{n=1}^{\infty} c_n \sin \dfrac{n\pi x}{2}$ for all x, then $f(x + 4) = f(x)$: The function f is periodic, and 4 is a period of f. Also show that $f(-x) = -f(x)$: The function f is odd.

(b) We saw in Exercise 5 that for $0 \leq x \leq 2$,

$$2x - x^2 = \sum_{k=0}^{\infty} \frac{32}{(2k+1)^3 \pi^3} \sin \frac{(2k+1)\pi x}{2}.$$

Show that the series on the right actually converges for *all* x; let $f(x)$ be the sum of the series. Use part (a) to sketch the graph of this function. (Note that for $0 \leq x \leq 2$, $f(x) = 2x - x^2$, but that outside the interval $[0, 2]$ this will no longer be true.)

7. (a) Show that if $f(x) = d_0 + \sum_{n=1}^{\infty} d_n \cos \dfrac{n\pi x}{2}$ for all x, then $f(x + 4) = f(x)$ and $f(-x) = f(x)$: The function f is even and periodic, with 4 as a period.

It turns out that *any* continuous function f that is even and periodic, with 4 as a period, can be written as the sum of a "cosine series" as above.

*__(b)__ Show that the coefficients in this cosine series are given by

$$d_0 = \frac{1}{2}\int_0^2 f(x)\,dx; \qquad d_m = \int_0^2 \cos \frac{m\pi x}{2} f(x)\,dx \qquad (m > 0).$$

[*Hint*: See Exercise 4 for an analogous situation.]

(c) Show that there is a continuous function f that is even and periodic, with 4 as a period, and such that $f(x) = x^2$ for $0 \leq x \leq 2$. Sketch the graph of this function.

(d) Show that for $0 \leq x \leq 2$,

$$x^2 = \frac{4}{3} + \sum_{n=1}^{\infty} \left[\int_0^2 x^2 \cos \frac{n\pi x}{2}\,dx\right] \cos \frac{n\pi x}{2}.$$

(e) Using integration by parts, show that for $0 \leq x \leq 2$,

$$x^2 = \frac{4}{3} + \frac{16}{\pi^2} \sum_{n=1}^{\infty} \frac{(-1)^n}{n^2} \cos \frac{n\pi x}{2}.$$

(f) By taking $x = 2$ in this formula, show that $\sum_{n=1}^{\infty} \dfrac{1}{n^2} = \dfrac{\pi^2}{6}$.

8. (a) Show that the solutions of $\dfrac{F''(x)}{F(x)} = \gamma$, γ constant, are given by

$$F(x) = C_1 e^{\sqrt{\gamma}x} + C_2 e^{-\sqrt{\gamma}x} \quad (\gamma > 0), \qquad F(x) = C_1 + C_2 x \quad (\gamma = 0),$$
$$F(x) = C_1 \cos(\sqrt{-\gamma}\,x) + C_2 \sin(\sqrt{-\gamma}\,x) \quad (\gamma < 0),$$

with C_1, C_2 arbitrary constants in each case.

(b) Find all solutions of the form $u(x, t) = F(x)G(t)$ to the homogeneous wave equation $\dfrac{\partial^2 u}{\partial x^2} - \dfrac{\partial^2 u}{\partial t^2} = 0$.

(c) Show that the only solutions from part (b) that satisfy the conditions $u(0, t) = u(2, t) = 0$ are of the form $\sin \dfrac{n\pi x}{2} \left(C_1 \cos \dfrac{n\pi t}{2} + C_2 \sin \dfrac{n\pi t}{2} \right)$.

(d) Show that the only solutions from part (c) that also satisfy $\dfrac{\partial u}{\partial t}(x, 0) = 0$ are of the form $C \sin \dfrac{n\pi x}{2} \cos \dfrac{n\pi t}{2}$.

9. The temperature $u(x, t)$ in a thin insulated wire (along the x-axis) without heat sources or sinks can be shown to satisfy the so-called **heat equation**
$$\dfrac{\partial^2 u}{\partial x^2} - \dfrac{\partial u}{\partial t} = 0,$$
provided that suitable units are chosen for x and t.

(a) Find all solutions of the form $u(x, t) = F(x)G(t)$ to the heat equation.

(b) Solve the initial value/boundary value problem $\dfrac{\partial^2 u}{\partial x^2} - \dfrac{\partial u}{\partial t} = 0$, $u(0, t) = 0$, $u(2, t) = 0$, $u(x, 0) = 1 - \cos \pi x$.

[*Hint*: Use Exercise 4(b). In computing the integrals, the identity
$$\sin p \cos q = \dfrac{1}{2} [\sin(p + q) + \sin(p - q)]$$
will help.]

10. The electrical potential $u(x, y)$ in a region of the x,y-plane without electrical charges can be shown to satisfy the (two-dimensional) **Laplace equation**
$$\dfrac{\partial^2 u}{\partial x^2} + \dfrac{\partial^2 u}{\partial y^2} = 0.$$

(a) Find all solutions of the form $u(x, y) = F(x)G(y)$ of the Laplace equation.

(b) Solve the boundary value problem $\dfrac{\partial^2 u}{\partial x^2} + \dfrac{\partial^2 u}{\partial y^2} = 0$, $u(x, 0) = 0$, $u(x, 2) = 0$, $u(0, y) = 0$, $u(2, y) = 2y - y^2$ for the electrical potential in the square $0 \le x \le 2$, $0 \le y \le 2$.

[*Hint*: Show (formally) that any function of the form
$$u(x, y) = \sum_{n=1}^{\infty} a_n (e^{n\pi x/2} - e^{-n\pi x/2}) \sin \dfrac{n\pi y}{2}$$
satisfies the first three boundary conditions. Then use Exercise 4(b) or Exercise 5(a), with x replaced by y, to find the coefficients a_n.]

11. (a) Given that $u(x, t) = \psi(x) F(t)$, that $F(t) = Ce^{-iEt/\hbar}$, and that $|u(x, t)|^2$ "represents" the probability that the particle is at position x at time t, show that E cannot have positive imaginary part.

(b) In the situation in part (a), show that if E has negative imaginary part, then for any given x the probability of finding the particle there decreases exponentially as $t \to \infty$.

(c) A more precise statement about the probability of finding the particle is that for any interval $a \leq x \leq b$, the probability that the particle is within that interval at time t is $\int_a^b |u(x,t)|^2 \, dx$. Explain why $\int_{-\infty}^{\infty} |u(x,t)|^2 \, dx = 1$ for all t.

*(d) Show, using the given from part (c), that if E has negative imaginary part, then for any finite interval the probability of finding the particle within that interval approaches zero as $t \to \infty$. (This means that the particle is "escaping" to ∞ or to $-\infty$.)

12. (a) Assuming that $E > 0$, solve the equation $\psi''(x) + \dfrac{2\mu E}{\hbar^2} \psi(x) = 0$ derived on p. 579 for a free particle.

(b) Still assuming that $E > 0$, show that $|u(x,t)|^2$ does not depend either on x or on t: At any given time, a free particle is just as likely to be anywhere as anywhere else.

*(c) Show that the assumption $E < 0$ leads to physically unacceptable solutions to the Schrödinger equation.

13. (See p. 580.) Show that the change of variable $\xi = \sqrt{\dfrac{\mu \omega}{\hbar}}\, x$ transforms the equation for the quantum-mechanical harmonic oscillator to

$$\frac{d^2\psi}{d\xi^2} + \left(\frac{2E}{\hbar\omega} - \xi^2\right)\psi = 0.$$

14. (See p. 580.)

(a) Show that the substitution $\psi(\xi) = e^{f(\xi)}$ transforms the equation

$$\frac{d^2\psi}{d\xi^2} + \left(\frac{2E}{\hbar\omega} - \xi^2\right)\psi = 0 \quad \text{into} \quad f''(\xi) + [f'(\xi)]^2 + \left(\frac{2E}{\hbar\omega} - \xi^2\right) = 0.$$

(b) Show that $[f'(\xi)]^2 = \xi^2$ implies $f(\xi) = \pm\dfrac{1}{2}\xi^2 + C$.

(c) What is the connection between part (a) of this exercise and Exercise 32 from Section 2.2 (p. 97)?

15. (See pp. 580–581.) Show that the substitution $\psi(\xi) = e^{-\xi^2/2} v(\xi)$ transforms the equation $\dfrac{d^2\psi}{d\xi^2} + \left(\dfrac{2E}{\hbar\omega} - \xi^2\right)\psi = 0$ into $\dfrac{d^2 v}{d\xi^2} - 2\xi\dfrac{dv}{d\xi} + \left(\dfrac{2E}{\hbar\omega} - 1\right)v = 0$.

16. It can be shown that the three-dimensional Laplace equation expressed in spherical coordinates (for which $x = \rho \sin\varphi \cos\theta$, $y = \rho \sin\varphi \sin\theta$, $z = \rho \cos\varphi$) has the form

$$\frac{\partial}{\partial \rho}\left(\rho^2 \frac{\partial u}{\partial \rho}\right) + \frac{1}{\sin\varphi} \frac{\partial}{\partial \varphi}\left(\sin\varphi \frac{\partial u}{\partial \varphi}\right) + \frac{1}{\sin^2\varphi} \frac{\partial^2 u}{\partial \theta^2} = 0.$$

(**Warning:** Although our notation for spherical coordinates is the standard one from most calculus books, in much of the technical literature the names of the spherical angles θ and φ are switched!)

(a) Show that the equation above can be rewritten as

$$\rho^2 \frac{\partial^2 u}{\partial \rho^2} + 2\rho \frac{\partial u}{\partial \rho} + \frac{\partial^2 u}{\partial \varphi^2} + \frac{\cos \varphi}{\sin \varphi} \frac{\partial u}{\partial \varphi} + \frac{1}{\sin^2 \varphi} \frac{\partial^2 u}{\partial \theta^2} = 0.$$

(b) Show that if $u(\rho, \varphi, \theta) = v(\rho, \varphi) w(\theta)$ is a solution to this equation, then there is a constant k such that

$$\frac{\rho^2}{v} \frac{\partial^2 v}{\partial \rho^2} + \frac{2\rho}{v} \frac{\partial v}{\partial \rho} + \frac{1}{v} \frac{\partial^2 v}{\partial \varphi^2} + \frac{\cos \varphi}{v \sin \varphi} \frac{\partial v}{\partial \varphi} = \frac{k}{\sin^2 \varphi}, \quad \frac{1}{w} \frac{\partial^2 w}{\partial \theta^2} = -k.$$

*(c) Explain why, if the function u is to be a solution throughout 3-space, $w(\theta)$ should be a periodic function of θ, and show why k should be the square of an integer. We'll write $k = m^2$.

(d) Show that if $v(\rho, \varphi) = F(\rho) G(\varphi)$ is a solution to the first equation in part (b), then there is a constant q such that

$$\rho^2 F''(\rho) + 2\rho F'(\rho) - q F(\rho) = 0,$$

$$G''(\varphi) + \frac{\cos \varphi}{\sin \varphi} G'(\varphi) + \left(q - \frac{m^2}{\sin^2 \varphi} \right) G(\varphi) = 0.$$

(e) Note that the first equation in part (d) is an Euler equation; show that $F(\rho) = \rho^\nu$ is a solution if $q = \nu(\nu + 1)$.

*(f) Assume that $G(\varphi)$ satisfies the second equation in part (d). Define a function H on $[-1, 1]$ by $H(x) = G(\arccos x)$. (Note that $0 \le \arccos x \le \pi$, as required for the spherical coordinate φ.) Show that H satisfies the **associated Legendre equation**

$$(1 - x^2) H''(x) - 2x H'(x) + \left(q - \frac{m^2}{1 - x^2} \right) H(x) = 0.$$

[*Hint:* $G(\varphi) = H(\cos \varphi)$, so $G'(\varphi) = H'(\cos \varphi) \cdot (-\sin \varphi)$.]
Note that for $q = \nu(\nu + 1)$ [from part (e)] and $m = 0$, this is the "ordinary" Legendre equation from Section 5.3.

17. Consider a uniform heavy chain, or rope, which is suspended from the ceiling, and which is swinging back and forth slightly in a vertical plane. (The chain is different from an "ordinary" pendulum in that not all the mass is concentrated at the free end.) Let $u(y, t)$ be the horizontal displacement of the chain at position y (taking $y = 0$ to be at the free end of the chain) and time t, as shown in Figure 7.4.4. It can be shown that if gravity, with constant acceleration g, is the only outside force on the chain, then the displacement satisfies $y \dfrac{\partial^2 u}{\partial y^2} + \dfrac{\partial u}{\partial y} = \dfrac{1}{g} \dfrac{\partial^2 u}{\partial t^2}$.

(a) Show that if $u(y, t) = F(y) G(t)$ is a solution to this equation, then there is a constant k such that

$$y F''(y) + F'(y) + k F(y) = 0, \qquad G''(t) = -kg G(t).$$

SEC. 7.4 PARTIAL DIFFERENTIAL EQUATIONS

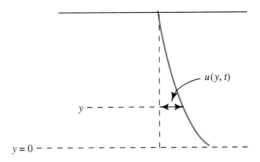

Figure 7.4.4. Suspended chain.

*(b) Assume that $F(y)$ satisfies the first equation in part (a) and that $k \neq 0$. Define a function $f(x)$ by $f(x) = F(\frac{x^2}{4k})$. Show that f satisfies the **Bessel equation of order** 0 (see Example 5.4.3 and Section 5.6):

$$xf''(x) + f'(x) + xf(x) = 0.$$

(c) What can you say about $f(2k\sqrt{L})$, where L is the length of the chain?

(d) Explain why we should expect $f(x)$ to remain bounded as $x \to 0^+$, which rules out most of the solutions to the Bessel equation of order 0 (see p. 436).

18. One might wonder why the equation $\frac{d^2\psi}{d\xi^2} + (\frac{2E}{\hbar\omega} - \xi^2)\psi = 0$ can't be solved more easily by looking for a power series solution $\psi(\xi) = \sum_{n=0}^{\infty} b_n \xi^n$.

    (a) Find the recurrence relation for the b_n.
    (b) Compare your result from part (a) with the recurrence relation for the Hermite equation (found on p. 398). Why is the latter preferable?

19. Show that one can get all solutions of the nonhomogeneous wave equation $\frac{\partial^2 u}{\partial x^2} - k\frac{\partial^2 u}{\partial t^2} = \frac{1}{\rho} F(x, t)$ by adding one fixed ("particular") solution of this equation to each of the solutions of the homogeneous wave equation $\frac{\partial^2 u}{\partial x^2} - k\frac{\partial^2 u}{\partial t^2} = 0$.

    [*Hint:* Compare Section 1.6, Exercise 23 and Section 2.6, Exercise 37.]

FURTHER READING

For more information on Fourier series and related topics, see Seeley, *An Introduction to Fourier Series and Integrals* (Benjamin, 1966) or Walker, *Fourier Analysis* (Oxford University Press, 1988).

APPENDIX A

Complex Numbers

A.1 INTRODUCTION

Although we are now used to having all the real numbers at our disposal, this has not always been the case—either in the history of mathematics or in our own mathematical development. We started with just the natural numbers 1, 2, 3, ..., which are needed for counting. This number system was expanded several times when it proved too limited to allow certain computations or measurements to be performed. For instance, negative numbers were introduced to make subtraction possible, fractions were needed for division, and square roots were required for geometric measurements. Finally, to enable the measurement of any length along a straight line, the real numbers were introduced.

While the real number system is satisfactory for most computations, it is sometimes unpleasant that not all quadratic equations can be solved in the system. To give the simplest example, $z^2 + 1 = 0$ has no solution. We will now remedy this—what could be more natural?—by expanding the number system still further. Fortunately, it will then turn out that *all* quadratic equations in the new system can be solved; the new number system, to be called the system of **complex numbers**, will also have many other convenient properties.

So: We wish to construct a number system that contains the real number system $\mathbb{R}$, in which addition, subtraction, multiplication, and division are defined consistent with addition, and so forth, in $\mathbb{R}$, and in which we can solve the quadratic equation $z^2 + 1 = 0$. Of course, we also want addition, and so forth, in this new number system to satisfy the usual laws of arithmetic.

To find such a number system, we fix a solution i to the equation above. That is, i is a "new number" with $i^2 + 1 = 0$; in other words, $i^2 = -1$.

If a and b are arbitrary real numbers, then we'll be forced to introduce a "new number" $a + b \cdot i$ as well, because we still want to be able to add and multiply in our new number system. However, once all the numbers $a + bi$ have been introduced, we have no need for more new numbers:

$$(a + bi) + (c + di) = (a + c) + (b + d)i;$$

$$(a + bi) - (c + di) = (a - c) + (b - d)i;$$

$$(a + bi) \cdot (c + di) = ac + (bc + ad)i + bdi^2 = (ac - bd) + (bc + ad)i;$$

$$\frac{a + bi}{c + di} = \frac{(a + bi)(c - di)}{(c + di)(c - di)} = \frac{(ac + bd) + (bc - ad)i}{c^2 + d^2}$$

$$= \frac{ac + bd}{c^2 + d^2} + \frac{bc - ad}{c^2 + d^2} i \quad \text{(provided } c \text{ and } d \text{ are not both 0)}.$$

It turns out that addition, and so forth, of these new numbers still satisfy the laws of arithmetic (commutativity, associativity, distributivity, etc.). The "old" real numbers are included in the new numbers, because for any real number a, $a = a + 0 \cdot i$.

Definition An expression $a + bi$, where a and b are real numbers, is called a **complex number**. a is called the **real part** and b is called the **imaginary part** of the complex number $z = a + bi$. [Notation: $a = \text{Re}(z); b = \text{Im}(z)$.]

We have addition, multiplication, and so forth, of complex numbers as indicated above. By our definition, two complex numbers are equal if and only if their real parts as well as their imaginary parts are equal.

The set of all complex numbers is denoted by $\mathbb{C}$.

NOTE: You may wonder why the notation $\sqrt{-1}$ isn't used for i. It is customary to reserve the square root sign for cases in which there is a positive square root, in order to distinguish that root from its opposite (which has the same square). For instance, $\sqrt{9} = 3$ and not -3, because 3 is positive and -3 is negative. We can't make such a distinction between i and $-i$ in a sensible way; if we insist, for instance, that i is "positive" and $-i$ is "negative," then how can their product be $i \cdot (-i) = 1$, which is certainly positive? It's better to think of i and $-i$ as identical twins: At birth, they can't be distinguished (they are the two new numbers whose square is -1), so their parents—the scientific community—must dress them differently to tell them apart. After that, the one in the red shirt is called i, the one in the blue shirt, $-i$.

EXERCISES

Compute the following.

1. $(2 + 3i)(5 - 2i)$ **2.** $\dfrac{1}{1 + i}$

SEC. A.1 INTRODUCTION

3. (a) Re($8i$) (b) Im($8i$)

4. $\dfrac{3 + 4i}{2 - i}$

5. $\dfrac{1}{i}$

6. $(-i)^2$

7. Show that $z = 1 + i$ satisfies the equation $z^2 - 2z + 2 = 0$. Find another complex solution to this equation.

8. Show that if a, b, c are real numbers such that $b^2 - 4ac < 0$, then
$$z_{1,2} = \frac{-b \pm i\sqrt{4ac - b^2}}{2a}$$
gives two solutions to the equation $az^2 + bz + c = 0$.

9. Show that if the product of two complex numbers is zero, (at least) one of the two numbers is zero.

*10. Show, using Exercise 9, that the two solutions to the equation $az^2 + bz + c = 0$ (a, b, c real numbers; $b^2 - 4ac < 0$) found in Exercise 8 are the *only* (complex) solutions.

Solve the following complex equations.

11. $\dfrac{z + i}{z - i} = 2$.

12. $z^2 + 8z + 20 = 0$.

13. $z^2 + 19 = 0$.

14. $z^2 - 6z + 10 = 0$.

15. $z^3 + 1 = 0$.

A.2 THE COMPLEX PLANE

As mentioned in Section A.1, it is a very useful fact that the real numbers correspond to the points of a straight line, the real line or number line. We will find a similar one-to-one correspondence for $\mathbb{C}$, except that we will have a "number plane" instead of a number line. A complex number, after all, is given by two real numbers: its real and its imaginary part. A point in the coordinate plane is also given by two real numbers: its horizontal and vertical coordinates (see Figure A.2.1). It is thus reasonable to associate to each complex number $a + bi$ the corresponding point (a, b) in the coordinate plane. The coordinate plane is called the **complex plane** (sometimes the **Argand diagram**) if its points are labeled by complex numbers in this way (see Figure A.2.2), just as the real line is a

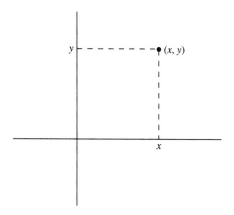

Figure A.2.1. Determining a point in the coordinate plane.

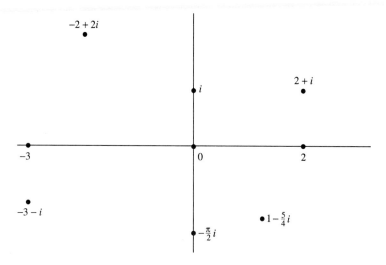

Figure A.2.2. Some points in the complex plane.

coordinate axis whose points are labeled by real numbers. The horizontal axis of the complex plane consists of those numbers whose imaginary part b is zero, that is, the real numbers. So it is nothing but the old number line; it is called the **real axis** in this context.

The vertical axis is called the **imaginary axis**[1] (and i is sometimes called the imaginary unit). Since the first coordinate a of any point on the vertical axis is zero, this axis consists of the numbers bi for various b; these are called **purely imaginary** numbers.

Addition of complex numbers corresponds to vector addition (by the parallelogram rule; see Appendix B, p. 606) in the coordinate plane, since

$$(a + bi) + (c + di) = (a + c) + (b + d)i,$$

while

$$(a, b) + (c, d) = (a + c, b + d)$$

for vector addition. See Figure A.2.3. Similarly for subtraction. Multiplication and division of complex numbers also have a geometric interpretation, as we will see below.

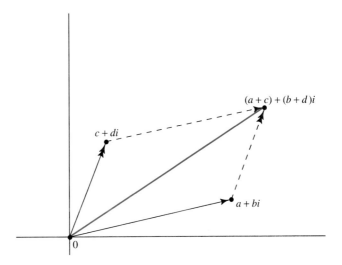

Figure A.2.3. Addition in the complex plane by the parallelogram rule.

The **complex conjugate** of a complex number $z = a + bi$ is defined to be $\bar{z} = a - bi$. For example, $\overline{3 - 8i} = 3 + 8i$. To pursue the analogy of the identical twins, each complex number z has its twin $\bar{z}$. In the complex plane, they are mirror images of each other in the real axis (see Figure A.2.4, p. 594). Some simple but important facts about complex conjugates, which will be proved in the exercises, follow.

---

[1] There is nothing insubstantial about this axis; it is "there" just as the real axis is. The terms "imaginary" and "complex" were coined when mathematicians were still very wary of complex numbers, much as you may be now.

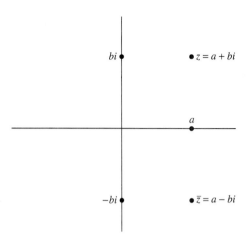

Figure A.2.4. Complex conjugates.

For any two complex numbers z and w, $\overline{z+w} = \overline{z} + \overline{w}$ and $\overline{zw} = \overline{z}\,\overline{w}$;

$$\text{Re}(z) = \frac{1}{2}(z + \overline{z}), \qquad \text{Im}(z) = \frac{1}{2i}(z - \overline{z});$$

$\overline{\overline{z}} = z;\quad \overline{z} = z$ if and only if z is real.

Note that the product $z\overline{z} = (a + bi)(a - bi) = a^2 + b^2$ is a nonnegative real number (which is zero if and only if $z = 0$). The **absolute value** (also called **length** or **modulus**) of the complex number $z = a + bi$ is defined to be $|z| = \sqrt{a^2 + b^2}$.

(**Careful:** *Not* $\sqrt{a^2 + (bi)^2}$!)
For example, $|3 - 8i| = \sqrt{9 + 64} = \sqrt{73}$. We have just seen that $z\overline{z} = |z|^2$ for any z.

NOTE: $|z|$ is the distance from z to the origin in the complex plane, that is, the length of the vector corresponding to z, as computed by Pythagoras (see Figure A.2.5). Of course, if z is a real number a, then $|z| = \sqrt{a^2} = |a|$ is its ordinary absolute value.

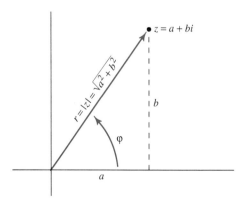

Figure A.2.5. $|z|$ as length of the vector; polar coordinates.

More basic facts, to be proved in the exercises: For any two complex numbers z and w,

$$|zw| = |z| \cdot |w|\,;$$
$$|z + w| \le |z| + |w|\,;$$
$$|z| = |\bar{z}|\,.$$

One can describe any complex number $z = a + bi$ by its rectangular coordinates a, b in the complex plane, but also by its **polar coordinates** r, φ. r is just the length $|z|$ defined above. The polar angle φ is called the **argument** of the complex number z; it is sometimes denoted by Arg z. We will think of φ as being defined up to multiples of 2π (though it sometimes pays to be more precise).

Recall that polar and rectangular coordinates are related by the formulas

$$\begin{cases} a = r \cos \varphi \\ b = r \sin \varphi \end{cases} \quad \text{and} \quad \begin{cases} r = \sqrt{a^2 + b^2} \\ \tan \varphi = \dfrac{b}{a} \end{cases}\,.\quad {}^2$$

We can therefore rewrite our complex number as

$$z = a + bi = r \cos \varphi + (r \sin \varphi)i$$
$$= r(\cos \varphi + i \sin \varphi)\,.$$

Conversely, if $z = r(\cos \varphi + i \sin \varphi)$, then r and φ are polar coordinates of z. A complex number is said to be in **polar form** if it is written as $r(\cos \varphi + i \sin \varphi)$.

If we multiply two complex numbers that are given in polar form, say

$$z = r(\cos \varphi + i \sin \varphi) \quad \text{and} \quad w = r_1(\cos \varphi_1 + i \sin \varphi_1),$$

the answer is

$$zw = rr_1(\cos \varphi + i \sin \varphi)(\cos \varphi_1 + i \sin \varphi_1)$$
$$= rr_1[\cos \varphi \cos \varphi_1 - \sin \varphi \sin \varphi_1 + i(\cos \varphi \sin \varphi_1 + \sin \varphi \cos \varphi_1)]$$
$$= rr_1[\cos(\varphi + \varphi_1) + i \sin(\varphi + \varphi_1)]\,.$$

In other words: The length of the product is the product of the lengths; the argument (polar angle) of the product is the *sum* of the arguments (up to multiples of 2π).[3] See Figure A.2.6, p. 596.

In particular, since powers of z are obtained by multiplying z by itself repeatedly, we have $|z^n| = |z|^n$, and Arg $(z^n) = n$ Arg z up to multiples of 2π.

EXAMPLE A.2.1 Find all complex solutions of $z^3 = 1$.

If z is such a solution, we can write $z = r(\cos \varphi + i \sin \varphi)$; we then have

$$r^3(\cos 3\varphi + i \sin 3\varphi) = 1 = 1(\cos 0 + i \sin 0)\,.$$

---

[2] Since $\tan \varphi = \dfrac{b}{a}$ determines φ only up to multiples of π, not 2π, this is not quite enough to find φ when a, b are given. See Exercise 12.

[3] See Exercises 17 and 18 for a geometric interpretation of division of complex numbers.

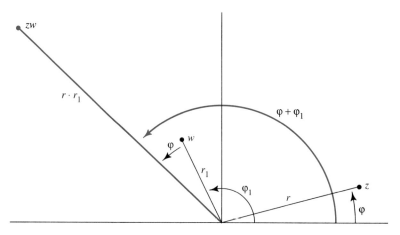

Figure A.2.6. Multiplication in the complex plane.

Therefore, $r^3 = 1$, and $3\varphi = 0$ up to multiples of 2π, since $r^3, 3\varphi$ on the one hand and 1, 0 on the other are polar coordinates for the same complex number 1. Now r is a *real, positive* number, so $r^3 = 1$ implies $r = 1$. What about φ? We have $3\varphi = 0$ up to multiples of 2π, so

$$3\varphi = 0 \quad \text{or} \quad 3\varphi = 2\pi \quad \text{or} \quad 3\varphi = 4\pi \quad \text{or} \quad 3\varphi = 6\pi \quad \text{etc.}$$

$$\varphi = 0 \quad \text{or} \quad \varphi = \frac{2\pi}{3} \quad \text{or} \quad \varphi = \frac{4\pi}{3} \quad \text{or} \quad \varphi = 2\pi \quad \text{etc.}$$

Up to multiples of 2π there are three possibilities for φ: $\varphi = 0, \frac{2\pi}{3}, \frac{4\pi}{3}$. So there are three solutions to $z^3 = 1$, namely,

$$1(\cos 0 + i \sin 0) = 1,$$

$$1\left(\cos \frac{2\pi}{3} + i \sin \frac{2\pi}{3}\right) = -\frac{1}{2} + \frac{1}{2}i\sqrt{3},$$

$$1\left(\cos \frac{4\pi}{3} + i \sin \frac{4\pi}{3}\right) = -\frac{1}{2} - \frac{1}{2}i\sqrt{3}.$$

In other words, $z^3 - 1$ factors completely, as

$$z^3 - 1 = (z - 1)\left(z + \frac{1}{2} - \frac{1}{2}i\sqrt{3}\right)\left(z + \frac{1}{2} + \frac{1}{2}i\sqrt{3}\right). \quad \blacksquare$$

It can be shown (although not easily) that *any* polynomial can be factored completely if complex roots are allowed; this even remains true if the polynomial has complex coefficients. As a result, it won't ever be necessary to expand the complex numbers further for the purpose of solving polynomial equations. Let us record this theorem:

Theorem

("**Fundamental Theorem of Algebra**," a bit of a misnomer). Every nth-degree polynomial with complex coefficients can be written as the product of n linear factors. That is, if $a_0, \ldots, a_n$ are complex numbers with $a_n \neq 0$, then we can write

$$a_n z^n + a_{n-1} z^{n-1} + \cdots + a_1 z + a_0 = a_n(z - z_1)(z - z_2) \cdots (z - z_n)$$

for suitably chosen complex numbers $z_1, \ldots, z_n$. [The numbers $z_1, \ldots, z_n$ (the roots of the polynomial) are not necessarily distinct.]

EXERCISES

1. Draw the points $-4 + 5i$, $8 - 3i$, $-12 + 8i$, $4 + 2i$ in the complex plane. Illustrate the facts that $(-4 + 5i) - (8 - 3i) = -12 + 8i$ and $(-4 + 5i) + (8 - 3i) = 4 + 2i$ in your diagram.

2. Show that for any complex number z,
$$\text{Re}(z) = \frac{1}{2}(z + \bar{z}) \quad \text{and} \quad \text{Im}(z) = \frac{1}{2i}(z - \bar{z}).$$

3. Show that for any complex numbers z and w,
$$\overline{z + w} = \bar{z} + \bar{w} \quad \text{and} \quad \overline{z \cdot w} = \bar{z} \cdot \bar{w}.$$

4. Show that $\bar{\bar{z}}$, the complex conjugate of $\bar{z}$, equals z; show that $\bar{z} = z$ if and only if z is real.

5. Consider the circle of radius 1, the so-called **unit circle**, whose center is the origin in the complex plane. Give conditions on a and b for the point $z = a + bi$ to be inside the circle, for the point to be on the circle, and for the point to be outside the circle. Then rephrase these conditions in terms of $|z|$.

6. Find the set of all complex numbers z such that $z + \bar{z} = 0$.

7. Prove directly from the definition of absolute value that for any complex numbers $z = a + bi$, $w = c + di$, we have $|z \cdot w| = |z| \cdot |w|$.

8. Show that $|z + w| \leq |z| + |w|$ for all z, w. (You may be able to find a geometric proof.)

9. Compute: (a) $|i|$; (b) $|3 + 4i|$; (c) $|(1 + i)^{10}|$; (d) $|5 - 6i|$.

10. Show that if z and w are complex numbers, then $|z - w|$ is the distance between them in the complex plane.

11. Find the set of all complex numbers z such that $|z + 1| = |z - 1|$.
    [*Hint*: Square both sides and write $|z + 1|^2 = (z + 1)(\bar{z} + 1)$, etc.]
    Can you explain the form of the answer?
    [*Hint*: Use Exercise 10.]

12. For $a = 3$, $b = -4$, find r and the two angles φ such that $r = \sqrt{a^2 + b^2}$, $\tan \varphi = \frac{b}{a}$. Which of the angles is $\text{Arg}(3 - 4i)$?

13. Show that $|\cos \varphi + i \sin \varphi| = 1$ for any angle φ.
14. Write the complex numbers $2 + 2i$, $-3i$, and $1 - \sqrt{3}\,i$ in polar form.
15. Draw the points corresponding to $2(\cos \frac{5\pi}{4} + i \sin \frac{5\pi}{4})$ and $3(\cos \frac{\pi}{8} - i \sin \frac{\pi}{8})$ in the complex plane.
16. Given $z = 2(\cos \frac{\pi}{3} + i \sin \frac{\pi}{3})$ and $w = 5(\cos \frac{\pi}{4} + i \sin \frac{\pi}{4})$, compute $z \cdot w$, $z^3 w$, and $z^3 w^4$.
17. Show that $\left|\frac{z}{w}\right| = \frac{|z|}{|w|}$, and $\text{Arg}(\frac{z}{w}) = \text{Arg } z - \text{Arg } w$ up to multiples of 2π, for any two complex numbers z, w, $w \neq 0$.
18. Draw a figure similar to Figure A.2.6, for division.
19. Find all complex solutions of $z^4 = 1$.
20. Find all complex solutions of $z^n = 1$, for n a positive integer.
21. (a) Show that if z is a solution to $z^3 - 18z^2 + 55z - 183 = 0$, then $\bar{z}$ is, also. [*Hint*: Use Exercise 3.]
    (b) Show that $z^3 - 18z^2 + 55z - 183$ has at least one real root, without using the Intermediate Value Theorem.
    (c) Show that if f is any polynomial with *real* coefficients, then $f(z) = 0$ implies $f(\bar{z}) = 0$.
    (d) Show that any polynomial of *odd* degree with real coefficients has at least one real root, without using the Intermediate Value Theorem.

FURTHER READING

If you're curious why "Fundamental Theorem of Algebra" is a bit of a misnomer (p. 597), see Chapter 4, "The Fundamental Theorem of Algebra" (by R. Remmert), in the fascinating book *Numbers* by Ebbinghaus et al. (Springer-Verlag, 1991).

A.3 POWER SERIES AND COMPLEX EXPONENTIALS

The main goal of this section is to extend the exponential function e^z to allow for complex values of z. This is very useful in many parts of mathematics; in this book it is used in solving certain constant-coefficient differential equations (see Section 2.5). Meanwhile, we'll start with some preliminary material on sequences and series of complex numbers.

By definition, a sequence $(z_n) = z_1, z_2, z_3, \ldots$ of complex numbers has a **limit** $a + bi$ if the real parts $\text{Re}(z_1), \text{Re}(z_2), \ldots$ have the limit $\lim_{n \to \infty} \text{Re}(z_n) = a$ and the imaginary parts have the limit $\lim_{n \to \infty} \text{Im}(z_n) = b$.

EXAMPLE A.3.1 Let $z_n = \dfrac{1}{1 + ni}$. Then $\text{Re}(z_n) = \dfrac{1}{n^2 + 1}$, $\text{Im}(z_n) = \dfrac{-n}{n^2 + 1}$ (why?), so $\lim_{n \to \infty} \text{Re}(z_n) = 0$, $\lim_{n \to \infty} \text{Im}(z_n) = 0$, and hence $\lim_{n \to \infty} z_n = 0$. ∎

EXAMPLE A.3.2 Let $z_n = i^n$. Then the sequence is $i, -1, -i, 1, i, -1, -i, 1, \ldots$. The real parts are $0, -1, 0, 1, 0, -1, 0, 1, \ldots$; this sequence of real numbers has no limit, so $\lim_{n \to \infty} i^n$ does not exist, either. ∎

EXAMPLE A.3.3 $\lim_{n \to \infty} \dfrac{2 + ni}{1 - 2ni} = -\dfrac{1}{2}$. (See Exercise 1.) ∎

It can be shown that the usual rules for the computation of limits apply.

Just as for real numbers, an infinite series $\sum_{n=1}^{\infty} w_n = w_1 + w_2 + w_3 + \cdots$ with complex terms is called **convergent** if the limit $\lim_{N \to \infty} \sum_{n=1}^{N} w_n$ of its partial sums exists; if so, this limit is called the **sum** of the infinite series. It follows that convergent infinite series with complex terms can be summed by summing the real and imaginary parts of the terms separately.

It can be shown that the ratio test for convergence applies: If $\lim_{n \to \infty} \left| \dfrac{w_{n+1}}{w_n} \right|$ exists and is less than 1, the series $\sum_{n=1}^{\infty} w_n$ is convergent; if $\lim_{n \to \infty} \left| \dfrac{w_{n+1}}{w_n} \right|$ exists and is greater than 1, the series is divergent.

In particular, we can study power series $\sum_{n=0}^{\infty} a_n z^n$ with a complex variable z and complex coefficients a_n (often, the coefficients will in fact be real).

The ratio test shows that $\sum_{n=0}^{\infty} a_n z^n$ converges for $|z| < R$ and diverges for $|z| > R$, where $R = \lim_{n \to \infty} \left| \frac{a_n}{a_{n+1}} \right|$, provided that this limit exists[4] (see Exercise 8). R is called the **radius of convergence**. Note that if a circle of radius R is drawn with its center at the origin (in the complex plane), then the power series converges for all z inside the circle and diverges for all z outside the circle, while the issue is still in doubt for $|z| = R$ (see Figure A.3.1).

For example, $\sum_{n=0}^{\infty} z^n = 1 + z + z^2 + \cdots$ converges for $|z| < 1$ and diverges for $|z| > 1$. (Exercise 9 shows that $\sum_{n=0}^{\infty} z^n = \frac{1}{1-z}$ for $|z| < 1$.)

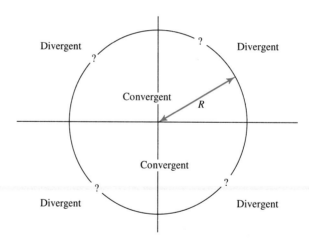

Figure A.3.1. Radius of convergence.

In the important special case that $R = \infty$, the "circle of convergence" consists of the whole complex plane, and the power series converges for all z.

We are now ready to define e^z for any complex z. Recall from calculus that e^x is given by the Taylor series expansion (about $x = 0$)

$$e^x = 1 + x + \frac{x^2}{2!} + \cdots = \sum_{n=0}^{\infty} \frac{x^n}{n!}.$$

---

[4] One common case in which the limit fails to exist occurs when $a_n = 0$ for infinitely many n. This case, of **gapped** power series, is discussed in Section 5.2.

Thus it seems natural to *define* e^z using the same power series:

$$e^z = \sum_{n=0}^{\infty} \frac{z^n}{n!} \quad \text{for all complex } z.$$

We should then check whether this series always converges. Since the coefficients are $a_n = \frac{1}{n!}$, we have

$$R = \lim_{n \to \infty} \left| \frac{a_n}{a_{n+1}} \right| = \lim_{n \to \infty} \frac{(n+1)!}{n!} = \lim_{n \to \infty} (n+1) = \infty,$$

and the series does indeed converge for all z.

Of course, it is not clear that the exponential $e^z = \sum_{n=0}^{\infty} \frac{z^n}{n!}$ will have the same properties that e^x has for real x. However, it can be shown that $e^{z+w} = e^z \cdot e^w$ for complex numbers z and w (see Exercise 10).

Let us see, in particular, what happens if z is on the imaginary axis, say $z = ix$ with x real. If we substitute this in the power series, we get

$$e^{ix} = \sum_{n=0}^{\infty} \frac{(ix)^n}{n!}$$

$$= 1 + ix + \frac{(ix)^2}{2!} + \frac{(ix)^3}{3!} + \cdots$$

$$= 1 + ix - \frac{x^2}{2!} - \frac{ix^3}{3!} + \cdots.$$

Remember that we are allowed to sum this infinite series by summing the real and imaginary parts separately. The real parts yield:

$$1 - \frac{x^2}{2!} + \frac{x^4}{4!} - \frac{x^6}{6!} + \cdots,$$

which is just the Taylor series for $\cos x$. Similarly, the terms with i yield

$$i\left(x - \frac{x^3}{3!} + \frac{x^5}{5!} - \frac{x^7}{7!} + \cdots \right) = i \sin x.$$

Putting this together, we have the basic identity

$$e^{ix} = \cos x + i \sin x,$$

which is known as **Euler's formula**.[5]

---

[5] After the great Swiss mathematician Leonhard Euler (1707–1783).

NOTE: We can use Euler's formula to shorten the expression for a complex number in polar form: $r(\cos \varphi + i \sin \varphi) = re^{i\varphi}$.

Now we'll look at the exponential e^z for an arbitrary complex number $z = a + bi$. As mentioned above, we can write $e^z = e^{a+bi} = e^a e^{bi}$, and so we have

$$e^z = e^a(\cos b + i \sin b) = e^a \cos b + i e^a \sin b.$$

In particular, $|e^z| = e^a = e^{\text{Re}(z)}$, while $\text{Arg}(e^z) = b = \text{Im}(z)$ up to multiples of 2π. Also, the exponential $e^{\bar z}$ of the conjugate $\bar z$ of z is the conjugate of e^z (see Exercise 5). As an important special case of this, for real x we have

$$e^{-ix} = \cos x - i \sin x = \overline{\cos x + i \sin x} = \overline{e^{ix}}.$$

Now we can express the trigonometric functions as combinations of complex exponentials. For instance, since $e^{ix} = \cos x + i \sin x$, $\cos x$ is the real part of e^{ix}, so

$$\cos x = \text{Re}(e^{ix}) = \frac{1}{2}(e^{ix} + \overline{e^{ix}}) \quad \text{(see Section A.2)}$$

$$= \frac{1}{2}(e^{ix} + e^{-ix}).$$

Similarly, $\sin x = \text{Im}(e^{ix}) = \dfrac{1}{2i}(e^{ix} - e^{-ix})$.

EXERCISES

1. Find: (a) $\lim\limits_{n \to \infty} \dfrac{2 + ni}{1 - 2ni}$ and (b) $\lim\limits_{n \to \infty} \dfrac{n}{1 + ni}$.

2. Compute: (a) $e^{2+(\pi/2)i}$; (b) $e^{\pi i}$; (c) $e^{\log 3 - (\pi/4)i}$; (d) $e^{2\pi i}$.

3. Show that $e^z = 1$ if and only if $z = 2k\pi i$ for some integer k.

4. Show that for real x, $\tan x = \dfrac{i(1 - e^{2ix})}{1 + e^{2ix}}$.

5. Prove that for any complex number z, we have $\overline{e^z} = e^{\bar z}$.

6. Find all complex numbers z such that $e^z = -1$.

7. Show that if (w_n) is a sequence of complex numbers such that $\sum\limits_{n=1}^{\infty} w_n$ converges, then $\lim\limits_{n \to \infty} w_n = 0$.

   [Hint: Consider the limit of $w_N = \sum\limits_{n=1}^{N} w_n - \sum\limits_{n=1}^{N-1} w_n$ as $N \to \infty$.]

8. Use the ratio test to show that if $R = \lim\limits_{n \to \infty} \left| \dfrac{a_n}{a_{n+1}} \right|$ exists, then $\sum\limits_{n=0}^{\infty} a_n z^n$ converges for $|z| < R$ and diverges for $|z| > R$.

SEC. A.3 POWER SERIES AND COMPLEX EXPONENTIALS

9. (a) Show that $\sum_{n=0}^{\infty} z^n$ converges for $|z| < 1$ and diverges for $|z| > 1$.

(b) Show that if $\sum_{n=0}^{\infty} z^n$ converges, then $(1-z)\left(\sum_{n=0}^{\infty} z^n\right) = 1$ and therefore,

$$\sum_{n=0}^{\infty} z^n = \frac{1}{1-z}.$$

*(c) Show that $\sum_{n=0}^{\infty} z^n$ diverges for $|z| = 1$.

*10. (a) Show that $\dfrac{(z+w)^n}{n!} = \sum_{k=0}^{n} \dfrac{1}{k!(n-k)!} z^k w^{n-k}$.

[*Hint:* Look up the binomial theorem.]

(b) Show that $\dfrac{(z+w)^n}{n!} = \sum_{\substack{k,l \\ k+l=n}} \dfrac{z^k}{k!} \dfrac{w^l}{l!}$, where the sum is taken over all pairs of integers $k, l \geq 0$ such that $k + l = n$.

(c) Show that $e^{z+w} = \sum_{n=0}^{\infty} \sum_{\substack{k,l \\ k+l=n}} \dfrac{z^k}{k!} \dfrac{w^l}{l!}$.

(d) Assuming that the terms of this infinite series for e^{z+w} may be rearranged in any order, and that the infinite series for e^z and e^w may be multiplied together using the "infinite" distributive law, show that $e^{z+w} = e^z \cdot e^w$.

[N O T E : It can be shown that the assumptions are valid because the series are absolutely convergent—same definition as in the real case.]

APPENDIX B

Basic Linear Algebra

B.1 VECTORS

In physics and mathematics, entities often come up which are not given by their numerical size alone, but which can be specified by giving both their size and their direction—for example, velocities and forces. Geometrically, such **vectors** can be indicated by arrows, with the understanding that two arrows with the same length and direction represent the same vector (see Figure B.1.1). Vectors are often denoted by boldface letters such as **v**, and we will do so.[1]

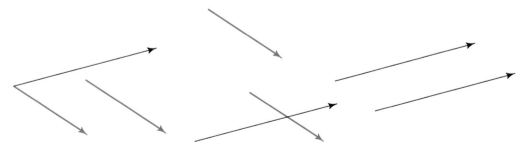

Figure B.1.1. The black arrows all represent the same vector; all the colored arrows represent a second vector.

---

[1] When writing by hand, $\underline{v}$ or $\vec{v}$ can be used instead of **v**.

605

A vector **v** can be multiplied by a number k; if k is positive, the resulting vector $k\mathbf{v}$ will have the same direction and be k times as long as **v**, while if k is negative, $k\mathbf{v}$ will have the opposite direction and be $(-k)$ times as long as **v** (see Figure B.1.2). In the special case when $k = 0$, $k\mathbf{v}$ will be the zero vector **0** whose length is 0 (and whose direction is undefined).

Figure B.1.2. Multiplication of a vector by numbers (**scalar multiplication**).

Vectors can also be added together; the sum of two vectors is formed in the same way that two forces acting on the same object combine to yield a resulting force. Specifically, to get a geometric construction of a sum $\mathbf{v} + \mathbf{w}$ of vectors, arrows representing **v** and **w** are drawn from the same starting point. The angle obtained in this way is completed to a parallelogram; a new arrow is drawn along the diagonal of this parallelogram from the common starting point of **v** and **w**, and this new arrow represents $\mathbf{v} + \mathbf{w}$. See Figure B.1.3.

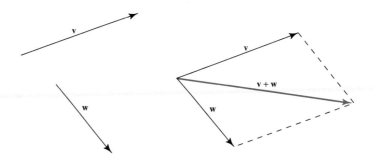

Figure B.1.3. Addition of vectors using the parallelogram rule.

In computations using vectors, it is often convenient to fix the starting point of each arrow to be the origin; this way, the vector is determined by the endpoint of the arrow, which will be a point (x, y) if the vector is in the plane and a point (x, y, z) if the vector is in 3-space.

[N O T E : The geometric discussion above is valid in both cases.] See Figure B.1.4.

In fact, the notation $\mathbf{v} = (x, y)$ or $\mathbf{v} = (x, y, z)$ is used in these cases. Notice that it is not clear from the notation (x, y) by itself when this should be interpreted as a point and when it should be interpreted, instead, as the vector given by the distance and direction from the origin to that point. Fortunately, the context will usually make this clear.

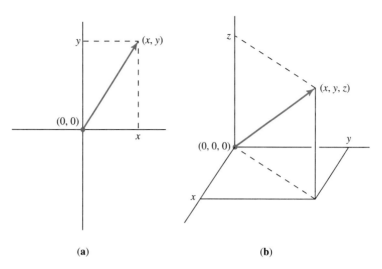

Figure B.1.4. A vector represented by an arrow starting at the origin: (a) in the plane; (b) in 3-space.

The coordinates x, y (or x, y, z) of the endpoint of the arrow, as described above, are called the **components** or **coordinates** of the vector $\mathbf{v} = (x, y)$ [or $\mathbf{v} = (x, y, z)$]. Scalar multiplication (multiplying vectors by numbers) and addition of vectors become easier when components are used: If $\mathbf{v} = (x_1, y_1)$ and $\mathbf{w} = (x_2, y_2)$, then $k\mathbf{v} = (kx_1, ky_1)$ and $\mathbf{v} + \mathbf{w} = (x_1 + x_2, y_1 + y_2)$ (see Exercise 2). The **length** of a vector $\mathbf{v}$ is usually denoted by $|\mathbf{v}|$; it is not hard to show that $|\mathbf{v}| = \sqrt{x^2 + y^2}$ if $\mathbf{v} = (x, y)$ and $|\mathbf{v}| = \sqrt{x^2 + y^2 + z^2}$ if $\mathbf{v} = (x, y, z)$ (see Exercise 4).

When a point (or a particle) moves in the x,y-plane, so that its position at time t is $(x(t), y(t))$, then its **velocity vector**[2] at time t is $(\frac{dx}{dt}, \frac{dy}{dt})$. The length of the velocity vector is the **speed**; it equals $\sqrt{(\frac{dx}{dt})^2 + (\frac{dy}{dt})^2}$ (why?). The direction of the velocity vector at time t is the direction in which the point is moving at that moment; when the velocity vector is drawn, it is usually *not* shown starting at the origin, but rather starting at the position of the moving point (see Figure B.1.5, p. 608).

It is not hard to define **limits**, **derivatives**, and **integrals** of vectors; in practice, they are found by looking at the different components separately. For instance, for a moving particle in the plane, the derivative of the position vector $(x(t), y(t))$ is the velocity vector $(\frac{dx}{dt}, \frac{dy}{dt})$. If we take the derivative again, we get $\frac{d}{dt}(\frac{dx}{dt}, \frac{dy}{dt}) = (\frac{d^2x}{dt^2}, \frac{d^2y}{dt^2})$, which is the **acceleration** vector $\mathbf{a}(t)$. This vector is related (in Newtonian mechanics) to the total force

---

[2] In this context the velocity vector is often denoted by $\mathbf{v}(t)$. A different notation, such as $\mathbf{r}(t)$, is then used for the position vector $(x(t), y(t))$.

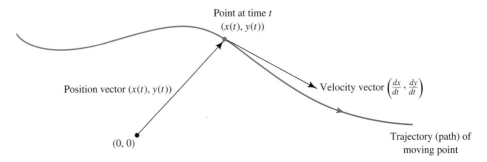

Figure B.1.5. Motion of a point in the plane.

$F(t)$ on the particle by Newton's law $F(t) = ma(t)$. Similar formulas, with the additional coordinate z, hold when a particle moves in 3-space.

By analogy with vectors in the coordinate plane and in 3-space, one can define vectors in n-dimensional space to be sequences of n numbers. Since it is not clear what the nth letter in the sequence $x, y, z, ?, ??, \ldots$ would be, the notation for such a vector is usually something like $\mathbf{v} = (v_1, v_2, \ldots, v_n)$ [or even $\mathbf{x} = (x_1, x_2, \ldots, x_n)$]. The geometric significance of such vectors is less clear, but they are still very useful in computations. As you might expect, addition and scalar multiplication are defined as follows: If

$$\mathbf{v} = (v_1, v_2, \ldots, v_n) \quad \text{and} \quad \mathbf{w} = (w_1, w_2, \ldots, w_n),$$

then

$$\mathbf{v} + \mathbf{w} = (v_1 + w_1, v_2 + w_2, \ldots, v_n + w_n) \quad \text{and} \quad k\mathbf{v} = (kv_1, kv_2, \ldots, kv_n).$$

The zero vector is $\mathbf{0} = (0, 0, \ldots, 0)$. If $v_1, v_2, \ldots, v_n$ are functions of t, then the vector $\mathbf{v} = (v_1, v_2, \ldots, v_n)$ is also written $\mathbf{v}(t) = (v_1(t), v_2(t), \ldots, v_n(t))$ and sometimes called a **vector function** (of t). The **derivative** of the vector (or vector function) $\mathbf{v}$ is given by

$$\frac{d\mathbf{v}}{dt} = (v_1'(t), v_2'(t), \ldots, v_n'(t)).$$

EXERCISES

1. If $\mathbf{v} = (2, -1)$ and $\mathbf{w} = (3, 4)$, find $2\mathbf{v} + \frac{1}{2}\mathbf{w}$ and $-3\mathbf{v} - \mathbf{w}$.

2. Show from the geometric definitions of $k\mathbf{v}$ and $\mathbf{v} + \mathbf{w}$ that if $\mathbf{v} = (x_1, y_1)$ and $\mathbf{w} = (x_2, y_2)$, then $k\mathbf{v} = (kx_1, ky_1)$ and $\mathbf{v} + \mathbf{w} = (x_1 + x_2, y_1 + y_2)$.

3. Give a geometric definition of the difference $\mathbf{v} - \mathbf{w}$ of two vectors $\mathbf{v}$ and $\mathbf{w}$. Show that $\mathbf{w} + (\mathbf{v} - \mathbf{w}) = \mathbf{v}$.

4. (a) Show that if $\mathbf{v} = (x, y)$, then $|\mathbf{v}| = \sqrt{x^2 + y^2}$.
   (b) Show that if $\mathbf{v} = (x, y, z)$, then $|\mathbf{v}| = \sqrt{x^2 + y^2 + z^2}$.

5. **(a)** Suppose that the position of a particle at time t is $(\cos t, \sin t)$. Show that the speed of the particle is constant.
   (b) As part (a), but let the position be $(\cos 5t, \sin 5t)$. How is the speed affected by this change? Can you explain why this should be so?

6. Suppose that the position of a particle at time t is $(2\cos 3t, 2\sin 3t)$.
   (a) Sketch the path of the particle.
   (b) Indicate the velocity vectors for $t = 0$, for $t = \dfrac{\pi}{12}$, for $t = \dfrac{\pi}{6}$, and for $t = \dfrac{\pi}{2}$ in your sketch.

7. Repeat Exercise 6 for the position vector $(3\cos 2t, -3\sin 2t)$.

8. Show that if $\mathbf{v}$ and $\mathbf{w}$ are vectors in n-dimensional space whose components are functions of t, then
$$\frac{d(\mathbf{v}+\mathbf{w})}{dt} = \frac{d\mathbf{v}}{dt} + \frac{d\mathbf{w}}{dt} \quad \text{and} \quad \frac{d(k\mathbf{v})}{dt} = k \cdot \frac{d\mathbf{v}}{dt}.$$

9. Define the length $|\mathbf{v}|$ of a vector $\mathbf{v} = (v_1, v_2, \ldots, v_n)$ in n-dimensional space.

10. The **dot product** of the vectors $\mathbf{v} = (v_1, v_2, \ldots, v_n)$ and $\mathbf{w} = (w_1, w_2, \ldots, w_n)$ in n-dimensional space is defined to be the number
$$\mathbf{v} \cdot \mathbf{w} = v_1 w_1 + v_2 w_2 + \cdots + v_n w_n.$$
Show that the dot product has the following properties.
    (a) $(\mathbf{u} + \mathbf{v}) \cdot \mathbf{w} = \mathbf{u} \cdot \mathbf{w} + \mathbf{v} \cdot \mathbf{w}$ (for any three vectors $\mathbf{u}, \mathbf{v}, \mathbf{w}$).
    (b) $(k\mathbf{v}) \cdot \mathbf{w} = k(\mathbf{v} \cdot \mathbf{w})$.
    (c) $\mathbf{v} \cdot \mathbf{v} = |\mathbf{v}|^2$ (see Exercise 9).

B.2 MATRICES AND COLUMN VECTORS

A rectangular array of numbers is called a **matrix**; if it has m rows and n columns, it is an $m \times n$ matrix (pronounced "m by n matrix"). For instance, $\begin{bmatrix} 1 & 0 & 5 \\ 2 & 3 & -\pi \end{bmatrix}$ is a 2×3 matrix. The numbers in the matrix are called its **entries**; the entry in the ith row and jth column is called the i,j entry. For example, the 1,3 entry in the matrix above is 5; the 2,1 entry is 2. Matrices[3] are usually indicated by boldface[4] capital[5] letters; their entries are often indicated by the corresponding lower-case letters with double subscripts indicating the row and column. For instance, an $m \times n$ matrix $\mathbf{A}$ whose i,j entry is a_{ij} (for any i,j) will look like

$$\mathbf{A} = \begin{bmatrix} a_{11} & a_{12} & \cdots & a_{1n} \\ a_{21} & a_{22} & \cdots & a_{2n} \\ \vdots & \vdots & & \vdots \\ a_{m1} & a_{m2} & \cdots & a_{mn} \end{bmatrix}$$

and can be written in shorthand as $\mathbf{A} = (a_{ij})_{1 \leq i \leq m, 1 \leq j \leq n}$ or simply $\mathbf{A} = (a_{ij})$ if the size of the matrix is clear from the context.

A matrix with only one column is usually referred to as a **column vector** and is interpreted as the vector in m-dimensional space whose components are the given entries. For example, the matrix $\mathbf{v} = \begin{bmatrix} 4 \\ 1 \\ -1 \end{bmatrix}$ corresponds to the vector $\mathbf{v} = (4, 1, -1)$ in 3-space [or to the displacement from the origin to the point $(4, 1, -1)$].

Just as vectors can be multiplied by numbers, any matrix $\mathbf{A}$ can be multiplied by a number k by multiplying each entry of $\mathbf{A}$ by k. For instance, if $\mathbf{A} = \begin{bmatrix} 1 & 0 & -1 \\ 2 & 3 & 5 \end{bmatrix}$ and $k = 10$, then

$$k\mathbf{A} = 10 \begin{bmatrix} 1 & 0 & -1 \\ 2 & 3 & 5 \end{bmatrix} = \begin{bmatrix} 10 & 0 & -10 \\ 20 & 30 & 50 \end{bmatrix}.$$

If two matrices have the same number of rows and also the same number of columns, they can be **added** by adding the entries in each position, for example:

$$\begin{bmatrix} 2 & 1 \\ 0 & -1 \\ 3 & 4 \end{bmatrix} + \begin{bmatrix} 1 & 5 \\ 2 & 0 \\ -1 & 1 \end{bmatrix} = \begin{bmatrix} 2+1 & 1+5 \\ 0+2 & -1+0 \\ 3-1 & 4+1 \end{bmatrix} = \begin{bmatrix} 3 & 6 \\ 2 & -1 \\ 2 & 5 \end{bmatrix}.$$

For column vectors of the same size this is the usual addition of vectors.

Multiplication of matrices by each other is more complicated (the motivation for the definition will appear below) but also more important. To form the product $\mathbf{AB}$ of the

---

[3] *Matrices* is the plural of *matrix*.
[4] When writing by hand, boldface can be shown by underlining, as in $\underline{A}$ for $\mathbf{A}$.
[5] Except lower-case for column vectors—see below.

SEC. B.2 MATRICES AND COLUMN VECTORS

matrices **A** and **B** (in that order) the number of columns of **A** must be equal to the number of rows of **B**. That is, if **A** is an $m \times n$ matrix, then **B** must be an $n \times p$ matrix for some p. The product **AB** will then be an $m \times p$ matrix. To find the i,j entry of **AB**, the ith row of **A** and the jth column of **B** are used; the entry in **AB** is obtained by adding the products of corresponding entries in that row of **A** and that column of **B** (that is, the first entry in the ith row of **A** times the first entry in the jth column of **B**, etc.)

EXAMPLE B.2.1

$$\mathbf{A} = \begin{bmatrix} 1 & 0 & 2 \\ 4 & -1 & 3 \end{bmatrix}, \quad \mathbf{B} = \begin{bmatrix} -2 & 6 & 8 & 7 \\ 0 & -3 & 1 & 5 \\ 0 & 0 & -4 & 2 \end{bmatrix}.$$

A is a 2×3 matrix, **B** is a 3×4 matrix; since the number of columns of **A** (three) equals the number of rows of **B**, we can form **AB**, which will be a 2×4 matrix. To get, say, the 2,3 entry of **AB**, we use the second row $(4 \quad -1 \quad 3)$ of **A** and the third column $\begin{bmatrix} 8 \\ 1 \\ -4 \end{bmatrix}$ of **B**, obtaining $4 \cdot 8 + (-1) \cdot 1 + 3 \cdot (-4) = 19$ for the 2,3 entry of **AB**. Using the same process for each of the other entries yields

$$\mathbf{AB} = \begin{bmatrix} 1 & 0 & 2 \\ 4 & -1 & 3 \end{bmatrix} \begin{bmatrix} -2 & 6 & 8 & 7 \\ 0 & -3 & 1 & 5 \\ 0 & 0 & -4 & 2 \end{bmatrix}$$

$$= \begin{bmatrix} 1 \cdot (-2) + 0 \cdot 0 + 2 \cdot 0 & 1 \cdot 6 + 0 \cdot (-3) + 2 \cdot 0 & \cdots & \cdots \\ 4 \cdot (-2) + (-1) \cdot 0 + 3 \cdot 0 & 4 \cdot 6 + (-1) \cdot (-3) + 3 \cdot 0 & \cdots & \cdots \end{bmatrix}$$

$$= \begin{bmatrix} -2 & 6 & 0 & 11 \\ -8 & 27 & 19 & 29 \end{bmatrix}.$$

Note that in this example the product **BA** is undefined, since **B** has four columns and **A** has two, not four, rows. Even when **AB** and **BA** are both defined, they are usually not equal! (See Exercises 1 and 2.) ∎

The matrix products that come up most often are products of a square matrix **A** (that is, a matrix with an equal number of rows and columns) with a square matrix **B** or a column vector **v** of the same size.

EXAMPLE B.2.2

$$\underbrace{\begin{bmatrix} 1 & 0 & 2 \\ 0 & 3 & 1 \\ 2 & 0 & 5 \end{bmatrix}}_{\mathbf{A}} \underbrace{\begin{bmatrix} 1 & -1 & 1 \\ 0 & 0 & 2 \\ 4 & 0 & 0 \end{bmatrix}}_{\mathbf{B}} = \underbrace{\begin{bmatrix} 9 & -1 & 1 \\ 4 & 0 & 6 \\ 22 & -2 & 2 \end{bmatrix}}_{\mathbf{AB}};$$

$$\underbrace{\begin{bmatrix} 1 & 0 & 2 \\ 0 & 3 & 1 \\ 2 & 0 & 5 \end{bmatrix}}_{\mathbf{A}} \underbrace{\begin{bmatrix} 1 \\ 0 \\ 4 \end{bmatrix}}_{\mathbf{v}} = \underbrace{\begin{bmatrix} 9 \\ 4 \\ 22 \end{bmatrix}}_{\mathbf{Av}}. \quad \blacksquare$$

Note that **Av** will again be a column vector, of the same size as **v**; thus **an $n \times n$ matrix A transforms any vector v in n-space to a new vector Av.**

EXAMPLE The matrix $\begin{bmatrix} 1 & 0 & 2 \\ 0 & 3 & 1 \\ 2 & 0 & 5 \end{bmatrix}$ transforms any vector (x, y, z) in 3-space to the new vector $(x + 2z, 3y + z, 2x + 5z)$, because

$$\underbrace{\begin{bmatrix} 1 & 0 & 2 \\ 0 & 3 & 1 \\ 2 & 0 & 5 \end{bmatrix}}_{A} \underbrace{\begin{bmatrix} x \\ y \\ z \end{bmatrix}}_{v} = \underbrace{\begin{bmatrix} x + 2z \\ 3y + z \\ 2x + 5z \end{bmatrix}}_{Av}.$$ ∎

Conversely, many transformations (of vectors) that come up in geometry, such as rotations and reflections that leave the origin fixed, can be described using matrices. It is shown in linear algebra textbooks that the composition of two such transformations is then described by the product of their matrices. This is the reason that matrix multiplication is defined the way it is, and it shows that **matrix multiplication is associative**: For any three matrices **A, B, C** for which the products are defined, $A(BC) = (AB)C$.

A particularly important $n \times n$ matrix is the **identity matrix** I_n,[6] whose entries along the diagonal from the upper left to the lower right of the matrix (the **main diagonal**) are 1 while all other entries are 0.

EXAMPLE $I_4 = \begin{bmatrix} 1 & 0 & 0 & 0 \\ 0 & 1 & 0 & 0 \\ 0 & 0 & 1 & 0 \\ 0 & 0 & 0 & 1 \end{bmatrix}.$ ∎

It can be shown (see Exercise 5) that $I_n B = B$ for any matrix **B** with n rows. In particular, $I_n v = v$ for any column vector **v** of size n, so I_n transforms every vector to itself.

Two (square) $n \times n$ matrices **A** and **B** are said to be **inverses** of each other if their product is I_n. It can be shown that if $AB = I_n$, then also $BA = I_n$, so the order doesn't matter in this case.

EXAMPLE B.2.3 $\begin{bmatrix} 3 & 2 \\ 1 & 1 \end{bmatrix} \begin{bmatrix} 1 & -2 \\ -1 & 3 \end{bmatrix} = \begin{bmatrix} 1 & 0 \\ 0 & 1 \end{bmatrix} = I_2$, so $\begin{bmatrix} 3 & 2 \\ 1 & 1 \end{bmatrix}$ and $\begin{bmatrix} 1 & -2 \\ -1 & 3 \end{bmatrix}$ are inverses of each other. ∎

---

[6] When n is clear from the context, the identity matrix is often simply denoted by **I**.

SEC. B.2 MATRICES AND COLUMN VECTORS

If $\mathbf{A}$ and $\mathbf{B}$ are inverses of each other, we write $\mathbf{B} = \mathbf{A}^{-1}$ (and $\mathbf{A} = \mathbf{B}^{-1}$). Any matrix $\mathbf{A}$ can have only one inverse (see Exercise 8).

Inverse matrices give rise to inverse transformations of vectors, that is, if a vector $\mathbf{v}$ is transformed first by a matrix $\mathbf{A}$ (yielding $\mathbf{Av}$) and the result is transformed by the inverse matrix $\mathbf{A}^{-1}$, the final result will be the original vector $\mathbf{v}$:

$$\mathbf{A}^{-1}(\mathbf{Av}) = (\mathbf{A}^{-1}\mathbf{A})\mathbf{v} = \mathbf{I}_n \mathbf{v} = \mathbf{v}.$$

Note that inverses are only defined for square matrices. However, not every square matrix has an inverse: If the matrix $\begin{bmatrix} 0 & 1 \\ 0 & 0 \end{bmatrix}$ had an inverse, there would be a matrix $\begin{bmatrix} a & b \\ c & d \end{bmatrix}$ such that $\begin{bmatrix} 0 & 1 \\ 0 & 0 \end{bmatrix} \begin{bmatrix} a & b \\ c & d \end{bmatrix} = \begin{bmatrix} 1 & 0 \\ 0 & 1 \end{bmatrix}$, that is, $\begin{bmatrix} c & d \\ 0 & 0 \end{bmatrix} = \begin{bmatrix} 1 & 0 \\ 0 & 1 \end{bmatrix}$.

Looking at the 2,2 entries on both sides, we see that we would have $0 = 1$, which is clearly impossible. In the next sections we will see which square matrices have an inverse and how these inverses can be computed.

EXERCISES

1. Let $\mathbf{A} = \begin{bmatrix} 1 & 2 & -4 \\ 3 & 0 & 1 \end{bmatrix}$, $\mathbf{B} = \begin{bmatrix} 2 & -1 \\ 1 & 3 \\ 0 & 1 \end{bmatrix}$. Compute $\mathbf{AB}$ and $\mathbf{BA}$.

2. Let $\mathbf{A} = \begin{bmatrix} 1 & 2 \\ 0 & 1 \end{bmatrix}$, $\mathbf{B} = \begin{bmatrix} 1 & 0 \\ 3 & 5 \end{bmatrix}$. Compute $\mathbf{AB}$ and $\mathbf{BA}$.

3. Let $\mathbf{A} = \begin{bmatrix} 4 & 0 & 1 \\ -1 & 1 & 3 \\ 5 & 1 & 2 \end{bmatrix}$, $\mathbf{v} = \begin{bmatrix} 2 \\ -1 \\ 3 \end{bmatrix}$. Compute $\mathbf{Av}$.

4. Let $\mathbf{A} = \begin{bmatrix} 6 & 2 \\ 2 & 6 \end{bmatrix}$, $\mathbf{v} = \begin{bmatrix} 1 \\ -1 \end{bmatrix}$. Show that $\mathbf{Av} = 4\mathbf{v}$.

5. Show that $\mathbf{I}_n \mathbf{B} = \mathbf{B}$ for any matrix $\mathbf{B}$ with n rows.

6. Are the following matrices inverses of each other?

   (a) $\begin{bmatrix} 1 & 2 & 3 \\ 0 & 1 & 0 \\ 0 & 0 & 1 \end{bmatrix}$ and $\begin{bmatrix} 1 & -2 & -3 \\ 0 & 1 & 0 \\ 0 & 0 & 1 \end{bmatrix}$.

   (b) $\begin{bmatrix} 1 & 4 \\ 0 & 1 \end{bmatrix}$ and $\begin{bmatrix} 1 & \frac{1}{4} \\ 0 & 1 \end{bmatrix}$.

   (c) $\begin{bmatrix} 5 & 0 \\ 0 & 1 \end{bmatrix}$ and $\begin{bmatrix} \frac{1}{5} & 0 \\ 0 & 1 \end{bmatrix}$.

7. Show that the matrix $\begin{bmatrix} 2 & 1 \\ 4 & 2 \end{bmatrix}$ has no inverse.

8. Show that if **B** and **C** are both inverses of the (square) matrix **A**, then **B** = **C**.

9. It is possible to define the **exponential** $e^{\mathbf{A}}$ of a square matrix **A** as the infinite sum

$$e^{\mathbf{A}} = \sum_{n=0}^{\infty} \frac{1}{n!} \mathbf{A}^n = \mathbf{I} + \mathbf{A} + \frac{1}{2}\mathbf{A}^2 + \frac{1}{6}\mathbf{A}^3 + \cdots.$$

Note that $\mathbf{A}^0$ is interpreted as **I**; as you would expect, $\mathbf{A}^2 = \mathbf{AA}$, $\mathbf{A}^3 = \mathbf{A}^2\mathbf{A}$, and so on. Recall that to find a term such as $\frac{1}{6}\mathbf{A}^3$, each entry in $\mathbf{A}^3$ is multiplied by $\frac{1}{6}$.

(a) Show that for $\mathbf{A} = \begin{bmatrix} 1 & 0 \\ 0 & 2 \end{bmatrix}$, $e^{\mathbf{A}} = \begin{bmatrix} e & 0 \\ 0 & e^2 \end{bmatrix}$.

(b) Show that for $\mathbf{A} = \begin{bmatrix} 1 & 2 \\ 0 & 1 \end{bmatrix}$, $e^{\mathbf{A}} = \begin{bmatrix} e & 2e \\ 0 & e \end{bmatrix}$.

10. (a) Show that if **A** is a square matrix and **v**, **w** are column vectors of the same size, then $\mathbf{A}(\mathbf{v} + \mathbf{w}) = \mathbf{Av} + \mathbf{Aw}$.

(b) Show that if **A** is a square matrix, **v** is a column vector of the same size, and k is a number, then $\mathbf{A}(k\mathbf{v}) = k(\mathbf{Av})$.

These two properties show that the transformation given by the matrix **A** (which transforms each vector **v** to the new vector **Av**) is a **linear transformation**.

11. Let $\mathbf{A} = \begin{bmatrix} 1 & 0 & 0 \\ 0 & 0 & 1 \end{bmatrix}$ and $\mathbf{B} = \begin{bmatrix} 1 & 0 \\ 0 & 0 \\ 0 & 1 \end{bmatrix}$. Show that $\mathbf{AB} = \mathbf{I}$, but that $\mathbf{BA} \neq \mathbf{I}$. Why doesn't this contradict the statement on p. 612 immediately after the definition of inverses?

B.3 SYSTEMS OF LINEAR EQUATIONS; INVERSES OF MATRICES

Consider a system of simultaneous linear equations, such as

$$\begin{cases} 2x + 3y + z = 1 \\ 4x - 3z = 9 \\ x + 4y + 5z = 0. \end{cases}$$

This system can be rewritten as follows: First form two vectors whose components are, respectively, the left- and right-hand sides of the equations. These vectors must be equal:

$$\begin{bmatrix} 2x + 3y + z \\ 4x - 3z \\ x + 4y + 5z \end{bmatrix} = \begin{bmatrix} 1 \\ 9 \\ 0 \end{bmatrix}.$$

Then interpret the left-hand side as a product of a known matrix with a vector whose components are the unknowns:

$$\begin{bmatrix} 2x + 3y + z \\ 4x - 3z \\ x + 4y + 5z \end{bmatrix} = \begin{bmatrix} 2 & 3 & 1 \\ 4 & 0 & -3 \\ 1 & 4 & 5 \end{bmatrix} \begin{bmatrix} x \\ y \\ z \end{bmatrix}.$$

The system then becomes

$$\begin{bmatrix} 2 & 3 & 1 \\ 4 & 0 & -3 \\ 1 & 4 & 5 \end{bmatrix} \begin{bmatrix} x \\ y \\ z \end{bmatrix} = \begin{bmatrix} 1 \\ 9 \\ 0 \end{bmatrix}.$$

Note that the entries of the matrix on the left are just the coefficients of the unknowns in the original system.

If we abbreviate this matrix by $\mathbf{A}$ and the vector of unknowns by $\mathbf{v}$, then the system has the form $\mathbf{Av} = \mathbf{b}$, where $\mathbf{b}$ is the known vector whose components are the right-hand sides of the original equations. If there were m equations in n unknowns, then $\mathbf{A}$ will be an $m \times n$ matrix; in particular, if the number of equations is equal to the number of unknowns, then $\mathbf{A}$ will be a square matrix. From now on we'll assume that this is the case.

If $\mathbf{A}$ has an inverse, then we can multiply both sides of $\mathbf{Av} = \mathbf{b}$ on the left by $\mathbf{A}^{-1}$ to obtain

$$\mathbf{A}^{-1}\mathbf{Av} = \mathbf{A}^{-1}\mathbf{b}$$

$$\mathbf{Iv} = \mathbf{A}^{-1}\mathbf{b}$$

$$\mathbf{v} = \mathbf{A}^{-1}\mathbf{b}.$$

Conversely, if $\mathbf{v} = \mathbf{A}^{-1}\mathbf{b}$, then $\mathbf{Av} = \mathbf{b}$. Conclusion:

If $\mathbf{A}$ has an inverse,[7] then the system $\mathbf{Av} = \mathbf{b}$ has a unique solution $\mathbf{v} = \mathbf{A}^{-1}\mathbf{b}$.

---

[7] A square matrix with an inverse is often called **invertible** or **nonsingular**; if $\mathbf{A}$ has no inverse, it is a **singular** matrix.

EXAMPLE B.3.1 The system $\begin{cases} 3x + 2y = 7 \\ x + y = 11 \end{cases}$ can be rewritten $\mathbf{Av} = \mathbf{b}$ with

$$\mathbf{A} = \begin{bmatrix} 3 & 2 \\ 1 & 1 \end{bmatrix}, \quad \mathbf{v} = \begin{bmatrix} x \\ y \end{bmatrix}, \quad \mathbf{b} = \begin{bmatrix} 7 \\ 11 \end{bmatrix}.$$

Since we saw in Example B.2.3 that $\mathbf{A}^{-1} = \begin{bmatrix} 1 & -2 \\ -1 & 3 \end{bmatrix}$, the system has a unique solution, given by

$$\begin{bmatrix} x \\ y \end{bmatrix} = \begin{bmatrix} 1 & -2 \\ -1 & 3 \end{bmatrix} \begin{bmatrix} 7 \\ 11 \end{bmatrix} = \begin{bmatrix} -15 \\ 26 \end{bmatrix}, \quad \text{that is,} \quad x = -15, \, y = 26. \quad \blacksquare$$

It can be shown that if a square matrix $\mathbf{A}$ has no inverse, then any system $\mathbf{Av} = \mathbf{b}$ has either infinitely many solutions or no solution at all.

EXAMPLE B.3.2 For the system $\begin{cases} 2x + y = 0 \\ 4x + 2y = 0 \end{cases}$, we get the matrix $\mathbf{A} = \begin{bmatrix} 2 & 1 \\ 4 & 2 \end{bmatrix}$, which has no inverse (see Section B.2, Exercise 7). Since the system obviously has the solution $x = 0, y = 0$, it must have infinitely many solutions (see also Exercise 14). $\quad \blacksquare$

A system of linear equations of the form $\mathbf{Av} = \mathbf{0}$, such as the system in the example above, is called **homogeneous**. A homogeneous system always has the **trivial solution** $\mathbf{v} = \mathbf{0}$; if $\mathbf{A}$ has an inverse, the system $\mathbf{Av} = \mathbf{0}$ has only the trivial solution, while if $\mathbf{A}$ has no inverse, that homogeneous system will have infinitely many solutions.

To find the inverse (if it exists) of a square matrix $\mathbf{A}$, the following technique, which is called **row reduction**, is used.[8] If $\mathbf{A}$ is an $n \times n$ matrix, one first forms the $n \times 2n$ matrix $[\mathbf{A}|\mathbf{I}]$, where $\mathbf{I}$ is the $n \times n$ identity matrix. For instance, if

$$\mathbf{A} = \begin{bmatrix} 2 & 4 & 1 \\ 3 & -1 & 2 \\ 4 & 0 & 1 \end{bmatrix},$$

then

$$[\mathbf{A}|\mathbf{I}] = \begin{bmatrix} 2 & 4 & 1 & | & 1 & 0 & 0 \\ 3 & -1 & 2 & | & 0 & 1 & 0 \\ 4 & 0 & 1 & | & 0 & 0 & 1 \end{bmatrix}.$$

(The vertical bar between $\mathbf{A}$ and $\mathbf{I}$ is not strictly necessary, but is a convenient way to keep track of the "middle" of the matrix.) Now arithmetic operations called row operations, to

---

[8] This technique is justified in books on linear algebra.

SEC. B.3 SYSTEMS OF LINEAR EQUATIONS; INVERSES OF MATRICES

be explained below, are used to change the rows of the matrix until it has the form [**I**|?] with the identity matrix to the left of the vertical bar; as a result, a new matrix (shown as ?) will appear to the right of the vertical bar. This new matrix is $\mathbf{A}^{-1}$.

The following three types of operations can be used as row operations:

1. Adding a multiple of one row to another. For instance, adding $(-\frac{3}{2})$ times the first row to the second row changes

$$\begin{bmatrix} 2 & 4 & 1 & | & 1 & 0 & 0 \\ 3 & -1 & 2 & | & 0 & 1 & 0 \\ 4 & 0 & 1 & | & 0 & 0 & 1 \end{bmatrix} \text{ to } \begin{bmatrix} 2 & 4 & 1 & | & 1 & 0 & 0 \\ 0 & -7 & \frac{1}{2} & | & -\frac{3}{2} & 1 & 0 \\ 4 & 0 & 1 & | & 0 & 0 & 1 \end{bmatrix}.$$

2. Multiplying a row by a nonzero constant. For instance, multiplying the second row by (-2) changes

$$\begin{bmatrix} 2 & 4 & 1 & | & 1 & 0 & 0 \\ 0 & -7 & \frac{1}{2} & | & -\frac{3}{2} & 1 & 0 \\ 4 & 0 & 1 & | & 0 & 0 & 1 \end{bmatrix} \text{ to } \begin{bmatrix} 2 & 4 & 1 & | & 1 & 0 & 0 \\ 0 & 14 & 1 & | & 3 & -2 & 0 \\ 4 & 0 & 1 & | & 0 & 0 & 1 \end{bmatrix}.$$

3. Interchanging two rows. For instance, interchanging the last two rows changes

$$\begin{bmatrix} 2 & 4 & 1 & | & 1 & 0 & 0 \\ 0 & 14 & 1 & | & 3 & -2 & 0 \\ 4 & 0 & 1 & | & 0 & 0 & 1 \end{bmatrix} \text{ to } \begin{bmatrix} 2 & 4 & 1 & | & 1 & 0 & 0 \\ 4 & 0 & 1 & | & 0 & 0 & 1 \\ 0 & 14 & 1 & | & 3 & -2 & 0 \end{bmatrix}.$$

Note that each of these operations is applied equally to the parts of the rows before and after the vertical bar.

EXAMPLE B.3.3 Find the inverse of $\mathbf{A} = \begin{bmatrix} 0 & 2 & 6 \\ 1 & -1 & 4 \\ 3 & 4 & 1 \end{bmatrix}$.

Solution We apply successive row operations to [**A**|**I**] until we obtain [**I**|$\mathbf{A}^{-1}$]:

$$\begin{bmatrix} 0 & 2 & 6 & | & 1 & 0 & 0 \\ 1 & -1 & 4 & | & 0 & 1 & 0 \\ 3 & 4 & 1 & | & 0 & 0 & 1 \end{bmatrix} \xrightarrow{R_1 \leftrightarrow R_2} \begin{bmatrix} 1 & -1 & 4 & | & 0 & 1 & 0 \\ 0 & 2 & 6 & | & 1 & 0 & 0 \\ 3 & 4 & 1 & | & 0 & 0 & 1 \end{bmatrix} \xrightarrow{R_3 - 3R_1}$$

$$\begin{bmatrix} 1 & -1 & 4 & | & 0 & 1 & 0 \\ 0 & 2 & 6 & | & 1 & 0 & 0 \\ 0 & 7 & -11 & | & 0 & -3 & 1 \end{bmatrix} \xrightarrow{R_2 \cdot \frac{1}{2}} \begin{bmatrix} 1 & -1 & 4 & | & 0 & 1 & 0 \\ 0 & 1 & 3 & | & \frac{1}{2} & 0 & 0 \\ 0 & 7 & -11 & | & 0 & -3 & 1 \end{bmatrix} \xrightarrow{R_1 + R_2}$$

$$
\begin{bmatrix} 1 & 0 & 7 & | & \frac{1}{2} & 1 & 0 \\ 0 & 1 & 3 & | & \frac{1}{2} & 0 & 0 \\ 0 & 7 & -11 & | & 0 & -3 & 1 \end{bmatrix} \xrightarrow{R_3 - 7R_2} \begin{bmatrix} 1 & 0 & 7 & | & \frac{1}{2} & 1 & 0 \\ 0 & 1 & 3 & | & \frac{1}{2} & 0 & 0 \\ 0 & 0 & -32 & | & -\frac{7}{2} & -3 & 1 \end{bmatrix} \xrightarrow{R_3 \cdot -\frac{1}{32}}
$$

$$
\begin{bmatrix} 1 & 0 & 7 & | & \frac{1}{2} & 1 & 0 \\ 0 & 1 & 3 & | & \frac{1}{2} & 0 & 0 \\ 0 & 0 & 1 & | & \frac{7}{64} & \frac{3}{32} & -\frac{1}{32} \end{bmatrix} \xrightarrow{R_1 - 7R_3} \begin{bmatrix} 1 & 0 & 0 & | & -\frac{17}{64} & \frac{11}{32} & \frac{7}{32} \\ 0 & 1 & 3 & | & \frac{1}{2} & 0 & 0 \\ 0 & 0 & 1 & | & \frac{7}{64} & \frac{3}{32} & -\frac{1}{32} \end{bmatrix}
$$

$$
\xrightarrow{R_2 - 3R_3} \begin{bmatrix} 1 & 0 & 0 & | & -\frac{17}{64} & \frac{11}{32} & \frac{7}{32} \\ 0 & 1 & 0 & | & \frac{11}{64} & -\frac{9}{32} & \frac{3}{32} \\ 0 & 0 & 1 & | & \frac{7}{64} & \frac{3}{32} & -\frac{1}{32} \end{bmatrix}.
$$

So we find that

$$
\mathbf{A}^{-1} = \begin{bmatrix} -\frac{17}{64} & \frac{11}{32} & \frac{7}{32} \\ \frac{11}{64} & -\frac{9}{32} & \frac{3}{32} \\ \frac{7}{64} & \frac{3}{32} & -\frac{1}{32} \end{bmatrix}.
$$

NOTES:

1. It would be wise to check this result by computing $\mathbf{AA}^{-1}$ (which should equal $\mathbf{I}$).

2. It is also a good habit to indicate which row operation is used at each step; the shorthand used above will work well. The notation $R_3 - 3R_1$ indicates that three times the first row is subtracted from the third; the first row itself remains the same. If you try to carry out several steps at once, make sure that the result of the first step doesn't influence the second. For instance, one cannot carry out $R_1 + 2R_2$ and $R_2 - R_1$ at once since the first of these steps changes R_1, which is used in the second step. ∎

SEC. B.3 SYSTEMS OF LINEAR EQUATIONS; INVERSES OF MATRICES

The procedure used in the example above is actually quite systematic; the identity matrix to the left of the vertical bar is obtained one column at a time, by getting a 1 in the correct position and then using multiples of the row containing it to get 0 elsewhere in the column. (Shortcuts may be possible for specific matrices.) The procedure will fail if $\mathbf{A}$ does not have an inverse; in this case a row of zeros will appear to the left of the vertical bar at some point in the row reduction. Conversely, if this happens, $\mathbf{A}$ does not have an inverse.

EXAMPLE B.3.4 Find the inverse (if it exists) of $\mathbf{A} = \begin{bmatrix} 2 & 4 \\ 3 & 6 \end{bmatrix}$.

$$[\mathbf{A}|\mathbf{I}] = \begin{bmatrix} 2 & 4 & | & 1 & 0 \\ 3 & 6 & | & 0 & 1 \end{bmatrix} \xrightarrow{R_1 \cdot \frac{1}{2}} \begin{bmatrix} 1 & 2 & | & \frac{1}{2} & 0 \\ 3 & 6 & | & 0 & 1 \end{bmatrix} \xrightarrow{R_2 - 3R_1} \begin{bmatrix} 1 & 2 & | & 1 & 0 \\ 0 & 0 & | & -\frac{3}{2} & 1 \end{bmatrix}.$$

Since the second row to the left of the vertical bar consists only of zeros, we can stop; $\mathbf{A}$ does not have an inverse. ∎

Since finding the inverse of a matrix can be laborious, it may be worthwhile to memorize the general result in the case of a 2×2 matrix:

$$\text{If} \quad \mathbf{A} = \begin{bmatrix} a & b \\ c & d \end{bmatrix}, \quad \text{then} \quad \mathbf{A}^{-1} = \frac{1}{ad - bc} \begin{bmatrix} d & -b \\ -c & a \end{bmatrix},$$

provided $ad - bc \neq 0$. If $ad - bc = 0$, $\mathbf{A}^{-1}$ does not exist.

EXAMPLE B.3.5 For $\mathbf{A} = \begin{bmatrix} 4 & 2 \\ 1 & 7 \end{bmatrix}$, $ad - bc = 4 \cdot 7 - 2 \cdot 1 = 26$, so

$$\mathbf{A}^{-1} = \frac{1}{26} \begin{bmatrix} 7 & -2 \\ -1 & 4 \end{bmatrix} = \begin{bmatrix} \frac{7}{26} & -\frac{2}{26} \\ -\frac{1}{26} & \frac{4}{26} \end{bmatrix}. \quad \blacksquare$$

The quantity $ad - bc$, which "determines" whether $\mathbf{A}$ has an inverse, is called the **determinant** of $\mathbf{A}$. In the next section, we'll consider determinants in general.

EXERCISES

Rewrite the following systems in the form $\mathbf{Av} = \mathbf{b}$.

1. $\begin{cases} 3x - 4y = 2 \\ -5x + y = 3. \end{cases}$

2. $\begin{cases} 4x - y = 8 \\ 3x + y + z = 6 \\ 2x - y - 3z = 10. \end{cases}$

3. Show that the formula on p. 619 for the inverse of a 2 × 2 matrix
$$\mathbf{A} = \begin{bmatrix} a & b \\ c & d \end{bmatrix} \text{ with } ad - bc \neq 0 \text{ is correct.}$$
[*Hint*: If $\mathbf{AB} = \mathbf{I}_2$, then $\mathbf{B}$ must be the inverse of $\mathbf{A}$.]

Find the inverses, if they exist, of the following matrices.

4. $\begin{bmatrix} 1 & 3 \\ 4 & 0 \end{bmatrix}$ 5. $\begin{bmatrix} 5 & 1 \\ 6 & 2 \end{bmatrix}$ 6. $\begin{bmatrix} 4 & 2 \\ 6 & 3 \end{bmatrix}$

7. $\begin{bmatrix} 4 & 0 & 0 \\ 0 & 2 & 3 \\ 0 & 1 & 1 \end{bmatrix}$ 8. $\begin{bmatrix} 1 & 5 & 3 \\ -3 & -10 & -7 \\ 2 & 12 & 7 \end{bmatrix}$ 9. $\begin{bmatrix} -17 & 14 & 18 \\ -1 & 1 & 1 \\ -4 & 3 & 4 \end{bmatrix}$

10. $\begin{bmatrix} 2 & 3 & 1 \\ 1 & 0 & 4 \\ 0 & 3 & -7 \end{bmatrix}$

Use the results of the exercises above to solve the following systems.

11. $\begin{cases} x + 5y + 3z = 4 \\ -3x - 10y - 7z = 2 \\ 2x + 12y + 7z = 0. \end{cases}$ 12. $\begin{cases} -17x + 14y + 18z = 0 \\ -x + y + z = 0 \\ -4x + 3y + 4z = 1. \end{cases}$

13. $\begin{cases} -17x + 14y + 18z = 0 \\ -x + y + z = 0 \\ -4x + 3y + 4z = 0. \end{cases}$

14. (See Example B.3.2.) Show directly that the system $\begin{cases} 2x + y = 0 \\ 4x + 2y = 0 \end{cases}$ has infinitely many solutions.

15. Use Exercise 10 to determine how many solutions the system
$$\begin{cases} 2x + 3y + z = 0 \\ x \phantom{{}+3y} + 4z = 0 \\ \phantom{x+{}} 3y - 7z = 0 \end{cases}$$
has. Then find all solutions.

16. Show that if $\mathbf{A}$ and $\mathbf{B}$ are square matrices of the same size such that $\mathbf{A}^{-1}$ and $\mathbf{B}^{-1}$ exist, then $(\mathbf{AB})^{-1}$ exists and equals $\mathbf{B}^{-1}\mathbf{A}^{-1}$.
[*Hint*: What does it mean if, indeed, $(\mathbf{AB})^{-1} = \mathbf{B}^{-1}\mathbf{A}^{-1}$?]

B.4 DETERMINANTS

As we saw in Section B.3, a system of linear equations $\mathbf{Av} = \mathbf{b}$ with a square matrix $\mathbf{A}$ has a unique solution if and only if $\mathbf{A}$ has an inverse. However, actually having to compute the inverse of $\mathbf{A}$ can be a nuisance; it is often helpful to be able to tell whether $\mathbf{A}$ has an inverse without having to find it. This can be done by computing the **determinant** $\det(\mathbf{A})$ of $\mathbf{A}$. The determinant is a number with, among other things, the property that

$\mathbf{A}$ has an inverse if and only if $\det(\mathbf{A}) \neq 0$

and therefore

$\mathbf{A}$ has no inverse if and only if $\det(\mathbf{A}) = 0$.[9]

Various equivalent definitions of the determinant can be given. All these definitions, though, involve a fair amount of work, either to show that the determinant is really well defined or to show that it has the desired property. See a textbook on linear algebra for a thorough treatment. Here one method of computing determinants and a list of some of their properties will be given, without motivation or proof.

To begin with, for a 1×1 matrix $\mathbf{A} = [a]$, the determinant is the entry itself, that is, $\det(\mathbf{A}) = a$. Next, for a 2×2 matrix $\mathbf{A} = \begin{bmatrix} a & b \\ c & d \end{bmatrix}$, the determinant is given by $\det(\mathbf{A}) = ad - bc$.

NOTE: When $\mathbf{A}$ is a given (2×2 or larger) matrix, a common way of writing $\det(\mathbf{A})$ is to replace the brackets around the matrix by vertical bars. For instance, $\begin{vmatrix} 3 & 5 \\ 2 & 6 \end{vmatrix}$ stands for $\det \begin{bmatrix} 3 & 5 \\ 2 & 6 \end{bmatrix} = 3 \cdot 6 - 5 \cdot 2 = 8$.

In the formula $\begin{vmatrix} a & b \\ c & d \end{vmatrix} = ad - bc$, note that there are two terms ad and bc on the right, which can be thought of as corresponding to the two entries a and b of the top row of the matrix. Each of these entries is multiplied by the entry (d or c, respectively) diagonally across from it, yielding ad and bc, and then the products are added with alternating signs: $+ad - bc$. Now one way to get the entry d which is diagonally across from a is to delete the row and the column that a is in from the matrix, as shown in Figure B.4.1; if we start with b instead of a, we get the entry c in this same way.

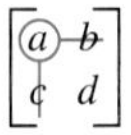

$$\begin{bmatrix} \cancel{a} & \cancel{b} \\ \cancel{c} & d \end{bmatrix}$$

Figure B.4.1. Deleting the row and column of a from the matrix.

---

[9] Thus $\mathbf{A}$ is **nonsingular** if and only if $\det(\mathbf{A}) \neq 0$ (the "usual" case), and $\mathbf{A}$ is **singular** when $\det(\mathbf{A}) = 0$. (See p. 615, footnote 7.)

For determinants of larger matrices, there will be a similar pattern: Each entry in the top row is multiplied by the *determinant* of the "matrix diagonally across" and the products are added *with alternating signs*. The "matrix diagonally across" or **minor** of an entry is defined to be the matrix[10] that remains when the row and column of that entry are deleted from the original matrix (see Figure B.4.2).

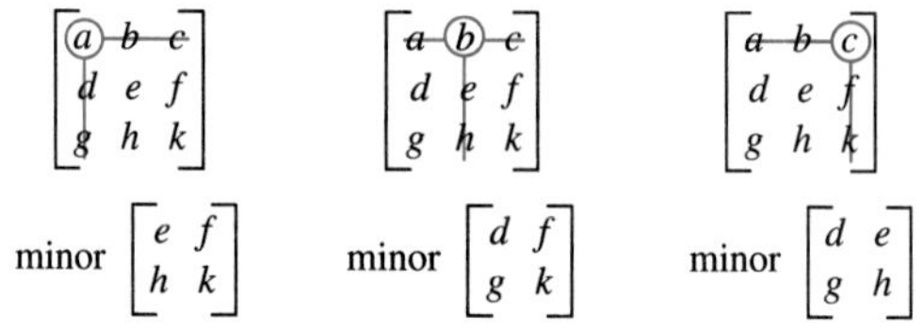

Figure B.4.2. Minors for the top-row entries of $\mathbf{A} = \begin{bmatrix} a & b & c \\ d & e & f \\ g & h & k \end{bmatrix}$.

Putting all this together, the definition of the determinant of a 3×3 matrix will be

$$\begin{vmatrix} a & b & c \\ d & e & f \\ g & h & k \end{vmatrix} = a \cdot \begin{vmatrix} e & f \\ h & k \end{vmatrix} - b \cdot \begin{vmatrix} d & f \\ g & k \end{vmatrix} + c \cdot \begin{vmatrix} d & e \\ g & h \end{vmatrix}.$$

Note that only 2×2 determinants appear on the right, and we already know how to compute these.

EXAMPLE

$$\begin{vmatrix} 3 & 4 & 5 \\ 1 & 0 & 2 \\ -1 & 1 & 3 \end{vmatrix} = 3 \cdot \begin{vmatrix} 0 & 2 \\ 1 & 3 \end{vmatrix} - 4 \cdot \begin{vmatrix} 1 & 2 \\ -1 & 3 \end{vmatrix} + 5 \cdot \begin{vmatrix} 1 & 0 \\ -1 & 1 \end{vmatrix}$$
$$= 3 \cdot [0(3) - 2(1)] - 4 \cdot [1(3) - 2(-1)] + 5 \cdot [1(1) - 0(-1)]$$
$$= -6 - 20 + 5 = -21. \quad \blacksquare$$

The same process is used for larger square matrices; each time, the determinant is "expanded" in terms of determinants of smaller size. Specifically, for an $n \times n$ matrix

$$\mathbf{A} = \begin{bmatrix} a_{11} & \cdots & a_{1n} \\ \vdots & & \vdots \\ a_{n1} & \cdots & a_{nn} \end{bmatrix},$$

we have

$$\det(\mathbf{A}) = a_{11} \det(\mathbf{M}_{11}) - a_{12} \det(\mathbf{M}_{12}) + a_{13} \det(\mathbf{M}_{13}) \cdots \pm a_{1n} \det(\mathbf{M}_{1n}),$$

---

[10] Many authors use the term *minor* for the *determinant* of this matrix, instead.

where $\mathbf{M}_{ij}$ is the minor of the entry a_{ij}, obtained by deleting the ith row and jth column of $\mathbf{A}$. This expression for $\det(\mathbf{A})$ is known as the **expansion** of the determinant **along the first row**.[11]

Properties of determinants include:

1. If two rows or two columns of a matrix are interchanged, the determinant changes sign.

EXAMPLE
$$\begin{vmatrix} 1 & 3 & 5 \\ 2 & 4 & 6 \\ 0 & 0 & 1 \end{vmatrix} = - \begin{vmatrix} 0 & 0 & 1 \\ 2 & 4 & 6 \\ 1 & 3 & 5 \end{vmatrix}.$$

Note that the determinant on the right is easier to compute using our definition. ∎

2. If a multiple of a row (or column) of a matrix is added to another row (or column), the determinant remains unchanged.

EXAMPLE
$$\begin{vmatrix} 4 & 1 & 2 & 7 \\ 2 & 0 & 1 & 4 \\ 3 & 2 & 6 & 5 \\ 1 & 3 & 8 & -1 \end{vmatrix} \underset{R_1 - 2R_2}{=} \begin{vmatrix} 0 & 1 & 0 & -1 \\ 2 & 0 & 1 & 4 \\ 3 & 2 & 6 & 5 \\ 1 & 3 & 8 & -1 \end{vmatrix}$$

$$= - \begin{vmatrix} 2 & 1 & 4 \\ 3 & 6 & 5 \\ 1 & 8 & -1 \end{vmatrix} - (-1) \cdot \begin{vmatrix} 2 & 0 & 1 \\ 3 & 2 & 6 \\ 1 & 3 & 8 \end{vmatrix} \quad \text{(why?)}.\quad ∎$$

3. If a matrix has a row or column consisting entirely of zeros, its determinant is 0.

EXAMPLE
$$\begin{vmatrix} 3 & 0 & -1 \\ 1 & 0 & 2 \\ 5 & 0 & 4 \end{vmatrix} = 0.\quad ∎$$

4. The determinant of a product is equal to the product of the determinants: $\det(\mathbf{AB}) = \det(\mathbf{A}) \cdot \det(\mathbf{B})$ for square matrices $\mathbf{A}, \mathbf{B}$ of the same size.

Warning: The determinant of the sum of two matrices is usually *not* equal to the sum of the determinants (see Exercise 17).

5. If a row (or column) of a matrix is multiplied (or divided) by a nonzero constant, the determinant is multiplied (or divided) by that same constant.

EXAMPLES
$$\begin{vmatrix} a & b & c \\ 5d & 5e & 5f \\ g & h & k \end{vmatrix} = 5 \cdot \begin{vmatrix} a & b & c \\ d & e & f \\ g & h & k \end{vmatrix}; \quad \begin{vmatrix} 100a & b \\ 100c & d \end{vmatrix} = 100 \cdot \begin{vmatrix} a & b \\ c & d \end{vmatrix}.\quad ∎$$

---

[11] See a linear algebra textbook for expansions of determinants along other rows, and along columns.

This property is sometimes expressed by saying that a constant can be "factored out" of any row or column of a determinant.

6. (Once more:) A square matrix **A** has an inverse if and only if $\det(\mathbf{A}) \neq 0$. Therefore, by Section B.3, a system of linear equations $\mathbf{Av} = \mathbf{b}$ has a unique solution if and only if $\det(\mathbf{A}) \neq 0$. In particular, a homogeneous system $\mathbf{Av} = \mathbf{0}$ has only the trivial solution if $\det(\mathbf{A}) \neq 0$ and has infinitely many solutions if $\det(\mathbf{A}) = 0$.

EXERCISES

Compute the following determinants.

1. $\begin{vmatrix} 2 & 5 \\ 3 & 7 \end{vmatrix}.$

2. $\begin{vmatrix} 4 & 0 \\ -1 & 1 \end{vmatrix}.$

3. $\begin{vmatrix} 1 & 5 & 0 \\ 2 & 4 & 6 \\ 1 & -1 & 2 \end{vmatrix}.$

4. $\begin{vmatrix} 2 & 4 & 6 \\ 1 & 5 & 0 \\ 1 & -1 & 2 \end{vmatrix}.$

5. $\begin{vmatrix} 3 & 2 & 3 \\ 1 & 0 & 1 \\ 6 & 7 & 8 \end{vmatrix}.$

6. $\begin{vmatrix} 2 & 1 & 3 \\ -1 & 1 & 1 \\ 6 & 0 & 4 \end{vmatrix}.$

7. $\begin{vmatrix} 3 & 0 & 5 & 2 \\ 0 & 0 & 0 & 0 \\ 1 & -1 & 4 & 6 \\ 2 & 3 & 0 & 5 \end{vmatrix}.$

8. $\begin{vmatrix} 1 & 0 & 2 & 4 \\ 3 & 0 & -1 & 6 \\ 1 & 0 & 5 & 8 \\ 2 & 0 & 1 & 3 \end{vmatrix}.$

9. $\begin{vmatrix} 3 & 0 & 0 \\ 0 & 2 & 0 \\ 0 & 0 & 5 \end{vmatrix}.$

10. $\begin{vmatrix} 1 & 2 & 0 & 3 \\ 4 & 5 & 6 & 7 \\ 4 & 6 & 7 & 8 \\ 4 & 6 & 8 & 9 \end{vmatrix}.$

11. $\begin{vmatrix} 0 & 2 & 1 & 0 \\ 3 & 1 & 2 & 4 \\ 1 & -1 & 1 & -1 \\ 2 & -1 & 3 & -1 \end{vmatrix}.$

12. (a) Show that $\det(\mathbf{I}_n) = 1$ for any n.

    (b) Show that for any invertible matrix **A**, $\det(\mathbf{A}^{-1}) = \dfrac{1}{\det(\mathbf{A})}$.

13. Does the system
$$\begin{cases} 2x + y + 3z - 4t = 0 \\ x - y - z + 2t = 0 \\ 3x + y + z - t = 0 \\ -y + z - t = 0 \end{cases}$$
have a nontrivial solution? Why?

14. Let **A** be a **diagonal** matrix, that is, a matrix all of whose entries off the main diagonal are zero:
$$\mathbf{A} = \begin{bmatrix} a_{11} & & & 0 \\ & a_{22} & & \\ & & \ddots & \\ 0 & & & a_{nn} \end{bmatrix}.$$
Show that $\det(\mathbf{A})$ is the product $a_{11} a_{22} \cdots a_{nn}$ of the entries on the main diagonal.

15. (a) Show that the result of Exercise 14 is still true if all entries *above* the main diagonal are zero (**A** is then called **lower triangular**) but the entries on and below the diagonal are arbitrary.

(b) As part (a), but now all entries *below* the main diagonal are zero (**A** is **upper triangular**).

16. Use the results of Exercise 15 to compute the following determinants.

(a) $\begin{vmatrix} 2 & 0 & 0 & 0 \\ 0 & 3 & 0 & 0 \\ 0 & 0 & 4 & 0 \\ 0 & 0 & 0 & 5 \end{vmatrix}.$
(b) $\begin{vmatrix} 2 & 0 & 0 & 0 \\ 6 & 3 & 0 & 0 \\ 7 & 0 & 5 & 0 \\ -1 & 1 & 2 & 1 \end{vmatrix}.$

(c) $\begin{vmatrix} 1 & 3 & 5 & 0 \\ 0 & 2 & 1 & 6 \\ 0 & 0 & -1 & 7 \\ 0 & 0 & 0 & 3 \end{vmatrix}.$
(d) $\begin{vmatrix} 0 & 0 & 1 & 3 \\ 0 & 1 & 5 & 2 \\ 1 & 0 & -1 & 4 \\ 0 & 0 & 0 & 5 \end{vmatrix}.$

[*Hint* for part (d): Interchange rows suitably.]

17. Let $\mathbf{A} = \begin{bmatrix} 1 & 0 \\ 0 & 2 \end{bmatrix}$, $\mathbf{B} = \begin{bmatrix} -1 & 0 \\ 0 & -2 \end{bmatrix}$. Compute det(**A**), det(**B**), and det(**A** + **B**).

18. (a) Does it follow from property 5 on p. 623 that

$$\begin{vmatrix} 100 & 1 & 0 \\ 200 & 0 & 1 \\ 300 & 400 & 500 \end{vmatrix} = 100^2 \cdot \begin{vmatrix} 1 & 1 & 0 \\ 2 & 0 & 1 \\ 3 & 4 & 5 \end{vmatrix}?$$

Why, or why not?

(b) Compute both sides of the "equality" in part (a) directly.

19. (a) Show that for any numbers $\lambda_1, \lambda_2, \lambda_3$,

$$\begin{vmatrix} 1 & 1 \\ \lambda_1 & \lambda_2 \end{vmatrix} = (\lambda_2 - \lambda_1) \quad \text{and}$$

$$\begin{vmatrix} 1 & 1 & 1 \\ \lambda_1 & \lambda_2 & \lambda_3 \\ \lambda_1^2 & \lambda_2^2 & \lambda_3^2 \end{vmatrix} = (\lambda_2 - \lambda_1)(\lambda_3 - \lambda_1)(\lambda_3 - \lambda_2).$$

(b) Show that for any numbers $\lambda_1, \ldots, \lambda_n$,

$$\begin{vmatrix} 1 & 1 & \cdots & 1 \\ \lambda_1 & \lambda_2 & \cdots & \lambda_n \\ \lambda_1^2 & \lambda_2^2 & \cdots & \lambda_n^2 \\ \vdots & \vdots & & \vdots \\ \lambda_1^{n-1} & \lambda_2^{n-1} & \cdots & \lambda_n^{n-1} \end{vmatrix} = \begin{vmatrix} \lambda_2 - \lambda_1 & \cdots & \lambda_n - \lambda_1 \\ \lambda_2^2 - \lambda_1^2 & \cdots & \lambda_n^2 - \lambda_1^2 \\ \vdots & & \vdots \\ \lambda_2^{n-1} - \lambda_1^{n-1} & \cdots & \lambda_n^{n-1} - \lambda_1^{n-1} \end{vmatrix}.$$

[*Hint*: Subtract the first column from each of the others.]

The determinant on the left is called a **Vandermonde determinant**; it is used in Section 4.1.

**(c) Show by induction on n that this Vandermonde determinant equals the product $(\lambda_2 - \lambda_1)(\lambda_3 - \lambda_1)(\lambda_3 - \lambda_2)(\lambda_4 - \lambda_1) \cdots (\lambda_n - \lambda_{n-1})$ of all differences $\lambda_j - \lambda_i$ with $j > i$ and $1 \leq i \leq n$, $1 \leq j \leq n$.
[*Hint*: Factor $\lambda_2 - \lambda_1$ out of the first column of the determinant on the right; similarly for the other columns. Then use row operations (subtracting multiples of rows from other rows) until the determinant simplifies to a Vandermonde determinant of size $(n-1)$.]

20. Show that if one row (or column) of a matrix is a multiple of another row (or column), then that matrix is not invertible.

Answers to Odd-Numbered Exercises

Section 1.1 (pp. 10–12)

1. $x(t) \approx 12 + 7(t - 3)$

5. **(a)** $x(t) = 15 + \int_2^t e^{-s^2}\, ds$

7.

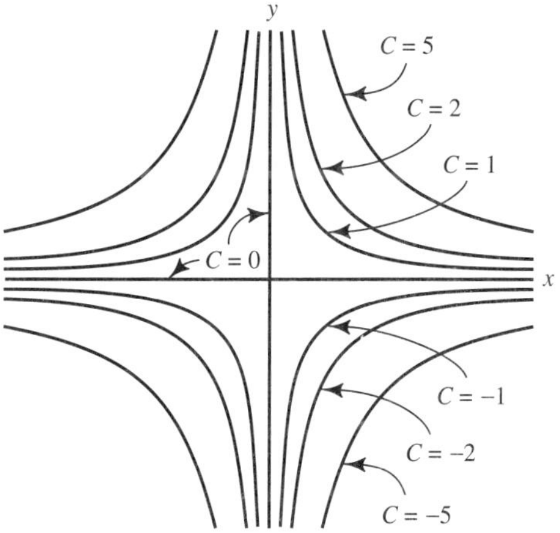

9.

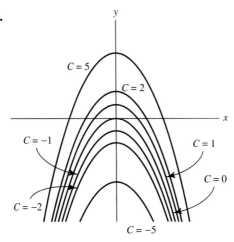

11. (a) $\dfrac{\partial f}{\partial x} = 2x$; $\dfrac{\partial f}{\partial y} = -\sin y$

(b) $\dfrac{\partial^2 f}{\partial x^2} = 2$; $\dfrac{\partial^2 f}{\partial x \, \partial y} = \dfrac{\partial^2 f}{\partial y \, \partial x} = 0$; $\dfrac{\partial^2 f}{\partial y^2} = -\cos y$

13. (a) $\dfrac{\partial f}{\partial x} = -2y \sin x \cos x$; $\dfrac{\partial f}{\partial y} = \cos^2 x$

(b) $\dfrac{\partial^2 f}{\partial x^2} = -2y(\cos^2 x - \sin^2 x)$; $\dfrac{\partial^2 f}{\partial x \, \partial y} = \dfrac{\partial^2 f}{\partial y \, \partial x} = -2 \sin x \cos x$; $\dfrac{\partial^2 f}{\partial y^2} = 0$

15. $z = -(x-1) - 2(y-1)$

17. $6e^{6t} - 8e^{4t} + 5e^{2t} + 2te^{2t} - 1$

19. $x \cos y = 1$; $y - \dfrac{\pi}{3} = \dfrac{1}{2\sqrt{3}}(x - 2)$

21. (b) If $\dfrac{dy}{dx}$ is defined and nonzero for all x

(c) (i) $x = \sqrt{y}$ is a function of y ;
(ii) x is not a function of y.

Section 1.2 (pp. 19–20)

1. (a) $x = \sqrt[3]{\tfrac{3}{2}t^2 + 3C}$

3. (a) $x = \sin(\tfrac{1}{2}t^2 - t + C)$

5. (a) $y = \tan(-x + C)$

(b) For $x \neq C + (2k+1)\dfrac{\pi}{2}$, $k = 0, \pm 1, \pm 2, \ldots$

ANSWERS TO ODD-NUMBERED EXERCISES

7. (a) $x = \dfrac{-1}{\arctan t + C}$

   (b) If $-\dfrac{\pi}{2} < C < \dfrac{\pi}{2}$, for $t \neq -\tan C$; otherwise, for all t.

9. (a) $y = -\tfrac{1}{2}e^{-x^2} + C$

11. (a) $x = \pm\sqrt[6]{6(t+C)}$
    (b) For $t > -C$

13. (a) $x = \tan(\tfrac{1}{3}t^3 + C)$
    (b) For $t \neq \sqrt[3]{3[(2k+1)\dfrac{\pi}{2} - C]}$, $k = 0, \pm 1, \pm 2, \ldots$

15. (a) $x = \sqrt[3]{3(t \log t - t + C)}$

17. (a) $x = -\tfrac{1}{3}\log(\tfrac{3}{2}e^{-2t} - 3C)$
    (b) If $C > 0$, for $t < -\tfrac{1}{2}\log(2C)$; if $C \leq 0$, for all t.

19. (a) $x = \pm\sqrt{2C - t^2}$
    (b) If $C > 0$, for $|t| < \sqrt{2C}$; if $C \leq 0$, never

25. (c) $(y^4 - 2)\,dy = x^2\,dx$ (for both equations!)

27. $\pm\sqrt[4]{10}$

Section 1.3 (pp. 26–28)

1. $5e^{7/2\log(3/5)} = 5 \cdot (\tfrac{3}{5})^{7/2} \approx 0.84$ kg; after $\dfrac{2\log(1/5)}{\log(3/5)} = \dfrac{2\log 5}{\log 5 - \log 3} \approx 6.3$ years; never (according to our model)

3. Half a year; half a year

5. $x(t) = (-5e^{-6})e^{3t} = -5e^{3(t-2)}$

9. (Population t years after 1900) $= 76e^{t\log(106/76)/20}$ million; in 1940, ≈ 148 million; in 1960, ≈ 206 million

11. (a) No (b) No
    (c) The *combined* population is growing exponentially, while the city population is declining at a constant rate.

13. $\dfrac{50[(\tfrac{3}{5})^{1/2} - 1]}{\log(3/5)} \approx 22.06$ kilograms; $\dfrac{50(e^{t\log(3/5)/2} - 1)}{\log(3/5)}$ kilograms

15. After $\dfrac{\log(3/110)}{\log(17/22)} \approx 14$ years

17. (a) $\dfrac{dx}{dt} = .05x$ (b) $1000e^{.1}$ dollars $\approx \$1105.17$ (c) By $\approx 5.13\%$

19. Yes; now bank B should get your money, with an APR (see Exercise 17) of $\approx 6.22\%$, as compared to $\approx 6.18\%$ for bank A

21. (a) $x(t) = \dfrac{AMe^{Mkt}}{1 + Ae^{Mkt}}$, where $A = \dfrac{x(0)}{M - x(0)}$ (provided $x(0) \neq M$)

(b) If $x(0) = M$, then $x(t) = M$ for all t; the population remains constant.

Section 1.4 (pp. 33–34)

1. Second-order PDE

3. First-order ODE, nonlinear

5. Second-order ODE, linear, homogeneous

7. Second-order ODE, nonlinear

9. Second-order ODE, nonlinear

11. Third-order ODE, linear, homogeneous

13. Third-order ODE, linear, homogeneous

Section 1.5 (pp. 46–48)

1. $x = \dfrac{1}{1-t}$

3. $x = \sin\left(t + \dfrac{\pi}{6}\right)$ (for $-\dfrac{2\pi}{3} \leq t \leq \dfrac{\pi}{3}$, to ensure $\dfrac{dx}{dt} \geq 0$)

5. $x = \dfrac{1}{\frac{3}{2}t^2 - \frac{1}{2}} = \dfrac{2}{3t^2 - 1}$

7. $x = \dfrac{2}{1 - t^2}$

9. $x = -3$

11. $y = -\frac{1}{3}\log(3e^{-x} - 3e^{-2} + 1)$

13. $x = -\sqrt[4]{\frac{4}{3}\sin(3t+1) - \frac{4}{3}\sin 1 + 1}$

15. $x = 0$

17. (a) Yes   **(b)** $x = \sqrt{2t + 9}$

(c)

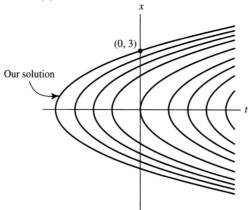

19. (a) Yes (b) $x = 5 - 3e^{t-1}$
(c)

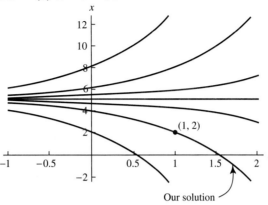

21. (a) No
(b) For t near 0, possibilities are
$$x(t) = 0 \quad \text{and} \quad x(t) = \begin{cases} 0 & (t \leq 0) \\ \pm t^{5/4} & (t \geq 0) \end{cases}.$$
(c)

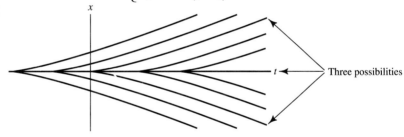

23. (a) Yes (b) $y = \frac{1}{27}(x+3)^3$
(c)

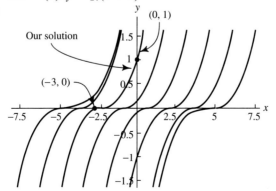

25. (a) No   (b) $x = At$, A arbitrary constant
(c)
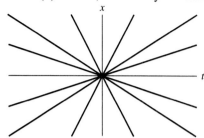
All nonvertical lines shown are possible.

27.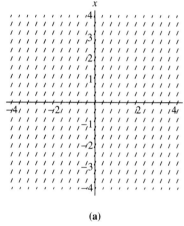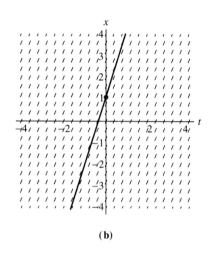
(a)   (b)

(c) $x = 3t + 1$

29.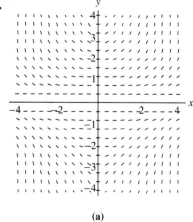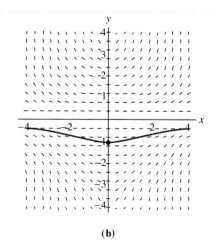
(a)   (b)

(c) $y = \dfrac{-8}{x^2 + 8}$

31.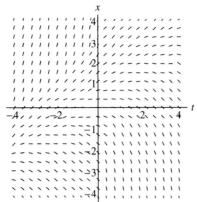

33.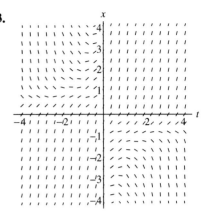

35.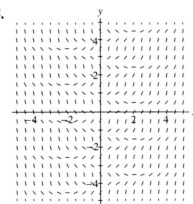

37.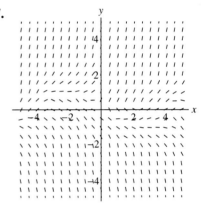

39. $x(t) = \begin{cases} 0 & (t \le -\frac{3}{2}x_0^{2/3}) \\ -(\frac{2}{3}t + x_0^{2/3})^{3/2} & (t \ge -\frac{3}{2}x_0^{2/3}) \end{cases}$

41. The curves for $C = -2$ and $C = -3$; yes.

43. (a) $\dfrac{dx}{dt} = Kx^{\alpha+1}$, $x(0) = x_0$

(b) $x(t) = \dfrac{x_0}{(1 - \alpha K x_0^\alpha t)^{1/\alpha}}$

(c) $x(t) \to \infty$

Section 1.6 (pp. 57–59)

3. $x = \dfrac{1}{3}t - \dfrac{1}{9} + Ce^{-3t}$

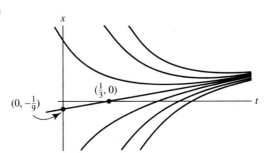

5. $y = 8 + \dfrac{C}{x}$

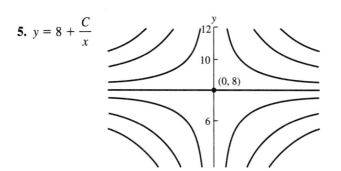

7. $y = 2x + \dfrac{C}{(x+3)^2}$

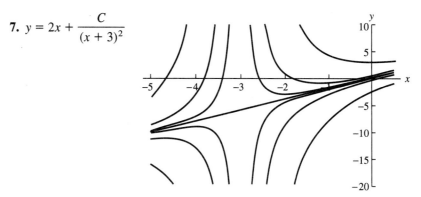

9. $y = 3 + Ce^{1/x}$

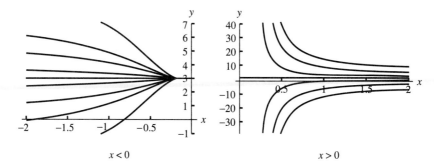

11. $x = \dfrac{2}{5}(1 - e^{5t})$ **13.** $x = t + 1$

15. $y = 2e^{e^{2x}/2} - 1$

17. $y = e^x - x - 1$; $y = 0$; two combinations of these:
$\begin{cases} y = 0 & (x \leq 0) \\ y = e^x - x - 1 & (x \geq 0) \end{cases}$, $\begin{cases} y = e^x - x - 1 & (x \leq 0) \\ y = 0 & (x \geq 0) \end{cases}$

19. $x = -e^{t^3} \displaystyle\int_0^t s^3 e^{-s^3}\, ds$

21. $x = 2t^2 + 4 - 3e^{t^2/2}$

23. (d) $x = \dfrac{33}{7} + Ae^{-7t}$; $x = -\dfrac{25}{18} + Ae^{18t}$

ANSWERS TO ODD-NUMBERED EXERCISES

Section 1.7 (pp. 69–75)

1. (a) $E(t, x) = tx - e^t + x^3$ (b) $tx - e^t + x^3 = C$
3. $x^5 y^2 - \frac{5}{2} x^4 y^4 = C$
5. Exact; $-\cos x - \frac{1}{3} x^3 y^3 + e^y = C$
7. Exact; $\frac{3}{2} x^2 + 2xy - 3y = C$
$$[y = \frac{C - \frac{3}{2} x^2}{2x - 3} \ (x \neq \tfrac{3}{2}), \text{ with one solution } y = -\tfrac{3}{4} x - \tfrac{9}{8}$$
(for $C = \frac{27}{8}$) defined for all x]
9. Exact; $\frac{3}{2} t^2 + 2tx - 3x = C$
$$[x = \frac{C - \frac{3}{2} t^2}{2t - 3} \ (t \neq \tfrac{3}{2}), \text{ with one solution } x = -\tfrac{3}{4} t - \tfrac{9}{8}$$
(for $C = \frac{27}{8}$) defined for all t]
11. $\mu(x) = e^{2x}$; $xe^{2x} \sin y = C$
13. $\mu(x) = x^2$; $tx^3 - 4t^3 x^4 = C$
15. $\mu(x) = e^{3x}$; $y = \sqrt{e^{-3x} - 2x}$
19. (c) $4x^2 y^2 + x^3 y^4 = C$
23. $(3x - y) e^{x^2 y^2} = C$
27. (a) $x \dfrac{dv}{dx} = \dfrac{2 - 2v^2}{3 + 2v}$

   (c) $\mu(x, y) = \dfrac{2}{(x - y)^6}$

(The factor 2 is unnecessary, but you're likely to find it.)

29. (a) $y = \dfrac{1}{x + Ce^{x^2/2}}$; $y = 0$

31. (a) $y = \dfrac{2}{\sqrt{1 - x^2}}$ (b) $\mu = \dfrac{1}{\sqrt{1 - x^2}}$

(c) $1 - x^2$ would be negative, and we would get
$$\int \frac{x \, dx}{1 - x^2} = -\frac{1}{2} \log(x^2 - 1) + C \text{ in part (a); } \mu(x) = \frac{1}{\sqrt{x^2 - 1}} \text{ in part (b).}$$
(Answer: $y = \dfrac{3\sqrt{3}}{\sqrt{x^2 - 1}}$.)

33. $\frac{1}{3} x^3 - 3xy + 2e^y - 5y = C$ 35. $x(y - 1)^2 = A$

37. $x = \dfrac{3}{t} + \dfrac{C}{t^2}$ 39. $2xy - 2x^2 + \frac{5}{2} y^2 = C$

41. $\frac{1}{2} x^4 y^2 - xy^3 = C$ 43. $y = \dfrac{1 + \sqrt{1 + 4(x^3 + x)(2x + 1)}}{4x + 2}$

45. $x = \dfrac{5}{3} t + \dfrac{C}{t^2}$ 47. $\begin{cases} x = 3t + 4 \log t + C & (t > 0) \\ x = 3t + 4 \log(-t) + C & (t < 0) \end{cases}$

49. $y = 2x - 1 + Ce^{-2x}$

51. $x = -\frac{4}{3}y + \frac{4}{9} + Ae^{-3y}$ (the equation is linear in x as a function of y)

53. $y = \frac{1}{2}\log(\sin 2x) + 1$

55. $\sin t + \frac{1}{2}e^{2t} + 3x^2 t + \cos x = \frac{1}{2}$

57. $xy^2 + \frac{4}{3}y^3 = C$

59. $y = \dfrac{-2}{x^2 + 2\log x - 3}$

In the figures for Exercises 63–71, the level curves are solid and the orthogonal trajectories are dashed.

63. $y = 3x + D$

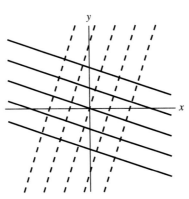

65. $x = Ay^4$ and $y = 0$

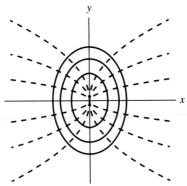

67. $y^2 - x^2 = D$

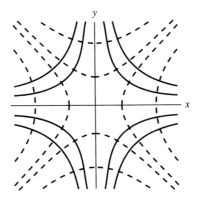

69. (b) $2x^3 - 6xy^2 + 5x = D$

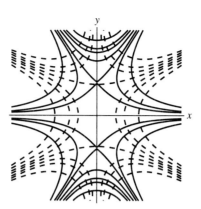

71. (b) $y = Ae^{2/3x} - 1$

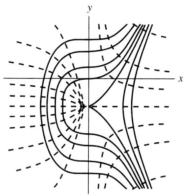

Section 2.1 (pp. 83–84)

3. $mv\dfrac{dv}{dx} + \gamma v + kx = 0$. For $\gamma \neq 0$, this equation is not separable, and there is no integrating factor of the form $\mu(x)$ or $\mu(v)$.

5. $x = 32t^5$; more generally, $x = (2t + C)^5$

7. $x = \dfrac{t^4}{144}$; more generally, $x = \left(\dfrac{t}{2\sqrt{3}} + C\right)^4$

9. $x = C_1 e^{2t} + C_2 e^{-2t}$, C_1, C_2 arbitrary

11. (b) $v = \sqrt{2\left(\dfrac{gR^2}{x+R} + C\right)}$

    (c) $C \geq 0$ **(d)** $\sqrt{2gR}$

    (e) $x(t) = (\tfrac{3}{2}R\sqrt{2g} \cdot t + R^{3/2})^{2/3} - R$

Section 2.2 (pp. 94–97)

1. No; $f(t, x, v) = \dfrac{t - 3v}{x + 1}$ is undefined for $x = -1$

3. Yes, by Theorem 2.2.2 (or Theorem 2.2.1)

5. No; although $f(t, x, v) = -v^{1/3} - x$ is defined for $x = 2$, $v = 0$, $\dfrac{\partial f}{\partial v}$ is not

7. $C = \pm \dfrac{1}{6\sqrt{6}}$; no, since for $f(t, x, v) = x^{1/3}$, $\dfrac{\partial f}{\partial x}$ is undefined for $x = 0$.

9. (b) $x_3(t)$ is not a solution; yes, since the equation isn't linear, the theorem doesn't apply.

11. $x(t) = \tfrac{3}{2} \sin 2t + \cos 2t$

13. $x(t) = -3e^t + 3e^{3t}$

15. (b) $x(t) = \alpha t^3 + \beta t$, α, β arbitrary
    (c) $x(t) = t^3 + 2t$
    (d) There are infinitely many solutions, including $x(t) = \alpha t^3 + t$,
    α arbitrary, and even $x(t) = \begin{cases} \alpha_1 t^3 + t & (t \leq 0) \\ \alpha_2 t^3 + t & (t \geq 0) \end{cases}$,
    α_1, α_2 arbitrary.

19. $x_1(t) = 2e^{\pi/4} x_2(t)$

21. $x'' + \dfrac{2 - 3t}{t^2 - t} x' + \dfrac{3t - 2}{t^3 - t^2} x = 0$; for $t \neq 0, 1$

23. $p(t) = \dfrac{x_1'' x_2 - x_1 x_2''}{W(x_1, x_2)}$, $q(t) = \dfrac{x_1' x_2'' - x_1'' x_2'}{W(x_1, x_2)}$; where $W(x_1, x_2) \neq 0$

25. $x = Ct + D$; $v = C$

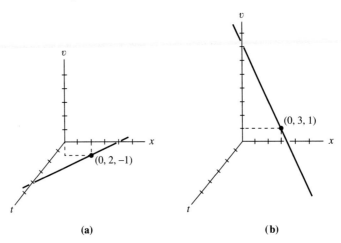

(a)   (b)

27. (b) $W = e^{-\int p(t)\, dt}$

33. $x = A \cos(t + C)$

35. $x = Ae^{2t}(1 - Be^t)$ or $x = Ce^{3t}$; in short, $x = \alpha e^{2t} + \beta e^{3t}$, α, β arbitrary

ANSWERS TO ODD-NUMBERED EXERCISES

Section 2.3 (pp. 100–101)

3. $x = \alpha + \beta e^{3t}$ (α, β arbitrary)
5. $y = \alpha e^{(-1+\sqrt{2})x} + \beta e^{(-1-\sqrt{2})x}$
7. $x = \frac{1}{7}(e^{2t} - e^{-5t})$
9. $y = -\frac{2}{3}e^{2x} - \frac{1}{3}e^{-10x}$
11. $x = -\frac{5}{2}e^3 \cdot e^{-t} + \frac{5}{2}e^9 \cdot e^{-3t} = \frac{5}{2}[e^{-3(t-3)} - e^{-(t-3)}]$
13. (c) $x = te^{2t}$; yes
    (d) $x = \alpha e^{2t} + \beta t e^{2t} = (\alpha + \beta t)e^{2t}$, α, β arbitrary

Section 2.4 (pp. 105–106)

In the answers to Exercises 1–11, α and β are arbitrary constants.

1. $x(t) = \begin{cases} \alpha t + \beta t \log t & (t > 0) \\ \alpha t + \beta t \log(-t) & (t < 0) \end{cases}$
3. $x(t) = \alpha \cos 2t + \beta \sin 2t$
5. $x(t) = e^{-t}(\alpha \sin t + \beta \cos t)$
7. $y(x) = \dfrac{\alpha \cos 2x + \beta \sin 2x}{e^x}$
9. $x(t) = \alpha \cos(\log t) + \beta \sin(\log t)$
11. $y(x) = \begin{cases} \alpha x + \beta\left(\dfrac{1}{2}x \log \dfrac{x-1}{x+1} + 1\right) & (x < -1; x > 1) \\ \alpha x + \beta\left(\dfrac{1}{2}x \log \dfrac{1-x}{x+1} + 1\right) & (-1 < x < 1) \end{cases}$
13. $x(t) = (2 - 6t)e^{3t}$
15. $x(t) = \dfrac{4}{e^2}e^{2t} - \dfrac{2}{e^4}e^{4t} = 4e^{2(t-1)} - 2e^{4(t-1)}$
17. $y(x) = (9 - 21x)e^{2x}$
19. No solutions (the initial conditions can't be satisfied)
21. $x(t) = \frac{4}{7}t^2 - \frac{3}{7}t^{-5}$
23. $y(x) = e^{2x}(\cos x - \sin x)$

Section 2.5 (pp. 110–111)

1. $\operatorname{Re}(z) = \dfrac{2 - t^3}{1 + t^4}$, $\operatorname{Im}(z) = \dfrac{t + 2t^2}{1 + t^4}$
3. $\operatorname{Re}(z) = e^{3t} \cos 2t$, $\operatorname{Im}(z) = e^{3t} \sin 2t$
5. $\operatorname{Re}(z) = \cos 4t$, $\operatorname{Im}(z) = -\sin 4t$
7. $\operatorname{Re}(z) = \cos 95t$, $\operatorname{Im}(z) = \sin 95t$

In the answers to Exercises 9–17, α and β are arbitrary constants.

9. $x = e^{3t}(\alpha \cos 2t + \beta \sin 2t)$
11. $x = \alpha e^{(-1+\sqrt{11})t} + \beta e^{(-1-\sqrt{11})t}$
13. $y = \alpha \cos x\sqrt{3} + \beta \sin x\sqrt{3}$
15. $x = \alpha + \beta e^{-7t}$
17. $x = (\alpha + \beta t)e^{-3t}$

19. $x = e^{-3t}(\cos t + 6 \sin t)$
21. $x = \left(-\dfrac{19}{e^3} + \dfrac{6t}{e^3}\right)e^t = (6t - 19)e^{t-3}$
23. $x = e^t(6 \sin t - \cos t)$
25. $x = 8$
29. **(b)** No (why?)
   (c) $x = Ce^{it} + De^{(-1-2i)t}$, C, D arbitrary *complex* constants
   (d) $x = 0$ (only)

Section 2.6 (pp. 117–118)

1. $x = 3e^t$
3. $x = -\tfrac{1}{2} \sin t + \tfrac{1}{2} \cos t$
5. $x = -te^{2t}$
7. $x = -\tfrac{2}{5} \sin t + \tfrac{9}{5} \cos t$
9. $x = -t^2 + t + 2$
11. $x = \tfrac{19}{6}$
13. $y = -\tfrac{1}{5}xe^x - 2$
15. $y = -\tfrac{1}{7} \sin 5x + \tfrac{1}{21} \cos 5x$
17. $x = \tfrac{1}{13}e^{2t} - \sin t - 2 \cos t$
19. $x = t^3 - 5t^2 + 4t + 6 + e^{-t/2}\left(\alpha \cos \dfrac{t\sqrt{3}}{2} + \beta \sin \dfrac{t\sqrt{3}}{2}\right)$, α, β arbitrary
21. $x = -\tfrac{1}{3}t + \tfrac{7}{9} - e^{-t} + \tfrac{2}{9}e^{-3t}$
23. $x = -te^t - 3e^t + 2e^{2t}$
25. $y = \tfrac{3}{2} \cos x - \tfrac{5}{2} \sin x - 3e^x + 2e^{4x}$
27. $x = \tfrac{1}{2}e^{-3t} - \tfrac{5}{2}e^t + 2e^{2t}$
29. $x = (\tfrac{1}{2}t^2 + 5t + 3)e^{-2t}$
31. $y = -xe^x - \tfrac{1}{2}e^{-2x} + e^x - \tfrac{1}{2}e^{2x}$
33. $x = \tfrac{1}{2} \cos t + \tfrac{3}{2} \sin t - \tfrac{1}{6}e^t - e^{-t} + \tfrac{2}{3}e^{-2t}$
35. $y = (5 - \tfrac{3}{4}x) \cos 2x + (\tfrac{7}{8} - \tfrac{1}{4}x) \sin 2x$

Section 2.7 (pp. 128–130)

1. **(a)** Underdamped
   (b) $x(0) = 0$, $x'(0) = 5$
   (c) $k = 68$, $\gamma = 12$
   (d) To $k = 340$, $\gamma = 60$; that is, k and γ must be multiplied by the same factor 5 that m is.

5.

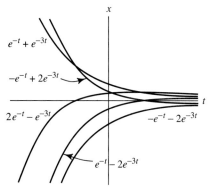

$x'' + 4x' + 3x = 0$; overdamped

7. $x = \dfrac{mg}{\gamma}t - \dfrac{m^2 g}{\gamma^2}(1 - e^{-\gamma t/m})$

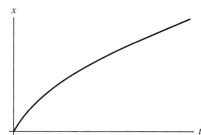

9. 436 Hz and 444 Hz

11.

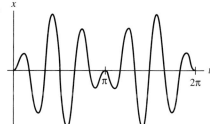

15. $\sqrt{2}\cos\left(4t + \dfrac{\pi}{4}\right)$

19. (c) "Excessive" damping will make it more difficult for the spring to return to its equilibrium position.

Section 2.8 (pp. 138–141)

1. $y = \tfrac{1}{150}e^{5x}\cos 3x + \tfrac{7}{150}e^{5x}\sin 3x - 2e^x + e^{2x}$

3. $x = \begin{cases} -\tfrac{1}{2}t\log t + Ct + Dt^3 & (t > 0) \\ -\tfrac{1}{2}t\log(-t) + Ct + Dt^3 & (t < 0) \end{cases}$

(If you're "missing" a term $-\tfrac{1}{4}t$, note that it can be absorbed into Ct.)

5. $x = e^{-2t}(\tfrac{1}{2}\int \sqrt{t}\, e^t\, dt) - e^{-4t}(\tfrac{1}{2}\int \sqrt{t}\, e^{3t}\, dt)$

7. $x = (\frac{4}{3}t^{3/2} - 3t + 1)e^{3t}$

9. $x = e^{-t}[-\log t - e - 1 + (2e + 1)t]$

11. $x = 3\cos t + 5\sin t + Ce^{-t} + De^{-4t}$

13. $x = \dfrac{31}{32}\sin 4t + \dfrac{1}{32\sin 4t}$

15. $y = -\frac{3}{11}x - \frac{5}{11} + Ce^{x\sqrt{11}} + De^{-x\sqrt{11}}$

17. $x = (8t + 32)e^{t/2} + (Ct + D)e^t$

19. $y = -x\cos x + (\sin x)\log(\sin x) + C\cos x + D\sin x$

21. $x = \frac{4}{35}t^{3/2} + \frac{8}{5}t^{-1} - \frac{5}{7}t^{-2}$

23. $x = e^{-3t}(-\frac{3}{10}\cos t - \frac{1}{10}\sin t) + Ce^{-t} + De^{-4t}$

25. $y = \begin{cases} \frac{3}{8} + x^2\log x + Cx^2 + Dx^{-4} & (x > 0) \\ \frac{3}{8} + x^2\log(-x) + Cx^2 + Dx^{-4} & (x < 0) \end{cases}$

(If you're "missing" a term $-\frac{1}{6}x^2$, note that it can be absorbed into Cx^2.)

27. $x = \frac{1}{2}e^{-2t}[\int h(t)e^{2t}\,dt] - \frac{1}{2}e^{-4t}[\int h(t)e^{4t}\,dt]$

29. $x = \frac{1}{4}[\int h(t)\cos 4t\,dt]\sin 4t - \frac{1}{4}[\int h(t)\sin 4t\,dt]\cos 4t$

31. $y = \dfrac{x^2}{6}\left[\int \dfrac{h(x)}{x^3}\,dx\right] - \dfrac{1}{6x^4}[\int h(x)x^3\,dx]$

33. (c) By the method of Section 2.1, considering $v = x'$ as an unknown function of x and using $x'' = v\,\dfrac{dv}{dx}$ to get a separable equation for v.

35. (b) $y = \alpha e^{-x^2} + \beta e^{x^2}$
    (c) $y = -\frac{1}{4}x^2 - \frac{1}{4} + Ce^{-x^2} + De^{x^2}$

37. (b) $x = \alpha t + \beta \sin t$
    (c) $x = t\displaystyle\int \dfrac{h(t)\sin t}{(\sin t - t\cos t)^2}\,dt - (\sin t)\displaystyle\int \dfrac{h(t)t}{(\sin t - t\cos t)^2}\,dt$

43. (d) $x(0) = x'(0) = 0$ for all solutions; no (why?)

Section 3.1 (pp. 151–154)

1. $x' = v$, $v' = 3tv - (\sin t)x + t^2$

3. $x' = v$, $v' = 17 - 4v^2 + 6xv$

5. Several possible models lead to systems of the form $\dfrac{dx}{dt} = x\,f(x)$, $\dfrac{dy}{dt} = Ky - Lxy$.

Here x, y are the sizes of the predator, prey population, respectively; K, L are positive constants; $f(x)$ is the excess birth rate of the predators, about which several assumptions are possible. Any such system can be solved by first solving the differential equation (which is separable) for x. Substituting the result into the other equation then yields a separable equation for y.

7. One possible model leads to $\dfrac{dx}{dt} = Kx + Lxy$, $\dfrac{dy}{dt} = My + Nxy$, K, L, M, N positive constants.

9. In the following systems (which are not the only possibilities), a, b, c represent the population sizes of species A, B, C, respectively, and $K_1, K_2, \ldots$ are positive constants.

(a) $\dfrac{da}{dt} = a(K_1 + K_2 b)$, $\dfrac{db}{dt} = b(-K_3 - K_4 a + K_5 c)$, $\dfrac{dc}{dt} = c(K_6 - K_7 b)$

(b) $\dfrac{da}{dt} = a(K_1 - K_2 a + K_3 b - K_4 c)$, $\dfrac{db}{dt} = b(-K_5 - K_6 a + K_7 c)$,

$\dfrac{dc}{dt} = c(K_8 - K_9 a - K_{10} b - K_{11} c)$

11. (b) Eliminating I, I_1, and V_C reduces the system to
$$RLC \dfrac{d^2 I_2}{dt^2} + L \dfrac{dI_2}{dt} + RI_2 = E(t).$$

13. $R_1 C_1 \dfrac{dV_1}{dt} + V_1 = R_2 C_2 \dfrac{dV_2}{dt} + V_2$, $(R_1 + R_3) C_1 \dfrac{dV_1}{dt} + R_3 C_2 \dfrac{dV_2}{dt} + V_1 = E(t)$

(V_1 and V_2 are the voltages across capacitors C_1 and C_2, respectively.)

15. (b) $L, \dfrac{1}{C}, R$

(c) (i) $R < 2\sqrt{\dfrac{L}{C}}$ (ii) $R = 2\sqrt{\dfrac{L}{C}}$ (iii) $R > 2\sqrt{\dfrac{L}{C}}$

(d) $\Omega = \dfrac{1}{\sqrt{LC}}$

Section 3.2 (pp. 172–177)

1. 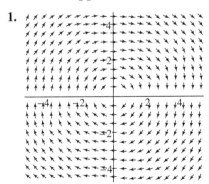   **3.**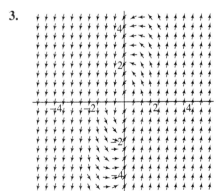

5.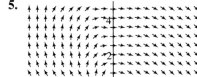

7.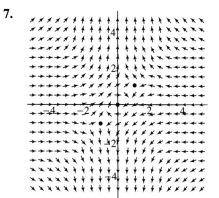

9. $(1, -1)$, $(0, 0)$

11. $(0, 0)$

13. (a)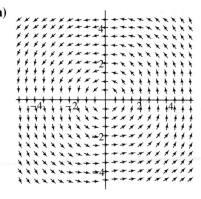

(b) $(0, 0)$
(c) $x^2 + 2y^2 = C$
(d)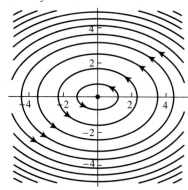

15. (a)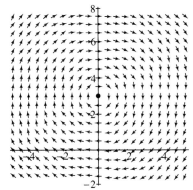
(b) $(0, 3)$
(c) $x^2 + y^2 - 6y = C$
(d)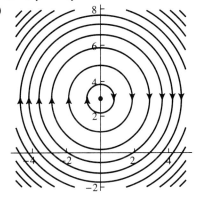

17. (a)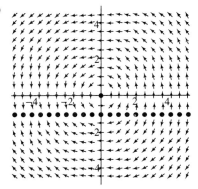

(b) $(0, 0)$, $(x, -1)$ (x arbitrary)
(c) $x^2 + 2y^2 = C$

(d)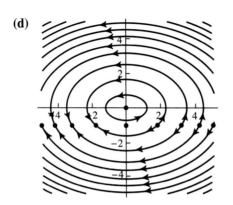

19. $(1 + x^2)y\,dx - dy = 0$; $\dfrac{dx}{dy} = \dfrac{1}{(1 + x^2)y}$; everywhere *except* along the x-axis

21. $(2x + y + 3)dx - dy = 0$; $\dfrac{dx}{dy} = \dfrac{1}{2x + y + 3}$; everywhere *except* along the line $2x + y + 3 = 0$

23. $\dfrac{dx}{dt} = 3x$, $\dfrac{dy}{dt} = x + y$

25. $\dfrac{dx}{dt} = 2x + y$, $\dfrac{dy}{dt} = 0$

27. (a) $(\tfrac{1}{2}, 2)$, $(-\tfrac{1}{2}, -2)$
  (b) $\dfrac{dx}{dt} = 2(x - \tfrac{1}{2}) + \tfrac{1}{2}(y - 2)$, $\dfrac{dy}{dt} = 4(x - \tfrac{1}{2}) - (y - 2) \,(= 4x - y)$;
  $\dfrac{dx}{dt} = -2(x + \tfrac{1}{2}) - \tfrac{1}{2}(y + 2)$, $\dfrac{dy}{dt} = 4(x + \tfrac{1}{2}) - (y + 2) \,(= 4x - y)$

29. (a) $(1, -1)$
  (b) $\dfrac{dx}{dt} = 2(x - 1) - 3(y + 1) \,(= 2x - 3y - 5)$,
  $\dfrac{dy}{dt} = 4(x - 1) + (y + 1) \,(= 4x + y - 3)$

31. $E(x,y) = xy$

33. $E(x,y) = ye^{-x}$

35. $E(x,y) = \dfrac{1}{2}\gamma x^2 - \alpha xy - \dfrac{1}{2}\beta y^2$

37. $E(x,y) = \cos x + 2x^3 + 2xy + y^3 - e^y$

39. $E(x,y) = \sin x - y\cos x + \sin y$

41. $E(x,y) = x^5 y^4 + x^3 y^2 - x^5$

43. (a) $E(x, y) = y^2 - 3x$
(b)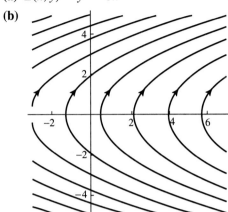

45. (a) $E(x, y) = xy$
(b)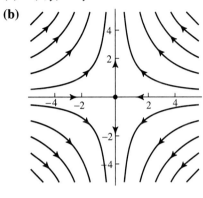

47. (a) $E(x, y) = x^2 - 4x + y^2$
(b)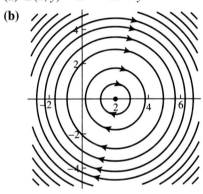

49. (a) $E(x, y) = y^2 - x^3$
(b)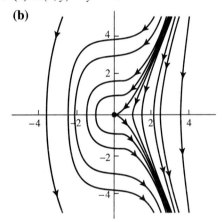

51. (a) $E(x, y) = \dfrac{y}{x^2 + 1}$
(b)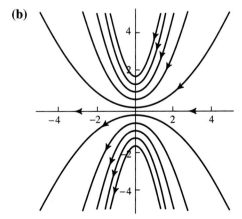

55. (a) $(0, 0)$, $\left(\dfrac{K}{L}, \dfrac{M}{P}\right)$

(c) $E(x, y) = Lx + Py - K \log x - M \log y$ $(x > 0, y > 0)$
(There are three other, similar, cases for other signs of x, y.)

57. (a) $(0, 0)$, $(-1, y)$ (y arbitrary)

(d)

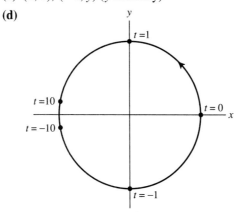

As $t \to \infty$ or $t \to -\infty$, $(x(t), y(t)) \to (-1, 0)$.

59. $x(t) = 1$, $y(t) = 1$ (for all t)

63. (a) $(x(t), y(t))$ goes to infinity (near the positive y-axis)
(b) $(x(t), y(t)) \to (0, 0)$
(c) $(x(t), y(t))$ goes to infinity (near the positive x-axis)
(d) $(x(t), y(t)) \to (0, 0)$
(e) $(x(t), y(t))$ goes to infinity (near the positive y-axis)
(f) $(x(t), y(t)) \to (0, 0)$

65. (a) $(-2, 0)$, $(0, 2)$, $(0, 3)$, $(3, 0)$

(b)

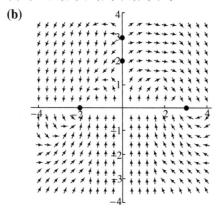

(c) $(x(t), y(t)) \to (3, 0)$

67. (a) $(1, 1)$, $(-1, 1)$

(b)

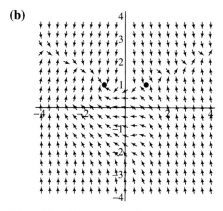

(c) $(x(t), y(t)) \to (-1, 1)$

69. (a) $(2, 0)$, $(-1, -3)$, $(\frac{1}{2}, \frac{3}{2})$

(b)

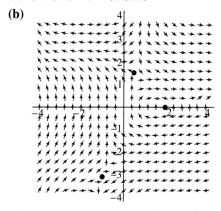

(c) $(x(t), y(t))$ goes to infinity (to the "northwest")

Section 3.3 (pp. 183–184)

1. $x = \alpha e^t + \beta e^{7t}$, $y = -\alpha e^t + \beta e^{7t}$ (α, β arbitrary)

3. $x = 2 + \alpha \sin 2t + \beta \cos 2t$, $y = 5 + 2\beta \sin 2t - 2\alpha \cos 2t$

5. $x = e^{2t}$, $y = e^{2t} + 2e^{-t}$

7. $x = (t + 9)e^t - 4$, $y = e^t + 1$

9. $x = \frac{1}{2}\alpha e^{3t} + \beta e^t$, $y = \alpha e^{3t}$

15. $x = 2t + C$, $y = -2t + 2 - C + De^t$ (C, D arbitrary)

17. $x = -\frac{1}{3}(C + 7) + De^{3t}$, $y = \frac{1}{3}(C - 14) + 2De^{3t}$

19. $x = -2t - \frac{1}{4}(C + 1) + De^{4t}$, $y = 2t + \frac{1}{4}(C - 3) + 3De^{4t}$

Section 3.4 (pp. 197–199)

1. (a), (b) $\lambda = 2$, eigenvector $(1, 0)$ (and nonzero multiples)
$\lambda = 3$, eigenvector $(5, 1)$

650 ANSWERS TO ODD-NUMBERED EXERCISES

(c) $\mathbf{T} = \begin{bmatrix} 1 & 5 \\ 0 & 1 \end{bmatrix}$

3. (a), (b) $\lambda = 1$, eigenvector $(1, -1)$
$\lambda = 3$, eigenvector $(1, 1)$

(c) $\mathbf{T} = \begin{bmatrix} 1 & 1 \\ -1 & 1 \end{bmatrix}$

5. (a), (b) $\lambda = 4$, eigenvector $(1, 1)$
$\lambda = -3$, eigenvector $(1, -6)$

(c) $\mathbf{T} = \begin{bmatrix} 1 & 1 \\ 1 & -6 \end{bmatrix}$

7. (a), (b) $\lambda = 1$, eigenvector $(1, 0)$
$\lambda = 0$, eigenvector $(2, -1)$

(c) $\mathbf{T} = \begin{bmatrix} 1 & 2 \\ 0 & -1 \end{bmatrix}$

9. (a) No real eigenvalues exist.
(b) No real eigenvectors exist.
(c) There is no such $\mathbf{T}$ (for real λ_1, λ_2)

11. (a) $x = Ce^t + De^{3t}$, $y = -Ce^t + De^{3t}$ (C, D arbitrary)
(b)

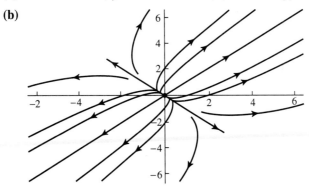

13. (a) $x = Ce^{-2t} + De^{-4t}$, $y = 2De^{-4t}$
(b)

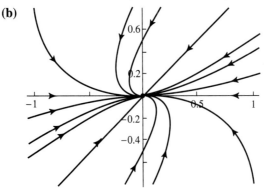

ANSWERS TO ODD-NUMBERED EXERCISES 651

15. (a) $x = 10e^{3t} - 10e^{-9t}$, $y = 10e^{3t} + 5e^{-9t}$
(b)

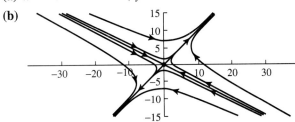

17. (a) $x = 4Ce^{3t} + 3De^{-t}$, $y = Ce^{3t} + De^{-t}$
(b)

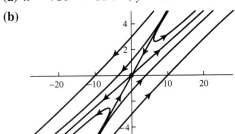

19. (a) $x = 2Ce^{4t} + 6De^{8t} - 2$, $y = -Ce^{4t} - De^{8t} + 1$
(b)

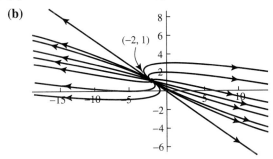

23.

Section 3.5 (pp. 207–209)

1. $\lambda = 0$ (double eigenvalue); eigenvector $(1, 0)$ (and nonzero multiples)

3. $\lambda = i\sqrt{3}$, eigenvector $(i\sqrt{3}, -1)$; $\lambda = -i\sqrt{3}$, eigenvector $(-i\sqrt{3}, -1)$

652 ANSWERS TO ODD-NUMBERED EXERCISES

5. $\lambda = 1 + i\sqrt{3}$, eigenvector $(1, i\sqrt{3})$; $\lambda = 1 - i\sqrt{3}$, eigenvector $(1, -i\sqrt{3})$
7. $\lambda = 4$ (double eigenvalue); eigenvector $(1, -1)$
9. $\lambda = 1$ (double eigenvalue); eigenvector $(1, -1)$
11. $x = (-Ct + D)e^{4t}$, $y = (-Ct + C + D)e^{4t}$ (C, D arbitrary)
13. $x = C_1 \cos 4t - C_2 \sin 4t$, $y = -2C_2 \cos 4t - 2C_1 \sin 4t$
15. $x = (t + 2)e^{-t/2}$, $y = (t + 4)e^{-t/2}$
17. $x = 5C_1 e^{2t} \cos 2t - 5C_2 e^{2t} \sin 2t - 5$,
    $y = (2C_2 - C_1)e^{2t} \cos 2t + (2C_1 + C_2)e^{2t} \sin 2t - 1$
19. $x = (3Ct + D)e^{4t}$, $y = (-3Ct + C - D)e^{4t} - 2$
21. $x = 5e^t \sin 4t$, $y = 5e^t \cos 4t + \frac{5}{2}e^t \sin 4t$
23. $x = C_1 \cos 3t - C_2 \sin 3t$, $y = C_2 \cos 3t + C_1 \sin 3t$

Section 3.6 (pp. 228–231)

1. (a) $(1, 1)$ is a stable spiral point.
   (b) The eigenvector $(2, 3 + i)$ for $\lambda = -2 + i$ can be used.

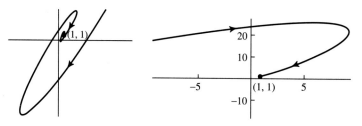

Direction of spiraling   Actual trajectory to scale

3. (a) $(2, 1)$ is a saddle point.
   (b) The eigenvectors $(1, 2)$ for $\lambda = 2$, $(1, 3)$ for $\lambda = -3$ can be used.

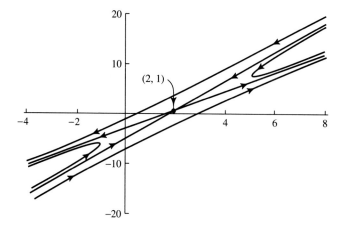

ANSWERS TO ODD-NUMBERED EXERCISES   **653**

5. (a) $(-2, -6)$ is an unstable improper node.
    (b) The eigenvector $(1, 2)$ for $\lambda = 1$ can be used.

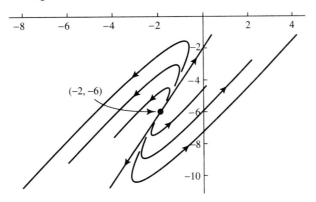

7. (a) $(-1, 1)$ is an unstable node.
    (b) The eigenvectors $(3, 1)$ for $\lambda = 1$, $(4, 1)$ for $\lambda = 2$ can be used.

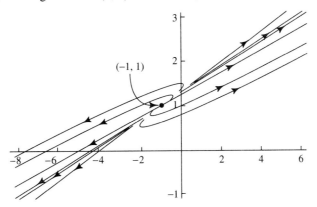

9. (a) All points on the line $y = \tfrac{1}{2}x$ are stationary points, which are "none of these."
    (b) The eigenvector $(2, 1)$ for the double eigenvalue $\lambda = 0$ can be used.

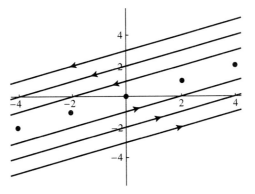

11. (a) $(31, -22)$ is a center.
   (b) The eigenvector $(10, -7 + i)$ for $\lambda = i$ can be used.

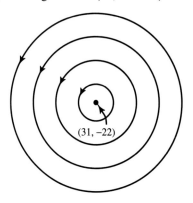

13. (a) $(-1, -1)$ is a stable spiral point.
   (b) The eigenvector $(2, 3 - i)$ for $\lambda = -3 + i$ can be used.

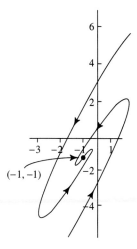

15. (a) All points on the line $y = \frac{1}{2}x$ are stationary points, which are "none of these."
   (b) The eigenvectors $(2, 1)$ for $\lambda = 0$, $(3, 1)$ for $\lambda = -3$ can be used.

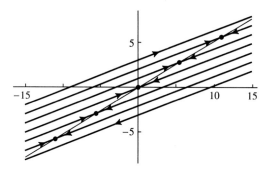

17. (a) $x = 2C_1 + 2C_2 e^{4t}$, $y = -C_1 + C_2 e^{4t}$ (C_1, C_2 arbitrary)

(b)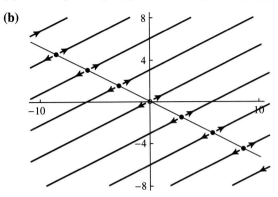

19. (a) $x = Ce^{2t} + De^{4t} + 2$, $y = Ce^{2t} - De^{4t} - 1$

(b)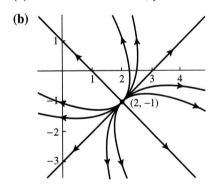

25. (e)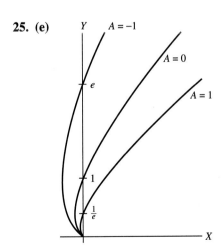

29. $k < -5$: stable node; $-5 < k < 7$: saddle point; $k > 7$: unstable node

31. $k < 0$: center; $k > 0$: saddle point

33. $k < -4$: stable spiral point; $-4 < k < 0$: stable node; $k > 0$: saddle point

Section 3.7 (pp. 244–246)

1. (a), (b) $(\frac{1}{2}, 2)$, saddle point; $(-\frac{1}{2}, -2)$, stable spiral point

(c)

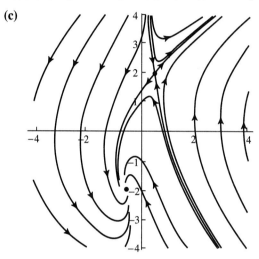

3. (a), (b) $(1, -1)$, unstable spiral point

(c)

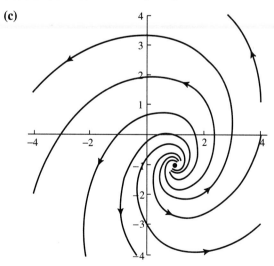

5. (a), (b) $(0, 0)$, saddle point; $(4, 0)$, unstable node; $(0, \frac{2}{3})$, stable node; $(1, 1)$, saddle point

(c)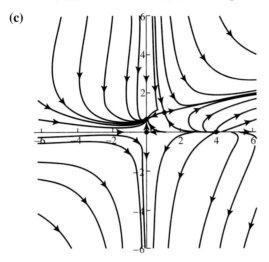

7. (a), (b) $(0, 0)$, ? (linear approximation has a center); $(1, -1)$, saddle point

(c)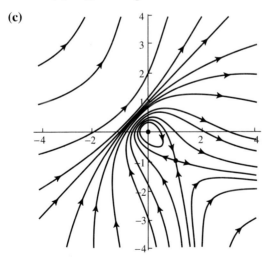

9. (a), (b) $(0, 0)$, ? (linear approximation has a center); $(1, 1)$, saddle point; $(-1, -1)$, saddle point

(c)

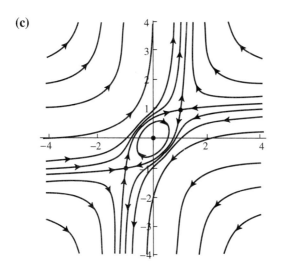

11. (a) $x = 0$ or $x = \dfrac{-1}{t + C}$, $y = De^t$, C, D arbitrary

13.

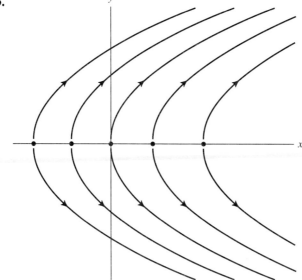

17. (d)

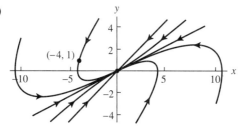

19. (b) $\lambda = -2, \lambda = -8$; $\lambda = -2, \lambda = 4$
(d) $y = -x$; $y = x - \sqrt{2}$; $y = x + \sqrt{2}$; see Figure 3.7.6 for directions.

Section 3.8 (pp. 259–262)

1. $V(x, y) = x^2 + 4y^2$; the stationary point $(0, 0)$ is stable.

3. $V(x, y) = x^2 + \frac{1}{3}y^2$; the stationary point $(0, 0)$ is asymptotically stable.

5. $V(x, y) = x^2 + y^2$; the stationary point $(0, 0)$ is asymptotically stable.

7. Although $V(x, y)$ is positive definite, $\dfrac{d(V(x, y))}{dt} = 4x^4 - 2y^4$ is *positive* for points $(x, 0)$ arbitrarily close to $(0, 0)$.

9. $V(x, y)$ is not positive definite.

11. Not positive definite, even on a domain containing $(0, 0)$, so can't be used.

13. Positive definite.

15. Not positive definite ($V(x, y) = 0$ along a line through $(0, 0)$); can't be used.

17. Not positive definite, but can be used on the domain D defined by $x^2 + y^2 < 4$.

19. Not positive definite; $V(x, y)$ is smaller near $(0, 0)$ than at $(0, 0)$, so can't be used.

27. (a) Note that if $\dfrac{\partial^2 V}{\partial x^2} \cdot \dfrac{\partial^2 V}{\partial y^2} - \left(\dfrac{\partial^2 V}{\partial x \partial y}\right)^2 > 0$, then certainly $\dfrac{\partial^2 V}{\partial x^2} \cdot \dfrac{\partial^2 V}{\partial y^2} > 0$.

31. For one thing, the function V defined by $V(x, y) = 0$ for all x, y would then qualify. (Why would this be undesirable?)

Section 3.9 (pp. 274–277)

5. (b) $(0, 0)$ is an unstable node; $\left(\dfrac{K}{L}, \dfrac{M}{P}\right)$ is a saddle point.

11. (b) $(0, 0)$, unstable node; $\left(\dfrac{A_1}{B_1}, 0\right)$, saddle point; $\left(0, \dfrac{A_2}{B_2}\right)$, stable node (except possibly if $A_2 C_1 = A_1 B_2 + A_2 B_2$, when the linear approximation has an improper node)

13. (a) $A_2 B_1 < A_1 C_2$, $A_1 B_2 < A_2 C_1$
(b) $(0, 0)$, unstable node;

$\left(\dfrac{A_1}{B_1}, 0\right)$, stable node (except possibly if $A_1 C_2 = A_1 B_1 + A_2 B_1$, when the linear approximation has an improper node);

$\left(0, \dfrac{A_2}{B_2}\right)$, stable node (except possibly if $A_2 C_1 = A_1 B_2 + A_2 B_2$)

17. (a) $(0, 0)$; saddle points at $(1, 1)$ and $(-1, -1)$
(c)

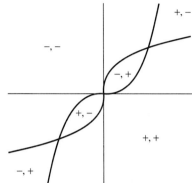

(d), (g)

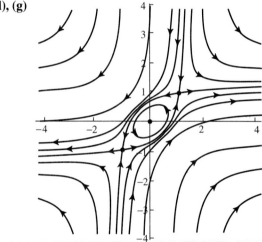

Section 4.1 (pp. 287–290)

1. $x = \alpha_1 e^t + \alpha_2 \cos 2t + \alpha_3 \sin 2t$ ($\alpha_1, \alpha_2, \alpha_3$ arbitrary)
3. $y = \alpha_1 e^{3x} + \alpha_2 \cos x + \alpha_3 \sin x$
5. $x = (\alpha_1 + \alpha_2 t) \cos 2t + (\alpha_3 + \alpha_4 t) \sin 2t$
7. $x = \alpha_1 e^{4t} + e^{t/2}(\alpha_2 \cos \frac{t\sqrt{3}}{2} + \alpha_3 \sin \frac{t\sqrt{3}}{2})$
9. $y = e^{3x} + 2e^{-3x} - e^x - 3e^{-x}$
11. $x^{(4)} - 6x^{(3)} + 13x'' - 24x' + 36x = 0$

Section 4.2 (pp. 301–303)

1. $x = t^4 - 2t^2 + 2t + 5t \log t - t^3$
3. **(b)** $x = \frac{1}{26} \cos 3t - \frac{1}{10} \cos t + \frac{1}{4}(e^{2t} - e^{-2t})$

ANSWERS TO ODD-NUMBERED EXERCISES

5. $\dfrac{1}{2i}e^{(-1+3i)t} - \dfrac{1}{2i}e^{(-1-3i)t}$

7. $\dfrac{1}{4i}t^2 e^{4it} - \dfrac{1}{4i}t^2 e^{2it} + \dfrac{1}{4i}t^2 e^{-2it} - \dfrac{1}{4i}t^2 e^{-4it}$

9. $\dfrac{1}{4i}e^{(1+2i)t} - \dfrac{1}{4i}e^{(1-2i)t} - \tfrac{1}{2}e^{3t} + \tfrac{1}{4}e^{(3+4i)t} + \tfrac{1}{4}e^{(3-4i)t}$

11. $x = (At^2 + Bt)e^{2t} + Ce^{-2t} + D\cos 3t + E\sin 3t$

13. $y = Ax\cos x + Bx\sin x + (Cx + D)e^{-x}$

15. $x = e^{-t}(At\cos 2t + Bt\sin 2t) + Cte^{t}$

17. $x = \tfrac{1}{15}e^{3t} + \tfrac{1}{8}t^2 + \alpha + \beta e^{2t} + \gamma e^{-2t}$

19. $x = -\tfrac{1}{6}te^{t} + \tfrac{1}{10}\sin t + \alpha e^{t} + \beta e^{-t} + \gamma e^{2t} + \delta e^{-2t}$

21. $x = \tfrac{1}{40}e^{2t} + \tfrac{1}{6}t\sin t + \alpha\cos t + \beta\sin t + \gamma\cos 2t + \delta\sin 2t$

23. $x = -\tfrac{1}{3}\int h(t)\,dt + \tfrac{1}{4}e^{t}\int h(t)e^{-t}\,dt + \tfrac{1}{12}e^{-3t}\int h(t)e^{3t}\,dt$

25. $x = \tfrac{1}{5}e^{2t}\int h(t)e^{-2t}\,dt + (\tfrac{2}{5}\cos t - \tfrac{1}{5}\sin t)\int h(t)\sin t\,dt$
$\qquad\qquad - (\tfrac{1}{5}\cos t + \tfrac{2}{5}\sin t)\int h(t)\cos t\,dt$

27. $x = -\dfrac{2}{t^2} + \dfrac{1}{t} - 3 + 4t - t^2$

29. $y = -\dfrac{2}{x} + x^3 + 4 - 2\log x$

33. (a) $W(x_1, x_2, x_3)(t) = 4t^2$

Section 4.3 (pp. 318–321)

1. $\begin{cases} x' = y_1 \\ y_1' = y_2 \\ y_2' = 3t^2 y_2 - (\sin t)x \end{cases}$

3. $\begin{cases} \dfrac{dx}{dt} = u \\ \dfrac{dy}{dt} = v \\ \dfrac{du}{dt} = 3v - 4y + t \\ \dfrac{dv}{dt} = 4u - t^2 x \end{cases}$

5. $\begin{cases} x' = u_1 \\ u_1' = u_2 \\ u_2' = 2tu_2 - x \\ y' = v \\ v' = 4y \end{cases}$

7. $\lambda = 2$, $\xi_1(1, 0, 0, 0)$, $\xi_1 \neq 0$; $\lambda = 0$, $\xi_4(0, 0, 2, 1)$, $\xi_4 \neq 0$;
$\lambda = 4$, $\xi_3(0, -9, 1, -2)$, $\xi_3 \neq 0$; $\lambda = -5$, $\xi_3(0, 0, 1, -2)$, $\xi_3 \neq 0$

9. $\lambda = 3$, $\xi_2(0, 1, -1)$, $\xi_2 \neq 0$; $\lambda = 1$, $\xi_3(-2, 1, 1)$, $\xi_3 \neq 0$;
$\lambda = 7$, $\xi_1(1, 1, 1)$, $\xi_1 \neq 0$

11. $\lambda = 2$, $\xi_4(-1, -1, 1, 1)$, $\xi_4 \neq 0$; $\lambda = -1$, $\xi_3(0, -2, 1, 4)$, $\xi_3 \neq 0$;
$\lambda = 3$, $\xi_3(0, -2, 1, 0)$, $\xi_3 \neq 0$; $\lambda = 5$, $\xi_2(0, 1, 1, 1)$, $\xi_2 \neq 0$

13. $(x_1, x_2, x_3) = C_1 e^{8t}(1, 1, 1) + C_2 e^{2t}(1, 1, -2) + C_3 e^{4t}(1, -1, 0)$

15. $(x_1, x_2, x_3) = C_1 e^t(0, 1, 1) + C_2 e^{-3t}(1, 1, 0) + C_3 e^{3t}(1, 1, -1)$

17. $x_1 = 16e^{3t}$, $x_2 = -\frac{1}{2}e^{3t} + e^t - \frac{1}{2}e^{-5t} + 4e^{-t}$,
$x_3 = 3e^{3t} + 2e^t - e^{-5t} - 8e^{-t}$, $x_4 = -\frac{5}{2}e^{3t} + \frac{3}{2}e^{-5t} + 4e^{-t}$

21. $X = x$, $Y = y - 2$, $Z = z + 3$

27. (a) $\dfrac{dx}{dt} = y_1$, $\dfrac{dy_1}{dt} = y_2$, ..., $\dfrac{dy_{n-2}}{dt} = y_{n-1}$,

$\dfrac{dy_{n-1}}{dt} = h(t) - p_{n-1}(t)y_{n-1} - \cdots - p_1(t)y_1 - p_0(t)x$

Section 4.4 (pp. 331–334)

1. (a) $\xi_1(1, 1, -1)$, $\xi_1 \neq 0$ (b) $\xi_3(-2, 1, 1)$, $\xi_3 \neq 0$

3. (a) $\lambda = 5$, $\lambda = \pm i$

(b) $\mathbf{T} = \begin{bmatrix} 1 & 2+i & 2-i \\ 0 & 1+5i & 1-5i \\ 0 & -5+i & -5-i \end{bmatrix}$, $\mathbf{T}^{-1}\mathbf{AT} = \begin{bmatrix} 5 & 0 & 0 \\ 0 & i & 0 \\ 0 & 0 & -i \end{bmatrix}$

5. (a) $\lambda = 1$ (double eigenvalue), $\lambda = -4$
   (b) No such $\mathbf{T}$ exists: Up to constant multiples, the only eigenvector for $\lambda = 1$ is $(2, 1, 3)$.

7. (a) $\lambda = 4$ (double eigenvalue), $\lambda = -2$

(b) $\mathbf{T} = \begin{bmatrix} 1 & 1 & 0 \\ -6 & 0 & 0 \\ 3 & 0 & 1 \end{bmatrix}$, $\mathbf{T}^{-1}\mathbf{AT} = \begin{bmatrix} -2 & 0 & 0 \\ 0 & 4 & 0 \\ 0 & 0 & 4 \end{bmatrix}$

9. (a) $\lambda = 0$, $\lambda = 1$, $\lambda = -1 \pm 3i$

(b) $\mathbf{T} = \begin{bmatrix} 1 & -22 & 1 & 1 \\ -4 & 16 & -1+i & -1-i \\ 0 & -13 & 0 & 0 \\ 10 & 0 & 0 & 0 \end{bmatrix}$, $\mathbf{T}^{-1}\mathbf{AT} = \begin{bmatrix} 0 & 0 & 0 & 0 \\ 0 & 1 & 0 & 0 \\ 0 & 0 & -1+3i & 0 \\ 0 & 0 & 0 & -1-3i \end{bmatrix}$

11. (a) $\lambda = 1$ (triple eigenvalue), $\lambda = 4$
    (b) No such **T** exists: There are only two linearly independent eigenvectors for $\lambda = 1$.

13. $x_1 = -\frac{3}{2}e^{5t} + \frac{3}{2}\cos t + \frac{1}{2}\sin t$, $x_2 = 3\cos t - 2\sin t$,
    $x_3 = -2\cos t - 3\sin t$

15. $x_1 = (2Ct + C + 2D)e^t + Ee^{-4t}$, $x_2 = (Ct + D)e^t$,
    $x_3 = (3Ct + 2C + 3D)e^t + Ee^{-4t}$ (C, D, E arbitrary)

17. $x_1 = 2C_1 e^{-2t}\cos t - 2C_2 e^{-2t}\sin t$,
    $x_2 = (-3C_1 - C_2)e^{-2t}\cos t - (C_1 - 3C_2)e^{-2t}\sin t$,
    $x_3 = D_1 \cos t - D_2 \sin t$,
    $x_4 = (-8C_1 - 2C_2)e^{-2t}\cos t - (2C_1 - 8C_2)e^{-2t}\sin t - (D_1 + D_2)\cos t - (D_1 - D_2)\sin t$

19. $x_1 = 7e^t - 8e^{2t}$, $x_2 = (-7t + 2)e^t + \frac{16}{3}e^{2t} + \frac{2}{3}e^{-t}$,
    $x_3 = -\frac{8}{3}e^{2t} - \frac{1}{3}e^{-t}$, $x_4 = (7t + 12)e^t - \frac{88}{3}e^{2t} - \frac{2}{3}e^{-t}$

21. $x_1 = [Ct^2 + (2D - C)t + E]e^{2t}$, $x_2 = (Ct + D)e^{2t}$, $x_3 = Ce^{2t}$, $x_4 = Fe^{-3t}$

25. (b) $\text{Re}(e^{\bar{\lambda}t}\bar{\xi}) = \text{Re}(e^{\lambda t}\xi)$; $\text{Im}(e^{\bar{\lambda}t}\bar{\xi}) = -\text{Im}(e^{\lambda t}\xi)$

Section 4.5 (pp. 348–353)

1. $\begin{bmatrix} \frac{1}{e} & 0 & 0 \\ 0 & 1 & 0 \\ 0 & 0 & e^2 \end{bmatrix}$ 3. $\begin{bmatrix} 1 & 1 & \frac{3}{2} \\ 0 & 1 & -1 \\ 0 & 0 & 1 \end{bmatrix}$

5. $\begin{bmatrix} (6 + e^{7t})/7 & (-2 + 2e^{7t})/7 \\ (-3 + 3e^{7t})/7 & (1 + 6e^{7t})/7 \end{bmatrix}$ 7. $\begin{bmatrix} \cos 2 & \frac{1}{2}\sin 2 \\ -2\sin 2 & \cos 2 \end{bmatrix}$

9. $\begin{bmatrix} 1 & 0 & 0 \\ 1 & 1 & 0 \\ \frac{1}{2} & 3 & 0 \end{bmatrix}$

11. $\begin{bmatrix} e^3 & 0 & 0 \\ \frac{1}{4}(e^3 - e^{-1}) & e^{-1} & 0 \\ \frac{1}{4}(5e^3 + e^{-1} - 6e) & e - e^{-1} & e \end{bmatrix}$

13. $\begin{bmatrix} e^{-3} & 0 & 0 \\ \frac{11}{40}e^{-3} - \frac{5}{8}e + \frac{7}{20}e^7 & \frac{5}{6}e + \frac{1}{6}e^7 & -\frac{5}{6}e + \frac{5}{6}e^7 \\ -\frac{19}{40}e^{-3} + \frac{1}{8}e + \frac{7}{20}e^7 & -\frac{1}{6}e + \frac{1}{6}e^7 & \frac{1}{6}e + \frac{5}{6}e^7 \end{bmatrix}$

15. $\begin{bmatrix} \frac{1}{3}(e^2 + 2e^{-4}) & \frac{1}{3}(e^2 - e^{-4}) & \frac{1}{3}(e^2 - e^{-4}) \\ \frac{1}{3}(e^2 - e^{-4}) & \frac{1}{6}(2e^2 + e^{-4} + 3e^{-2}) & \frac{1}{6}(2e^2 + e^{-4} - 3e^{-2}) \\ \frac{1}{3}(e^2 - e^{-4}) & \frac{1}{6}(2e^2 + e^{-4} - 3e^{-2}) & \frac{1}{6}(2e^2 + e^{-4} + 3e^{-2}) \end{bmatrix}$

17. $\begin{bmatrix} \cos 2t & 0 & -2\sin 2t \\ 2e^{3t} - 2\cos 2t & e^{3t} & 4\sin 2t \\ \frac{1}{2}\sin 2t & 0 & \cos 2t \end{bmatrix}$

19. $x = 20e^t - 12e^{2t}$, $y = 5e^t - 6e^{2t}$

21. $x_1 = -5e^{2t}$, $x_2 = -5e^{2t} + e^{5t} + 4e^{3t} + 2e^{-t}$,
    $x_3 = 5e^{2t} + e^{5t} - 2e^{3t} - e^{-t}$, $x_4 = 5e^{2t} + e^{5t} - 4e^{-t}$

23. $x_1 = -10 \cos 2t - 2 \sin 2t$, $x_2 = 4 \cos 2t - 20 \sin 2t - 4e^{-t}$,
    $x_3 = \cos 2t - 5 \sin 2t$

25. $x_1 = -1 - 6t + \frac{5}{2}t^2$, $x_2 = 4 + 5t$, $x_3 = -5$

27. $x_1 = (6t + 5)e^{4t}$, $x_2 = (6t - 1)e^{4t}$

29. $x_1 = -\frac{1}{2}e^{-2t} - \frac{1}{2}e^{-4t} + 6e^t$, $x_2 = e^{-4t} - 12e^t$, $x_3 = -\frac{1}{2}e^{-2t} + \frac{1}{2}e^{-4t} - 11e^t$

31. $x_1 = 20t^2 - 11t + 4$, $x_2 = 40t - 1$, $x_3 = 10$

45. (a) $\mathbf{B}^2 = \begin{bmatrix} 9t^2 & 12t^2 \\ 0 & 9t^2 \end{bmatrix}$, $\mathbf{B}^3 = \begin{bmatrix} 27t^3 & 54t^3 \\ 0 & 27t^3 \end{bmatrix}$, $\mathbf{B}^4 = \begin{bmatrix} 81t^4 & 216t^4 \\ 0 & 81t^4 \end{bmatrix}$

(b) $\begin{bmatrix} 3^m t^m & 2m \, 3^{m-1} t^m \\ 0 & 3^m t^m \end{bmatrix}$

Section 4.6 (pp. 362–365)

1. $x_1 = \frac{15}{2} \cos t + \frac{3}{2} \sin t + C(6 - e^{-5t}) + D(-3 + 3e^{-5t})$,
   $x_2 = 2 \cos t + 3 \sin t + C(2 - 2e^{-5t}) + D(-1 + 6e^{-5t})$

3. $x_1 = C \cos 2t + \frac{1}{2}D \sin 2t + e^t$, $x_2 = -2C \sin 2t + D \cos 2t - 2e^t$

5. $x_1 = 2 + 2 \sin t \cos t + C \cos t + D \sin t$,
   $x_2 = -1 - \sin^2 t - C \sin t + D \cos t$

7. $x_1 = e^t + \frac{2}{5}e^{3t} + C(e^{5t} + 6e^{-2t}) + D(2e^{5t} - 2e^{-2t})$,
   $x_2 = -e^t + \frac{4}{5}e^{3t} + C(3e^{5t} - 3e^{-2t}) + D(6e^{5t} + e^{-2t})$

9. $x_1 = -\frac{1}{4}t^4 + \frac{1}{3}t^3 + 5t^2 + C(1 - 3t) + Dt$,
   $x_2 = -\frac{3}{4}t^4 + 15t^2 + 10t + D(1 + 3t) - 9Ct$

11. $\mathbf{v} = (x_1, v_1, x_2, v_2)$,

$$\mathbf{A} = \begin{bmatrix} 0 & 1 & 0 & 0 \\ -\dfrac{k_1 + k_2}{m_1} & 0 & \dfrac{k_2}{m_1} & 0 \\ 0 & 0 & 0 & 1 \\ \dfrac{k_2}{m_2} & 0 & -\dfrac{k_2}{m_2} & 0 \end{bmatrix},$$

$\mathbf{h}(t) = \left(0, g, 0, \dfrac{F(t)}{m_2} + g\right)$

13. $\mathbf{v} = (x_1, v_1, x_2, v_2, x_3, v_3)$,

$$\mathbf{A} = \begin{bmatrix} 0 & 1 & 0 & 0 & 0 & 0 \\ -\dfrac{k_1+k_2}{m_1} & 0 & \dfrac{k_2}{m_1} & 0 & 0 & 0 \\ 0 & 0 & 0 & 1 & 0 & 0 \\ \dfrac{k_2}{m_2} & 0 & -\dfrac{k_2+k_3}{m_2} & 0 & \dfrac{k_3}{m_2} & 0 \\ 0 & 0 & 0 & 0 & 0 & 1 \\ 0 & 0 & \dfrac{k_3}{m_3} & 0 & -\dfrac{k_3}{m_3} & 0 \end{bmatrix},$$

$$\mathbf{h}(t) = \left(0, g, 0, g, 0, \dfrac{F(t)}{m_3} + g\right)$$

17. (a) $\mathbf{v} = (I_L, V_C)$, $\mathbf{A} = \begin{bmatrix} 0 & \dfrac{1}{L} \\ -\dfrac{1}{C} & -\dfrac{1}{RC} \end{bmatrix}$, $\mathbf{h}(t) = \begin{bmatrix} 0 \\ \dfrac{E(t)}{RC} \end{bmatrix}$

(b) $I_L = 4(3e^{-2t} - e^{-6t})\int (e^{2t} - e^{6t})E(t)\, dt$
$\qquad + 4(e^{-2t} - e^{-6t})\int(-e^{2t} + 3e^{6t})E(t)\, dt,$
$V_C = 6(-e^{-2t} + e^{-6t})\int(e^{2t} - e^{6t})E(t)\, dt$
$\qquad + 2(-e^{-2t} + 3e^{-6t})\int(-e^{2t} + 3e^{6t})E(t)\, dt$

Section 5.1 (pp. 374–376)

1. $y''(0) = -1$, $y^{(3)}(0) = 2$ **3.** $y''(0) = 0$, $y^{(3)}(0) = 3$

5. (a) $y''(0) = 4$, $y^{(3)}(0) = 8$, $y^{(4)}(0) = 16$
(b) $y^{(n)}(0) = 2^n$

(c) $1 + 2x + \dfrac{4}{2!}x^2 + \dfrac{8}{3!}x^3 + \cdots = \sum_{n=0}^{\infty} \dfrac{2^n}{n!}x^n$. This is the Taylor series for e^{2x}.

7. (a) $a_2 = 9$, $a_3 = 27$, $a_4 = 81$; $a_n = 3^n$

9. (a) $a_2 = 4$, $a_3 = 8$, $a_4 = 16$; $a_n = 2^n$

11. (a) $a_2 = 2$, $a_3 = 6$, $a_4 = 24$; $a_n = n!$

15. (b) $a_2 = -1$, $a_3 = 1$, $a_4 = -1$, $a_5 = 1$; $a_n = (-1)^{n+1}$

(d) $y = -\dfrac{1}{1+x}$, defined for all $x \neq -1$; the power series converges for $|x| < 1$

Section 5.2 (pp. 391–394)

1. $-x + \dfrac{x^2}{4} - \dfrac{x^3}{9} + \cdots$; $R = 1$ **3.** $1 + 10x + 50x^2 + \cdots$; $R = \infty$

5. $\dfrac{x^2}{4} - \dfrac{x^4}{64} + \dfrac{x^6}{576} + \cdots$; $R = 2$

7. $\sum\limits_{k=1}^{\infty} \dfrac{(-1)^{k+1} x^{2k}}{8k}$; $R = 1$

9. $\sum\limits_{n=2}^{\infty} (-1)^n \dfrac{2(n-1)x^n}{3^{n-1}}$; $R = 3$

11. $\sum\limits_{n=0}^{\infty} (n+1)(n+6) x^n$

13. $\sum\limits_{n=1}^{\infty} (n-1)^2 x^n$

15. $\sum\limits_{n=0}^{\infty} \dfrac{(n+2)(n+1)}{3^{n+2}} x^n$

17. $\sum\limits_{n=2}^{\infty} \dfrac{2^{n-2}}{n+1} x^n$

19. $(a_0 + b_1) + \sum\limits_{n=1}^{\infty} (a_n + b_{n+1} - 3c_{n-1}) x^n$

21. $\sum\limits_{n=0}^{\infty} \dfrac{(-1)^{n+1}}{n+1} x^n$

23. Analytic at $x = 0$ 25. Not analytic at $x = 0$

27. Not analytic at $x = 0$ [$f^{(4)}(0)$ does not exist.]

29. Analytic at $x = 0$

31. Not analytic at $x = 0$ (Radius of convergence is 0.)

33. $1 + 3x - \dfrac{3^2 x^2}{2!} + \cdots = \sum\limits_{n=0}^{\infty} \dfrac{3^n x^n}{n!}$ (all x)

35. $5 + 5x + 5x^2 + 4x^3 + 4x^4 + \cdots = 5 + 5x + 5x^2 + 4\sum\limits_{n=3}^{\infty} x^n$ ($|x| < 1$)

37. $2x - \tfrac{8}{3}x^3 + \tfrac{32}{5}x^5 - \tfrac{128}{7}x^7 + \cdots = \sum\limits_{k=0}^{\infty} \dfrac{(-1)^k 2^{2k+1}}{2k+1} x^{2k+1}$ ($|x| < \tfrac{1}{2}$)

39. $\log 3 + \dfrac{x}{3} - \dfrac{x^2}{18} + \dfrac{x^3}{81} - \dfrac{x^4}{324} + \cdots = \log 3 + \sum\limits_{n=0}^{\infty} \dfrac{(-1)^n x^{n+1}}{(n+1) 3^{n+1}}$ ($|x| < 3$)

41. $\sqrt{3}$ 43. 3

45. (a) $1 - x^2 + x^4 - x^6 + \cdots = \sum\limits_{k=0}^{\infty} (-1)^k x^{2k}$; for $|x| < 1$

(b) See p. 386.

47. (a) $2x - \dfrac{8x^3}{3!} + \dfrac{32x^5}{5!} - \dfrac{128x^7}{7!} + \cdots = \displaystyle\sum_{k=0}^{\infty} (-1)^k \dfrac{2^{2k+1} x^{2k+1}}{(2k+1)!}$ (all x)

(b) $\dfrac{dy}{dx} = \displaystyle\sum_{k=0}^{\infty} \dfrac{(-1)^k 2^{2k+1}}{(2k)!} x^{2k}$, $\dfrac{d^2y}{dx^2} = \displaystyle\sum_{k=1}^{\infty} \dfrac{(-1)^k 2^{2k+1}}{(2k-1)!} x^{2k-1}$

49. $-7 + 30x - 113x^2 + 422x^3 (+ \cdots)$

51. $x + \tfrac{1}{3}x^3 + \tfrac{2}{15}x^5 + \tfrac{17}{315}x^7 (+ \cdots)$

53. $1 - \dfrac{1}{2!}\left(x - \dfrac{\pi}{2}\right)^2 + \dfrac{1}{4!}\left(x - \dfrac{\pi}{2}\right)^4 - \dfrac{1}{6!}\left(x - \dfrac{\pi}{2}\right)^6 + \cdots = \displaystyle\sum_{k=0}^{\infty} \dfrac{(-1)^k}{(2k)!}\left(x - \dfrac{\pi}{2}\right)^{2k}$

(all x)

55. $\dfrac{1}{5}\left[1 - \dfrac{x-5}{5} + \dfrac{(x-5)^2}{5^2} - \dfrac{(x-5)^3}{5^3} + \cdots\right] = \displaystyle\sum_{k=0}^{\infty} (-1)^k \dfrac{(x-5)^k}{5^{k+1}}$

($|x - 5| < 5$)

57. $1 + \dfrac{x-e}{e} - \dfrac{(x-e)^2}{2e^2} + \dfrac{(x-e)^3}{3e^3} - \dfrac{(x-e)^4}{4e^4} + \cdots =$

$1 + \displaystyle\sum_{k=1}^{\infty} (-1)^{k+1} \dfrac{(x-e)^k}{ke^k}$ ($|x - e| < e$)

59. Converges; $-\tfrac{77}{192} \approx -0.40$ is within 0.01

61. Diverges

Section 5.3 (pp. 405–407)

1. (a) $R = 1$
(b) $2(n+1)(n+2)a_{n+2} = (n^2 - n + 1)a_n + (n^2 - 1)a_{n+1}$
(c) $y_1 = 1 + \tfrac{1}{4}x^2 + \tfrac{1}{32}x^4 + \cdots$, $y_2 = x - \tfrac{1}{4}x^2 + \tfrac{1}{12}x^3 + \cdots$

3. (a) $R = 3$
(b) $3(n+1)(n+2)a_{n+2} + (n^2 - 1)a_{n+1} + 2a_n = 0$
(c) $y_1 = 1 - \tfrac{1}{3}x^2 + \tfrac{1}{54}x^4 + \cdots$, $y_2 = x + \tfrac{1}{6}x^2 - \tfrac{1}{9}x^3 + \cdots$

5. (a) $R = \infty$
(b) $a_2 = 0$; $(n+1)(n+2)a_{n+2} - na_n + 3a_{n-1} = 0$ ($n \geq 1$)
(c) $y_1 = 1 - \tfrac{1}{2}x^3 - \tfrac{3}{40}x^5 + \cdots$, $y_2 = x + \tfrac{1}{6}x^3 - \tfrac{1}{4}x^4 + \cdots$

7. (a) $R = 2$
(b) $2(n+1)(n+2)a_{n+2} = (n+1)^2 a_{n+1} + (2n-1)a_n$
(c) $y_1 = 1 - \tfrac{1}{4}x^2 - \tfrac{1}{12}x^3 + \cdots$, $y_2 = x + \tfrac{1}{4}x^2 + \tfrac{1}{6}x^3 + \cdots$

9. (a) $R = \infty$
(b) $a_{n+2} = \dfrac{5n - 3}{(n+1)(n+2)} a_n$

(c) $y_1 = 1 - \frac{3}{2}x^2 + \frac{(-3)7}{4!}x^4 + \frac{(-3)7(17)}{6!}x^6 + \cdots$,

$y_2 = x + \frac{2}{3!}x^3 + \frac{2(12)}{5!}x^5 + \frac{2(12)22}{7!}x^7 + \cdots$

11. (a) $R = \sqrt{3}$

(b) $a_{n+2} = \frac{n-1}{3(n+2)} a_n$

(c) $y_1 = 1 - \frac{x^2}{3 \cdot 2} - \frac{x^4}{3^2 \cdot 2 \cdot 4} - \frac{3x^6}{3^3 \cdot 2 \cdot 4 \cdot 6} - \frac{3 \cdot 5 x^8}{3^4 \cdot 2 \cdot 4 \cdot 6 \cdot 8} + \cdots$, $y_2 = x$

13. (a) $R = \infty$

(b) $a_2 = 0$; $(n+1)(n+2)a_{n+2} + 3a_{n-1} = 0 \quad (n \geq 1)$

(c) $y_1 = 1 - \frac{3}{2 \cdot 3}x^3 + \frac{3^2}{2 \cdot 3 \cdot 5 \cdot 6}x^6 - \frac{3^3}{2 \cdot 3 \cdot 5 \cdot 6 \cdot 8 \cdot 9}x^9 + \cdots$,

$y_2 = x - \frac{3}{3 \cdot 4}x^4 + \frac{3^2}{3 \cdot 4 \cdot 6 \cdot 7}x^7 - \frac{3^3}{3 \cdot 4 \cdot 6 \cdot 7 \cdot 9 \cdot 10}x^{10} + \cdots$

15. $a_5 = \frac{319}{40}$

17. $\lambda = 0, 4, 8, 12, 16, \ldots$: the nonnegative multiples of 4

19. $\lambda = 0, 1, 4, 9, 16, \ldots$: the perfect squares

21. (a) $a_{n+2} = \frac{n^2 - \nu^2}{(n+1)(n+2)} a_n$

(b) $y_1 = 1 - \frac{\nu^2}{2}x^2 + \frac{(4-\nu^2)(-\nu^2)}{4!}x^4 + \frac{(16-\nu^2)(4-\nu^2)(-\nu^2)}{6!}x^6 + \cdots$,

$y_2 = x + \frac{1-\nu^2}{3!}x^3 + \frac{(9-\nu^2)(1-\nu^2)}{5!}x^5 + \cdots$;

$R \geq 1$ [actually, $R = 1$ except for the special cases in part (c)]

(c) For integers ν; $1, x, 1 - 2x^2, x - \frac{4}{3}x^3, 1 - 8x^2 + 8x^4$

23. (b) $y = \frac{1}{1-2x}$

(c) $y = \frac{Ce^{-4x} + D}{1-2x}$, C, D arbitrary

25. (a) $y = \frac{1}{1-x}$

(b) $y = \frac{1}{1-3x}$

(c) $y = \dfrac{C}{1-x} + \dfrac{D}{1-3x}$, C, D arbitrary

Section 5.4 (pp. 415–416)

1. $x = \pm\sqrt{3}$

3. $x = 0,\; x = -1$

5. (a) $y = a_1(x + \dfrac{4}{2}x^2 + \dfrac{4\cdot 7}{2\cdot 2\cdot 3}x^3 + \dfrac{4\cdot 7\cdot 10}{2\cdot 2\cdot 3\cdot 3\cdot 4}x^4 + \cdots)$, a_1 arbitrary
   (b) All these converge for all x, so $R = \infty$.

7. (a) $y = a_3 x^3$, a_3 arbitrary
   (b) All these converge for all x, so $R = \infty$.

9. (a) $y = a_1(x + \tfrac{3}{4}x^2 + \tfrac{5}{16}x^3 + \cdots)$, a_1 arbitrary
   (b) $R = 2$

11. (a) $y = a_0(1 + x + \tfrac{1}{2}x^2 + \tfrac{1}{2}x^3 + \tfrac{7}{8}x^4 + \cdots)$, a_0 arbitrary
    (b) $R = 0$: except if $a_0 = 0$, the "solution" only converges for $x = 0$

13. (a) $y = a_1 x$, a_1 arbitrary
    (b) $R = \infty$

15. (a) The series starts off $y = x^4 + \tfrac{16}{5}x^5 + \tfrac{27}{4}x^6 + \cdots$.
    (b) $R = 1$ [*Hint:* First show that $a_{n+1} > a_n$ for all $n > 3$, then use this to show that $\lim\limits_{n\to\infty} \dfrac{a_{n+1}}{a_n} = \lim\limits_{n\to\infty} \dfrac{n^2}{(n-3)(n+1)}$.]

17. (a) $y = \displaystyle\int \dfrac{e^x}{x}\, dx$

19. Including the integration constant has the effect of adding constant multiples of $y_1(x)$ to a particular $y_2(x)$. Since $y_1(x)$ is a solution and the equation is linear (and homogeneous), this will indeed give us other solutions.

Section 5.5 (pp. 425–427)

1. $y = -7x + 3x^3$

3. $y = 2x^2 - 4x^2 \log(-x)$ $(x < 0)$

5. $y = x^2[\alpha \cos(\log(-x)) + \beta \sin(\log(-x))]$, α, β arbitrary

7. $y = \dfrac{7}{2} - \dfrac{8}{x^4}$

9. $y = \alpha x + \beta x^2 + \dfrac{\gamma}{x}$, α, β, γ arbitrary

11. $y = x^3 \log x + \alpha x^2 + \beta x^3$, α, β arbitrary

17. (a) The graph of $y = x \log x$ has a vertical (one-sided) tangent line at $x = 0$.
    (b) The graph of $y = x^2 \log x$ has a horizontal tangent line at $x = 0$.

21.

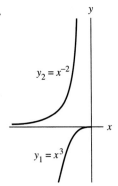

23.

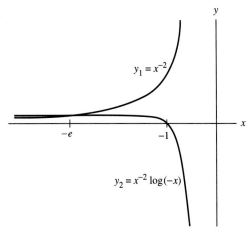

25.

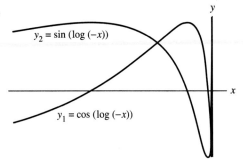

29. Yes; for $\lambda > 2$

Section 5.6 (pp. 437–441)

7. Including the integration constant has the effect of adding constant multiples of $y_1(x)$ to a particular $y_2(x)$. Since $y_1(x)$ is a solution and the equation is linear (and homogeneous), this will indeed give us other solutions. (Compare Section 5.4, Exercises 18 and 19.)

ANSWERS TO ODD-NUMBERED EXERCISES **671**

13. (c) $b_3 = -\frac{1}{6}$, $b_5 = \frac{1}{120}$
    (d) $b_5 = -\frac{1}{10}$
    (e) $b_{2\nu+2} = -\dfrac{1}{2(2\nu+2)}$; the solution is the same.

Section 5.7 (pp. 455–459)

1. (a) Regular singular point   (b) $x^2 y'' + 5xy' + 3y = 0$
   (c) There will be basic solutions of the form $y_1 = x^{-1} \sum_{n=0}^{\infty} a_n x^n$,

   $y_2 = By_1 \log x + x^{-3} \sum_{n=0}^{\infty} b_n x^n$; they "blow up" as $x \to 0^+$.

3. (a) Regular singular point   (b) $x^2 y'' + \frac{5}{2}xy' - y = 0$
   (c) There will be basic solutions of the form $y_1 = x^{1/2} \sum_{n=0}^{\infty} a_n x^n$,

   $y_2 = x^{-2} \sum_{n=0}^{\infty} b_n x^n$; $y_1 \to 0$ as $x \to 0^+$, while y_2 "blows up" as $x \to 0^+$.

5. (a) Irregular singular point
7. (a) Regular singular point   (b) $x^2 y'' + 5xy' + 4y = 0$
   (c) There will be basic solutions of the form $y_1 = x^{-2} \sum_{n=0}^{\infty} a_n x^n$,

   $y_2 = y_1 \log x + x^{-2} \sum_{n=0}^{\infty} b_n x^n$; they "blow up" as $x \to 0^+$.

9. (a) Ordinary point
11. (a) Regular singular point   (b) $x^2 y'' + 2xy' = 0$
    (c) There will be basic solutions of the form $y_1 = \sum_{n=0}^{\infty} a_n x^n$,

    $y_2 = By_1 \log x + x^{-1} \sum_{n=0}^{\infty} b_n x^n$; as $x \to 0^+$, $y_1 \to a_0$, while y_2 "blows up."

13. (a) Regular singular point   (b) $x^2 y'' + xy' + 2y = 0$
    (c) There will be basic solutions of the form

    $y_1 = \text{Re}\left(x^{i\sqrt{2}} \sum_{n=0}^{\infty} a_n x^n\right)$, $y_2 = \text{Im}\left(x^{i\sqrt{2}} \sum_{n=0}^{\infty} a_n x^n\right)$, where $a_0 = 1$ and a_1, a_2, ... are complex coefficients; these basic solutions will fluctuate more and more rapidly as $x \to 0^+$, with "amplitude" approaching 1.

17. $y = e^{-2x}(\alpha\sqrt{x} + \dfrac{\beta}{x^2})$ (α, β arbitrary)

19. $y = \sqrt{x}(\alpha e^{2x} + \beta)$ **21.** $y = x^{1/3}e^{-x}(\alpha + \beta \log x)$

23. $y_1 = \dfrac{e^x}{x}$, $y_2 = x^{7/2}\left(1 + \dfrac{2}{11}x + \dfrac{4}{11\cdot 13}x^2 + \dfrac{8}{11\cdot 13\cdot 15}x^3 + \cdots\right)$

25. $y_1 = \left(x - \dfrac{2}{33}x^2 - \dfrac{5}{99}x^3 + \cdots\right)\cos(2\sqrt{2}\log x) +$
$\left(-\dfrac{8\sqrt{2}}{33}x^2 + \dfrac{2\sqrt{2}}{99}x^3 + \cdots\right)\sin(2\sqrt{2}\log x),$

$y_2 = \left(\dfrac{8\sqrt{2}}{33}x^2 - \dfrac{2\sqrt{2}}{99}x^3 + \cdots\right)\cos(2\sqrt{2}\log x) +$
$\left(x - \dfrac{2}{33}x^2 - \dfrac{5}{99}x^3 + \cdots\right)\sin(2\sqrt{2}\log x)$

27. $y = x^4 - \frac{2}{5}x^5 + \frac{1}{15}x^6 + \cdots$

29. $y = x - 3x^2 + \frac{9}{4}x^3 + \cdots = \displaystyle\sum_{n=1}^{\infty} \dfrac{(-1)^{n+1}3^{n-1}}{[(n-1)!]^2}x^n$

31. $y = x^3 - \dfrac{x^4}{1\cdot 6} + \dfrac{x^5}{1\cdot 2\cdot 6\cdot 7} - \dfrac{x^6}{1\cdot 2\cdot 3\cdot 6\cdot 7\cdot 8} + \cdots$

33. **(b)** Not necessarily; although $\lambda = 0$ will be a root of the indicial equation, it might be the smaller of two integer roots. For an example, see Exercise 27.

Section 6.1 (pp. 475–478)

1. $\dfrac{24}{s^4} - \dfrac{4}{s^3} + \dfrac{5}{s^2} + \dfrac{10}{s}$ **3.** $\dfrac{4s}{s^2+1} - \dfrac{6}{s^2+4}$

5. $\dfrac{2}{(s+2)^2+4} + \dfrac{2}{(s+2)^3} - \dfrac{3}{s+2}$ **9.** Converges

11. Converges (compare Example 6.1.7)   **13.** Diverges

15. Converges (note that since $t \geq 2$, $\dfrac{1}{t^2-1} \leq \dfrac{2}{t^2}$)

17. $\dfrac{6}{s^4}$; for $s > 0$

21. $\dfrac{s}{s^2+b^2}, \dfrac{b}{s^2+b^2}$

27. $x(t) = \begin{cases} \cos 3t + \frac{1}{3}\sin 3t & \left(t \leq \frac{\pi}{2}\right) \\ \frac{1}{8}\sin t + \cos 3t + \frac{11}{24}\sin 3t & \left(\frac{\pi}{2} \leq t \leq \frac{2\pi}{3}\right) \\ (\frac{1}{16}\sqrt{3} + 1)\cos 3t + \frac{7}{16}\sin 3t & \left(t \geq \frac{2\pi}{3}\right) \end{cases}$

29. $x(t) = \begin{cases} e^t + e^{3t} & (t \leq 3) \\ \frac{2}{3} + (1 - e^{-3})e^t + (1 + \frac{1}{3}e^{-9})e^{3t} & (3 \leq t \leq 5) \\ (1 - e^{-3} + e^{-5})e^t + (1 + \frac{1}{3}e^{-9} - \frac{1}{3}e^{-15})e^{3t} & (t \geq 5) \end{cases}$

37. (f) $\mathcal{L}\{t\sqrt{t}\} = \frac{3}{4}\sqrt{\pi}\, s^{-5/2}$; $\mathcal{L}\{t^2\sqrt{t}\} = \frac{15}{8}\sqrt{\pi}\, s^{-7/2}$

Section 6.2 (pp. 487–489)

3. $\dfrac{s^2 + 5s + 3}{s(s^2 - 4)}$

5. $\dfrac{s^2 + 2s + 2}{(s+1)^3}$

7. $\dfrac{s^4 - 2s^3 - s - 2}{(s+1)^2(s^2+1)(s-3)} \left(= \dfrac{1}{s+1} + \dfrac{s}{(s+1)(s^2+1)(s-3)} + \dfrac{1}{(s+1)^2(s-3)} \right)$

9. $1 - 3e^{-t}$

11. $\frac{1}{2}t^2 e^{2t} - \frac{5}{2}\sin 2t + 2e^{3t}$

13. $\frac{1}{4}(e^{4t} - 1)$

15. $\frac{5}{8}e^{2t} + \frac{9}{8}e^{-2t} - \frac{3}{4}$

17. $\frac{1}{2}e^{4t}\sin 2t$

19. $1 - 2e^{-4t}\sin 2t$

21. $\frac{1}{8}(\cos t - \cos 3t)$

23. $5 - \cos 2t + \frac{3}{2}\sin 2t$

25. $x(t) = 3e^{4t}$

27. $x(t) = 7e^{2t} - 5e^{3t}$

29. $x(t) = (2 - 6t)e^{3t}$

31. $x(t) = e^{-3t}(\cos t + 6\sin t)$

33. $x(t) = \frac{7}{9} - \frac{1}{3}t - e^{-t} + \frac{2}{9}e^{-3t}$

35. $x(t) = (\frac{1}{2}t^2 + 5t + 3)e^{-2t}$

37. $x(t) = \frac{1}{2}\cos t + \frac{3}{2}\sin t - \frac{1}{6}e^t + \frac{2}{3}e^{-2t} - e^{-t}$

39. $x(t) = \frac{1}{2}t - e^t + \frac{13}{4}e^{2t} - \frac{3}{4}e^{-2t}$

41. $x(t) = (t + 9)e^t - 4$, $y(t) = e^t + 1$

43. $x(t) = \cos 2t + \frac{1}{2}\sin 2t$, $y(t) = \frac{5}{2}\sin 2t$

49. (b) $\dfrac{6s^2 - 2}{(s^2 + 1)^3}$

Section 6.3 (pp. 500–503)

1. (a) $h(t) = u_1(t) - u_2(t)$
(b) $\dfrac{1}{s}(e^{-s} - e^{-2s})$

3. (a) $h(t) = \sin t + (\cos t - \sin t)u_0(t) + (1 - \cos t)u_\pi(t)$
(b) $\dfrac{s}{s^2 + 1} + e^{-\pi s}\left(\dfrac{1}{s} + \dfrac{s}{s^2 + 1}\right)$

5. (a) $h(t) = 2tu_2(t) + (2e^t - 2t)u_3(t)$
(b) $e^{-2s}\left(\dfrac{2}{s^2} + \dfrac{4}{s}\right) + e^{-3s}\left(\dfrac{2e^3}{s - 1} - \dfrac{2}{s^2} - \dfrac{6}{s}\right)$

7. (a) $h(t) = 1 - e^{2t} + (e^{2t} - 1)u_\pi(t) + (\cos t)u_{2\pi}(t)$
(b) $\dfrac{1}{s} - \dfrac{1}{s - 2} + e^{-\pi s}\left(\dfrac{e^{2\pi}}{s - 2} - \dfrac{1}{s}\right) + \dfrac{se^{-2\pi s}}{s^2 + 1}$

9. $\dfrac{e^{-s} - e^{-2s} + 2s^2 - 9s}{s(s^2 - 4s + 3)}$

11. $\dfrac{3}{s} + \dfrac{1 - e^{-\pi s}}{s(s^2 - 4s)} + \dfrac{e^{-\pi(s-2)} - 1}{(s - 2)(s^2 - 4s)} + \dfrac{e^{-2\pi s}}{(s^2 + 1)(s - 4)}$

13. $x(t) = \begin{cases} 4e^{-3t} & (t \le 2) \\ \left(4 + \dfrac{e^6}{3}\right)e^{-3t} - \dfrac{1}{3} & (t \ge 2) \end{cases}$

15. $x(t) = \begin{cases} e^t + 2e^{3t} & (t \le 2) \\ -\tfrac{1}{3} + (1 + \tfrac{1}{2}e^{-2})e^t + (2 - \tfrac{1}{6}e^{-6})e^{3t} & (2 \le t \le 4) \\ (1 + \tfrac{1}{2}e^{-2} - \tfrac{1}{2}e^{-4})e^t + (2 - \tfrac{1}{6}e^{-6} + \tfrac{1}{6}e^{-12})e^{3t} & (t \ge 4) \end{cases}$

17. $x(t) = \begin{cases} 2e^{2t} - e^{-2t} & (t \le 2) \\ -\dfrac{1}{4}t + \left(2 + \dfrac{5}{16e^4}\right)e^{2t} + \left(\dfrac{3e^4}{16} - 1\right)e^{-2t} & (t \ge 2) \end{cases}$

19. $x(t) = \begin{cases} -e^{-3t} & (t \le 2) \\ -\dfrac{1}{3}t + \dfrac{4}{9} - \left(\dfrac{5e^6}{18} + 1\right)e^{-3t} + \dfrac{e^2}{2}e^{-t} & (2 \le t \le 4) \\ \left(\dfrac{11}{18}e^{12} - \dfrac{5e^6}{18} - 1\right)e^{-3t} + \left(\dfrac{e^2}{2} - \dfrac{3e^4}{2}\right)e^{-t} & (t \ge 4) \end{cases}$

21. (a) $\dfrac{2}{s - 2} - \dfrac{4}{(s - 2)^2}$; $\dfrac{3/2}{s} - \dfrac{3/2}{s - 2} + \dfrac{3}{(s - 2)^2}$

23. $x(t) = \begin{cases} \cos 2t + \sin 2t & (t \le \pi) \\ \tfrac{1}{3}\cos t + \tfrac{4}{3}\cos 2t + \sin 2t & (t \ge \pi) \end{cases}$

25. $x(t) = \begin{cases} \cos 2t + \sin 2t & (t \leq \pi) \\ \frac{1}{3}\cos t + \frac{4}{3}\cos 2t + \sin 2t & (\pi \leq t \leq 2\pi) \\ \frac{1}{2} + \frac{7}{6}\cos 2t + \sin 2t & (2\pi \leq t \leq 4\pi) \\ \frac{5}{3}\cos 2t + \sin 2t & (t \geq 4\pi) \end{cases}$

27. (a) $x(t) = \begin{cases} 0 & (t \leq \pi) \\ -\sin t & (t \geq \pi) \end{cases}$; yes

(b) $x(t) = \begin{cases} 0 & (t < \pi) \\ -\cos t & (t > \pi) \end{cases}$; no ($x(t)$ is not continuous for $t = \pi$)

29. $\dfrac{1}{s(1 + e^{-s})} = \dfrac{e^s}{s(e^s + 1)}$

31. $\dfrac{e^s - 1 - s}{s^2(e^s - 1)}$

33. $\dfrac{s}{s^2 + 1}$

Section 6.4 (pp. 512–514)

1. $\frac{1}{12}t^4 + \frac{5}{3}t^3$

3. $\frac{1}{3}(\cos 2t - \cos 4t)$

5. $\frac{1}{4}(e^{3t} - e^{-t})$

7. $\frac{1}{2}(\sin t - t \cos t)$

9. $\dfrac{1}{2}t \cos bt + \dfrac{1}{2b}\sin bt$

11. $t \cos bt$

13. $\frac{1}{9}(e^{5t} - e^{-4t})$

15. $\frac{1}{16}t - \frac{1}{64}\sin 4t$

17. te^{at}

19. $\begin{cases} 0 & (t \leq \pi) \\ -\sin t & (t \geq \pi) \end{cases}$

21. $\begin{cases} 0 & (t \leq 4) \\ \frac{1}{2} + \frac{1}{2}e^{2(t-4)} - e^{t-4} & (t \geq 4) \end{cases}$

23. $x(t) = -\frac{1}{3}t \sin 3t + 4 \cos 3t$

25. $x(t) = -\frac{1}{8}t^2 \cos 2t + \frac{1}{16}t \sin 2t$

29. (a) $\displaystyle\int_0^t f(\tau)\,d\tau$

35. $x(t) = \begin{cases} 5 \sin t - 2 \cos t & (t \leq \pi) \\ \dfrac{11}{2}\sin t + \left(\dfrac{\pi}{2} - 2\right)\cos t - \dfrac{1}{2}t \cos t & (\pi \leq t \leq 2\pi) \\ 5 \sin t - \left(2 + \dfrac{\pi}{2}\right)\cos t & (t \geq 2\pi) \end{cases}$

Section 6.5 (pp. 526–530)

1. $x(t) = \begin{cases} \cos 2t - 2 \sin 2t & (t \leq 2\pi) \\ \cos 2t - \frac{1}{2}\sin 2t & (t \geq 2\pi) \end{cases}$

3. $x(t) = \begin{cases} -\dfrac{10}{3} + 2t + 4e^{-t} - \dfrac{2}{3}e^{-3t} & (t \leq 1) \\ -\dfrac{10}{3} + 2t + \left(4 + \dfrac{e}{2}\right)e^{-t} - \left(\dfrac{2}{3} + \dfrac{e^3}{2}\right)e^{-3t} & (t \geq 1) \end{cases}$

5. $x(t) = e^{3t} - e^{2t} \quad (t \geq 0)$

7. $x(t) = \begin{cases} (\frac{1}{2}t^2 - 4t + 2)e^{2t} & (t \leq 1) \\ [\frac{1}{2}t^2 + (e^{-2} - 4)t + (2 - e^{-2})]e^{2t} & (t \geq 1) \end{cases}$

9. $x(t) = \frac{5}{2}e^t - \frac{9}{2}e^{-t} + 5 - 10t \quad (t \geq 0)$

13. (a) $v_1 = 0, v_2 = v_0$; $v_1 = \frac{2}{5}v_0, v_2 = \frac{3}{5}v_0$
    (b) After the collision, the first particle should no longer be at rest.

15. (a) $h_2(t) = \left(-\dfrac{2}{\epsilon^2}t + \dfrac{2}{\epsilon}\right)u_0(t) + \left(\dfrac{2}{\epsilon^2}t - \dfrac{2}{\epsilon}\right)u_\epsilon(t)$

17. (a) $h_4(t) = \left(\dfrac{\pi}{2\epsilon}\sin\dfrac{\pi}{\epsilon}t\right)[u_0(t) - u_\epsilon(t)]$

19. (c) To match the jump of $12u_2(t)$.

21. (a) $2\delta(t) - \cos 3t - \frac{17}{3}\sin 3t$

Section 7.1 (pp. 541–544)

1. $(\Delta_t x)(t, u) = x(t + 1, u) - x(t, u)$; $(\Delta_u x)(t, u) = x(t, u + 1) - x(t, u)$.
   A partial difference equation for x is an equation in which one or more partial differences of x (of first or higher order) occur.

3. 341, 532

5. (a) $\Delta x = 4x + 2$
   (b) $\Delta^2 x - \Delta x - x = 4$
   (c) $\Delta^2 x + \Delta x - x = 0$

7. For the particular sequence (x_t) given by $x_t = x^t$, we have $(\Delta^n x)_t = x^t(x - 1)^n$.

9. A homogeneous linear difference equation for the unknown sequence (x_t) is an equation of the form $f_n(t)(\Delta^n x)_t + f_{n-1}(t)(\Delta^{n-1}x)_t + \cdots + f_1(t)(\Delta x)_t + f_0(t)x_t = 0$.
   A homogeneous linear recurrence relation for (x_t) has the form
   $x_{t+n} = g_{n-1}(t)x_{t+n-1} + g_{n-2}(t)x_{t+n-2} + \cdots + g_1(t)x_{t+1} + g_0(t)x_t$.

13. (a) $r = 3$
    (b) $x_t = (\alpha + \beta t)3^t$, α, β arbitrary

15. $x_t = \begin{cases} \dfrac{b(a^t - 1)}{a - 1} + Ca^t, & C \text{ arbitrary } (a \neq 1) \\ bt + C, & C \text{ arbitrary} \quad (a = 1) \end{cases}$

17. (a) $\lambda = 1$, eigenvectors $\xi_2(2, 1), \xi_2 \neq 0$;
       $\lambda = -0.2$, eigenvectors $\xi_1(1, -1), \xi_1 \neq 0$

(b) $T = \begin{bmatrix} 2 & 1 \\ 1 & -1 \end{bmatrix}$, $D = \begin{bmatrix} 1 & 0 \\ 0 & -0.2 \end{bmatrix}$

(c) $\begin{bmatrix} \frac{2}{3} - \frac{1}{3}(0.2)^9 & \frac{2}{3} + \frac{2}{3}(0.2)^9 \\ \frac{1}{3} + \frac{1}{3}(0.2)^9 & \frac{1}{3} - \frac{2}{3}(0.2)^9 \end{bmatrix}$

(d) $\frac{200}{3} - \frac{100}{3}(0.2)^9 \approx 66.66665\%$. (Obviously, this is unrealistic "accuracy." What else do you notice?)

Section 7.2 (pp. 555–557)

1. $x(t) = \sqrt{t^3 + 1}$

5. Approximate value for $n = 2$: 1.031250; for $n = 8$: 1.159597; for $n = 32$: 1.227488; for $n = 128$: 1.249340; for $n = 512$: 1.255210

9. Approximate value for $n = 2$: 1.095703; for $n = 8$: 1.233510; for $n = 32$: 1.255324; for $n = 128$: 1.257082; for $n = 512$: 1.257200

11. (a) $h = \dfrac{\partial^2 f}{\partial t^2} + \dfrac{\partial f}{\partial x}\dfrac{\partial f}{\partial t} + \left[2\dfrac{\partial^2 f}{\partial x \partial t} + \left(\dfrac{\partial f}{\partial x}\right)^2\right] f + \dfrac{\partial^2 f}{\partial x^2} f^2$

(c) There is often a great deal of work involved in computing values of $h(t, x)$.

15. Approximate value for $n = 2$: 1.307268; for $n = 8$: 1.261181; for $n = 32$: 1.257506; for $n = 128$: 1.257227; for $n = 512$: 1.257208

19. Approximate value for $n = 2$: 0.746855; for $n = 8$, $n = 32$, $n = 128$, and $n = 512$: 0.746824

Section 7.3 (pp. 567–571)

1. (c) $\left(\dfrac{5}{x}\right)^2 = 5$, and $\dfrac{1}{2}\left(x + \dfrac{5}{x}\right) = x$.

(d) The statements in (a) and (b) are still correct, but now $x^2 < 5$ when $x > -\sqrt{5}$ and $x^2 > 5$ when $x < -\sqrt{5}$.

7. (d) If $m = 0$, then f is identically zero on the rectangle and all functions x_n will be identically zero there; the function $x(t) = 0$ will be a solution to the initial value problem.

9. (a) $f_2(x) = \begin{cases} 0 & (x < \frac{1}{2}) \\ 2x - 1 & (\frac{1}{2} \le x \le 1) \\ 1 & (x > 1) \end{cases}$, $f_3(x) = \begin{cases} 0 & (x < \frac{2}{3}) \\ 3x - 2 & (\frac{2}{3} \le x \le 1) \\ 1 & (x > 1) \end{cases}$,

$f_4(x) = \begin{cases} 0 & (x < \frac{3}{4}) \\ 4x - 3 & (\frac{3}{4} \le x \le 1) \\ 1 & (x > 1) \end{cases}$

(b) $f_n(x) = \begin{cases} 0 & (x < \frac{n-1}{n}) \\ nx - n + 1 & (\frac{n-1}{n} \le x \le 1) \\ 1 & (x > 1) \end{cases}$

(c) $L(x) = \begin{cases} 0 & (x < 1) \\ 1 & (x \geq 1) \end{cases}$

11. (a) $x_5(t) = \frac{1}{12}t^{12} + \frac{1}{3}t^9 + t^6 + 2t^3$,
$x_6(t) = \frac{1}{60}t^{15} + \frac{1}{12}t^{12} + \frac{1}{3}t^9 + t^6 + 2t^3$

(b) $x_n(t) = 2t^3 + t^6 + \frac{1}{3}t^9 + \cdots + 2\frac{t^{3(n-1)}}{(n-1)!} = \sum_{k=1}^{n-1} 2\frac{t^{3k}}{k!}$

15. (b) $\dfrac{n}{n+1}$

Section 7.4 (pp. 581–587)

1. (a) $k = \dfrac{1}{\alpha^2}$

(b) When $\beta \neq 0$, for $k = \dfrac{\alpha^2}{\beta^2}$; when $\alpha = 0$, for every k.

9. (a) $u(x,t) = C_1 e^{(x\sqrt{\gamma} + \gamma t)} + C_2 e^{(-x\sqrt{\gamma} + \gamma t)}$ $(\gamma > 0)$;
$u(x,t) = C_1 + C_2 x$;
$u(x,t) = [C_1 \cos(\sqrt{-\gamma}\, x) + C_2 \sin(\sqrt{-\gamma}\, x)]e^{\gamma t}$ $(\gamma < 0)$.
(C_1, C_2, and γ are arbitrary constants.)

(b) $u(x,t) = \sum_{n=1}^{\infty} c_n (\sin \dfrac{n\pi x}{2}) e^{-n^2\pi^2 t/4}$, where

$c_n = \displaystyle\int_0^2 (1 - \cos \pi x) \sin \dfrac{n\pi x}{2}\, dx$

$= \begin{cases} 0 & (n \text{ even}) \\ \dfrac{16}{\pi n(4 - n^2)} & (n \text{ odd}) \end{cases}$, so

$u(x,t) = \displaystyle\sum_{k=0}^{\infty} \dfrac{16}{\pi(2k+1)(3 - 4k - 4k^2)} (\sin \dfrac{(2k+1)\pi x}{2}) e^{-(2k+1)^2\pi^2 t/4}$.

11. (c) The particle is always *somewhere*.
17. (c) Since we want $u(L,t) = 0$, we'll look for solutions with $F(L) = 0$, that is, $f(2k\sqrt{L}) = 0$.
    (d) Otherwise, the displacement of the chain will presumably be unbounded (by the same sort of "fishy" argument as on p. 579).

Appendix A.1 (pp. 589–591)

1. $16 + 11i$ 3. (a) 0   (b) 8
5. $-i$ 7. $z = 1 - i$

11. $z = 3i$

13. $z = \pm i\sqrt{19}$

15. $z = -1$; $z = \tfrac{1}{2} \pm \tfrac{1}{2}i\sqrt{3}$

Appendix A.2 (pp. 592–598)

1.
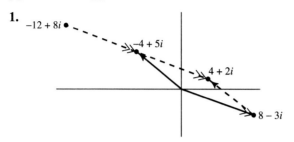

5. $a^2 + b^2 < 1$, $a^2 + b^2 = 1$, $a^2 + b^2 > 1$; $|z| < 1$, $|z| = 1$, $|z| > 1$

9. (a) 1   (b) 5   (c) 32   (d) $\sqrt{61}$

11. The set of all numbers of the form $z = bi$ (the imaginary axis, which is the perpendicular bisector of the line segment from -1 to 1).

15.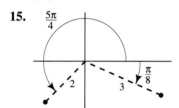

19. $z = \pm 1$; $z = \pm i$

Appendix A.3 (pp. 599–603)

1. (a) $-\tfrac{1}{2}$ (b) $-i$

Appendix B.1 (pp. 605–609)

1. $(\tfrac{11}{2}, 0)$; $(-9, -1)$

3.
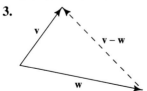

5. (b) The speed is multiplied by 5, since the particle gets to each position in a fifth of the original time.

7.

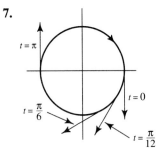

(Vectors scaled down)

9. $|\mathbf{v}| = \sqrt{v_1^2 + v_2^2 + \cdots + v_n^2}$

Appendix B.2 (pp. 610–614)

1. $\mathbf{AB} = \begin{bmatrix} 4 & 1 \\ 6 & -2 \end{bmatrix}$; $\mathbf{BA} = \begin{bmatrix} -1 & 4 & -9 \\ 10 & 2 & -1 \\ 3 & 0 & 1 \end{bmatrix}$

3. $\begin{bmatrix} 11 \\ 6 \\ 15 \end{bmatrix}$

11. $\mathbf{BA} = \begin{bmatrix} 1 & 0 & 0 \\ 0 & 0 & 0 \\ 0 & 0 & 1 \end{bmatrix}$; **A** and **B** aren't square matrices.

Appendix B.3 (pp. 615–620)

1. $\underset{\mathbf{A}}{\begin{bmatrix} 3 & -4 \\ -5 & 1 \end{bmatrix}} \underset{\mathbf{v}}{\begin{bmatrix} x \\ y \end{bmatrix}} = \underset{\mathbf{b}}{\begin{bmatrix} 2 \\ 3 \end{bmatrix}}$

5. $\begin{bmatrix} \frac{1}{2} & -\frac{1}{4} \\ -\frac{3}{2} & \frac{5}{4} \end{bmatrix}$

7. $\begin{bmatrix} \frac{1}{4} & 0 & 0 \\ 0 & -1 & 3 \\ 0 & 1 & -2 \end{bmatrix}$

9. $\begin{bmatrix} 1 & -2 & -4 \\ 0 & 4 & -1 \\ 1 & -5 & -3 \end{bmatrix}$

11. $x = 58$, $y = 30$, $z = -68$

13. $x = 0$, $y = 0$, $z = 0$

15. Infinitely many solutions; $x = -4z$, $y = \frac{7}{3}z$, $z = z$, z arbitrary.

Appendix B.4 (pp. 621–626)

1. -1 **3.** 24   **5.** -4

7. 0   **9.** 30   **11.** -13

13. Yes; $\begin{vmatrix} 2 & 1 & 3 & -4 \\ 1 & -1 & -1 & 2 \\ 3 & 1 & 1 & -1 \\ 0 & -1 & 1 & -1 \end{vmatrix} = 0$

17. $\det(\mathbf{A}) = 2$; $\det(\mathbf{B}) = 2$; $\det(\mathbf{A} + \mathbf{B}) = 0$

Index

A

AC, DC, 461–462
Abel's identity, 96
Absolute value (of complex number), 594
Acceleration (for particle along line), 77
   vector, 607
Addition of vectors, 606
Airy equation, 402
Airy functions, 402*
Analytic function(s), 385
   at $x = c$, 387
   list of, 386
Annual percentage rate, 28
Antiderivative, 3
Approximating Euler equation, 442
Approximation
   by Taylor polynomials, 368
   of value of power series, 379
Approximation, successive
   for solution to initial value problem, 562
   for square roots, 559–560
Argand diagram, 592
Argument (of complex number), 595
Arithmetic
   of complex numbers, 590
   of matrices, 610
Arithmetic sequence, 536
Associated Legendre equation, 586
Asymptotically stable, 237
Attractor basin, 240
Autonomous system, 156, 306
   linear, 178, 306
   linear homogeneous, 179, 306
Axes, locating new, 193, 219

B

Basic solutions (for homogeneous linear ODE), 89, 280
   in power series, 403, 447
   for systems, 193, 209, 315, 325*
Basic step function, 490
   derivative of, 521
Beats, 127
Bernoulli equations, 72
Bessel equation, 428
   of order 0, 433, 587
   of order $\frac{1}{2}$, 431
   of order 1, 440
Bessel functions, 430*
Big game hunting, 151
Binet's formula, 538*
Boundary value problem, 36
   for PDE, 573
Branching, 40

C

C^1 functions, 8
Capacitance, 147
Capacitor, 147
Cauchy–Euler equation, *see* Euler equation
Center (type of stationary point), 217
Certificates of deposit, 534
Chain rule
   one variable, 3
   two variables, 9
Chain suspended from ceiling, 409, 586
Change of coordinates
   to shift stationary point, 178
   to simplify linear system, 185, 308

*This indicates that the reference is to a footnote.

Characteristic equation
  of nth-order linear ODE, 283
  of linear system, 187
  of $n \times n$ matrix, 313
  of second-order linear ODE, 99
Characteristic polynomial, 313
Chebyshev equation, 406
Chebyshev polynomials, 406
Circle of convergence, 600
Circuit, electrical, 146, 461
Classification
  of critical points of $V(x, y)$, 257
  of differential equations, 29
  of stationary points of system, 214, 217–218
Coefficients
  of linear ODE, 31, 87
  of power series, 376
Collision of particles, 514
Column vector, 610
Commuting matrices, 347
Companion matrix, 321
Comparison test (for improper integrals), 468
Competition (of populations), 145, 268
Complementary probabilities, 541
Complementary solutions, 131*
Complex conjugate, 593
Complex eigenvalues, 202, 322
Complex exponentials, 601
  trigonometric functions in terms of, 296, 602
Complex number(s), 590
  absolute value of, 594
  argument of, 595
  arithmetic of, 590
  conjugate of, 593
  convergent series of, 599
  exponential for, 601
  imaginary part of, 590
  limit of, 599
  polar form of, 595, 602
  purely imaginary, 593
  real part of, 590
Complex plane, 592
Complex-valued function, 107
Complex-valued solution, 108
Components (of vector), 607
Conjugate (of complex number), 593
Conservation of energy, 169
Constant of integration, 3
  parametrization by, 17
Convergence (of improper integral), 467
Convergent series (with complex terms), 599
Convolution, 506
  properties of, 508

Coordinate axes, locating new, 193, 219
Coordinates, change of
  to shift stationary point, 178
  to simplify linear system, 185, 308
Critical point
  of system, 163
  of $V(x, y)$, 165
Critically damped (vibration), 120, 122
Cycle, 240
  population, 266

D

DC, AC, 461–462
Definite integral, 3
Delta "function," 518
  as derivative of step function, 521
  as identity for convolution, 519
  integrals involving, 519
Derivative
  definition, 2
  of complex-valued function, 108
  of power series, 380
  of vector function, 608
Determinant, 621
  expansion along first row, 622–623
  properties of, 623
  Vandermonde, 284, 625–626
  Wronskian, 92, 133, 282
Diagonal matrix, 191, 308, 624
Diagonalizable (matrix), 191
Diagonalizing (a matrix), 191
Difference equation(s), 534
  system of, 539
Difference function, 534
Differentiable, 2
  for complex-valued function, 108
Differential equation(s), 12. See also First-order ODE; Second-order ODE; Higher-order ODE; Partial differential equations
  classification of, 29
  solutions of (definition), 13
Differential form (of first-order ODE), 15*, 162
Differential operator, 477
Differentiating power series, 380
Differentiation operator, 477
Direction field, 42
  of system, 158
Distributions, 519
Division of power series, 389
Domain, 252
Dot product, 609
Double eigenvalue, 200

INDEX

Doubly periodic functions, 82*
Dummy variable, 4

E

Eigenvalues, 190, 310
   complex, 202, 322
   double, 200
Eigenvectors, 190, 310
   as columns of **T**, 191, 309
   along new axes, 193
Elastic beam, 279
Elastic collision, 514
Electrical circuit, 146, 461
Electrical network, 152, 359
Elimination method (for system), 180
Elliptic integrals, 82
Energy, 169, 251
   levels, 581
Enterprise, 26
Entries of matrix, 610
Equilibrium
   of particle in general, 166
   of particle on spring, 119
Equilibrium point (of system), 163
Error function (erfc), 562
Escape velocity, 84
Euler equation(s), 417
   approximating, 442
   behavior near $x = 0$, 422
   solutions to, 421
Euler's formula, 601
Euler's method, 546
Exact (first-order ODE), 60
   test for, 62
Excess birth rate, 24
Existence and Uniqueness Theorem, 40, 559
Existence of power series solutions, 396, 447
Expansion of determinant, 622–623
Exponential(s)
   for complex numbers, 601
      trigonometric functions in terms of, 296, 602
   for matrices, 336, 614
      how to compute, 340
      solving system using, 342

F

Fibonacci numbers, 537
Fibonacci sequence, 537
First integral (for system), 167
First-order ODE, 30
   exact, 60
   homogeneous $dy/dx = f(y/x)$, 72
   in differential form, 15*, 162
   integrating factor for, 64
   linear, 32
      homogeneous, how to solve, 49
      inhomogeneous, how to solve, 51
   separable, 15
Forced vibrations, 124
Forcing term, 31, 87, 356
   with jumps, 463, 493
Formal solution (of ODE), 412
Fourier series, 576
Free particle, 77
Frequency, 124*
Frictional force, 78
Frobenius's method, 444
Functional, 530
Fundamental matrix, 354
Fundamental Theorem of Algebra, 597
Fundamental Theorem of Calculus, 4

G

Gamma function, 438–439
Gapped power series, 377
Generator (electrical), 147
Geometric sequence, 536
Geometric series, 372
Gradient system, 166, 254
Gravity, acceleration of, 78

H

Half-life, 26
Harmonic oscillator, 580
Heat equation, 584
Hermite equation, 395, 398, 581
Hermite polynomials, 400*, 407
Heun's method, 552
Higher-order ODE, 279
   basic solutions for, 280
   constant-coefficient
      homogeneous, 283
      inhomogeneous, 291
   undetermined coefficients for, 296
   variation of constants for, 291
Homogeneous equation $dy/dx = f(y/x)$, 72
Homogeneous linear ODE, 31
   basic solutions for, 89, 280
   constructing solutions from others, 88
   first-order (how to solve), 49
   higher-order, constant-coefficient, 283
   reduction of order for, 102

Homogeneous linear ODE *(Cont.)*
   second-order constant-coefficient, 98, 104, 108
      connection with Euler equation, 418
   solving using power series, 370, 396
   trivial solution to, 88
Homogeneous system
   of linear (algebraic) equations, 616
   of n differential equations, 306, 354
   of two differential equations, 179
Hooke's law, 78

I

Identical twins, 590, 593
Identity matrix, 612
Imaginary axis, 593
Imaginary part (of complex number), 590
Implicit function theorem, 35
Improper integral, 466
   comparison test for, 468
Improper node, 214
Indefinite integral, 3
Index of summation, shifting, 381
Indicial equation, 448
Inductance, 147
Inductor, 147
Inhomogeneous linear ODE, 31
   first-order (how to solve), 51
   higher-order, 291
   obtaining all solutions from one, 52, 112, 131, 291
   second-order
      by undetermined coefficients, 114
      by variation of constants, 134
Initial condition, 36
Initial population size, 25
Initial value/boundary value problem, 573
Initial value problem, 36, 85
   existence, uniqueness of solution, 40, 86, 150, 280, 559
   for linear first-order ODE, 56
   for linear second-order ODE, 87
   for higher-order DE, 279–280
   for PDE, 573
   for system of two differential equations, 150
   how to find solution, 43
   solving using Laplace transform, 479
   successive approximation to solution, 562
   with jumps in forcing term, 463, 493
Integral, 3
   convolution, 506
   improper, 466
   involving delta function, 519
   of a matrix, 354*
   of a power series, 382

Integral curve, 39
   for system, 149
Integral equation, 561
Integral test, 467
Integrating factor, 64
Integrating power series, 382
Integration constant, parametrization by, 17
Interest rate, 28
Interval (open, closed), 3*
Interval-by-interval method, 463
Inverse Laplace transform, 482
   of product, 506
   of rational function, 483
Inverse (matrix), 612
   how to find, 616
   of 2×2 matrix, 619
Invertible matrix, 615*
Irregular singular point, 443
Isolated stationary point, 233

J

Jordan form, 330
Jump (of function), 464

K

Kinetic energy, 169, 251
Kirchhoff's laws, 147

L

Laguerre equation, 409, 412
Laguerre polynomials, 412*
Laplace equation, 584
   in spherical coordinates, 585
Laplace transform(s), 469
   of basic step function, 491
   of derivative, 479
   of function with jumps, 492, 502
   of higher derivatives, 479
   inverse, 482
   linearity of, 472
   solving initial value problem with, 479, 493
   solving system with, 485
   tables of, 474, 511
Legendre equation, 396, 400
   associated, 586
Legendre polynomials, 402*
Length
   of complex number, 594
   of vector, 607
Lerch's theorem, 481

INDEX

Level curves, 4, 255
  slope of, 10
Limit of sequence
  of complex numbers, 599
  of continuous functions, 563
Linear approximation
  to $f(x, y)$, 8
  to system, 167, 210
  to $x(t)$, 2
Linear combination, 89, 280
Linear equations, systems of, 615
Linear independence, 89, 281
Linear ODE, 31. *See also* Homogeneous linear ODE; Inhomogeneous linear ODE
Linear operator, 472
Linear PDE, 572
Linear resistor, 146
Linear system(s) (of ODE), 178, 306
  characteristic equation of, 187
  complex eigenvalue case, 202
  double eigenvalue case, 200
  eigenvalue method for, 186
  elimination method for, 180
  homogeneous, 179
  phase portrait for, 213, 215, 218
  stability for, 238
Linear transform(ation), 472, 614
Linearly independent, 89, 281
Local maximum/minimum, 165*, 257
Locating new axes, 193, 219
Logarithm, 17*
Lower triangular matrix, 625
Lucas sequence, 542
Lyapunov function, 249, 254
Lyapunov's theorem, 250, 254

M

Main diagonal (of matrix), 612
Markov chain, 541
Markov process, 541
Matrix/matrices, 610
  commuting, 347
  determinant of, 621
  diagonal, 191, 308, 624
  diagonalizing, 191
  eigenvalues and eigenvectors of, 190, 310
  entries of, 610
  exponential of, 336, 614
    how to compute, 340
    solving system using, 342
  fundamental, 354
  identity, 612
  inverse of, 612
    for 2×2 matrix, 619
    how to find, 616
  invertible, 615*
  lower triangular, 625
  main diagonal of, 612
  minors of, 622
  multiplication of, 610–611
    associativity of, 612
  nilpotent, 337
  nonsingular, 615*, 621*
  row operations on, 616–617
  singular, 615*, 621*
  symmetric, 257*
  transforming vectors by, 612
  upper triangular, 329, 347, 625
Mechanical energy, 251
Mechanical vibration, 119
  with periodic external force, 124
Method of Frobenius, 444
Method of undetermined coefficients, 114, 296
Method of variation of constants, 51–52, 134, 291, 357
Minor (of matrix entry), 622
Model
  competition, 271
  population growth, 24
  predator–prey, 145, 262
  Verhulst, 28
Modulus (of complex number), 594
Mowing lawns, 540
Multiplication
  of matrices, 610–611
    associativity of, 612
  of power series, 384

N

Newton's second law, 77, 607–608
Nilpotent (matrix), 337
Node, 214
Nonlinear differential equation, 31
Nonlinear systems, stability for, 239
Nonsingular matrix, 615*, 621*
Numbers, complex, *see* Complex numbers
Numerical analysis, 545

O

Ohm's law, 146
Open set, 252*
Operator
  differential, 477
  differentiation, 477

Operator *(Cont.)*
   linear, 472
Order
   of differential equation, 30
   of difference equation, 534
Ordinary differential equation(s) (ODE), 30. *See also* First-order ODE; Second-order ODE; Higher-order ODE
Ordinary point (for an ODE), 395, 409
Orthogonal trajectories, 74, 255*
Overdamped (vibration), 120, 123

P

Parallelogram rule, 593, 606
Partial derivatives, 5
Partial differential equation(s) (PDE), 30
   initial value/boundary value problem for, 573
   separation of variables for, 577
Partial fractions, 483
Particle, moving
   on spring(s), 78, 119, 355, 514, 579
   along straight line, 77
Particular solution, 131*
Path (of autonomous system), 157
Periodic
   forcing term, 124
   function with jumps, 502
Phase portrait, 159
   for linear system, 213, 215, 218
   for nonlinear system, 232
Piecewise continuous (function), 464
Polar coordinates, 595
Polar form (of complex number), 595, 602
Population cycles, 266
Population growth, 24
   two populations, 145, 262
   Verhulst model, 28
Porcupine, 42
Positive definite, 250
Potential energy, 169, 251, 579
Potential function, 61
Power series, 376
   approximating a value of, 379
   with complex variable, 600
   differentiating, 380
   dividing, 389
   expansion(s)
     in $x - c$, 386–387
     list of, 386
   gapped, 377
   geometric, 372
   integrating, 382
   multiplying, 384
   "new" function defined by, 378
   radius of convergence of, 376, 600
   shifting index of summation, 381
   solutions, existence of, 396, 447
   solving second-order ODE using, 370, 396
   Taylor series, 370, 385
Predator–prey model, 145, 262
Purely imaginary numbers, 593

Q

Quadratic forms, 254

R

RLC circuit, 147, 461
Radioactive decay, 21
Radius of convergence, 376, 600
   for rational function, 390
Rate of change, 3
Ratio test, 599
Rational functions, 367*
   inverse Laplace transform of, 483
   radius of convergence for, 390
Real axis, 593
Real part (of complex number), 590
Recurrence relation, 372, 535
Reduction of order, 102
   for inhomogeneous equation, 140
Regular singular point, 443
Related rate problems, 143
Remainder term (for Taylor polynomial), 369
Resistance, 147
Resistor, 146
Resonance, 125
Riccati equation, 97
Round-off error, 551
Row operations (on matrix), 616–617
Row reduction, 616
Runge–Kutta method, 554, 556–557

S

Saddle point
   for system, 214
   for $V(x, y)$, 257
Sawtooth wave, 503
Scalar multiplication, 606
Schrödinger equation, 578
   time-independent, 579
Science fiction, 424
Second Derivative Test, 258
Second partials, 8
Second-order ODE, 30
   existence, uniqueness of solutions, 86

$x'' - f(x) = 0$ (how to solve), 79
homogeneous linear
  constant-coefficient, 98, 104, 108
  solving using power series, 370, 396
inhomogeneous linear
  by undetermined coefficients, 114
  by variation of constants, 134
rewriting as system, 145
Separable differential equation, 15
Separation of variables (for PDE), 577
Separatrix, 241
Sequence, 534
  arithmetic, 536
  of complex numbers, 599
  of continuous functions, 563
  geometric, 536
Series. *See also* Power series
  with complex terms, 599
Shifting
  index of summation, 381
  stationary point (to origin), 178–179
Simultaneous linear equations, 615
Singular matrix, 615*, 621*
Singular point(s) (for an ODE), 409, 442
  indicial equation for, 448
  irregular, 443
  regular, 443
Slap shot, 520
Solution curve, 39
Solution(s) (of a differential equation), 13
  complex-valued, 108
  formal, 412
  power series, 370, 396
  of systems, 149, 305
Speed, 607
Spherical coordinates, 585
"Spike" functions, 516
Spiral point, 217–218
Spring(s) (with particle attached), 78, 119, 355, 514, 579
Square wave, 502
Stability for linear systems, 238
Stability for nonlinear systems, 239
Stable
  definition of, 236
  node, 214
  spiral point, 218
Standard basis vectors, 333, 364–365
Stationary point (of system), 163
  attractor basin of, 240
  classification of, 214, 217–218
  isolated, 233
  shifting to the origin of, 178–179
  stability of, 236
Steady-state (equilibrium) population, 25

Step function, basic (unit), 490
  derivative of, 519
Stockpile, 27
Strict Lyapunov function, 250
Successive approximation
  for solution to initial value problem, 562
  for square roots, 559–560
Sum (of series with complex terms), 599
Summation, shifting index of, 381
Superposition principle, 131, 297
Symbiosis (of populations), 145
Symmetric matrix, 257*
System of difference equations, 539
System of simultaneous linear equations, 615
System(s) of differential equations
  autonomous, 156, 306
  "basic solutions" for, 193, 209, 315, 325*
  direction field of, 158
  existence, uniqueness of solutions, 150
  first integrals for, 167
  gradient, 166, 254
  linear, 178, 306
  linear approximation to, 167, 210
  linear homogeneous, 179
  phase portrait of, 159
  rewriting second-order ODE as, 145
  rewriting higher-order ODE as, 304
  solutions of, 149, 305
  solving using Laplace transform, 485
  solving using matrix exponential, 342
  stability for, 238, 239
  stationary point of, 163
  variation of constants for, 357

T

T (transformation matrix), 188, 200, 308
Table of differences, 535
Tangent line, 2
Tangent plane, 4
Taylor expansion near $x = c$, 387
Taylor polynomials, 368
  remainder term for, 369
Taylor series, 370
  near $x = c$, 387
Terms of sequence, 534
Test for exactness, 62
Test for linear independence, 281
Three-term (Taylor series) method, 550
Time reversal, 176, 212
Time series, 534*
Total energy, 169
Trajectory (of autonomous system), 157
Transform(ation), linear, 472, 614

Transformation matrix **T**, 188, 200, 308
Transformation of vectors (by matrix), 612
Triangular matrix, 329, 347, 625
Trigonometric functions
   in terms of complex exponentials, 296, 602
Trigonometric series, 576
Trivial solution
   to homogeneous linear ODE, 88
   to homogeneous system of equations, 616

U

Uncoupled (equations), 185, 309
Undamped (vibration), 120
Underdamped (vibration), 120
Undetermined coefficients, 114, 296
Unstable
   definition of, 236
   node, 214
   spiral point, 218
Upper triangular matrix, 329, 347, 625

V

Vandermonde determinant, 284, 625–626
Variation of constants
   for first-order linear ODE, 51–52
   for higher-order ODE, 291
   for second-order linear ODE, 134
   for system, 357

Variation of parameters, *see* Variation of constants
Vector differential equation, 189, 307, 354
Vector field, 61
Vector function, 189,* 307, 608
Vector(s), 605
   addition of, 606
   column, 610
   components of, 607
   length of, 607
   in n-dimensional space, 608
   zero, 606
Velocity
   of particle along line, 77
   vector, 157, 607
Verhulst model, 28
Vibrating string, 572
Vibration (of particle on spring), 119
   with periodic external force, 124
Voltage, 146
Volterra–Lotka equations, 146, 262

W

Water tank, 143
Wave equations, 572
Wave function, 578
Wronskian, 92, 133, 282

Z

Zero vector, 606

$x'' - f(x) = 0$ (how to solve), 79
homogeneous linear
    constant-coefficient, 98, 104, 108
    solving using power series, 370, 396
inhomogeneous linear
    by undetermined coefficients, 114
    by variation of constants, 134
    rewriting as system, 145
Separable differential equation, 15
Separation of variables (for PDE), 577
Separatrix, 241
Sequence, 534
    arithmetic, 536
    of complex numbers, 599
    of continuous functions, 563
    geometric, 536
Series. *See also* Power series
    with complex terms, 599
Shifting
    index of summation, 381
    stationary point (to origin), 178–179
Simultaneous linear equations, 615
Singular matrix, 615*, 621*
Singular point(s) (for an ODE), 409, 442
    indicial equation for, 448
    irregular, 443
    regular, 443
Slap shot, 520
Solution curve, 39
Solution(s) (of a differential equation), 13
    complex-valued, 108
    formal, 412
    power series, 370, 396
    of systems, 149, 305
Speed, 607
Spherical coordinates, 585
"Spike" functions, 516
Spiral point, 217–218
Spring(s) (with particle attached), 78, 119, 355, 514, 579
Square wave, 502
Stability for linear systems, 238
Stability for nonlinear systems, 239
Stable
    definition of, 236
    node, 214
    spiral point, 218
Standard basis vectors, 333, 364–365
Stationary point (of system), 163
    attractor basin of, 240
    classification of, 214, 217–218
    isolated, 233
    shifting to the origin of, 178–179
    stability of, 236
Steady-state (equilibrium) population, 25

Step function, basic (unit), 490
    derivative of, 519
Stockpile, 27
Strict Lyapunov function, 250
Successive approximation
    for solution to initial value problem, 562
    for square roots, 559–560
Sum (of series with complex terms), 599
Summation, shifting index of, 381
Superposition principle, 131, 297
Symbiosis (of populations), 145
Symmetric matrix, 257*
System of difference equations, 539
System of simultaneous linear equations, 615
System(s) of differential equations
    autonomous, 156, 306
    "basic solutions" for, 193, 209, 315, 325*
    direction field of, 158
    existence, uniqueness of solutions, 150
    first integrals for, 167
    gradient, 166, 254
    linear, 178, 306
    linear approximation to, 167, 210
    linear homogeneous, 179
    phase portrait of, 159
    rewriting second-order ODE as, 145
    rewriting higher-order ODE as, 304
    solutions of, 149, 305
    solving using Laplace transform, 485
    solving using matrix exponential, 342
    stability for, 238, 239
    stationary point of, 163
    variation of constants for, 357

T

T (transformation matrix), 188, 200, 308
Table of differences, 535
Tangent line, 2
Tangent plane, 4
Taylor expansion near $x = c$, 387
Taylor polynomials, 368
    remainder term for, 369
Taylor series, 370
    near $x = c$, 387
Terms of sequence, 534
Test for exactness, 62
Test for linear independence, 281
Three-term (Taylor series) method, 550
Time reversal, 176, 212
Time series, 534*
Total energy, 169
Trajectory (of autonomous system), 157
Transform(ation), linear, 472, 614

Transformation matrix **T**, 188, 200, 308
Transformation of vectors (by matrix), 612
Triangular matrix, 329, 347, 625
Trigonometric functions
    in terms of complex exponentials, 296, 602
Trigonometric series, 576
Trivial solution
    to homogeneous linear ODE, 88
    to homogeneous system of equations, 616

U

Uncoupled (equations), 185, 309
Undamped (vibration), 120
Underdamped (vibration), 120
Undetermined coefficients, 114, 296
Unstable
    definition of, 236
    node, 214
    spiral point, 218
Upper triangular matrix, 329, 347, 625

V

Vandermonde determinant, 284, 625–626
Variation of constants
    for first-order linear ODE, 51–52
    for higher-order ODE, 291
    for second-order linear ODE, 134
    for system, 357

Variation of parameters, *see* Variation of constants
Vector differential equation, 189, 307, 354
Vector field, 61
Vector function, 189,* 307, 608
Vector(s), 605
    addition of, 606
    column, 610
    components of, 607
    length of, 607
    in n-dimensional space, 608
    zero, 606
Velocity
    of particle along line, 77
    vector, 157, 607
Verhulst model, 28
Vibrating string, 572
Vibration (of particle on spring), 119
    with periodic external force, 124
Voltage, 146
Volterra–Lotka equations, 146, 262

W

Water tank, 143
Wave equations, 572
Wave function, 578
Wronskian, 92, 133, 282

Z

Zero vector, 606